A Study on Next-Generation Materials and Devices

Dr. M. S. Vijaya Kumar, Department of Polymer Science and Technology / Chemistry, JSS Science and Technology University, Mysore, Karnataka, India. Dr. M S Vijaya Kumar obtained his Doctoral Degree from Tokyo Metropolitan University, Tokyo, JAPAN. He did his post-doctoral research work at Tufts University, Boston, USA. Dr. M S Vijaya Kumar has the vision of fostering academic learning and research among students and research scholars for the development of innovative concepts and products to the betterment of society. He published 46 research articles, 6 Conference Proceedings.
E-mail address: vijayakumarms@jssstuniv.in
Mobile number: +91-72593 66827

Dr. K. Srujan Raju is currently working as Dean Research and Development at CMR Technical Campus. He obtained his Doctorate in Computer Science in the area of Network Security. He has more than 20 years of experience in academics and research. His research interest areas include Data Mining, Cognitive Radio Networks, and Image Processing.
E-mail address: ksrujanraju@cmrtc.ac.in
Mobile number: +91 79899 29663

Dr. K. Rajakumar is presently working on 2 projects funded by the Government of India, has filed 7 patents at the Indian Patent Office, edited 10 books from Springer publications – AISC series, authored 4 books, contributed chapters in various books and published 25 papers at reputed and peer-reviewed International Conferences and Journals. His involvement with students is very conducive for solving their day-to-day problems. He mentored more than 100 students in incubating cutting-edge solutions. Dr. Rajakumar has acted as a reviewer and Technical Member for many conferences.
E-mail address: kantkhapazhamr@susu.ru
Mobile number: +91-81240 11542

Dr. Subramanian Saravanakumar is serving as the Head of the Department of Physics at Kalasalingam Academy of Research and Education, Tamil Nadu. With over 17 years of research experience and 10 years of post-Ph.D. teaching expertise, his scholarly contributions span luminescent and perovskite materials, and nanomaterials. He has authored more than 60 high-impact international journal publications and actively reviews for reputed journals. His advanced proficiency in crystallographic analysis and charge density mapping underpins his significant research output. Dr. Saravanakumar also serves as an editorial board member and has guided multiple Ph.D, and postgraduate students, demonstrating his leadership in both research and education.
E-mail address: s.saravanakumar@klu.ac.in
Mobile number: (+ 91) 99438 12370

A Study on Next-Generation Materials and Devices

International Conference on A Study on Next-Generation Materials and Devices

Edited by

M. S. Vijaya Kumar
K. Srujan Raju
K. Rajakumar
S. Saravanakumar

CRC Press
Taylor & Francis Group
Boca Raton London New York

CRC Press is an imprint of the
Taylor & Francis Group, an **informa** business

First edition published 2026
by CRC Press
4 Park Square, Milton Park, Abingdon, Oxon, OX14 4RN

and by CRC Press
2385 NW Executive Center Drive, Suite 320, Boca Raton FL 33431

CRC Press is an imprint of Informa UK Limited

British Library Cataloguing-in-Publication Data
A catalogue record for this book is available from the British Library

ISBN: 9781041146100 (hbk)
ISBN: 9781041146117 (pbk)
ISBN: 9781003675259 (ebk)

DOI: 10.1201/9781003675259

Typeset in Times New Roman
by HBK Digital

Contents

List of figures x
List of tables xvi
Foreword xviii
Preface xix
Acknowledgements xx

Chapter 1 Development of plant health monitoring device 1
B. Perumal, Pallikonda Rajasekaran, A. Lakshmi, Rajesh V., Krishna Priya R., and Saravanan Velusamy

Chapter 2 Development of MEMS-based sensors for industrial applications 6
Aakansha Soy and Ahilya Dubey

Chapter 3 An analysis of deep learning techniques used for semantic segmentation of remote sensing images 9
Roshni Rajendran and P. Nagaraj

Chapter 4 Managing distribution transformers using GSM-based remote monitoring systems 15
Aakansha Soy and Ahilya Dubey

Chapter 5 Bio-innovations in drug packaging: A novel approach using microcrystalline cellulose and modified starch based bio-composite materials 23
Sakthivel sankaran, Marimuthu Chandran, Krishnan Selvaraj, Arumugaprabhu Veerasimman, and S. Vijiyalakshmi

Chapter 6 Modelling and design analysis of a ripple-free high step-up converter for fuel cell application 29
Abhijeet Madhukar Haval and Dhablia Dharmesh Kirit

Chapter 7 Transcutaneous electrical nerve stimulator for cervical cancer 33
S. Shanmugapriya, S. Mahathi, and K. Vishaly

Chapter 8 Advanced power electronics for electric vehicle charging stations 37
Anu G. Pillai and Shital Kewte

Chapter 9 Smart gloves using SpO2, temperature and ECG sensor 40
Kalimuthukumar S., Ramkarthick B., and Manoj Kumar S.

Chapter 10 A comprehensive overview of internet-of-things-based water monitoring systems 46
Anupa Sinha and Pooja Sharma

Chapter 11 A detailed review of the concepts, technologies, and industrial applications of Digital Twin 54
J. Loyola Jasmine, M. Carmel Sobia, and M. Jayalakshmi

Chapter 12 PSO-based optimization of EVCS and DG placement for improved grid durability and efficiency in distribution systems 60
Ashu Nayak and Ankita Tiwari

Chapter 13 Tumor extraction from MR brain images using modified seeded region growing algorithm 65
D. B. Shanmugam, Arun Francis G., Manikandan S., KottaimalaiR., Thilagaraj M., and Ambika B.

Chapter 14 Threats to the IOT technologies from security point of view 71
Debarghya Biswas and Balasubramaniam Kumaraswamy

Chapter 15 Blockchain-powered smart meter billing: Enabling transparency and security 75
Rakhee M., M. Sudheep Elayidom, and Baiju Karun

Chapter 16 Sensor based emergency locating module (ELM) 81
F. Rahman and Priti Sharma

Chapter 17 Predictive analytics for renal health: Machine learning in chronic kidney disease prediction 85
Radha M. and Muthukumar A.

Chapter 18 Analysing the efficiency of a panel of monocrystalline cells utilizing a nanofluid cooling technique based on Aluminium-Metal Oxide (Al2O3) 89
Gaurav Tamrakar and Shailesh Singh Thakur

Chapter 19 Bandwidth improvement using C stub in simple compact monopole antenna for wireless applications 93
Sathyamoorthy Sellapillai and Kavitha Thandapani

Chapter 20 Transforming Ayurvedic medicine access: A prototype on healthcare systems 97
Kamlesh Kumar Yadav and Md Afzal

Chapter 21 Multiscale information fusion to segment white blood cells for early stage detection of acute lymphoblastic leukemia 101
M. Thilagaraj, S. Vaira Prakash, G. Petchinathan, Kottaimalai Ramaraj, C. S. Sundar Ganesh, and S. Krishnanarayanan

Chapter 22 Augmented Reality (AR) and Virtual Reality (VR) for educational applications 107
Manish Nandy and Lalnunthari

Chapter 23 Mammography images: Visual Geometric Group (VGG) model and customization for breast cancer detection 111
M. Shanmugapriya and J. Pradeepkandhasamy

Chapter 24 Innovative 3D printing technologies for biomedical application 118
Priya Vij and Ghorpade Bipin Shivaji

Chapter 25 Effective human heart disease prognosis through machine learning 124
Gayathiri Jeyachandra Gandhi, Vijaya Kumar K., Arunprasath T., Pallikonda Rajasekaran M., Kottaimalai R., and Thiruppathy Kesavan V.

Chapter 26 Edge AI-integrated photonic crystals for real-time and adaptive optical sensing applications 128
F. Rahman and Priti Sharma

Chapter 27 Unveiling the challenges in extracting cerebral tumour on MRI utilizing Glowworm Swarm Optimization based PCM Clustering 134
Anitta D., Anu Joy, Krishna Narayanan S., Padmavathi M., Kottaimalai Ramaraj, and M. Thilagaraj

Chapter 28 Analysis of PV penetration with smart inverter on DG system at grid system 140
Shailesh Madhavrao Deshmukh and Ikhar Avinash Khemraj

Chapter 29 Unveiling interdependencies between Alzheimer's disease and lung cancer through integrated CGNN modelling 144
G. Akiladevi, M. Arun, and J. Pradeepkandhasamy

Chapter 30 Low-calcium calcined clay or mortar in an acidic environment 151
Subrata Majee and Nasar Ali R.

Chapter 31 Indian sign language recognition system facilitating communication for individuals with special needs 156
Nishadha S. G., and R. Murugeswari

Chapter 32 Next-generation soft matter-infused metasurfaces for adaptive and reconfigurable optoelectronic devices 161
Madhu Sahu and Cherry

Chapter 33 Hybrid deep learning model for brain tumour segmentation 167
Chelli N. Devi

Chapter 34 Biodegradable optical sensors with quantum-assisted signal processing for sustainable and high-precision sensing 171
Ashu Nayak and Ankita Tiwari

Chapter 35 Implement door unlocking by using an RFID enabled ATM card system 177
B. Perumal, A. Lakshmi, Rajesh V., Ramalingam H. M., Saravanan Velusamy, and Krishna Priya R.

Chapter 36 Enhance performance the particle Swarm optimization based optimizing load balancing technique in cloud computing 182
Manish Nandy and Lalnunthari

Chapter 37 Implementing an RSSI-based localization algorithm for Alzheimer's disease in smart home systems using wireless sensor networks 187
Senthilnathan N., Sarojini R., Kulasekarapandian S., Adaikalam A., Kavitha T., and Thilakavathi B.

Chapter 38 A comprehensive optimization strategy for ZVS and ZCS-enabled half-bridge LLC resonant conversion devices 193
Kamlesh Kumar Yadav and Md Afzal

Chapter 39 Combining firefly optimization and Mahalanobis distance based fuzzy C-means clustering for earlier breast cancer diagnosis 197
Farha Kowser, C. Arul Murugan, G. Vijaykumar, Kottaimalai Ramaraj, Thilagaraj M., and Petchinathan G.

Chapter 40 Modeling and intelligent control of hybrid system for microgrid operation 203
Priya Vij and Ghorpade Bipin Shivaji

Chapter 41 Decrypting theft suspects in low-resolution snapshots 207
Mervin Jerel D., Moneshwar C., Naveenkumar S., and D. Menaka

Chapter 42 RFID technology for supply chain management 213
Anupa Sinha and Pooja Sharma

Chapter 43 A comprehensive survey on OFDM autoencoders: Integrating machine learning for enhanced communication systems 216
P. Rajarajan and Madona B. Sahaai

Chapter 44 IoT-enhanced smart vending cart revolutionizing retail with connected technology 222
F. Rahman and Priti Sharma

Chapter 45 Exploration of blockchain integrated IoT devices in healthcare application for secured remote patient monitoring 228
Varsha P. Hotur, Ramani U., Menakadevi N., R. Krishna Kumar, Kottaimalai R., and Thilagaraj M.

Chapter 46 A study of fourth-generation communication systems' multiple-access techniques and their relative effectiveness 234
Ashu Nayak and Ankita Tiwari

Chapter 47 Transfer learning with vision transformers for Alzheimer's disease detection: An ablation study 240
Sathvik Rajampalli, Kaviya Dharshini, and Jeeva JB

Chapter 48 Hybrid energy storage systems for smart homes 245
Kamlesh Kumar Yadav and Md Afzal

Chapter 49 Advances in artificial pancreas technology: A comprehensive review and future prospects 249
Radha M., Muthukumar A., Friska J., Uma M., Krishnaveni G., and Saranya M.

Chapter 50 IoT-based appropriate crop Identification Method after Soil Analysis and Local Weather Prediction 256
Abhijeet Madhukar Haval and Dhablia Dharmesh Kirit

Chapter 51 5G network based spectrum analysis for mm-wave and sub 6 GHz 261
Supraja C. and Kavitha T.

Chapter 52 AI-driven crystalline defect engineering for tunable light emission: A machine learning approach to optoelectronic material design 267
Swati Agrawal and Abhay Dahiya

Chapter 53 Reduction of noise in multimodal brain images using adaptive filtering techniques 272
N. Thenmoezhi, B. Perumal, A. Lakshmi, Pallikonda Rajasekaran, Kottaimalai Ramaraj, and Arunprasath Thiyagarajan

Chapter 54 AI-Directed Synthesis of Bio-Optical Nanomaterials for Medical Imaging 278
Shailesh Madhavrao and Ikhar Avinash Khemraj

Chapter 55 Deep learning approaches for classification of abnormal respiratory signals 284
Kaleeswari P., Ramalakshmi R., Arunachalam Muthukumar, and Thanga Raj M.

Chapter 56 Advanced irrigation systems employing recent technologies 289
Anupa Sinha and Pooja Sharma

Chapter 57 Plant care app crop disease detection using machine learning 292
G. K. Jayaprakash and Pandiaraj Kadarkarai

Chapter 58 Fault classification and monitoring in induction machine using artificial neural network 298
Manish Nandy and Lalnunthari

Chapter 59 A Unique Stochastic Slime Mould Converter Control (SMC) model with Quadratic High Gain Converter (QHGC) for PV-EV System 302
Sarath S., and K. Vijayakumar

Chapter 60 Intelligent interface of solid oxide fuel cell for micro grid operation 308
Abhijeet Madhukar Haval and Dhablia Dharmesh Kirit

Chapter 61 Development and characterization of polyvinyl alcohol-activated charcoal composites for advanced wound dressings 312
S. Shanmugapriya, K. Ruth Esther, Marimuthu Chandran, Konda Mahesh, and Aman Kumar

Chapter 62 Infrastructural study of smart cities 319
Aakansha Soy and Ahilya Dubey

Chapter 63 Dehazing the Hazed images for buildings and underwater objects using KWM algorithm 322
Buvanesh Pandian V., Arunprasath T., Pallikonda Rajasekaran M., Kottaimalai R., and Krishna Priya R.

Chapter 64 AI-powered metamaterial-based dynamic light steering for Next-Gen displays 327
Kamlesh Kumar Yadav and Md Afzal

Chapter 65 Design and implementation of heart patient health monitoring system using IoT 332
B. Santhikiran and Kavitha T.

Chapter 66 AI-enhanced image sensors for autonomous vehicles 338
Rahul Mishra and Vinay Chandra Jha

Chapter 67 An explorative analysis of T cell activating drugs 341
Salins S. S., Muthukumar A., Thanga Raj M., and Kaleeswari P.

Chapter 68 Bioinformatics: Using machine learning for genome analysis 347
Ashu Nayak and Ankita Tiwari

Chapter 69 Diagnosis of PCOS using the optimal bald eagle based SVM model 351
Oviya Graselin S., Arunprasath T., Pallikonda Rajasekaran M., Ramalakshmi R., Kottaimalai R., and Thiruppathy Kesavan V.

Chapter 70 AI-driven self-adaptive crystalline polymers for next-generation smart lenses: a novel approach to real-time optical adjustment and wearable vision enhancement 357
Nidhi Mishra and Patil Manisha Prashant

Chapter 71 An Efficient medical decision-making system for skin cancer classification using SENet 362
Muthuselvi S and Sumathi R.

Chapter 72 A smart adaptive neuro fuzzy inference system (ANFIS) model integrated with DAB converter for EV charging systems 368
Revathy K. P. and K. Vijayakumar

Chapter 73 Mathematical analysis to enhance the frequency stability of interconnected power system 375
Shailesh Madhavrao Deshmukh and Ikhar Avinash Khemraj

Chapter 74 Optimizing the scheduling of electric vehicles in a static G2V system, incorporating grid stability 380
Nidhi Mishra and Patil Manisha Prashant

Chapter 75 Adaptive contextual emotion-infused transfer learning network for respiratory surveillance 385
Kaleeswari P., Ramalakshmi R., Muthukumar A. and Thanga Raj M.

Chapter 76 HoloNeuroNet: A nano-holography-based AI optical computing framework 391
Anupa Sinha and Pooja Sharma

Chapter 77 Utilizing DBSCAN clustering for Alzheimer's disease identification in MR brain images 397
Anu Joy, Anitta D., Karthikeyan K., Vidyalakshmi R., Kottaimalai Ramaraj, and M. Thilagaraj

Chapter 78 Electric vehicle modelling for grid integration and charging analysis 403
F. Rahman and Priti Sharma

Lists of figures

Figure 1.1	Circuit diagram of plant health monitoring device	3
Figure 1.2	Plant health monitoring device	4
Figure 2.1	MEMS-based sensors	6
Figure 2.2	System architecture MEMS-based sensors	7
Figure 3.1	CNN-based image classification	10
Figure 3.2	Basic block diagram of semantic segmentation of remote sensing images	10
Figure 4.1	Remote monitoring system for distribution transformers	17
Figure 4.2	Data transmission process	18
Figure 4.3	Remote monitoring system setup	20
Figure 4.4	GSM based monitoring system for transformers	21
Figure 5.1	Block diagram of newly developed biocomposite sample material	26
Figure 5.2	Cellulose powder	26
Figure 5.3	Starch powder	26
Figure 5.4	Mixture of both the cellulose and modified starch	27
Figure 5.5	Developed biocomposite material	27
Figure 6.1	Circuit diagram of proposed converter	30
Figure 6.2	Graph between voltage gain corresponding to various turns ratio and duty cycle	31
Figure 6.3	Graph between voltage gain corresponding to duty cycle for all the converters	31
Figure 7.1	Symptoms of cervical cancer	34
Figure 7.2	Building block of TENS unit	34
Figure 7.3	TENS Hardware for cervical cancer system	35
Figure 7.4	Coding for Transcutaneous electrical nerve stimulator	35
Figure 8.1	System architecture	38
Figure 8.2	Working of advanced power electronics for electric vehicle charging stations	38
Figure 10.1	Block diagram of proposed system	47
Figure 10.2	IoT system component	51
Figure 10.3	IoT based water monitoring system	52
Figure 11.1	Digital Twin	55
Figure 11.2	DT network	57
Figure 11.3	Digital twin migration	58
Figure 12.1	Profile for the voltage distribution network following the installation of distributed generating sources (DGs) and electric vehicle charging stations (EVCS)	62
Figure 12.2	Actual loss of power in the bus system following the best possible placement of distributed generating sources (DGs) and electric vehicle charging stations (EVCS)	63
Figure 12.3	Reactive power outages when distributed generating sources (DGs) and electric vehicle charging stations (EVCS) are installed	63
Figure 12.4	Total power loss on the distribution network following the integration of distributed generating sources (DGs) and electric vehicle charging stations (EVCS)	63
Figure 13.1	Benign tumour segmented outputs using proposed algorithm	68
Figure 13.2	Malignant tumour segmented outputs using proposed algorithm	69
Figure 14.1	Interfacing of firewall	72
Figure 14.2	Steps included in cybersecurity	73
Figure 15.1	Energy resources in Kerala	76
Figure 15.2	Smart energy meter with Blockchain	77
Figure 15.3	The billing process	77
Figure 15.4	Snippet of billing process	78
Figure 15.5	Billing processing time	78
Figure 15.6	Data update frequency	78
Figure 15.7	Operational costs	79
Figure 15.8	Error reduction costs	79
Figure 15.9	Security improvements 1	79
Figure 15.10	Security improvements 2	79

Figure 16.1 Working of sensors inbuilt in vehicle 82
Figure 16.2 Employment of sensors in vehicles 82
Figure 16.3 Steps included in sensor-based emergency locating 83
Figure 17.1 Block diagram 86
Figure 17.2 Different diagnostic parameters plots 87
Figure 17.3 CKD result 87
Figure 18.1 Equivalent circuit diagram of Solar cell 90
Figure 18.2 Circuit diagram of Boost Converter 90
Figure 19.1 Design structure of proposed C shaped monopole antenna 94
Figure 19.2 |S11| Comparison of monopole and proposed antenna 94
Figure 19.3 The suggested antenna's impedance characteristics 94
Figure 19.4 VSWR of the antenna 95
Figure 19.5 Current distribution of the suggested antenna (a) 2.42 GHz and (b) 5.24GHz 95
Figure 19.6 Two dimensional Radiation pattern of the antenna 95
Figure 19.7 3-dimensional Radiation pattern of antenna 96
Figure 20.1 RFID sensor 98
Figure 20.2 PIR motion sensor 98
Figure 20.3 Stepper motor 98
Figure 20.4 Microcontroller 99
Figure 20.5 Temperature & humidity sensor 99
Figure 20.6 Wi-Fi module 99
Figure 20.7 RGB liquid crystal display 99
Figure 20.8 LED indicator 100
Figure 20.9 Emergency stop button 100
Figure 20.10 Functional block diagram 100
Figure 21.1 Types of Leukemia 102
Figure 21.2 Types of WBCs 103
Figure 21.3 Proposed workflow 103
Figure 21.4 MIF-Net 104
Figure 21.5 MIF-Net Segmentation outcomes 105
Figure 22.1 Augmented reality (AR) and virtual reality (VR) for educational applications 107
Figure 22.2 System architecture augmented reality (AR) and virtual reality (VR) for educational applications 109
Figure 23.1 Sample mammogram images 114
Figure 23.2 Workflow of the proposed methodology 115
Figure 23.3 Comparative analysis 115
Figure 24.1 3d printing and their application 119
Figure 24.2 Various 3D printing technologies 119
Figure 25.1 Class distribution 125
Figure 25.2 Distribution of age (a), Distribution of sex (b) 126
Figure 25.3 Distribution of chest discomfort (a), Distribution of trestbps (b) 126
Figure 25.4 Function of ML classifier 126
Figure 25.5 Function of DL classifier 126
Figure 26.1 Flow diagram of the proposed methodology 130
Figure 26.2 Graphical representation of sensitivity (RIU) 131
Figure 26.3 Graphical representation of adaptive optical response comparison 131
Figure 26.4 Graphical representation of processing speed 132
Figure 27.1 BraTS tumour dissection results 137
Figure 27.2 Comparison of DBSCAN's accuracy values with SOTA 138
Figure 27.3 Comparison of DS value with SOTA 138
Figure 28.1 Block diagram of grid integrated PV array 141
Figure 28.2 Block diagram of converter 141
Figure 28.3 Frequency of renewable energy with various location 141
Figure 28.4 VSC controller 142
Figure 28.5 Comparison of consumption of pv power as down stream and upstream 142
Figure 28.6 Consumption of pv power as upstream 142
Figure 28.7 Consumption of pv power as down stream 142

Figure 29.1	Pre-processed image	147
Figure 29.2	Features extracted from CNN	148
Figure 29.3	Training and validation Loss	148
Figure 29.4	Visualization of lung cancer caused by Alzheimer disease	149
Figure 29.5	Heatmap visualization of feature prediction–lung cancer caused by Alzheimer disease	149
Figure 29.6	Comparison model-CGNN	149
Figure 30.1	Calcined clay powder	152
Figure 30.2	Research of the properties of fly ash and calcined clay mortar, namely its consistency	153
Figure 30.3	Initiation setting times for fly ash and calcined clay are measured	153
Figure 30.4	Examining the compressive strength of fly ash mortars	153
Figure 30.5	Examining the compressive strength of fly ash mortars	154
Figure 31.1	Working of ISL recognition model 1	158
Figure 31.2	Sample recognized hand sign codes	158
Figure 31.3	Accuracy and Loss estimated for 14 epochs of training	159
Figure 32.1	Graphical representation of optical modulation efficiency	164
Figure 32.2	Graphical representation of performance retention at 100°C (%)	165
Figure 32.3	Graphical representation of real-time tunability	165
Figure 33.1	Brain MRI scans	169
Figure 34.1	System architecture	173
Figure 34.2	Graphical representation of sensitivity and accuracy comparison	174
Figure 34.3	Graphical representation of real-time processing efficiency (%)	175
Figure 34.4	Graphical representation of energy consumption and sustainability	175
Figure 34.5	Graphical representation of data integrity	176
Figure 35.1	Block diagram Implement door unlocking by using an RFID-enabled ATM card system	179
Figure 35.2	Door unlocking by using an RFID-enabled ATM card system output model	179
Figure 35.3	RFID-enabled ATM card system output Model–Arduino code	180
Figure 35.4	RFID-enabled ATM card system output model	180
Figure 36.1	Delivery model of cloud	183
Figure 36.2	Meta heuristic group-based search	184
Figure 36.3	Show the successful tasks	185
Figure 36.4	Average scheduled length	185
Figure 36.5	Ratio of successful execution	185
Figure 37.1	General block diagram for Alzheimer's patients with WSN	188
Figure 37.2	Mean Estimation error Vs No. of Nodes respectively	190
Figure 37.3	Estimated and actual locations of Alzheimer's patient	190
Figure 38.1	Energy description for generalized harmonic tanks	194
Figure 38.2	Graph how the Gain vs load curve	194
Figure 38.3	Graph show the angle of ZVS with load for LLC	195
Figure 38.4	Graph show the angle of ZVS with load for CLL	195
Figure 38.5	Efficiency measurement	195
Figure 39.1	Proposed model workflow	200
Figure 39.2	Dissection outcomes of FO_mFCM method	200
Figure 39.3	Comparison of sensitivity values of FO_mFCM with SOTA	201
Figure 39.4	Comparison of accuracy values of FO_mFCM with SOTA	201
Figure 39.5	Comparison of dice score values of FO_mFCM with SOTA	201
Figure 39.6	Comparison of computational time values of FO_mFCM with SOTA	201
Figure 40.1	Typical MG structure	204
Figure 40.2	Architecture of the Hybrid system	204
Figure 40.3	SOFC module	205
Figure 40.4	Equivalent circuit of Battery	205
Figure 40.5	Response of BSS as part of Hybrid system	205
Figure 40.6	Response of SOFC as part of Hybrid system	206
Figure 41.1	Architecture of object detection	208
Figure 41.2	Architecture of a siamese network	209
Figure 41.3	Training loss and accuracy plot	209

Figure 41.4	Criminal detected	211
Figure 41.5	Mail alert and movement mapping	211
Figure 42.1	RFID technology for supply chain management	213
Figure 42.2	System architecture RFID technology for supply chain management	214
Figure 43.1	Principles of OFDM	217
Figure 43.2	Compensation for frequency and timing offsets	217
Figure 44.1	Vending cart	223
Figure 44.2	Flow chart of purchase order	224
Figure 44.3	Design of smart vending cart	225
Figure 45.1	Process of BC	229
Figure 45.2	Key facilities of BC	229
Figure 45.3	BC healthcare companies	230
Figure 45.4	Benefits of IoT in healthcare	232
Figure 46.1	IDMA scheme receiver structures with K concurrent users	237
Figure 46.2	An analysis contrasting CDMA and IDMA	238
Figure 47.1	No dementia (left) vs mild dementia (right)	240
Figure 47.2	ViT architecture flow (Top to down)	242
Figure 47.3	Training parameters	243
Figure 47.4	((a) head-only, (b) full-fine tuning, (c) two-stage fine-tuning, (d) layer-by-layer fine-tuning) are provided with an input image (mild-demented) and the corresponding prediction made	243
Figure 48.1	Hybrid energy storage systems for smart homes	*246*
Figure 48.2	Hybrid energy storage systems for smart homes	247
Figure 49.1	Flow of artificial pancreas	251
Figure 49.2	Block diagram	251
Figure 49.3	Profiles of Glucose levels and hormonal deliveries during Artificial pancreas visits	252
Figure 49.4	Post-meal levels during visits	254
Figure 50.1	Dataflow diagram	257
Figure 50.2	ESP32	257
Figure 50.3	Soil NPK sensor	257
Figure 50.4	pH sensor probe with transmitter	257
Figure 50.5	Soil temperature probe sensor	258
Figure 50.6	Soil moisture solution	258
Figure 50.7	Block Diagram of proposed work	258
Figure 50.8	Corresponding data from sensors	259
Figure 50.9	Working model	259
Figure 50.10	Dashboard of our application	259
Figure 51.1	Analog beam-forming	264
Figure 51.2	Hybrid beam-forming	264
Figure 51.3	Spectral efficiency for Sub-6 GHz (3 GHz)	265
Figure 51.4	Spectral efficiency for mm-wave (60 GHz)	265
Figure 51.5	Heat map of spectral efficiency for mm-wave (60 GHz) and Sub 6 GHz (3 GHz)	265
Figure 51.6	Beam forming gain comparison for mm wave (60 GHz)	265
Figure 51.7	Beam forming gain comparison for Sub-6 GHz (3 GHz)	265
Figure 52.1	Flow diagram of the proposed methodology	268
Figure 52.2	Performance comparison of emission wavelength tuning	269
Figure 52.3	Stability of engineered defects	270
Figure 53.1	Tumour cell in brain	274
Figure 53.2	Examples of multi modal brain medical image fusion	275
Figure 53.3	Proposed method	275
Figure 53.4	Output	275
Figure 54.1	Proposed AI-directed synthesis framework	280
Figure 54.2	Flow diagram of the methodology	280
Figure 54.3	Graphical representation of comparison of traditional vs. AI-directed synthesis in bio-optical nanomaterials	281
Figure 54.4	Comparison of cytotoxicity and cellular uptake. traditional vs. AI-directed synthesis	281

Figure 55.1 Proposed method of abnormal respiratory signals classification 286
Figure 55.2 Comparative analysis of classification performance 287
Figure 56.1 (a) surface irrigation, (b) drip irrigation, (c) sprinkler system 290
Figure 56.2 Smart field watering system 290
Figure 57.1 Rust disease 293
Figure 57.2 Yellow leaf disease 293
Figure 57.3 Eye spot disease 293
Figure 57.4 Red dot disease 293
Figure 57.5 Image processing 295
Figure 57.6 Flow chart of crop disease identification 295
Figure 57.7 CNN for disease prediction 296
Figure 57.8 Community support 296
Figure 57.9 Weather forecast 297
Figure 57.10 Crop disease detection 297
Figure 58.1 Flow chart for diagnosis of fault and monitoring the condition 299
Figure 58.2 Flow chart of an ANN based fault diagnosis system for induction motor 299
Figure 58.3 Performance of neural network with different numbers of neuron in hidden layers with all features as an input 299
Figure 58.4 Validation performance curve for scheme 1 with all features as an input 300
Figure 58.5 Performance of neural network with different numbers of neuron in hidden layers with reduced features as an input 300
Figure 59.1 Schematic representation of the PV-integrated EV system 303
Figure 59.2 Schematic model of the proposed QHGC-S^2MC^2 based PV-EV system 303
Figure 59.3 Schematic model of the proposed QHGC model 304
Figure 59.4 Voltage, power and current of PV system 305
Figure 59.5 Evaluation of the QHGC and standard model based on the voltage stress between diodes 306
Figure 59.6 Comparison of the proposed QHGC and conventional design using normalized inductor time 306
Figure 59.7 Comparison of the proposed QHGC with the standard one using switching stress 306
Figure 59.8 Comparison of efficiency between the input voltage 75V and 85V 306
Figure 59.9 Inverter voltage 306
Figure 59.10 Inverter current 306
Figure 59.11 Harmonic distortion 306
Figure 60.1 Schematic of fuzzy system 309
Figure 60.2 Fuzzy rule based models 309
Figure 60.3 Processes in the fuzzy controller 309
Figure 60.4 Grid side control for SOFC MG operation 310
Figure 60.5 MV network 310
Figure 61.1 Flow chart of the developed methodology 314
Figure 61.2 PVA in powder form 315
Figure 61.3 Activated charcoal in powder form 315
Figure 61.4 Calendula officinalis in powder form 315
Figure 61.5 Chromolaena Odorata in powder form 315
Figure 61.6 Sample vs tensile strength graph 317
Figure 61.7 Sample vs stiffness graph 317
Figure 62.1 Model of smart city 320
Figure 62.2 Emergency services in smart cities 321
Figure 64.1 Proposed framework 329
Figure 64.2 Comparison of power consumption and energy efficiency of different display technologies 329
Figure 64.3 Comparison of resolution and clarity improvement across display technologies 330
Figure 64.4 Response time and refresh rate 330
Figure 65.1 Proposed work block diagram 334
Figure 65.2 Data transmission and storage 334
Figure 65.3 Real-time monitoring and Alerts user interface 334
Figure 65.4 User interface 335
Figure 65.5 Prototype 335
Figure 65.6 LCD output values 335
Figure 65.7 SMS output 336

Figure 65.8	Temperature, blood pressure, oxygen quality, ECG checking	336
Figure 65.9	ECG monitoring using arduino plot	336
Figure 66.1	AI-enhanced image sensors for autonomous vehicles	339
Figure 66.2	System architecture AI-enhanced image sensors for autonomous vehicles	339
Figure 67.1	Align result	345
Figure 68.1	Bioinformatics using machine learning for genome analysis	348
Figure 68.2	System architecture Bioinformatics. Using machine learning for genome analysis	349
Figure 69.1	Work frame of the suggested PCOS classification model	352
Figure 69.2	Flowchart of the optimal feature selection process	353
Figure 69.3	Performance of the proposed BESO-SVM by varying the k-fold	354
Figure 69.4	ROC performance of the proposed BESO-SVM	355
Figure 70.1	Proposed system: AI-based self-adaptive crystalline polymers	359
Figure 70.2	Adaptation speed across different smart lens technologies	360
Figure 70.3	Chart representing the power consumption (mW)	360
Figure 71.1	(a) Benign and (b) malignant dermatological images	364
Figure 71.2	Optimization of SENet model	365
Figure 71.3	SENet mobile pre-training: Epoch vs accuracy	366
Figure 71.4	SENet large pre-training: Epoch vs accuracy	366
Figure 71.5	SENet mobile Epoch vs loss	366
Figure 71.6	SENet large: Epoch vs loss	366
Figure 71.7	Adam optimizer's ROC curve for SENet mobile.	366
Figure 72.1	Block diagram of the proposed DAB-ANFIS model	370
Figure 72.2	Schematic model of DAB converter	370
Figure 72.3	DAB-ANFIS integrated model for EV charging system	371
Figure 72.4	Input voltage and training data set of ANFIS	371
Figure 72.5	Generation of FIS	372
Figure 72.6	Plot of training vs testing data set of ANFIS	372
Figure 72.7	Output voltage and current of DAB converter	372
Figure 72.8	Output voltage of inverter	372
Figure 72.9	Voltage gain and efficiency of converter	372
Figure 72.10	THD of DAB-ANFIS system	372
Figure 72.11	DAB-Fuzzy system	373
Figure 72.12	Comparison among the classic and proposed controlling techniques based on Efficiency & Settling time	373
Figure 73.1	Multi area power system without SSSC	377
Figure 73.2	Multi area power system with SSSC	377
Figure 73.3	Frequency response with SSSC	378
Figure 73.4	Deviated tie-line power response with SSSC	378
Figure 74.1	Schematic design of electric vehicle charging networks	381
Figure 74.2	Comparative analysis between the performance parameters of EVA and the number of EV	382
Figure 74.3	Comparative analysis between the performance parameters of EVA and charging rate limit	383
Figure 75.1	Proposed methodology	387
Figure 75.2	Accuracy comparison of different models	389
Figure 75.3	Proportion of normal and abnormal breathing patterns detected	389
Figure 76.1	Proposed HoloNeuroNet framework	393
Figure 76.2	System design and implementation	394
Figure 76.3	Representation of computation time (ms) and speed improvement (%)	394
Figure 76.4	Pie chart representing the power consumption	395
Figure 76.5	Graphical representation of model accuracy (%) and accuracy improvement (%)	395
Figure 77.1	Dissection outcomes of AD on MR images using DBSCAN	400
Figure 77.2	Comparison of DBSCAN's accuracy values with SOTA	401
Figure 78.1	Graph showing energy demand and distance curve for EV	404
Figure 78.2	Charging schedules (Dumb charging scheme)	405
Figure 78.3	Charging schedules (smart charging scheme)	406
Figure 78.4	Power loss for charging hours	406

Lists of tables

Table 3.1 Summary of methods used for semantically segmenting the RS imagery based on FCN 12
Table 6.1 Comparative performance analysis of proposed converter 31
Table 11.1 Digital design pattern catalog 55
Table 13.1 Performance measurements of the proposed method on tumor detection 70
Table 13.2 Comparison of metrics 70
Table 15.1 Efficiency metrics 78
Table 15.2 Cost savings 78
Table 18.1 Temperature variation of PV panel 91
Table 18.2 Performance parameters of panel after application of Aluminium-metal oxide (Al2O3) Nanofluid cooling 91
Table 21.1 Comparison of various network models 104
Table 21.2 Segmentation of cytoplasm using MIF-Net 105
Table 21.3 Segmentation of nucleus using MIF-Net 105
Table 23.1 Distribution of full-mammography picture data 115
Table 23.2 Performance metrics for assessment 116
Table 24.1 Comparative analysis 122
Table 25.1 Comparison of accuracy using different classifiers 125
Table 26.1 Performance comparison of AI-driven inverse design 131
Table 26.2 Adaptive optical response comparison 131
Table 26.3 Signal processing performance 132
Table 26.4 Sensitivity enhancement using hybrid structures 132
Table 30.1 Show the quantity of compound that is chemical composition of opc, fly ash, and calcined clay 152
Table 33.1 Overview of recent works 168
Table 33.2 Segmentation results 169
Table 33.3 Comparison of results 169
Table 34.1 Sensitivity and accuracy comparison 174
Table 34.2 Response time comparison 175
Table 34.3 Energy consumption and sustainability 175
Table 34.4 Security and data integrity comparison 176
Table 37.1 Model evaluation: actual vs. predicted X and Y 191
Table 37.2 RSSI measurements by distance and antenna 191
Table 41.1 Comparison of literature and proposed model for decrypting thief suspect in low resolution snapshots 210
Table 41.2 Comparison of Siamese convolutional neural network on the Omniglot verification task 211
Table 43.1 Types of Autoencoders 218
Table 43.2 Summary of recent advancements in deep learning integration with wireless communication systems 219
Table 45.1 Blockchain applications 231
Table 46.1 Comparing IDMA with other current MA technologies 237
Table 46.2 Comparisons of the IDMA and CDMA MAC Protocols 238
Table 47.1 Training protocol 242
Table 49.1 Comparisons of insulin-alone artificial pancreas, rapid insulin-and-pramlintide artificial pancreas, and regular insulin-and-pramlintide artificial pancreas 253
Table 52.1 Performance comparison of emission wavelength tuning 269
Table 52.2 Enhancement of quantum efficiency: comparison of traditional, alloying-based, and AI-optimized defect engineering methods 270
Table 52.3 Stability of engineered defects 270
Table 52.4 Efficiency comparison of defect engineering methods: processing time, computational cost, and reduction percentage 270
Table 53.1 SNR value and extraction 276

Table 54.1	Enhanced optical properties of nanomaterials: traditional vs. AI-directed synthesis	281
Table 54.2	Biocompatibility and safety comparison: traditional vs. AI-directed synthesis	281
Table 55.1	Comparative analysis	288
Table 56.1	Table showing conditions on which the model operates	290
Table 59.1	Components study	304
Table 59.2	Simulation parameter setting	305
Table 59.3	Loss value analysis of QHGC	306
Table 60.1	THD for PI, HCC and fuzzy based intelligent control	310
Table 61.1	Mechanical properties of the films	316
Table 63.1	Qualitative analysis of the SOTS dataset	324
Table 63.2	FSIM value for the SOTS dataset for the proposed work	325
Table 63.3	The average time taken in each step	325
Table 63.4	SSIM value for the SOTS dataset for proposed work	325
Table 63.5	F1 score for proposed work	325
Table 64.1	Comparison of power consumption and energy efficiency improvement across display technologies	329
Table 64.2	Comparison of power consumption and energy efficiency improvement across display technologies	330
Table 64.3	Comparison of viewing angle and glare reduction across display technologies	330
Table 64.4	Comparison of response time, refresh rate, and performance improvement across display technologies	330
Table 67.1	Antigen vs antibody	343
Table 67.2	PCD1 vs CTLA 1 identity matrix from align	345
Table 67.3	Keytruda vs Ipilimumab identity matrix from align	345
Table 69.1	Performance measures	354
Table 69.2	Confusion matrix of the proposed BESO-SVM	355
Table 69.3	Comparative analysis	355
Table 70.1	Comparison of adaptation speed across different smart lens technologies	360
Table 70.2	Comparison of lens and power consumption (mW)	360
Table 70.3	Visual clarity and adaptive contrast	360
Table 70.4	User comfort and long-term wearability	361
Table 71.1	Distribution of images	363
Table 71.2	Image enhancement methods	364
Table 71.3	Lists the SENet mobile characteristics	365
Table 71.4	SENet mobile confusing matrix	365
Table 71.5	SENet large confusing matrix	365
Table 71.6	Comparing the recommended strategy with different approaches	366
Table 72.1	Different types of converters and application	370
Table 72.2	Simulation parameters	371
Table 72.3	Comparison of proposed system with conventional methods	373
Table 73.1	Variables of the multi are pawer system with and without SSSC	377
Table 74.1	Parameters used for simulation	382
Table 75.1	Performance metrics of soldier BreathWaveNet vs Benchmark algorithms	388
Table 75.2	Comparative analysis of soldier BreathWaveNet with Benchmark algorithms	389
Table 76.1	Comparison of computation time (ms) and speed improvement (%)	394
Table 76.2	Differentiation of power consumption (W) and energy reduction (%)	395
Table 76.3	AI model accuracy enhancement	395
Table 76.4	Distinguishing the network throughput (Gbps) and scalability improvement(x)	395
Table 78.1	Parameters of new fitted distribution	404

Foreword

On behalf of the organizing committee of the International Conference on A Study on Next-Generation Materials and Devices (ICNMD, 2024), held from August 01-03, 2024, in Virudhunagar, India, we are pleased to present *A Study on Next-Generation Materials and Devices*. This volume captures the cutting-edge research presented at the conference, focusing on state-of-the-art developments for sustainable progress. The topics span critical areas including energy solutions, environmental challenges, advanced sensors, biomaterials, next-generation semiconductors, artificial intelligence, and more. The conference aimed to foster interdisciplinary collaboration and exchange of knowledge. We believe this handbook will serve as a valuable resource, offering key insights into emerging trends in materials science and engineering.

Preface

A Study on Next-Generation Materials and Devices proudly presents the proceedings of the International Conference on Next-Generation Materials and Devices (ICNMD, 2024) held from August 01–03, 2024, in Virudhunagar, India.

ICNMD 2024 served as a crucial platform, focusing on state-of-the-art research and development in A Study on Next-Generation Materials and Devices for sustainable development. The diverse program explored major topics such as energy solutions, environmental concerns, advanced sensors, the role of artificial intelligence, and computational approaches for materials design. It also delved into biomaterials for medical applications, alongside discussions on next-generation semiconductors, and flexible electronics poised to revolutionize the electronics industry. The event covered all the significant verticals related to materials and devices, featuring pioneers who shed light on uncharted domains.

Intended for a broad audience, including students, researchers, and industry professionals, the conference successfully fostered interdisciplinary collaboration and meaningful knowledge exchange. This handbook captures the breadth and depth of those discussions, serving as a valuable resource for academics and professionals in materials science and engineering. It offers insights into future trends and potential applications while promoting a collaborative environment for innovative solutions to complex material challenges.

Acknowledgements

We extend our sincere gratitude to all who contributed to *A Study on Next-Generation Materials and Devices*, representing the proceedings of the *International Conference on A Study on Next-Generation Materials and Devices (ICNMD, 2024)*. Special thanks to *Kalasalingam Academy of Research and Education* for hosting the event. We deeply appreciate the **authors** for their valuable contributions, as well as the *peer reviewers* and *Programme Committee members for their rigorous efforts* in maintaining academic quality. Finally, we acknowledge *T&F Conference Proceedings* for facilitating this publication.

1 Development of plant health monitoring device

B. Perumal[1,a], *Pallikonda Rajasekaran*[1,b], *A. Lakshmi*[2,c], *Rajesh V.*[3,d], *Krishna Priya R.*[4,e], *and Saravanan Velusamy*[5,f]

[1]Department of Electronics and Communication Engineering, Kalasalingam Academy of Research and Education, Krishnankoil, Tamil Nadu, India
[2]Department of Electronics and Communication Engineering, Ramco Institute of Technology, Rajapalayam, Tamil Nadu, India
[3]Department of Electronics and Communication Engineering, SRM Institute of Science and Technology, Tiruchirappalli, Tamil Nadu, India
[4]Head of Research and Consultancy Department, University of Technology and Applied Sciences, Musandam, Sultanate of Oman
[5]Department of Electrical and Electronics Engineering, University of Technology and Applied Sciences, Muscat, Oman

Abstract: The promise of precision agriculture to maximize resource use and boost agricultural yields has drawn a lot of interest. Monitoring plant health is crucial for effective farm management, yet traditional methods often lack precision and real-time data. In response, this study presents the development of a novel Plant Health Monitoring Device (PHMD) designed to address these limitations. The PHMD integrates advanced sensing technologies such as multispectral imaging, infrared thermography, and moisture sensors to provide comprehensive and real-time assessments of plant health indicators. Through the fusion of these data streams, the device offers insights into various aspects of plant physiology, including nutrient deficiencies, water stress, pest infestations, and disease outbreaks. Key features of the PHMD include its portability, ease of deployment, and compatibility with existing farm machinery and unmanned aerial vehicles (UAVs). Additionally, leveraging wireless connectivity and cloud-based platforms, the device facilitates smooth data processing and transfer, enabling farmers to make informed decisions promptly. Field trials conducted across diverse agricultural settings demonstrate the efficacy of the PHMD in identifying initial indications of abnormalities, thus enabling timely interventions to mitigate potential yield losses. Furthermore, its scalability and adaptability make it suitable for deployment in small-scale farms as well as large commercial operations.

Keywords: IoT, DC-to-DC converter, resistors, transistors, PHMD (plant health management division), smart farming, smart power management

1. Introduction

The agricultural sector faces numerous challenges in ensuring crop productivity and sustainability, with plant health being a critical factor influencing yields and overall farm profitability. Traditional methods of assessing [1] plant health often rely on visual inspection or sporadic sampling, which can be labour-intensive, high consumption of time and subjected to inaccuracies. Furthermore, the demand for improved and more precise methods of tracking is developing due to the effects of climate change and the intricate nature of contemporary agricultural practices [2]. In response to these challenges [3], there has been a surge in the development of advanced technologies for precision agriculture. In order to achieve optimal yields, eliminate input waste, and optimize resource allocation, these innovations seek to give farmers access to real-time data and relevant knowledge [4].

Central to this endeavor is the development of Plant Health Monitoring Devices (PHMDs) [5], which integrate cutting-edge sensing technologies to assess various indicators of plant health and vigor. This paper presents an overview of the development of a novel PHMD tailored for the needs of modern agriculture. The device combines state-of-the-art sensors, data analytics algorithms, and wireless communication capabilities to offer farmers a comprehensive and timely assessment of crop health status [6]. The PHMD offers vital data for enhancing irrigation plans, nutrient administration methods, and insect prevention tactics by constantly monitoring important biological variables like soil moisture levels, leaf temperature, and the amount of chlorophyll. The motivation behind the development of this PHMD stems from the recognition of the limitations of existing monitoring methods and the pressing need for more effective tools to support sustainable agricultural methods.

[a]perumal@klu.ac.in, [b]m.p.raja@klu.ac.in, [c]lakshmi@ritrjpm.ac.in, [d]rajeshv@srmist.edu.in, [e]krishna.priya@utas.edu.om, [f]saranhct@gmail.com

DOI: 10.1201/9781003675259-1

The PHMD seeks to strengthen decision-making, increase the effectiveness of resources, and eventually boost farming operation' robustness and revenue by providing farmers with real-time insights into plant health and stress levels [7].

In the following sections, we will delve into the design principles, technological components, and operational Through field trials and case studies [21].

2. Literature Review

The prior establishment of PHMDs, which aims to overcome the issues with conventional techniques of monitoring crop health and vigor, marks an important step forward in precision farming [8]. This review delivers an idea about existing research and developments in the field of PHMDs, focusing on key technologies, applications, and challenges. Sensor Technologies: Various sensing technologies have been integrated into PHMDs to capture different aspects of plant physiology. Multispectral imaging systems, for example, allow for the assessment of chlorophyll content and vegetation indices, providing insights into plant stress levels and nutrient deficiencies [3]. Infrared thermography sensors offer non-invasive methods for monitoring leaf temperature, which correlates with water stress and disease susceptibility [1]. Additionally, soil moisture sensors and nutrient analysers enable the monitoring of soil conditions, facilitating targeted irrigation and fertilization practices [4].

Processing and analyzing the enormous volumes of data produced by PHMDs requires the use of sophisticated data analytics methods, such as machine learning techniques [9]. ML models can detect patterns suggestive of particular plant illnesses or stressors by examining temperature patterns, spectral signatures, and other sensor data [10]. These predictive models enable early detection and timely interventions, thereby minimizing yield losses and optimizing resource utilization [9]. Wireless Communication and IoT Integration: PHMDs are often equipped with wireless communication capabilities [11], delivering real-time data transfer and remote access. Incorporation of Internet of Things (IoT) platforms allows for seamless connectivity with farm management systems and decision support tools [12]. Cloud-based storage and assessment platforms further enhance the scalability and accessibility of PHMD data, facilitating collaboration and knowledge sharing among stakeholders [7]. Field Validation and Performance Evaluation [12]: Field trials and validation studies are necessary for calculating the accuracy and effectiveness of PHMDs under real-world conditions. Researchers have conducted experiments across various crop types [13] and agricultural settings to evaluate the performance of PHMDs in detecting plant stress, nutrient deficiencies, and disease outbreaks [2]. These studies highlight the potential of PHMDs to improve crop management practices and enhance farm productivity. Challenges and Future Directions: Despite their promising potential, PHMDs face several challenges, including cost constraints, data interpretation complexities, and interoperability issues. Future research efforts should [14] focus on addressing these challenges through advancements in sensor technology, data analytics algorithms, and user interface design. Additionally, there is a need for interdisciplinary collaborations between agronomists, engineers [15].

3. Data and Variables

The DC to DC converter can be used in various applications, from low-power devices like small batteries to high-power systems for things like power transmission. The primary function of this converter is to change the input voltage to a different output voltage while maintaining the power efficiency of the system.

A resistor is a component in an electrical circuit designed to provide resistance, regulating the current flow. Fixed resistors are pre-determined during the design of the circuit and their resistance values should not be altered. They can be made from materials like carbon or metal and come in various sizes depending on their power rating [16]. Variable resistors, such as potentiometers, rheostats, and trimmers, have an adjustable resistance. They include a sliding mechanism that allows for changing the resistance value between two fixed points in the circuit.

BC547 is commonly used for tasks such as signal amplification, switching, and other processing applications. The BC547 is a general-purpose transistor, often chosen for low- to medium-power circuits due to its reliability and efficiency in various electronic systems [17].

RESISTOR: Fixed resistors are designed to set the right conditions in a circuit. Their values should never be changed to adjust the circuit since those were determined during the design phase. It can have a carbon composition or chip-and-wire wound type. It can also be made with a mixture of finely ground carbon or be very small in size and for high power rating [11]. Examples of this are potentiometers, rheostats, trimmers, and so on.

TRANSISTOR: BC547 is frequently used general-purpose NPN BJT in electronic devices for processing signals, conversion, and amplifying purposes [18]. Here's overview of its key characteristics and typical applications.

4. Methodology and Model Specifications

The working principle of a PHMD involves the integration of various sensors to measure variables related to plant physiology, coupled with data processing and analysis to evaluate the plant's condition. Sensor Integration: The PHMD incorporates multiple sensors designed to capture different aspects of plant health and environmental conditions. These sensors may include: Multispectral Imaging Sensor: Captures images of plants across different wavelengths of light, providing information on chlorophyll content, leaf area, and stress levels.

Infrared (IR) Thermography Sensor: Measures the temperature of plant leaves, which can indicate stress caused by water deficiency, disease, or pest infestation. Soil Moisture Sensor [19]: Determines the soil moisture, helping to optimize watering schedule and prevent overwatering or underwatering [8]. Nutrient Sensors: Measure levels of essential nutrients in the soil or plant tissues, aiding in fertilizer management and nutrient uptake analysis.

Data Acquisition: The sensors collect raw data corresponding to the measured parameters, such as spectral reflectance, temperature readings, moisture levels, and nutrient concentrations Data Processing and Analysis: The raw sensor data is processed and analysed to extract meaningful insights regarding plant health and environmental conditions. This may involve: Feature Extraction: Identifying relevant features or patterns from the sensor data, such as spectral signatures indicative of nutrient deficiencies or temperature differentials associated with water stress [14].

The development of algorithms performed an essential part in interpreting sensor data and determining the health status of plants according to set criteria [20]. Machine learning methods can be applied to train models using labelled data, helping to predict plant health based on various indicators. One key aspect is setting thresholds for different plant health metrics, allowing the system to find problems like lack of nutrition, pest swarms, or occurrences of disease when conditions fall outside of normal ranges.

Decision support systems then offer practical recommendations to farmers or agronomists based on the analyzed data.

Data visualization is another important component, where the analyzed data is displayed in an easy-to-understand format—such as graphs, charts, or maps. This helps farmers to quickly assess the health of their crops and deliver informed decisions. Real-time monitoring tools, including dashboards or mobile apps, enhance this process by enabling remote access and collaboration with other stakeholders.

Additionally, the system incorporates feedback and adaptation mechanisms. Continuous monitoring allows the PHMD to adjust its recommendations based on changing environmental conditions and improve over time. Farmers can provide input based on their observations, which helps refine the system's predictive accuracy and overall effectiveness.

5. Empirical Results

Development of the PHMD reached a significant milestone with successful field trials, offering important insights into its effectiveness and potential impact on farming practices. During the testing phase, the PHMD proved its ability to accurately measure various indicators of plant health and stress, providing real-time data that is essential for making informed decisions. In trials conducted in different agricultural environments, the device consistently delivered reliable

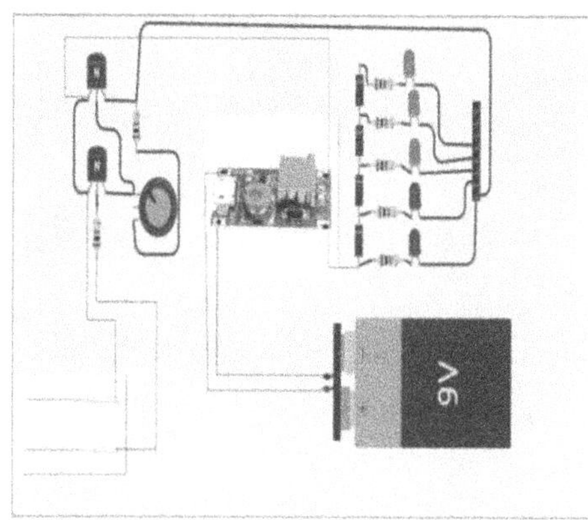

Figure 1.1. Circuit diagram of plant health monitoring device.

Source: Author.

results in detecting early signs of stress, nutrient imbalances, and diseases. Figure 1.1 depicts the circuit diagram of plant health monitoring device.

The use of multispectral imaging sensors allowed for detailed analysis of chlorophyll content and leaf area, helping monitor plant health and stress levels precisely. Additionally, infrared thermography sensors provided valuable data on leaf temperature, which helped identify water stress and potential disease risks.

A key finding from the trials was the PHMD's ability to enhance resource management, particularly in irrigation. By continuously monitoring soil moisture, the device enabled farmers to apply targeted irrigation, reducing water wastage and preventing overwatering or underwatering. This not only improved water efficiency but also contributed to healthier crops and better yields. The integration of nutrient sensors allowed the PHMD to evaluate soil fertility and nutrient levels, helping farmers adjust fertilizer use and improve nutrient management. By identifying deficiencies early, the device prevented yield loss and enhanced overall crop productivity.

The PHMD's wireless communication capabilities were crucial in ensuring smooth data transmission and enabling remote monitoring. This allowed farmers to access real-time information and make timely interventions. Cloud-based platforms for data storage and analysis further facilitated scalability and collaboration, allowing stakeholders to share knowledge and insights. The development of the PHMD culminated in successful field trials, providing valuable insights into its efficacy and potential impact on agricultural practices. Throughout the testing phase, the PHMD demonstrated its capability to accurately assess various indicators of plant health and stress, offering real-time data important for informed decision-making. In field trials conducted across diverse agricultural settings, the PHMD consistently delivered reliable results in identifying prompt indications of stress, nutrient deficiencies, and disease outbreaks. Multispectral imaging sensors captured detailed information about chlorophyll content and leaf area, enabling precise

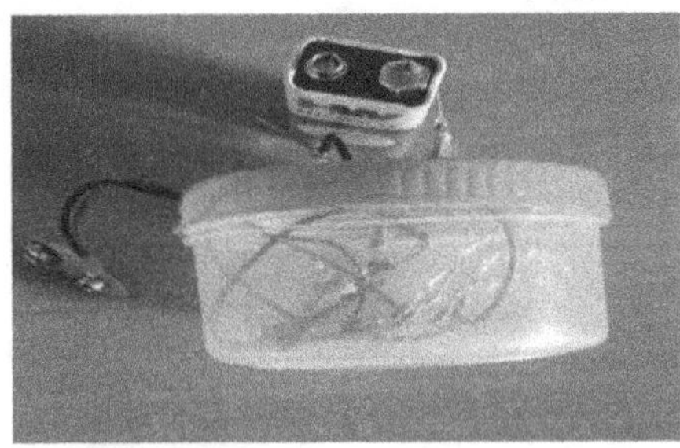

Figure 1.2. Plant health monitoring device.

Source: Author.

monitoring of plant vigor and stress levels. Additionally, infrared thermography sensors provided crucial data on leaf temperature, aiding in the identification of water stress and disease susceptibility.

The field trials demonstrated the PHMD's potential to transform plant health monitoring and management in agriculture. By providing actionable insights and supporting data-driven decisions, the device can boost crop productivity, optimize resource use, and encourage sustainable farming practices. Ongoing research and development are needed to refine its capabilities and ensure its widespread adoption in the agricultural industry.

6. Conclusion

The creation of a PHMD marks a major step forward in precision agriculture, providing farmers with an advanced tool to boost crop yield, improve resource efficiency, and support sustainable farming practices. By incorporating cutting-edge sensors, data analysis techniques, and wireless communication, the PHMD offers real-time information on critical factors affecting plant health and environmental conditions. The need for more accurate and efficient monitoring solutions motivated the development of this device, as traditional methods of assessing plant health have limitations. The PHMD tracks key indicators while field testing, showing the device's effectiveness in improving farming practices and reducing crop losses. With timely interventions and precise treatments, the PHMD helps farmers make better-informed decisions that optimize yields, reduce waste, and lessen environmental impact. Moving forward, additional research and development are necessary to eradicate the challenges such as cost, data analysis complexities, and device compatibility. Collaboration among agronomists, engineers, and data experts will be crucial for advancing the technology and ensuring its successful integration into agricultural decision-making. Overall, the PHMD represents a promising innovation in agriculture, offering farmers a valuable tool to enhance crop production, promote environmental sustainability, and contribute to global food security efforts. The developed plant health monitoring device is illustrated in Figure 1.2.

7. Acknowledgement

The authors thank the International Research Centre of Kalasalingam Academy of Research and Education, Tamil Nadu, India, for permitting the use of the computational facilities available in the Centre for Biomedical Research and Diagnostic Techniques Development.

References

[1] Ranjan, R., Jha, P., Kumar, V., & Lal, B. (2018). Role of infrared thermography for crop water stress detection: A review. *Agricultural Reviews*, *39*(4), 263–270.

[2] Sankaran, S., Mishra, A., Ehsani, R., & Davis, C. (2015). A review of advanced techniques for detecting plant diseases. *Computers and Electronics in Agriculture*, *72*, 1–13.

[3] Dash, S., Yadav, R., Mishra, S., & Yadav, G. (2017). Plant Disease Detection Using Image Processing: A Review. Xiv preprint arXiv:1708.03466.

[4] Gautam, N., Singh, R., Pandey, R., & Singh, S. K. (2019). Wireless sensor networks for precision agriculture: A review. *Computers and Electronics in Agriculture*, *157*, 436–448.

[5] Singh, A., Ganapathy Subramanian, B., & Sarkar, S. (2016). Machine learning for high-throughput stress phenotyping in plants. *Trends in Plant Science*, *21*(2), 110–124.

[6] Pandian, S. R., & Ravindran, A. D. (2017). Precision agriculture based on Internet of Things. *Procedia Computer Science*, *133*, 295–302.

[7] Kamble, S. S., Patil, S. S., & Patil, A. V. (2020). Cloud Based IoT Platform for Precision Agriculture: A Review. *IJERT*, *9*(8), 706–712.

[8] Mahalenin, A. K. (2016). Plant disease detection by imaging sensors–parallels and specific demands for precision agriculture and plant phenotyping. *Plant Disease*, *100*(2), 241–251.

[9] Liu, C., Zhou, X., Zhou, Y., & Xie, X. (2019). Hyperspectral imaging for plant disease detection: From sensor systems to practical applications. *Remote Sensing*, *11*(8), 812.

[10] Rathore, A., Kar, S., Kala, R., & Kumar, S. (2020). Machine Learning for Precision Agriculture: A Survey. arXiv preprint arXiv:2004.12537.

[11] Wang, W., & Yang, J. (2019). Precision agriculture technology for crop farming. *Precision Agriculture Technology for Crop Farming*, 1–336.

[12] Noh, H. R., Hong, S. G., Yoon, Y. M., & Hong, S. J. (2021). An IoT-based plant disease prediction and monitoring system using deep learning. *Computers and Electronics in Agriculture*, *184*, 106122.

[13] Tadesse, A. W., & Alharthi, A. I. (2019). Advances and future trends in precision agriculture technologies for crop monitoring. In *Emerging Technologies and Their Applications in Agriculture* (pp. 217–239). Springer.

[14] Nagarajan, V. K., Raj, R., Laha, P., & Laha, D. K. (2018). Precision agriculture technologies and factors affecting their adoption: A review. *Computers and Electronics in Agriculture*, *155*, 26–35.

[15] Patel, J., & Kar, A. (2016). A review on methods for detection of plant diseases. *International Journal of Computer Applications*, *134*(1), 1–6.

[16] Feng, Z., & Krebs, M. (2017). Plant sensors for disease detection: Review and future perspectives. *Trends in Biotechnology*, *35*(4), 335–341.

[17] Lobos, G. A., Poblete-Echeverría, C., & Ahumada, M. (2017). Precision agriculture: A review of methods for data acquisition and analysis on field crops. *Chilean Journal of Agricultural Research*, *77*(2), 172–184.

[18] Mahlein, A. K., Oerke, E. C., Steiner, U., & Dehne, H. W. (2012). Recent advances in sensing plant diseases for precision crop protection. *European Journal of Plant Pathology*, *133*(1), 197–209.

[19] Perumal, B., Nagarai, P., Venkatesh, R., Muneeswaran, V., GopiShankar, Y., SaiKumar, A., Koushik, A., & Anil, B. (2022). Real time transformer health monitoring system using IoT in r. In *2022 International Conference on Computer Communication and Informatics (ICCCI)* (pp. 1–5).

[20] Singh, A., Seres, G., & Wolpert, T. (2016). Application of spectral reflectance for detection of grapevine leafroll disease in two red-berried grapevine cultivars. *Precision Agriculture*, *17*(3), 269–284.

2 Development of MEMS-based sensors for industrial applications

Aakansha Soy[1,a] *and Ahilya Dubey*[2,b]

[1]Assistant Professor, Department of CS & IT, Kalinga University, Raipur, India
[2]Research Scholar, Department of CS & IT, Kalinga University, Raipur, India

Abstract: MEMS-based gadgets are very useful in modern industry because they are small, accurate, and use little power. These monitors let you keep an eye on important things like pressure, temperature, and movements in real time. This makes things run more smoothly and safely. Recent progress in both making sensors and processing data has directly led to better metal-organic semiconductor (MEMS) sensor performance in a wide range of uses. The main goal of this paper is to find out how adding MEMS sensors to industrial systems affects tracking and making decisions. The study also tries to guess what problems and new developments will happen in MEMS technology in the future. This should be used as a plan for how it will grow and get better in the future.

Keywords: MEMS, sensors, industrial applications, microfabrication, automation

1. Introduction

Industry likes MEMS sensors because they are small, work well, and don't cost a lot. Because they accurately measure pressure, temperature, acceleration, and chemical concentrations, these sensors are very important in industry, healthcare, transportation, and aeroplanes [9]. Because they are small, they are easy to add to complex systems, which improves performance [6]. Even though sensor technology is useful, it has problems with sensitivity, accuracy, and reliability in many situations. Real-time, high-resolution, low-power monitors are needed because technology changes so quickly. MEMS technology gets around these problems by making it possible to make things smaller without lowering their performance. Design and manufacturing improvements that make MEMS devices faster and more sensitive are making them more useful in industrial settings [1]. The unique mechanical and electrical features of MEMS sensors can improve production and automation, changing enterprises. Improved sensor performance and applications from MEMS technology research will improve industrial processes [5].

2. Methodology

2.1. MEMS technology

MEMS sensors are microfabricated using photolithography, etching, and depositing. These methods enable precise micro-scale structure creation and physical event detection. Surface and bulk micromachining produce accurate and reliable accelerometers, gyroscopes, and pressure sensors. Motion detectors, orientation sensors, pressure sensors, microphones, and fluid dynamics monitors are some MEMS sensors utilized in diverse industries. Figure 2.1 shows various types of MEMS sensors and their respective applications across domains. Each form of MEMS technology has specialist uses, from automotive safety systems to healthcare diagnostics [2].

2.2. Sensor integration

MEMS sensors can connect with microcontrollers or PLCs in embedded applications, allowing them to be used in industrial systems. Integration enables real-time control and monitoring, improving operational efficiency. UART, SPI, and I2C are used to communicate sensor data to control systems [7]. MEMS sensors must be calibrated for accuracy and reliability. For calibration, sensor outputs are compared to standards and measurements altered. Sensors undergo multiple tests to ensure they can endure temperature and mechanical stress [3].

Figure 2.1. MEMS-based sensors.

Source: https://www.analogictips.com/mems-pressure-sensors-feature-fast-response-high-resolution-long-term-stability/

[a]ku.aakanshasoy@kalingauniversity.ac.in, [b]ahilya.dubey@kalingauniversity.ac.in

DOI: 10.1201/9781003675259-2

2.3. *System architecture*

MEMS sensors are installed across a facility to collect data on numerous parameters for an industrial monitoring system. These sensors send data to a CPU or cloud-based system for analysis. Dashboards and mobile apps give operators real-time insights and alarms [4]. MEMS sensor data is preprocessed to remove noise and extract useful features. Trend analysis and system performance prediction can be done with machine learning algorithms. This data-driven strategy improves operational efficiency and predictive maintenance by improving decision-making.

2.4. *Uses*

- Improved accuracy and reliability in industrial monitoring.
- Enhanced automation and control.
- Reduced maintenance costs.
- Configure the feedback system to react to sensor data in real time and adjust operational parameters from analysis.
- Allow operators to monitor system performance and make decisions using dashboards or applications' data and alarms.

3. Working

3.1. *Data acquisition and signal processing*

MEMS sensors are installed across a facility to collect data on numerous parameters for an industrial monitoring system. These sensors send data to a CPU or cloud-based system for analysis. Dashboards and mobile apps give operators real-time insights and alarms [8]. Figure 2.2 shows the architecture of a MEMS-enabled industrial monitoring system with data flow from sensors to analytics platforms. MEMS sensor data is preprocessed to remove noise and extract useful features. Trend analysis and system performance prediction can be done with machine learning algorithms. This data-driven strategy improves operational efficiency and predictive maintenance by improving decision-making.

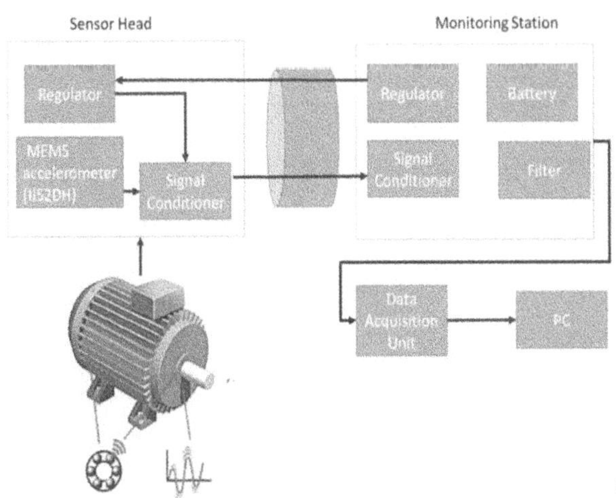

Figure 2.2. System architecture MEMS-based sensors.

Source: Author.

3.2. *Real-time monitoring and feedback*

The system processes data to monitor industrial conditions in real time. Operators can quickly check stats with dashboards and mobile apps. The system can use automated feedback mechanisms to warn or adjust at specified criteria. If a pressure sensor reads too high, the system might turn off machines or alert for repair. The feedback loop and real-time monitoring make industrial settings safer, more efficient, and less downtime-prone.

4. Algorithm

Collect raw MEMS sensor data continuously or periodically.

- Reduce noise and improve signal quality with low-pass and high-pass digital filtering.
- Normalize or scale data to ensure consistency among sensors.
- Select the most important peaks, trends, and averages from the filtered data for study.
- Analyse retrieved features using statistical or machine learning methods for insights, anomalies, or predictive maintenance.
- Send processed sensor data to the control system via I2C, SPI, or MQTT.
- Processed data should be used in system control algorithms. Temperatures above a specific level can activate cooling systems or alarms.

5. Conclusion

New MEMS sensors have transformed industrial automation and monitoring. These miniaturized, high-sensitivity, low-power sensors improve data collection in many applications. MEMS technology increases industrial system efficiency, safety, and dependability by providing real-time insights into key parameters. There's hope for MEMS sensors. Advanced sensing capabilities, such as multi-modal sensors that monitor several parameters, need more investigation. MEMS sensor deployment for remote monitoring may benefit from wireless communication technology advances. Predictive analytics and autonomous decision-making enabled by MEMS technology, AI, and ML may soon make industrial surroundings smarter.

References

[1] Eckhart, M., Brenner, B., Ekelhart, A., & Weippl, E. (2019). Quantitative security risk assessment for industrial control systems: Research opportunities and challenges. *Journal of Internet Services and Information Security*, *9*(3), 52–73.

[2] Bhatt, G., Manoharan, K., Chauhan, P. S., & Bhattacharya, S. (2019). MEMS sensors for automotive applications: A review. *Sensors for Automotive and Aerospace Applications*, 223–239.

[3] Bordel Sánchez, B., Alcarria Garrido, R. P., Sánchez de Rivera, D., & Sánchez Picot, Á. (2016). Enhancing process control in industry 4.0 scenarios using cyber-physical systems. *Journal of Wireless Mobile Networks, Ubiquitous Computing, and Dependable Applications*, *7*(4), 41–64.

[4] Hajare, R., Reddy, V., & Srikanth, R. (2022). MEMS based sensors–A comprehensive review of commonly used fabrication techniques. *Materials Today: Proceedings*, *49*, 720–730.

[5] Surendar, A., Saravanakumar, V., Sindhu, S., & Arvinth, N. (2024). A bibliometric study of publication-citations in a range of journal articles. *Indian Journal of Information Sources and Services*, *14*(2), 97–103. https://doi.org/10.51983/ijiss-2024.14.2.14

[6] Algamili, A. S., Khir, M. H. M., Dennis, J. O., Ahmed, A. Y., Alabsi, S. S., Ba Hashwan, S. S., & Junaid, M. M. (2021). A review of actuation and sensing mechanisms in MEMS-based sensor devices. *Nanoscale Research Letters*, *16*, 1–21.

[7] Cao, Y., Dhahad, H. A., Alsharif, S., Sharma, K., Shafy, A. S. E., Farhang, B., & Mohammed, A. H. (2022). Multi-objective optimizations and exergoeconomic analyses of a high-efficient bi-evaporator multigeneration system with freshwater unit. *Renewable Energy*, *191*, 699–714.

[8] Cao, Y., Dhahad, H. A., Sharma, K., ABo-Khalil, A. G., El-Shafay, A. S., & Ibrahim, B. F. (2022). Comparative thermoeconomic and thermodynamic analyses and optimization of an innovative solar-driven trigeneration system with carbon dioxide and nitrous oxide working fluids. *Journal of Building Engineering*, *45*, 103486.

[9] Veera Boopathy, E., Peer Mohamed Appa, M. A. Y., Pragadeswaran, S., Karthick Raja, D., Gowtham, M., Kishore, R., Vimalraj, P., & Vissnuvardhan, K. (2024). A Data Driven Approach through IOMT based Patient Healthcare Monitoring System. *Archives for Technical Sciences*, *2*(31), 9–15.

3 An analysis of deep learning techniques used for semantic segmentation of remote sensing images

Roshni Rajendran[1,a] *and P. Nagaraj*[2,b]

[1]Research Scholar, Department of Computer Science and Engineering, Kalasalingam Academy of Research and Education, Krishnankoil, Virudhunagar, India
[2]Associate Professor, Department of Computer Science and Engineering, Kalasalingam Academy of Research and Education, Krishnankoil, Virudhunagar, India

Abstract: Remote sensing imagery has significant roles in various domains, such as environmental monitoring, law enforcement, and disaster response. In any case, the customary techniques for distinguishing proof of items in these pictures are frequently unreasonable because of the broad geographic region and predetermined number of accessible experts. Thusly, there is a requirement for robotization. Conventional techniques for object recognition and arrangement miss the mark with regards to precision and dependability. Utilizing a few AI strategies this challenge can be tended to. Profound learning has displayed wonderful abilities in picture grasping, especially through semantic division calculations. This examination study centers around the acknowledgment of items and designs in remote detecting symbolism. The proposed framework uses progressed semantic division calculations like Deeplab, Refinet, and Pspnet to accomplish exact and precise semantic division in and remote detecting pictures. By utilizing the force of profound learning, the point is to robotize and improve the proficiency of article acknowledgment in remote detecting pictures, empowering more successful applications in misfortune reaction, policing, ecological checking.

Keywords: Remote sensing images, deep learning, semantic segmentation

1. Introduction

Image segmentation is pointed toward separating a picture into unmistakable and homogeneous areas in view of its inside qualities. One famous methodology, semantic segmentation, utilizes profound learning calculations to mark or sort every pixel in each picture. This method empowers the separation of different pixel bunches having a place with various classifications. For instance, in independent vehicles, semantic division is used to distinguish vehicles, walkers on road/path, traffic signs, other street components. Semantic segmentation remembers applications for different fields including driving, medical imaging. Image segmentation plays an urgent part in computer vision and digital image processing by relegating names to individual pixels inside an image. Its primary goal is to accurately assign a particular class to a set of pixels based on their discernible characteristics. There are two main types of image segmentation: instance segmentation, which aims to identify each instance or object with a unique label, and semantic segmentation, which aims to identify and label all instances and objects belonging to a specific category as a whole. Some image classification techniques based on CNN are widely studied. It includes CNN to classify images. The CNN consists of a lot of layers and finally produces predicted probability as a result that leads to knowing and classifying the input images.

While image semantic segmentation can effectively guide human activities in remote sensing applications, current techniques often struggle to meet the high-precision demands required for complex images. This motivates the exploration of a novel approach to enhance the precision of segmentation by leveraging Deeplab, Refinet, and PSPNet algorithms.

The CNN consists of a lot of layers and finally produces predicted probability as a result that leads to knowing and classifying the input images.

While image semantic segmentation can effectively guide human activities in remote sensing applications, current techniques often struggle to meet the high-precision demands required for complex images. This motivates the exploration of a novel approach to enhance the precision of segmentation by leveraging Deeplab, Refinet, and PSPNet algorithms. The CNN consists of a lot of layers and finally produces predicted probability as a result that leads to knowing and classifying the input images. While image semantic segmentation can effectively guide human activities in remote sensing applications, current techniques often struggle to meet the high-precision demands required for complex images. This motivates the exploration of a novel approach to enhance the precision of segmentation by leveraging Deeplab, Refinet, and PSPNet algorithms.

[a]roshnirajendran24@gmail.com, [b]nagaraj.p@klu.ac.in

DOI: 10.1201/9781003675259-3

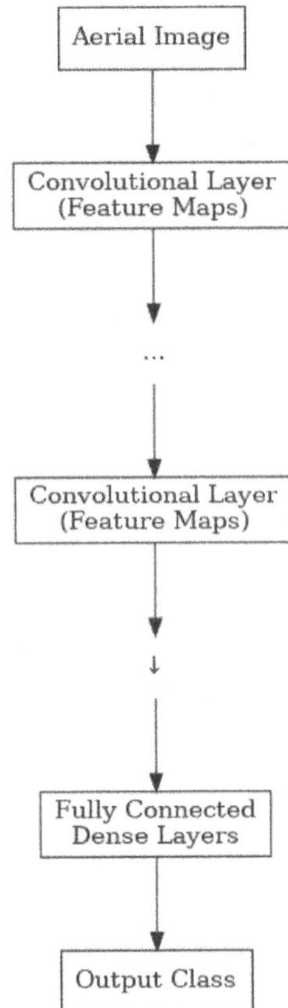

Figure 3.1. CNN-based image classification.

Source: Author.

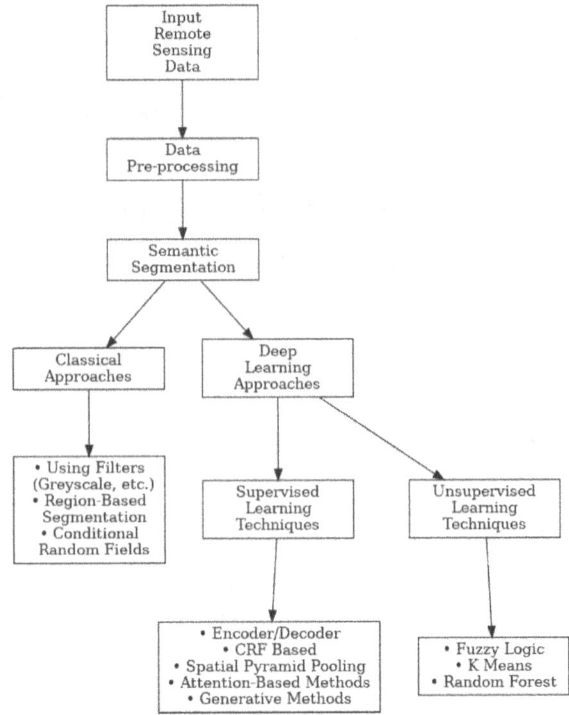

Figure 3.2. Basic block diagram of semantic segmentation of remote sensing images.

Source: Author.

Figure 3.1 depicts the process in CNN-based image classification. Deeplab, one of the considered algorithms, employs multiple atrous rates in its modules and uses atrous convolution in a cascade or parallel manner to collect information at different scales. This addresses the challenge of segmenting objects of varying sizes. Additionally, Deeplab incorporates image-level features to improve efficiency and include global context. PSPNet, also known as Pyramid Scene Parsing Network, is a semantic segmentation method that utilizes a pyramid parsing module to collect global context information from different regions. By combining local and global cues, PSPNet achieves more accurate predictions. Deeplab, one of the considered algorithms, employs multiple atrous rates in its modules and uses atrous convolution in a cascade or parallel manner to collect information at different scales. This addresses the challenge of segmenting objects of varying sizes. Additionally, Deeplab incorporates image-level features to improve efficiency and include global context.

PSPNet, also known as Pyramid Scene Parsing Network, is a semantic segmentation method that utilizes a pyramid parsing module to collect global context information from different regions. By combining local and global cues, PSPNet achieves more accurate predictions. Figure 3.2 represents the basic block diagram of semantic segmentation of remote sensing images

Refinet, the third algorithm, excels in enhancing the precision of fine-grained spatial details in saliency detection. Its compatibility with deep learning-based models allows it to refine object regions identified by these models, which often lack precise spatial details or produce blurry object boundaries.

However, Refinet's performance improves the accuracy of the saliency maps. If the target object cannot be accurately located in the saliency map, then the provision of additional spatial details becomes futile, resulting in limited enhancement of saliency map quality.

The rest of the paper is organized as follows. Section I gives an introduction to the subject area. Section II comprises related studies and section III consists of a conclusion.

2. Related Works

Several deep learning methods are used for doing semantic segmentation in remote sensing images. Some of them are discussed below.

From the study of Thitisiriwech et al. [1], they observed that transferring pre-trained weights from the Tiramisu model on cityscapes images to an unlabeled dataset leads to poor segmentation results. To overcome this challenge, the study introduces two solutions. Firstly, they propose an enhanced version of DeepLab-V3+ called DeepLab-V3-A1, which incorporates the Xception module. By adding 11 convolution kernels to the decoder side, significant performance improvements are achieved over the original DeepLab-V3+. The effectiveness of these techniques is evaluated by the validation set of the cityscapes dataset, and their applicability to other 2D semantic segmentation datasets is demonstrated.

Jingchun Zhou et al. [2] say that the proposed enhancement for PSPnet involves the integration of the global pyramid pooling feature. PSPnet is originally designed with an FCN module that incorporates a pyramid structure to gather information and then give valuable characteristics used for pixel classification. In the context of multi-focus image fusion, accurately determining whether a pixel belongs to the focused or non-focused area is crucial for achieving optimal fusion results. To address this task, a scene parsing-based image segmentation method utilizing PSPnet is employed.

By leveraging the hierarchical global context provided by PSPnet, the focused area in the source image can be precisely classified, preventing the loss of contextual information across different regions. The pyramid pooling module of PSPnet generates four pooling scales, creating representations for different sub-regions. These representations are then combined to reconstruct the final feature representation, encompassing both local and global contextual information.

Liang Zhang et al. [3] discuss about in the aerospace field, the image-collecting module produces image data, necessitating the exploration of efficient and rapid processing techniques. Deep learning advancements have made remarkable contributions to image classification tasks. This study applies a deep learning approach to classify images generated by optical measurement equipment.

Initially, a binary image classification network is constructed using a residual network, which serves as a versatile framework for image classification. The network is specifically tailored to distinguish rocket images from other types of images. To enhance generalization performance, especially for challenging images, a modified loss function based on binary cross-entropy is employed.

Charming Shen et al. [4] discuss that cloud cover percentage is a crucial indicator in the analysis of satellite imagery. Currently, cloud cover assessment is predominantly performed manually at ground stations, which can be a time-consuming process.

Various temporal information is considered and the dataset is labelled. The labels are assigned based on predefined criteria, where A represents no clouds, B indicates a minimal number of clouds, and C, D, and E correspond to increasing levels of cloud coverage. The * label is used to signify no-data instances, which can occur when there is a temporary interruption in data collection due to sensor switching. Manual assessment has drawbacks, including its labour-intensive nature and the potential for subjective judgments leading to inaccuracies. The algorithm utilizes parallel VGG-16 networks. By combining state-of-the-art techniques, this algorithm demonstrates reasonable results in cloud cover assessment.

Postadjian et al. [5] say that classification plays a crucial role in the creation of land cover maps. Deep neural networks, known for their exceptional performance in tasks such as semantic segmentation and speech recognition, have surpassed traditional classifiers in various machine learning challenges. As a result, these techniques have become popular in land-cover mapping.

These difficulties incorporate the expanded degree of detail because of upgraded goal, the presence of items with differing scales and directions inside a similar scene, and the intricacy of high-goal pictures containing numerous semantic classes. To address these difficulties, a powerful way to deal with include portrayal is vital. These difficulties incorporate the expanded degree of detail because of upgraded goal, the presence of items with differing scales and directions inside a similar scene, and the intricacy of high-goal pictures containing numerous semantic classes. To address these difficulties, a powerful way to deal with include portrayal is vital.

The first satellite picture is distorted into different scales, and each scale is utilized to prepare a DCNN. To accelerate the preparation cycle, a DCNN with SPP is thought of. Then, the multiscale satellite pictures are taken care of into their separate SPP nets to acquire features at various scales.

These features catch important data at various scales. To make a complete element portrayal, a various bit learning strategy is proposed. This technique thinks about these highlights. By utilizing SPP nets and multiple kernel learning, it works on the productivity and viability of element extraction and order.

Postadjian et al. [6] discuss the performance of DCNN in various tasks, including object classification, and semantic labeling. The architecture culminates with a full connected layer that combines information from all features in the last convolutional layer. The model can handle the unique characteristics and variations in land-cover classes across different regions, enabling accurate classification at a larger scale.

Patrick Helber et al. [7] discuss the task of classifying Sentinel-2 satellite images presents a significant challenge. The dataset consists of Sentinel-2 satellite images. This dataset covers 13 spectral bands, providing rich and comprehensive information for classification.

The availability of accurate and efficient classification systems based on deep CNNs can contribute to better understanding and management of land resources, as well as support decision-making processes in diverse fields related to Earth observation.

Keyang Cai et al. [8] discuss the classification of satellite images plays a crucial role in meteorological forecasting. However, the advent of CNN has revolutionized multi-layer learning algorithms. In a CNN, local image regions serve as inputs at the bottom layer of the hierarchical structure, and the extracted features are progressively transferred to higher layers. Each layer applies digital filters to identify data characteristics. This approach effectively handles observations with variations in scaling, rotation, and other factors.

A dedicated convolutional neural network is constructed for cloud classification, enabling automatic feature learning and accurate classification results. This method demonstrates high precision and robustness in cloud classification tasks. By leveraging the power of deep CNNs, cloud classification in satellite images can be significantly enhanced. The automated feature learning capabilities of the CNNs facilitate improved performance and more accurate cloud classification, ultimately benefiting meteorological forecasting and related applications.

Ulku [9] explains in this study about various segmentation models for pixel-wise tree classification are considered. For that classification using multispectral imagery is considered.

Also, a comparison with various algorithms is considered to analyze the precision of each algorithm. The author considers geometrical complexity in the dataset. Due to their expensive nature, the satellite images are not considered here but the aerial images are considered. Even it is necessary to utilize a very high-resolution image data set for training because of the architecture used here. For properly discriminating trees NIR reflectance is used. Pretraining is done with Resnet-34.

Mykola Lavreniuk et al. [10] consider the production of crop classification maps using high-resolution images. However, several challenges arise in their creation, such as the collection of input images, validation datasets, and the presence of clouds. To enhance the efficiency of ground data utilization, classifiers are developed that can be trained using data collected in past years. In this study, a deep learning method is explored for generating crop classification maps. The aim is to employ a deep learning approach based on sparse autoencoders.

Table 3.1 summarizes the methods used for semantically segmenting the RS imagery based on FCN. Initially, the autoencoder is trained solely on satellite data, capturing the underlying patterns and features. Subsequently, fine-tuning of the neural network is performed using in-situ data from the previous year. By utilizing this approach, the need for collecting annual in-situ data for the same territory can be avoided, considering the time-consuming and challenging nature of ground truth data collection. Overall, this work demonstrates the effectiveness of using a deep learning approach, specifically employing sparse autoencoders, to generate crop classification maps. By utilizing in-situ information from earlier years, the proposed methodology offers a proficient arrangement that beats the difficulties related with ground information assortment, at last supporting reasonable land the executives rehearse.

Lior Bragilevsky et al. [17] says that RS image analysis plays an important part in different domains, including checking cataclysmic events, predicting quakes and torrents, following boats and route, and concentrating on the impacts of environmental change. The Amazon locale, known for its enormous size and far off regions, requires a high-level satellite imaging framework to acquire significant data. Extracting useful insights from RS images can contribute to observing and comprehending the dynamic nature of the Amazon basin, aiding in the management of deforestation and its far-reaching consequences. The primary objective of this competition was to develop well-trained models that could enhance the efficiency of classifying the particular land and provide insights for more effective management strategies.

Z. Nong et al. [18] introduced Encoder-decoder structures like U-Net often struggle to preserve local details due to the downsampling operations involved. To address this, attention mechanisms (AMs) can be used to enhance low-level features, aiding in the recovery of local details that are lost. Additionally, incorporating Boundary Attention Modules (BAM) can improve the segmentation process by providing strong boundary information. Experiments conducted on very high-resolution (VHR) data have demonstrated the

Table 3.1. Summary of methods used for semantically segmenting the RS imagery based on FCN

References	*Dataset*	*Data*	*Features extracted*	*Results*
Z. Shao et al. [11]	UC Merced archive	Land cover	Colour histogram,Gabor texture, GIST, BoVW	Accuracy: 71.93% Precision: 81.78% Recall: 88.70% F1 Score: 0.8150
L. Li et al. [12]	Landsat-8 Biome Cloud Validation Masks	Cloud images	Infrared features	Pixel accuracy: 85.14 Mean IoU: 69.19
N. Zang et al. [13]	ISPRS Potsdam	Vehicle Instance segmentation	Colour histogram, Multilevel contextual features	Accuracy: 99.79 F1 Score: 0.9343
D. Pan et al. [14]	VHR datasets	Roads	Textures, colours, shape etc.	Completeness:97.6% Correctness: 98.2% Quality:95.8%
M. Helleis et al. [15]	Sentinel-1 dataset	Flood images	Multitemporal intensity	IoU: 73.1
B. Guo et al. [16]	Pascal VOC2012 Dataset	River images	Edge features	MIoU: 82.88% Pixel accuracy: 84.91% Precision: 84.06%

Source: Author.

effectiveness of this approach, particularly in recovering object boundaries, such as those of man-made structures with well-defined edges. However, the reliance on ground-truth images with finely annotated boundaries poses a significant challenge. Possible solutions to this limitation include utilizing a combination of colour, texture, and shape features to reduce the dependency on such high-quality annotations.

3. Evaluation Metrics

Liu et al. [19] suggested some measures that are widely used for segmentation in images are considered.

a. Pixel accuracy: One of the simplest metrics is pixel accuracy. It gives the relation between total pixels and classified pixels.
b. mIoU: The mean intersection over union is the ratio of the intersection of pixels to be categorized and the ground truth pixels and their union.
c. Recall (Sensitivity): It is a ratio of true positives to the instances which are negatively predicted. It tells about the correctly classified remote sensing images.

$$Recall = \frac{TP}{TP+FN}$$

d. Precision: It is the ratio of correctly predicted and wrongly predicted instances.

$$Precision = \frac{TP}{TP+FP}$$

e. F2 Score: It is derived from F1 score. It is obtained from precision and recall. By giving more importance to recall, the F1 score increases the chances of correctly predicting the images.
f. Accuracy: It is based on the accurate classifications made.

$$Accuracy = \frac{TP+TN}{TP+FP+TN+FN}$$

4 Conclusion and Future work

This paper reviews the semantic segmentation of remote sensing images. In this study, several machine learning systems that classify images in remote sensing imagery are considered. The remote-sensing image contains a vast amount of information; however, the sample size is highly uneven. As a result, while conventional networks can partially segment remote sensing images, there is significant room for improving the segmentation accuracy.

The future work includes a semantic segmentation system by improving the segmentation accuracy even if the sample size is uneven. For that very effective segmentation algorithms like Deeplab, Refinet, and Pspnet are expected to be considered.

References

[1] Thitisiriwech, K., Panboonyuen, T., Kantavat, P., Iwahori, Y., & Kijsirikul, B. (2022). The Bangkok Urbanscapes Dataset for semantic urban scene understanding using enhanced encoder-decoder with atrous depthwise separable A1 convolutional neural networks. *IEEE Access, 10*, 59327–59349.

[2] Zhou, J., Hao, M., Zhang, D., Zou, P., & Zhang, W. (2019). Fusion PSPNet image segmentation-based method for multi-focus image fusion. *IEEE Photonics J, 11*(6), 1–12.

[3] Zhang, L., Chen, Z., Wang, J., & Huang, Z. (2018). Rocket image classification based on deep convolutional neural network. *Paper presented at 10th Int. Conf. Commun., Circuits, Syst. (ICCCAS)* (pp. 383–386). Chengdu, China.

[4] Shen, C., Zhao, C., Yu, M., & Peng, Y. (2018). Cloud cover assessment in satellite images via deep ordinal classification. *IGARSS–IEEE Int. Geosci. Remote Sens. Symp* (pp. 3509–3512). Valencia.

[5] Postadjian, T., Bris, A. L., Mallet, C., & Sahbi, H. (2018). Superpixel partitioning of very high-resolution satellite images for large-scale classification perspectives with deep convolutional neural networks. *IGARSS–IEEE Int. Geosci. Remote Sens. Symp* (pp. 1328–1331). Valencia.

[6] Postadjian, T., Bris, A. L., Sahbi, H., & Mallet, C. (2018). Domain adaptation for large-scale classification of very high-resolution satellite images with deep convolutional neural networks. *IGARSS–IEEE Int. Geosci. Remote Sens. Symp* (pp. 3623–3626). Valencia.

[7] Helber, P., Bischke, B., Dengel, A., & Borth, D. (2018). Introducing EuroSAT: A novel dataset and deep learning benchmark for land use and land cover classification. *IGARSS–IEEE Int. Geosci. Remote Sens. Symp* (pp. 204–207). Valencia.

[8] Cai, K., & Wang, H. (2017). Cloud classification of satellite images based on convolutional neural networks. *Paper presented at 8th IEEE Int. Conf. Softw. Eng. Serv. Sci. (ICSESS)* (pp. 874–877). Beijing.

[9] Ulku, E., Akagündüz, P., & Ghamisi, P. (2022). Deep semantic segmentation of trees using multispectral images. *IEEE J Sel Top Appl Earth Obs Remote Sens, 15*, 7589–7604.

[10] Lavreniuk, M., Kussul, N., & Novikov, A. (2018). Deep learning crop classification approach based on sparse coding of time series of satellite data. *IGARSS–IEEE Int. Geosci. Remote Sens. Symp* (pp. 4812–4815). Valencia.

[11] Shao, Z., Zhou, W., Deng, X., Zhang, M., & Cheng, Q. (2020). Multilabel remote sensing image retrieval based on fully convolutional network. *IEEE J Sel Top Appl Earth Obs Remote Sens, 13*, 318–328.

[12] Li, L., Li, X., Liu, X., Huang, W., Hu, Z., & Chen, F. (2021). Attention mechanism cloud detection with modified FCN for infrared remote sensing images. *IEEE Access, 9*, 150975–150983.

[13] Zang, N., Cao, Y., Wang, Y., Huang, B., Zhang, L., & Mathiopoulos, P. T. (2021). Land-use mapping for high-spatial resolution remote sensing image via deep learning: A review. *IEEE J Sel Top Appl Earth Obs Remote Sens, 14*, 5372–5391.

[14] Pan, D., Zhang, M., & Zhang, B. (2021). A generic FCN-based approach for the road-network extraction from VHR remote sensing images—using OpenStreetMap as benchmarks. *IEEE J Sel Top Appl Earth Obs Remote Sens, 14*, 2662–2673.

[15] Helleis, M., Wieland, M., Krullikowski, C., Martinis, S., & Plank, S. (2022). Sentinel-1-based water and flood mapping: Benchmarking convolutional neural networks against an operational rule-based processing chain. *IEEE J Sel Top Appl Earth Obs Remote Sens, 15*, 2023–2036.

[16] Guo, B., Zhang, J., & Li, X. (2023). River extraction method of remote sensing image based on edge feature fusion. *IEEE Access*, *11*, 73340–73351.

[17] Bragilevsky, L., & Bajić, I. V. (2017). Deep learning for Amazon satellite image analysis. *Paper presented at IEEE Pacific Rim Conf. Commun., Comput. Signal Process (PACRIM)* (pp. 1–5). Victoria, BC.

[18] Nong, Z., Su, X., Liu, Y., Zhan, Z., & Yuan, Q. (2021). Boundary-aware dual-stream network for VHR remote sensing images semantic segmentation. *IEEE J Sel Top Appl Earth Obs Remote Sens*, *14*, 5260–5268.

[19] Liu, J., Xiong, X., Li, J., Wu, C., & Song, R. (2020). Dilated residual network based on dual expectation maximization attention for semantic segmentation of remote sensing images. *IGARSS–IEEE Int. Geosci. Remote Sens. Symp* (pp. 1825–1828). Waikoloa, HI, USA.

[20] Zhan, Z., Zhang, X., Liu, Y., Sun, X., Pang, C., and Zhao, C. (2020). Vegetation land use/land cover extraction from high-resolution satellite images based on adaptive context inference. IEEE Access, 8:21036–21051.

[21] Hu, H., Li, Z., Li, L., Yang, H., and Zhu, H. (2020). Classification of very high-resolution remote sensing imagery using a fully convolutional network with global and local context information enhancements. IEEE Access, 8:14606–14619.

[22] Liu, Q., Hang, R., Song, H., and Li, Z. (2018). Learning multiscale deep features for high-resolution satellite image scene classification. IEEE Trans. Geosci. Remote Sens., 56(1):117–126.

[23] Peng, C., Zhang, K., Ma, Y., and Ma, J. (2022). Cross fusion net: A fast semantic segmentation network for small-scale semantic information capturing in aerial scenes. IEEE Trans. Geosci. Remote Sens., 60:1–13.

4 Managing distribution transformers using GSM-based remote monitoring systems

Aakansha Soy[1,a] and Ahilya Dubey[2,b]

[1]Assistant Professor, Department of CS & IT, Kalinga University, Raipur, India
[2]Research Scholar, Department of CS & IT, Kalinga University, Raipur, India

Abstract: Distribution transformers play a pivotal role in the electrical grid by facilitating the efficient transmission of electricity from high-voltage transmission lines to lower voltage levels suitable for distribution to homes, businesses, and industrial facilities. However, the reliability and efficiency of distribution transformers have not yet been fully explored. In this article, we explore the integration of GSM technology into a remote monitoring system designed specifically for transformers. The system enables early detection of potential faults, allowing for proactive maintenance and prevention of costly failures. By leveraging this architecture, utilities and operators can enhance reliability, optimize maintenance schedules, and mitigate risks associated with transformer failures, thus contributing to the resilience of electrical grid infrastructure. GSM-based monitoring systems enable utility and operator to proactively manage transformer assets, optimize asset management practices, and uphold high standards of service delivery in the dynamic landscape of modern electrical distribution networks. In addition, the system's predictive maintenance capabilities, supported by advanced data analytics, further enhance grid resilience by allowing timely interventions and reducing the risk of unexpected outages.

Keywords: Remote monitoring systems, GSM, transformer

1. Introduction

Distribution transformers play a pivotal role in the electrical grid by facilitating the efficient transmission of electricity from high-voltage transmission lines to lower voltage levels suitable for distribution to homes, businesses, and industrial facilities [1]. These transformers are ubiquitous in the infrastructure that powers our modern society, silently ensuring that electricity reaches us reliably and safely. They are integral components in substations and are often located at various points within the distribution network, where they step down voltages to levels suitable for local consumption [18]. Without these transformers, the functioning of our electrical infrastructure would be severely compromised, highlighting their critical importance. As essential as distribution transformers are, ensuring their optimal performance and reliability is equally crucial. Efficient monitoring systems are essential for this purpose, as they provide real-time insights into the operational status of transformers [3]. Traditionally, monitoring has been performed through periodic manual inspections or automated systems that relay information via wired connections. However, these methods often have limitations in terms of frequency, accessibility, and the ability to detect early signs of potential issues [19]. Advances in telecommunications technology have revolutionized the field of remote monitoring, offering solutions that overcome many traditional limitations. One such technology that has proven highly effective in remote monitoring applications is GSM (Global System for Mobile Communications) [4, 5]. Originally developed for mobile phone communication, GSM technology has found diverse applications beyond its initial scope, including machine-to-machine (M2M) communication and Internet of Things (IoT) devices [2]. Its robustness, widespread coverage, and cost-effectiveness make it an ideal choice for implementing remote monitoring systems for distribution transformers. GSM technology allows data to be transmitted wirelessly over cellular networks, enabling real-time monitoring of critical parameters such as voltage levels, current loads, temperature, and operational status [20]. This capability is particularly valuable in the context of distribution transformers, where early detection of anomalies can prevent costly failures, reduce downtime, and enhance overall grid reliability [7] [8]. By leveraging GSM-based remote monitoring systems, utilities and operators gain the ability to monitor transformer health continuously, irrespective of their geographical location [21]. In this article, we explore the integration of GSM technology into a remote monitoring system designed specifically for distribution transformers [9]. The architecture, working principles, implementation considerations, and benefits of such systems. By doing so, we aim to highlight the transformative impact of GSM technology on the reliability and efficiency of electrical distribution networks. This discussion will provide valuable insights

[a]ku.aakanshasoy@kalingauniversity.ac.in, [b]ahilya.dubey@kalingauniversity.ac.in

DOI: 10.1201/9781003675259-4

for engineers, utilities, and stakeholders involved in the management and maintenance of distribution transformer assets [4]. Traditional methods of monitoring distribution transformers have predominantly relied on periodic manual inspections or automated systems that use wired communication protocols. Manual inspections involve on-site visits by technicians to visually inspect transformers, record data such as temperature and oil level, and perform routine maintenance checks [11]. While these inspections provide direct observation of transformer conditions, they are inherently limited by their infrequency and the potential for human error in data collection and interpretation. Moreover, manual inspections do not provide continuous real-time monitoring, making it challenging to detect subtle changes or early signs of impending issues [12]. Automated monitoring systems, on the other hand, have improved the frequency and reliability of data collection compared to manual methods. These systems often use wired communication protocols such as Modbus or DNP3 to transmit data from sensors installed on transformers to centralized monitoring stations. Sensors typically measure parameters like temperature, oil level, winding currents, and voltage fluctuations. While automated systems offer advantages in terms of data accuracy and consistency, they still face limitations related to accessibility and scalability [13]. Wired communication requires infrastructure investments and may not be feasible in remote or expansive distribution networks. Traditional monitoring systems, whether manual or automated, exhibit several limitations that compromise their effectiveness in ensuring optimal transformer performance. One major drawback is the inability to provide real-time data continuously [14]. Manual inspections are periodic and therefore miss transient faults or deteriorating conditions that can lead to unexpected failures. Automated systems, while more frequent in data collection, still rely on wired connections that are susceptible to physical damage, require maintenance, and are costly to deploy over large geographic areas. Another limitation is the lack of predictive analytics capabilities in traditional systems. These systems often focus on data collection and basic threshold alarms, reacting to problems after they occur rather than proactively predicting potential issues [15]. This reactive approach can result in increased downtime, higher maintenance costs, and disruptions to the electrical supply. Furthermore, traditional monitoring systems may not be equipped to handle the growing complexity and demands of modern electrical grids. With the integration of renewable energy sources, increased load variability, and the need for grid resilience, monitoring systems must evolve to provide more comprehensive insights into transformer health and performance. GSM technology offers significant advantages over traditional monitoring systems, making it an attractive choice for remote monitoring of distribution transformers. One key advantage is its ability to provide real-time data transmission over cellular networks. By utilizing GSM modules installed on transformers, data from sensors can be transmitted continuously to a central monitoring station without the need for physical wired connections. This capability enables operators to monitor transformer parameters remotely and respond promptly to any anomalies or critical events. Moreover, GSM technology enhances accessibility and scalability in monitoring infrastructure. Cellular networks have extensive coverage across urban, suburban, and rural areas, ensuring that even remote transformers can be monitored effectively [16]. This ubiquitous coverage reduces the need for extensive infrastructure investments associated with deploying wired communication networks over long distances. Another advantage of GSM-based monitoring systems is their cost-effectiveness and ease of deployment. Compared to laying down new wired communication lines, installing GSM modules is relatively straightforward and does not require significant capital expenditure. This makes GSM technology particularly suitable for retrofitting existing transformer installations with remote monitoring capabilities, thereby extending the lifespan and improving the efficiency of aging infrastructure [10, 17]. Furthermore, GSM technology supports data security and encryption protocols, ensuring that sensitive information transmitted from transformers to monitoring stations remains protected against unauthorized access or cyber threats. This aspect is crucial in safeguarding critical infrastructure assets and maintaining the integrity of the electrical grid. The adoption of GSM technology in transformer monitoring systems represents a paradigm shift towards more efficient, reliable, and scalable solutions. By leveraging real-time data transmission capabilities, operators can enhance predictive maintenance strategies, reduce operational costs, and improve the overall resilience of distribution networks [6].

2. System Architecture

2.1. Description of the overall system design

The system architecture for remote monitoring of distribution transformers using GSM technology comprises a well-integrated framework designed to capture, process, and transmit essential operational data in real-time. At its core, the architecture ensures continuous monitoring of critical parameters to preemptively identify potential faults, optimize maintenance schedules, and enhance overall grid reliability.

2.2. Key components

2.2.1. Sensors for monitoring voltage, current, temperature

Sensors are pivotal components installed on distribution transformers to capture vital operational data. These sensors include voltage and current sensors to monitor electrical parameters, temperature sensors to track the thermal conditions of the transformer, and oil level sensors to assess the health of transformer insulation. The data collected by these sensors provide insights into the transformer's operational status and enable early detection of anomalies or potential failures. Sensor technologies have evolved to offer high precision, durability, and compatibility with various transformer configurations, ensuring accurate data acquisition across diverse environmental conditions.

2.2.2. GSM module for communication

The GSM module acts as the communication gateway between the distribution transformers and the centralized monitoring system. It utilizes GSM (Global System for Mobile Communications) technology to establish wireless connectivity over cellular networks. The module facilitates the transmission of real-time data collected by sensors to a remote monitoring center or cloud-based platform. GSM technology ensures robust and reliable data transfer, enabling operators to receive timely updates on transformer performance regardless of their geographical location. Moreover, GSM modules support secure communication protocols, safeguarding sensitive data transmitted from transformers against unauthorized access or cyber threats.

2.2.3. Microcontroller for data processing

A microcontroller serves as the central processing unit within the monitoring system, responsible for managing data acquisition, processing sensor inputs, executing control algorithms, and interfacing with the GSM module for data transmission. The microcontroller processes raw sensor data to derive actionable insights and performs local computations to assess the operational health of the transformer. It also coordinates diagnostic routines and fault detection algorithms, enabling the system to autonomously respond to critical events or trigger alerts based on predefined thresholds. Modern microcontrollers offer high computational capabilities, low power consumption, and robust integration with sensor interfaces, making them suitable for real-time monitoring applications.

2.3. Block diagram of the system

The system architecture can be visualized through a block diagram that illustrates the interconnectivity and functional components of the remote monitoring system for distribution transformers using GSM technology:

- These components include voltage, current, temperature, and oil level sensors installed on the distribution transformer. They continuously monitor the operational parameters and send analog or digital signals to the microcontroller.
- At the heart of the system, the microcontroller interfaces with the sensors to receive data inputs. It processes these inputs through embedded software algorithms to analyze the transformer's operational condition. Themicrocontroller also manages the operation of the GSM module for data transmission.
- This module establishes communication with the cellular network infrastructure using GSM protocols. It converts the processed data from the microcontroller into a format suitable for wireless transmission. The GSM module securely transmits the data packets over the cellular network to a designated server or cloud-based platform.
- The transmitted data packets are received and stored on a centralized server or cloud-based platform hosted by the utility or monitoring service provider. Here, the data is further processed, analyzed, and visualized through user interfaces or dashboards accessible to authorized personnel.
- Operators and maintenance personnel can access real-time data and analytics through a user-friendly interface provided by the monitoring center or cloud platform. Alerts, notifications, and historical trends are displayed, enabling informed decision-making and proactive maintenance strategies.

This block diagram encapsulates the flow of information and interactions among the key components within the remote monitoring system. It emphasizes the seamless integration of sensors, microcontroller, and GSM technology to enable continuous monitoring, data-driven insights, and operational efficiency improvements in distribution transformer management (Figure 4.1). By leveraging this architecture, utilities can enhance reliability, optimize maintenance schedules, and mitigate risks associated with transformer failures, thereby contributing to the resilience of the electrical grid.

3. Working Principle

Sensors deployed on distribution transformers are designed to capture various operational parameters critical to assessing the transformer's health and performance. These sensors include voltage sensors, current sensors, temperature

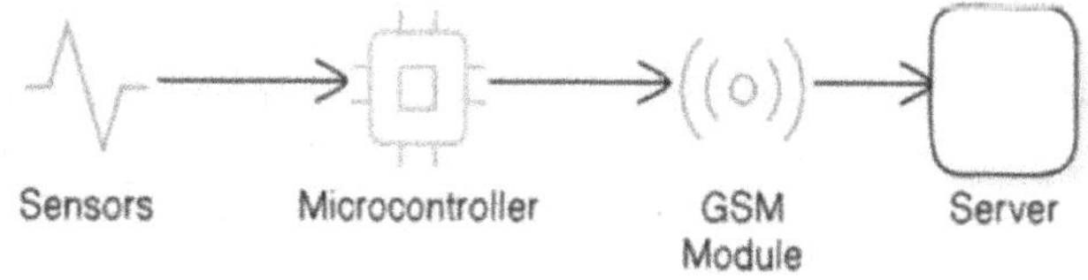

Figure 4.1. Remote monitoring system for distribution transformers.

Source: Author.

sensors, and oil level sensors, among others. Each sensor is strategically placed at key points on the transformer to ensure comprehensive monitoring of its operational condition. For instance, voltage and current sensors are typically connected to the primary and secondary windings to measure electrical parameters, providing insights into power consumption and load variations. Temperature sensors are placed in critical areas to monitor thermal conditions, ensuring that the transformer operates within safe temperature limits. Oil level sensors, located within the transformer's oil reservoir, monitor oil levels and quality, which are crucial indicators of insulation integrity and potential overheating.

The sensors operate by converting physical quantities such as voltage, current, temperature, or oil level into electrical signals (analog or digital). These signals are then transmitted to the microcontroller for further processing and analysis. The accuracy and reliability of sensor data collection are essential for detecting anomalies early, allowing operators to take preventive actions and avoid costly transformer failures.

3.1. Role of the microcontroller in data acquisition and processing

The microcontroller serves as the central processing unit within the remote monitoring system, responsible for managing data acquisition, processing sensor inputs, and executing control algorithms. Upon receiving signals from sensors, the microcontroller interfaces with analog-to-digital converters (ADCs) to convert analog sensor readings into digital data that can be processed digitally. This digital data undergoes real-time processing within the microcontroller, where embedded software algorithms analyze the information to assess the transformer's operational status. The microcontroller plays a critical role in executing predefined control logic and diagnostic routines tailored to the specific requirements of transformer monitoring. It monitors trends in sensor data, compares current readings with historical norms or threshold values, and triggers alerts or alarms when deviations exceed acceptable limits. Additionally, the microcontroller manages power consumption, ensuring efficient operation while maintaining continuous data acquisition and processing capabilities. Moreover, the microcontroller coordinates communication with the GSM module for data transmission to the remote monitoring center. It formats the processed data into packets suitable for transmission, manages data integrity and encryption protocols to secure sensitive information, and initiates communication cycles based on predefined schedules or event-driven triggers. Overall, the microcontroller acts as the intelligent hub of the monitoring system, enabling real-time decision-making and proactive maintenance strategies.

3.2. Data transmission process via GSM module

Once the microcontroller processes and formats the sensor data, it initiates the data transmission process through the GSM module (Figure 4.2). The GSM module utilizes Global System for Mobile Communications (GSM) technology to establish wireless communication over cellular networks. It interfaces with the microcontroller to receive formatted data packets and modulates them into signals compatible with GSM protocols for transmission.

By leveraging GSM technology, the data transmission process is streamlined, enabling real-time monitoring

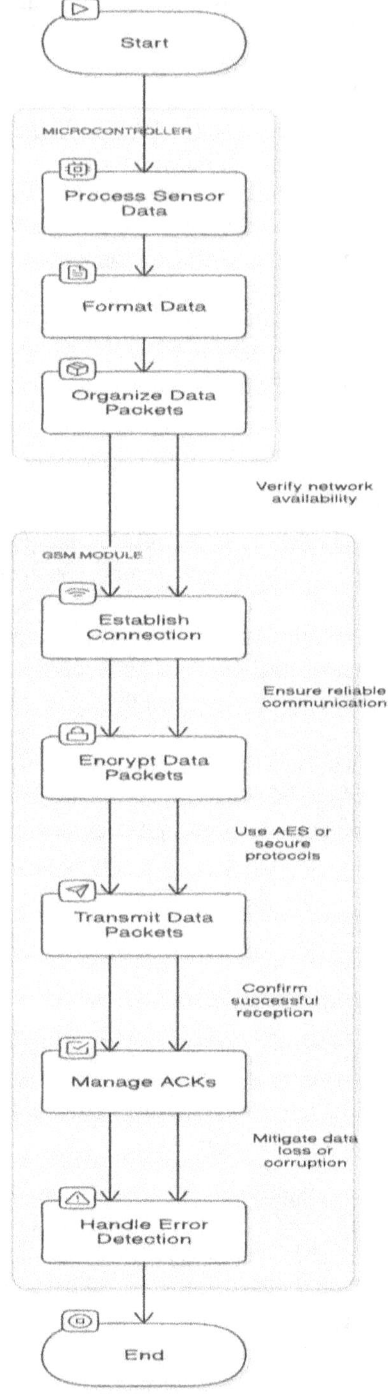

Figure 4.2. Data transmission process.
Source: Author.

capabilities across distributed transformer assets without relying on physical wired connections. This wireless approach enhances operational flexibility, scalability, and cost-effectiveness, particularly in remote or inaccessible locations where deploying wired infrastructure would be impractical or economically prohibitive.

3.3. Data Reception and Interpretation at the Control Center

At the control center or monitoring facility, the transmitted data packets are received and processed by a centralized server or cloud-based platform. This platform serves as the interface for operators, providing real-time visibility into transformer performance and operational metrics. Upon reception, the data undergoes several stages of interpretation and analysis:

- Received data packets are parsed to extract sensor readings, timestamps, and metadata. Parsed data is then stored in a structured database or data repository for historical analysis and trend monitoring.
- Processed data is visualized through intuitive dashboards or graphical interfaces accessible to operators and maintenance personnel. Real-time displays include parameter trends, alarm notifications, and operational status indicators, enabling users to monitor transformer health and performance at a glance.
- The monitoring platform incorporates predefined thresholds and alert mechanisms to detect abnormal conditions or critical events in real-time. Operators receive immediate notifications via email, SMS, or dashboard alerts when sensor readings exceed set thresholds, allowing for prompt intervention and preventive maintenance actions.
- Historical data analysis tools enable deeper insights into transformer behaviour over time. Operators can generate performance reports, conduct root cause analyses of past incidents, and optimize maintenance schedules based on data-driven insights.

The seamless integration of sensor data collection, microcontroller processing, GSM-based communication, and centralized data interpretation forms the foundation of an efficient remote monitoring system for distribution transformers. This integrated approach enhances operational visibility, enables proactive maintenance strategies, and improves overall grid reliability by minimizing downtime and optimizing asset performance.

4. Implementation

4.1. Planning and requirements gathering

Begin by conducting a thorough assessment of monitoring requirements and objectives. Identify critical parameters to monitor (e.g., voltage, current, temperature, oil level), determine communication needs, and define system scalability requirements based on the size and complexity of the distribution network.

4.2. Selection of components

Choose appropriate components based on the identified requirements. This includes selecting sensors capable of accurately measuring desired parameters under varying environmental conditions. Common sensor types include voltage and current sensors (e.g., Hall effect sensors), temperature sensors (e.g., thermistors or RTDs), and oil level sensors (e.g., capacitive or float-based sensors). For communication, select a GSM module compatible with the chosen cellular network and capable of secure data transmission (Figure 4.3). Choose a microcontroller platform with sufficient processing power, analog-to-digital conversion capabilities, and support for communication interfaces (e.g., UART, SPI) needed to interface with sensors and the GSM module.

4.3. Programming the microcontroller

Develop or customize firmware for the microcontroller to manage sensor data acquisition, processing, and communication tasks. The firmware should include algorithms for data parsing, error handling, encryption (if applicable), and periodic transmission scheduling. Implement protocols for interfacing with sensors and configuring communication with the GSM module. Test the firmware thoroughly to ensure compatibility with selected sensors and GSM module functionalities.

4.4. Integrating the system components

Physically integrate sensors, microcontroller, and GSM module onto the distribution transformer or within a suitable enclosure. Ensure sensors are securely mounted in optimal locations to facilitate accurate data collection. Establish electrical connections between sensors and the microcontroller, ensuring proper signal integrity and power supply. Install the GSM module and configure it to establish a reliable cellular connection, verifying network coverage and signal strength at the installation site.

4.5. Testing and calibration

Conduct comprehensive testing of the integrated system to validate functionality and performance under simulated operating conditions. Test sensor readings against known benchmarks or calibrated instruments to verify accuracy and consistency. Perform communication tests to ensure data transmission reliability over the GSM network, checking for latency, packet loss, and encryption integrity. Calibrate sensors as necessary to align measurements with expected operational ranges and environmental factors.

5. Deployment and Optimization

Deploy the monitoring system across targeted distribution transformer assets within the electrical grid. Monitor system performance during initial deployment to identify any operational issues or optimization opportunities. Optimize data transmission schedules, sensor placement, and firmware parameters based on real-world performance data and feedback from initial deployment phases.

5.1. Training and maintenance

Provide training to operational staff responsible for monitoring and managing the system. Educate personnel on system operation, troubleshooting procedures, and interpreting data analytics provided by the monitoring platform. Establish a proactive maintenance schedule to inspect sensors, update firmware, and perform periodic system checks to ensure continued reliability and performance.

5.2. Programming the microcontroller

Programming the microcontroller involves several key steps to ensure seamless operation of the monitoring system:

- Implement firmware routines to read analog sensor inputs using ADCs integrated into the microcontroller. Configure sampling rates and resolution settings to optimize sensor data accuracy and minimize noise interference.
- Develop algorithms to process sensor data in real-time, including data filtering, averaging, and threshold monitoring. Implement diagnostic routines to detect anomalies or deviations from normal operating conditions based on predefined thresholds.
- Integrate protocols for establishing and maintaining communication with the GSM module. Develop routines for formatting sensor data into packets suitable for transmission over the cellular network. Implement error handling mechanisms to ensure reliable data delivery and retransmission strategies in case of communication failures.
- Incorporate encryption algorithms (e.g., AES) to secure sensor data before transmission over the GSM network. Implement authentication protocols to verify data integrity and prevent unauthorized access to sensitive information.
- Optimize power consumption by implementing sleep modes and power-saving strategies for both the microcontroller and peripheral components. Ensure continuous operation while minimizing energy consumption to extend system longevity and reliability.

5.3 Improved Reliability and Efficiency

The adoption of GSM-based monitoring systems contributes to improved reliability and efficiency of distribution transformers and the broader electrical grid infrastructure. By continuously monitoring operational parameters, these systems provide comprehensive visibility into transformer performance and health status. Operators can proactively

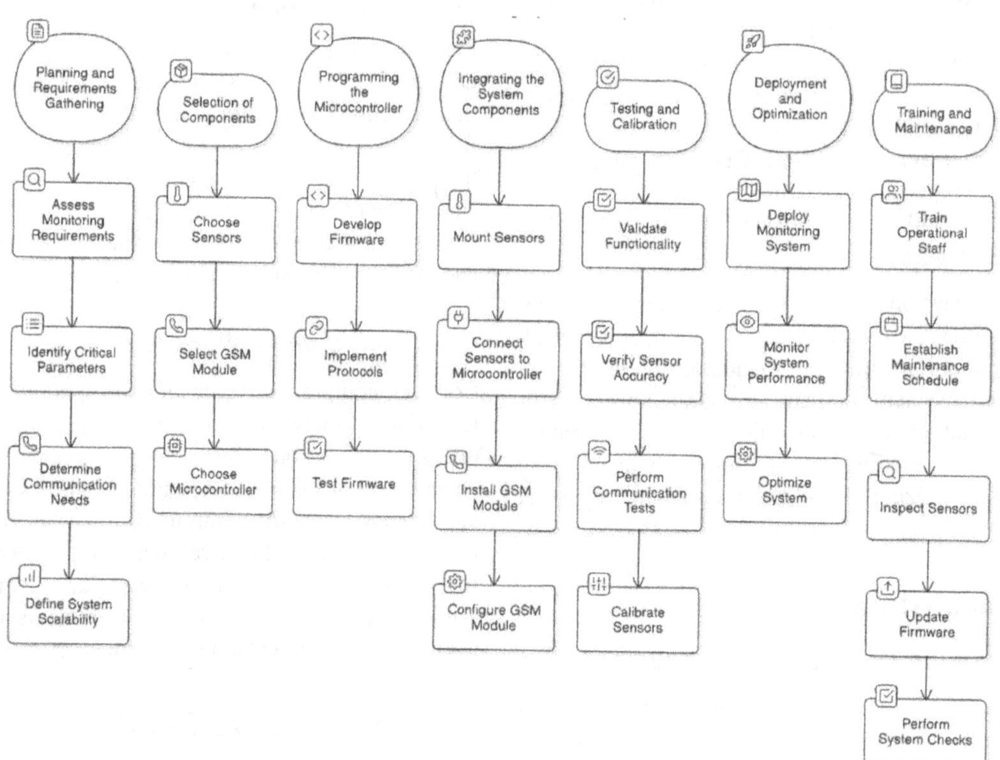

Figure 4.3. Remote monitoring system setup.

Source: Author.

schedule maintenance activities based on actual asset condition rather than predefined intervals, optimizing resource utilization and minimizing downtime. Enhanced reliability is further supported by the system's ability to facilitate predictive maintenance strategies. By analyzing historical data trends and performance metrics, operators can anticipate potential issues, preemptively replace aging components, and mitigate risks associated with unexpected failures. This predictive approach not only enhances operational reliability but also improves overall grid efficiency by reducing energy losses, enhancing voltage regulation, and ensuring consistent delivery of electricity to consumers.

The adoption of GSM-based monitoring systems for distribution transformers offers multifaceted benefits that enhance operational capabilities, improve asset management practices, and strengthen grid resilience. By enabling real-time monitoring, early fault detection, cost-effective scalability, and enhanced reliability, these systems empower utilities and operators to proactively manage transformer assets, optimize maintenance strategies, and uphold high standards of service delivery in the dynamic landscape of modern electrical distribution networks (Figure 4.4).

6. Conclusion

The deployment of a GSM-based monitoring system for distribution transformers represents a significant advancement in the management and operation of electrical grid infrastructure. This technology leverages real-time data acquisition and transmission, providing operators with immediate insights into transformer performance and health. By integrating sensors to monitor critical parameters such as voltage, current, temperature, and oil levels, the system enables early detection of potential faults, allowing for proactive maintenance and prevention of costly failures.

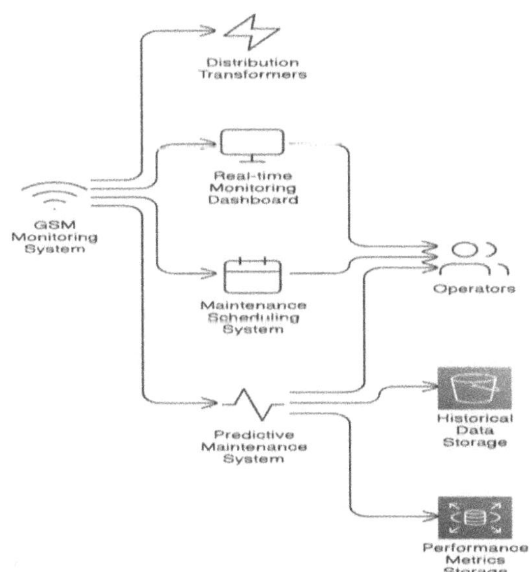

Figure 4.4. GSM based monitoring system for transformers.

Source: Author.

The microcontroller at the core of the system efficiently processes sensor data and manages communication with the GSM module, ensuring reliable data transmission over cellular networks. This wireless communication capability not only enhances operational flexibility but also reduces the need for extensive physical infrastructure, making the system both cost-effective and scalable. The ease of deploying GSM modules, coupled with their secure and robust communication protocols, ensures that even remote or geographically dispersed transformers can be monitored effectively.

Moreover, the system's ability to provide continuous, real-time monitoring improves the reliability and efficiency of distribution transformers. Operators can optimize maintenance schedules based on actual asset conditions, minimizing downtime and extending the lifespan of transformer components. The system's predictive maintenance capabilities, supported by advanced data analytics, further enhance grid resilience by allowing for timely interventions and reducing the risk of unexpected outages.

Overall, GSM-based monitoring systems offer a comprehensive solution for modernizing transformer management practices. They provide utilities and operators with the tools needed to enhance grid reliability, optimize resource utilization, and ensure the consistent delivery of electricity. By embracing this technology, the electrical industry can better meet the demands of an evolving energy landscape, characterized by increased load variability, the integration of renewable energy sources, and the need for greater grid resilience.

References

[1] Ojo, T. P., Akinwumi, A. O., Ehiagwina, F. O., Ambali, J. M., & Olatinwo, I. S. (2022). Design and implementation of a GSM-based monitoring system for a distribution transformer. *European Journal of Engineering and Technology Research*, *7*(2), 22–28.

[2] Surendar, A., Saravanakumar, V., Sindhu, S., & Arvinth, N. (2024). A bibliometric study of publication-citations in a range of journal articles. *Indian Journal of Information Sources and Services*, *14*(2), 97–103. https://doi.org/10.51983/ijiss-2024.14.2.14

[3] Kepa, W., Luhach, A. K., Kavi, M., Fisher, J., & Luhach, R. (2020, December). GSM based remote distribution transformer condition monitoring system. In *International Conference on Advanced Informatics for Computing Research* (pp. 59–68). Singapore: Springer Singapore.

[4] Chaturvedi, R., Islam, A., & Sharma, K. (2021). A review on the applications of PCM in thermal storage of solar energy. *Materials Today: Proceedings*, *43*, 293–297.

[5] Hayati, M. A. E. A. E., & Babiker, S. F. (2016, February). Design and implementation of low-cost SMS based monitoring system of distribution transformers. In *2016 conference of basic sciences and engineering studies (SGCAC)* (pp. 152–157). IEEE.

[6] Kitana, A., Traore, I., & Woungang, I. (2020). Towards an epidemic SMS-based cellular botnet. *Journal of Internet Services and Information Security*, *10*(4), 38–58.

[7] Al-Muntaser, A. A., Pashameah, R. A., Sharma, K., Alzahrani, E., & Tarabiah, A. E. (2022). Reinforcement of structural, optical, electrical, and dielectric characteristics of CMC/PVA based on GNP/ZnO hybrid nanofiller: nanocomposites materials for energy-storage applications. *International Journal of Energy Research, 46*(15), 23984–23995.

[8] Daehyeon, S., Youngshin, P., Bonam, K., & Ilsun, Y. (2024). A study on the implementation of a network function for real-time false base station detection for the next generation mobile communication environment. *Journal of Wireless Mobile Networks, Ubiquitous Computing, and Dependable Applications (JoWUA), 15*(1), 184–201.

[9] Mahesar, S., Ali, A., Aslam, M., Tunio, N. A., Akhtar, R., & Osama, A. (2020). Monitoring and control of 500 kva transformer through GSM based Raspberry Pi controller. *International Journal of Electrical Engineering & Emerging Technology, 3*(2), 40–44.

[10] Ando, R., Takahashi, K., & Suzaki, K. (2012). Inter-domain communication protocol for real-time file access monitor of virtual machine. *Journal of Wireless Mobile Networks, Ubiquitous Computing, and Dependable Applications, 3*(1/2), 120–137.

[11] Kumar, T. A., & Ajitha, A. (2017, July). Development of IOT based solution for monitoring and controlling of distribution transformers. In *2017 international conference on intelligent computing, instrumentation and control technologies (ICICICT)* (pp. 1457–1461). IEEE.

[12] Sun, J., Yan, G., Abed, A. M., Sharma, A., Gangadevi, R., Eldin, S. M., & Taghavi, M. (2022). Evaluation and optimization of a new energy cycle based on geothermal wells, liquefied natural gas and solar thermal energy. *Process Safety and Environmental Protection, 168*, 544–557.

[13] Satter, M. D. A. (2023). *Power transformer monitoring and controlling using GSM* (Doctoral dissertation, Daffodil International University).

[14] Sharma, A., Sharma, K., Islam, A., & Roy, D. (2020). Effect of welding parameters on automated robotic arc welding process. *Materials Today: Proceedings, 26*, 2363–2367.

[15] Narejo, G. B., Bharucha, S. P., & Pohwala, D. Z. (2015). Remote microcontroller based monitoring of substation and control system through GSM modem. *International Journal of Scientific Engineering Research, 6*(1), 714.

[16] Sachan, A. (2013). GSM based automated embedded system for monitoring and controlling of smart grid. *International Journal of Electrical Robotics, Electronics and Communications Engineering, 7*(12), 1273–1277.

[17] Hai, T., Ali, M. A., Dhahad, H. A., Alizadeh, A. A., Sharma, A., Almojil, S. F., ... & Wang, D. (2023). Optimal design and transient simulation next to environmental consideration of net-zero energy buildings with green hydrogen production and energy storage system. *Fuel, 336*, 127126.

[18] Jalilian, M., Sariri, H., Parandin, F., Karkhanehchi, M. M., Hookari, M., Jirdehi, M. A., & Hemmati, R. (2016). Design and implementation of the monitoring and control systems for distribution transformer by using GSM network. *International Journal of Electrical Power & Energy Systems, 74*, 36–41.

[19] Chen, Y., Feng, L., Mansir, I. B., Taghavi, M., & Sharma, K. (2022). A new coupled energy system consisting of fuel cell, solar thermal collector, and organic Rankine cycle; generation and storing of electrical energy. *Sustainable Cities and Society, 81*, 103824.

[20] Amesimenu, D. K., Chang, K. C., Sung, T. W., Zhou, Y., Gakiza, J., Omer, A. A. I., ... & Haque, S. M. O. (2020, March). Study of smart monitoring and protection of remote transformers and transmission lines using GSM technology. In *2020 IEEE International Conference on Artificial Intelligence and Information Systems (ICAIIS)* (pp. 297–301). IEEE.

[21] Mahesar, S., Ali, A., Aslam, M., Tunio, N. A., Akhtar, R., & Osama, A. (2020). Monitoring and control of 500 kva transformer through GSM based Raspberry Pi controller. *International Journal of Electrical Engineering & Emerging Technology, 3*(2), 40–44.

5 Bio-innovations in drug packaging: A novel approach using microcrystalline cellulose and modified starch based bio-composite materials

Sakthivel sankaran[1,a], Marimuthu Chandran[1,b], Krishnan Selvaraj[1,c], Arumugaprabhu Veerasimman[2,d], and S. Vijiyalakshmi[3,e]

[1]Department of Biomedical Engineering, Kalasalingam Academy of Research and Education, Tamil Nadu, India
[2]Department of Mechanical Engineering, Kalasalingam Academy of Research and Education, Tamil Nadu, India
[3]Department of Electronics and Communication Engineering, Vel Tech Rangarajan Dr. Sagunthala R&D Institute of Science and Technology, Tamil Nadu, India

Abstract: Synthetic materials, while offering superior properties, pose significant environmental risks. Bio composite materials, derived from natural resources such as plant fibres and bio-based polymers, present sustainable alternatives. This study builds upon previous research, focusing on the development of bio composite materials reinforced with modified starch and microcrystalline cellulose. Our objective was to create materials suitable for food packaging applications in the Consumables industry, demonstrating superior tensile strength compared to conventional packaging materials. We explored a novel approach using modified starch and microcrystalline cellulose. The starch was modified using specific quantities of sodium hydroxide, adipic acid, and mustard oil relative to its weight. Microcrystalline cellulose, derived from α-cellulose precursor, was chosen for its high biodegradability, minimal environmental impact, improved mechanical properties, reduced energy consumption, enhanced thermal and moisture resistance, and superior barrier properties compared to synthetic materials. By focusing on these eco-friendly materials, we aim to contribute to sustainable practices in the packaging industry, addressing environmental concerns associated with synthetic alternatives.

Keywords: Cellulose, microcrystalline cellulose, bio composite, modified starch

1. Introduction

In an era marked by escalating environmental concerns and increasing need for environmentally friendly alternatives, development of biocomposite materials has risen to prominence in the realm of scientific exploration. These materials, derived from renewable sources and equipped with biodegradable properties, provide a compelling avenue to mitigate the ecological impact of conventional synthetic materials. This research addresses the pressing demand for eco-friendly materials within various industries, with a specialized emphasis on their aptness for drug delivery in the medical field. Through the utilization of natural resources and innovative fabrication techniques, we endeavour to provide a comprehensive evaluation of these biocomposite materials. This evaluation will underscore their eco-friendly characteristics, mechanical performance, and their potential to excel in the realm of pharmaceutical applications. By combining the biodegradability, mechanical strength, and renewable sourcing of microcrystalline cellulose and modified starch, these bio composites emerge as promising candidates for sustainable drug transport solutions. This research represents a pivotal stride toward environmentally-conscious materials in the dynamic landscape of scientific innovation, with the potential to positively impact both the pharmaceutical industry and environmental conservation efforts.

2. Starch-Based Eco-Friendly Materials

They have described about a bio composite material using modified starch and sugarcane bagasse [1]. They conducted various tests, including density, absorbency, thermal analysis, impact resistance, tensile and flexural strength. The composite showed higher impact and flexural strength than polystyrene foam, with lower density compared to cardboard. However, it's important to note the substance displayed some limitations in specific tests [2]. They have described about the creating biodegradable and bio-based polymers as alternatives to conventional petroleum-based ones. Their approach involves modifying existing materials and crafting new polymer composites derived from organic origins like starch and plant fibers. This process entails complexity

[a]drsakthivelsankaran2024@gmail.com, [b]marimuthu98300@gmail.com, [c]krishnansk2003@gmail.com, [d]v.arumugaprabhu@klu.ac.in, [e]Drvijayalakshmis@veltech.edu.in

DOI: 10.1201/9781003675259-5

and requires specialized equipment. Achieving mechanical and thermal properties comparable to traditional plastics poses a challenge [3]. They have described about the biocomposite fabrication techniques like electrospinning, filament winding, injection molding, extrusion, nanotech, and advanced printing. They highlighted filament winding for high-strength aerospace and automotive applications with high fibre fractions, and the versatility of extrusion for blending and shaping. However, scalability issues could result in elevated production costs, necessitating further property optimization [4]. They have described about the starch as an eco-friendly alternative to non-biodegradable materials. They enhance starch's properties through blending, composites, and nanocomposites, aiming for diverse applications. Challenges include mechanical traits, processing complexity, and varying biodegradation rates [5]. They have described about the complex nanoprecipitation to create a pair of nanoparticles from starch-palmitic acid complexes. These were integrated into starch composite films, enhancing tensile strength, wettability, and thermal stability, particularly with amylose-palmitic acid complexes. However, the latter's larger size might pose challenges for uniform dispersion in the composite material [6]. They have described about developing starch-based blends and composites as eco-friendly alternatives to single-use plastics, employing plasticization for improved process ability. Promising in diverse packaging, additives enhance properties, but achieving petroleum-product parity remains a challenge [7]. They have described about the starch based biocomposite films with synthesized Nano cellulose from cotton linters. Glycerol and polyvinyl alcohol aided processing. The study showed significant improvements in mechanical strength and barrier properties compared to untreated films. However, limited dimensional stability and moisture sensitivity require further consideration for practical applications [8]. They have described about the utilize jackfruit seed starch and tamarind kernel xyloglucan to create eco-friendly food packaging. They blend these with zinc oxide nanoparticles, resulting in improved properties and antimicrobial activity. Coating tomatoes with these materials extends shelf-life, but scalability, cost, and nanoparticle disposal pose challenges in large-scale production [9]. They have described about a strong biocomposite using thermoplastic corn starch and bacterial cellulose, showing a 3.6-fold increase in tensile strength. While promising for eco-friendly materials, challenges in complexity and cost exist [10]. They have described about a biocomposite made of cassava starch and Cymbopogon citratus fiber (CCF). The CCF was extracted using the water retting method and processed further. The thermoplastic cassava starch (TPCS) was prepared by combining starch, glycerol, and palm wax. However, bio-composites can be sensitive to moisture, causing dimensional instability and reduced mechanical properties over time [11]. They have described about the impact of rice husk and sawdust fillers on biocomposites made from sago starch. The process involved precise measurements, blending of components, and heating to create thermoplastic starch, which was subsequently merged with polypropylene (PP). However, biodegradable materials like these can degrade faster when exposed to moisture, limiting certain applications.

3. Cellulose and Biocomposite Development

They have described about the investigated Syagrus Romanzoffiana rachis-derived cellulose fibers for eco-friendly composites [12]. Extracted using the water retting method, these fibers show promise as lightweight, sustainable reinforcements. While potential limitations were not specified, further research is recommended for a comprehensive understanding [13]. They have described about employed cellulose nanocrystals (CNC) and cellulose micro fibrils (CMF) in flat sheet membrane fabrication for filtration. These enhanced membrane properties like strength, permeability, and wetting-spreading time due to CNC's hydroxyl groups aiding water transport. However, higher CNC concentration increased porosity and reduced salt rejection. Yet, it also stiffened the membrane, compromising elasticity and break limit [14]. They have described about the cellulose microfibers (CMFs) and cellulose nanocrystals (CNCs) as reinforcing agents for starch biopolymer, creating versatile biocomposites for food packaging. While CMFs enhanced stiffness but reduced ductility, CNCs showed improved properties at specific concentrations. The study highlighted trade-offs between stiffness, ductility, and transparency in the composite materials [15]. They have described about the colleagues utilized cellulose nanocrystals (CNCs) and cellulose microfibers (MFCs) to reinforce polylactic acid (PLA) biocomposites. Incorporating 1–5% CNCs notably improved thermal stability, crystallization processes, and mechanical properties, with a 40% increase in flexural strength and a 35% rise in storage modulus at room temperature. However, specific drawbacks of this method were not mentioned in the provided information [16]. They have studied about carboxymethyl cellulose (CMC)/poly(3hydroxybutyrate-co-3-hydroxyvalerate) (PHBV) biocomposites using coupling agents (MA and VTMS). They discovered that these substances formed covalent bonds, enhancing CMC distribution and interfacial bonding. Tensile and flexural properties increased significantly. However, drawbacks like cost and environmental impact related to the coupling agents' use were not discussed in the provided information [17]. They have described about the carrageenan-based films with cellulose fillers, highlighting carboxymethyl cellulose's effectiveness. Ana González Moreno's research created pectin-cellulose nanocrystal composites, impacting optical and thermal properties. Despite promise in food packaging, weak interactions in components could limit certain uses, underscoring challenges in biocomposite applications [18]. They have described about to created pectin-cellulose nanocrystal (CNC) biocomposites, altering material properties significantly. Pectin reduced crystallinity, modified CNC film structure, and affected optical properties. CNC-rich samples reflected visible light, while pectin-containing samples showed high UV absorbance. Despite promise in food packaging and optics, weak interactions between components might limit specific applications, despite improved thermal stability and seawater biodegradability [19]. They have described about to developed muffin liners with enhanced water resistance and thermal stability using cellulose nanofibers and starch. Challenges in production scaling and long-term biodegradability assessment need consideration for practical use [20]. They have described about the modified micro fibrillated cellulose (MFC) using an aqueous-phase approach, enhancing its ability to work well with polylactic acid (PLA) composites. The modified MFC, specifically vinyl laurate modified MFC (VL-MFC), significantly improved tensile strength and Young's modulus in oven-dried PLA composites. While promising, potential industrial drawbacks indicate the necessity for further research in this area [21]. They have described about the interface between short flax fiber and polypropylene (PP) thermoplastic composites using a dual approach: chemical coupling with Maleic Anhydride Polypropylene (MAPP) and nanoscale structuring via cellulose nanocrystals (CNC) and xyloglucan (XG). This modification improved interfacial properties and mechanical performance.

4. Multidisciplinary Applications of Biocomposite Materials

They have described about sustainable packaging using starch-based composite films reinforced with modified bamboo cellulose [22]. They detailed the cellulose extraction process and analyzed its characteristics. The researchers reached the conclusion that the material exhibited enhanced mechanical strength and hydrophobicity. However, it's worth noting that it showed some limitations in relation to flexibility [23]. They have described about the investigated films blended with ST-GE for packaging applications. Their study covered characteristics of pure ST and GE films, blending processes, additive effects, and overall performance enhancements. The researchers inferred that these films serve as highly biodegradable alternatives for food packaging, with improved mechanical and barrier properties. However, it's important to highlight that the material has a shorter lifespan due to its rapid biodegradation rate [24]. They have described about the developed Composite Phase Change Materials (CPCMs) for efficient renewable energy use. They combined ternary chloride with starch to form a SiC ceramic skeleton, optimizing the process. The composite demonstrated impressive thermal properties, but challenges include complex fabrication, limited temperature range, and potential environmental concerns [25]. They have described about to develop starch/hemp agro-composites for eco-friendly construction, targeting reduced carbon emissions. They assess physical and mechanical traits with varying hemp shiver sizes and Hemp/Starch ratios. Promising for low emissions and energy efficiency, these materials face challenges in achieving high density levels and non-linear mechanical properties [26]. They have described about the developed eco-friendly biodegradable starch films for food packaging. By incorporating date seed extracts, cellulose, and chitin, the films gained improved mechanical properties and antioxidation. They effectively prolonged fruit freshness, but their hydrophilic nature may limit use in specific food packaging applications due to water-vapour permeability [27]. They have described about biocompatible trans femoral amputee sockets using an innovative composite material comprising jute fibers and epoxy. The composite exhibits exceptional mechanical properties, surpassing other materials. It also demonstrates an 82% strength improvement, alongside good moisture resistance. The prototype socket offers a biocompatible, strong, and moisture-resistant solution for prosthetic applications, promising improved comfort and durability for above-knee amputee patients [28]. They have described about a bio-composite film from keratin and turmeric starch, derived from chicken feather waste. This sustainable alternative to fossil-based plastics offers high water solubility for drug delivery. The film exhibits notable stability, desirable mechanical properties, and capacity for diverse industrial uses, although thermal stability is a limiting factor [29]. They have described about the biodegradable packaging using chitosan, alginate, and cellulose nanocrystals. They examined structure, morphology, and thermal stability. While chitosan-based films showed higher initial biodegradation, they didn't completely degrade. Alginate-based films fully degraded in 107–112 hours. Further improvements are needed due to incomplete chitosan-based film degradation [30]. They have described about to meet the demand for eco-friendly products by creating Kombucha leather for fashion, footwear, bags, and interior coverings. They combine Kombucha-derived bacterial cellulose with polyurethane elastomer and polylactic acid, achieving good elasticity, hydrophobicity, and moderate biodegradability. However, achieving strong bond interactions between components remains a challenge, potentially affecting tear strength and durability [31]. They have described about the cellulose biocomposites as promising materials in biomedical engineering owing to their biocompatibility, biodegradability, and versatility. These green composites, especially those from bacterial sources, find applications in wound care, the field of tissue engineering, and pharmaceutical transport. Despite limitations in mechanical strength and scalability, cellulose based biocomposites offer sustainable solutions in biomedical engineering, minimizing environmental impact [32]. They have described about the materials for prosthetic limbs. Their methodology involves using a natural fiber-reinforced bio-composite material for sockets, aiming to reduce skin sores and enhance mechanical strength.

Silicone liners are recommended for stability, adhesion, and comfort, outperforming copolymer liners. The study suggests combining silicone liners with glass and nylon fibers for sockets to create ideal prosthetic limbs. However, drawbacks and limitations of this approach are not discussed in the abstract.

5. Methodology

Microcrystalline cellulose and modified starch were employed in the creation of biocomposite materials. The blending process involved a fusion of hand layup method and the compression method (Figure 5.1). An extensive study was conducted to assess the biodegradable properties of these substances with a comprehensive analysis performed under both laboratory and composting conditions. In the laboratory, controlled environments were established to assess the biodegradability of the biocomposite materials, simulating real-world scenarios to understand their behaviour during degradation.

Various mechanical property tests were conducted to analyse the structural integrity and durability of the biocomposite materials. Impact resistance tests measured the materials ability to withstand sudden forces, flexural strength tests assessed their bending properties, and tensile durability tests determined their ability to withstand tension. This assessment encompassed a thorough evaluation of the extensive features of biocomposite material, combining findings from the studies on biodegradable properties and mechanical properties. The results were compiled into a final report, presenting a comprehensive assessment of the biocomposite materials. The report highlighted their suitability for biodegradable applications and provided insights into their mechanical performance. It served as a valuable resource for understanding the attributes of the substances their potential use in various applications.

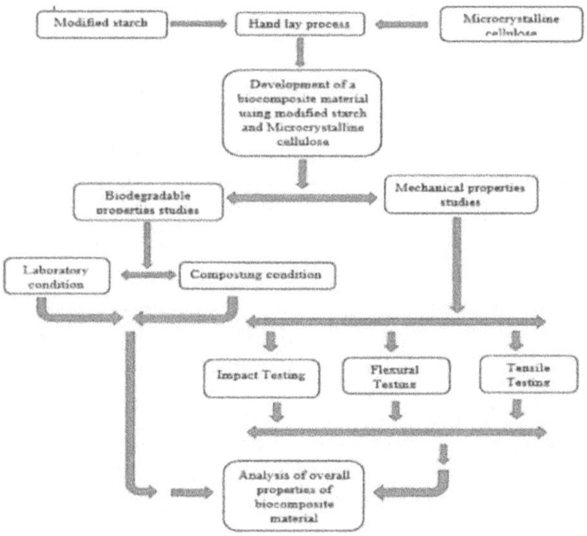

Figure 5.1. Block diagram of newly developed biocomposite sample material.

Source: Author.

6. Implementation

Starch and cellulose are both polysaccharides, but they differ in their structures and functions. Starch, derived from plants, consists of glucose units linked by alpha-1,4 and alpha-1,6 glycosidic bonds. It serves as an energy storage molecule in plants. Cellulose, also plant-derived, is composed of glucose units linked by beta-1,4 glycosidic bonds. Its linear structure and intermolecular hydrogen bonding create a strong, rigid polymer. In biocomposite materials, starch is often used for its biodegradability and renewable sourcing, while cellulose provides strength and stability due to its fibrous nature. Blending these polymers can yield biocomposites with a balance of properties suitable for various applications.

When starch (Figure 5.3) and cellulose (Figure 5.2) are mixed with oil in the context of biocomposite materials, it typically involves creating blends or composite to achieve specific properties. The oil components can act as a plasticizer, enhancing the flexibility and process capability of the composite material. It may also improve compatibility between the hydrophilic starch/cellulose and hydrophobic oil phases. The resulting mixture (Figure 5.4) can exhibit improved mechanical properties, such as flexibility and elongation, making it suitable for applications like biodegradable packaging or coating. The choice of oil and its concentration

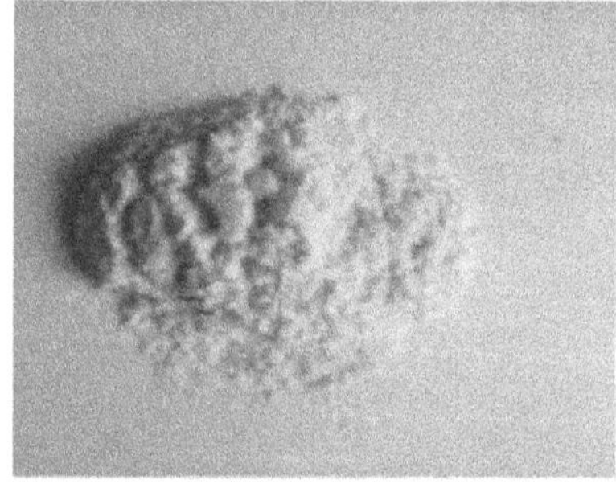

Figure 5.2. Cellulose powder.

Source: Author.

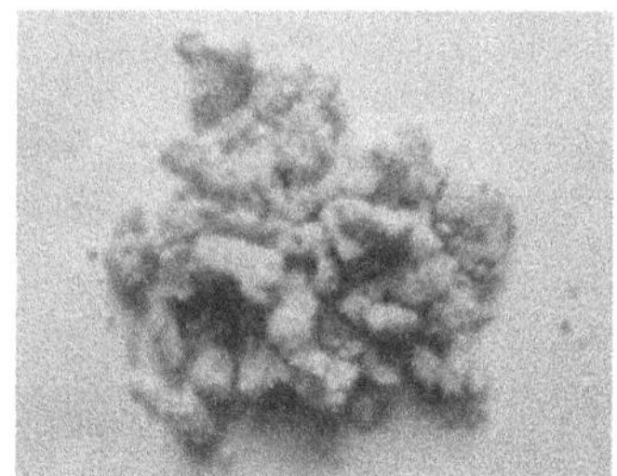

Figure 5.3. Starch powder.

Source: Author.

in the mixture can influence the final material's characteristics, including its thermal stability, moisture resistance, and overall behaviour in various contexts environments. This approach leverages the unique properties of each component to generate a composite material with a tailored set of features. Following this blending process, the composite is placed in a mould, exposed to sunlight for 2 to 3 days, and subsequently dried (Figure 5.5). The outcome is a bio-composite material with a tailored set of features, leveraging the unique properties of each component, and ready for application.

7. Conclusion

Creating bio-composite materials using modified starch and microcrystalline cellulose through compression holds significant promise for medical applications. These materials offer distinct advantages in the medical field, meeting the growing demand for eco-friendly and biodegradable alternatives. Their use of sustainable and biodegradable natural materials makes them particularly suitable for medical devices and implants, reducing the risk of long-term environmental contamination compared to traditional plastics. The compression method employed in this study ensures a strong bond between the components, resulting in a durable material critical for implantable devices and prosthetics, thereby enhancing their longevity and performance in medical settings.

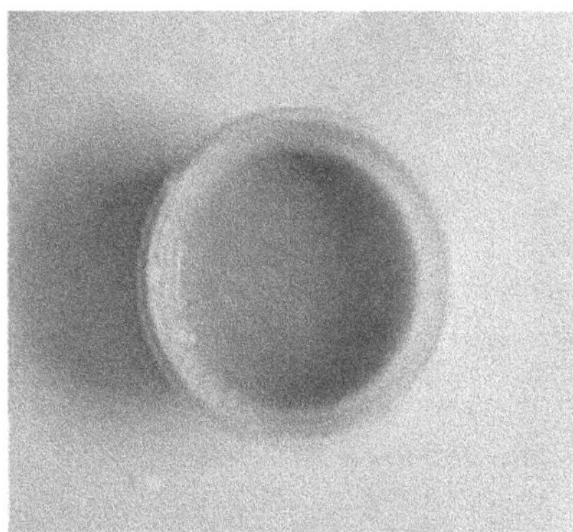

Figure 5.4. Mixture of both the cellulose and modified starch.

Source: Author.

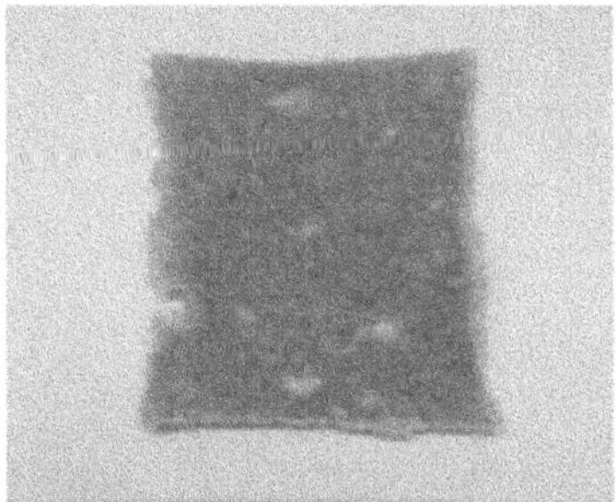

Figure 5.5. Developed biocomposite material.

Source: Author.

The development of this bio-composite material represents a promising advancement in sustainable materials for medical applications. Its renewable, biodegradable, and durable properties make it an environmentally conscious choice that has the potential to improve patient outcomes while reducing the industry's ecological footprint.

References

[1] Mahmud, M. A., Belal, S. A., & Gafur, M. A. (2023). Development of a biocomposite material using sugarcane bagasse and modified starch for packaging purposes. *Journal of Materials Research and Technology*, *24*, 1856–1874.

[2] Islam, M. S., Islam, M. M., Islam, K. N., & Sobuz, M. H. R. (2022). Biodegradable and bio-based environmentally friendly polymers. *Encyclopaedia of Materials: Plastics and Polymers*, Elsevier, 820–836.

[3] Siraj, M. T., Alshybani, I., Payel, S. B., Shahadat, M. R. B., & Rahman, M. Z. (2023). Advances in biocomposite fabrication: Emerging technologies and their potential applications. *Reference Module in Materials Science and Materials Engineering*, Elsevier.

[4] Zanela, J., Reis, M. O., & Shirai, M. A. (2023). Chapter 4 Recent studies on starch-based materials: Blends, composites, and nanocomposites. *Handbook of Natural Polymers, 1*.

[5] Yan, X., Liu, R., Bai, J., Wang, Y., & Fu, J. (2023). Preparation of starch-palmitic acid complex nanoparticles and their effect on properties of the starch composite film. *International Journal of Biological Macromolecules*, Elsevier.

[6] Surendren, A., Mohanty, A. K., Liu, Q., & Misra, M. (2022).A review of biodegradable thermoplastic starches, their blends and composites: Recent developments and opportunities for single-use plastic packaging alternatives. *Green Chemistry*, Elsevier.

[7] Patil, S., Bharimalla, A. K., Nadanathangam, V., Dhakane-Lad, J., Mahapatra, A., Jagajanantha, P., & Saxena, S. (2022). Nano cellulose reinforced corn starch-based biocomposite films: Composite optimization, characterization and storage studies. *Food Packaging and Shelf Life*, Elsevier.

[8] Santhosh, R., & Sarkar, P. (2022). Jackfruit seed starch/tamarind kernel xyloglucan/zinc oxide nanoparticles-based composite films: Preparation, characterization, and application on tomato (Solanum lycopersicum) fruits. *Food Hydrocolloids*, Elsevier.

[9] Santos, T. A., & Spinace, M. A. (2021). Sandwich panel biocomposite of thermoplastic corn starch and bacterial cellulose. *International Journal of Biological Macromolecules*, Elsevier.

[10] Kamaruddin, Z. H., Jumaidin, R., Ilyas, R. A., Selamat, M. Z., Alamjuri, R. H., & Yusof, F. A. M. (2022). Biò composite of cassava starch-cymbopogon citratus fibre: Mechanical, thermal and biodegradation properties. *Polymers*.

[11] Dewi, R., Sylvia, N., & Riza, M. (2021). The effect of rice husk and saw dusk filler on mechanical property of bio composite from Sago Starch. *International Journal of Engineering, Science and Information Technology*, *1*(3), 98–103.

[12] Ferfari, O., Belaadi, A., Bedjaoui, A., Alshahrani, H., & Khan, M. K. (2023). Characterization of a new cellulose fiber extracted from Syagrus Romanzoffiana rachis as a potential reinforcement in biocomposites materials. *Materials Today Communications*, *36*, 106576.

[13] Rijal, M. S., Nasir, M., Purwasasmita, B. S., & Asri, L. A. (2023). Cellulose nanocrystals-microfibrils biocomposite with improved membrane performance. *Carbohydrate Polymer Technologies and Applications*, *5*, 100326.

[14] Khalili, H., Bahloul, A., Ablouh, E. H., Sehaqui, H., Kassab, Z., Hassani, F. Z. S. A., & El Achaby, M. (2023). Starch biocomposites based on cellulose microfibers and nanocrystals extracted from alfa fibers (Stipa tenacissima). *International Journal of Biological Macromolecules*, *226*, 345–356.

[15] Ruz-Cruz, M. A., Herrera-Franco, P. J., Flores-Johnson, E. A., Moreno-Chulim, M. V., Galera-Manzano, L. M., & Valadez-González, A. (2022). Thermal and mechanical properties of PLA-based multiscale cellulosic biocomposites. *journal of materials research and technology*, *18*, 485–495.

[16] Zhou, Y., Katsou, E., & Fan, M. (2021). Interfacial structure and property of eco-friendly carboxymethyl cellulose/poly (3-hydroxybutyrate-co-3-hydroxyvalerate) biocomposites. *International Journal of Biological Macromolecules*, *179*, 550–556.

[17] Hamdan, M. A., Ramli, N. A., Othman, N. A., Amin, K. N. M., & Adam, F. (2021). Characterization and property investigation of microcrystalline cellulose (MCC) and carboxymethyl cellulose (CMC) filler on the carrageenan-based biocomposite film. *Materials Today: Proceedings*, *42*, 56–62.

[18] Moreno, A. G., Guzman-Puyol, S., Domínguez, E., Benítez, J. J., Segado, P., Lauciello, S., ... & Heredia-Guerrero, J. A. (2021). Pectin-cellulose nanocrystal biocomposites: Tuning of physical properties and biodegradability. *International journal of biological macromolecules*, *180*, 709–717.

[19] Shih, Y. T., & Zhao, Y. (2021). Development, characterization and validation of starch based biocomposite films reinforced by cellulose nanofiber as edible muffin liner. *Food Packaging and Shelf Life*, *28*, 100655.

[20] Li, K., Mcgrady, D., Zhao, X., Ker, D., Tekinalp, H., He, X., ... & Ozcan, S. (2021). Surface-modified and oven-dried microfibrillated cellulose reinforced biocomposites: Cellulose network enabled high performance. *Carbohydrate Polymers*, *256*, 117525.

[21] Doineau, E., Coqueugniot, G., Pucci, M. F., Caro, A. S., Cathala, B., Bénézet, J. C., ... & Le Moigne, N. (2021). Hierarchical thermoplastic biocomposites reinforced with flax fibres modified by xyloglucan and cellulose nanocrystals. *Carbohydrate Polymers*, *254*, 117403.

[22] Omar, F. N., Hafid, H. S., Zhu, J., Bahrin, E. K., Nadzri, F. Z. M., & Wakisaka, M. (2022). Starch-based composite film reinforcement with modified cellulose from bamboo for sustainable packaging application. *Materials Today Communications*, *33*, 104392.

[23] Zhang, W., Azizi-Lalabadi, M., Jafarzadeh, S., & Jafari, S. M. (2023). Starch-gelatin blend films: A promising approach for high-performance degradable food packaging. *Carbohydrate polymers*, *320*, 121266.

[24] He, X., Wang, W., Qiu, J., Hou, Y., & Shuai, Y. (2023). Controllable preparation method and thermal properties of composite phase change materials based on starch pore formation. *Solar Energy Materials and Solar Cells*, *253*, 112255.

[25] Gacoin, A., & Li, A. (2023). Optimal composition of a starch-hemp agro-composite materials. *Construction and Building Materials*, *400*, 132711.

[26] Thakwani, Y., Karwa, A., Bg, P. K., Purkait, M. K., & Changmai, M. (2023). A composite starch-date seeds extract based biodegradable film for food packaging application. *Food Bioscience*, *54*, 102818.

[27] Sankaran, S., Murugan, P. R., & Simman, A. (2022). Statistical and Experimental Analysis of the Mechanical Properties of Jute Fiber Reinforced Epoxy Composite for the Transfemoral Prosthesis Biocomposite Socket Applications. *Journal of Natural Fibers*, *19*(16), 13840–13851.

[28] Oluba, O. M., Osayame, E., & Shoyombo, A. O. (2021). Production and characterization of keratin-starch bio-composite film from chicken feather waste and turmeric starch. *Biocatalysis and Agricultural Biotechnology*, *33*, 101996.

[29] Vidmar, B., Oberlintner, A., Stres, B., Likozar, B., & Novak, U. (2023). Biodegradation of polysaccharide-based biocomposites with acetylated cellulose nanocrystals, alginate and chitosan in aqueous environment. *International Journal of Biological Macromolecules*, *252*, 126433.

[30] Nguyen, H. T., Saha, N., Ngwabebhoh, F. A., Zandraa, O., Saha, T., & Saha, P. (2023). Silane-modified kombucha-derived cellulose/polyurethane/polylactic acid biocomposites for prospective application as leather alternative. *Sustainable Materials and Technologies*, *36*, e00611.

[31] Hoque, M. E., Sharif, A., & Jawaid, M. (Eds.). (2021). *Green Biocomposites for Biomedical Engineering: Design, Properties, and Applications*. Woodhead Publishing.

[32] Sankaran, S., Murugan, P. R., Johnson, J. C., Abdullah, H. J. S., Raj, C. M. N., & Ashokan, D. (2019, April). Prevention of skin problems in patients using prosthetic limb: A review of current technologies and limitations. In *2019 International Conference on Communication and Signal Processing (ICCSP)* (pp. 0077–0081). IEEE.

6 Modelling and design analysis of a ripple-free high step-up converter for fuel cell application

Abhijeet Madhukar Haval[1,a] *and Dhablia Dharmesh Kirit*[2,b]

[1]Assistant Professor, Department of CS & IT, Kalinga University, Raipur, India
[2]Research Scholar, Department of CS & IT, Kalinga University, Raipur, India

Abstract: As a source of energy that is both high-power density and environmentally friendly, fuel cells are becoming increasingly widespread across a wide range of industries. Fuel cells are a source of energy that is becoming increasingly widespread. Direct current (dc) to direct current (dc–dc) converters are utilized in the beginning stages of fuel cell systems. These converters are utilized to convert the variable, low-level voltage output from the fuel cell into a stable, high-level voltage. For the purpose of ensuring that the fuel cell continues to exhibit its high power density characteristic, it is necessary to incorporate dc–dc converters that possess a significant amount of power density. Within the scope of this investigation, a DC-DC converter that makes use of high-voltage gain is presented. This converter is able to effectively extract smooth current from the source by utilising duty cycles that can be adjusted. The significance of this converter topology is especially noteworthy in the context of renewable energy systems that are characterized by low-voltage authorities. It is extremely likely that these systems will experience current ripples as a result of conventional converters. As part of the investigation, the fundamentals of the converter's operation are investigated, with a particular emphasis on modelling and design suggestions. It has been demonstrated that the proposed topology is effective by means of computer-based PSIM.

Keywords: Converter, ripple, voltage, power density, high power, PSIM

1. Introduction

The use of fuel cells as a source of energy that is both high-power density and environmentally friendly is becoming increasingly widespread across a wide range of industries. Within the initial stages of fuel cell systems, direct current (dc) to direct current (dc–dc) converters are employed to transform a variable, low-level voltage output from the fuel cell into a stable, high-level voltage [1]. For the purpose of preserving the high power density characteristic of the fuel cell, it is necessary to incorporate dc–dc converters that possess a significant power density. In addition, these converters need to demonstrate a minimal fluctuation in the input current in order to prevent any potential adverse impact on the operational lifespan of the fuel cell [13].

Conventional non-isolated buck-boost DC-DC converters are widely used because they are simple to design and because they are cost-effective. The voltage conversion ratio is restricted due to the presence of parasitic elements in the circuit as well as an extreme duty cycle. In addition, when it comes to high voltage applications, its power switch is not suitable because of the strains that are caused by high voltage. Over the course of the last few decades, numerous improved non-isolated DC-DC converters have been introduced [3] in order to address these deficiencies [2].

Generally speaking, photovoltaic (PV) systems generate low-level output voltages, which are required to be increased in order to be utilized in practical applications [14]. However, these systems experience fluctuations in the maximum power point as a result of inherent oscillations. These oscillations are made worse by the input current ripples that originate from the series-connected dc-dc converter stage that is connected to the PV source. An increase in the output voltage of a fuel cell was achieved through the utilisation of a high-voltage gain single-input single-output converter throughout the course of a previous research study [5, 17]. The existing body of literature [7] provides documentation of a wide range of high-voltage-gain direct current to direct current converters. The following types of converters are included in this category: full-bridge isolated converters, non-isolated converters that use switched capacitor (SC) circuits, non-isolated interleaved converters with multi-winding coupled inductors, non-isolated converters that incorporate diode-capacitorinductor (D-C-L) circuits, and non-isolated interleaved converters that feature diode-capacitor settings. One thing that should be brought to your attention is that some of these designs involve a relatively high number of components, which results in losses associated with them. Utilising multiple voltage multiplier (VM) cells is a method that is comparable to the approach that is utilized in certain converters in [9, 16]. This method can be utilized to achieve the goal of increasing the voltage gain. Higher voltage gain is achieved through the utilisation of these VM cells, which

[a]ku.abhijeetmadhukarhaval@kalingauniversity.ac.in, [b]dhablia.dharmesh@kalingauniversity.ac.in

DOI: 10.1201/9781003675259-6

are made up of diode-capacitor pairs. The performance of some configurations is improved by the elimination of input current ripples, while others have multi-input single-output features [4].

There have been a number of dc-dc converters introduced that have the capability to cancel out ripples in the input current. These converters offer voltage gains that are comparable to those of conventional boost converters. A few examples of specific techniques include the utilisation of three-winding coupled inductors for the purpose of suppressing input current ripple, the utilisation of filter blocks for the purpose of achieving ripple-free input/output currents, and the incorporation of ripple-cancelation networks within interleaved converters [10]. With regard to photovoltaic (PV) systems, these innovations are especially appealing. Furthermore, bidirectional direct current-direct current converters have been proposed. These converters make use of coupled inductors and interleaved structures in order to improve their conversion ratios and reduce the negative effects of ripples in the input current [15]. In conclusion, the research landscape presents a number of different strategies that can be utilized to improve the voltage output of photovoltaic (PV) systems while simultaneously reducing the ripples in the input current. These advancements have the potential to improve the efficiency of photovoltaic applications as well as their operational viability [11, 12].

Using the findings of the literature review, the purpose of this paper is to develop a converter that is free of ripples and has a high gain. Because of its emphasis on high voltage gain, low input currency ripple, and low voltage stress on power devices, it is also suitable for use in fuel cell applications, battery storage applications, and other similar applications. A conventional boost converter is included in addition to the two stages that are present. The parasitic resistor power dissipations of the magnetic and semiconductor components are taken into consideration in order to improve the efficiency of the system. Furthermore, the low voltage stress that is placed on the output diode is yet another advantage of this converter device [6].

2. Modelling of Proposed Converter

The suggested converter design involves integrating two boost stages, as shown in Figure 6.1. This configuration results in a converter that operates across four distinct modes [8].

In first mode, switch is on and diode *D*2 is conducting. The input, magnetizing, output inductor (*Lin*, *Lm*, *L*0) receives energy from the input voltage source, input capacitance *Cin*, capacitor *C*2 respectively and then their current increases linearly expressed by the following expressions:

$$V_{Lin} = V_s - V_{Cin}$$

$$V_{Lm} = V_{Cin}$$

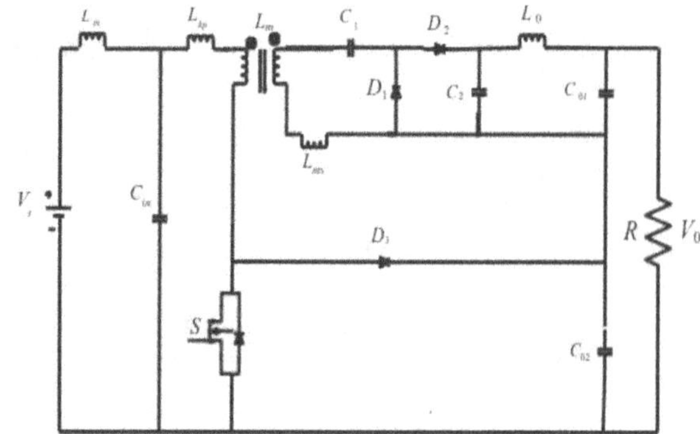

Figure 6.1. Circuit diagram of proposed converter.

Source: Author.

In second mode, the switch is turned off and diode, *D*1 is on and the expressions for this mode are as follows:

$$V_{Lin}^{off} = V_s - V_{Cin}$$

$$V_0 = V_{C01} + V_{C02}$$

In mode third, *D*3 is also turned off. For the fourth mode of operation, all the power devices is turned off, Input capacitor *Cin* is storing the energy from the input source and output capacitors are giving the energy to the load.

3. Results and Discussion

3.1. *Efficiency analysis*

The parasitic resistors in the circuit components can be used to estimate the theoretical efficiency of the converter. The following is an outline of the main parasitic elements of the suggested converter elements:

- *rLin*, *rLm*, *rL*0 are the equivalent series resistance of input, coupled and output inductor.
- *rCi*, *rC*1, *rC*2 are the equivalent series resistance of the capacitors.
- *VD*1, *VD*2, *VD*3 are the threshold voltages of the diodes.
- RDS is the static drain-to-source resistance of the power switch.

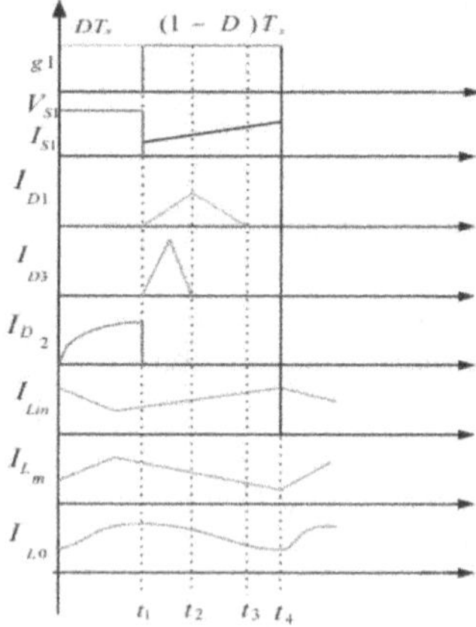

The operation of the converter is shown by the waveforms in the Figure 6.1. Voltage gain depends on the turn-ratio and duty cycle, as shown in the Figure 6.2. It is clear that there is a lot of scope for improvement in the voltage conversion ratio of the suggested converter. When d = 0.5 and n = 2, the converter achieves an impressive ideal voltage gain of MCCM = 6.5.

The input voltage, which is typically drawn from renewable energy sources like fuel cells and battery stations, is assumed to be 30 volts. This converter topology's gain is 6.5 times on D = 0.5, indicating that 200V is the output voltage that is utilized for renewable energy sources. Based on the simulation results, it is clear that the power switch and output diode are under less voltage stress than the output voltage.

Table 6.1 represents the comparative analysis of the proposed converter along with earlier published converter topologies. Comparison is done on the basis of number of components (switches, diodes, capacitors, inductor, coupled-inductor) voltage gain, input current ripple, utilization factor, voltage and current stress across the MOSFET devices. Figure 6.3 illustrates the graph of the voltage gain for various duty cycles.

4. Conclusion

A high voltage ripple-free step-up converter is examined in this paper used in fuel cells, battery energy storage and photovoltaic arrays. Coupled-inductor is implemented for achieving the high gain, this topology only used a single switch which simplifies the controlling pulse. This topology is advantages for getting high output voltage, making it well suited for applications like DC microgrids, EVs, and renewable sources. While using the coupled inductor, the proposed configuration achieves a substantial increase in voltage conversion efficiency for the low-duty cycle. this is achieved by reducing the turns ratio of the transformer. In addition to offering a high voltage gain with minimal normalized voltage stress across its power switch and diodes, the new topology carries over the advantages of the non-isolated transformer-type ripple-free converter, as demonstrated in the comparative section. To validate its performance, the converter's precision is confirmed through the simulation platform PSIM. In the Future, the hardware implementation will be done.

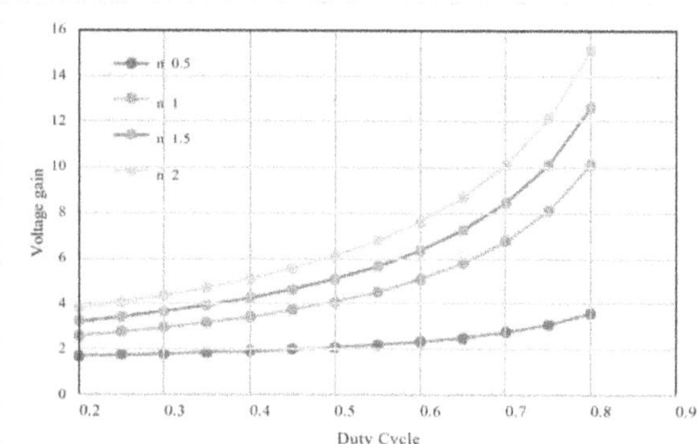

Figure 6.2. Graph between voltage gain corresponding to various turns ratio and duty cycle.

Source: Author.

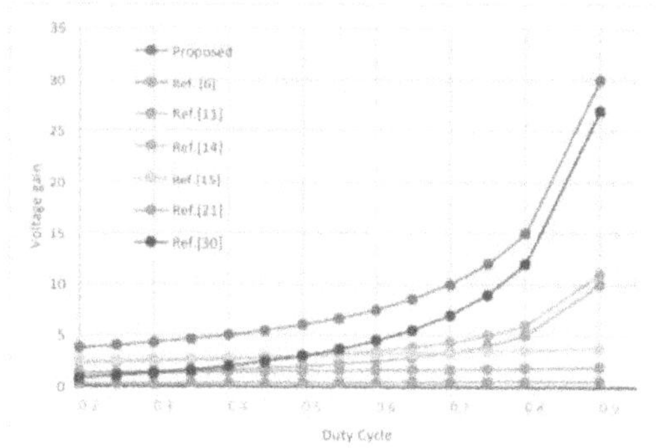

Figure 6.3. Graph between voltage gain corresponding to duty cycle for all the converters.

Source: Author.

Table 6.1. Comparative performance analysis of proposed converter

No. of components S/D/C/L=T	*Voltage gain*	*Voltage stress on MOSFETs*	*Voltage stress on output diodes*	*Input current ripple*	*Utilization factor*	*Efficiency*	*Converter*
11	$n1-d$	$Vin1-d$	$(2n-1)Vin1-d$	High	0.0064	95	[6]
7	$1+d$	Vin	Vin	Low	0.9	98.18%	[11]
13	$2+d1-d)$	$(1+M)Uhigh\ 3$	-	Low	0.16	91.12%	[14]
15	$2+nd\ 1-d$	$Vin1-d$	$2Vin1-d$	High	0.12	92.5%	[15]
15	$N+2\ 4d+1$	$4Vin\ 4d+1$	-	Low	0.099	95.4%	[21]
12	$2n2\ 1-d$	$Vin1-d$	$2nVin1-d$	Low	0.25	93%	[32]
12	$1+n1-d$	$Vo-nVin1+d$	$nV0\ 1+d$	Very Low	0.002	95%	Proposed

Source: https://www.researchgate.net/publication/343363115_Design_and_Development_of_Non-Isolated_Modified_SEPIC_DC-DC_Converter_Topology_for_High-Step-Up_Applications_Investigation_and_Hardware_Implementation/figures?lo=1&utm_source=google&utm_medium=organic.

References

[1] Bi, H., Mu, Z., & Chen, Y. (2022). Common grounded wide voltage-gain range DC–DC converter with zero input current ripple and reduced voltage stresses for fuel cell vehicles. *IEEE Transactions on Industrial Electronics*, *70*(3), 2607–2616.

[2] Babenko, V., Danilov, A., Vasenin, D., & Krysanov, V. (2021). Parametric optimization of the structure of controlled high-voltage capacitor batteries. *Archives for Technical Sciences*, *1*(24), 9–16.

[3] Chaturvedi, R., Sharma, A., & Islam, A. (2021). Design analysis of BRB energy dissipated devices in commercial building structures. *Materials Today: Proceedings*, *45*, 2949–2952.

[4] Rajesh, D., Giji Kiruba, D., & Ramesh, D. (2023). Energy proficient secure clustered protocol in mobile wireless sensor network utilizing blue brain technology. *Indian Journal of Information Sources and Services*, *13*(2), 30–38.

[5] Lee, S. W., & Do, H. L. (2017). Isolated SEPIC DC–DC converter with ripple-free input current and lossless snubber. *IEEE transactions on industrial electronics*, *65*(2), 1254–1262.

[6] Enokido, T., Aikebaier, A., & Takizawa, M. (2011). Computation and transmission rate based algorithm for reducing the total power consumption. *Journal of Wireless Mobile Networks, Ubiquitous Computing, and Dependable Applications*, *2*(2), 1–18.

[7] Chaturvedi, R., & Singh, P. K. (2021). A practicable learning under conversion of plastic waste and building material waste keen on concrete tiles. *Materials Today: Proceedings*, *45*, 2938–2942.

[8] Santhosh, G., & Prasad, K. V. (2023). Energy saving scheme for compressed data sensing towards improving network lifetime for cluster based WSN. *Journal of Internet Services and Information Security*, *13*(1), 64–77.

[9] Andres, B., Faistel, T. M. K., Andrade, A. M. S. S., Roggia, L., & Schuch, L. (2023). Analysis and comparison of high step-up converters based on Greinacher voltage multiplier cells. *Int. J. Circuit Theory Appl*, *51*(1), 115–146.

[10] Sharma, A., Sharma, K., Islam, A., & Roy, D. (2020). Effect of welding parameters on automated robotic arc welding process. *Materials Today: Proceedings*, *26*, 2363–2367.

[11] Akin, O., Gulmez, U. C., Sazak, O., Yagmur, O. U., & Angin, P. (2022). GreenSlice: An energy-efficient secure network slicing framework. *Journal of Internet Services and Information Security*, *12*(1), 57–71.

[12] Islam, A., Sharma, S., Sharma, K., Sharma, R., Sharma, A., & Roy, D. (2020). Real-time data monitoring through sensors in robotized shielded metal arc welding. *Materials Today: Proceedings*, *26*, 2368–2373.

[13] Goudarzian, A., Khosravi, A., & Ali Raeisi, H. (2022). Modeling, design and control of a modified flyback converter with ability of righthalf-plane zero alleviation in continuous conduction mode. *Eng. Sci. Technol. an Int. J*, *26*(xxxx).

[14] Chaturvedi, R., Islam, A., & Sharma, A. (2022). Analysis on manufacturing automated guided vehicle for MSME Projects and its fabrication. In *Computational and Experimental Methods in Mechanical Engineering: Proceedings of ICCEMME 2021* (pp. 357–366). Springer Singapore.

[15] Yan, Z., et al. (2022). Battery-friendly DC-DC converter with ripple free current modulation strategy. *IEEE J. Emerg. Sel. Top. Power Electron*, *11*(2), 2300–2310.

[16] Hasanpour, S., Siwakoti, Y., & Blaabjerg, F. (2020). Hybrid cascaded high step-up DC/DC converter with continuous input current forrenewable energy applications. *IET Power Electron*, *13*(15), 3487–3495.

[17] Yan, Z., et al. (2022). Ripple-free bidirectional DC-DC converter with wide ZVS range for battery charging/discharging system. *IEEE Trans. Ind. Electron*, *70*(10), 9992–10002.

7 Transcutaneous electrical nerve stimulator for cervical cancer

S. Shanmugapriya[a], S. Mahathi[b], and K. Vishaly[c]

Department of Biomedical Engineering, Kalasalingam Academy of Research and Education, Krishnankoil, Tamil Nadu, India

Abstract: Transcutaneous Electrical Nerve Stimulation (TENS) is a non-invasive treatment that applies low-voltage electrical currents to particular body parts using a tiny, battery-powered device. It has been investigated in several medical settings and is frequently used for pain management, including TENS, which reduces acute and chronic pain. It works by stimulating nerves to block or reduce the perception of pain signals. TENS can help with muscle rehabilitation after injury or surgery by promoting muscle contractions and reducing muscle atrophy. TENS has been used as a non-pharmacological pain relief method during labour, providing women with a drug-free option for pain management during childbirth. TENS is sometimes used to manage neuropathic pain conditions, such as diabetic neuropathy or phantom limb pain. Some individuals with fibromyalgia have found relief from pain and muscle stiffness through TENS therapy. TENS may be used to manage the pain associated with osteoarthritis, particularly in the knees and hips. TENS units can be applied to the head and neck to help alleviate certain types of headaches and migraines. The exact mechanism of how TENS works isn't fully understood, but it's believed to modulate pain perception through the stimulation of sensory nerves. Because the right electrode placement and settings are essential to TENS's effectiveness, it should only be administered under a healthcare provider's supervision. Because the right electrode placement and settings are essential to TENS's effectiveness, it should only be administered under a healthcare provider's supervision. If you have a specific question or need more information about TENS in a particular context, please feel free to ask.

Keywords: TENS, cervical cancer, human papillomavirus, Pap test

1. Introduction

Cervical cancer is common in women nowadays. It starts with cancer in the cervix in the lower parts of the urinary bladder, which may be caused due to human papillomavirus [3]. All over the country, there are many women affected by the many that have led to death due to lack of awareness. Cervical cancer is when an abnormality is found by treating the cervix region. According to WHO, a pap test was taken and coloscopic images were taken to find the proper image view of the cervix [8]. HPV vaccination is used to detect cancer across countries [9]. Artificial intelligence and Pap test is helping to detect cervical cancer across the country. To reduce the pain of this cancer TENS is used to reduce mainly neck pain which is caused by this cancer and to reduce lower back pain [10]. CT scans and MRI scans are taken for the picture quality of cervical cancer behind the vagina they use techniques like 3D images for better scanning and providing clinical treatment for infected persons [11]. Acupuncture point was determined with several people's database for curing cancer pain in a particular region using TENS which is a noninvasive method that can be placed in any part of the body [12]. Acetic acid test is a method that they use in this paper in which they take an image of the portion of the affected area in the urinary bladder [14]. Irrespective of diagnosis protocol 381 uses TENS to reduce pain in the pain. Figure 7.1 represents the symptoms of cervical cancer.

2. Literature Review

TENS use invasive to stimulate nerves in neck region pain this may include techniques like cervical epidural steroid injection, radio frequency, ablation, or spinal cord stimulation. To ascertain the precise area and possible advantages of TENS therapy, we must speak with a qualified physician [1]. Figure 7.1 represents symptoms of cervical cancer.

TENS can be used to treat pain in place of medication therapy such as immunotherapy, hormone therapy, or chemotherapy [2]. People with cervical cancer after surgery with depression, and anxiety to reduce this clinical trial is used in this particular paper the persons are suffering they have taken a report and analyze the literature with Revman 5.3 software. Figure 7.2 represents the building blocks of TENS unit [5].

The administrator of TENS provides acupuncture-like TENS to produce non–painful senses in the pain region. For blind people with cervical cancer researchers have to find controlling devices and other pain relief to provide the results

[a]shanmugapriya.s@klu.ac.in, [b]9921020014@klu.ac.in, [c]9921020012@klu.ac.in

DOI: 10.1201/9781003675259-7

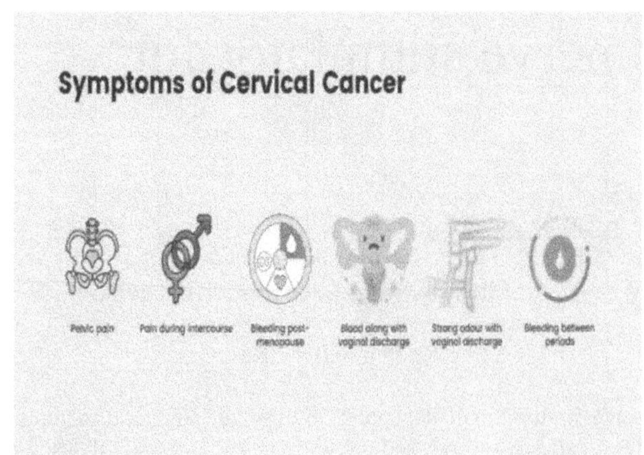

Figure 7.1. Symptoms of cervical cancer.

Source: Regency medical centre.

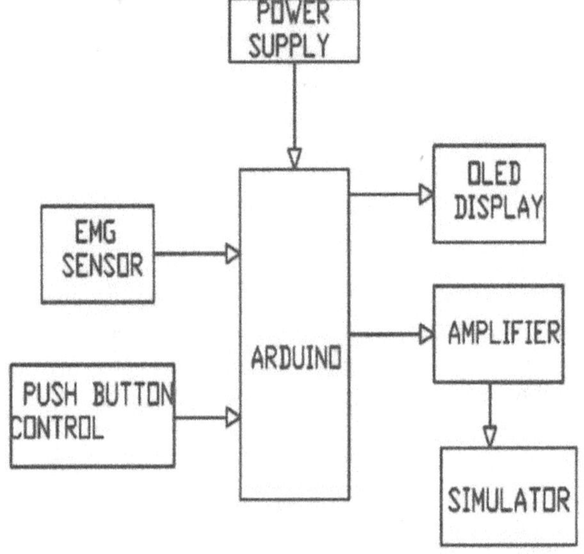

Figure 7.2. Building block of TENS unit.

Source: Author.

which have been found in many databases and the records have been recorded in an assessment based on research peoples [6]. Muscle contraction using low frequency to reduce neck pain by using TENS, clinical evidence and analgesic effects of TENS for persons is listed and analgesic effects have taken place in pain region [13]. For women with labor pain and cervical cancer, this TENS is used in the lower back to reduce the pain temporarily [7]. The paralytic ileus was evaluated in persons who have undergone gynecologic operation based on cancer they use TENS for pain relief during post-surgery [11]. Severe urinary pain with cervical cancer uses TENS In the lower hip to reduce pain these record is taken by doctors to review every year [12]. CT scans and MRI scans are taken for the picture quality of cervical cancer behind the vagina they use techniques like 3D images for better scanning and providing clinical treatment for infected persons. Acupuncture point was determined with several people's [13]. Database for curing cancer pain in a particular region using TENS which is a noninvasive method that can be placed in any part of the body [7]. Acetic acid test is a method that they use in this paper in which they take an image of the portion of the affected area in the urinary bladder [14]. Irrespective of diagnosis protocol 381 uses TENS to reduce pain in the pain region [4].

3. Material and Methods

We go into great length on the elements used in this approach. The design of the system was based on the elements that are somewhat common in this field. Usually using low-voltage electrical currents to stimulate the nerves across the skin, TENS units help to relieve pain [15]. The design of the unit evaluated elements including the type of electrode location, the frequency and intensity of electrical current. The block diagram Figure 7.2 represents the blocking of TENS unit.

3.1. EMG sensor

The EMG sensor uses a conductive pad applied to the skin to identify electrical activity coming from a muscle. Individual muscle fibres get electrical impulses upon activation, which causes the muscles to contract each time. Numerous research labs studying EMG sensors and signals find application in biomechanics, movement disorders, motor control, postural control, neuromuscular physiology, and physical therapy as well. Electrical signals transmitted by motor neurons produce muscle contraction. Using small devices called electrodes, an EMG turns these impulses into graphs, sounds, or numerical data. Which are subsequently analyzed by a professional. The analogue output signal uses 1.5 V as its reference voltage. The range of the output voltage is 0 to 3.0 V. The degree of muscle activity determines the strength of the signal. The signal output.

3.2. Power supply

A power supply is any piece of electrical equipment used to supply electricity to a load. The main purposes of a power supply are converting electric current from a source into the correct voltage, current, and frequency needed to run a load. Power supplies are hence sometimes known as electric power converters. While some power sources are standalone devices, others are included into the load appliances they run. In a hybrid series parallel resonant architecture (PRCLCC) high-voltage power converter uses a capacitor as the output filter. This construction helps to minimize the transformer radio's secondary volume as well.

3.3. Arduino UNO

Arduino UNO microcontroller uses the ATmega328 as controller. Among novices, electronics projects find great popularity with the Arduino UNO board. Arduino.cc created the open-source Arduino Uno microcontroller board depending on the Microchip ATmega328P CPU. It first appeared in 2010. Both digital and analog, the board comprises input/output (I/O) pins that can be coupled to various expansion boards and other circuitry. Six PWM-capable digital I/O pins and six analogue I/O pins comprise the board. One may program it using a type B USB connector and the Arduino IDE, or Integrated Development Environment. A barrel connection or a USB cable allowing voltages between 7 and 20 will power it.

3.4. *OLED display*

OLED, or Organic Light-Emitting Diode, is a technology that uses organic molecules to generate light through the use of LEDs. These organic LEDs are used to create the best display screens in the world. Compared to traditional displays, which only use red, green, and blue light to make images, OLEDs add an additional white light to create even more color. This leads to richer, more varied, and more realistic images. If you care about image quality, OLED can really deliver the best.

3.5. *Amplifier*

Electrical amplifier, also known as an electronic amplifier, is a circuit that generates a higher output signal than its input by using an external power supply. A use that is easily recognized is an audio amplifier, which is used to boost a speaker's voice loudness for greater audibility in vast spaces. The sections of an audio input, amplifier, driver, output, and power supply make up a power amplifier. Most of the voltage gain is produced by the amplifier portion. Between the amplifier portion and the output stage, the driver stage serves as a buffer. Typically, a low-impedance load like a loudspeaker needs to be driven by the output stage.

3.6. *Push button control*

An internal switching mechanism of a push button switch is activated by the user manually pressing a button on the mechanical device, which controls an electrical circuit. Depending on the needs of the design, they are available in a wide range of sizes, forms, and configurations. A push button switch's primary purpose is to turn anything on or off. Nevertheless, there are various kinds of push button switches, and each kind has a unique purpose.

4. System Implementation

4.1. *Hardware implementation*

The design of a TENS device for patients with cervical cancer required taking into account the particular requirements and difficulties faced by this demographic, including the location and intensity the Figure 7.3 represents the hardware of TENS.

5. Result and Discussion

An intriguing application is transcutaneous electrical nerve stimulation (TENS) for cervical cancer. Although TENS is frequently used to relieve pain, treating cancer directly with it is uncommon. It is plausible, therefore, that TENS is being investigated as an adjunctive therapy for the management of pain associated with cancer or medication side effects. This may entail analysing the data you've gathered statistically. You can explore the ramifications of your findings, how they stack up against the body of literature, and possible future directions or uses for TENS in the treatment of cervical cancer throughout the conversation. is a treatment that relieves pain by using low-voltage electrical currents. If you require assistance with the findings and analysis after conducting a TENS study or experiment. The Gate Control idea is one idea that explains the efficiency of TENS. It implies that the electrical impulses emitted by the TENS machine have the ability to 'gate' or obstruct the brain's receipt of pain signals (Figure 7.4).

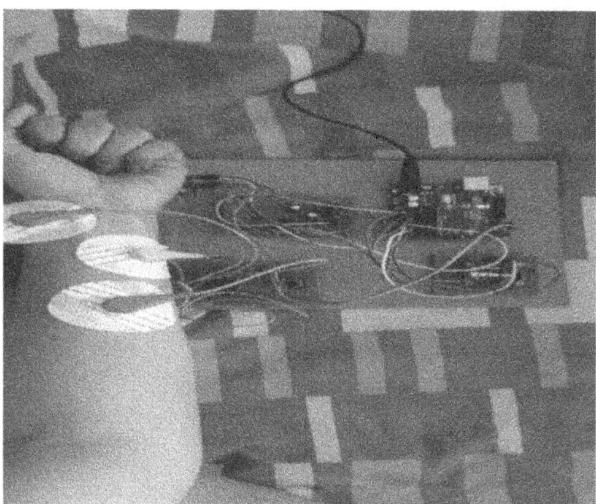

Figure 7.3. TENS Hardware for cervical cancer system.

Source: Author.

```
#include <SoftwareSerial.h>

SoftwareSerial bluetooth(2, 3); // RX, TX

int smokePin = A0;       // Analog pin for the smoke sensor (MQ-135 or similar)
int ledPin = 8;          // Digital pin for the LED
int buzzerPin = 9;       // Digital pin for the buzzer
int threshold = 200;     // Smoke level threshold (adjust as needed)
unsigned long lastMessageTime = 0;
const unsigned long messageInterval = 10000; // Interval for sending messages (10 seconds)

void setup() {
  Serial.begin(9600);  // Initialize Serial Monitor
  pinMode(ledPin, OUTPUT);
  pinMode(buzzerPin, OUTPUT);
  bluetooth.begin(9600);  // Initialize Bluetooth communication
  lastMessageTime = millis(); // Initialize last message time
}

void loop() {
  int smokeLevel = analogRead(smokePin);
  float voltage = (smokeLevel / 1024.0) * 5.0; // Convert analog reading to voltage
  float smokePPM = (voltage - 1.4) / 0.01;     // Calculate smoke concentration in ppm
```

Figure 7.4. Coding for Transcutaneous electrical nerve stimulator.

Source: Author.

6. Conclusion

In summary, transcutaneous electrical nerve stimulation, or TENS, shows promise as an adjuvant treatment for improving cervical cancer symptoms. TENS provides patients with a potentially useful technique for pain relief, lowering pharmaceutical dependence, and enhancing general quality of life due to its non-invasive nature and capacity to modify pain perception. Even though the results of current research are optimistic, more carefully planned clinical trials are necessary to completely clarify its effectiveness, ideal parameters, and long-term advantages in the context of treating cervical cancer. However, the available data indicates that TENS has a great deal of promise as an important part of all-encompassing care plans for patients with cervical cancer, providing a route to better symptom control and wellbeing.

References

[1] Inamdar, M. U., & Mehendale, N. (2021). A review on transcutaneous electrical nerve stimulation and its applications. *SN Comprehensive Clinical Medicine*, 3(12), 2548–2557.

[2] Paolucci, T., Agostini, F., Paoloni, M., de Sire, A., Verna, S., Pesce, M., Ribecco, L., Mangone, M., Bernetti, A., & Saggini, R. (2021). Efficacy of TENS in cervical pain syndromes: An umbrella review of systematic reviews. *Applied Sciences*, 11(8), 3423.

[3] Matikas, A., Foukakis, T., & Bergh, J. (2017). Dose intense, dose dense and tailored dose adjuvant chemotherapy for early breast cancer: An evolution of concepts. *Acta Oncologica*, 56(9), 1143–1151.

[4] Patrono, M. G., Calvo, M. F., Franco, J. V. A., Garrote, V., & Vietto, V. (2021). A systematic review and meta-analysis of the prevalence of therapeutic targets in cervical cancer. *ecancermedicalscience*, 15, 1200.

[5] Johnson, M. I., Paley, C. A., Jones, G., Mulvey, M. R., & Wittkopf, P. G. (2022). Efficacy and safety of transcutaneous electrical nerve stimulation (TENS) for acute and chronic pain in adults: A systematic review and meta-analysis of 381 studies (the meta-TENS study). *BMJ Open*, 12(2), e051073.

[6] Nirmal, R., Yun, C., Le, M., Paripoonnanonda, P., & Yi, J. (2013, September). Digital health game on cervical health and its effect on American women's cervical cancer knowledge. In *2013 IEEE International Games Innovation Conference (IGIC)* (pp. 191–198). IEEE.

[7] Lee, J., Lee, H., Eizad, A., & Yoon, J. (2021, October). Effects of using TENS as electro-tactile feedback for postural balance under muscle fatigue condition. In *2021 21st International Conference on Control, Automation and Systems (ICCAS)* (pp. 1410–1413). IEEE.

[8] Anand, M. S., Manimozhi, S., Walid, M. A. A., Kumar, A. N., Kumar, P. D., & Suryanarayana, N. V. S. (2023, August). CSO—VGG-19: Prediction of Cervical Cancer Using Cuckoo Search-Based Deep VGG-19. In *2023 5th International Conference on Inventive Research in Computing Applications (ICIRCA)* (pp. 932–937). IEEE.

[9] Reeve, J., Menon, D., & Corabian, P. (1996). Transcutaneous electrical nerve stimulation (TENS): A technology assessment. *International Journal of Technology Assessment in Health Care*, 12(2), 299–324.

[10] Rahman, M. L., Alam, M. J., Rashid, N., Tithy, L. H., Ahmed, A. U., & Arafat, M. T. (2020, June). Iot based cost efficient muscle stimulator for biomedical application. In *2020 IEEE Region 10 Symposium (TENSYMP)* (pp. 634–637). IEEE.

[11] Vij, N., Tolson, H., Kiernan, H., Agusala, V., Viswanath, O., & Urits, I. (2022). Pathoanatomy, biomechanics, and treatment of upper cervical ligamentous instability: A literature review. *Orthopedic Reviews*, 14(3), 37099.

[12] Li, Q. Y., Yang, W. X., Yao, L. Q., Chen, H., Li, Z. R., Gong, Y. B., & Shi, J. (2023). Exploring the rules of related parameters in transcutaneous electrical nerve stimulation for cancer pain based on data mining. *Pain and Therapy*, 12(6), 1355–1374.

[13] Wang, S., Sun, X., Cheng, W., Zhang, J., & Wang, J. (2019). Pilot in vitro and in vivo study on a mouse model to evaluate the safety of transcutaneous low-frequency electrical nerve stimulation on cervical cancer patients. *International Urogynecology Journal*, 30, 71–80.

[14] Sun, X. L., Wang, H. B., Wang, Z. Q., Cao, T. T., Yang, X., Han, J. S., Wu, Y. F., Reilly, K. H., & Wang, J. L. (2017). Effect of transcutaneous electrical stimulation treatment on lower urinary tract symptoms after class III radical hysterectomy in cervical cancer patients: Study protocol for a multicentre, randomized controlled trial. *BMC Cancer*, 17, 1–7.

[15] Johnson, M. I., Paley, C. A., Wittkopf, P. G., Mulvey, M. R., & Jones, G. (2022). Characterising the features of 381 clinical studies evaluating transcutaneous electrical nerve stimulation (TENS) for pain relief: A secondary analysis of the meta-TENS study to improve future research. *Medicina*, 58(6), 803.

8 Advanced power electronics for electric vehicle charging stations

Anu G. Pillai[1,a] *and Shital Kewte*[2,b]

[1]Assistant Professor, Department of Electrical, Kalinga University, Raipur, India
[2]Department of Electrical and Electronics Engineering, Kalinga University, Raipur, India

Abstract: Because electric cars (EVs) are becoming more popular so quickly, better charging infrastructure needs to be built to make sure that energy use is efficient and sustainable. This article looks into how power electronics can be used to make EV charging stations better. It talks about current problems like charging speed, grid integration, and energy management. Modern power electronics parts (like inverters and converters) can be combined with smart control programmes to make the charging process run more smoothly and efficiently. Techniques like demand response and load predictions make it even easier for the system to respond to changes in the grid in real time. This paper focuses on how better charging methods could help the growing market for electric vehicles while also being better for the environment and encouraging the use of renewable energy sources.

Keywords: Power electronics, electric vehicles, charging stations, efficiency, renewable energy

1. Introduction

To minimize greenhouse gas emissions and increase energy sustainability, electric automobiles (EVs) and charging infrastructure must be widely embraced. Reliable charging infrastructure meets the demand for user-friendly, affordable charging options while growing the electric vehicle sector. Increasing electric car sales require a broad and accessible charging infrastructure. Despite its relevance, electric car charging is difficult [1]. Chargers are scarce, charging speeds are poor, and the electrical system cannot satisfy demand. EV model and protocol interoperability and urban and rural charging station accessibility slow charging infrastructure expansion. These concerns must be addressed to boost customer trust and electric mobility [2].

Power electronics must be enhanced to make EV charging systems more efficient and effective. Modern power electronics enable grid integration and faster charging, enhancing charging station efficiency. They help create smart charging solutions that respond to grid situations, energy prices, and customer preferences. Through bidirectional charging, electric vehicles (EVs) can improve energy management and grid stability. Modern power electronics may help electric vehicle charging infrastructure meet sustainable transportation needs.

2. Methodology

2.1. Power electronics components

Controllers, transformers, and converters are used by chargers. For energy to be sent from the grid to the batteries of electric cars, converters are needed to change AC to DC and DC to AC. Inverters convert battery DC electricity to grid-integrated AC power for bidirectional energy flow. Controllers monitor these elements to maximize charging efficiency and safety [11]. Integration of power electronics components dramatically improves charging station functionality and efficiency. Inverters enable vehicle-to-grid (V2G) technology, which enables EVs return electricity to the grid when demand is high, and smart converters can adjust charging rates at grid conditions [4].

2.2. Charging technologies

There are two basic charging technologies: quick and slow. Because they supply up to 350 kW of power and reduce charging time, public charging networks are ideal for fast charging stations. Slower Level 1 and Level 2 chargers take hours to fully charge an electric vehicle and are better for household use [3]. Each technology has pros and cons depending on infrastructure, cost, and speed. Wireless charging is a promising alternative to plug-in charging. Electric vehicles can be charged automatically without physical connections using inductive charging stations in the ground. Users can charge more easily with this technology in smart city infrastructure [6, 7].

2.3. System architecture

A modern electric vehicle charging station has several subsystems. Power converter unit, power management software, user interface, and communication systems. The user

[a]ku.anugpillai@kalingauniversity.ac.in, [b]shital.kewte@kalingauniversity.ac.in

DOI: 10.1201/9781003675259-8

interface simplifies customer communication and payment processing, while the power conversion unit contains all power electronics [12]. Communication technologies ensure electric vehicle and charging standard interoperability. Power flow at the charging station is carefully monitored for efficiency and safety. Efficient energy management systems monitor grid and user demand in real time. Change billing rates and scheduling on the fly to reduce peak load. This automated power flow management aids the shift to sustainable energy by increasing user experience and grid stability (Figure 8.1).

2.4. *Uses*

- Enhanced efficiency and reliability of charging stations.
- Reduced charging time.
- Integration with renewable energy sources.

3. Working

Modern electric vehicle charging stations use complex control and power regulation algorithms to maximize charging efficiency.

3.1. *Power conversion and regulation processes*

First, the charging station converts grid AC into DC for electric vehicle batteries. Power converters' load-dependent voltage amplitude allows this. After conversion [8], DC power is regulated to meet battery needs to prevent harm and extend battery life (Figure 8.2). The regulation dynamically adjusts output voltage and current in response to real-time feedback from the electric vehicle's battery management system to ensure a smooth charging operation [10].

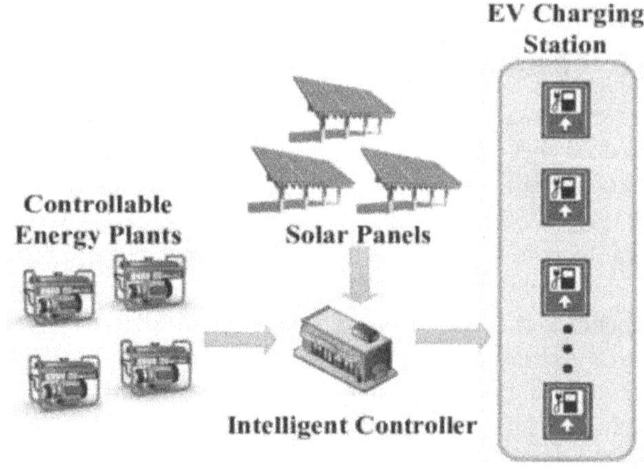

Figure 8.1. System architecture.

Source: https://www.google.com/imgres?h=647&w=850&tbnh=196&tbnw=257&osm=1&lns_uv=1&source=lens-native&usg=AI4_-kSomjP_CAdTzIiyZCBXlA_wV89DUA&imgurl=https://www.researchgate.net/publication/280621606/figure/fig1/AS:613929009025029@1523383490437/The-architecture-of-the-EV-charging-station.png&imgrefurl=https://www.researchgate.net/figure/The-architecture-of-the-EV-charging-station_fig1_280621606&tbnid=2Y58k20k6ygiEM&docid=mg7165cZ508esM.

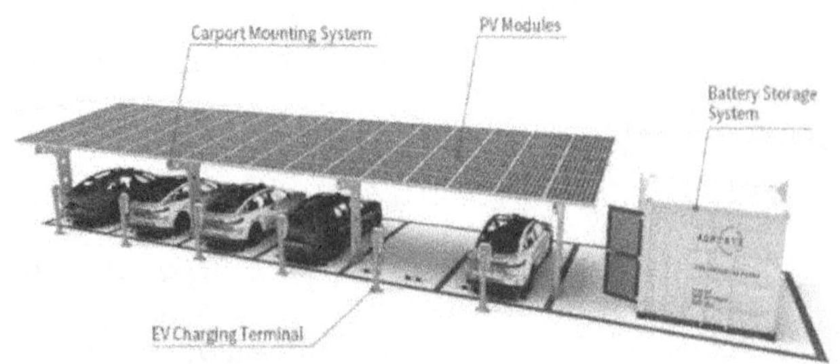

Figure 8.2. Working of advanced power electronics for electric vehicle charging stations.

Source: https://www.google.com/imgres?h=440&w=1000&tbnh=149&tbnw=339&osm=1&lns_uv=1&source=lens-native&usg=AI4_-kQANmMP7orRCgSFc3m3HkOOp10diA&imgurl=https://agreate.com/wp-content/uploads/2022/06/PBC-1.jpg&imgrefurl=https://agreate.com/pbc-pv-bess-ev-charging/&tbnid=ET-RvmzsNVzYMM&docid=d-EedR_r2FqvJM

3.2. *Control algorithms for efficient charging*

Advanced control algorithms automatically manage charging. These algorithms monitor battery temperature, SOC, and grid demand. By predicting optimal charging times and rates using predictive analytics, the algorithms cut costs and optimize renewable energy utilisation. They enable smart charging, which reduces peak demand by altering charging rates based on client requests or grid conditions [5]. Overall, these control systems stabilize and sustain the electricity grid and improve user experience [9].

4. Algorithm

- Set up the charging station, check grid status, and check battery SOC to start the system.
- Monitor weather, grid load, and electric vehicle battery management system (BMS) data in real time. Temperature, current, and voltage are included.
- Compare data to thresholds like the maximum charging current or optimal state of charge range to determine if charging is safe.
- Depending on grid circumstances and charge state, choose speedy, slow, or delayed charging to avoid peak demand hours.
- Activate the right power converters to convert grid AC electricity to DC to protect the electric vehicle battery.
- Adjust charging rates based on real-time BMS and grid data to maximize efficiency.
- The interface should display charge status, anticipated time to full charge, and cost implications in real time. Turn off the charger and put the system back in standby when the battery temperature or charge is safe.

5. Conclusion

Modern power electronics in EV charging infrastructure have greatly improved charging efficiency, costs, and renewable energy consumption. Through improved power conversion and clever control algorithms, these technologies enhance the electrical grid and improve charging efficiency. Demand response and load forecasting make electric vehicle charging more sustainable and responsive. More investment in cutting-edge charging infrastructure is needed to fulfil rising EV demand. Future advances may include smart grid integration, more efficient charging, and wireless charging. These improvements will eventually make transportation more ecologically friendly, which benefits everyone.

References

[1] Amry, Y., Elbouchikhi, E., Le Gall, F., Ghogho, M., & El Hani, S. (2022). Electric vehicle traction drives and charging station power electronics: Current status and challenges. *Energies*, *15*(16), 6037.

[2] Giliberto, M., Arena, F., & Pau, G. (2019). A fuzzy-based solution for optimized management of energy consumption in e-bikes. *Journal of Wireless Mobile Networks, Ubiquitous Computing, and Dependable Applications*, *10*(3), 45–64.

[3] Islam, A., Sharma, A., Chaturvedi, R., & Singh, P. K. (2021). Synthesis and structural analysis of zinc oxide nano particle by chemical method. *Materials Today: Proceedings*, *45*, 3670–3673.

[4] Rajesh, D., Giji Kiruba, D., & Ramesh, D. (2023). Energy proficient secure clustered protocol in mobile wireless sensor network utilizing blue brain technology. *Indian Journal of Information Sources and Services*, *13*(2), 30–38.

[5] Sharma, A., Chaturvedi, R., Saraswat, M., & Kalra, R. (2022). Weld reliability characteristics of AISI 304L steels welded with MPAW (Micro Plasma Arc Welding). *Materials Today: Proceedings*, *60*, 1966–1972.

[6] Laith, A. A. R., Ahmed, A. A., & Ali, K. L. A. (2023). IoT cloud system based dual axis solar tracker using arduino. *Journal of Internet Services and Information Security*, *13*(2), 193–202.

[7] Waheed, A., Rehman, S.U., Alsaif, F., Rauf, S., Hossain, I., Pushkarna, M., & Gebru, F. M. (2024). Hybrid multimodule DC–DC converters accelerated by wide bandgap devices for electric vehicle systems. *Scientific Reports*, *14*(1), 4746.

[8] Mausam, K., Sharma, A., & Singh, P. K. (2021). Calculating stress, temperature in brake pad using ANSYS composite materials. *Materials Today: Proceedings*, *45*, 3547–3550.

[9] Sharma, A., Chaturvedi, R., Singh, P. K., & Sharma, K. (2021). AristoTM robot welding performance and analysis of mechanical and microstructural characteristics of the weld. *Materials Today: Proceedings*, *43*, 614–622.

[10] Sawant, D., Mehta, J., & Khatri, R. (2022). Electric vehicle supply unit reconfigurable through HMI. In *2022 2nd International Conference on Power Electronics & IoT Applications in Renewable Energy and its Control (PARC)* (pp. 1–5). IEEE.

[11] Saadaoui, A., Ouassaid, M., & Maaroufi, M. (2023). Overview of integration of power electronic topologies and advanced control techniques of ultra-fast EV charging stations in standalone microgrids. *Energies*, *16*(3), 1031.

[12] Rajendran, G., Vaithilingam, C. A., Naidu, K., & Oruganti, K. S. P. (2020). Energy-efficient converters for electric vehicle charging stations. *SN Applied Sciences*, *2*, 1–15.

[13] Jadoun, V. K. (2022). Investigation of economic feasibility of a virtual power plant and electric vehicle charging station in grid tied DC microgrid.

9 Smart gloves using SpO2, temperature and ECG sensor

Kalimuthukumar S.[a], *Ramkarthick B.*[b], *and Manoj Kumar S.*[c]

Department of Biomedical Engineering, Kalasalingam Academy of Research and Education, Krishnankoil, Srivilliputhur, Tamil Nadu, India

Abstract: Smart gloves using SpO2, temperature, and ECG sensors are wearable devices which can be used to monitor a range of health parameters, including oxygen rate, body temperature, and heart rate. For example, smart fingerprints can be used to monitor patients with cardiovascular, respiratory, or other chronic diseases. Athletes and fitness enthusiasts can use smart wristbands to track their efforts and progress. For example, smart wristbands can be used to monitor pulse rate and oxygen saturation during exercise, as well as subsequent recovery time. Smart gloves can be used by operators in construction and manufacturing industries to watch for hand and finger movements and signs of ergonomic dangers.

Keywords: Smart gloves, SpO2 sensor, temperature sensor, ECG sensor, wearable devices, ergonomics

1. Introduction

The healthcare landscape is constantly evolving, driven by technological advances that allow individuals to take control of their health and wellbeing. One such groundbreaking innovation is a smart wristband that is equipped with SP02 (blood oxygen saturation), temperature, and ECG (electrocardiogram) sensors This smart wristband represents an amazing pulse has remote health monitoring and the potential to change the way to manage ourselves and monitor our vital signs.

In recent years, wearable technology has become increasingly popular due to its ability to provide real-time health information and improve overall health literacy through devices such as fitness trackers and smartwatches paving the way for a health-conscious public because of the greatness of the. Nevertheless, they are usually inadequate for tracking vital signs like oxygen saturation, temperature, and cardiac output.

This is where smart fingerprints enter the picture, offering a more complete solution than just activity tracking. As our society turns into an increasing number of health-conscious, there is a developing call for personal fitness information that goes beyond the primary metrics supplied by using traditional wearable gadgets. While fitness trackers and smartwatches have carved a gap for themselves in tracking physical interest and sleep styles, they fall brief in taking pictures crucial symptoms critical for a comprehensive information of our health. The capabilities bestowed upon these smart gloves are nothing short of outstanding. In a world where data is paramount, smart gloves seamlessly combo actual-time tracking with consumer-friendly interfaces and wireless connectivity.

The destiny of smart gloves is very promising. The value of the technology is predicted to come down inside the coming years, and the devices are expected to become more comfortable and simpler to use. As a result, smart gloves are expected to become greater broadly adopted in a wide variety of applications.

2. Literature Review

Smart gloves are continually evolving, with advancements in materials, sensor technology, and miniaturization, promising even more versatile and intuitive human-computer interactions in the future. This literature review synthesizes key findings and insights from existing research on enhancing different type of biosensors and their applications used in the medical field for different problem statements.

This paper offers a comprehensive study the patents and research papers on the improvement of automation glove for remote control and healthcare applications. [1]. The main purpose is to evaluate existing robotic gloves and determine the competition of robotic gloves. Discover the smartest gloves and get insight into future developments. The review shows a lot of promise, including the use of advanced data filtering algorithms, the integration of AIML techniques for signature identification, and expansion in the wide range of indicators. Most importantly, the smartest gloves are made using muscledriven mechanisms or soft actuators.

[a]kali.12eee@gmail.com, [b]ramkarthick312@gmail.com, [c]manojpandian1404@gmail.com

DOI: 10.1201/9781003675259-9

This review focuses on comprehensive efforts in academia and industry to study and develop an advanced wearable device for healthcare [2]. The support comes mainly from the rise in medical costs and current technology with advances at the micron level. and nanotechnology, miniaturization ofsensors, and creation of smart fabrics will continue to advance SWS, allowing men or women to manage and monitor patients on a regular basis. People's health will gradually change the healthcare landscape. Consisting of different components and 2 components, such as detectors and controllers, to digital media, these models simplify sophisticated medical procedures and enable the use of affortable, painless devices.

Paralysis is a debilitating circumstance that can rob human beings in their mobility, independence, and communique [3]. This review is focused on the smart glove which is a innovative device which can assist paralyzed sufferers triumph over those demanding situations. Using plenty ofsensors, the Smart Glove can track body temperature, oxygen saturation, pulse fee, coronary heart function, and emotional nation. It can also recognize hand gestures and convert them into textual content or instructions. This lets in paralyzed sufferers to talk with others, manipulate their surroundings, or even manipulate their personal fitness. The Smart Glove is still below improvement, however it has the capability to noticeably enhance the first-class of existence for humans having paralysis. It is a handy and green manner to monitor their health, and it represents a higher breakthrough on scientific technology.

In this research paper titled 'Utilization of microcontroller technology using Arduino board for Internet of Things (a systematic review)'. To examine the results of studies regarding integrating embedded processor with IoT technology [4]. This research review is designed explicitly to discover the integration of microcontrollers era of IoT in the fitness quarter. The studies technique employs a comprehensive evaluation with four (four) levels of work: identity, evaluation, qualification, and entry. The embedded system is based on the Arduino uno design and features temperature detectors, moisture detectors, fuel detectors, gas detectors, image sensors, sensors for light, and movement sensor. IoT controllers could be utilized across a variety of applications, including infrastructure and business. The uses of microcontroller technology in the exercise of IoT is to analyze the situation of affected people.

The purpose of this article is to review existing physical and exercise equipment, describe their operation, and compare their ability to understand [5]. Such a generation could be specifically could be beneficial for specific groups of people whose tend to change moods rapidly, including individuals having autism along with those with impairments. Portable detectors offer problem-solving capabilities that can supplement and enhance existing psychological therapies. This article provides an overview of current and new technologies on the industry. Evaluates the research on current models, examines their effectiveness, clinical effectiveness, and discusses scientific considerations. A modest amount of items provide accurate domestic products and might be appropriate with those having autism spectrum disorder. Better solutions are probably to emerge in the distant. Thus, carers ought to be cautious to choose equipment that meets the needs of the person receiving care and has reliable and valid scientific recommendations.

This project is a survey measuring current student engagement in education [6]. This review describes existing commercial and noncommercial tools for monitoring student engagement and identifies key physical indicatorsrelevant to the research. Their results show that physiologic indicators considered for evaluating students' involvement comprise cardiovascular activity, temperature on the skin, breathing rate, saturation of oxygen, BP, and ECG. Likewise, Anxiety and confusion remain key characteristics of student involvement. This method may assist to optimize the method of instruction by identifying characteristics such as tension, anxiety, stress, emotion, especially discomfort at the moment, help teachers create suggested new lessons or improve existing lessons.

This review aims to offer a full understanding of fabric technology in the context of the resources and production techniques utilized in fabric production [7]. The emphasis is on competition and advanced solutions for gas sensors for collecting different physical data, some expansions of textiles used to obtain bioelectricity, breathlessness, heat and sweat are presented.

It is an in-depth analysis of data and methods developed for structure-wide textile sensors in the production process (fibers, yarns, fabrics and garments) [8]. It then reviews the performance existing parameters utilized for those detectors and stresses the necessity for evaluation models in several areas: biologic compatibility, heat as well as physical ease, longevity, and medical monitoring processes based on the human yawn model. It also analyses the importance of textile form factor preprocessing and conditioning reporting based on the impact of textile form factor on the electrical and electronic properties of textile sensors.

The present study was conducted to narrow the gap by offering a current overview of the extant design, manufacturing and analytics literature on smart wearables in one location to aid subsequent research and work in this fast-emerging field [9]. The innovation of sensors and device manufacturing, combined with advances in big data technologies, has enabled the creation of smart wearables for a variety of health care applications. The devices are useful in the remote monitoring and management of different diseases, as well as in the rehabilitation of patients. Smart wearables are a nonintrusive and cost-effective way of tackling heavy and costly healthcare approaches, including hospitalization and the development of late-diagnosed diseases.

A study revealed that a new body analysis device for the purpose of the continuous monitoring of biophysical as well as psychophysical illnesses, the current methods are not suitable for integration with wearable or portable devices [10]. Smart technologies symbolize the foundation of IoT-based medical platforms that enable capillary and instantaneous monitoring of patients. Additionally, the modern architecture and support services of IoT platforms for healthcare are examined, providing insight into future developments in healthcare. All decision makers use tools that can be used to record patient 3 parameters and share them with the cloud platform for processing and providing instant feedback.

This review provides insight into the opportunities and limitations of renewable energy generation [11]. The most recent innovations in portable devices focus on three primary topics: (1) personal health monitoring, which can record a variety of physical and biochemical markers, (2) assistive robots and artificial devices to assist the disabled, used for movement and pain or sensation. Providing information and communications to patients during daily operations, including infrared (IR) sensing and holograms (III). Considering different situations, this review allows doctors, engineers, scientists, and researchers to investigate the potential of wearable electronic devices and make recommended statements for future developments.

This paper review focus on the use of sensing and biological applications data rather than business [12]. Additionally last conversation. on identifying solutions to demonstrate the advantages of physical care in recovery and to demonstrate future advancements in this field is published at the end of this article. The main conclusions of this article are: (1) an introduction to the expanding body of work in the biomedical field since 2016, focusing mainly on physical therapy and healthcare services and systems, and (2) to sensing and biological applications.

Three categories serve as the foundation for this 2020 review: Assistive Systems; as well as sensors used in medical care, personal medical care, and constant monitoring of health [13]. The rapid expansion of portable technology and detecting prototypes for monitoring active efforts. has been fueled by advancements in sensing techniques, more affordable ICs, and the growth of networking innovations. These devices can be used for medical rehabilitation, sports monitoring, or general well-being. The expansion of connected devices and the numerous health and behaviour monitoring systems are made possible by the ongoing advancements in information technology and electronic device technology, which has led to the prediction of an abrupt increase in financial investment in these fields.

Their review centers on a number of prerequisites for the creation of these kinds of environments, including the implementation of systems for monitoring physiological signs, methods for identifying daily activities, and indoor air quality monitoring systems [14]. Additionally, the most popular machine learning techniques in movement identification and physical movement monitoring in the literature are mentioned. Additionally, the significance of providing the elderly population with physical and cognitive training through the use of exergames and immersive environments is discussed. Implementing the Internet of Things (IoT) in the healthcare ecosystem is one of the most effective ways to handle these difficulties, and thereby prevent and identify potential health impairments in people.

The advancements and therapeutic value of smart wearable body sensors are reviewed in this paper [15]. The integration of intelligent wearable sensors into daily patient care has the potential to improve patient-physician relationships, increase patient autonomy and involvement, and enable innovative remote monitoring methods that will transform healthcare spending and management. For preventative measures in a variety of medical fields, including cardiovascular, hormonal, neurological health, and rehabilitation therapy, smart wearable detectors are efficient and trustworthy. Additionally, perioperative monitoring and rehabilitation therapy have demonstrated the accuracy and use of these sensors.

In their assessment, they emphasize how crucial it is for patients and society at large to enhance the standard and effectiveness of medical care, both in hospitals and at home [16]. With the availability of numerous technologies such as nano technology, and low-power applications design, novel materials, and retractable detectors, it is currently possible to create simple to use gadgets that improve patient security and comfort. Nearly 90% of the skin's surface is in direct touch with textiles and clothing, making non-invasive smart clothing and smart sensors an appealing option for at-home and ambulatory health monitoring. Additionally, wearable technology and smart houses equipped with exosensors are viable options. These systems can all offer a secure and cozy setting for athome medical care.

With the goal of achieving more effective and accurate mental health prediction as well as early detection, prevention, and rehabilitation [17]. To give an example for enhancing the degree that people's psychological well-being, their research offers a thorough function classifications, implementation systems, similar indications, common approaches, and outcomes of wearable technology for detecting mental wellness issues. One useful strategy to stop mental health illnesses from happening is to anticipate a person's mental health in real time and with accuracy. Smart wearables have been increasingly significant in the monitoring of psychological wellness in the past few years.

This study presents a summary on the current state of the knowledge on adjustable pressure detectors, covering three basic types of absorption systems, and their corresponding advancements in science, and applicationsin wearable technology and E-skin [18]. There are still difficulties in creating flexible, sensitive, and highly extensible multi-function equipment. The main directions in the field of sensors going forward will be investigating new sensing mechanisms, looking for new functional materials, and creating innovative flexible device integration technologies. Scientists are exhibiting a lot of enthusiasmin flexible pressure sensors, which are being used extensively 4 in new electronic equipment due to their unique qualities of high flexibility, high sensitivity, and light weight; wearable flexible sensing devices and electronic skin (E-skin) are two examples.

A discussion of present and upcoming sensor and tracking techniques appropirate to accurate wellness is provided in this document, with a focus on high-demand technologies like wearables, implanted sensors, and mobile and portable devices [19]. The rapidly developing discipline of precision health monitors each person's health and wellbeing continually in order to provide early identification and individualized treatment of medical disorders. Continuous, portable, and implantable health monitoring devices can be used to accomplish active monitoring and assessment, giving a thorough view of an individual's wellbeing and lifestyle. Therefore, it's imperative to obtain a better grasp of the present and developing solutions which may support gathering information for personalized care.

This review focuses on the larger scientific community, which will expedite the development of flexible electronics and support the logical design of devices for monitoring human health [20]. One of the upcoming categories wearable technology health tracking devices are one example of personal portable gadgets used in telemedicine practice. The basis of these systems is the monitoring of several biological signals that humans release through their breath, saliva, urine, and excessive perspiration on their skin. However, because the building process of conventional semiconductor equipment is no longer suitable, wearable healthcare equipment development and commercialization are still proceeding at a rather slow rate. Notwithstanding these obstacles, developments in the fields of material research, chemical testing, design of devices, and assembly have opened the way for radically new functional technologies, resulting in the ongoing evolution of wearable devices across time.

In order to discuss and present the main issues of the development of the field in order to provide a market with stable and valuable products, this paper includes information about user needs, technologies, research and development of integrated systems, and upcoming challenges [21]. The study in new sensing and monitoring tools, such as wearable wireless devices and sensor networks, for personal use in safety, protection, well-being and healthcare has increased significantly in the last ten years. Sensor based combined systems on body worn platforms, or smart wearable systems (SWS) offer ubiquitous and personalized answers for continuous external tracking of the body and its surroundings, and feedback to the user.

Regarded as a hot topic in the field of technology, automated health surveillance is the subject of their study [22]. Even though the population of senior citizens is growing, there is no doubt that there is an immediate need for a distributed medical treatment system with distant monitoring with the goal of decreasing the increasing expenses of treatment. Sixty percent of lives can be saved with early detection and continuous health monitoring. These characteristics make a wireless, wearable, reasonably priced, and automated health monitoring system a good choice. Checking vital signslike temperature, heart rate, gyro, oxygen level, etc., when necessary, can be difficult. It is possible to develop and build an IOT-based patient health monitoring system with Arduino and a standard ESP8266.

The aim of their study was to design and implement a cost-effective embedded technologies for advanced military support [23]. When the modules are combined, they assist in the development of a network that is used to monitor the whereabouts of fellow soldiers and send alert messages based on health and gesture parameters. The health monitoring devices include temperature, PPG, and an ECG; the gesture detection component is a handwear made of flex and IMU detectors; and indoor localization and alert messaging are handled by UWB-based communication.

This article presents an summary of the cutting-edge micro/nano systems being suggested for wearable and implantable medical diagnostic, therapeutic, and therapy purposes [24]. Apart from the adaptability or conformability of the transducers (sensors and actuators) and the distinctiveness of their construction materials, the special feature of these technologies in their integration with different kinds of power collectors, which enables the technology to function independently or without the need for batteries. Mart materials respond environmental stimulus such as stress, and temperature, light, humidity, pH, and magnetic or electrical fields, which makes them useful for a wide range of sensing and actuation applications in healthcare. These materials can also be used to shape biofuel powered devices or to capture mechanical energy driven from individual's motion, the surroundings, or heat from the body.

This paper compares various present-day configurations methods and identifies their practical inadequacies to present a thorough survey of study and innovation on wearable biosensor systems for health-monitoring [25]. To create a comprehensive study, a set of important features that best capture the features and functionality of wearable biosensor systems have been chosen. This survey is not meant to be critical; rather, it is meant to be a guide for future research improvements and to serve as a benchmark for current accomplishments and their intellectual stage. An emerging trend 5 in health monitoring is wearable biosensor systems, which are expected to improve treatment of various medical conditions and allow proactive personal health management.

3. Research Findings

Healthcare: Smart gloves can be used to display sufferers' essential signs and symptoms, together with heart price, blood oxygen tiers, and body temperature. These statistics can be used to locate early symptoms of clinical troubles, together with coronary heart assault, stroke, and respiration misery. Smart gloves can also be used to monitor sufferers' recovery from surgical operation or other clinical procedures.

Fitness and well-being: Smart gloves may be used to song fitness metrics, which includes heart price, calorie burn, and steps taken. This data can assist people to achieve their health desires and live wholesome. Smart gloves can also be used to monitor pressure tiers and sleep first-rate.

Sports and recreation: Smart gloves may be used to improve athletic overall performance and save you injuries. For instance, smart gloves may be used to music the motion of the hand and wrist, and to offer comments on how to improve approach. Smart gloves also can be used to hit upon early symptoms of fatigue or muscle strain.

Occupational safety: Smart gloves can be used to defend workers from hazards in the place of work. For instance, smart gloves may be used to discover electric hazards, chemical compounds, and heat. Smart gloves also can be used to monitor the health and protection of people in risky environments, together with creation websites and mines.

4. Conclusion

Smart gloves the usage of SpO2, temperature, and ECG sensors have the capability to be a treasured device for lots of programs in healthcare, fitness and well-being, sports activities and pastime, and occupational protection. In healthcare, smart gloves can be used to display sufferers' vital signs and locate early signs and symptoms of clinical problems. In fitness and well-being, clever gloves may be used to track fitness metrics and help people obtain their fitness goals. In sports activities and recreation, smart gloves can be used to improve athletic performance and save you accidents. In occupational protection, smart gloves may be used to shield workers from hazards in the administrative center. Overall, smart gloves using SpO2, temperature, and ECG sensors have the capacity to make a sizable effect on our lives by using helping us to live healthful, safe, and efficient.

5. Future Scope of the Work

In future work, we are going to serve our community like to build a low-cost smart glove consist of SpO2, ECG, and temperature sensor to regularly monitor their health and Smart gloves could also be used to regulate the wellness of the aged and individuals with disabilities. Overall, the future scope of smart gloves using SpO2, temperature, and ECG sensors is very bright. These technologies have an opportunity to transform the way we monitor our health, stay safe, and interact with the world around us.

References

[1] Saypulaev, G. R., Merkuryev, I. V., Saypulaev, M. R., Shestakov, V. K., Glazkov, N. V., & Andreev, D. R. (2023, March). A review of robotic gloves applied for remote control in various systems. In *2023 5th International Youth Conference on Radio Electronics, Electrical and Power Engineering (REEPE)* (Vol. 5, pp. 1–6). IEEE.

[2] Chan, M., Estève, D., Fourniols, J. Y., Escriba, C., & Campo, E. (2012). Smart wearable systems: Current status and future challenges. *Artificial intelligence in medicine*, *56*(3), 137–156.

[3] Kalpana, H. M., Kulkarni, A. B., Tiwari, M., Chitmalwar, O. R., & TR, R. S. (2023, July). Smart Glove with Gesture Based Communication and Monitoring of Paralyzed Patient. In *2023 International Conference on Smart Systems for applications in Electrical Sciences (ICSSES)* (pp. 1–6). IEEE.

[4] Yusro, M., Guntoro, N. A., & Rikawarastuti. (2021, April). Utilization of microcontroller technology using Arduino board for internet of things (A systematic review). In *AIP conference proceedings* (Vol. 2331, No. 1, p. 060004). AIP Publishing LLC.

[5] Taj-Eldin, M., Ryan, C., O'Flynn, B., & Galvin, P. (2018). A review of wearable solutions for physiological and emotional monitoring for use by people with autism spectrum disorder and their caregivers. *Sensors*, *18*(12), 4271.

[6] Bustos-Lopez, M., Cruz-Ramirez, N., Guerra-Hernandez, A., Sánchez-Morales, L. N., Cruz-Ramos, N. A., & Alor-Hernandez, G. (2022). Wearables for engagement detection in learning environments: A review. *Biosensors*, *12*(7), 509.

[7] Angelucci, A., Cavicchioli, M., Cintorrino, I. A., Lauricella, G., Rossi, C., Strati, S., & Aliverti, A. (2021). Smart textiles and sensorized garments for physiological monitoring: A review of available solutions and techniques. *Sensors*, *21*(3), 814.

[8] Lopez, X., Afrin, K., & Nepal, B. (2020). Examining the design, manufacturing and analytics of smart wearables. *Medical Devices & Sensors*, *3*(3), e10087.

[9] Shuvo, I. I., Shah, A., & Dagdeviren, C. (2022). Electronic textile sensors for decoding vital body signals: state-of-the-art review on characterizations and recommendations. *Advanced Intelligent Systems*, *4*(4), 2100223.

[10] De Fazio, R., De Vittorio, M., & Visconti, P. (2021). Innovative IoT solutions and wearable sensing systems for monitoring human biophysical parameters: A review. *Electronics*, *10*(14), 1660.

[11] Farooq, A. S., & Zhang, P. (2022). A comprehensive review on the prospects of next-generation wearable electronics for individualized health monitoring, assistive robotics, and communication. *Sensors and Actuators A: Physical*, *344*, 113715.

[12] Palumbo, A., Vizza, P., Calabrese, B., & Ielpo, N. (2021). Biopotential signal monitoring systems in rehabilitation: A review. *Sensors*, *21*(21), 7172.

[13] Nascimento, L. M. S. D., Bonfati, L. V., Freitas, M. L. B., Mendes Junior, J. J. A., Siqueira, H. V., & Stevan Jr, S. L. (2020). Sensors and systems for physical rehabilitation and health monitoring—A review. *Sensors*, *20*(15), 4063.

[14] Jacob Rodrigues, M., Postolache, O., & Cercas, F. (2020). Physiological and behavior monitoring systems for smart healthcare environments: A review. *Sensors*, *20*(8), 2186.

[15] Appelboom, G., Camacho, E., Abraham, M. E., Bruce, S. S., Dumont, E. L., Zacharia, B. E., … & Connolly, E. S. (2014). Smart wearable body sensors for patient self-assessment and monitoring. *Archives of public health*, *72*, 1–9.

[16] Axisa, F., Schmitt, P. M., Gehin, C., Delhomme, G., McAdams, E., & Dittmar, A. (2005). Flexible technologies and smart clothing for citizen medicine, home healthcare, and disease prevention. *IEEE Transactions on information technology in biomedicine*, *9*(3), 325–336.

[17] Long, N., Lei, Y., Peng, L., Xu, P., & Mao, P. (2022). A scoping review on monitoring mental health using smart wearable devices. *Math. Biosci. Eng*, *19*(8), 7899–7919.

[18] Xu, F., Li, X., Shi, Y., Li, L., Wang, W., He, L., & Liu, R. (2018). Recent developments for flexible pressure sensors: A review. *Micromachines*, *9*(11), 580.

[19] Silvera-Tawil, D., Hussain, M. S., & Li, J. (2020). Emerging technologies for precision health: An insight into sensing technologies for health and wellbeing. *Smart Health*, *15*, 100100.

[20] Lou, Z., Wang, L., Jiang, K., Wei, Z., & Shen, G. (2020). Reviews of wearable healthcare systems: Materials, devices and system integration. *Materials Science and Engineering: R: Reports*, *140*, 100523.

[21] Lymberis, A. (2010). Advanced wearable sensors and systems enabling personal applications. *Wearable and autonomous biomedical devices and systems for smart environment: issues and characterization*, 237–257.

[22] Dinesh, E., Poovitha, K., Pranikaa, V., & Rosini, M. (2023, February). A Real Time System to Analyze Patient's Health Condition using Second Layer Computing. In *2023 7th International Conference on Computing Methodologies and Communication (ICCMC)* (pp. 653–661). IEEE.

[23] Hota, G., Sharma, S., Rathore, A., Joshi, S., & Shah, H. (2019, February). An integrated visual signalling, localisation & health monitoring system for soldier assistance. In *2019 IEEE International Conference on Electrical, Computer and Communication Technologies (ICECCT)* (pp. 1–6). IEEE.

[24] Hassani, F. A., Shi, Q., Wen, F., He, T., Haroun, A., Yang, Y., … & Lee, C. (2020). Smart materials for smart healthcare–moving from sensors and actuators to self-sustained nanoenergy nanosystems. *Smart Materials in Medicine*, *1*, 92–124.

[25] Pantelopoulos, A., & Bourbakis, N. (2008, August). A survey on wearable biosensor systems for health monitoring. In *2008 30th Annual International Conference of the IEEE Engineering in Medicine and Biology Society* (pp. 4887–4890). IEEE.

10 A comprehensive overview of internet-of-things-based water monitoring systems

Anupa Sinha[1,a] *and Pooja Sharma*[2,b]

[1]Assistant Professor, Department of CS & IT, Kalinga University, Raipur, India
[2]Research Scholar, Department of CS & IT, Kalinga University, Raipur, India

Abstract: The Internet of Things (IoT) represents a paradigm shift in how devices interact and communicate over the internet, enabling unprecedented levels of connectivity, automation, and data-driven decision-making across diverse sectors. Water monitoring plays a crucial role in safeguarding public health, ensuring environmental sustainability, and supporting economic activities worldwide. By harnessing the power of IoT, water management authorities, utilities, industries, and environmental agencies can achieve operational efficiencies, optimize resource utilization, and enhance overall environmental stewardship in the realm of water management. In this paper, we provide an in-depth overview of the functionalities, features, and features of the Internet-of-Things-based water monitoring systems, emphasizing their capabilities in real-time monitoring, remote accessibility and control, and a data analytics for enhanced operational efficiency, decision making, and sustainability in water management practices.

Keywords: IoT, water monitoring, environmental sustainability, utilities, industries

1. Introduction

Water monitoring plays a crucial role in safeguarding public health, ensuring environmental sustainability, and supporting economic activities worldwide. The availability of clean water is fundamental for drinking, agriculture, industrial processes, and ecosystem health. Monitoring water quality and quantity helps identify pollutants, assess the impact of human activities, and manage water resources effectively [1]. Inaccurate or inadequate monitoring can lead to contamination risks, ecosystem degradation, and scarcity issues, highlighting the critical need for reliable and timely data on water conditions [15]. Governments, industries, and environmental organizations rely on comprehensive water monitoring systems to make informed decisions, implement regulatory measures, and protect water sources for future generations.

The Internet of Things (IoT) represents a paradigm shift in how devices interact and communicate over the internet, enabling unprecedented levels of connectivity, automation, and data-driven decision-making across diverse sectors [3]. IoT encompasses a network of interconnected physical devices, sensors, actuators, and software that collect and exchange data. This technology has revolutionized industries such as healthcare, transportation, agriculture, and manufacturing by enhancing operational efficiency, reducing costs, and enabling innovative services. In IoT systems, devices equipped with sensors gather real-time data, which is then processed, analyzed, and utilized to automate tasks, monitor environments, and optimize processes [16]. The scalability and versatility of IoT have paved the way for transformative applications in smart cities, energy management, logistics, and notably, environmental monitoring systems like water quality and quantity assessment [2, 5].

An IoT-based water monitoring system refers to a network of interconnected devices, sensors, and technologies designed to collect, transmit, and analyze data related to water quality and quantity in real-time. This system integrates IoT principles with water management practices to enable continuous monitoring and remote management capabilities [17]. The scope of IoT-based water monitoring extends across various applications, including urban water supply networks, industrial wastewater management, agricultural irrigation systems, and environmental monitoring in natural water bodies [7]. By leveraging IoT, these systems offer enhanced visibility into water parameters such as pH levels, turbidity, dissolved oxygen, temperature, and water flow rates. This real-time data acquisition facilitates proactive decision-making, early detection of anomalies, and efficient resource allocation, thereby improving water resource management strategies and ensuring sustainable use of water resources [4].

2. Components Involved: Sensors, IoT Devices, Data Communication Technologies

The IoT-based water monitoring system comprises several key components that work synergistically to gather, transmit, and interpret data from diverse water sources. Central to this system are sensors, which are specialized devices designed

[a]ku.anupasinha@kalingauniversity.ac.in, [b]pooja.sharma@kalingauniversity.ac.in

DOI: 10.1201/9781003675259-10

to detect and measure specific water parameters [18]. These sensors may include pH sensors, turbidity sensors, conductivity sensors, dissolved oxygen sensors, and flow meters, among others. These sensors are deployed at strategic locations within the water infrastructure or natural water bodies to continuously monitor and collect data [6].

IoT devices serve as the interface between sensors and the digital ecosystem, equipped with microcontrollers, processors, and communication modules that facilitate data aggregation, processing, and transmission [9]. These devices may range from embedded systems and microcontrollers to specialized IoT gateways capable of aggregating data from multiple sensors and transmitting it to centralized servers or cloud-based platforms [8].

Data communication technologies form the backbone of IoT-based water monitoring systems, enabling seamless connectivity and data transmission over various networks. Common communication protocols used include Wi-Fi, cellular networks (3G/4G/5G), Bluetooth Low Energy (BLE), Zigbee, LoRa (Long Range), and NB-IoT (Narrowband IoT). These technologies ensure reliable and secure data transfer from remote sensor nodes to centralized data repositories or control centers, where the data is processed, analyzed, and visualized in user-friendly interfaces [10].

The integration of these components facilitates real-time monitoring of water quality and quantity, enabling stakeholders to respond promptly to changes in water conditions, identify potential threats to water resources, and implement preventive measures or corrective actions as necessary (Figure 10.1). By harnessing the power of IoT, water management authorities, utilities, industries, and environmental agencies can achieve operational efficiencies, optimize resource utilization, and enhance overall environmental stewardship in the realm of water management [19].

3. Detailed Description of Sensors Used for Water Quality and Quantity Monitoring

Sensors play a pivotal role in IoT-based water monitoring systems by collecting precise and real-time data on various water parameters critical to assessing water quality and quantity. These sensors are specialized devices designed to detect and measure specific attributes of water, ensuring accurate monitoring and analysis.

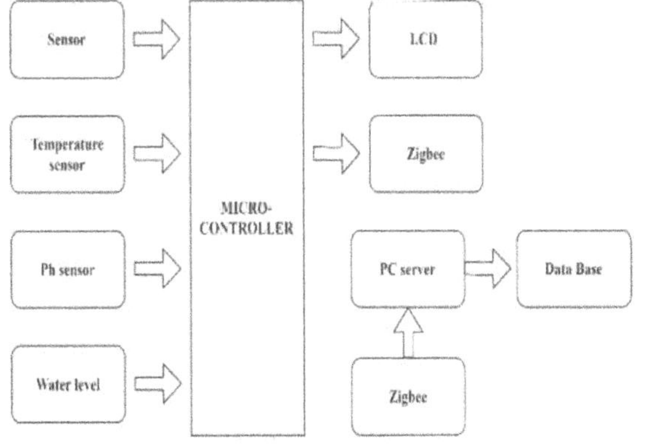

Figure 10.1. Block diagram of proposed system.

Source: Author.

Water Quality Sensors: These sensors measure parameters such as pH levels, turbidity, conductivity, dissolved oxygen (DO), chemical concentrations (e.g., nutrients, heavy metals), and biological indicators (e.g., algae, bacteria). pH sensors detect the acidity or alkalinity of water, crucial for understanding its chemical balance. Turbidity sensors quantify the clarity of water by measuring suspended particles, providing insights into sediment levels or pollutants. Conductivity sensors gauge the ability of water to conduct electrical current, correlating with its dissolved ion content and salinity. DO sensors assess oxygen levels dissolved in water, vital for aquatic life support [11]. Chemical and biological sensors detect specific pollutants or organisms, aiding in pollution detection and ecological health assessment. These sensors are typically deployed in water bodies, treatment plants, or distribution networks to monitor and ensure compliance with regulatory standards.

Water Quantity Sensors: Flow meters are primary sensors used to measure the volume or flow rate of water moving through pipes, channels, or natural water bodies. They employ various technologies such as electromagnetic, ultrasonic, or mechanical methods to accurately quantify water flow. Flow meters provide critical data for water distribution, irrigation management, leakage detection, and flood monitoring applications. Additionally, water level sensors measure the height or depth of water in reservoirs, tanks, or rivers, offering insights into water storage capacity, flood risk assessment, and drought monitoring [12]. These sensors enable efficient water resource management by optimizing water distribution, preventing overuse, and ensuring adequate supply during periods of high demand or environmental stress.

The integration of diverse water quality and quantity sensors within IoT-based monitoring systems enhances operational efficiency, facilitates timely decision-making, and supports sustainable water management practices across urban, industrial, agricultural, and environmental domains.

4. IoT Devices and Their Role in Data Collection and Transmission

IoT devices serve as the backbone of water monitoring systems, facilitating the seamless integration of sensor data into digital networks and enabling remote monitoring and management capabilities. Figure 10.2 shows IoT system component. These devices encompass a range of hardware components and communication technologies tailored to specific operational requirements and environmental conditions.

In practice, IoT devices enable continuous data acquisition, real-time monitoring, and proactive decision-making in water management scenarios. By leveraging these devices, stakeholders can optimize resource allocation, mitigate risks of water contamination or scarcity, and enhance overall operational efficiency within water distribution networks, treatment facilities, and environmental monitoring systems.

5. Communication Protocols Utilized (e.g., Wi-Fi, LoRa, Cellular Networks)

Communication protocols are fundamental to IoT-based water monitoring systems, enabling reliable data transmission and connectivity between sensors, IoT devices, and centralized data repositories. The selection of communication protocols depends on factors such as data bandwidth requirements, transmission range, power efficiency, and network infrastructure availability.

- *Wi-Fi (802.11):* Wi-Fi offers high data transfer rates and operates over short to medium distances within local area networks (LANs). It is suitable for applications where continuous power supply and high-speed data transmission are required, such as urban water distribution networks, wastewater treatment plants, and smart building management systems.
- *LoRa (Long Range):* LoRa is a low-power, wide-area network (LPWAN) protocol designed for long-range communication over several kilometers in urban, suburban, or rural environments. It provides extended battery life for IoT devices and is ideal for remote monitoring applications such as groundwater level monitoring, agricultural irrigation systems, and environmental monitoring in remote locations.
- *Cellular Networks (3G/4G/5G):* Cellular networks utilize existing telecommunications infrastructure to provide ubiquitous coverage and high-speed data connectivity for IoT deployments. They are suitable for mobile applications and urban environments where IoT devices require continuous connectivity, real-time data transmission, and integration with cloud-based platforms.
- *NB-IoT (Narrowband IoT):* NB-IoT is a cellular LPWAN technology optimized for low-power IoT applications requiring long battery life and deep indoor penetration. It supports a large number of devices per cell and is well-suited for applications such as smart metering, leak detection, and asset tracking within water distribution networks.

In conclusion, the choice of communication protocols in IoT-based water monitoring systems influences data transmission efficiency, network reliability, and overall system performance. By leveraging these protocols, organizations can establish robust IoT infrastructures capable of supporting scalable, interoperable, and cost-effective water monitoring solutions across diverse operational environments.

6. Real-Time Monitoring Capabilities

Real-time monitoring capabilities are essential features of IoT-based water monitoring systems, enabling continuous observation and assessment of water quality and quantity parameters. These systems utilize interconnected sensors and IoT devices to gather data in real-time from various monitoring points such as water treatment plants, distribution networks, reservoirs, and natural water bodies [13]. By leveraging advanced sensing technologies and communication protocols, real-time data acquisition facilitates immediate detection of anomalies, changes in water conditions, or potential threats to water resources.

For instance, sensors deployed in water distribution networks can monitor parameters like pressure, flow rate, and chlorine levels continuously. Any deviation from preset thresholds triggers automated alerts or notifications, enabling operators to promptly respond to leaks, contamination incidents, or operational inefficiencies. Real-time monitoring also supports proactive decision-making by providing timely insights into water usage patterns, demand fluctuations, and environmental impacts [14]. This capability is crucial for optimizing resource allocation, ensuring compliance with regulatory standards, and maintaining water quality throughout the distribution network. Overall, real-time monitoring enhances operational efficiency, reduces response times to critical events, and improves the overall resilience of water management systems.

6.1. Remote accessibility and control

Remote accessibility and control are integral functionalities enabled by IoT-based water monitoring systems, empowering stakeholders to manage and oversee water infrastructure from anywhere at any time. These systems leverage cloud-based platforms, IoT gateways, and mobile applications to provide secure remote access to real-time data, operational parameters, and system performance metrics. Authorized personnel, such as water utility operators, engineers, and environmental regulators, can remotely monitor sensor readings, receive alerts, and implement control actions via web interfaces or dedicated mobile applications.

Remote accessibility allows for proactive maintenance and troubleshooting, as operators can remotely diagnose equipment faults, adjust operational parameters, and schedule maintenance activities without the need for on-site visits. This capability minimizes downtime, optimizes resource utilization, and enhances overall system reliability. Furthermore, remote control functionalities enable rapid response to emergency situations, such as sudden changes in water quality or supply disruptions, by initiating automated protocols or manual interventions remotely.

In addition to operational benefits, remote accessibility fosters transparency and accountability in water management practices by facilitating real-time data sharing with stakeholders, policymakers, and the public. It supports data-driven decision-making, collaborative efforts in water conservation initiatives, and compliance with regulatory requirements. Overall, remote accessibility and control empower organizations to enhance operational efficiency, ensure continuous service delivery, and mitigate risks associated with water resource management.

6.2. Data analytics and insights generation

Data analytics and insights generation are critical functionalities of IoT-based water monitoring systems, transforming raw sensor data into actionable intelligence for informed decision-making and strategic planning. Figure 10.3 shows IoT based water monitoring system. These systems employ advanced data analytics techniques, such as machine learning algorithms, statistical analysis, and predictive modeling, to derive meaningful insights from large volumes of sensor data collected over time. Data analytics capabilities enable trend analysis, anomaly detection, and correlation of multiple variables to identify patterns, trends, and potential risks in water quality and quantity dynamics. Furthermore, real-time data processing facilitates immediate detection of abnormal conditions or deviations from expected norms, triggering proactive alerts or automated responses to mitigate risks. Data visualization tools and dashboards provide intuitive interfaces for visualizing trends, generating reports, and communicating key findings to stakeholders effectively. This facilitates data-driven decision-making, supports evidence-based policymaking, and enhances operational efficiency across water management domains.

Moreover, data analytics contribute to continuous improvement in water management practices by optimizing resource allocation, prioritizing infrastructure investments, and implementing targeted interventions for environmental sustainability. By harnessing the power of data analytics, organizations can enhance resilience to climate change impacts, improve water quality standards, and achieve long-term sustainability goals. Overall, data analytics and insights generation are pivotal functionalities that empower stakeholders to leverage data as a strategic asset in optimizing water resource management, enhancing environmental stewardship, and ensuring the resilience of water infrastructure in the face of evolving challenges.

7. Challenges and Solutions in Deploying IoT Water Monitoring Systems

While IoT-based water monitoring systems offer significant benefits, their deployment presents various challenges that require careful consideration and strategic solutions:

- IoT devices collect and transmit sensitive data, raising concerns about data security breaches and unauthorized access. To mitigate these risks, stakeholders must implement robust encryption protocols, authentication mechanisms, and data access controls. Regular cybersecurity audits and compliance with data protection regulations (e.g., GDPR) are essential to safeguarding sensitive information and maintaining stakeholder trust.
- Integrating diverse IoT devices, sensors, and communication protocols from different vendors can pose interoperability challenges. Standardization of communication protocols and adoption of open-source platforms facilitate seamless integration and data exchange between heterogeneous systems. Additionally, IoT gateways play a crucial role in translating data formats and ensuring compatibility across interconnected devices.
- Many IoT devices operate in remote or inaccessible locations, relying on battery power for extended periods. Optimizing energy-efficient designs, deploying low-power consumption sensors, and implementing energy harvesting techniques (e.g., solar, kinetic) prolong battery life and reduce maintenance costs. Remote monitoring of battery levels and predictive maintenance practices help preemptively replace or recharge batteries to ensure uninterrupted operation.
- Scaling IoT deployments to accommodate growing sensor networks and expanding operational environments requires robust network infrastructure and reliable connectivity solutions. Utilizing LPWAN technologies like LoRa, NB-IoT, or satellite communications extends network coverage in remote areas with limited cellular connectivity. Cloud-based platforms offer scalability by accommodating increasing data volumes and supporting distributed computing for real-time data processing and analytics.
- Initial deployment costs, including sensor procurement, infrastructure setup, and software development, can be substantial. Conducting thorough cost-benefit analyses, evaluating long-term operational savings, and quantifying potential ROI are essential steps in securing investment and funding for IoT projects. Leveraging cloud-based subscription models and scalable solutions helps manage upfront costs and optimize resource allocation over the project lifecycle.

These challenges through proactive planning, stakeholder collaboration, and technological innovation is essential to realizing the full potential of IoT-based water monitoring systems. By overcoming these hurdles, stakeholders can harness the transformative power of IoT to enhance water management practices, promote environmental sustainability, and ensure reliable access to clean water resources for future generations.

7.1. Improved accuracy and efficiency in monitoring

IoT-based water monitoring systems revolutionize traditional monitoring practices by providing enhanced accuracy and efficiency in data collection, analysis, and reporting. These systems integrate advanced sensors and IoT devices to continuously monitor key water parameters such as pH levels, turbidity, dissolved oxygen, and flow rates in real-time. Unlike manual sampling methods, which are labor-intensive and prone to human error, IoT sensors ensure consistent data quality and reliability by automating data acquisition processes.

Real-time monitoring capabilities enable rapid detection of water quality anomalies or operational deviations, triggering immediate alerts and notifications to stakeholders.

This proactive approach facilitates timely intervention and corrective actions to prevent potential water contamination events, optimize treatment processes, and ensure compliance with regulatory standards. Moreover, IoT-enabled data analytics and predictive modeling techniques identify trends, patterns, and correlations in water quality, dynamics, enhancing predictive maintenance strategies and operational decision-making.

By improving accuracy and efficiency in monitoring, IoT-based systems empower water utilities, environmental agencies, and industrial facilities to enhance operational transparency, streamline regulatory compliance, and uphold water quality standards. These capabilities support sustainable water management practices and contribute to safeguarding public health, ecosystem integrity, and overall water resource sustainability.

7.2. Environmental impact and sustainability benefits

IoT-based water monitoring systems play a pivotal role in promoting environmental sustainability and mitigating the impacts of water pollution, resource depletion, and climate change. These systems contribute to environmental conservation efforts through several key mechanisms:

- Real-time monitoring of water quality parameters allows for early detection of pollutant sources, contamination events, and ecological stressors in natural water bodies. Timely intervention and remediation measures mitigate the environmental impact of pollutants, safeguarding aquatic ecosystems, biodiversity, and public health.
- IoT sensors facilitate precise monitoring of water usage patterns, enabling efficient water conservation strategies and sustainable water management practices. By optimizing irrigation scheduling, reducing water losses, and promoting water reuse initiatives, these systems minimize freshwater demand, alleviate pressure on water resources, and support drought resilience in agricultural and urban settings.
- IoT-enabled environmental monitoring programs support ecosystem restoration initiatives by assessing habitat quality, monitoring water flow regimes, and implementing adaptive management strategies. These efforts enhance ecosystem resilience to climate change impacts, preserve natural habitats, and promote biodiversity conservation in freshwater ecosystems.
- IoT-based data analytics and reporting capabilities improve transparency, accountability, and regulatory compliance in water management practices. By providing stakeholders with real-time access to accurate data and performance metrics, these systems facilitate evidence-based decision-making, policy formulation, and stakeholder engagement in environmental governance processes.

In summary, the environmental impact and sustainability benefits of IoT-based water monitoring systems extend beyond operational efficiencies to encompass ecosystem protection, water conservation, and resilience to environmental challenges. By leveraging IoT technologies, stakeholders can achieve a balance between socio-economic development and environmental stewardship, ensuring the long-term sustainability of water resources for future generations.

8. Conclusion

IoT-based water monitoring systems offer unparalleled advantages in modernizing water management practices. These systems integrate advanced sensors, IoT devices, and data analytics to provide real-time monitoring and actionable insights into water quality and quantity parameters. By automating data collection and analysis, IoT technologies ensure high accuracy in detecting water quality anomalies, which is critical for maintaining regulatory compliance and safeguarding public health. Moreover, these systems optimize resource allocation, reduce operational costs through proactive maintenance, and enhance environmental sustainability by minimizing water losses and promoting efficient water use practices. IoT-based water monitoring enhances operational efficiency, resilience of water infrastructure, and contributes significantly to sustainable water management practices globally.

8.1. Future prospects and importance of continued research and development

The future of IoT-based water monitoring systems holds immense promise driven by ongoing advancements in sensor technologies, artificial intelligence, and connectivity solutions. Continued research and development efforts are crucial for enhancing the precision, scalability, and interoperability of IoT systems in water management. Future innovations will focus on integrating emerging technologies such as blockchain for secure data management, 5G networks for enhanced real-time communication, and satellite-based remote sensing for comprehensive water resource monitoring. These advancements will enable more effective management of water resources, improved resilience to climate

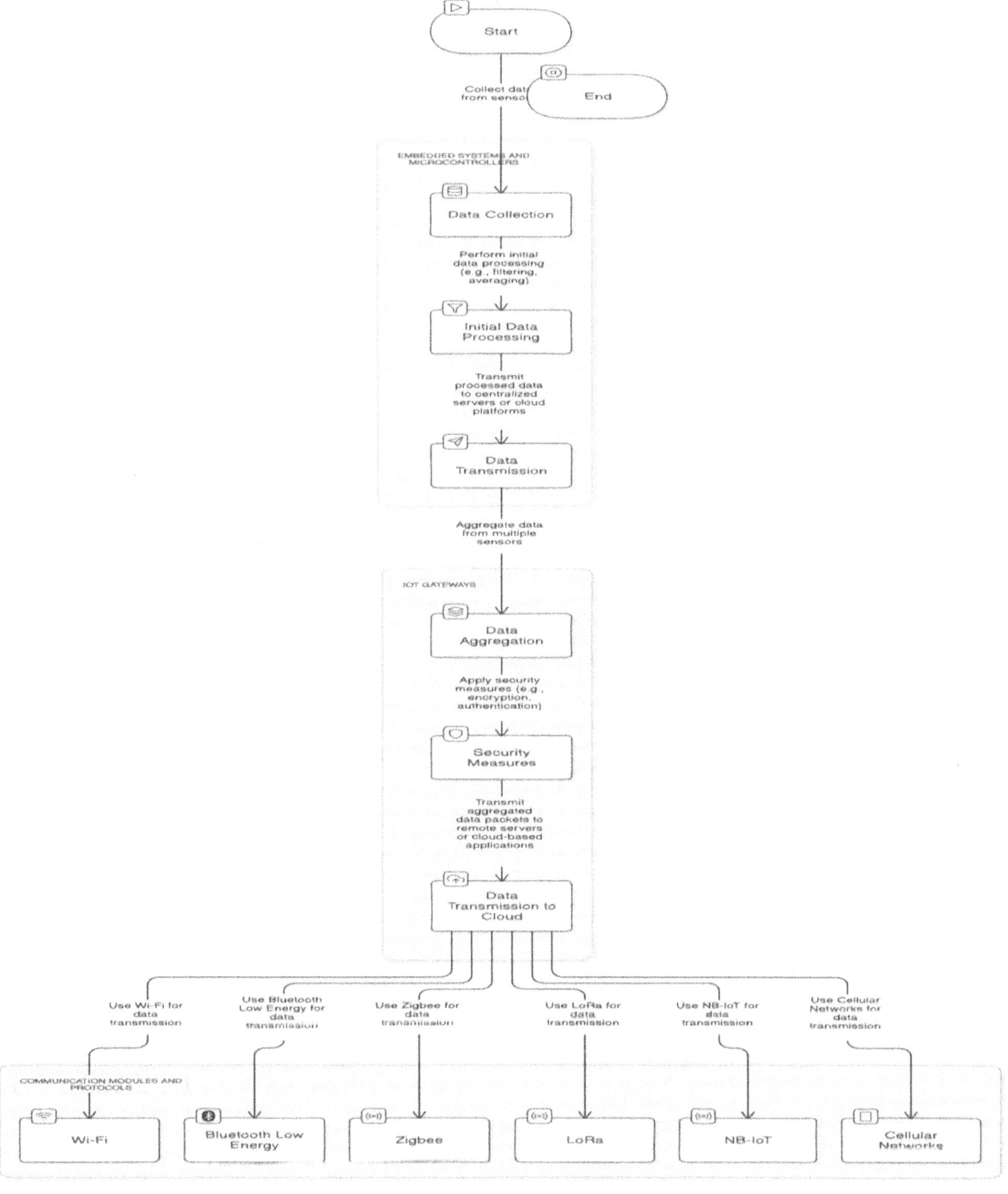

Figure 10.2. IoT system component.

Source: Author.

change impacts, and support for sustainable development goals. Investing in R&D will also foster innovation ecosystems, promote technology transfer, and ensure the ethical and equitable deployment of IoT solutions in water governance. Global collaboration and knowledge sharing will further accelerate progress, enabling stakeholders to address complex water challenges and secure access to clean water for future generations.

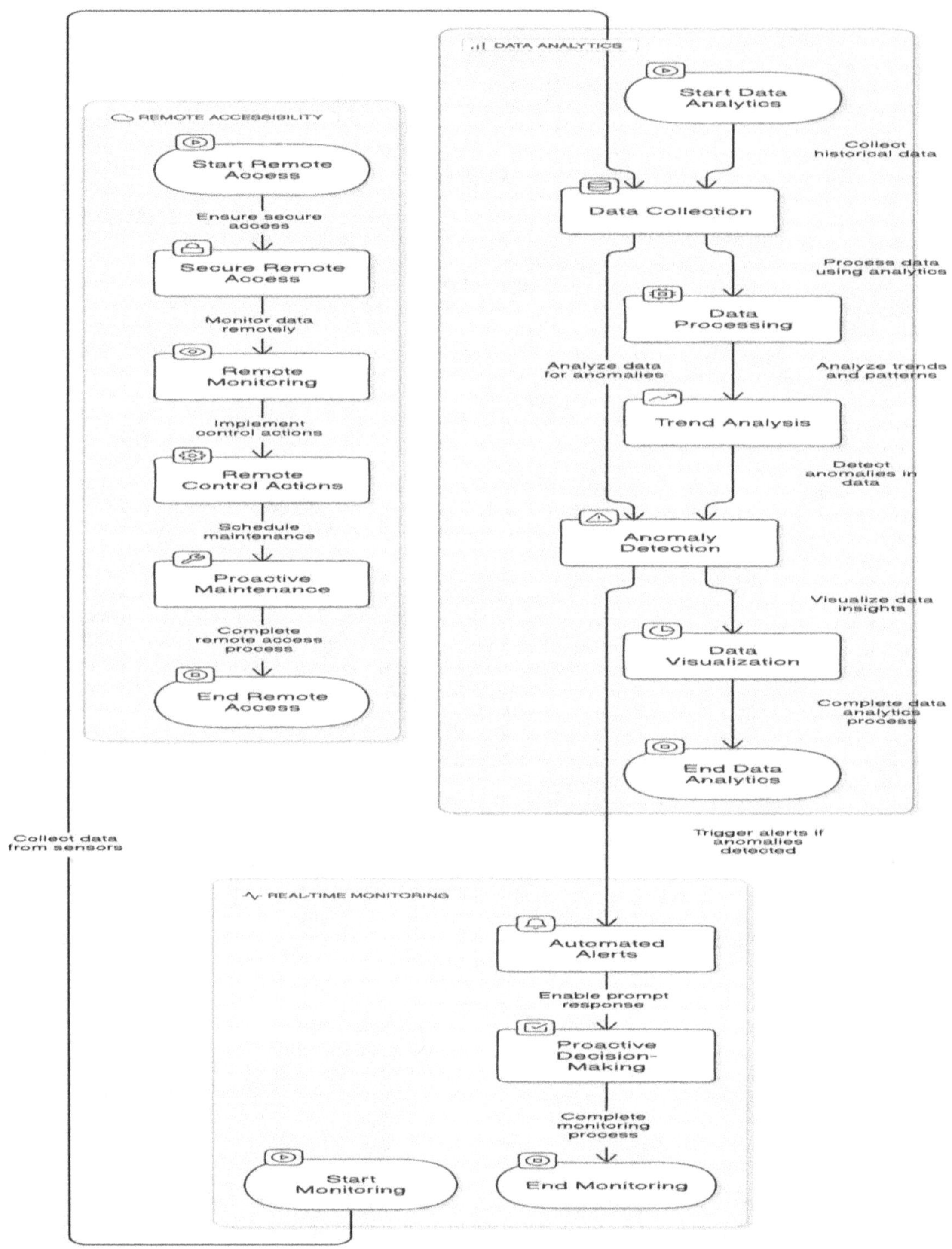

Figure 10.3. IoT based water monitoring system.

Source: Author.

References

[1] Ighalo, J. O., Adeniyi, A. G., & Marques, G. (2021). Internet of things for water quality monitoring and assessment: a comprehensive review. *Artificial Intelligence for Sustainable Development: Theory, Practice and Future Applications*, 245–259.

[2] Buljubašić, S. (2020). Application of new technologies in the water supply system. *Archives for Technical Sciences*, *1*(22), 27–34.

[3] Villamil, S., Hernández, C., & Tarazona, G. (2020). An overview of internet of things. *Telkomnika (Telecommunication Computing Electronics and Control)*, *18*(5), 2320–2327.

[4] Jayapriya, R. (2021). Scientometrics analysis on water treatment during 2011 to 2020. *Indian Journal of Information Sources and Services*, *11*(2), 58–63.

[5] Chaturvedi, R., & Singh, P. K. (2021). A practicable learning under conversion of plastic waste and building material waste keen on concrete tiles. *Materials Today: Proceedings*, *45*, 2938–2942.

[6] Robles, T., Alcarria, R., De Andrés, D.M., De la Cruz, M.N., Calero, R., Iglesias, S., & Lopez, M. (2015). An IoT based reference architecture for smart water management processes. *Journal of Wireless Mobile Networks, Ubiquitous Computing, and Dependable Applications*, *6*(1), 4–23.

[7] Dogo, E. M., Salami, A. F., Nwulu, N. I., & Aigbavboa, C. O. (2019). Blockchain and internet of things-based technologies for intelligent water management system. *Artificial Intelligence in IoT*, 129–150.

[8] Badii, A., Carboni, D., Pintus, A., Piras, A., Serra, A., Tiemann, M., & Viswanathan, N. (2013). CityScripts: Unifying Web, IoT and Smart City Services in a Smart Citizen Workspace. *Journal of Wireless Mobile Networks, Ubiquitous Computing, and Dependable Applications*, *4*(3), 58–78.

[9] Vo, D. T., Nguyen, X. P., Nguyen, T. D., Hidayat, R., Huynh, T. T., & Nguyen, D. T. (2021). A review on the internet of thing (IoT) technologies in controlling ocean environment. *Energy Sources, Part A: Recovery, Utilization, and Environmental Effects*, 1–19.

[10] Llopiz-Guerra, K., Daline, U. R., Ronald, M. H., Valia, L. V. M., Jadira, D. R. J. N., & Karla, R. S. (2024). Importance of environmental education in the context of natural sustainability. *Natural and Engineering Sciences*, *9*(1), 57–71.

[11] Durga, B. G., Kumaran, T. H. M., Devika, I. V., & Akshaya, M. J. (2022, August). Internet of Things based Weather and Water Quality Monitoring System. In *2022 3rd International Conference on Electronics and Sustainable Communication Systems (ICESC)* (pp. 998–1002). IEEE.

[12] Guo, J., & Nazir, S. (2021). Internet of things based intelligent techniques in workable computing: An overview. *Scientific Programming*, *2021*(1), 6805104.

[13] Marques, G., & Pitarma, R. (2020). A cost-effective real-time monitoring system for water quality management based on internet of things. In *Science and Technologies for Smart Cities: 5th EAI International Summit, SmartCity360, Braga, Portugal, December 4–6, 2019, Proceedings* (pp. 312–323). Springer International Publishing.

[14] Li, J., Abdulghani, Z. R., Alghamdi, M. N., Sharma, K., Niyas, H., Moria, H., & Arsalanloo, A. (2023). Effect of twisted fins on the melting performance of PCM in a latent heat thermal energy storage system in vertical and horizontal orientations: Energy and exergy analysis. *Applied Thermal Engineering*, *219*, 119489.

[15] Prapti, D. R., Mohamed Shariff, A. R., Che Man, H., Ramli, N. M., Perumal, T., & Shariff, M. (2022). Internet of Things (IoT)-based aquaculture: An overview of IoT application on water quality monitoring. *Reviews in Aquaculture*, *14*(2), 979–992

[16] Biqing, L., Xiaomei, Y., & Shiyong, Z. (2018). An Internet of Things–based simulation study on Lijiang River water environment monitoring. *Journal of Coastal Research*, (82), 106–113.

[17] Robles, T., Alcarria, R., Martín, D., Morales, A., Navarro, M., Calero, R., ... & López, M. (2014, May). An internet of things-based model for smart water management. In *2014 28th international conference on advanced information networking and applications workshops* (pp. 821–826). IEEE.

[18] Chaturvedi, R., Singh, P. K., & Sharma, V. K. (2021). Analysis and the impact of polypropylene fiber and steel on reinforced concrete. *Materials Today: Proceedings*, *45*, 2755–2758.

[19] Hai, T., Chaturvedi, R., Mostafa, L., Kh, T. I., Soliman, N. F., & El-Shafai, W. (2024). Designing g-C3N4/ZnCo2O4 nanocoposite as a promising photocatalyst for photodegradation of MB under visible-light excitation: response surface methodology (RSM) optimization and modeling. *Journal of Physics and Chemistry of Solids*, *185*, 111747.

11 A detailed review of the concepts, technologies, and industrial applications of Digital Twin

J. Loyola Jasmine[1,a], *M. Carmel Sobia*[2,b], *and M. Jayalakshmi*[1,c]

[1]Assistant Professor, Kalasalingam Academy of Research and Education, Srivilliputtur, Tamil Nadu, India
[2]Associate Professor, P.S.R. Engineering College, Sivakasi, Tamil Nadu, India

Abstract: The fast advancement of industrial digitalization and intelligence has made it imperative to have a digitized model of the physical world which is both accurate and up-to-date. Consequently, this opens the door for the optimisation and control of physical entity operations via the analysis of large amounts of data. Due to its ability of bridging the gaps between the digital and physical realms, the digital twin technology has garnered considerable interest as a means to address the aforementioned demands. Researchers from academia and industry have taken an interest in digital twins due to their numerous possible applications. Various points of view have informed a great deal of the study. Articles summarizing the existing research are a consequence. The articles provide only one viewpoint, and the introduction and summary are disorganized and incomplete. The extensive evaluation of digital twin research in this paper is built around four primary pillars: data, models, networks, and applications. Finding out how the field evolves when it transitions between the virtual and real worlds and vice versa is the main objective. In addition to thoroughly assessing the present research obstacles, the report goes above and beyond by describing potential future paths for advancing digital twin technology.

Keywords: Digital twin, twin model, twin network, twin data

1. Introduction

The fast expansion of digitized industries, informatization, and intelligence across a wide range of industries and fields has been a major factor in the increasing significance and frequency of data exchange between the real and virtual worlds. An urgent need exists for comprehensive collection of data and evaluation that can faithfully portray the physical world in the virtualized world, foretell its future evolution, and regulate the actions of physical entities [1]. Hence, this demand is skyrocketing across a number of sectors, including aerospace, smart cities, intelligent transportation, and smart manufacturing [2]. Because of this, scientists have been on the lookout for a new technology that can meet this need. Digital twins hold the most promise as a means to fulfil these demands. It can detect, recognize, and predict physical things' current states. Although digital twins as a concept have been around for more than 20 years, the technologies that support them have only just received the attention they deserve. Digital twins are electronic copies of a physical item that can automatically and seamlessly exchange data in both directions with the real thing [3]. According to [4], a physical object, physical object's digital twin, and a mapping between the physical object and its digital twin are the three main parts of a digital twin. The digital twin of a product represents the product that is created in real-time using data from the physical, virtual, and interactive parts of the product's lifecycle. This concept was defined by [5] as part of 'Product Lifecycle Management (PLM)'. Digital Twins are virtualized, dynamic model which faithfully mimics a real-life item in a digital setting; this concept was defined by [6]. Barricelli et al. [7] states that the DT is a smart, evolving virtual representation of the real-world objects or processes. They have been the subject of much study and practical application, and a wide variety of definitions here reflects that.

Digital shadows aren't quite as interactive or self-developing as digital twins, but they better depict tangible things digitally [8]. Furthermore, digital shadows, which are digital representations of the state of physical items at a certain instant in time, may be seen as an antecedent of digital twins, as mentioned by [9]. To improve the in-vehicle network's data collecting, prediction, capability of validation and assessment, [10] present the idea of a digital twin. A mapping network that links several digital twins (DTs) is called a digital twin network (DTN) according to [4]. In this network, physical and digital things may work together to complete complex tasks. Building a digitalized twin network in the digital realm is shown in Figure 11.1.

2. Construction of Digital Twins

2.1. *The emergence of digital twins*

The aircraft industry is where the concept of digital transformation first emerged. Thanks to innovations in next-gen information technology, DT has progressed through four

[a]loyolajasminej@gmail.com, [b]carmelsobia@psr.edu.in, [c]jayalacsmi@gmail.com

DOI: 10.1201/9781003675259-11

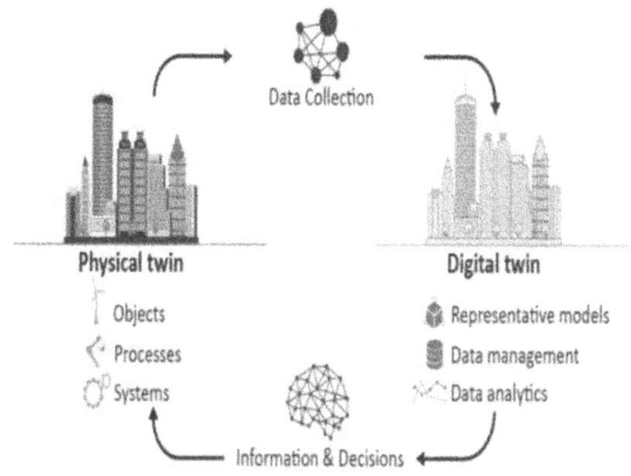

Figure 11.1. Digital Twin.

Source: Medium.

stages of development: technological discovery, concept formulation, application germination, and industry penetration.

DT contains a variety of features, such as the specific instance of a physical structure, its performance, maintenance, repair history, health status, and other characteristics. It establishes preventive maintenance schedules by utilizing historical data regarding the physical structure. The virtual model is employed to predict future performance and maintenance patterns and to monitor and comprehend the performance of the physical asset. The performance of the system can be monitored by developers/facilities administrators, who can then make any necessary adjustments to ensure that the anticipated requirements are met. The digital strands' connectivity offers the potential to trace the life cycle aspects of physical assets. Predictive analytics data obtained from the physical structure can be used to forecast future system performance after assumptions are refined. The potential exists to conduct remote maintenance by troubleshooting remote apparatus that is malfunctioning. By integrating data from the tangible asset with data from the IoT, it enhances and optimises services and operations. The age of the physical system can be reflected in the operational and maintenance data simulated by a DT. An abstracted catalog of design patterns is listed in Table 11.1.

A virtual and real object has a causal relationship and is coordinated in the Digital Twin. The layout template library also included designs without two-way coordination along with the 'real' Digital Twin designs.

2.2. *Concept of DT*

As a general rule, digital twin implementations include making digital representations, or 'twin entities', of physical items. To do this, the four layers of the model—geometric, physical, behavioural, and rule-based—are combined and integrated. Because of this, each twin has skills including assessment, optimisation, evaluation, and prediction. Thus, the concept of representing an analogous thing is vital to digital twin technologies.

Table 11.1. Digital design pattern catalog [11]

Name	*Lifecycle stage*	*Description*
Digital Model	Concept, Development, and Production	A blueprint for manually developing a physical object
Digital Generator	Concept, Development, and Production	A blueprint for automatically developing a physical object
Digital Shadow	Concept, Development, and Production	Modeling an existing physical object
Digital Matching	Utilization, Support	Finding physical objects that match the features of the digital model
Digital Proxy	Utilization, Support, and Retirement	Acting as a proxy for the physical object
Digital Restoration	Utilization and Support	Restoring a physical object to its earlier state
Digital Monitor	Utilization and Support	Monitoring a physical object
Digital Control	Utilization and Support	Monitoring and controlling a physical object
Digital Autonomy	Utilization and Support	Monitoring and controlling a physical object without manual, human intervention

Source: Author.

2.2.1. *Modelling of Twins*

Building a high-fidelity twin entity model is an essential first step in creating a digital twin. What this implies is that the model has to be an accurate and fast reflection of the physical entity's real state. Many versions of the design are often caused by the reliance on numerical parameters for digital modelling in conventional modelling methodologies. With the advent of digital twin technology, a new paradigm for modelling has developed [12] proposed a digital twin modelling strategy to solve the problems of inaccurate models and long feedback loops. This method aims to attain consciousness of oneself, self-prediction, and self-sustaining through the use of digital twin cognitive designs, digital twin mapping designs, and digital twin descriptive structures. Mo et al. [13] developed a smart system to address the challenge of accurately modelling the intricate behaviour of airplanes employing real-time multimodal sensors along with data inputs. This framework might continuously upgrade and enhance its representations to better represent the actual state of tangible items. Yet, persistent worries about the quantity and caliber of data accessible for data-driven modeling restrict the models' reliability [14]. A data-driven dynamic modelling technique was presented by [15] for building virtual models using data, models, and algorithms. In an attempt to guarantee model correctness and enhance the integrity of data, they use deep learning techniques to manage massive amounts of data. Besides data-driven twin modelling, it is critical to build a model that includes the interaction and interdependence of physical elements. Hence, digital twin systems can accurately portray the dynamic behaviour of physical things. A very realistic motion system including gaps and friction was simulated by [16] utilizing the digital twin. To address the problem of vehicle trajectory recovery within the framework of the Internet of Vehicles (IoV), Ji et al. [17] utilized the 'Spatio-Temporal Tracker (STT)'. They helped with the recovery of vehicle trajectories. They were able to represent intricate and unpredictable traffic situations by modelling the relationship between cars and associated worldwide movement characteristics. For examining product quality at several scales during processing, Liu et al. [18] came up with a technique for building a knowledge model for product quality utilizing a digital twin approach. In order to investigate the link between quality indicators and to create a multi-scale representation of product quality aspects, a quality knowledge map was developed. In order to combine data from multi-dimensional context machining processes, Liu et al. [19] suggested a DT modelling technique that uses bio simulation. In addition to enhancing the accuracy and efficiency of component processing, this strategy seeks to create a high level of digital twin system fidelity and flexibility.

2.2.2. *Update of Twin model*

For keeping the twin model updated, reduce the gap between it and the real thing, and make re-creating the model easier in terms of synchronisation, it must be updated often while in use. The majority of current twin modelling upgrade study has been on updating frameworks with virtual or real variables either from mathematical frameworks or real objects. A method for updating models was proposed by [20]. It entails first getting the model's internal parameters via simulation, and then updating the model using actual data from a small number of physical entities. Improving the model's precision and consistency is the goal of this procedure [21]. Employed an iterative learning technique to constantly change the variables in the model, retraining and updating the framework frequently using information gathered in actual time from the physical device. This approach is unable to guarantee the accuracy of the latest version because the internal variables do not depend on real data. The goal is to keep the model's accuracy high while taking ageing into consideration. Still, we need further studies to find out how often to update the model for continuous model change. In their proposal, [22] combine physical models with machine learning classifiers to create a digital twin perception method. This technique educates an interpreter utilizing the optimal decision tree and modifies the DT model by replicating the training information provided by actual individuals under various conditions. Uncertainty in model upgradations is heightened since device functionality is susceptible to several disturbances. As a result, it becomes very tough to upgrade the framework in a way that stays truest to the real thing. Using libraries grounded in physics, [23] lay forth a protocol for creating and overseeing virtual representations of dynamic systems. The latest perturbations factor in the current DT paradigm is derived utilizing the candidate functional libraries and input and output data from system design or outcome inspection. Then, the perturbation term is fine-tuned for model update using sparse Bayesian regression. Using the results of the simulations, this method may successfully infer additional perturbation terms for the DT framework. Also, it updates the model quite well and provides clear and accurate explanations of these disturbances. Improving the model with new information gleaned from simulations or real-world objects is the main emphasis of this research. It may also identify areas of uncertainty while upgrading the quantitative model to make it more accurate. System engineers often struggle with data scarcity, which makes it difficult for model simulation techniques to provide accurate results. While there's no pressing reason to postpone the model update, it may be even more accurate if we include expert knowledge and historical data.

3. Fundamentals of DT

This paper concludes that DTs have four unique technical traits: interdependent development, actual time synchronization, virtual-real mapping, and closed-loop optimisation, after reviewing their development history and underlying concepts. Researchers have looked at a plethora of potential applications for DT systems for a variety of tasks.

3.1. *Data acquisition and transmission technology*

A dynamic simulation is a real-time, dynamic model of a physical system that makes use of surreal mapping methods. An integral part of digital transformations is the actual-time collection, transfer, and up-gradation of data. Numerous scattered, extremely accurate sensors of different types that provide an essential sensory role form the core of the whole twin network. The principles of speed, security, and accuracy govern the installation of sensors and the creation of sensor networks. In this procedure, several types of physical data related to the system are collected by scattered sensors in order to characterize its state [17].

3.2. *Lifecycle data management*

The backbone of DT systems is data storage and administration. Fast data acceptance and safe redundant backup are made possible by cloud servers, which also enable dispersed administration of large-scale operational data. Important for keeping the whole DT system running well, sophisticated data analysis algorithms rely on this availability of trustworthy data sources [18].

3.3. Advanced computing systems

The computing system is essential in the efficient running of Digital Twin systems, which include complex procedures. One important measure for DT system performance is how well they work in real-time. Improving the data and algorithm structures to speed up task execution is important to ensure the network's real-time behaviour.

3.4. Virtual modelling and simulation technology

'DTs are built on high-fidelity virtual modelling technologies. Since DT is an ongoing process that encompasses the complete product lifetime, an adaptive model faithfully portrays such fact. Accurately simulating the physical properties and geometric principles of connected things is the goal of DT's high-fidelity virtual modelling and dynamic simulations. Several domains, dimensions, and timelines of model data must be combined with great precision to accomplish high-fidelity dynamic simulation and virtual modelling. In order to accomplish thorough self-updating and optimisation, the system should also be able to actively monitor the simulation process in real-time and receive feedback data. When physical systems from several domains are combined in a multidomain model, both healthy and unhealthy operating conditions are included. In order to fully understand and provide a thorough mechanical simulation of the fusion architecture, multidomain modelling is used starting with the initial stage of conceptual design [19].

3.5. Other key technologies

A DT system is characterized by a number of parameters, substantial data duplication, and complex and unavoidable noise. Data quality is an important consideration when building DT models, and the parameters show strong coupling, nonlinearity, and temporal fluctuation that affect this aspect. Therefore, developing efficient means of handling massive amounts of data is crucial. The best way to understand important data and make better decisions is via the use of visualisation techniques for decision tree systems.

4. Digital Twin Networks

Because of the growing complexity of software scenarios, an individual DT unit cannot satisfy the requirements of an application. This has led to the emergence of twin structures, which are composed of several twin units. Several physical entities must be able to communicate and exchange information effectively in the physical space domain. To accomplish complex system tasks, it is necessary for physical entities to interact with one another, share information, and learn from one another; this includes twins and other similar entities. In order to make information exchange easier, a digital twin network is built using several networked twins, as shown in Figure 11.2. A synopsis of the work done to apply physical and virtual state synchronisation is given in this chapter. State synchronisation and rectification, twin network deployment, and computing job offloading are all covered.

4.1. Physical–virtual state synchronisation

If digital twin systems want their virtual models to be accurate, they need to synchronize the physical and virtual states via real-time data transmission [24]. It enables the virtualized devices to replicate the real devices' operation and anticipate its actions before they happen, that result in very precise information and analysis. Numerous new approaches to guaranteeing physical and virtual state synchrony have been developed in recent years.

4.2. Twins deployment optimization

In order to ensure that each digital twin and its associated physical entity may communicate in real-time, it is essential to think about their deployment while setting up digital twin networks.

Twin adoption necessitates the adaptable positioning of twin units due to the flexibility of actual groups, the accessibility of processing resources at the edges of the networks, and the requirement for the smallest possible latency between physical entities and their twins. The dilemma of where to place the network edges of digital twins that are socially connected to the Internet of Things was optimally solved by [25], taking into account the edge cloud's restricted processing capacity and the social interactions amongst IoT gadgets. After explaining the placement optimization problem utilizing a mixed-integer linear programming framework, the problem was converted into a mixed-integer linear programming issue by the application of linear relaxation.

The constant movement of real objects necessitates frequent server changes to keep the twins in sync with the most up-to-date locations. If you want to move a twin model from one server to another, you'll need to do triple migration.

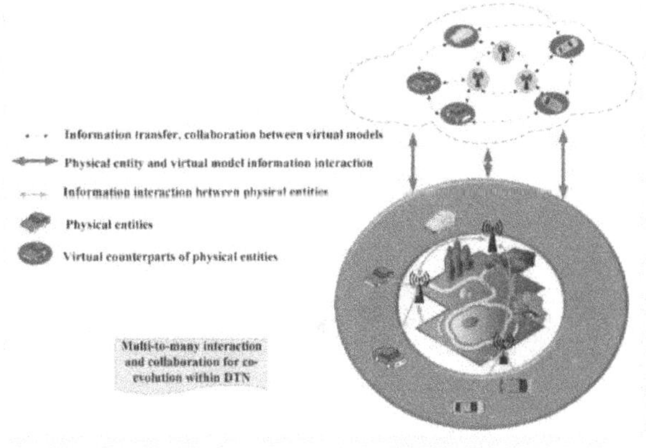

Figure 11.2. DT network.
Source: Wu et al. 2023.

For both the real and virtual equivalents, this ensures effective communication and satisfies the need for low latency. Figure 11.3 shows how the digital doppelganger follows the physical object's path and communicates with a nearby target edge server.

By using migration learning to reconfigure the twin model on the targeted server, the aforementioned research achieves beneficial effects. However, optimisation of system loads, management of storage, and allocation of computing power are all crucial aspects of digital twin transfer that must be carefully considered because to the high demand for computer resources. In order to make twin migration more efficient, accurate, and easy to implement, further research is needed.

4.3. Computational offloading within Twin networks

Several computing tasks, each with its own time constraints and performance needs, must be handled by the DT network in a complex DT implementation scenario. Hardware infrastructure is used to execute calculations and allow data transfer between the twin entities in order for actions to be executed. The complex nature of the tasks and the need for real-time performance in twin systems make it clear that a single twin entity cannot do these tasks alone. This means that several sets of identical twins must work together. Therefore, a digital twin network relies heavily on the procedure of offloading computing activities to external devices.

5. Conclusion

After careful analysis, the below definitions of the phrase 'digital twin' may be drawn:

- Individualized, meaning that each person's digital twin is closely related to their physical twin. Regarding construction, operation, and upkeep, a DT is indistinguishable from its physical analogue. To be considered high-fidelity, a DT should faithfully simulate the actions of its physical equivalent in a computer simulation. Using multi-physics modelling and continuously updating the model throughout its lifetime is necessary to achieve high realism.
- The term 'real-time' describes how fast and accurate a digital duplicate can respond to its physical counterpart. Modern innovations in mobile communication and the Internet of Things have made this feasible by drastically reducing latency and delays. When one twin (digital or physical) is changed, it immediately affects the other (controllable).
- This is the final step in achieving digital-physical convergence, which involves linking the DT with the real twin. Typically, the focus of each research was on improving certain aspects of digital twins, and the methods used to do this varied greatly. Using a wide range of instruments, each one was tailored to meet the unique demands of its designated application field. In an effort to faithfully reproduce physical behaviour, some researchers adopt a modelling-oriented viewpoint that develops out of technical engineering considerations. From the vantage point of information management, some people put an emphasis on semantic linkages and effective information flow. Despite the advancement of several DT frameworks and reference models, none of them have achieved significant industry recognition. This paper offers a brief history, current understanding, and application of DTs before defining them precisely and offering an Industry-based DT model. More references and insights into the development and use of DT technology should be forthcoming from the research's analysis and summary.

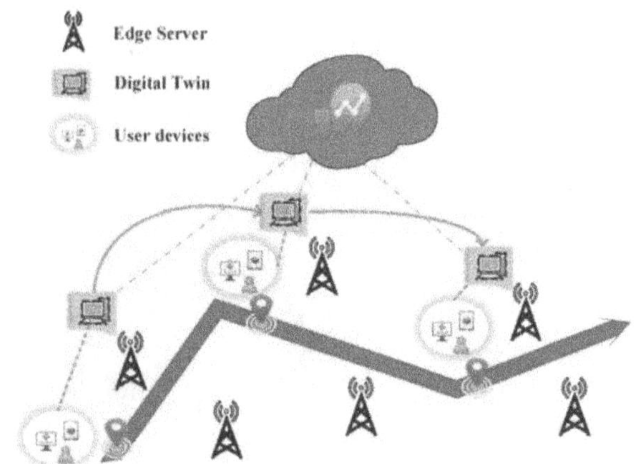

Figure 11.3. Digital twin migration.
Source: Wu et al. 2023.

References

[1] Tao, F., Zhang, H., Liu, A., & Nee, A. Y. (2018). Digital twin in industry: State-of-the-art. *IEEE Trans Ind Inform, 15*, 2405–2415.

[2] Mylonas, G., Kalogeras, A., Kalogeras, G., Anagnostopoulos, C., Alexakos, C., & Muñoz, L. (2021). Digital twins from smart manufacturing to smart cities: A survey. *IEEE Access, 9*, 143222–143249.

[3] Kritzinger, W., Karner, M., Traar, G., Henjes, J., & Sihn, W. (2018). Digital Twin in manufacturing: A categorical literature review and classification. *Ifac-Papers Online, 51*, 1016–1022.

[4] Wu, Y., Zhang, K., & Zhang, Y. (2021). Digital twin networks: A survey. *IEEE Internet Things J, 8*, 13789–13804.

[5] Tao, F., Cheng, J., Qi, Q., Zhang, M., Zhang, H., & Sui, F. (2018). Digital twin-driven product design, manufacturing and service with big data. *Int J Adv Manuf Technol, 94*, 3563–3576.

[6] Zhuang, C., Liu, J., & Xiong, H. (2018). Digital twin-based smart production management and control framework for the complex product assembly shop-floor. *Int J Adv Manuf Technol, 96*, 1149–1163.

[7] Barricelli, B. R., Casiraghi, E., & Fogli, D. (2019). A survey on digital twin: Definitions, characteristics, applications, and design implications. *IEEE Access*, *7*, 167653–167671.

[8] Eleftheriou, O. T., & Anagnostopoulos, C. N. (2022). Digital twins: A brief overview of applications, challenges and enabling technologies in the last decade. *Digit. Twin*, *2*, 2.

[9] Bergs, T., Gierlings, S., Auerbach, T., Klink, A., Schraknepper, D., & Augspurger, T. (2021). The concept of digital twin and digital shadow in manufacturing. *Procedia CIRP*, *101*, 81–84.

[10] Zhao, L., Han, G., Li, Z., & Shu, L. (2020). Intelligent digital twin-based software-defined vehicular networks. *IEEE Netw*, *34*, 178–184.

[11] Qi, Q., Tao, F., Hu, T., Anwer, N., Liu, A., Wei, Y., Wang, L., & Nee, A. (2021). Enabling technologies and tools for digital twin. *J Manuf Syst*, *58*, 3–21.

[12] Luo, W., Hu, T., Zhu, W., & Tao, F. (2018). Digital twin modeling method for CNC machine tool. In *Proceedings of the 2018 IEEE 15th International Conference on Networking, Sensing and Control (ICNSC)*, Zhuhai, China, 27–29; pp. 1–4.

[13] Mo, Y., Ma, S., Gong, H., Chen, Z., Zhang, J., Tao, D. (2021). Terra: A smart and sensible digital twin framework for robust robot deployment in challenging environments. *IEEE Internet Things J*, *8*, 14039–14050.

[14] Hui, L., Wang, M., Zhang, L., Lu, L., & Cui, Y. (2022). Digital twin for networking: A data-driven performance modeling perspective. *IEEE Netw*, *37*, 202–209.

[15] Wang, D., Zhang, Z., Zhang, M., Fu, M., Li, J., Cai, S., Zhang, C., & Chen, X. (2021). The role of digital twin in optical communication: Fault management, hardware configuration, and transmission simulation. *IEEE Commun Mag*, *59*, 133–139.

[16] Guerra, R. H., Quiza, R., Villalonga, A., Arenas, J., & Castano, F. (2019). Digital twin-based optimization for ultraprecision motion systems with backlash and friction. *IEEE Access*, *7*, 93462–93472.

[17] Ji, Z., Shen, G., Wang, J., Collotta, M., Liu, Z., & Kong, X. (2023). Multi-vehicle trajectory tracking towards digital twin intersections for internet of vehicles. *Electronics*, *12*, 275.

[18] Liu, S., Lu, Y., Li, J., Song, D., Sun, X., & Bao, J. (2021). Multi-scale evolution mechanism and knowledge construction of a digital twin mimic model. *Robot Comput Integr Manuf*, *71*, 102123.

[19] Liu, S., Bao, J., Lu, Y., Li, J., Lu, S., & Sun, X. (2021). Digital twin modeling method based on biomimicry for machining aerospace components. *J Manuf Syst*, *58*, 180–195.

[20] Kang, J. S., Chung, K., & Hong, E. J. (2021). Multimedia knowledge-based bridge health monitoring using digital twin. *Multimed Tools Appl*, *80*, 34609–34624.

[21] Jafari, S., & Byun, Y. C. (2022). Prediction of the battery state using the digital twin framework based on the battery management system. *IEEE Access*, *10*, 124685–124696.

[22] Kapteyn, M. G., & Willcox, K. E. (2022). Design of digital twin sensing strategies via predictive modeling and interpretable machine learning. *J Mech Des*, *144*, 091710.

[23] Tripura, T., Desai, A. S., Adhikari, S., & Chakraborty, S. (2023). Probabilistic machine learning based predictive and interpretable digital twin for dynamical systems. *Comput Struct*, *281*, 107008.

[24] Zhang, M., Zuo, Y., & Tao, F. (2018). Equipment energy consumption management in digital twin shop-floor: A framework and potential applications. In *Proceedings of the 2018 IEEE 15th International Conference on Networking, Sensing and Control (ICNSC)*, Zhuhai, China, 27–29, pp. 1–5.

[25] Chukhno, O., Chukhno, N., Araniti, G., Campolo, C., Iera, A., & Molinaro, A. (2020). Optimal placement of social digital twins in edge IoT networks. *Sensors*, *20*, 6181.

[26] Wu, H., Ji, P., Ma, H. and Xing, L., 2023. A comprehensive review of digital twin from the perspective of total process: Data, models, networks and applications. *Sensors*, 23(19), p.8306.

12 PSO-based optimization of EVCS and DG placement for improved grid durability and efficiency in distribution systems

Ashu Nayak[1,a] *and Ankita Tiwari*[2,b]

[1]Assistant Professor, Department of CS & IT, Kalinga University, Raipur, India
[2]Research Scholar, Department of CS & IT, Kalinga University, Raipur, India

Abstract: To reduce pollution in the environment, traditional fossil fuel cars must give way to electric vehicles (EVs). Electricity networks and utility operators are greatly impacted by the location of electric vehicle charging stations (EVCS). Issues including elevated load, uneven power output, power outages, and decreased voltage stability are caused by improperly positioned EVCS. Increasing the use of electric vehicles (EVs) through distributed generation (DG) helps alleviate these problems. DG and EVCS placement optimization in radial distribution networks is the main goal of this effort. The particle swarm optimization (PSO) is used in the algorithm, the technique analyses weight flow utilizing a forward sweep method as well the backward, and it gives the formal placements and sizes for EVCS and DG. By employing a novel green filter compensator based on FACTS, the integrated scheme is completely stable. This enables low inrush currents, steady DC bus voltage, and load excursions, resulting in the most efficient functioning in the renewable energy sources integrated with the diesel engine. PSO algorithm performs 1–2.5 times better than other approaches overall, showing higher optimization efficacy and computational economy. This method is a potential tool for improving EVCS and DG placement in distribution networks since it produces much better outcomes when used.

Keywords: Electric vehicle, fossil fuel, harmonics distortion, particle swarm optimization, DG

1. Introduction

In recent years low fuel consumption and the emission outputs of car technologies have been enhanced by use of the plug-in hybrid electric vehicles (PHEVs) and electric vehicles (EVs). The depletion of fossil fuels and the emergence of concerns about the climate have also brought an increased interest in plug-in hybrid electric vehicles (PHEVs) and electric cars (EVs), mostly named internal combustion engine vehicles (ICEVs). Battery electric vehicles or BEVs are an option for internal combustion engines (ICEV). Batteries within BEVs require grid electricity to be recharged to start operating. Because the batteries in BEVs are refilled with grid power, it is evident that these vehicles represent a vital connection between the transportation and electrical sectors. Furthermore, because they consume little energy and produce no local emissions, BEVs can offer a perfect way to lessen the impact of transportation on the environment and reduce reliance on energy sources. BEVs are, to put it another way, zero-emission vehicles (ZEVs) [1–3]. These days, advancements in electric motors, battery systems, power electronics converters, and control algorithms have largely made the commercialization of BEVs conceivable. Electric motors power the wheels of a BEV powertrain. As such, the EM is a crucial element needed for high-performance BEV powertrains to arise and be accepted. Thus, a good agreement has been made recently in research on electric motors that maybe utilized to meet the needs of automotive applications and BEVs [4–7]. The main problem with this design, however, is that the count of independent voltage for DC sources at the input rises in proportion to the count of voltage ranges to be generated. As an example, the CHB inverter features independent DC-to-DC converters that supply each H-bridge unit's DC connections without relying on a common DC bus bar. With constant power sharing among Hbridge units, a modular structure, and fewer voltage levels than their asymmetric counterpart, symmetric CHB topologies are more efficient. A well-known PV array source shows a 27-level asymmetric CHB design [9], but big line-frequency transformers need to be installed at the output to separate single H-bridge units. As proposed in [13], the 'Level Doubling Network' (LDN) is a capacitor given to the half-bridge unit joined in cascade with a symmetric CHB architecture. The 27-level asymmetric CHB architecture of a well-known PV array source is presented in [9], however big line-frequency transformers need to be installed at the output to separate single H-bridge units. As proposed in [10], the 'Level Doubling Network' (LDN) is a capacitor-supplied half-bridge unit joined in cascade with a symmetric CHB architecture [2].

[a]ku.ashunayak@kalingauniversity.ac.in, [b]ankita.tiwari@kalingauniversity.ac.in

DOI: 10.1201/9781003675259-12

Steady AC output may be produced by the diesel-powered synchronous generator, but the PV source's output power is directly dependent on sunshine [11]. Consequently, the source-load power balance needs to be adjusted for variations in solar radiation levels. By keeping a steady supply of power, a DC/AC converter regulates the flow of the power from the hybrid PV-FC sources to the load side [12]. To assist in steady the flow of electric power, a fuel cell was incorporated. Natural selection and the genetic search process are the foundations of GA, an iterative search method. But GA is quite picky; it includes scenarios for crossing, mutation, and selection, among other things. Moreover, the encoding and decoding procedure affects the genetic algorithm's accuracy and adds to its complexity [4].

While driving, the EVs' batteries are charged, and choosing the best spot for the EVCS has a big impact on the methods that may be used. The position and capability of the EVCS have a major impact on how well the power system operates. The distribution network is poor due to the current and voltage restriction violations, and the unexpected deployment of EVs makes planning difficult. Road networks have a major impact on the best placement for EVCS at the same time. As a result, it is important to think about how best to distribute EVCS road and distribution networks [5]. Power loss and bus voltage are significant considerations that must be considered when the EVs are charging their batteries at the charging station. Here's a summary of the suggested work: A computationally efficient load flow method called backward and forward sweep (BFS) is used to determine the goal functions, which are voltage and power loss. Investigations are conducted on how EVCS affects the distribution network, and distributed generators are included to lessen these effects [8, 10].

2. Particle Swarm Optimization

Psoc is a technique of computation optimization approach based on a natural system with search strategy development by Kennedy and Eberhart [9]. In the first phase of the system, we have a population of randomly chosen solutions. As a particle, every possible resolution is mentioned. Each particle is moved over the issue space after having been given a random velocity. As the particles have memory, they support a record of prior quality position (the Pbest) along with the associated fitness. They mention the global best of the swarm as the one with the highest fitness. There are several Pbest for every particle of the swarm. The basic idea behind the PSO approach of accelerating each particle way to its Pbest and Gbest positions at every time step is coupled with a random weighted speed. The selection procedure and the particle swarm optimization method include these steps of the primary:

- Create a population of particles in the issue space's d dimensions, giving them random places and velocities, and then launch them.
- Assess the swarm's individual particles' fitness.
- Comparison was made between a value of each particle (and best fitness) and values of the same variables from the previous iteration. Then if the current value is greater than the value of Pbest, it sets Pset to the value and Pbest location to the current position in d dimensional space.
- By comparing the Pbest of the particles with one another To promote the swarm's global quality position with the highest fitness (Gbest).
- Modify the particle's location and speed. As per Equations (1) and (2), in that order.

$$V_{id} = \omega \times V_{id} + C_1 \times rand_1 \times (P_{id} - X_{id}) + \\ + C_2 \times rand_2 \times (P_{gd} - X_{id}) \quad (1)$$

$$X_{id} = X_{id} + V_{id} \quad (2)$$

The dynamic total error minimization technique is used by various important parameters in the PSO optimization search. The following are below and the ones I refer to here, ω is called the inertia weight, which controls the exploration and exploitation of search space by looking dynamically at velocity. The maximum velocity at which particles may move is known as Vmax; if a particle's velocity is higher than Vmax, it can only go up to Vmax. Vmax therefore establishes search fitness and resolution. Particles will transcend an appropriate resolution if Vmax is too high. In local minima, the particles will caught if Vmax is set to decrease. The variables C1 and C2, denoted as the social and cognitive components, respectively, in (1) and (2). The acceleration constants listed below are what shift a particle's velocity in the direction of Pbest and Gbest (usually in the middle). The decision made by PSO to prioritize cooperation over competition sets it apart from the other evolutionary algorithms. Similar to the survival of the fittest, other optimization techniques frequently employ decimation. On the other hand, there is no individual replacement or destruction in the PSO population, which is constant. The best work of their neighbors has an impact on individuals. People finally get together at the best locations inside the issue domain. Furthermore, the PSO does not often contain genetic operators such as individual-to-individual crossover or mutation, and additional people never replace particles throughout the run. Therefore, in contrast to GA, all of the particles in PSO tend to converge fast on the optimal solution.

3. Closed Loop Control

The suggested Multi-Objective Optimization (MO) search method makes use of the following definitions:

- Def.l: The following formulation of the generic MO problem, which calls for the optimization of N goals, is possible: Reduce- stations (EVCS).

$$\vec{y} = \vec{F}(\vec{x}) = \left[\vec{f}_1(\vec{x}), \vec{f}_2(\vec{x}), \vec{f}_3(\vec{x}), \ldots, \vec{f}_N(\vec{x})\right]^T \quad (3)$$

$$subject\ to\ g_j(\vec{x}) \leq 0 \qquad j = 1, 2, \ldots, M \quad (4)$$

$$\text{Where}: \vec{x}^* = \left[\vec{X}_1^*, \vec{X}_2^*, \ldots, \vec{X}_P^*\right]^T \in \Omega \tag{5}$$

- The decision variables are represented by $\vec{x}$ *, a P-dimensional vector inside a parameter space Ω, the target vector by $\vec{y}$, and the limitations by $\vec{g}$ I ($\vec{x}$). The region through which the goal vectors pass is known as the objective space. A subspace that satisfies the requirements among the objective vectors is called the feasible space.
- Concept 2: If the decision vector $\vec{x}_1$ does not rigitly outperform $\vec{x}_2$ in atleast single goal and does not outperform $\vec{x}_2$ in all objectives, then $\vec{x}$, ∈Ω dominates the decision vector $\vec{x}_2$∈Ω (represented by $\vec{x}_1 < \vec{x}_2$).
- Def 3: A decision vector is Pareto-optimal if it has nil other x_l 2∈Ω that overpowered it. It is claimed that two decision vectors are similar when they both match an objective vector and are Pareto-optimal.

According to definition 4, the Pareto optimal set is defined to be the non-dominated set of the whole suitable search space Ω. The collection of values in the objective space that is Pareto optimal is referred to as a parametric optimum front.

A group of particles are randomly initiated within the MOPSO decision space. Velocity Vi and position xi are assigned to each particle i in the decision space. The change in particles and their positions constitute moving toward the best-found solutions thus far. The non-dominated solutions from the previous generations are the archive. The archive is an external population whose solutions are the non-dominated solutions that have been identified so far. To go towards the optima, velocities are calculated in the following way. (DGs) and electric vehicle charging stations (EVCS).

$$\mathrm{V}_{id} = \omega \times V_{id} + C_1 \times rand_1 \times (P_{pd} - X_{id}) + C_2 \times rand_2 \times (P_{rd} - X_{id}) \tag{6}$$

In this above case, Pp,d, Pr,d are randomly picked from only the global Pareto archive; inertia factor ω influences the ability of the algorithm to act locally and globally; Vi,d is the particle i's velocity in d_th dimension; and values of c1 and c2 representing the weight on the social and cognitive element respectively. We have two equal and independent randoms in the [0, 1] interval named r1 and r2. The two particles, Pr,d and Pp,d, which the particle should move to according to Equation (6), should be taken from the edited set of non-dominated solutions kept in the archive. The particles shift places until a termination requirement is satisfied through generations. It is possible to run the MOPSO for multiple generations and obtain a decent-sized amount of Pareto optimal trade-off solutions.

4. Results and Discussion

A disruption in the distribution network occurs when a lot of EVs are charged at the same time, increasing the demand at the charging station (CS). This results in a drop in the system's voltage and an increase in power losses. When installing an EVCS, there are a few things to consider. Look for a charging station where electric vehicle owners can quickly and conveniently charge their cars with the least amount of power loss. The suggested study makes it abundantly evident that power reduction in the distribution network rises in tandem with the number of EVCSs. Numerous scholarly articles discuss how reducing I2R losses lowers distribution network power losses. However, in the suggested work, power losses are decreased by strategically placing distributed generators at various points across the system, causing the voltage of the distribution network to increase as power losses in the distribution system decrease.

The true power losses rise to 218 kW when the three EVCSs with fixed sizes are ideally positioned at 2, 3, and 4 in the distribution network, as shown in Figure 12.1. After that, EVCSs number two, three, four, nine, and thirteen. This results in an actual power loss of 256 KW. As the load on the charging station grows, the losses also increase. The distribution network's bus 30 is the best location for a single DG of size 1310.705 kW, which minimizes the losses to 112. 5737 KW. Reduce the actual power losses in the distribution system to 70.85 kW by installing two DGs: one at bus 6 with a capacity of 1311.805 kW and the other at bus 32 with a capacity of 874.8701 kW.

When EVCS are efficiently assigned at buses 2, 3, and 4 of the bus system given in Figure 12.2, reactive losses in the power in the distribution network increase to 143 kVar. Reactive low increases to 166 kbar when there are five EVCS at sites 2, 3, 4, 9, and 13 in the distribution network, which is the number of EVCS positioned ideally. Distributed generators are included in the distribution network to reduce power loss, allowing electric vehicle (EV) owners to conveniently charge their cars without interfering with the distribution system. With a capacity of 1311.805 kW, one DG is placed in the bus system at bus 30. To 71.9387 kVar,

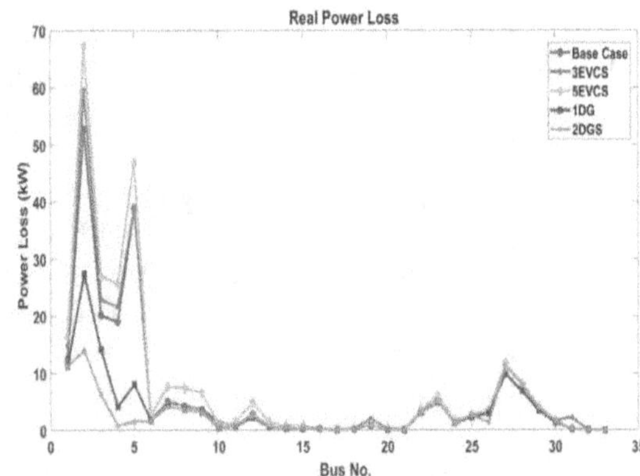

Figure 12.1. Profile for the voltage distribution network following the installation of distributed generating sources (DGs) and electric vehicle charging stations (EVCS).

Source: Author.

the reactive power loss drops. Ideal locations for two DGs with capacities of 1311.805 and 872.8701 kW are buses 6 and 32 in IEEE 33 bus systems. By installing two DGs, the reactive power loss is reduced to 49.8781 kVar. Figure 12.3 illustrates the position of the DGs and EVCS in the 33-bus network distribution and the actual and reactive power losses following their ideal arrangement. The table makes it evident that, after the best distribution of VCS, active power losses rise to 256 kW and reactive losses to 143 kVar, respectively, and then fall to 70.85 kW and 49.8781 kVar following the placement of the distributed generator. The difference in power losses and the voltage increase after the required configuration of EVCS and DGs. Through the ideal location of a single distributed generator, the suggested technique reduces actual and reactive power losses to 46.103% and 48.253%, respectively. The installation of two DGs has reduced power losses by 65.5278% and 63.2201%, respectively (Figure 12.4).

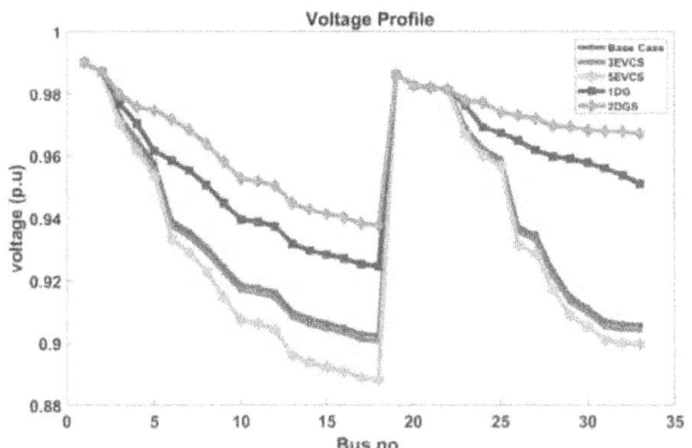

Figure 12.2. Actual loss of power in the bus system following the best possible placement of distributed generating sources (DGs) and electric vehicle charging stations (EVCS).

Source: Author.

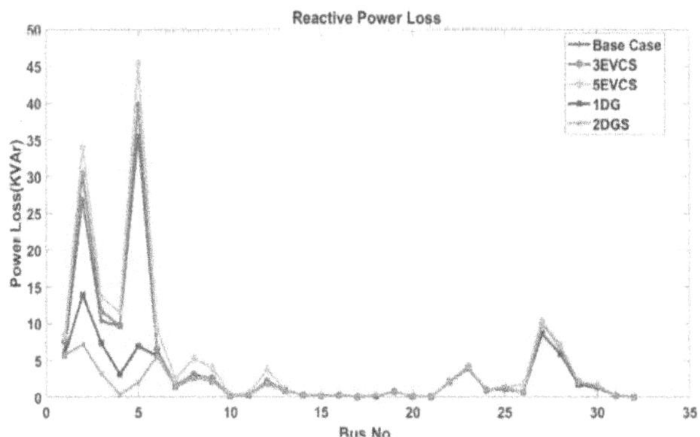

Figure 12.3. Reactive power outages when distributed generating sources (DGs) and electric vehicle charging stations (EVCS) are installed.

Source: Author.

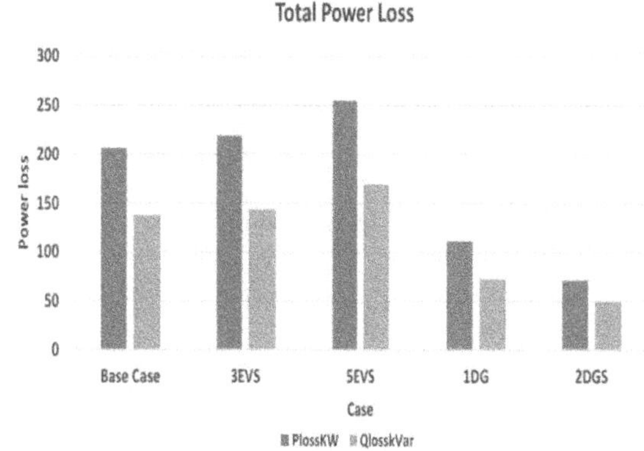

Figure 12.4. Total power loss on the distribution network following the integration of distributed generating sources (DGs) and electric vehicle charging stations (EVCS).

Source: Author.

5. Conclusions

Transportation with fossil fuels produces more pollution than when electric vehicles are used. EVCS installations resulted from an enhancement in the usage of electric cars. Thus, the power system is adversely affected by the EVCS installation. Here, the BFS load flow analysis is used to show how EVCS affects the power system. When several electric vehicles charge, the system's power losses rise and its voltage falls. Distributed generators are positioned to maximize voltage increases and offset such losses. The Multi-Objective Optimization (MOPSO) approach is employed to dynamically modify the add and settings of all controllers to reduce the overall absolute error of each controller. Under extreme load and excursion conditions, this control technique effectively ensures voltage stabilization while improving power/energy usage and source-load performance. These control techniques' benefits and downsides are discussed. It has been demonstrated that, particularly at low load and speed, IFOC founded on PSO and PWM voltage has higher effectiveness than alternative controlled motor schemes. DG and EVCS placement optimization is only one of the many uses for this study. It might be improved to help grid modernization initiatives, link energy storage devices to distribution networks, and manage electric vehicle fleets more effectively across a range of businesses. It may also be used to resolve cybersecurity concerns related to demand response programs, residential microgrid configurations, smart city planning, and the integration of EVCS and DG.

References

[1] Rosales-Tristancho, A., Brey, R., Carazo, A. F., & Brey, J. J. (2022). Analysis of the barriers to the adoption of zero-emission vehicles in Spain. *Transportation Research Part A: Policy and Practice*, *158*, 19–43.

[2] Giliberto, M., Arena, F., & Pau, G. (2019). A fuzzy-based Solution for Optimized Management of Energy Consumption

in e-bikes. *Journal of Wireless Mobile Networks, Ubiquitous Computing, and Dependable Applications*, *10*(3), 45–64.

[3] Bilal, M., Rizwan, M., Alsaidan, I., & Almasoudi, F. M. (2021). AI-based approach for optimal placement of EVCS and DG with reliability analysis. *IEEE Access*, *9*, 154204–154224. https://doi.org/10.1109/ACCESS. 2021.3125135

[4] Nair, J. G., Raja, S., & Devapattabiraman, P. (2019). A scientometric assessment of renewable biomass research output in India. *Indian Journal of Information Sources and Services*, *9*(S1), 72–76.

[5] Tomasov, M., Motyka, D., Kajanova, M., & Bracinik, P. (2019). Modelling effects of the distributed generation supporting e-mobility on the operation of the distribution power network. *Transp Res Procedia*, *40*, 556–563 https://doi.org/10.1016/J.TRPRO.2019.07.080

[6] Islam, A., Sharma, A., Chaturvedi, R., & Singh, P. K. (2021). Synthesis and structural analysis of zinc oxide nano particle by chemical method. *Materials Today: Proceedings*, *45*, 3670–3673.

[7] Fadel, S., Al-Awami, A. T., & Mohammed, O. A. (2018). Charge control and operation of electric vehicles in power grids: A review. *Energies*, *11*(4), 701. https://doi.org/10.3390/EN11040701

[8] Mohamed, S., Kumaran, U., & Rakesh, N. (2024). An approach towards forecasting time series air pollution data using LSTM-based auto-encoders. *Journal of Internet Services and Information Security*, *14*(2), 32–46.

[9] Patrao, I., Figueres, E., González-Espín, F., & Garcerá, G. (2011). Transformerless topologies for grid-connected single-phase photovoltaic inverters. *Renewable and Sustainable Energy Reviews*, *15*(7), 3423–3431.

[10] Praveenchandar, J., Venkatesh, K., Mohanraj, B., Prasad, M., & Udayakumar, R. (2024). Prediction of air pollution utilizing an adaptive network fuzzy inference system with the aid of genetic algorithm. *Natural and Engineering Sciences*, *9*(1), 46–56.

[11] Moniruzzaman, M. (2019). Design and simulation of photovoltaic, diesel hybrid AC mini-grid system for rural electrification.

[12] Nejabatkhah, F., Danyali, S., Hosseini, S. H., Sabahi, M., & Niapour, S. M. (2011). Modeling and control of a new three-input DC-DC boost converter for a hybrid PV/FC/battery power system. *IEEE Transactions on Power Electronics*, *27*(5), 2309–2324.

[13] Sharma, A., Islam, A., Sharma, K., & Singh, P. K. (2021). Optimization techniques to optimize the milling operation with different parameters for a composite of AA 3105. *Materials Today: Proceedings*, *43*, 224–230.

13 Tumor extraction from MR brain images using modified seeded region growing algorithm

D. B. Shanmugam[1,a], *Arun Francis G.*[2,b], *Manikandan S.*[3,c], *KottaimalaiR.*[4,d], *Thilagaraj M.*[5,e], *and Ambika B.*[6,f]

[1]Department of Computer Science and Applications, SRM Institute of Science and Technology Ramapuram Campus Chennai, Tamil Nadu, India
[2]Department of Electronics and Communication Engineering, Karpagam College of Engineering, Coimbatore, India
[3]Department of Electronics and Telecommunication Engineering, Karpagam College of Engineering, Coimbatore, India
[4]Department of Electronics and Communication Engineering, Kalasalingam Academy of Research and Education, Krishnankoil, Tamil Nadu, India
[5]Department of Industrial Internet of Things, MVJ College of Engineering, Bangalore, Karnataka, India
[6]Department of Electrical and Electronics Engineering, Pandiyan Saraswathi Yadhav Engineering College, Tamil Nadu, India

Abstract: An aberrant development of cells inside the cerebral cortex or adjacent tissues is called a brain tumour. One of the most important steps in the analysis of medical images for the diagnosis and treatment of numerous illnesses, especially in cancer, is tumour removal, sometimes referred to as tumour segmentation. The procedure entails separating the tumour area from surrounding medical pictures. One of the most important imaging techniques for identifying, classifying, and tracking brain malignancies is magnetic resonance imaging (MRI). To view various biological tissues and structures, several MRI sequences are employed. Preprocessing techniques are used to lower distortion and improve the clarity of the medical images. The preprocessed pictures can be used for extracting the tumour region using a variety of segmentation approaches. Using methods such as modified seeded region growth, pixels are grouped into regions according to spatial connection and resemblance criterion. The modified algorithm may incorporate advanced segmentation techniques that improve the accuracy of tumour delineation. This could include better handling of tumour boundaries, irregular shapes, and varying intensities within the tumour region. The precision of the retrieved tumour areas is evaluated by cross-referencing the results with information from skilled human annotations or histological analyses.

Keywords: Brain tumour, Tumour segmentation, magnetic resonance imaging (MRI), seeded region growing algorithm, therapy, diagnosis

1. Introduction

A cerebral tumour refers to a tumour that originates within the brain itself, specifically within the cerebrum [1]. These tumours can be either benign or malignant. The course of therapy for a cerebral tumour is determined by the individual's desires and general wellness along with the kind, position, dimension, and grade of the tumour. Tumour excision procedures, chemotherapy, radiation therapy, selective therapy, immunotherapy, or an assortment of these procedures are possible forms of treatment [2]. The complete removal of the tumour as can be securely removed, alleviating symptoms, reducing or halting tumour development, and enhancing general quality life span are feasible therapeutic objectives [3].

A region of interest (ROI) is a particular region or area in an image that has been chosen for additional evaluation, analysis, or explanation because it is thought to be relevant to a certain medical concern or scientific goal [4]. ROIs can be physically defined by utilizing specific software programs that create edges or contours surrounding the region of interest. Although manual segmentation offers exact control over the selection process, it can be arbitrary and taking time [5]. Artificial methods employ techniques that autonomously identify and divide up ROIs according to pre-established standards, including texture features, intensity thresholds, or deep learning approaches. Although computerized categorization might require confirmation and refinement, it can be more quickly accomplished objective [6]. ROIs can

[a]shanmugd@srmist.edu.in, [b]ja.arunji@gmail.com, [c]mailingtomani@gmail.com, [d]r.kottaimalai@klu.ac.in, [e]m.thilagaraj@gmail.com, [f]ambi_kab@yahoo.co.in

DOI: 10.1201/9781003675259-13

graphically reflect the targeted region by being superimposed or emphasized on the source image. ROIs can be examined continuously to observe variations in dimensions, form, or other aspects in long-term study or medical follow-up visits. This makes it possible to evaluate the course of the illness, the impact of therapy, or the reaction to treatments. Finding and correlating relevant portions among several picture acquisitions is one way that automated tracking algorithms can help with the longitudinal analysis of ROIs. In order to make informed decisions about diagnosis, therapy, and patient care, ROI analysis is essential. To arrive at well-informed decisions about evaluation, prognosis, and therapy plans, physicians employ data retrieved from ROIs. ROI examination is a crucial part of the understanding and evaluation of clinical images because it allows useful information to be extracted from complicated image data and is quantitatively assessed. In a variety of healthcare settings, it makes precise diagnosis, impartial assessment, and individualized treatment strategy easier.

Xilei et al. developed an automated method for aneurysm diagnosis with a sensitivity of 91% in aneurysm recognition by utilizing a DL approach on 2D local projection pictures [7]. When compared to current cutting-edge methods, the selected approach turned out to be more effective. Xin et al. created a method that uses an AI methodology and a 3D CNN to identify aneurysms in CTA and DSA images [8]. The determined values for sensitivity, specificity, and accuracy are 74%, 18%, and 84%, respectively. In order to help forecast brain disorders, Emmanuel et al. investigated the application of DL algorithms to identify aneurysms in a variety of medical photos [9]. The distinctions between a vein with an aneurysm and a healthy vein were emphasized by Guangchen et al. [10]. He investigated and examined aneurysm parameters like size, morphology, and location using the chi-squared test and multivariate logistic regression. The range of accuracy that was attained is 92% to 97%.

An image can be divided into regions using a technique called seeded region growth in image processing, which is based on pre-established seed locations [11]. When removing regions of interest (ROIs) from medical imaging, like MRI or CT scans, it is especially helpful. A flexible method, seeded region growing can be applied to various image forms and segmentation applications. It works very well for segmenting areas of medical imaging that have uniform features, such tumours. To get precise and significant segmentation outcomes, though, great consideration should go into choosing the seed points and fine-tuning the similarity metrics. Consequently, the Seeded region growth method helps with the automated recognition and dissection of the tumour area in medical imaging. By sharing the generated images with clinicians and patients to expedite the diagnosis procedure, including this technique additionally helps in earlier illness identification.

2. Methodology

2.1. Pre-processing method

The seed growing method, sometimes known as seeded region growing, is a popular technique for picture segmentation. To increase the accuracy and effectiveness of this procedure, preprocessing steps are crucial.

- To help keep the algorithm from becoming trapped in tiny, unimportant places, use a Gaussian filter to smooth out the image and decrease noise [12].
- Normalize the image's intensity values to a specified range such as 0 to 1 to guarantee consistency and enhance the algorithm's functionality.
- Use CLAHE or histogram equalization to enhance the image's contrast. By taking this action, the ROI may become easier to identify.
- Calculate the image's gradient to find locations with sharp contrasts, as they may represent possible region borders.
- Choose the starting seed points manually or by applying methods like thresholding, clustering, or choosing them manually based on domain expertise.
- For generating a binary or multi-level image that distinguishes the subject matter from the surrounding area, employ thresholding methods. Images with different illumination settings are able to processed using adaptive thresholding.

By putting these preprocessing stages into practice, the standard of the regions that the seed growing method segments can be greatly improved, producing greater precision and substantial outcomes.

2.2. Modified seeded region growing

In image processing and machine vision, a technique called seeded region growth is utilized to divide a picture into areas according to resemblance standards. The beginning areas of the method are represented by a set of seed points. Following that, through the addition of nearby pixels that satisfy specific matching requirements, repeatedly expands these areas [13]. Adams and Bischof developed seeded region growth (SRG), which is sturdy, fast, and parameter-free. These qualities enable the development of an excellent technique that can be used with a wide variety of image types. Because of its ability to incorporate high-level understanding of image features into the seed selection process, SRG is also particularly appealing for semantic image segmentation. Nevertheless, the SRG method also faces issues with pixel grouping orders for labeling and autonomous seed generation.

An overview of the seeded region expanding technique is provided below:

Initialization: Choose a few image-based seed spots. These starting places can be picked at random or automatically depending on predetermined standards. Create an initial blank space for every seed point.

Region Developing: Include a seed point to the appropriate section for every seed point. Verify whether each surrounding pixel of the seed point satisfies the requirements for being integrated into the area based on resemblance. Add the nearby pixel to the region and designate it as attended if the requirements are satisfied. Till the region stops growing—that is, until the last pixels can be added—continue this operation for all unvisited nearby pixels.

Dismissal: Once every region has expanded to its full potential, resume the region-growing procedure for each seed point.

Post-processing: It's possible to optionally carry out post-processing operations like minimizing noise or combining of too-small nearby areas.

The definition of the resemblance metrics that are employed to decide whether to add a pixel to a region is a crucial step in the seeded region expanding method. This criterion may be based on a mixture of these or on intensity, colour, or texture similarity, among other considerations.

The following are a few typical resemblance standards for seeded region growth [14].

Intensity or Colour Similarity: Assess how closely the present pixel matches the seed pixel in terms of intensity or colour. When the absolute difference is less than a specific threshold, the pixels are said to be comparable.

Restrict the expansion of regions according to their spatial proximity. As a result, regions are kept from expanding into irrelevant portions of the picture.

Measures of Homogeneity: To determine whether the pixel values within a region are homogeneous, use statistical measurements like variance or entropy.

Among its many uses are segmenting images, recognizing objects, and analysis of medical images. Seeded region growth is a flexible technique. It might, however, need precise parameter adjustment and be susceptible to noise and picture artifacts.

Let initial seeds be S1, S2, S3, ..., Sq could be substituted by the centroids of these generated homogeneous regions, R1, R2, R3, ..., Rq. The labels for the pixels in the identical region are exactly the same, but the labels for the pixels in the variant regions change. The remaining pixels are referred to as unused pixels, while all of these labeled pixels are referred to as assigned pixels.

Let H represent the collection of all unallocated pixels that border at least one identified region.

$$H = \{(x, y) \nexists \cup_{i=1}^{q} R_i | N(x, y) \cap \cup_{i=1}^{q} R_i \neq \emptyset\} \quad (1)$$

Where,

$N(x, y)$ = 2nd order neighborhood of the pixel.

Calculating the variation among the testing pixel and the identified region next to it is done as

$$\delta(x, y, R_i) = |g(x, y) - g(X_i^c, Y_i^c)| \quad (2)$$

Where,

$g(x, y)$ = three colour components values of (x, y)

$g(X_i^c, Y_i^c)$ = average values of three colour components of R_i

$$\varphi(x, y) = \min_{(x,y)\epsilon H} \{\delta(x, y, R_j) | j \epsilon \{1, \ldots, q\}\} \quad (3)$$

Unless all of the image's pixels are assigned to their appropriate areas, this seeded region growth process occurs again [15]. The remaining partition of the image is split into a collection of regions that are as homogenous as feasible based on the restrictions provided, according to the definitions of Equations (1) and (3). The SRG technique is a highly desirable option for semantic image segmentation since it is quick, efficient, and requires no parameter modification. But the issues with pixel grouping and autonomous seed selection also affect the SRG method.

The fundamental tenet of region-growing techniques is that adjacent pixels inside an area have comparable values. Comparing a single pixel with its neighbors is the standard technique. The pixel can be designated as belonging to the cluster together with one or more of its neighbors if a similarity condition is met. In every case, noise affects the findings, and the choice of the similarity criterion is important.

This technique accepts the image as input in addition to a set of seeds. Every object that is to be segmented is marked by a seed. By comparing all unallocated surrounding pixels to the regions, the regions are developed iteratively. A pixel's intensity value divided by the region's mean is used to calculate how similar two pixels are. The pixel belonging to the corresponding region is the one with the least difference, as determined by this method. Until all pixels have been allocated to a region, the procedure is repeated. The selection of seeds affects the outcomes of segmentation since seeded region growing necessitates seeds as an extra input, and noise in the image might lead to incorrectly positioned seeds.

The modified seeded growing method is particularly useful for brain tumor segmentation due to several key reasons:

- Accurate Boundary Detection
- Handling Heterogeneous Tumors
- Reducing Over-segmentation
- Improving Computational Efficiency
- Clinical Relevance and Interpretability
- Integration with Advanced Imaging Techniques

In summary, the need for modified seeded growing methods in brain tumor segmentation arises from their ability to address the challenges specific to medical image analysis, such as heterogeneous tumor characteristics, noise, and computational efficiency, while striving for clinically relevant and accurate results. These modifications ensure that the segmentation method is robust, reliable, and capable of supporting clinical decision-making processes effectively.

3. Implementation and Outcomes

Due to the magnetic field created throughout the test, there are noises in the input image that was collected from the testing area. The preprocessing step involves removing any undesirable noises from the supplied image. The image is resized during the pre-processing phase in accordance with how flexible the method is. For better visual comprehension, the image's contrast and brightness should also be improved.

Using an automated technique to determine which pixels are probably positioned in the given area, choose the initial seed points. Give each seed point a distinct label that identifies the starting regions. Look at the pixels surrounding each seed point. With regard to a resemblance criterion, decide if the adjacent pixel belongs in the region. This standard may be derived from colour, texture, intensity, or other characteristics of the pixels. A surrounding pixel has been added to the region and given the appropriate name if it satisfies the similarity criterion. Fresh pixels are included and those turn into the latest seed points.

Repeatedly, neighbours are looked at, similarity is tested, and pixels are added to the region. Until no more pixels satisfy the requirements for any region, the method adds identical pixels that are adjacent to expand the areas. When every pixel in the image has been allocated to a region or whenever no further pixels satisfy the similarity requirements for any region, the process comes to an end.

The dataset employed in this work is a publicly available one which is downloaded from Kaggle database. Figure 13.1 depicts the input, pre-processed, region of interest (ROI) and the segmented outputs of benign tumours.

Figure 13.2 depict the input, pre-processed, region of interest (ROI) and the segmented outputs of malignant tumors.

The performance or efficacy of the proposed technique is assessed by using the performance metric parameters like, Computational time, Recall, Specificity, Sensitivity, F1-Score, Accuracy, Precision, Dice overlap index (DOI) and Tanimoto Co-efficient (TC) [20].

$$Accuracy = \frac{TP + TN}{TP + FP + TN + FN}$$

$$Precision = \frac{TP}{TP + FP}$$

$$Recall = \frac{TP}{TP + FN}$$

$$F1 - Score = \frac{2 * (Recall * Precision)}{(Recall + Precision)}$$

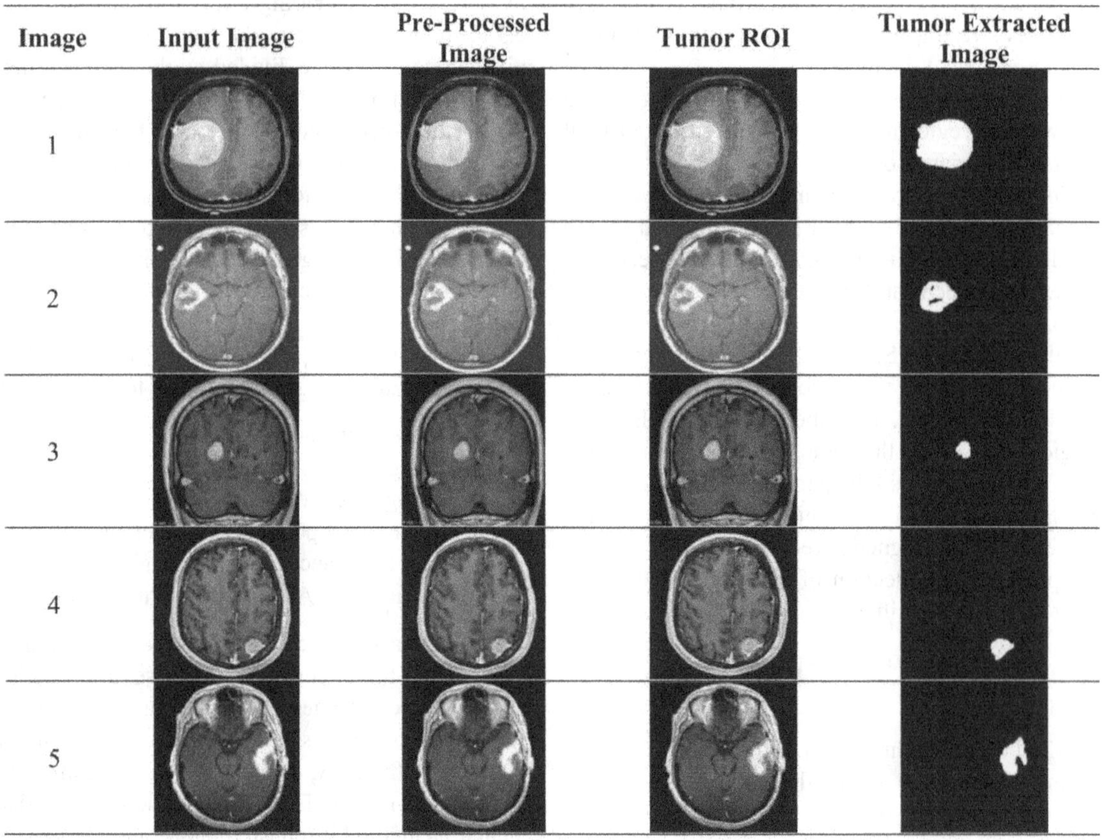

Image	Input Image	Pre-Processed Image	Tumor ROI	Tumor Extracted Image
1				
2				
3				
4				
5				

Figure 13.1. Benign tumour segmented outputs using proposed algorithm.

Source: Author.

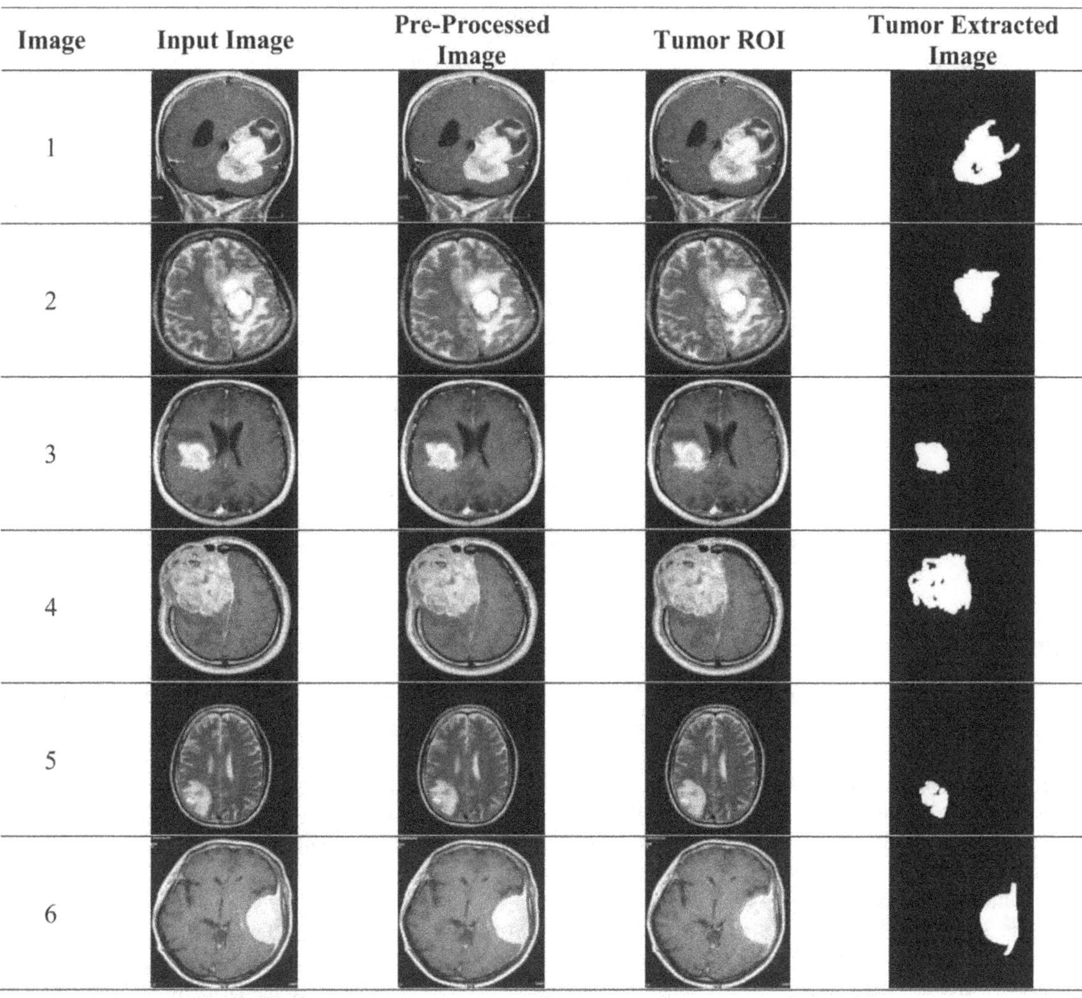

Figure 13.2. Malignant tumour segmented outputs using proposed algorithm.

Source: Author.

$$Specificity = \frac{TN}{TN + FP}$$

$$Sensitivity = \frac{TP}{TP + FN}$$

$$Tanimoto\ Co - efficient = \frac{|X \cap Y|}{|X \cup Y|}$$

$$Dice\ Overlap\ Index = \frac{2|X \cap Y|}{|X| + |Y|}$$

Where, TP = True Positive, TN = True Negative, FP = False Positive, FN = False Negative, X = Pixels in input image, Y = Pixels in segmented image.

Table 13.1 exemplifies the average performance measurement outcomes of the proposed technique on detecting and segmenting the benign and malignant tumors. Table 13.2 describes the performance metrics comparison. The highest values attained by the proposed method on all metrics implies it effectiveness on tumor segmentation.

4. Conclusion

The suggested method evaluated on a dataset and was designed to autonomously produce the seed point initialization for every input brain image. It can determine in the findings of the experiment as the suggested approach is capable of locating brain tumours and identifying the optimal ROIs. The suggested approach produced greater outcomes than conventional methodologies.

5. Acknowledgement

The authors thank the International Research Centre of Kalasalingam Academy of Research and Education, Tamil Nadu, India, for permitting the use of the computational facilities available in the Centre for Biomedical Research and Diagnostic Techniques Development.

Table 13.1. Performance measurements of the proposed method on tumor detection

Image	*Time (sec)*	*Recall*	*Precision*	*Sensitivity*	*Specificity*	*F1-Score*	*Accuracy*	*TC*	*DOI*
Benign	4.56	97.74	96.4	98.21	98.64	96.4	98.16	94.74	97.24
Malignant	5.12	96.98	95.12	97.16	97.28	94.6	97.42	93.86	95.21

Source: Author.

Table 13.2. Comparison of metrics

Algorithms	*Time (s)*	*TC (%)*	*DOI (%)*	*OF (%)*	*Specificity (%)*
Cuckoo-IT2FCM	15.40	22.03	36.09	71.43	93.75
PSO-FCM	22.36	22.54	36.79	87.01	79.65
BFOA-MFKM	10.36	23.35	37.86	90.32	98.21
BFOA-MFCM	7.93	25.23	40.29	90.48	98.25
Proposed method (Benign)	4.56	94.74	97.24	98.21	98.64
Proposed method (malignant)	5.12	93.86	95.21	97.16	97.28

Source: Author.

References

[1] Naydin, S., Marquez, B., & Liebman, K. M. (2022). Vascular Anatomy of the Brain. *Introduction to Vascular Neurosurgery*, 3–29.

[2] Ramaraj, K., Amiya, G., Rajasekaran, M. P., Govindaraj, V., Vasudevan, M., Thirumurugan, M., ... & Thiyagarajan, A. (2023). A robust multi-utility neural network technique integrated with discriminators for bone health decisioning to facilitate clinical-driven processes. *Research on Biomedical Engineering*, *39*(1), 139–157.

[3] Ramadan, O. M. A., & Lv, X. (2022). History of Endovascular Surgery of Cerebral Aneurysms. In *Endovascular Surgery of Cerebral Aneurysms* (pp. 29–40). Singapore: Springer Nature Singapore.

[4] Sengan, S., Arokia Jesu Prabhu, L., Ramachandran, V., Priya, V., Ravi, L., & Subramaniyaswamy, V. (2020). Images super-resolution by optimal deep AlexNet architecture for medical application: a novel DOCALN. *Journal of Intelligent & Fuzzy Systems*, *39*(6), 8259–8272.

[5] Yoon, N. K., McNally, S., Taussky, P., & Park, M. S. (2016). Imaging of cerebral aneurysms: a clinical perspective. *Neurovascular Imaging*, *2*(1), 1–7.

[6] Pandian, B., Arunprasath, T., Vishnuvarthanan, G. & Ramaraj, K. (2022). Object identification from dark/blurred image using WBWM and Gaussian pyramid techniques. In *2022 International Conference on Augmented Intelligence and Sustainable Systems (ICAISS)* (pp. 637–642). IEEE.

[7] Dai, X., Huang, L., Qian, Y., Xia, S., Chong, W., Liu, J., ... & Ou, C. (2020). Deep learning for automated cerebral aneurysm detection on computed tomography images. *International Journal of Computer Assisted Radiology and Surgery*, *15*, 715–723.

[8] Wei, X., Jiang, J., Cao, W., Yu, H., Deng, H., Chen, J., ... & Zhou, Z. (2022). Artificial intelligence assistance improves the accuracy and efficiency of intracranial aneurysm detection with CT angiography. *European Journal of Radiology*, *149*, 110169.

[9] Mensah, E., Pringle, C., Roberts, G., Gurusinghe, N., Golash, A., & Alalade, A. F. (2022). Deep learning in the management of intracranial aneurysms and cerebrovascular diseases: A review of the current literature. *World Neurosurgery*, *161*, 39–45.

[10] He, G., Wang, J., Zhang, Y., Li, M., Lu, H., Cheng, Y., & Zhu, Y. (2022). Diagnostic Performance of MRA for UnrupturedAneurysms at the Distal ICA. *Clinical Neuroradiology*, *32*(2), 507–515.

[11] Joy, A., Anitta, D., Ramaraj, K., & Thilagaraj, M. (2023, August). Smart Drug Delivering System in Hospitals Using RFID. In *2023 5th International Conference on Inventive Research in Computing Applications (ICIRCA)* (pp. 1645–1649). IEEE.

[12] Chen, Z., Yao, L., Liu, Y., Han, X., Gong, Z., Luo, J., Zhao, J., & Fang, G. (2024). Deep learning-aided 3D proxy-bridged region-growing framework for multi-organ segmentation. *Scientific Reports*, *14*(1), 9784.

[13] Ramaraj, K., Murugan, P. R., Amiya, G., et al. (2023). Emphatic information on bone mineral loss using quantitative ultrasound sonometer for expeditious prediction of osteoporosis. *Sci Rep*, *13*, 19407.

[14] Khan, Z. A., & Gostick, J. T. (2024). Enhancing pore network extraction performance via seed-based pore region growing segmentation. *Advances in Water Resources*, *183*, 104591.

[15] Prasad, P., & Mishra, A. (2023). Segmentation using AI for identifying tumors in brain MRI. *American Journal of Multidisciplinary Research & Development (AJMRD)*, *5*(02), 58–64.

14 Threats to the IOT technologies from security point of view

Debarghya Biswas[1,a] *and Balasubramaniam Kumaraswamy*[2,b]

[1]Assistant Professor, Department of CS & IT, Kalinga University, Raipur, India
[2]Research Scholar, Department of CS & IT, Kalinga University, Raipur, India

Abstract: The Internet of Things (IoTs) requires ubiquitous, stable, high-performing, efficient, and scalable infrastructure and facilities. IoT services are compatible with a vast array of devices, spanning from basic to sophisticated technology, and communication takes place over multiple networks. The Internet of Things' most important feature is cloud computing, which links servers and analyses vital data collected from sensors in addition to offering enough storage. Upcoming organisation and resale data will be needed to complete this attribution. Some uses of the IoT make it possible to automate processes and enable physical objects that are not living to function without human intervention. Applications of the IoT in the industrial sector have the potential to increase output and provide wider networks of communication between personnel and machinery. In the end, this would make it possible for more rival companies to join the market, which would enhance quality assurance and lower losses. The many applications of IoT, such as smart cities and towns, healthcare, and industry, are covered in this article along with the hazards and benefits of various technologies.

Keywords: Ransomware, IOT, cyber-attacks, cyber-warfare

1. Introduction

Now a days, the most concern topics between all academicians and research specialist is Internet of Things (IoT). It allows all objects /things available in our surrounding are to be connect with each other without involvement of mankind. IoT is a rapidly growing field of study with lots of untapped potential. Thanks to its boundless innovation, it is on the verge of converting the internet's existing shape into a modern and interconnected one. The number of internet-connected gadgets is increasing every day, and connecting all with IoT, whether by wire or wireless, will ensure a continuous flow of data [8].

2. Problems with IOT

IoT services can be used with a broad variety of devices, from simple to complicated machinery, and communication is done through a number of networks. IoT services work with a wide range of devices, from simple to complicated technologies, and communication occurs through a variety of networks. To ensure data security in the IoT environment, end-to-end (E2E) data security across the complete IoT service should be offered. Data is generated from a wide range of sources and disseminated at randomly through an open network such as the internet [1]. As a result, throughout the data life cycle, the data protection framework must be capable of controlling and measuring sensitive data protection information. The IoTs connects things in a natural manner, and the objects that are connected change with time. The linked gadgets should be capable of maintaining the required level of security in this case. Local sensing devices used in the home, for example, should communicate securely with one another and be kept safe to allow multi-thing cooperation. Furthermore, when interacting with mobile devices, they all should follow the same security policy [2]. There are several different sorts of devices and platforms in IoT environments, ranging from small sensors to smart phones. As previously noted, a security concern in one item can rapidly spread to certain other items, making multi-thing safety challenging to secure. As a consequence, each item must have a secure SW execution environment. Because everything has different abilities, such as computer power and memory size, the same infrastructure security cannot be applied to all. As a result, security mechanisms should be built that give the right degree of security based on object abilities and responsibilities [12]. Mis-configuration by users is the source of many security and privacy flaws [4]. On the other hand, ensuring that users are aware of complex security/privacy policies or processes will be incredibly difficult and unrealistic. As a result, technologies that make creating and enforcing security and privacy policies simple are essential [6].

3. Different Technologies

Some uses of the IoT make it possible to automate processes and enable physical objects that are not living to function

[a]ku.debarghyabiswas@kalingauniversity.ac.in, [b]Balasubramaniam.kumaraswamy@kalingauniversity.ac.in

DOI: 10.1201/9781003675259-14

without human intervention. Applications of the IoT in the industrial sector have the potential to increase output and provide wider networks of communication between personnel and machinery. In the end, this would make it possible for more rival companies to join the market, which would enhance quality assurance and lower losses. The many applications of IoT, such as smart cities and towns, healthcare, and industry, are covered in this article along with the hazards and benefits of various technologies.

A system called RFID (Radio Frequency Identification) uses an object's distinctive frequency to identify it. It can be integrated into any object because to its tiny size and low cost. Depending on the requirements, it is either an active or passive transmitter microchip in the shape of an adhesive sticker. Tags have batteries attached to them because they are always active and release energy [3]. Active tags are only active when they are prepared to transmit data, whereas passive tags are only engaged when signals are activated.

Compared to passive RFID tags, active RFID tags are more costly and widely utilised. A system made up of linked RFID tags and readers broadcasts the object's position, identity, and any other relevant parameters when any relevant signals are generated. The Readers receive the appropriate object-related supplied data. Wireless signals are used to verify the data before it is sent to the CPU.

WSNs are multi-hop, bi-directional sensors made up of multiple nodes connected to one or more information sensors, dispersed around the sensing region. The perception layer makes use of multi-hop communications. Every sensor is a transceiver that is equipped with an antenna, microprocessor, and sensor interface circuit. However, an additional proposal has been made [13] as a component for communication, actuation, and sensing with a power source for harvesting energy, which might be a battery or another kind of power generation. One kind of memory device that can hold data is a memory unit. The sensor network gathers information on temperature, humidity, speed, and other object-specific parameters. It's off to the processing facilities after that.

Recent advances in optical technology, such as Li-Fi and Cisco's BiDi optical technology, have the potential to revolutionise the IoT. For IoT-connected devices, Li-Fi, an epoch-making Visible Light Communication (VLC) technology, will offer better connectivity at a higher bandwidth. Conversely, BiDi technology offers a 40G Ethernet cable for transferring massive amounts of data from various IoTs devices [5].

4. Cyber Security

Data security is one of every government's top priorities, and more and more information is being handled digitally or through cyberspace these days. Additionally, a large number of social media provide a decent platform where people may communicate with users worldwide and feel protected at the same time. Social media is the primary source of information used by the offences under discussion to evaluate unauthorized content. Apart from social media, consumers also tried to exercise the necessary prudence when doing other types of transactions, like online bank transactions [14].

4.1. Cyber security techniques

By utilizing novel approaches, cyberattacks against cyberspace have the potential to expand. Most often, cybercriminals will alter the virus signatures in order to capitalize on newly discovered technical flaws. In other cases, they genuinely look for unique characteristics of cutting-edge technology to find vulnerabilities in malware insertion [7]. Cybercriminals are leveraging the millions and billions of daily users on the Internet and the rapidly developing technologies to their advantage in order to quickly and efficiently access a vast number of individuals.

4.2. Access control and password security

A straightforward method of protecting sensitive information and maintaining privacy is to employ usernames and passwords. One of the most important cyber security measures is this method of security.

- Authentication of Data: Prior to being communicated, the information must be verified as having originated from a reliable source and hasn't been altered. Frequently, the fighting virus software programme on PCs is used to authenticate these papers. A sincere anti-virus programme is more important to safeguard gadgets against viruses.
- Malware Scanners: A computer programme which occasionally checks every record and file for malicious code or destructive viruses inside the system. In this field, Trojan horses, worms, and viruses are typically used to identify and classify samples of hostile software systems.
- Firewall: A firewall is a piece of hardware or software that keeps viruses, worms, and hackers out of your computer when they try to connect to it online. Each message that enters the firewall is inspected, and those that don't match the security standards are blocked from being sent at all. A firewall is essential for detecting malware. Figure 14.1 shows interfacing of firewall.

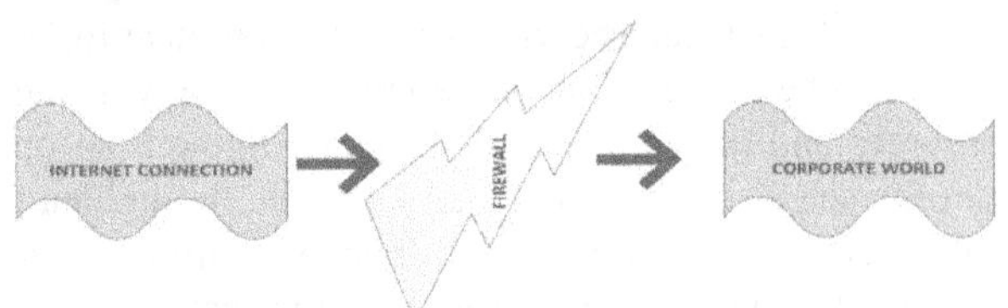

Figure 14.1. Interfacing of firewall.

Source: Author.

5. Role of Social Media in Cyber Security

Businesses that are interactive in today's world must devise novel strategies for safeguarding customer information in an increasingly complex setting. Social media is crucial to both personal intrusions and cyber safety. Since most employees use social media or social networking sites almost daily, there is an increasing concern of attack due to the increasing popularity of social media among them. This has created a massive platform for cybercriminals to breach private information and steal valuable data. These days, it's quite simple to disclose personal information, so companies need to be sure to identify security breaches early on, take immediate action to stop them, and stop them all. Social media has made it simple for users to publish their personal information, which hackers can then use. People therefore must take appropriate precautions to prevent information loss and misuse on these social media platforms [15]. Figure 14.2 shows steps included in cybersecurity process.

6. Attacks

In general terms, there are two categories of cybersecurity threats that are referred to be actioned targeted because they are thought to have detrimental impacts on digital media systems and certain courses of action that are necessary for taking advantage of the cyber infrastructure or for malicious purposes without causing any kind of affect that could immediately have an impact on the system's operation, such as the Trojan Horse, which is beneficial for computer security [9]. The misuse of digital technology includes using the internet and other digital platforms to engage in any kind of extortion, such as taking, planning mental tyrant, and possessing an infringement on copyrights, for disseminating dubious messages that combine hateful and political discourse, and for providing children's stimulating entertainment or unique but restricted resources. In close proximity to the proliferation of free hacking tools and low-level traps like as keyloggers and RF scanners, if someone is using their PC for email and other gaming purposes, at some point they are likely to be targeted.

6.1. Phishing attacks

In Verizon's most recent survey on data infractions, 32% of the confirmed violations were related to natural events. By tricking the victims into thinking they are communicating with a reliable individual via email or text, and increasingly over phone, the attacks aim to obtain sensitive data such as usernames, passwords, SS numbers, and card details [10].

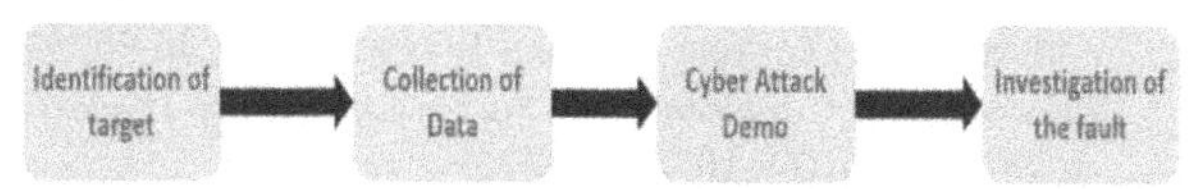

Figure 14.2. Steps included in cybersecurity.

Source: Author.

6.2. IoT Ransomware

A variety of network-connected devices, such as service sensors and residential appliances, are part of the worldwide IoTs. Refrigerators and climate control units may be held as hostages and are potential targets for hackers seeking to obtain information from backend systems, including those in power supply and communication facilities, even though they rarely hold sensitive information on their own.

6.3. Advanced persistent threats

Advanced Persistent Threats, or APTs, are typically novel platforms used by cybercriminals. For many years, the primary component for assessing target attacks has been the computer network system. Since many attacks employ more sophisticated techniques, network security should collaborate with other security solutions to detect attacks in a timely manner. For this reason, it is necessary to improve security measures in order to shield the network from new threats.

6.4. Through mobiles

Since people can now connect with one other anywhere in the world, network security has become very important. Various security tools are insufficient because individuals are utilizing the computing equipment such as laptops and cell phones that needed additional security measures in addition to those that came pre-installed. It is necessary to take into account the issues raised by these mobile devices. In addition, users should exercise caution when clicking on links and advertisements because the mobile network is more vulnerable to cybercrimes [11]. Smart cities are designed metropolitan settings where anyone, anywhere, at any time, can use any service. The IoTs is now a crucial component of smart city construction, with the goal of the issues with conventional urban development. The nature of IoT information exchange allows for the collection of vast volumes of data on people and other items in the smart city, especially between linked objects, or 'Things', and remote places for data to be processed and preserved. As a result, virus attacks and security flaws could affect this data. Bandwidths of data gathered and sent via an IoT infrastructure will thereby impact the security and privacy of the people living in smart cities.

7. Possible Impacts of New Technologies on Internet of Things Security

Emerging technologies have the ability to drastically change the IoT security landscape by posing new difficulties and providing creative solutions for those that already exist. By automated threat detection, strengthening intrusion prevention systems, and modifying security measures based on real-time data analysis, AI and ML can be utilised to improve IoT security. But hackers might also use these technologies to create more advanced attacks that target IoTs devices. Blockchain and distributed ledger technologies provide transparent, immutable, and decentralised systems for access control and authentication, making them viable options for protecting IoTs devices and data. For IoT security, more investigation and advancement are required to fully realise the promise of blockchain technology. It is anticipated that the extensive implementation of 5G networks would lead to a rise in the quantity of interconnected IoT gadgets, augment their functionalities, and present novel security obstacles. In 5G networks, safeguarding the safety and privacy of IoT gadgets will necessitate new Online safety IoT-Related Hygiene: Best Practices and Challenges methods and the modification of current security mechanisms.

8. Conclusion

In conclusion, a new era of networked devices has been ushered in by the IoT explosive expansion, presenting a host of advantages and conveniences to a wide range of businesses and people. But this unprecedented degree of connectedness has also brought about a number of intricate cybersecurity dangers and problems, emphasizing how crucial it is to uphold strong cybersecurity hygiene in the IoT era. Safeguarding IoT devices and the copious amounts of sensitive data they gather, store, and communicate requires cybersecurity care. Good IoT cybersecurity procedures may guard against unwanted access, data leaks, and cyberattacks, guaranteeing the privacy, availability, and integrity of IoT devices as well as the important data they handle. Furthermore, upholding solid IoT cybersecurity hygiene will help foster user confidence and encourage the broad use of IoT technologies, ultimately enabling the realization of their full potential to revolutionize sectors and enhance quality of life. A diversified approach is necessary to address the particular threats and problems related to cybersecurity in the IoTs. This entails creating thorough security frameworks, boosting end-user education and awareness, stimulating industrial, academic, and governmental collaboration, and encouraging innovation in IoT security solutions. Stakeholders may contribute to a safer, more secure IoT ecosystem that benefits all users and opens the door for ongoing innovation and progress in the IoT era by emphasizing IoT cybersecurity awareness and cooperating to develop and execute effective security solutions.

References

[1] Ghelani, D. (2022). Cyber security, cyber threats, implications and future perspectives: A Review. *Authorea Preprints*.

[2] Thevenon, P. H., Riou, S., Tran, D. M., Puys, M., Polychronou, N. F., El-Majihi, M., & Sivelle, C. (2022). iMRC: Integrated Monitoring & Recovery Component, a Solution to Guarantee the Security of Embedded Systems. *Journal of Internet Services and Information Security, 12*(2), 70–94.

[3] Singh, P. K., Singh, P. K., & Sharma, K. (2022). Electrochemical synthesis and characterization of thermally reduced graphene oxide: Influence of thermal annealing on microstructural features. *Materials Today Communications, 32*, 103950.

[4] Edgar, R. W., Min, T. A., & Gunther, P. (2011). Guest editorial: Advances in applied security. *Journal of Wireless Mobile Networks, Ubiquitous Computing, and Dependable Applications, 2*(4), 1–3.

[5] Kumar, R., Pandey, A. K., Samykano, M., Mishra, Y. N., Mohan, R. V., Sharma, K., & Tyagi, V. V. (2022). Effect of surfactant on functionalized multi-walled carbon nano tubes enhanced salt hydrate phase change material. *Journal of Energy Storage, 55*, 105654.

[6] Oleksandr, K., Viktoriya, G., Nataliia, A., Liliya, F., Oleh, O., & Maksym, M. (2024). Enhancing economic security through digital transformation in investment processes: theoretical perspectives and methodological approaches integrating environmental sustainability. *Natural and Engineering Sciences, 9*(1), 26–45.

[7] Hai, T., Aziz, K. H. H., Zhou, J., Dhahad, H. A., Sharma, K., Almojil, S. F., ... & Abdelrahman, A. (2023). Neural network-based optimization of hydrogen fuel production energy system with proton exchange electrolyzer supported nanomaterial. *Fuel, 332*, 125827.

[8] Rajesh, D., Giji Kiruba, D., & Ramesh, D. (2023). Energy proficient secure clustered protocol in mobile wireless sensor network utilizing blue brain technology. *Indian Journal of Information Sources and Services, 13*(2), 30–38.

[9] Aslan, Ömer, et al. (2023). A comprehensive review of cyber security vulnerabilities, threats, attacks, and solutions. *Electronics, 12*(6), 1333.

[10] Chen, Y., Feng, L., Jamal, S. S., Sharma, K., Mahariq, I., Jarad, F., & Arsalanloo, A. (2021). Compound usage of L shaped fin and Nano-particles for the acceleration of the solidification process inside a vertical enclosure (A comparison with ordinary double rectangular fin). *Case Studies in Thermal Engineering, 28*, 101415.

[11] Alzoubi, Haitham M., et al. (2022). Cyber security threats on digital banking. *2022 1st International Conference on AI in Cybersecurity (ICAIC)*. IEEE.

[12] Alfandi, O., Khanji, S., Ahmad, L. et al. (2021). A survey on boosting IoT security and privacy through blockchain. *Cluster Comput, 24*, 37–55. https://doi.org/10.1007/s10586-020-03137-8.

[13] Singh, J., Singh, G., & Negi, S. (2023, May). Evaluating security principals and technologies to overcome security threats in iot world. In *2023 2nd International Conference on Applied Artificial Intelligence and Computing (ICAAIC)* (pp. 1405–1410). IEEE.

[14] Obarafor, V., Qi, M., & Zhang, L. (2023, July). A Review of Privacy-Preserving Federated Learning, Deep Learning, and Machine Learning IIoT and IoTs Solutions. In *2023 8th International Conference on Signal and Image Processing (ICSIP)* (pp. 1074–1078). IEEE.

[15] Nižetić, S., Šolić, P., Gonzalez-De, D. L. D. I., & Patrono, L. (2020). Internet of Things (IoT): Opportunities, issues and challenges towards a smart and sustainable future. *Journal of Cleaner Production, 274*, 122877.

15 Blockchain-powered smart meter billing: Enabling transparency and security

Rakhee M.[1,a], *M. Sudheep Elayidom*[2,b], *and Baiju Karun*[3,c]

[1]Research Scholar, Department of Computer Science & Engineering, School of Engineering, Cochin University of Science and Technology, Cochin, India
[2]Professor, Department of Department of Computer Science & Engineering, School of Engineering, Cochin University of Science and Technology, Cochin, India
[3]Lecturer, University of Technology & Applied Sciences, Muscat, Oman

Abstract: Energy is vital for economic development and improving living standards. Addressing the rising demand efficiently is a key government priority. Renewable sources like solar, wind, and bioenergy are gaining momentum as sustainable alternatives. The surge in automated systems, electric vehicles, and smart cities has significantly increased power consumption, necessitating advanced energy management. Smart electricity meters, which track and transmit real-time energy data, are transforming consumption monitoring. However, challenges related to data security, transparency, and reliability persist. Blockchain technology offers a robust solution with its decentralized ledger, ensuring secure and transparent transactions. In our proposed model, smart contracts eliminate the need for third-party intermediaries, enabling reliable and tamper-proof operations. This system is set for deployment in Kerala, India, to improve power management efficiency. By integrating blockchain with smart metering, the proposed approach enhances energy distribution, security, and accountability, paving the way for a smarter power grid.

Keywords: Smart meter, smart contract, blockchain, power management, electricity usage monitoring, billing system

1. Introduction

Kerala is ranked second in NITI Aayog's 2022 State Energy and Climate Index (SECI), showcasing its significant strides in the energy sector. Unlike many other states where energy consumption is dominated by industry and agriculture, Kerala's electricity usage is primarily driven by the domestic and commercial sectors. Approximately 30% of the energy needs the state is met through renewable sources [1].

The power supply in Kerala is sourced from various providers, including central stations, private generators, power exchanges, and internal sources such as KSEBL-owned plants, Independent Power Producers (IPPs), captive power plants (CPPs), and prosumers. The peak demand for 2022–23 reached 4517 MW, reflecting a 3.12% increase from 4380.04 MW in 2021–22. The total installed capacity of 3514.81 MW is divided among the state sector (65.43%, or 2299.89 MW), the private sector (24.33%, or 855.34 MW), and the central sector (10.23%, or 359.58 MW). As per the database of kerenvis.nic.in Kerala's electrical energy consumption increased by 5.83% in 2021–22, totaling 25,383.77 MU.

Kerala frequently experiences power outages, especially during high demand periods. The state's growing population, urbanization, and industrial growth have steadily increased electricity demand, stressing the existing power systems. Kerala heavily relies on electricity imported from neighboring states, as its own Kerala's power distribution system faces daily challenges, including high transmission and distribution network costs, power theft, and inefficient accounting, which financially strain distribution companies.

Figure 15.1 illustrates the installed capacity from hydel, thermal, and renewable sources in Kerala.

Monitoring energy consumption in urban areas is essential, with rising demand for real-time energy usage insights. Smart grids offer macro-level energy management, but a micro-level perspective is equally important [2]. Smart meters, with two-way communication and real-time monitoring, revolutionize power grids and promote sustainable energy use. However, concerns about real-time personal data collection and privacy breaches persist, making data privacy and user trust critical [3].

Advanced Metering Infrastructure (AMI) enhances energy efficiency in smart grids. These bidirectional systems automate electricity generation, transmission, and consumption, forming the backbone of modern power infrastructure [4]. Smart meters collect and transmit consumption data via data collectors to the meter data management system, enabling real-time pricing, billing, and outage management [5]. However, data exchange over public networks exposes the system to cybersecurity threats, including data

[a]rakheebaiju@gmail.com, [b]sudheepelayidom@gmail.com, [c]baijukarun@gmail.com

DOI: 10.1201/9781003675259-15

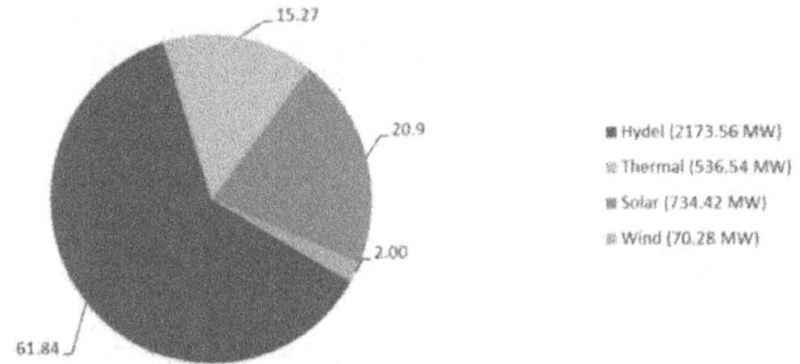

Figure 15.1. Energy resources in Kerala.

Source: KSEBL.

theft, unauthorized access, meter tampering, and false data injection, jeopardizing infrastructure security and energy management [6].

As essential elements of IoT-based energy systems, smart meters face growing cybersecurity challenges. Blockchain technology addresses these concerns by offering secure authentication via private-public encryption with controlled key access. It eliminates the reliance on a central authority by distributing the ledger across all participants, reducing risks associated with single points of failure. Merkle trees or hash trees further enhance transaction security by ensuring efficient and tamper-proof validation [7]. This research introduces a blockchain-integrated distributed AMI security framework, ensuring a cost-effective and robust communication protocol between smart meters and MDMS.

1.1. *Key contributions*

- Blockchain-based encryption for secure smart meter communication, replacing traditional third-party authorities.
- Decentralized architecture, preventing single points of failure and improving system reliability.
- Comprehensive security assessment, demonstrating resistance to common cyber threats, enhancing the integrity of smart grid networks.

2. Literature Review

Integrating blockchain technology with smart energy meters ensures the secure and timely transmission of power consumption data to users. To achieve this, technologies such as IoT, cloud computing, and sensors must be implemented to gather values from the customer end. Over the past decade, researchers have developed numerous approaches to perform this process.

This section reviews and summarizes some of these significant works.

Zanghi [8] introduced a centralized and sustainable method using blockchain to manage a broad spectrum of metering systems, optimizing existing private or public communications infrastructure while reducing costs, enhancing accuracy, and simplifying deployment and management and addressing privacy and information integrity issues.

To secure data transmission from publishers to subscribers, Ramyasri et al. [9] applied Elliptic Curve Cryptography (ECC) with public-key encryption, suitable for burst and continuous data communication. Agarwal et al. [10] introduced an IoT-based smart meter using Arduino for energy monitoring and theft detection. Razak et al. [11] developed an IoT smart meter that tracks usage and stores data in the cloud, promoting energy conservation. Mololoth et al. [12] explored blockchain and machine learning in smart grids, addressing grid safety and demand management. Waseem et al. [13] highlighted blockchain's role in mitigating cybersecurity risks in smart grids. Barman et al. [14] proposed an IoT smart meter with cloud integration to detect theft and provide consumers with real-time energy data. Avancini et al. [15] discussed the benefits and challenges of a novel IoT-based smart energy meter, featuring multiple communication interfaces for easy integration with tracking applications. The solution was confirmed and demonstrated in real-world scenarios.

Prathik et al. [16] developed an energy tracking device using Arduino and a GSM module to display power loss. This method helps users manage their electrical devices more effectively, reduce energy waste, and receive updates on power outage dates and bill payments.

Suciu et al. [17] explored security strategies for smart grids, emphasizing blockchain's role in ensuring privacy and system reliability. They highlighted blockchain's effectiveness in real-time energy monitoring and attack resistanc. Mukherjee et al. [18] proposed an e-Chain architecture using Ethereum blockchain and smart contracts, enhancing security and privacy in smart energy systems. Dorri et al. [19] introduced a secure private blockchain (SPB) framework for energy trading, enabling direct transactions between producers and consumers. Their approach, based on public key (PK) routing, minimizes overheads and eliminates the need for third-party verification.

3. Methodology

3.1. *Smart meter*

Smart meters provide real-time monitoring of electricity, water, and gas usage, eliminating the need for manual readings. These devices leverage Advanced Metering Infrastructure (AMI) to facilitate two-way communication with utility companies [20]. However, current smart meters transmit usage data directly to energy providers, raising concerns over security and privacy. Centralized data control creates a single point of failure, increasing the risk of hacking and ransomware attacks [21]. Implementing blockchain technology can enhance data security and decentralization, mitigating vulnerabilities while ensuring transparent and tamper-proof energy transactions.

Challenges with current energy meters also include vulnerabilities like susceptibility to spoofing and Denial of Service (DoS) attacks. These vulnerabilities allow for the mimicking of IDs and flooding systems with fake data, leading to inaccurate usage predictions. Consequently, voltage fluctuations may occur due to false data dissemination. If a meter inaccurately reports low consumption despite high usage, it results in reduced power distribution to the feeder, causing a voltage drop and increased current flow, risking critical systems. Conversely, if the meter falsely indicates higher consumption, it triggers excessive power dispatch, resulting in voltage spikes that may damage equipment if there isn't sufficient load to absorb the excess voltage.

Additionally, privacy concerns arise regarding user data stored in the Meter Data Management System (MDMS) of the Head End System (HES). Data mapped to consumer numbers and customer details may be exploited by hackers, leading to billing errors and disruptions in daily usage patterns, potentially causing equipment damage.

3.2. Proposed Blockchain enabled smart meter technology

Blockchain is a decentralized digital ledger shared across peer-to-peer (P2P) networks using secure consensus mechanisms.

Figure 15.2 illustrates the workflow of a blockchain-based smart meter.

By enabling direct energy trading among consumers, prosumers, and companies, blockchain eliminates intermediaries. It ensures data integrity, confidentiality, and reliability in smart grid systems, fostering transparency and economic benefits. Integrating blockchain in smart grids enhances security by mitigating cyber threats like Denial of Service (DoS), IP spoofing, and Man-in-the-Middle (MITM) attacks, which can disrupt power loads. It also protects sensitive utility data by anonymizing personal and business details. Unlike conventional databases, blockchain operates through a globally distributed network, removing single points of failure and ensuring secure, tamper-proof transactions.

By leveraging blockchain, existing infrastructure can adapt to modern regulations, allowing prosumers to seamlessly connect to power grids, ensuring secure, efficient, and transparent energy transactions.

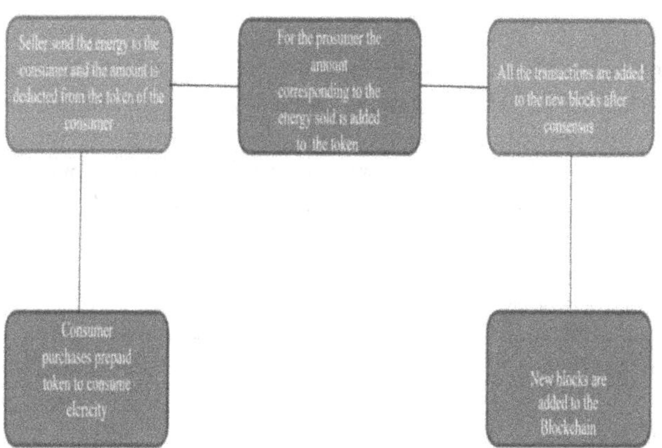

Figure 15.2. Smart energy meter with Blockchain.

Source: Author.

4. Implementation

The billing system outlined in this paper utilizes the Ethereum blockchain, for its implementation. The billing process is managed through smart contracts developed in the Solidity programming language. These smart contracts are initially deployed on a test network, allowing for comprehensive testing and refinement of the billing system's functionality before it is deployed on the main Ethereum network. In this system, each participant, or node, is provided with a private key and an address generated by the Ethereum network, ensuring secure and authenticated interactions within the blockchain environment.

The proposed billing strategy employs a prepaid model. The billing process dataflow diagram is shown in Figure 15.3.

In this process customers are issued prepaid tokens, which they use to pay for their energy consumption. When the smart energy meter records usage data, this information is transmitted to the smart contract. The smart contract then deducts the appropriate amount from the customer's prepaid token balance. Furthermore, customers who generate excess electricity, such as through solar energy, can sell this surplus back to the grid. In these instances, the earnings from selling the excess electricity are credited to the customer's prepaid token balance.

A snippet of the smart contract for billing is in Figure 15.4.

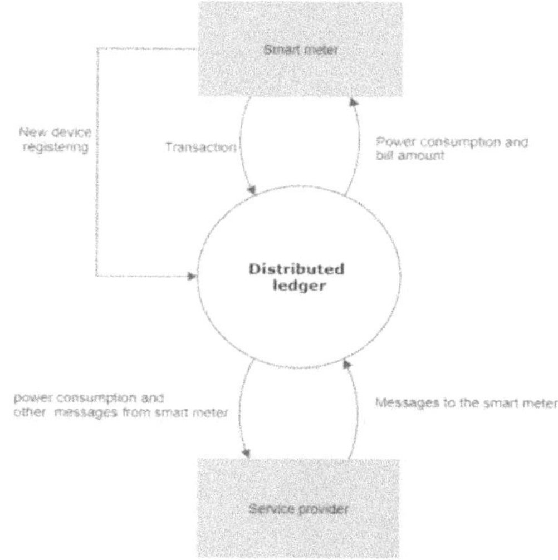

Figure 15.3. The billing process.

Source: Author.

```
// SPDX-License-Identifier: MIT
pragma solidity >=0.4.22 <0.6.0;

contract SMContract {

    address private utility;
    uint private counter;

    struct Reading{
      address uAddr;
      uint256 energy_consumed;
      uint256 readingNo;
    }

    event readingEvent(address _user, uint256 _reading);

    mapping (uint256=> Reading) public _readings;
    uint256[] public readingsArr;

    constructor() public {
        utility = msg.sender;
        counter = 1;
    }
```

Figure 15.4. Snippet of billing process.

Source: Author.

5. Results and Discussion

The smart contract developed for smart energy meter billing on the Ethereum blockchain successfully facilitated transparent and efficient billing processes.

5.1. *Efficiency and transparency*

The implementation of the smart contract ensured efficiency by automating the billing process, reducing the billing processing time from 48 hours to just 2 hours, as shown in Table 15.1 and Figure 15.5. The data update frequency also improved significantly, from every 24 hours to every 15 minutes, as depicted in Figure 15.6. This transparency allowed customers to verify their energy consumption and billing details in real-time, fostering trust and accountability in the billing system.

Table 15.1. Efficiency metrics

Metric	*Before block chain*	*After block chain*	*Improvement(%)*
Billing Processing Time	48 hours	2 hours	95.83%
Data Update Frequency	Every 24 hours	Every 15 minutes	96.88%

Source: Author.

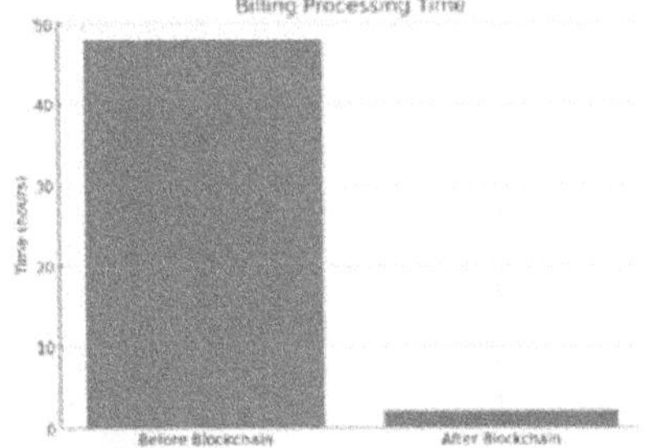

Figure 15.5. Billing processing time.

Source: Author.

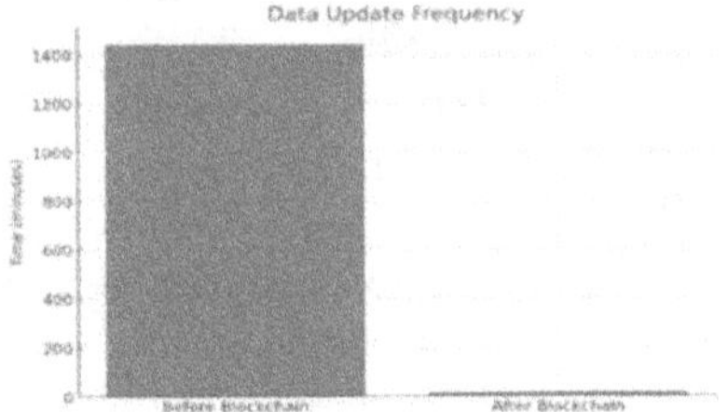

Figure 15.6. Data update frequency.

Source: Author.

5.2. *Cost savings and rrror reduction*

Smart contracts powered by blockchain technology significantly reduced operational costs by 80%, as detailed in Table 15.2 and Figures 15.7 and 15.8. Additionally, the reduction in billing errors led to a 90% decrease in error-related costs, as shown in Figure 15.8.

5.3. *Customer Empowerment*

Utilizing Ethereum blockchain technology gave customers greater control over their energy consumption and billing activities. With real-time data and transparent billing information, customers could make informed decisions about their energy usage, potentially leading to energy savings and environmental benefits. The ability to monitor usage in real-time empowers consumers to adjust their consumption habits, promoting more sustainable energy use.

5.4. *Security*

Blockchain's decentralized and immutable nature made it resistant to tampering and fraud. The number of unauthorized access attempts decreased by 93.33%, and fraud incidents were completely eliminated, as shown in Figures 15.9 and 15.10.

5.5. *Challenges and future prospects*

Despite its benefits, blockchain faces scalability issues and interoperability challenges in smart metering. Future research should enhance smart contract efficiency for energy billing and explore decentralized computing platforms to improve functionality. Advancing computational capabilities can further optimize smart grid solutions for seamless energy management.

Table 15.2. Cost savings

Cost type	*Before block chain*	*After block chain*	*Savings (%)*
Operational Costs	$10,000/month	$2,000/month	80%
Error Reduction Costs	$5,000/month	$500/month	90%

Source: Author.

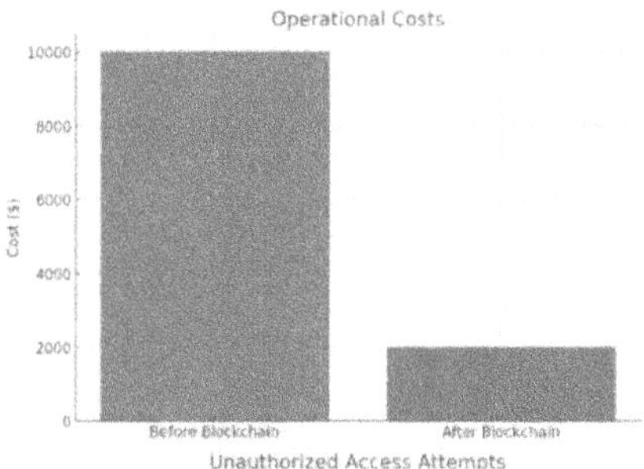

Figure 15.7. Operational costs.

Source: Author.

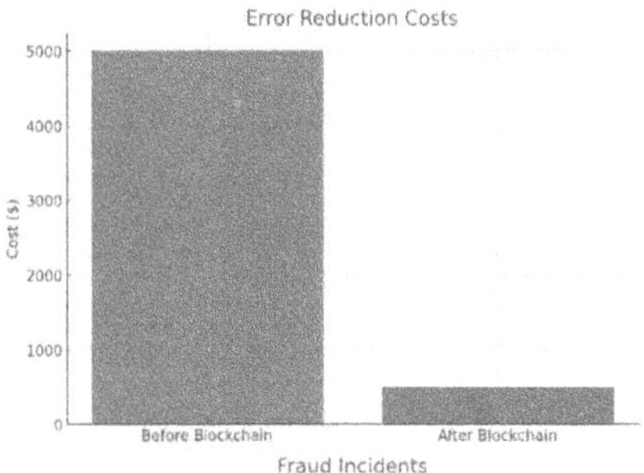

Figure 15.8. Error reduction costs.

Source: Author.

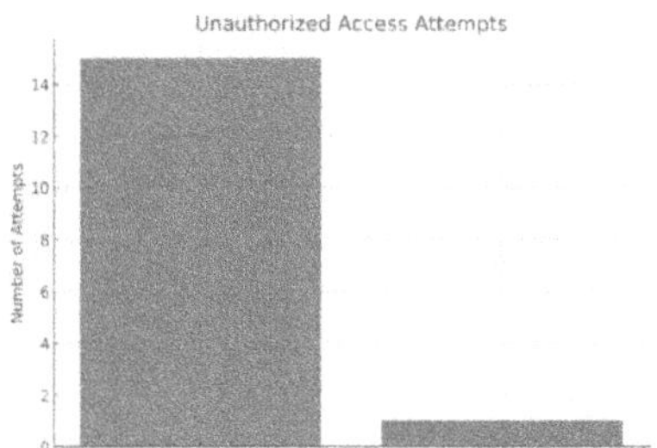

Figure 15.9. Security improvements 1.

Source: Author.

Figure 15.10. Security improvements 2.

Source: Author.

6. Conclusion

The smart energy meter efficiently manages energy supply and consumption, calculating costs for both conventional grid and solar energy. This system enhances transparency, reducing manpower, costs, and time compared to traditional methods.

Implementing smart contracts on the Ethereum blockchain strengthens data security, privacy, and integrity, ensuring efficient and transparent energy billing. It enhances cost-effectiveness and customer empowerment, fostering a more reliable energy infrastructure. While challenges remain, ongoing research can further refine blockchain integration, paving the way for a sustainable and advanced energy sector.

References

[1] Malar, J. S. J., Bisharathu Beevi, A., & Jayaraju, M. (2023). Efficient power flow management in hybrid renewable energy systems. *IETE Journal of Research, 69*(2), 1088–1100. http://kerenvis.nic.in/Database/ENERGY_811.aspx.

[2] Rehman S., Natarajan N., Vasudeva, M., Mohammed A. B., Mohandes M. A., Khan F., & Al-Sulaiman F. A. (2023). Performance evaluation of grid-connected photovoltaic system for Kuttiady village in Kerala, India. *Environmental Science and Pollution Research, 30*(44), 99147–99159.

[3] Seenath Beevi, P. T., Harikumar, R., & Aravind, M. L. (2022, May). Smart Grid and Utility Challenges. In *ISUW 2020: Proceedings of the 6th International Conference and Exhibition on Smart Grids and Smart Cities* (pp. 245–255). Singapore: Springer Nature Singapore.

[4] Sworna Kokila, M. L., Venkatarathinam, R., Manivasagam, M. A., & Kishore, K. H. (2024). Optimizing energy consumption in smart grids using demand response techniques. *Distributed Generation & Alternative Energy Journal*, 111–136.

[5] Koasidis K., Marinakis V., Doukas H., Doumouras N., Karamaneas A., & Nikas A. (2023). Equipment-and time-constrained data acquisition protocol for non-intrusive appliance load monitoring. *Energies, 16*(21), 7315.

[6] Steephen S., Sheeba R., & Sundar S. (2023, May). Demand forecasting in smart homes using deep learning techniques. In *2023 International Conference on Control, Communication and Computing (ICCC)* (pp. 1–6). IEEE.

[7] Condon, F., Franco, P., Martínez, J. M., Eltamaly, A. M., Kim, Y. C., & Ahmed, M. A. (2023). EnergyAuction: IoT-blockchain architecture for local peer-to-peer energy trading in a microgrid. *Sustainability, 15*(17), 13203.

[8] Zanghi, E., Brown Do Coutto Filho, M., & Stacchini de Souza, J. C. (2023). Collaborative smart energy metering system inspired by blockchain technology. *International Journal of Innovation Science.*

[9] Ramyasri, G., Murthy, G. R., Itapu, S., & Krishna, S. M. (2023). Data transmission using secure hybrid techniques for smart energy metering devices. *e-Prime-Advances in Electrical Engineering, Electronics and Energy, 4*, 100134.

[10] Agarwal, N. K., Kishore, M. P., Rastogi, P. K., Singh, S., Singh, S., & Raj, Y. (2023, March). Arduino Employed Power Theft Controller and IoT based Load Controlling for Smart Energy Meter System. In 2023 10th International Conference on Computing for Sustainable Global Development (INDIACom) (pp. 1021–1025). IEEE.

[11] Razak, N. A. A., Ramlee, R. A., Ibrahim, A. F. T., Andre, H., & Khmag, A. (2023, May). IoT-BSEM: Internet of things-based smart energy meter for domestic house. In *AIP Conference Proceedings* (Vol. 2592, No. 1). AIP Publishing.

[12] Mololoth, V. K., Saguna, S., & Åhlund, C. (2023). Blockchain and machine° learning for future smart grids: A review. *Energies*, *16*(1), 528.

[13] Waseem, M., Adnan Khan, M., Goudarzi, A., Fahad, S., Sajjad, I. A., & Siano, P. (2023). Incorporation of blockchain technology for different smart grid applications: Architecture, prospects, and challenges. *Energies*, *16*(2), 820.

[14] Barman, B. K., Yadav, S. N., Kumar, S., & Gope, S. (2018, June). IoT based smart energy meter for efficient energy utilization in smart grid. In *2018 2nd international conference on power, energy and environment: towards smart technology (ICEPE)* (pp. 1–5). IEEE.

[15] Avancini, D. B., Rodrigues, J. J., Rabêlo, R. A., Das, A. K., Kozlov, S., & Solic, P. (2021). A new IoT-based smart energy meter for smart grids. *International Journal of Energy Research*, *45*(1), 189–202.

[16] Prathik, M., Anitha, K., & Anitha, V. (2018, February). Smart energy meter surveillance using IoT. In *2018 International conference on power, energy, control and transmission systems (ICPECTS)* (pp. 186–189). IEEE.

[17] Suciu, G., Sachian, M. A., Dobrea, M., Istrate, C. I., Petrache, A. L., Vulpe, A., & Vochin, M. (2019, September). Securing the smart grid: A blockchain-based secure smart energy system. In *2019 54th International Universities Power Engineering Conference (UPEC)* (pp. 1–5). IEEE.

[18] Mukherjee, P., Barik, R. K., & Pradhan, C. (2021). echain: Leveraging toward blockchain technology for smart energy utilization. In *Applications of Advanced Computing in Systems: Proceedings of International Conference on Advances in Systems, Control and Computing* (pp. 73–81). Springer Singapore.

[19] Dorri, A., Luo, F., Kanhere, S. S., Jurdak, R., & Dong, Z. Y. (2019). SPB: A secure private blockchain-based solution for distributed energy trading. *IEEE Communications Magazine*, *57*(7), 120–126.

[20] Jemei, S., & Pahon, E. (2023). Encyclopedia of electrical and electronic power engineering. *Control of fuel cell systems*, *3*, 472–484.

[21] Suriyan, K., Ramalingam, N., Jayaraman, M. K., & Gunasekaran R. (2023). Recent developments of smart energy networks and challenges. *Smart Energy and Electric Power Systems*, 37–47.

16 Sensor based emergency locating module (ELM)

F. Rahman[1,a] *and Priti Sharma*[2,b]

[1]Assistant Professor, Department of CS & IT, Kalinga University, Raipur, India
[2]Research Scholar, Department of CS & IT, Kalinga University, Raipur, India

Abstract: Traffic accidents result in the deaths of individuals. Such issues stem from an insufficient level of cooperation among the involved organizations. The graph also gets bigger when the required steps and techniques are not correctly practiced. This research examines various approaches to accident prevention that have been previously proposed, taking into account the technologies used. A few studies also suggest methods other than accident tracking for the recovery of the injured passengers. Certain techniques are estimated and prepared with consideration for the present conditions around traffic accidents. A prototype of an accident detection module has been presented here that incorporates vibration sensors. Vibration sensors can identify when an unusual vibration occurs in an automobile due to a collision.

Keywords: Traffic accidents, piezoelectric sensor, HC-SR04, SONAR, GPS and GSM

1. Introduction

Due to the automobile's quick development and the rising number of vehicles on the road, traffic accidents have been on the rise. The form known as a scene drawing (SD), which describes the scene conditions with pictures and symbols when a traffic accident occurs, is the primary piece of evidence. Quick scene drawing is necessary to minimize traffic congestion [2]. The scene data must also be digitalized and visualized in order to supply information for accident reconstruction [1]. Rather of being seen as tragedies that can result in fatalities, traffic accidents are often dismissed as inconsequential mishaps. These days, there are a wide variety of transportation options available, which contributes to a decrease in the frequency of traffic accidents. As highways become congested, traffic seems to be getting worse. They harm not just the modes but also the lives.

The general public has to be more focused and aware of this. In India, 1.55 lakh people died in traffic accidents in 2021 compared to 1.33 lakh in 2020. The most cases that have been reported were from Tamil Nadu. The three-point seat belt is an inexpensive form of restraint that prevents passengers from being thrown forward in the event of an accident [13]. The seatbelt distributes the physical impact of a crash across the stronger parts of the body, like the chest and pelvis, slowing the passenger down at the same speed as the vehicle [6]. The alert system employs image processing techniques to identify driver fatigue through the use of the Eye Aspect Ratio (EAR). This method is precise and accurate in real-time analysis, and it sounds an alert with an alert buzzer when the driver is detected to be drowsy. This helps to partially prevent accidents from happening. Even in the unlikely event of an accident, the GPS Receiver Module provides the precise position of the incident, and the vibration sensor in use recognizes it. Thus, a Real Time Driver Fatigue Detection and Smart Rescue System is suggested as an effective way to address all of the aforementioned issues. It can identify drowsiness in real time, preventing accidents [3]. Additionally, if an accident were to occur, saving the victim would be my priority. When an accident is detected, it is located on an online system that ambulance drivers use to promptly provide medical attention to the victim. The victim's parents receive a notice message with geographical values including the same information.

2. Related Work

When it comes to edge detection and notifying when an automobile departs its lane, a number of strategies are employed in automotive lane identification [14]. Many of the modern transportation systems that leverage technology advancements use lane detection systems. Taking into account the fact that getting there could be challenging due to a variety of road conditions, particularly when travelling at night or during inclement weather during the day. An accident detection system has been prepared in [5] using IR sensors and Arduino Uno technology. IR sensors are employed to avoid collisions by notifying drivers of nearby automobiles when the gap between them surpasses a specified threshold.

To ensure life safety, [15] propose an automated, Internet of Things-based accident detection system [8]. After an incident, data is submitted to a web server, friends of the victim receive instant SMS messages, and the relevant authorities—such as the local police station, traffic control centre, and ambulance service—are notified [4, 10]. It is necessary to have a system that can seamlessly coordinate the

[a]ku.frahman@kalingauniversity.ac.in, [b]PRITI.SHARMA@kalingauniversity.ac.in

DOI: 10.1201/9781003675259-16

several actions that must be completed in order to respond quickly to the scene of the accident. As per reference [7], these detection systems employ an array of technologies, such as mobile phone applications and the Global Positioning System (GPS) and Global System for Mobile Communication (GSM). There are several contexts in which the visual accident elements detection technique described in [16] can be used. It is vital to have a system in place that can smoothly coordinate the various tasks that must be finished in order to react swiftly at the accident scene. [9] determines incident type (limited to pedestrian falls) through the examination of risk factors in pedestrian accident cases. In order to adjust the accident detection algorithm to real-world conditions, data should be collected and processed based on specific accident types identified from risk variables. The crash trial setting had been built up similarly to the real world. The procedure described in [17] immediately notifies the nearby hospital, police department, and insurance company, enabling them to promptly respond to the scene of the occurrence and complete their assigned duties. A detection phase is also included in [11] to address such situations and ensure that the driver and passengers have the required medical attention as soon as is practical. When a collision causes an anomalous vibration in the car, vibration sensors are utilized to identify the accident. In addition to vibration detection, the newest technology being developed, deep learning, has been included to help identify accidents and notify the responsible centres. By using the design, the victim's chances of survival can be increased by preventing the delay of emergency aid (Figure 16.1).

3. Sensors Employed for Emergency Situations

To find out what caused the accident, a few sensors are mounted to the cars to gather information on the position and direction of the vehicle (Figure 16.2). A piezoelectric sensor is a piece of equipment that uses the piezoelectric effect to convert changes in force, strain, temperature, acceleration, and pressure into an electrical current. To find out what caused the accident, a few sensors are mounted to the

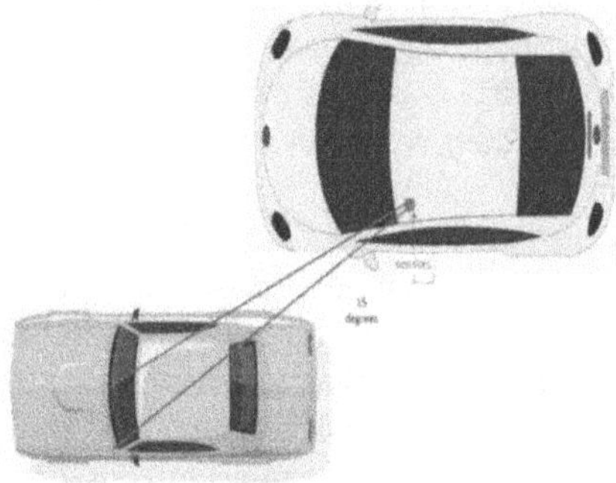

Figure 16.1. Working of sensors inbuilt in vehicle.

Source: Author.

Figure 16.2. Employment of sensors in vehicles.

Source: Author.

cars to gather information on the position and direction of the vehicle. The principle of operation for a piezoelectric sensor is the conversion of energy into mechanical and electrical energy kinds. An increase in electrical current results from the physical distortion that pressure creates in a polarized crystal. A piezo sensor might then be used to assess the mechanical deformation or electric charge that was produced. Vibration sensors are instruments that monitor and evaluate linear speed, displacement, proximity, and the starting force of different shocks.

The HC-SR04 is an ultrasonic detector that uses sound navigation via SONAR technology. The sensor is incredibly reliable and efficient. The form factor's length varies from 2 to 400 cm. It works best in direct sunlight or with dark objects, however garments and other soft materials could be difficult to identify. It consists of a receiver unit and an ultrasonic emitter. When there is an accident, this module determines whether there is a passenger in the car and notifies the microcontroller [12]. The circuit board can be designed to function as a wireless smartwatch that the user wears on their palm. An MSP430 microcontroller receives the recorded heartbeat as a spiking signal, which is then amplified, processed, and converted into a pulse signal. The beat is investigated if the vibration intensity is low. The area is recorded and notified if the heartbeat is likewise low. In addition, the condition is normal because the eye flicker sensor's output is significant. When it's low, it means the eyes are closed. After then, the signal is turned on. If it doesn't shut off physically and keeps going at a high yield, the area is monitored and sent as a notice.

To distinguish between different eye closures, one uses the eye-blink sensor. It works by shining infrared light into the area around the eye and eyelid, then measuring the variations in light that comes back because of a differentiator and phototransistor. It runs on a 5volt battery. A 5volt DC power source that is interface regulated is supplied to the eye blink sensor. The red wire represents guaranteed supply, the centre wire is yield, and the black wire is ground. Just the 5V and ground leads need to be connected in order to test the sensor's operation. Don't detach the yield wire. The yield is 0V and the LED goes out when the eye is closed. Lay an eye blink detector glass 15 mm from your eyes on your face, and watch how the LED frowns with each blink. At the point when the eye is opened, the result will be favourable. The microcontroller receives the output legitimately in order to link actions.

4. Steps Included in the Operation of the Model

A detection phase is included in Figure 16.3 to guarantee that the driver and passengers receive medical assistance as soon as possible. Vibration sensors can identify when an unusual vibration occurs in an automobile due to a collision. Vibration detection has been enhanced using deep learning, the most recent technical development that is now being expanded, in order to eventually recognize accidents and connect with the accountable centres.

5. Conclusion

The current research has addressed several accident detection tactics based on the technologies employed. A few techniques for accident detection include IOT-automation, GPS, GSM, and deep learning. A variety of sensors, including RFID, ultrasonic and piezoelectric sensors, are utilized to locate the accident and then enable prompt service delivery. A few research studies also offer various suggestions for rescuing the impacted passengers in addition to accident detection. Certain approaches are assessed and built with consideration for the current road-accident scenario. Herein, a vibration sensor-equipped accident detection module design is provided. When an automobile experiences an unusual vibration as a result of a collision, vibration sensors can detect it.

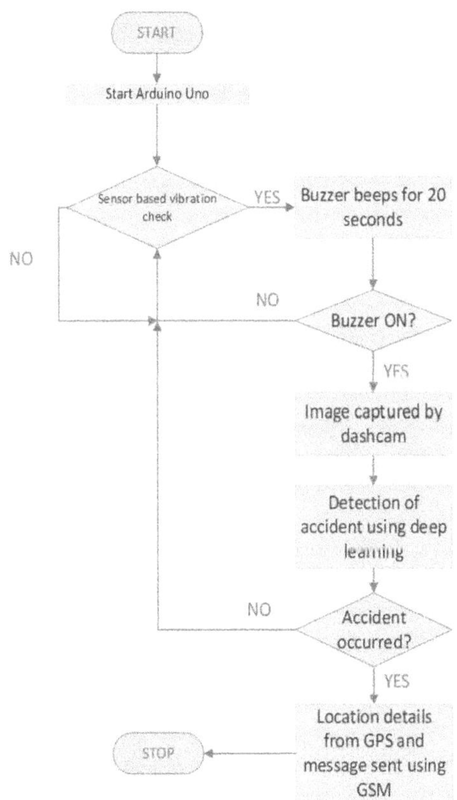

Figure 16.3. Steps included in sensor-based emergency locating.

Source: Author.

References

[1] Lakshmy, S., Gopan, R., Meenakshi, M. L., Adithya, V., & Elizabeth, M. R. (2022, March). Vehicle accident detection and prevention using iot and deep learning. In *2022 IEEE International Conference on Signal Processing, Informatics, Communication and Energy Systems (SPICES)* (Vol. 1, pp. 22–27). IEEE.

[2] Zahraa, T., & Laith, A. A. R. (2024). Smart traffic system using infrared sensors-based IoT. *Journal of Internet Services and Information Security*, *14*(3), 29–41.

[3] Li, J., Abdulghani, Z. R., Alghamdi, M. N., Sharma, K., Niyas, H., Moria, H., & Arsalanloo, A. (2023). Effect of twisted fins on the melting performance of PCM in a latent heat thermal energy storage system in vertical and horizontal orientations: Energy and exergy analysis. *Applied Thermal Engineering*, *219*, 119489.

[4] Jazem, M.A. (2023). Effective machine-learning based traffic surveillance moving vehicle detection. *Journal of Wireless Mobile Networks, Ubiquitous Computing, and Dependable Applications*, *14*(1), 95–105.

[5] Chen, Y., Feng, L., Mansir, I. B., Taghavi, M., & Sharma, K. (2022). A new coupled energy system consisting of fuel cell, solar thermal collector, and organic Rankine cycle; generation and storing of electrical energy. *Sustainable Cities and Society*, *81*, 103824.

[6] Trivedi, J., Devi, M. S., & Solanki, B. (2023). Step towards intelligent transportation system with vehicle classification and recognition using speeded-up robust features. *Archives for Technical Sciences*, *1*(28), 39–56.

[7] Sharma, H. K., Kumar, S., & Verma, S. K. (2022). Comparative performance analysis of flat plate solar collector having circular &trapezoidal corrugated absorber plate designs. *Energy*, *253*, 124137.

[8] Madhan, K., & Shanmugapriya, N. (2024). Efficient object detection and classification approach using an enhanced moving object detection algorithm in motion videos. *Indian Journal of Information Sources and Services*, *14*(1), 9–16.

[9] Narayanan, K. L., Ram, C. R. S., Subramanian, M., Krishnan, R. S., & Robinson, Y. H. (2021, February). IoT based smart accident detection & insurance claiming system. In *2021 Third international conference on intelligent communication technologies and virtual mobile networks (ICICV)* (pp. 306–311). IEEE.

[10] Kim, M. J., Go, Y. B., Choi, S. Y., Kim, N. S., Yoon, C. H., & Park, W. (2023). A Study on the Analysis of Law Violation Data for the Creation of Autonomous Vehicle Traffic Flow Evaluation Indicators. *Journal of Internet Services and Information Security*, *13*(3), 185–198.

[11] Chaturvedi, R., Sharma, A., Sharma, K., & Saraswat, M. (2022). Nanotech Science as well as its multifunctional implementations.

Recent Trends in Industrial and Production Engineering: Select Proceedings of ICCEMME 2021, 217–228.

[12] Jeong, M., Lee, S., Bae, M., & Lee, K. (2021, January). Accident Type Definition for Implementation of Pedestrian Accident Recognition Algorithm. In *2021 International Conference on Information Networking (ICOIN)* (pp. 735–737). IEEE.

[13] Patel, D. H., Sadatiya, P., Patel, D. K., & Barot, P. (2019, June). IoT based obligatory usage of safety equipment for alcohol and accident detection. In *2019 3rd International conference on Electronics, Communication and Aerospace Technology (ICECA)* (pp. 71–74). IEEE.

[14] Ravi, A., Phanigna, T. R., Lenina, Y., Ramcharan, P., & Teja, P. S. (2020, July). Real time driver fatigue detection and smart rescue system. In *2020 International Conference on Electronics and Sustainable Communication Systems (ICESC)* (pp. 434–439). IEEE.

[15] Kumar, M. N., Kumar, S. P., Premkumar, R., & Navaneethakrishnan, L. (2021, March). Smart characterization of vehicle impact and accident reporting system. In *2021 7th International Conference on Advanced Computing and Communication Systems (ICACCS)* (Vol. 1, pp. 964–968). IEEE.

[16] Hai, T., Aziz, K. H. H., Zhou, J., Dhahad, H. A., Sharma, K., Almojil, S. F., … & Abdelrahman, A. (2023). RETRACTED: Neural network-based optimization of hydrogen fuel production energy system with proton exchange electrolyzer supported nanomaterial. *Fuel*, *332*, 125827.

[17] Chen, Y., Feng, L., Jamal, S. S., Sharma, K., Mahariq, I., Jarad, F., & Arsalanloo, A. (2021). Compound usage of L shaped fin and Nano-particles for the acceleration of the solidification process inside a vertical enclosure (A comparison with ordinary double rectangular fin). *Case Studies in Thermal Engineering*, *28*, 101415.

17 Predictive analytics for renal health: Machine learning in chronic kidney disease prediction

Radha M.[1,a] *and Muthukumar A.*[2,b]

[1]Research Scholar, Department of Electronics and Communication Engineering, Kalasalingam Academy of Research and Education, Virdhunagar, Tamil Nadu, India
[2]Associate Professor, Department of Electronics and Communication Engineering, Kalasalingam Academy of Research and Education, Virdhunagar, Tamil Nadu, India

Abstract: Chronic kidney disease (CKD) is a serious and paralyzing state with profound implications for patient morbidity and mortality. Despite its widespread impact, early detection and risk stratification remain challenging. In this study, we present a comprehensive framework utilizing predictive analytics and machine learning algorithms to advance CKD prediction. By integrating diverse datasets encompassing patient demographics, clinical variables, and biomarker profiles, our model exhibits superior performance in identifying individuals at various stages of CKD progression. Furthermore, we elucidate the accountability of our model, offering many perspectives on the critical elements pertaining to predictive accuracy. Through the deployment of this predictive tool in clinical settings, healthcare providers can proactively identify high-risk individuals, facilitate timely interventions, and ultimately improve patient outcomes in the realm of renal health.

Keywords: Chronic kidney disease, predictive analytics, machine learning algorithms, early detection

1. Introduction

Chronic Kidney Disease (CKD) causes a major impact all over the world and imposing considerable economic expense on the healthcare department. Millions of individuals were affected by this disease [1]. In medical science, early detection and precise prediction of CKD progression remain formidable challenges. Timely identification of individuals at risk of CKD deterioration is crucial for implementing proactive interventions to deal with the complications and increase the patient outcomes. In the last several years, the emergence of predictive analytics and Machine Learning has become development of sophisticated system capable of vast amounts of patient data to forecast disease trajectories. In this paper, the aim of the research is to implement the power of predictive analytics and various algorithms of machine learning techniques to construct a robust predictive model for CKD. By integrating diverse patient data sources, including demographic information [2], clinical variables, and biomarker profiles, our research seeks to enhance the accuracy of CKD prediction and provide actionable insights for clinicians. Through this interdisciplinary approach, we endeavor to advance our understanding of CKD progression, optimize resource allocation, and ultimately improve the level of care for individuals affected by this debilitating condition.

2. Literature Review

The following are the different papers which are surveyed to get a knowledge of the existing work for the better analysis.

Khalid et al. [1], proposed the hybrid technique for the prediction. This model deals with the Pearson correlation for feature selection. This technique was implemented by combining the different algorithms such as base classifier and the meta classifier. The objective of the paper is to provide a better understanding of algorithms techniques among the different classifiers. As a result, overfitting is addressed and the maximum accuracy is attained. Four different models were used in the implementation for the prediction analysis.

Islam et al. [2] investigated the comparison between the data variables with the features selection in terms of target class. It deals with the predictive analysis of machine learning methods. It was implemented by using the variables with the addition of different class properties. In this study, twelve different algorithms were tested in a supervized learning environment. By analyzing the variables with the different techniques, we came to the result that the XgBoost classifier has the better performance. It was discovered that the variables such as haemoglobin, albumin and specific gravity had the major impact in predicting the CKD.

Debal et al. [4], proposed the binary and multi classification for stage prediction plays a major role. This model

[a]radha.murugan1@gmail.com, [b]muthuece.eng@gmail.com

DOI: 10.1201/9781003675259-17

includes different machine learning concepts. It is obtained by applying cross validation for the feature selection. It includes tenfold cross validation. It concluds the random forest is better than the other different algorithms.

Revathy et al. [3] proposed the study deals with the various types of Machine Learning algorithms. It processes with the preprocessing of different data, data identification and various techniques for the prediction of CKD and determine the best framework for the prediction. It helps in the detection of the early stages of CKD.

Nikhilesh Reddy et al. [8] proposed the different machine learning approaches with various attributes which consist of the term class which represent the person is affected with the disease or not.

3. Existing System

The existing system encompasses a spectrum of methodologies leveraging both traditional statistical methods and machine learning algorithms. These approaches typically utilize diverse data sources, including reports of the patient in the digital format, required test analysis in the laboratory, image processing, and patient demographics, to extract relevant features are serum creatinine levels [3], estimated glomerular filtration rate (eGFR), and urine protein levels. Machine learning algorithms commonly employed in CKD prediction studies include various concepts of machine learning algorithms, each offering unique potential and considerations. Evaluation parameters such as accuracy, sensitivity, specificity, AUC-ROC, etc. plays a important role in the analysis of the model analysis. However, challenges such as data scarcity, data inequality, interpretability, lack of variable implementation and generalizability to diverse populations persist, suggesting avenues for future research including the integration of multi-omics data, development of interpretable models, application of deep learning techniques, Debal et al. [4], and implementation of predictive models in clinical practice.

4. Proposed Methodology

The proposed system deals with the two different ML algorithms such as Decision tree algorithm and Support Vector machine algorithms. This Decision tree classifier algorithm recursively partitions the feature space by selecting the best parameters at each node by splitting the data into different subsets, using criteria like Gini impurity or information gain, until a certain parameter are implemented in the system. Decision trees are interpretable, handle non-linear relationships well, and require no assumptions about data distribution. They can tolerate missing values and outliers, are scalable, particularly when employed in ensemble methods like Random Forest or Gradient Boosting, and are adaptable for both classification and regression applications. Conversely, Support Vector Machines (SVM) are a type of supervized learning method utilized for classification and regression issues. which, following specific techniques, are utilized to determine the hyperplane and provide the data with the various classifications. Increasing the margin between the classes of the provided data is another usage for it. When it comes to complex decision boundaries, mapping data parameters into higher-dimensional spaces using better functions, and further processing to the linear condition – all of which are especially helpful when working with high-dimensional data – SVM performs well with medium-ranged datasets.

5. Block Diagram

Figure 17.1 shows the block diagram of Chronic Kidney Disease Prediction. In that first the dataset is pre-processing and then the dataset is training and also testing using the suitable machine learning algorithm.

5.1. Data collection

The datasets are the major step for the analysis of the research. Gather patient data including demographics, clinical history, laboratory results, and imaging data.

5.2. Data preprocessing

The scatter and density plots display the relationship between different diagnostic parameters, such as blood pressure and age, segmented by CKD classification as specified in Figure 17.2. Perform feature engineering, manage missing values, normalize or standardize features, and clean up the gathered data.

5.3. Training data

In the training process, the selected algorithm technique should be trained effectively with the system matched data, using procedures like cross-validation to tune hyper parameters and prevent over-fitting.

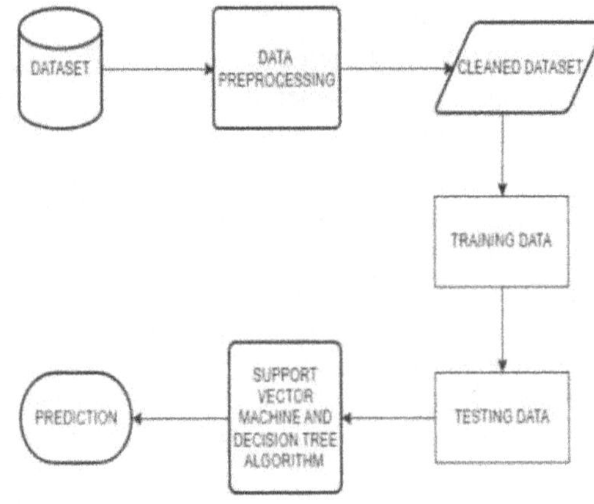

Figure 17.1. Block diagram.

Source: Author.

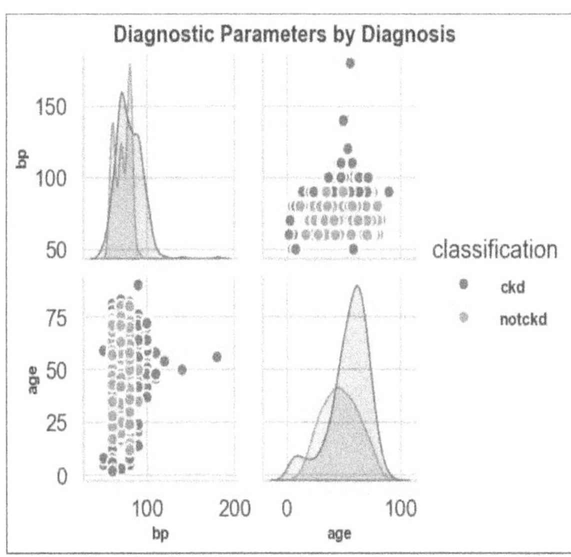

Figure 17.2. Different diagnostic parameters plots.

Source: Author.

5.4. Testing

A variety of evaluation metrics can be used to access the trained models' performance. It includes the following such as accuracy, sensitivity, specificity, AUC-ROC, and precision-recall curve. These parameters play a major part in terms of analysis.

5.5. Deployment

Deploy the trained models into clinical practice or healthcare systems, integrating them into existing work flows and ensuring interoperability with electronic health record systems.

5.6. Machine learning algorithm

Chronic renal illness can be predicted using a variety of machine learning techniques. It depends on the major factors such as the how much data is available, classification of data, the size, structure and the desired understanding capacity of the model.

5.7. Prediction

The key to ensuring the model's effectiveness in properly and consistently forecasting chronic renal disease is validation of its performance, interpretability, and generalizability.

6. Results and Discussion

The dataset which is used for our project implementation with various parameters. It consists of different variables. The datasets serve as the foundation upon which the predictive models are built and evaluated [4]. Typically, datasets used in chronic kidney disease prediction projects comprise a diverse range of demographic, clinical, and laboratory variables collected from patients with and without kidney disease. These variables may include age, gender, blood pressure, serum creatinine levels, presence of comorbidities (e.g., diabetes, hypertension), and urinalysis results, among others.

The datasets are often sourced from healthcare databases, clinical repositories, or public datasets such as the UCI Machine Learning Repository [6]. Additionally, it's essential to mention some of the predefined steps undertaken, such as data cleaning, implementation of data, feature training and testing and feature engineering, to ensure the datasets are suitable for analysis and modelling [3]. Providing this introductory information sets the stage for discussing the methodology and results of the project, giving readers a clear understanding of the data used and its relevance to the study objectives.

The following Figure 17.3 represents the number of persons predicted with the CKD or not. From the observation, we conclude that 250 persons are affected with CKD and 150 persons are not affected with CKD which is based on the analysis from the dataset.

6.1. Support vector machine

One supervized learning approach that can be used mostly for classification problems is the support vector machine, or SVM [9]. However, it is most commonly used for categorization problems. The goal of this approach is to generate a decision boundary or hyperplane that could separate datasets into different classifications. It addresses the fact that the data points that determine the hyperplane are referred to as an analysis support vector [10]. The different real-time applications are handled by this algorithm.

6.2. Decision tree algorithm

This algorithm is a subset of supervized learning algorithms that focuses mostly on problem classification. It also has the

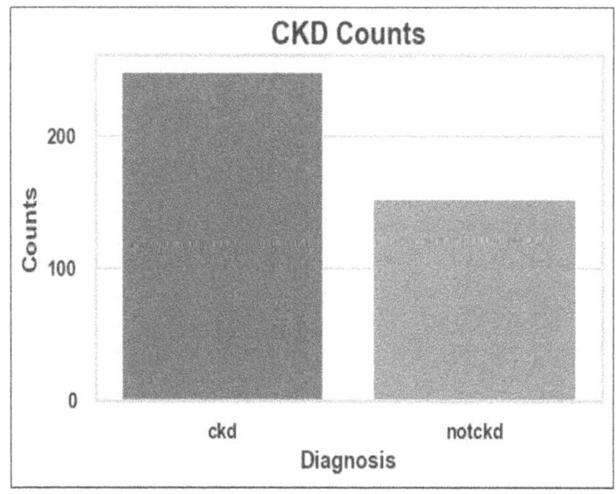

Figure 17.3. CKD result.

Source: Author.

property of regression type of problems in the system analysis. This algorithm is based on both the qualitative variables and the quantitative variables. It represents a hierarchical structure that includes nodes and branches. The structure starts from the decision node which is also called as root node. Then it divides into two different nodes called as sub tree which has a different leaf node. By analyzing the leaf nodes in the end, the outcome can be verified for the data analysis [5]. The datasets features are represented by the internal node, decision rules by the branches and problem outcome from the leaf nodes.

We can determine that the support vector machine learning algorithm outperforms the decision tree classifier algorithm by comparing the two algorithms. The outcomes demonstrate that the support vector machine learning algorithm and the decision tree classifier have 98% and 94% accuracy levels, respectively. The support vector machine learning approach is therefore chosen for improved analysis and prediction based on findings and computation.

7. Conclusion

To sum up, our initiative uses machine learning algorithms to forecast chronic kidney disease, which show promise in precisely identifying people who are at risk. The created model demonstrates resilience and reliability by utilizing sophisticated techniques like feature selection, model optimization, and cross-validation. Additionally, the successful implementation of this project underscores the importance of interdisciplinary collaboration between healthcare professionals and data scientists. By harnessing the power of machine learning, clinicians can potentially streamline diagnosis processes and allocate resources more effectively. Furthermore, ongoing refinement and adaptation of the model will be crucial to keep pace with evolving medical knowledge and technological advancements. Overall, this endeavor marks a significant step towards early detection and intervention, ultimately helping the patients in a limited duration and providing a good exposure in the healthcare department, while also signifying a noteworthy advancement in leveraging data-driven approaches for personalized medicine.

8. Future Work

In future endeavors, our project could benefit from expanding the dataset to various parameters of demographic and the clinical variables, enhancing with the model's accuracy and applicability. Additionally, integrating real-time data sources from wearable devices or electronic health records could enable ongoing monitoring and early intervention. Collaboration with healthcare institutions for prospective validation studies would validate the model's real-world effectiveness. Lastly, creating intuitive user interfaces and decision assistance tools would make it easier to incorporate into clinical practice, which would enhance patient care in the treatment of chronic renal disease.

References

[1] Khalid, H., Khan, A., Zahid Khan, M., Mehmood, G., & Shuaib Qureshi, M. (2023). Machine learning hybrid model for the prediction of chronic kidney disease. *Computational Intelligence and Neuroscience*, 2023(1), 9266889.

[2] Islam, M. A., Majumder, M. Z. H., & Hussein, M. A. (2023). Chronic kidney disease prediction based on machine learning algorithms. *Journal of Pathology Informatics*, 14, 100189.

[3] Chittora, P., Chaurasia, S., Chakrabarti, P., Kumawat, G., Chakrabarti, T., Leonowicz, Z., Jasiński, M., Jasiński, Ł., Gono, R., Jasińska, E., & Bolshev, V. (2021). Prediction of chronic kidney disease-a machine learning perspective. *IEEE Access*, 9, 17312–17334.

[4] Debal, D. A., & Sitote, T. M. (2022). Chronic kidney disease prediction using machine learning techniques. *Journal of Big Data*, 9(1), 109.

[5] Kaur, C., Kumar, M. S., Anjum, A., Binda, M. B., Mallu, M. R., & Al Ansari, M. S. (2023). Chronic kidney disease prediction using machine learning. *Journal of Advances in Information Technology*, 14(2), 384–391.

[6] Dritsas, E., & Trigka, M. (2022). Machine learning techniques for chronic kidney disease risk prediction. *Big Data and Cognitive Computing*, 6(3), 98.

[7] Revathy, S., Bharathi, B., Jeyanthi, P., & Ramesh, M. (2019). Chronic kidney disease prediction using machine learning models. *International Journal of Engineering and Advanced Technology*, 9(1), 6364–6367.

[8] Nikhilesh Reddy, O., Sai Gowtham, K., Abdul Sami, S., & Karthik, D. V. M. (2023). Chronic kidney disease prediction using machine learning. *International Journal for Multidisciplinary Research*, 5(5), 1.

[9] Radha, M., & Murugan, K. (2023, December). An efficient IoT device for chronic kidney disease affected Patient Data Collection. In *2023 International Conference on Energy, Materials and Communication Engineering (ICEMCE)* (pp. 1–5). IEEE.

[10] Radha, M., & Murugan, K. (2024, February). Monitoring and Storage of Health Data in Secured Cloud Environment: A Detailed Survey. In *2024 2nd International Conference on Computer, Communication and Control (IC4)* (pp. 1–6). IEEE.

18 Analysing the efficiency of a panel of monocrystalline cells utilizing a nanofluid cooling technique based on Aluminium-Metal Oxide (Al2O3)

Gaurav Tamrakar[a] *and Shailesh Singh Thakur*[b]

Assistant Professor, Department of Mechanical, Kalinga University, Raipur, India

Abstract: We examine the current state of the art in cooling and temperature regulation of monocrystalline cells (PV) panels in this study. In particular, the increased electrical efficiency that can be achieved by reducing the panel temperature has led to the development and experimental testing of numerous cooling solutions in recent years. The efficiency decreases with a temperature increase of about 0.5% per degree Celsius. Although various other cooling methods have been investigated, active water and air cooling have been found to be the most popular and easy to implement. Cooling with phase-change materials, conductive cooling, and other methods are available. Methods of cooling, module size and type, location, season, and other variables can affect the range of improvements in electrical efficiency, which usually fall between three and five percent. Finally, this paper assesses various temperature control methods, including concurrent cooling.

Keywords: Solar power, Al2O3, nanofluid cooling, SMPS

1. Introduction

Solar power uses the sun's rays and heat to generate clean, renewable energy. Photovoltaic cells, more commonly known as solar panels, are the most common way to harness the sun's energy. Solar power stations use panels that are almost edge-to-edge in order to collect as much sunlight as possible from large areas. Sometimes you can see them perched on top of houses and other buildings. In order to construct the cells, semiconductor materials are utilized. Radiation from the sun makes the electrons in atoms within a cell more mobile [2]. In turn, this allows electrons to flow freely inside the cell, which is what drives the generator [1].

Thousands of homes can be powered simultaneously by solar collectors [7]. Many different techniques are used by solar-thermal power plants to collect energy from the sun. Solar thermal collectors use the sun's heat to boil water, similar to how steam turbines in coal or nuclear power plants generate electricity. Reducing the monthly cost of electricity would be beneficial for most households, and solar power offers a simple solution for just that.

Renewable energy sources like solar could boost profits while cutting expenses. Certain solar energy installations may be eligible for federal and state tax credits, rebates, and incentives. To be eligible, your system must be certified by \the Solar Rating and Certification Corporation (SRCC) or an equivalently recognized organization. You can minimize the total cost of the project by a substantial amount by opting for a cleaner and more sustainable electricity source, which will qualify you for a tax break [9]. Solar energy produces far less carbon dioxide than traditional, non-renewable power sources that take their cues from fossil fuels. Solar power generates very little greenhouse gas because it does not involve the combustion of any fossil fuels [6]. As solar power increases in efficiency, emissions of greenhouse gases (CO2) and other dangerous pollutants (sulfur oxides, nitrogen oxides, particulate matter, etc.) decrease. More people using renewable energy sources will lead to lower air pollution levels [3–5, 10].

Solar energy system warranties, which can reach 25 years in length, are among the industries longest. When it comes to maintenance, solar thermal and PV components usually need to be changed every ten years, but the system itself only needs to be cleaned once a year. Due to the lack of moving parts, the possibility of malfunction is greatly diminished. Solar power installations require less maintenance and fewer expensive repairs. This is why solar energy systems are gaining popularity among those who aren't handy around the house and can't fix common household appliances.

Solar technologies can be roughly categorized as 'Active solar' technology includes systems that convert light into electricity, such as photovoltaics, and systems

[a]ku.gauravtamrakar@kalingauniversity.ac.in, [b]ku.shaileshsinghthakur@kalingauniversity.ac.in

DOI: 10.1201/9781003675259-18

that heat water using solar energy [12]. Activated sunlight has several direct energy applications, such as drying clothes and heating spaces. In passive solar design, strategies such as facing the sun directly, choosing materials with high thermal mass or light diffusion properties, and incorporating natural ventilation into interior spaces are all considered.

- The energy that is extracted from the sun's rays is known as solar energy (Figure 18.1). Using solar energy to generate electricity is called a 'photovoltaic' process. Doing so requires a substance that acts as a semiconductor. Another method of converting solar energy into usable heat is solar thermal technologies.
- The first is 'solar concentration', a method that uses concentrated sunlight to produce electricity through thermal turbines.
- One alternative is to use solar water heating and air conditioning systems, which combine heating and cooling (Figure 18.2).

2. Boost Converter

As a direct current (DC) to direct current (DC) power converter, a boost converter—sometimes called a step-up converter—raises the output voltage while lowering the load current. The energy storage components of this type of switched-mode power supply (SMPS) include a diode, a transistor, and either an inductor or a capacitor, or both. To reduce voltage ripple at the converter's input and output, capacitor filters are commonly used, sometimes in conjunction with inductors, to smooth out the voltage at the supply side. The capacity of various linear and nonlinear control strategies to keep voltage regulation good under changing loads has been studied for boost converters, which are inherently nonlinear systems [11]. The Voltage-Lift Type Boost Converter is a one-of-a-kind boost converter that finds application in solar photovoltaic (PV) systems. Incorporating the passive components (diode, inductor, and capacitor) of a typical boost-converter, these converters improve the overall performance and electricity quality of a photovoltaic system.

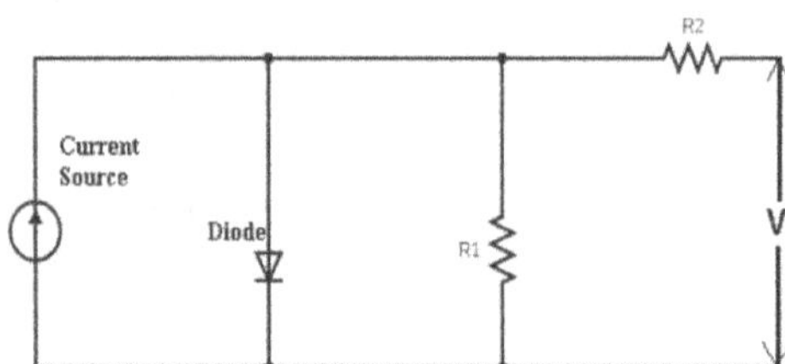

Figure 18.1. Equivalent circuit diagram of Solar cell.

Source: Author.

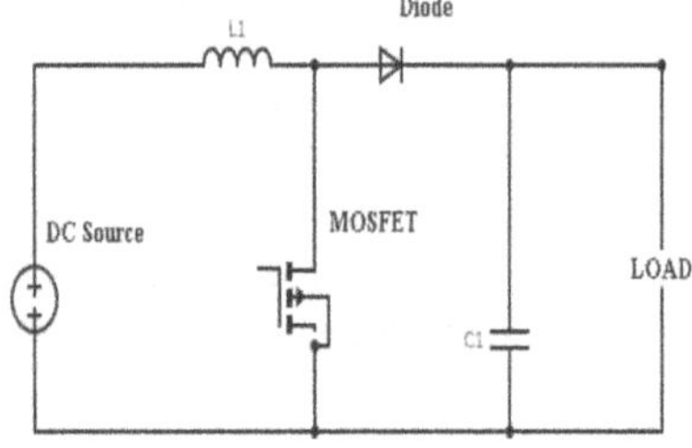

Figure 18.2. Circuit diagram of Boost Converter.

Source: Author.

3. Nanofluid Cooling

By combining the base fluid with nanoparticles, nanofluids enhance the thermal conduction properties of the base fluid. Metals like Cu, Al, and Fe, as well as metal oxides like CuO, A12O3, TiO2, and Fe2O3, might make up these nanoparticles.

Measurements or theoretical calculations of the nanofluid's thermophysical properties, such as pH, volume concentration, density, viscosity, specific heat, and thermal conductivity, could be done using a set of equations developed from models of earlier experimental data. There are two approaches to nanofluid synthesis; the first is a one-step process, while the second is a two-step process that is both easier and more cost-effective. Here, intense magnetic force agitation helps disperse the A12O3 nano powder into the base fluid (distillate water) after its preparation. The next step was to use a sonicator to create an emulsion of the nanoparticles and the base fluid, which improved the nanofluid's stability and made the nanoparticles more evenly dispersed. Because of its high thermal conductivity (40.0 W/m K), aluminum-metal oxide (A12O3) is considered to be one of the most popular nanoparticles for making an effective nanofluid. White alumina, in its prepared form, is nano-aluminium oxide. There are two main approaches to nanofluid synthesis: the one-step process and the more practical and cost-effective two-step process. In the two-step method, Al_2O_3 nanopowder is first synthesized and then mixed with a base fluid—typically distilled water—using a high-powered magnetic stirrer. To enhance the dispersion and stability of the nanoparticles within the fluid, a sonicator is used to create a uniform emulsion.

Aluminum oxide (Al_2O_3) is a popular choice for nanofluid applications due to its high thermal conductivity of 40.0 W/m·K. One of the promising applications of such nanofluids is in solar panel cooling. Overheating is a major issue that reduces solar panel efficiency, and lowering the temperature can significantly improve current, power output, and overall efficiency by reducing ohmic resistance.

Among the various cooling methods available, air cooling using back channels is one approach. Another method involves water spray cooling. Laboratory experiments using

Table 18.1. Temperature variation of PV panel

Sr. No.	I_R	T_{PV} *(Before Cooling)*	T_{PV} *(After Cooling)*	P_{max} *(Before Cooling)*	P_{max} *(After Cooling)*
1	810	76.48	58.3	146.30	158.76
2	815	66.35	58.95	148.05	164.99
3	825	52.15	38.47	157.78	171.72
4	843	69.3	41.2	165.35	184.58

Source: Author.

Table 18.2. Performance parameters of panel after application of Aluminium-metal oxide (Al2O3) Nanofluid cooling

Sr.No.	I_R	T_P	V_{pmax}	I_{pmax}	P_{max}	V_{oc}	I_{sc}	*F. F*	η
1	810	58.3	25.79 V	6.15 A	158.76	31.81 V	6.852 A	0.785	0.20
2	815	58.95	26.58 V	6.21 A	164.99	32.68 V	6.319 A	0.765	0.199
3	825	38. 47	34.43 V	6.16 A	171.72	33.7 V	6.983 A	0.729	0.191
4	843	41.2	34.15 V	6.78 A	184.58	33.23 V	7.252 A	0.77	0.197

Source: Author.

a sun simulator under controlled conditions and varying irradiance levels have demonstrated that water spray cooling can result in a temperature reduction of 5–22 °C, which corresponds to a 9–22% increase in power output [8].

$$\frac{d T_{PV}}{dt} = \left\{\frac{\varphi \cdot \alpha - \sigma\, \epsilon_P\{(T_{PV}^2 + T_S^2)(T_{PV} + T_S)\} - C_{FF}\left(\ln \frac{K_1\, \varphi}{T_{PV}}\right) - Q_H - Q_{cv}}{C_{PV}}\, Used\ control\ technique\right\} - \frac{Conduction}{C_{PCM}} \quad (1)$$

$$T_S = 0.037536\,[T_{amb}{}^{1.5}] + 0.32\,[T_{amb}] \quad (2)$$

$$L_{min} = \frac{D\,(1 - D^2)\,R}{2f} \quad (3)$$

$$C_{min} = \frac{D}{R\left(\frac{\Delta V_O}{V_O}\right) f} \quad (4)$$

$$L_{min} = \frac{D\,(1 - D^2)\,R}{2f} \quad (5)$$

$$C_{min} = \frac{D}{R\left(\frac{\Delta V_O}{V_O}\right) f} \quad (6)$$

4. Results and Discussion

Combining active and passive cooling methods with aluminum-metal oxide (Al2O3) nanofluid cooling leads to a significant temperature drop of approximately 16°C and is associated with enhanced efficiency and fill factor (Tables 18.1 and 18.2). The efficiency and fill factor are around 19.7 percent and 78.5 percent, respectively, at 58.3 degrees Celsius, and 41.2 degrees Celsius, respectively.

5. Conclusion

Methods for cooling using nanofluids containing aluminium metal oxide (Al2O3) have been studied. On the other hand, there are situations where nanofluid cooling with aluminium-metal oxide (Al2O3) is preferable than single cooling. Such rare cases include, for instance, the use of polycrystalline microcrystal (PCM) material and the deployment of tiny, extremely concentrated PV cells. Taking pump costs into account, passive methods, particularly those involving back-side cooling, may on occasion lead to a higher power increase than active methods. An Al2O3 nanofluid cooling system is the most effective means of reusing PV/thermal energy system waste heat. The use of aluminum-metal oxide (Al2O3) nanofluid cooling was shown to be the most effective method for optimizing electricity efficiency. Therefore, developing effective methods of cooling photovoltaic panels using nanofluids containing aluminum metal oxide (Al2O3) should be the focus of future research. The efficiency and temperature of the released water could be enhanced by installing an extra solar panel at the water exit. A more economical way to reduce pumping costs, especially in hot climates, is to use front surface cooling. Water evaporation is a concern with front-side cooling since it would require continuous water replacement.

References

[1] Thorat, P. (2023). Experimental investigation of change in performance parameters on cooling of lithium-ion battery pack using nanofluids. *Energy Storage*, *5*(6), e451.

[2] Gridnev, S., Podlesnykh, I., Skalko, Y., & Rezunov, A. (2020). Estimating the influence of solar radiation at different seasons on the mode of deformation of a span structure with an ortrotropic plate. *Archives for Technical Sciences*, *2*(23), 59–66.

[3] Verma, S. K., Sharma, K., Gupta, N. K., Soni, P., & Upadhyay, N. (2020). Performance comparison of innovative spiral shaped solar collector design with conventional flat plate solar collector. *Energy*, *194*, 116853.

[4] Nair, J. G., Raja, S., & Devapattabiraman, P. (2019). A scientometric assessment of renewable biomass research output in India. *Indian Journal of Information Sources and Services*, *9*(S1), 72–76.

[5] Sharma, R., Sharma, K., & Saraswat, B. K. (2023). A review of the mechanical and chemical properties of aluminium alloys AA6262 T6 and its composites for turning process in the CNC. *Materials Today: Proceedings*.

[6] Abduljaleel, A., Laith, A. A., Hassan, M. G., & Zahraa, E.F. (2024). IoT structure based supervisor and enquired the greenhouse parameters. *Journal of Internet Services and Information Security*, *14*(1), 138–152.

[7] Rahman, M. S., Basu, A., Nakamura, T., Takasaki, H., & Kiyomoto, S. (2018). PPM: Privacy policy manager for home energy management system. *Journal of Wireless Mobile Networks, Ubiquitous Computing, and Dependable Applications*, *9*(2), 42–56.

[8] Chaturvedi, R., & Singh, P. K. (2021). Synthesis and characterization of nano crystalline nitrogen doped titanium dioxide. *Materials Today: Proceedings*, *45*, 3666–3669.

[9] Chen, Y., Feng, L., Jamal, S. S., Sharma, K., Mahariq, I., Jarad, F., & Arsalanloo, A. (2021). Compound usage of L shaped fin and Nano-particles for the acceleration of the solidification process inside a vertical enclosure (A comparison with ordinary double rectangular fin). *Case Studies in Thermal Engineering*, *28*, 101415.

[10] Appadurai, M., Fantin Irudaya Raj, E., & Chithambara Thanu, M. (2023). Application of nanofluids in heat exchanger and its computational fluid dynamics. *Mathematics and Computer Science*, *1*, 505–524.

[11] Yan, G., Shawabkeh, A., Chaturvedi, R., Nur-Firyal, R., & Youshanlouei, M. M. (2022). Using MHD free convection to receive the generated heat by an elliptical porous media. *Case Studies in Thermal Engineering*, *36*, 102153.

[12] Majdi, A., Alqahtani, M. D., Almakytah, A., & Saleem, M. (2021). Fundamental study related to the development of modular solar panel for improved durability and repairability. *IET Renewable Power Generation*, *15*(7), 1382–1396.

19 Bandwidth improvement using C stub in simple compact monopole antenna for wireless applications

Sathyamoorthy Sellapillai[1,a] and Kavitha Thandapani[2,b]

[1]Research Scholar, Veltech Rangarajan Dr. Sagunthala R & D Institute of Science and Technology,Chennai, India
[2]Professor, Veltech Rangarajan Dr. Sagunthala R & D Institute of Science and Technology, Chennai, India

Abstract: A CPW fed C structured compact antenna is proposed in favour of wireless applications. The proposed antenna's dimensions are $26.5 \times 30 \times 1.6$ mm^3. The C shaped stub width is fixed based on the parametric study with the finite ground in order to achieve one narrow band 2.26–2.55 GHz and a broad band 3.49–6.28 GHz having an adequate bandwidth about 290 MHz as well as 2.79 GHz accordingly. It blankets Wireless Local Area Network (WLAN), Worldwide Interoperability (WiMAX) and Industrial, Scientific and Medical Purposes (ISM) and the other parameters like impedance is matching, Voltage standing wave ratio falls under 2 and produces a stable radiation pattern towards required bands.

Keywords: Coplanar waveguide, monopole antenna, broadband antenna, WLAN, WiMAX, wireless

1. Introduction

Antennas are essential for effective two way communication of electromagnetic signals in wireless communication systems. Antennas are crucial components that determine the dependability and efficiency of wireless communication lines because they act as the interface between electronic devices and their surroundings. Among the wide range of antenna designs that are accessible.

Microstrip patch antennas as well as co-planar waveguide (CPW) antennas are popular options because of their unique features and versatility in a range of communication contexts. Microstrip patch antennas are becoming increasingly common in wireless communication systems because of their low profile, small size, simplicity in construction. Using a more effective feeding method [1] also helps to reduce the antenna device's size. The layout of coplanar waveguides [2] offer good flexibility, and CPW antennas achieve superior antenna size reduction than microstrip antennas. Analyzed various shaped antennas [3] like square, circular and triangle shaped antennas in that the square shape is good at size and bandwidth, circular is slightly good at bandwidth but size is more triangular shaped antenna bandwidth and return loss is less hence to propose compact and wide bandwidth modified square like c shaped is proposed here, including asymmetric triangle slot in the patch [4], Metamaterial inspired OSRR based compact antenna [5] was designed for WiMAX and WLAN applications in which antenna occupy less space but failed to accommodate 3.5 GHz WiMAX frequency band and adding different slots in the patch and ground [6] are used to produce multiband performance, inverted l slot is etched from monopole antenna [7] to achieve multiband for wireless applications but it's not in compact size, techniques to be considered [8] in order to miniaturization, a ring shaped decagon [9] is used for multiband here inverted dual slot is used [10], C shape is etched from ground and semicircular based patch is used to radiate but the size is big, to achieve triple band c shaped slot is used with inverted triangle with two slots but occupied area is more [11]. It is made up of two C-shaped strips, one of which is inverted and has a bottom plane, This [12] work introduces the use of an integrated baluns, which achieves three frequency bands utilizing C-shaped resonators but not in compact size, modified rectangular and semicircular shape is used in patch with the shrinking ground [13]. In this work a simple monopole is used then a c shaped slot is introduced for dual band and to achieve wide bandwidth, fine impedance match and good radiation pattern for the required frequency bands.

2. Antenna Configuration

Since microstrip antenna possesses single band and it may not contain a broad bandwidth. Therefore in order to construct multiple resonance with a broad bandwidth and each to be operated individually we need combine many resonance bands, In order to get an multiple and broad bandwidth this design helps to achieve that.

[a]sathya1067@gmail.com, [b]drkavitha@veltech.edu.in

DOI: 10.1201/9781003675259-19

A CPW-fed dual broadband antenna was created upon FR4 substrate having dielectric constant of 4.4, fed by a 50 Ω transmission line, and a substrate height about 1.6 mm.

The compact antenna's structure is shown in Figure 19.1 its size remains 26.5 × 30 ×1.6 mm³. The substrate Width W = 26.5 mm and length L = 30 mm, the ground width GW = 11.75 mm and length GL = 9 mm, Feed width FW = 2.6 mm, Gap between ground and feed G1 = 0.2 mm, Gap between ground and C shaped stub G2 = 0.5 mm, Gap between C shaped stub and feed G3 = 0.9 mm, Stub gap AG = 2, The width of the C shaped stub which left unconnected AW1 = 11 mm and connected AW2 = 9.45 mm. The length of the C shape is AL = 11 mm.

3. Findings with Elaboration

3.1. The proposed antenna's S11

The suggested antenna layout is numerically investigated and optimized using the electromagnetic solver Ansoft HFSS. The computed S11 with respect to the suggested antenna and monopole antenna is displayed in Figure 19.2.

It shows the |S11| similarity upon monopole antenna as well as suggested antenna, initially monopole antenna resonating at 5.6 GHz with 1.14 GHz of bandwidth to enhance the bandwidth as well as to generate additional resonance c stub is introduced based on parametric study in monopole antenna.

The simulation findings in Figure 19.2 clearly indicate a widespread bandwidth spanning from 3.49. GHz to 6.28 GHz, it accommodates two resonant bands located at 2.42 GHz and 5.24 GHz, It appears that the bandwidth covers the necessary applications between 2.26–2.55 GHz and 3.49–6.28 GHz.

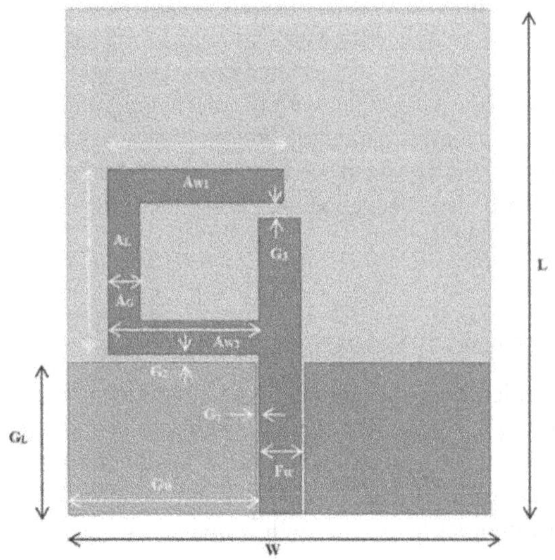

Figure 19.1. Design structure of proposed C shaped monopole antenna.

Source: Author.

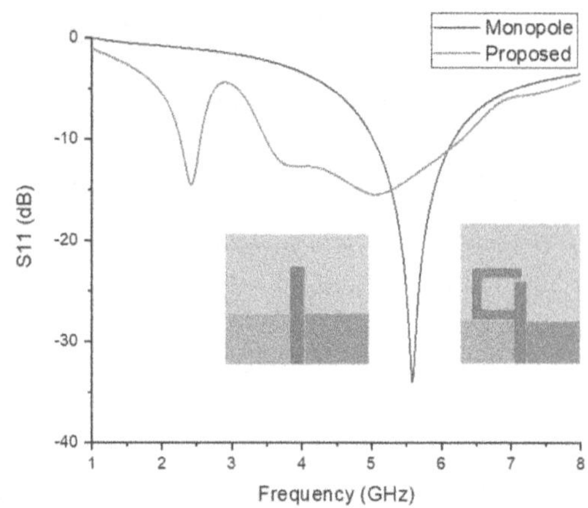

Figure 19.2. |S11| Comparison of monopole and proposed antenna.

Source: Author.

3.2. Impedance characteristics of the antenna

The suggested antenna's impedance value for both real and imaginary values is displayed in Figure 19.3. It may be observed that at the necessary frequencies, imaginary values are positioned near 0 ohm while real values are present near 50 ohm. Better impedance values for the antenna at the intended resonance frequencies are validated by the Figure 19.3.

3.3. VSWR of the Antenna

Figure 19.4 shows a plot of the suggested antenna's VSWR; we able to observe the values are less than 2 for the following required bands 2.26–2.55 GHz and 3.49–6.28 GHz it shows that the antenna works well for the particular frequencies.

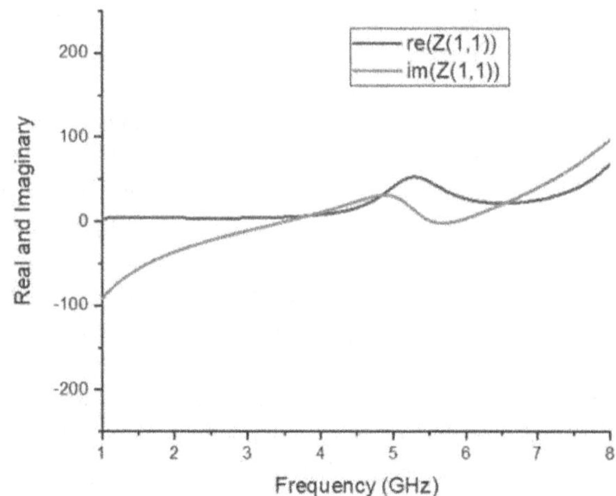

Figure 19.3. The suggested antenna's impedance characteristics.

Source: Author.

3.4. Antenna's current distribution

Figure 19.5 displays the simulated current distribution for both frequencies, 2.42 GHz and 5.24 GHz. in the Figure 19.5 (a) 2.42 GHz the concentration of the current is more at feed and parallel to both sides which nearer to the feed of the ground the less concentration at C shape. In Figure 19.5 (b) we can observe the current concentration in more at both the feed line as well as edge of the C shape hence both is responsible for the frequency 5.24 GHz.

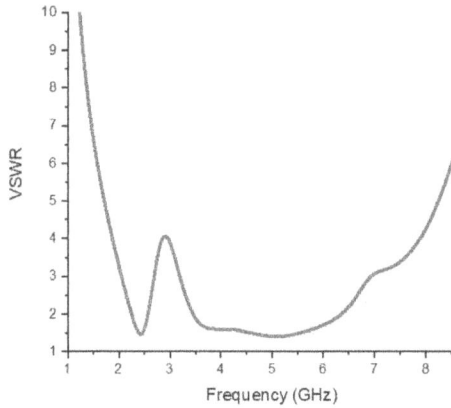

Figure 19.4. VSWR of the antenna.

Source: Author.

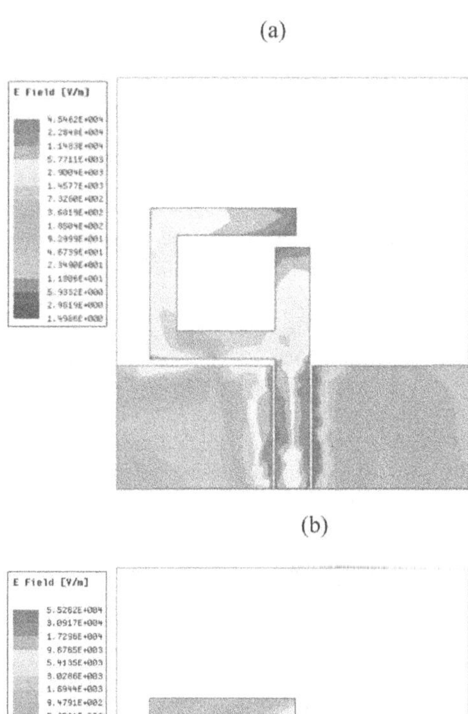

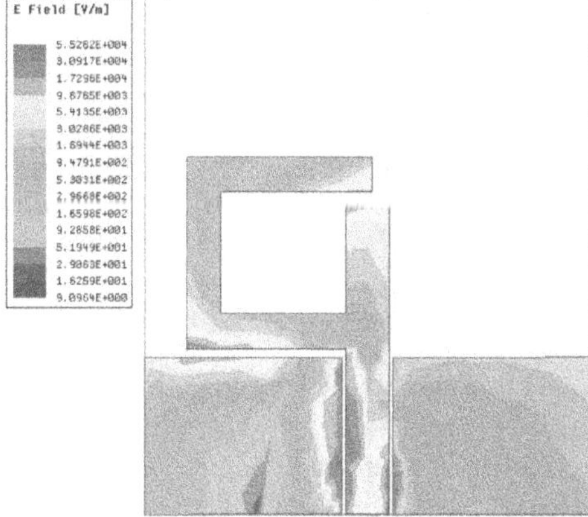

Figure 19.5. Current distribution of the suggested antenna (a) 2.42 GHz and (b) 5.24GHz.

Source: Author.

3.5. Antenna's Radiation pattern

The 2D radiation pattern of the antenna is plotted in Figure 19.6 (a and b) it shows that at both frequencies 2.42 GHz & 5.24 GHz it produces both omnidirectional and dumbbell shape for the desired frequencies.

In Figure 19.7 (a and b) shows the 3D radiation pattern towards following resonance frequencies 2.42 GHz and 5.24

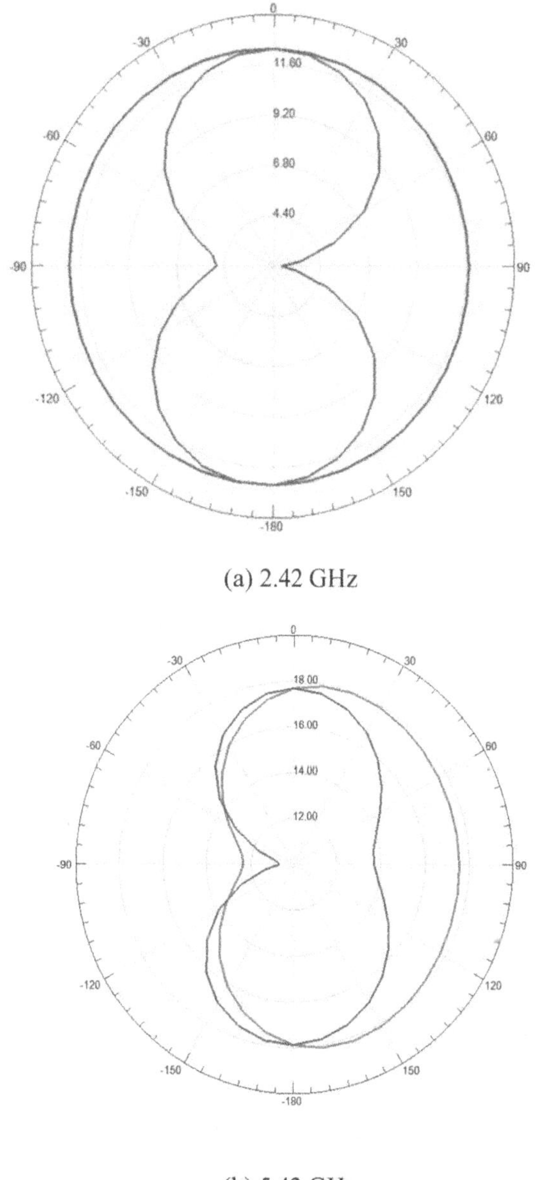

Figure 19.6. Two dimensional Radiation pattern of the antenna.

Source: Author.

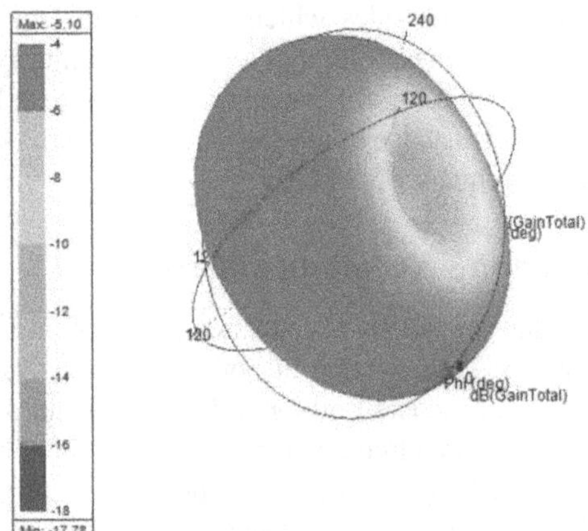

(a) 2.42 GHz

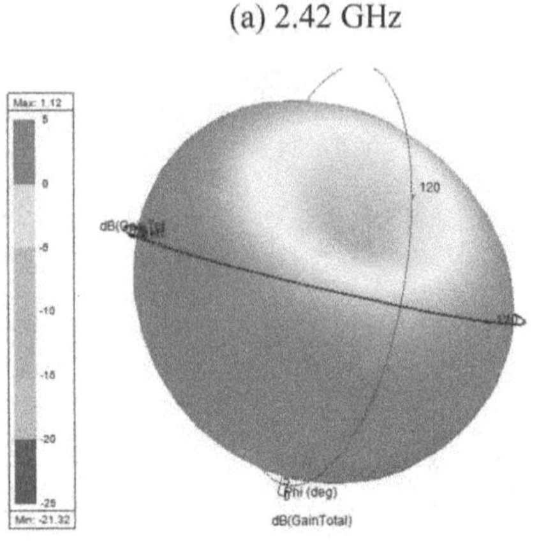

(b) 5.24 GHz

Figure 19.7. 3-dimensional Radiation pattern of antenna.

Source: Author.

GHz It is evident to us that the antenna possess good radiation characteristics for the required resonance bands.

4. Conclusion

An antenna with a small size has been created with the geometry of 26.5 × 30 × 1.6 mm^3 for wireless applications. A simple monopole is responsible to create resonance at 5.6 GHz then by adding C shape the frequency resonance are 2.42 GHz and 5.24 GHz with a narrow and a wide bandwidth. The proposed antenna is small in size of 795 mm^2 with this it covers the most wireless applications like WLAN, WiMAX, WiMAX and ISM simultaneously. It will be better candidate for modern wireless applications.

References

[1] Ameen, M., Mishra, A., & Chaudhary, R. K. (2020). Asymmetric CPW-fed electrically small metamaterial-inspired wideband antenna for 3.3/3.5/5.5 GHz WiMAX and 5.2/5.8 GHz WLAN applications. *AEU-International Journal of Electronics and Communications*, *119*, 153177.

[2] Wen, C. P. (1969). Coplanar waveguide: A surface strip transmission line suitable for nonreciprocal gyromagnetic device applications. *IEEE Transactions on Microwave Theory and Techniques*, *17*(12), 1087–1090.

[3] Sellapillai, S., Rengasamy, R., & Praveen Naidu, V. (2021, December). Study and Comparison of Various Metamaterial-Inspired Antennas. In *International Conference on Futuristic Communication and Network Technologies* (pp. 309–316). Singapore: Springer Nature Singapore.

[4] Azeez, Y. F., Abboud, M. K., & Qasim, S. R. (2023). Design of miniaturized multi-band hybrid-mode microstrip patch antenna for wireless communication. *Indonesian Journal of Electrical Engineering and Computer Science*, *31*(2), 794–801.

[5] Rajkumar, R., & Usha Kiran, K. (2017). A metamaterial inspired compact open split ring resonator antenna for multiband operation. *Wireless personal communications*, *97*, 951–965.

[6] Huang, H., Liu, Y., Zhang, S., & Gong, S. (2014). Multiband metamaterial-loaded monopole antenna for WLAN/WiMAX applications. *IEEE antennas and wireless propagation letters*, *14*, 662–665.

[7] Wheeler, H. A. (1947). Fundamental limitations of small antennas. *Proceedings of the IRE*, *35*(12), 1479–1484.

[8] Dhanasekaran, D., Somasundaram, N., & Rengasamy, R. (2021). A compact decagon ring-shaped multiband antenna for WLAN/WiMAX/WAVE/satellite applications. *Journal of Applied Science and Engineering*, *24*(5), 757–761.

[9] Thandapani, K., & Subramani, S. (2018). Inverted dual U slot loaded truncated microstrip patch antenna for wireless applications. *International Journal of Engineering and Technology (UAE)*, *7*, 151–155.

[10] Ding, K., Gao, C., Yu, T. B., & Qu, D. X. (2015). CPW-fed C-shaped slot antenna for broadband circularly polarized radiation. *International Journal of RF and Microwave Computer-Aided Engineering*, *25*(9), 739–746.

[11] Rezvani, M., Zehforoosh, Y., & Beigi, P. (2019). Circularly-polarized and high-efficiency microstrip antenna with C-shaped stub for WLAN and WiMAX applications. *Radioelectronics and Communications systems*, *62*(11), 604–608.

[12] Ali, A., & Shukla, S. M. (2023, June). A Dual Band Dual C Shaped Monopole Antenna with a Low Profile for Wi-Fi, WiMAX, and WLAN Applications. In *2023 International Conference on IoT, Communication and Automation Technology (ICICAT)* (pp. 1–5). IEEE.

[13] Thaher, Raad & Nori, Lina. (2022). Design and analysis of multiband circular microstrip patch antenna for wireless communication. Periodicals of Engineering and Natural Sciences (PEN). 10. 23. 10.21533/pen.v10i3.2996.

[14] Hatamian A, Ghobadi C, Nourinia J, Shokri M. A compact triple-band printed dipole antenna using C-shaped resonators with stable radiation for GSM/ISM/WLAN applications. Microw Opt Technol Lett. 2023; 65: 611–618.

[15] Bala, S., Bag, B., Sarkar, S., & Sarkar, P. P. (2024). A single feed circularly polarized dual band microstrip monopole antenna for wireless applications. *Microwave and Optical Technology Letters*, *66*(2), e34048.

20 Transforming Ayurvedic medicine access: A prototype on healthcare systems

Kamlesh Kumar Yadav[1,a] *and Md Afzal*[2,b]

[1]Assistant Professor, Department of CS & IT, Kalinga University, Raipur, India
[2]Research Scholar, Department of CS & IT, Kalinga University, Raipur, India

Abstract: Ayurvedic Medicine is an oldest medical system in the world. Now-a-days people give preference to the Ayurvedic medicines as the allopathic medicines are costlier and have side effects. Ayurvedic medicines are based on plants, animals extract and minerals both in single ingredient drugs and compound formulations, however, Ayurveda does not rule out any substances from being used as a potential source of medicine. Dispensing machines are critical components of an optimized inventory management solution as they are the equipment that securely store assets and enable convenient, point-of-use access through a human less transaction. Operation of efficient stock control appropriate to the needs of the dispensary with the objective of ensuring continuity of supply for patients and minimizing wastage. The Automated Ayurvedic Medicine Vending System represents an innovative approach to the distribution of traditional Ayurvedic remedies, aligning ancient healing practices with modern technology. This invention introduces a smart vending machine equipped with a user-friendly interface and precision dispensing technology. Users interact with the machine and select specific Ayurvedic medicines through the intuitive interface. The system employs advanced sensors for inventory management, ensuring optimal stock levels and expiration monitoring and their combination facilitates accurate medicine selection and dispensing. This invention enhances accessibility, efficiency, and reliability in obtaining Ayurvedic healthcare solutions, catering to the evolving needs of individuals seeking holistic wellness within contemporary healthcare settings.

Keywords: Ayurvedic, medicine dispenser, smart, user friendly, efficiency, reliability

1. Introduction

Ayurvedic medicine, one of the world's oldest medical systems, has long been esteemed for its holistic approach to health and wellness. With its roots deeply embedded in ancient traditions, Ayurveda utilizes natural ingredients such as plant extracts, animal products, and minerals in both single-ingredient and compound formulations [1]. As modern healthcare increasingly acknowledges the limitations and side effects associated with allopathic medicines—often characterized by higher costs and potential adverse reactions—there has been a notable shift towards more sustainable and natural alternatives [2]. This resurgence in interest in Ayurvedic remedies underscores the need for innovative solutions that bridge the gap between traditional practices and contemporary healthcare demands [8].

Dispensing machines play a crucial role in optimizing inventory management by securely storing assets and providing point-of-use access through automated transactions. They are instrumental in ensuring efficient stock control, which is essential for maintaining a steady supply of medications and minimizing waste [11]. This functionality is particularly valuable in settings that require precision and reliability, such as the distribution of specialized Ayurvedic medicines. The Automated Ayurvedic Medicine Vending System represents a groundbreaking advancement in the distribution of Ayurvedic remedies, merging traditional healing practices with modern technological innovation [3]. This system introduces a smart vending machine designed to offer an enhanced user experience through an intuitive interface and precision dispensing technology. By allowing users to interact directly with the machine, select specific Ayurvedic medicines, and receive accurate doses, the system ensures a high level of accessibility and convenience [4, 9].

Equipped with advanced sensors, the vending machine facilitates effective inventory management by monitoring stock levels and tracking expiration dates. This integration of technology not only streamlines the dispensing process but also aligns with the evolving needs of individuals seeking holistic healthcare solutions [5, 6]. By combining the time-honored principles of Ayurveda with contemporary technological advancements, this invention addresses the demand for efficient and reliable access to Ayurvedic medicines, making it a significant step forward in modernizing traditional healthcare practices [7, 10].

[a]ku.kamleshkumaryadav@kalingauniversity.ac.in, [b]md.afzal@kalingauniversity.ac.in

DOI: 10.1201/9781003675259-20

2. Proposed Prototype

In this Paper, the attempt is to create a prototype version of automatic medicine vending machine specifically for ayurvedic health care units. The automated Ayurvedic medicine vending machine operates through a systematic process, combining various components and technologies to provide users with a streamlined and efficient way to access Ayurvedic remedies. Block diagram below gives the short presentation on what the framework will do in this specific invention. Microcontroller is utilized for controlling entire procedure of this automated medicine dispenser. Users approach the vending machine and interact with the user interface, which includes a touchscreen display. The interface prompts users to input their selections, browse available medicines, or provide necessary information. RFID Sensor used to monitor the inventory levels of Ayurvedic medicines in storage units. The system ensures that the selected medicines are available and not expired. A stepper motor controls the dispensing mechanism, precisely selecting and retrieving the specified Ayurvedic medicines. The medicines move along a conveyance system, ensuring accurate positioning for dispensing. In summary, the automated Ayurvedic medicine vending machine seamlessly integrates user interaction, inventory management, precise dispensing mechanisms, and real-time communication to provide users with a modern, efficient, and reliable way to obtain Ayurvedic healthcare solutions.

3. Components and Functionality

3.1. *For medicine selection and dispensing*

3.1.1. RFID sensor

Each piece of Ayurvedic medicine can have an RFID tag put on it. Then, RFID readers inside the vending machine can identify the medicines and keep track of how they move in and out of stock, giving real-time updates. Different sensors are used in the inventory management system to keep an eye on the amount of Ayurvedic medicines in the storage units. These sensors keep track of how much of each medicine is in stock, so the machine knows how much is available (Figure 20.1).

3.1.2. PIR motion sensor (HC SR501)

The PIR Motion Sensor Detector Module HC SR501 is used here. It is almost always used to track the movement of a person within the sensor's range. 'PIR', 'Pyroelectric', 'Passive Infrared', and 'IR Motion' sensors are all terms that are often used to describe it. This invention checks to see if the person taking the medicine moves and lets the person know if the dispenser is working right. Motion sensors can tell when someone is getting close to the vending machine. With this information, you can turn on the user interface. When it's not being used, it stays in a low-power state, which saves energy. Motion sensors can help manage lines when there are a lot of people waiting to use the vending machine at the same time. The system can serve and prioritize users based on how close they are, making the experience more organized and quicker. The motion sensor in the fully automated Ayurvedic medicine vending machine makes the system smarter. It not only saves energy but also improves safety, the user experience, and the overall efficiency of operations (Figure 20.2).

3.1.3. Stepper motor

The stepper motor in the automated Ayurvedic medicine vending machine is a key part of the mechanism that makes sure the medicine is dispensed precisely. Its use makes it possible for the dispensing parts to move precisely and safely, which makes it possible to choose and get Ayurvedic medicines from the storage units. The stepper motor's step-by-step rotation makes controlled dispensing easier. This lowers the chance of mistakes and makes the vending process more reliable. This level of accuracy is necessary to make sure that users get the right medicines, which makes the vending machine more efficient and effective at distributing Ayurvedic health products (Figure 20.3).

Figure 20.1. RFID sensor.

Source: https://images.app.goo.gl/NYZrgMF1HFXmJnd18.

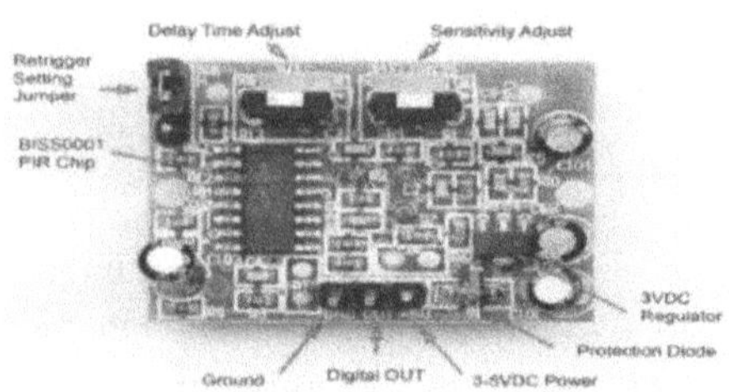

Figure 20.2. PIR motion sensor.

Source: https://www.google.com/imgres?h=500&w=666&tbnh=194&tbnw=259&osm=1&lns_uv=1&source=lens-native&usg=AI4_-kTtcd3XN0osWhYaRRDl7thhi-Jnwg&imgurl=https://europe1.discourse-cdn.com/arduino/optimized/4X/e/9/3/e93ace6878c4e1cfe13b2c941462c603268bd269_2_666x500.jpeg&imgrefurl=https://forum.arduino.cc/t/help-modifying-on-time-for-arduino-controlled-ir-sensor-lamp/551858&tbnid=6Q8bmc91NRIJ8M&docid=Ul5C8l6hXT8SMM.

Figure 20.3. Stepper motor.

Source: https://images.app.goo.gl/drGk5CrLrUSJyFcp8.

3.1.4. Micro controller (ESP 32)

The ESP32 family is a low-cost, low-power system-on-a-chip microcontroller that has built-in Wi-Fi and dual-mode Bluetooth. The ESP32 Board comes after the ESP 8266 Board. It has two cores of processing power and can connect to both Wi-Fi and Bluetooth networks. The board has a USB-to-UART interface, which makes it easy to program with popular development tools like the Arduino IDE. The Micro USB port, the 3.3V pin, or the Vin pin can all be used to power it. The ESP32, which was made by Espressif Systems, is famous for being fast and having both Wi-Fi and Bluetooth built in. A hall effect sensor is also built into the ESP32 so it can pick up on changes in the magnetic field around it. The sensor's output voltage goes up as the magnetic field gets stronger. You can see this in the serial monitor tools (Figure 20.4).

3.1.5. Temperature and humidity sensor

Temperature and humidity sensors are low-cost electronic devices that can find, measure, and report both temperature and humidity. The ratio of the amount of moisture that can be seen all around to the amount of moisture that is present at a certain air temperature. In the automated Ayurvedic medicine vending machine, the temperature and humidity sensor is very important for keeping the quality and effectiveness of the medicines that are given out. The sensor keeps an eye on the environment all the time to make sure that Ayurvedic products are kept in the right temperature and humidity ranges. This proactive approach helps keep the healing properties of sensitive herbal mixtures from breaking down or going bad. Adding this sensor makes the vending machine more reliable overall, giving users confidence in the quality and effectiveness of the Ayurvedic medicines it gives out. It also follows best practices for pharmaceutical storage to provide safe and effective medical solutions (Figure 20.5).

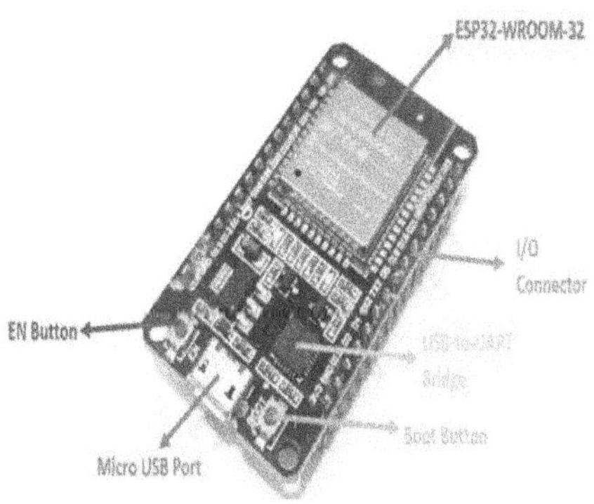

Figure 20.4. Microcontroller.

Source: https://images.app.goo.gl/MWJfTgttULH5n4zk6.

Figure 20.5. Temperature & humidity sensor.

Source: https://images.app.goo.gl/e8iX5gfjHAEiaU7t9.

3.2. For connectivity module

3.2.1. WIFI module

Wi-Fi lets the vending machine connect to the internet, which lets it talk to a central server in real time. This makes it easier to update data, process transactions, and keep an eye on things from afar (Figure 20.6).

3.2.2. Mobile networks

The vending machine can work in different places without Wi-Fi because it is connected to mobile networks. This is especially helpful for deployments in places with little or no internet access.

3.3. For control and navigation

3.3.1. RGB liquid crystal display

An RGB (Red, Green, Blue) Liquid Crystal Display (LCD) is an important part of the automated Ayurvedic medicine vending machine that makes the user experience immersive and educational. In the first place, it makes it easy for people to use by showing a visually appealing menu of Ayurvedic medicines. Users can easily move between the options, see information about the products, and make clear choices (Figure 20.7).

3.3.2. LED indicators

LED lights are used to show users what's going on with the machine, whether the medicine is working, or any other information need to know (Figure 20.8).

Figure 20.6. Wi-Fi module.

Source: https://images.app.goo.gl/hv2rgCat1xPLuAsp8.

Figure 20.7. RGB liquid crystal display.

Source: https://images.app.goo.gl/EEMAjrQEYmKLVDsk9.

3.3.3. *Emergency Stop Button*

For safety, there is a physical emergency stop button that lets users stop any dispensing or transaction process if there are problems or emergencies (Figure 20.9).

4. Design and Operation

Medicine dispensing devices are available with a variety of features depending on their design, purpose, and intended users. This invention has an additional feature, that it meant for use in Ayurvedic medicine dispensing unit. In this clinical healthcare, there are a variety of medication availability which may be available in the form of powder, tablets, pills, liquid, semi-solid etc. and are classified in different categories like arishta, asava, Rasa Rasayan Lauha, Bati, Churna, Avaleha, Ghrita, Parpati, Taila, Goggulu. In today's era, the contactless service is the need of an hour. The dispensing of these medications is a time-consuming process as well as extra care must be taken to ensure safety and accuracy. Good dispensing practices ensure that an effective form of the correct medicine is delivered to the right patient, in the correct dosage and quantity, with clear instructions, and in a package that maintains the potency of the medicine (Figure 20.10).

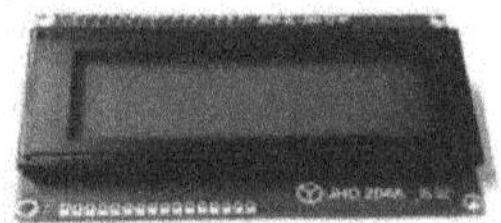

Figure 20.8. LED indicator.

Source: https://images.app.goo.gl/EEMAjrQEYmKLVDsk9.

Figure 20.9. Emergency stop button.

Source: https://images.app.goo.gl/AiXbto2hTAjXWtzH7.

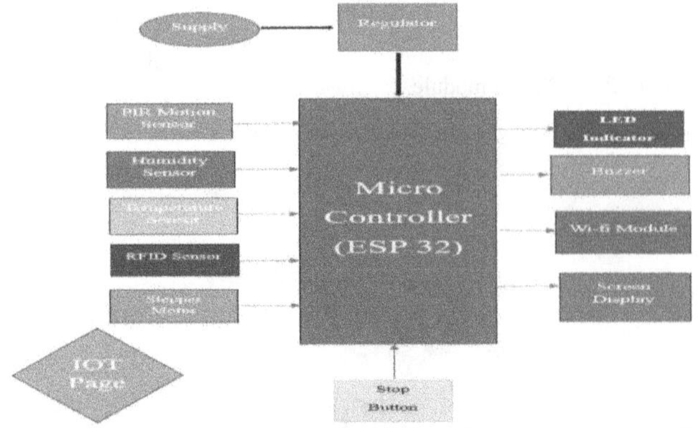

Figure 20.10. Functional block diagram.

Source: Author.

5. Conclusion

The purpose of this protoype is to simplify and modernize the process of dispensing Ayurvedic medicines, thereby providing users with a method that is both convenient and effective for gaining access to these traditionally effective treatments. Individuals who are looking for natural healthcare solutions will have easier access to Ayurvedic medicines as a result of the system's incorporation of technology and automation, which improves the distribution process. A precision dosing mechanism that is based on machine learning algorithms for personalized dosage recommendations is included in this dispenser, which also includes user-friendly elements such as voice-activated and augmented reality guidance. This dispenser combines traditional Ayurvedic medicine practices with modern technology. The invention intends to improve the accuracy of Ayurvedic medicine administration, as well as the user experience and adherence to the prescribed dosage.

References

[1] Nasir, Z., Asif, A., Nawaz, M., & Ali, M. (2023). Design of a smart medical box for automatic pill dispensing and health monitoring. *Engineering Proceedings*, *32*(1), 7.

[2] Vijayakumar, P., Sivasubramaniyan, G., & Saraswati Rao, M. (2019). Bibliometric analysis of Indian Journal of Nuclear Medicine (2014–2018). *Indian Journal of Information Sources and Services*, *9*(1), 122–127.

[3] Sharma, A., Yadav, R., & Sharma, K. (2021). Optimization and investigation of automotive wheel rim for efficient performance of vehicle. *Materials Today: Proceedings*, *45*, 3601–3604.

[4] Salman, R., & Banu, A. A. (2023). DeepQ Residue Analysis of Computer Vision Dataset using Support Vector Machine. *Journal of Internet Services and Information Security*, *13*(1), 78–84.

[5] Sharma, A., Islam, A., Sharma, K., & Singh, P. K. (2021). Optimization techniques to optimize the milling operation with different parameters for composite of AA 3105. *Materials Today: Proceedings*, *43*, 224–230.

[6] Malathi, K., Shruthi, S. N., Madhumitha, N., Sreelakshmi, S., Sathya, U., & Sangeetha, P. M. (2024). Medical data integration and interoperability through remote monitoring of healthcare devices. *Journal of Wireless Mobile Networks, Ubiquitous Computing, and Dependable Applications (JoWUA)*, *15*(2), 60–72. https://doi.org/10.58346/JOWUA.2024.I2.005

[7] Islam, A., Sharma, S., Sharma, K., Sharma, R., Sharma, A., & Roy, D. (2020). Real-time data monitoring through sensors in robotized shielded metal arc welding. *Materials Today: Proceedings*, *26*, 2368–2373.

[8] Uyan, A. (2022). A Review on the Potential Usage of Lionfishes (Pterois spp.) in Biomedical and Bioinspired Applications. *Natural and Engineering Sciences*, *7*(2), 214–227.

[9] Shibata, T., & Matsuura, T. (2020). U.S. Patent Application No. 16/610, 484.

[10] World Health Organization. (2022). *WHO benchmarks for the practice of ayurveda*. World Health Organization.

[11] Philip, J., Abraham, F. M., Giboy, K. K., Feslina, B. J., & Rajan, T. (2020). Automatic medicine dispenser using IoT. *International Journal of Engineering Research & Technology*, *9*(08), 342–349.

21 Multiscale information fusion to segment white blood cells for early stage detection of acute lymphoblastic leukemia

M. Thilagaraj[1,a], *S. Vaira Prakash*[2,b], *G. Petchinathan*[3,c], *Kottaimalai Ramaraj*[4,d], *C. S. Sundar Ganesh*[5,e], *and S. Krishnanarayanan*[6,f]

[1]Department of Industrial Internet of Things, MVJ College of Engineering, Bangalore, Karnataka, India
[2]Department of Electronics and Communications Engineering, Ramco Institute of Technology, Rajapalayam, Tamil Nadu, India
[3]Department of Biomedical Engineering, Sri Shanmugha College of Engineering and Technology, Tiruchengode, Tamil Nadu, India
[4]Department of Electronics and Communication Engineering, Kalasalingam Academy of Research and Education, Krishnankoil, Tamil Nadu, India
[5]Department of Electrical and Electronics Engineering, Karpagam College of Engineering, Coimbatore, Tamil Nadu, India
[6]Department of Computer Science and Engineering, MVJ College of Engineering, Bengaluru, India

Abstract: White blood cells (WBC) are most affected by the malignancy leukemia. Leukocytes, another name for WBCs, combat infestations. The squishy substance found within bones called the marrow of the bone is where RBC, WBC, and platelets are produced. The marrow in the bones produces ineffective WBCs when a patient has leukemia. The human system cannot be protected from pathogens by these aberrant cells. It overcrowds the bone marrow, get into the circulation of the blood, and may migrate to the cerebral cortex, liver, or lymph glands, among various regions of the human anatomy. If the immune system produces an excessive amount of lymphocytes, a kind of WBC, it can result in acute lymphoblastic leukemia (ALL). Among the deadliest forms of blood-based cancer, leukemia affects a large number of individuals annually. The detection of leukemia is significantly correlated with WBCs. Investigations have shown that leukemia alters the shape and quantity of WBCs. Effective WBC separation makes it possible to identify the quantity and morphological that subsequently aids in leukemia identification and treatment. Human WBC evaluation processes are laborious, arbitrary, and not as precise. To address challenges in White Blood Cell (WBC) categorization, researchers propose the utilization of a Multi-Scale Information Fusion Network (MIF-Net). This deep structure integrates both external and internal processes to fuse spatial information effectively. The difficulty in accurate segmentation of WBC images arises from the low contrast between the cytoplasm and the background, as well as the complex structure of nuclei with fuzzy borders. The early layers of the network focus on capturing precise boundary details as a spatial property. MIF-Net strategically di-vides and extends this boundary information across multiple scales to facilitate the fusion of external data. MIF-Net improves segmentation efficiency while protecting border details. Additionally, MIF-Net leverages internal information fusion at certain intervals to enhance features at various network phases.

Keywords: Leukemia, White blood cells (WBC), acute lymphoblastic leukemia (ALL), multiscale information fusion network (MIF-Net), diagnosis, therapy

1. Introduction

Cancer, a medical condition, arises when specific cells in the body undergo uncontrolled proliferation and invade other areas [1]. Given the vast number of cells constituting the human body, cancer can potentially originate in any location. The normal process of cell division, essential for the body's growth and replenishment, can be disrupted. Instead of orderly growth and multiplication to replace aging or damaged cells, occasional malfunctions lead to the unchecked proliferation of impaired or abnormal cells [2].

The fast development of aberrant blood cells is an indicator of leukemia, a type of blood cancer. The majority of our body's blood is produced from our bone marrow, which is the place where this unregulated development occurs [3]. The cell mutates and becomes a specific type of leukemia cell. After the marrow cell goes through a leukemic transformation, the leukemia cells are able to proliferate and survive

[a]m.thilagaraj@gmail.com, [b]vairaprakashklu@gmail.com, [c]gpetchi@gmail.com, [d]r.kottaimalai@klu.ac.in, [e]sundarganesh.cs@kce.ac.in, [f]krishna.narayanan08@gmail.com

DOI: 10.1201/9781003675259-21

longer than what is typical for normal cells. Regular cell growth is eventually suppressed or crowded out by the leukemia cells. Every kind of leukemia has a distinct progression pace and mechanism for replacing healthy blood and bone marrow cells.

Typically, leukemia cells are undeveloped WBC that are still growing. The cells that mature into platelets, red blood cells, and WBC are made in the bone marrow. Every one kind of cell performs a distinct function: Red blood cells carry the oxy-gen from the lungs to the tissues and parts; the platelets aid in the formation of clots that prevent blood loss; and aid in the body's defense against the spread of infection [4]. Modifications in the genetic material (DNA) of bone marrow cells result in leukemia. It is unidentified that which triggered such changes in genetics. The various types of leukemia is illustrated in Figure 21.1.

'Acute' refers to the leukemia's rapid progression and likelihood of death in a matter of months if treatment fails to be received [5]. Being 'lymphocytic' indicates that it originates from in advance, undeveloped lymphocytes, a subset of white blood cells. Initial lymphoid precursors multiply and substitute the bone marrow's typical hematopoietic cells in ALL, a cancerous (clonal) illness of the marrow. The most frequent form of leukemia and cancer among children in the US is called ALL. Kids under five years old have an elevated probability of getting ALL. Following subsequently, the likelihood gradually decreases till mid-twenties after which it gradually increases once more following age 50. In general, youngsters account for roughly four out of each ten instances of ALL. Less than 0.5 percent of cancers reported in the US are related to ALL, indicating that it is not a prevalent cancer. The probability of developing ALL is approximately 1 in 1,000 for a typical individual [6]. Males are somewhat more susceptible compared to females, and White people are more at threat instead of African Americans. While kids account for the majority of ALL scenarios, youngsters account for roughly 4 out of 5 mortalities from ALL.

Named for the blood cells impacted by ALL, the two primary types are B-cell ALL and T-cell ALL: Our B-cells, that assist in combating infection and produce antigens, are impacted by B-cell ALL. Between 75 and 80 percent of ALL cases are B-cell ALL. Our T-cells, that neutralize pathogens and assist other cells of the immune system, are impacted by T-cell ALL. Rarely does another type—natural killer ALL—occur [7]. A change in the DNA in the precursor cells leads to the huge production of WBC, which is the cause of ALL. Prior to that they become fully grown and capable of battling infection like competent WBCs, the bone marrow releases these immature white blood cells.

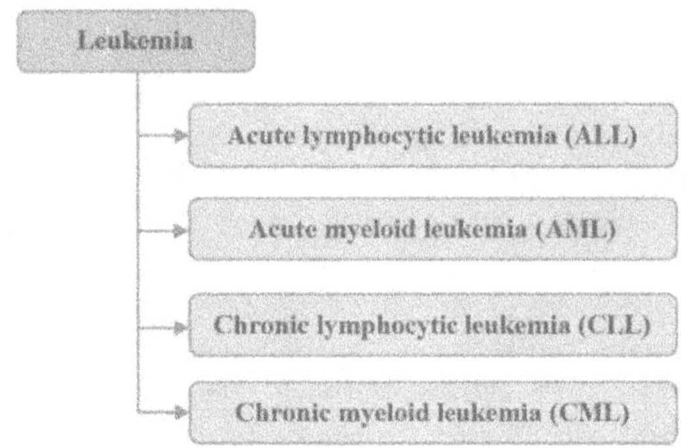

Figure 21.1. Types of Leukemia.
Source: Author.

The precise cause or reasons for ALL are unknown, however there are a number of established risk factors, including excessive radiation contact, benzene being exposed, chemotherapy treatment with drugs in previous years, certain chromosome abnormalities, getting a transplant of bone marrow, being exposed to toxic substances like benzene, and a substance utilized in production that is additionally present in tobacco products. The majority of ALL indications strike unexpectedly and have an analogous impact on children and adults alike. The condition of anemia, hemorrhage like bleeding from the nose or hefty periods with menstruation, scratches, coughing, lightheadedness, exhaustion, a high temperature, recurring infections caused by bacteria or viruses, Pain in the joints, loss of appetite, per-spires at midnight, Blisters, or red, pinhead-sized pimples on the surface of the skin, breathing difficulty, More pale or lighter-than-normal skin tone, enlarged lymphatic vessels Inexplicable reduction in size and fatigue constitute typical beginning indications [8].

Over 65 of every 100 individuals, or over 65%, are expected to live with leukemia for at least five years following receiving a diagnosis [9]. Individuals who may have leukemia are first recommended to have a complete blood count, that involves the WBC count. In addition, WBC appearance is examined to verify leukemia assessment results. The traditional process of determining WBC count and structure evaluation is difficult, lengthy, highly susceptible to mistake, and unreliable [10]. Therefore, computerized systems that rely on machine learning must be used as a substitute for human labour. Technological developments in computerized diagnosis could potentially have an effect on the screening market.

In particular, deep learning has been essential to many disease identification mechanisms. Contemporary medical diagnostics require automated detection. The diagnosis of leukemia is strongly correlated with WBC. Leukemia patients have altered WBC appearance and count. The human evaluation and investigation of WBCs forms the basis of conventional diagnosis [11]. The most effective way to close the substantial discrepancy in leukemia, identification accuracy and timeliness is to use effortless and reliable diagnosis systems based on deep learning. Figure 21.2 portrays the types of WBCs.

WBC separation is difficult due to certain instances, the cell nucleus possesses an indistinguishable boundary with a convoluted form, whereas the cytoplasm encounters minimal contrast in the surrounding tissue. The majority of WBC division methods currently in use fail to factor joint cytoplasm and nucleus classification into account. Furthermore, while joint segmentation is performed with certain techniques, their ability to segment is lacking. Finally, a lot of current networks need a lot of parameters that can be trained because they exhibit poor computational performance.

In order to tackle these challenges, we create a WBC classification system that can efficiently carry out joint segmentation of the nucleus and cytoplasm. We propagated and preserved object boundary data using multi-scale information fusion. Additionally, we employed image in-formation fusion to boost learning as well as obtaining features. Combining data from several network stages facilitates in-depth feature learning of images and yields precise forecasts. By employing innovative methods, we were able to enhance classification efficiency while maintaining computational cost. We demonstrated state-of-the-art classification efficiency by evaluating the structure we developed on a dataset that is publicly accessible. WBC count and morphology are altered by leukemia, while morphological identification primarily relies on precise boundary forecasts for targeted classes. Furthermore, there exist certain statistical methods that rely on the cytoplasmic and nucleic acid regions, thereby facilitating the confirmation of leukemia.

It could be summarize as the primary contributions to the research as ensues: In order to aid in the evaluation of leukemia, a novel network named MIF-Net was developed. This network integrates the segmentation of both cytoplasm and nucleus, placing emphasis on White Blood Cell (WBC) morphology and count. With its shallow structure and internal information fusion following intervals, MIF-Net enhances segmentation efficiency by empowering features.

Within the advanced stages of the network, MIF-Net extends to incorporate external information fusion by dividing low-level multi-scale fine boundary data originating from the initial layer. It facilitates precise boundary estimation for the nucleus and cytoplasm. The design suggested achieves state-of-the-art classification efficiency and is tested on freely accessible databases from WBC.

2. Multi-Scale Information Fusion

The shape and quantity of white blood cells are important indicators for leukemia diagnosis. As a result, we create a distinctive framework for the division of the nucleus and cytoplasm in WBC microscopic pictures. The WBC segmentation procedure may become difficult due to the lack of contrast of the cytoplasm and the blurry nucleus boundary for certain images [12]. To address such problems, we create a framework that utilizes multi-scale information fusion. The network's initial layers contain extracted features that may contain low-level spatial information. The workflow of the suggested system is displayed in Figure 21.3.

The spatial information splitter receives WBC pictures as inputs. This spatial data encompasses acceptable boundary data that has been divided using a spatial information splitter into multiple scales. For the purpose of information fusion, fine boundary-related data is transmitted to the network's advance levels. Boundary information has been combined with down sampled spatial information in the outside data fusion phase. The fusion of external data serves as a major aid in the forecasting of the cytoplasmic and nucleic boundaries. Splitter further transmits spatial data for internal fusion. Amplification of features throughout the network is ensured via internal information fusion. Upon going through several stages of processing, internal fused data is eventually downsampled and fused with external information fusion. Ultimately, prediction masks are produced based on fused characteristics. Our network's output, including separated cytoplasm and nucleus and WBC count, will assist with leukemia assessment and prediction.

Segmenting WBC microscopic pictures can be difficult because of their variations in dimensions, form, and colour state. We have created a multi-scale information fusion network that may succeed significantly in categorization under difficult circumstances. Utilizing the image input layer, the spatial information of the input image is retrieved.

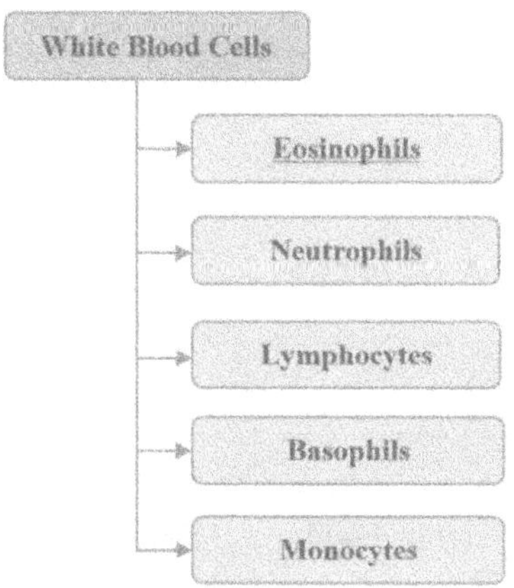

Figure 21.2. Types of WBCs.

Source: Author.

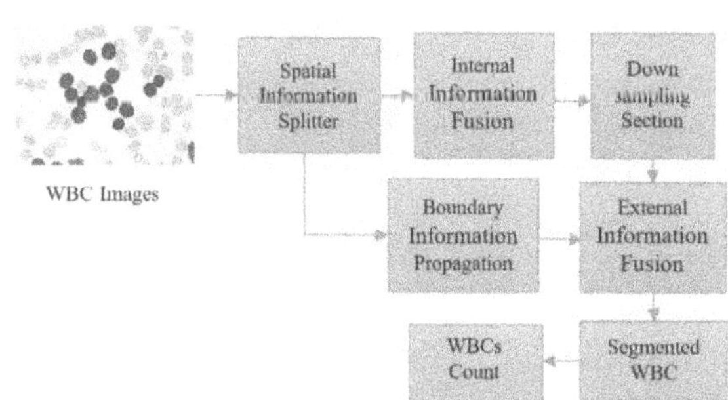

Figure 21.3. Proposed workflow.

Source: Author.

The spatial information splitter that relies on a convolution layer divides this spatial data. Employing strided convolution layers, the spatial information splitter divides the spatial information into three separate levels. Multi-scale picture plays a primary role in enhancing categorization power. Well-known encoder-decoder-based segregation designs are U-Net and SegNet. Pooling layers and un-pooling layers in encoder-decoder structures are employed to upsample and downsample spatial data, correspondingly. Data loss occurs when layers are pooled and unpooled in addition to during regular procedure. Since prediction masks are formed following decoding, this data lack becomes of greater importance at the decoder end.

MIF-Net utilizes a shallow structure to address this issue, and transpose convolution is used to maintain the position of un-pooling layers. In a comparable manner for multi-scale information downsampling and transmission, we employ strided convolution instead of pooling layers. Transpose and strided convolution layers are two kinds of learnable layers that aid in the best possible learning for the network [13]. Furthermore, MIF-Net offers a novel approach for the merging of internal and external data. In the boundary information propagation (BIP) phase, we disseminated the intricate boundary data associated with the items that are present in the first few layers of the network. BIP utilizes strided convolution to propagate spatial information across many scales. In multi-scale distribution of data, three strided convolution layers with the appropriate stride participated. Subsequently, the resulting data is fused with spatial information in external information fusion blocks. We employed two types of fusions, external information fusion (EIF) and internal information fusion (IIF). Three strided convolution layers with the appropriate stride were involved in multi-scale data distribution. The consequent residue data is combined with spatial information in the external information fusion blocks. We used external information fusion (EIF) and internal information fusion (IIF). As a result, EIF is the moniker given to all fusions in the decoder that receive external inputs from the encoder. Three EIFs are used in the suggested design to fuse external information on multiple scales [14]. Skip interconnections across every IIFs begin in the convolution layer that comes before the encoder and end in the convolution layer that comes after. As a result, IIF refers to all fusions in encoders that receive internal input from encoders. Four IIF are used in the suggested network to promote spatial features at periods. All network levels typically degrade certain features as part of its regular functioning. To counteract the aforementioned feature deterioration issue, MIF-Net created an internal information fusion technique. Four IIF blocks—IIF-1, IIF-2, IIF-3, and IIF-4—are used in the MIF-Net architecture to empower features.

The strided convolution with stride 1 is displayed in S-1. It receives input through the splitter and outputs the result to EIF-3. Following the same way, S-2 stands for strided convolution with stride 2 [15]. It receives an input from the splitter, halve the dimension of the feature map, and feeds the result to EIF-2. Figure 21.4 depicts the MIF-Net. Table 21.1 describes the comparison of developed method with various models.

Table 21.1 describes the comparison of developed method with various models.

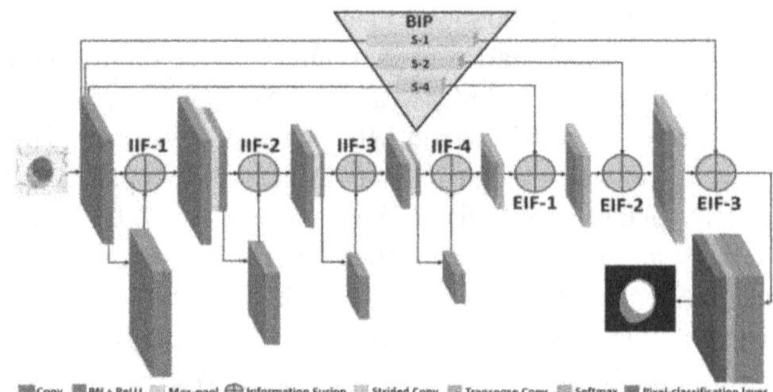

Figure 21.4. MIF-Net.

Source: Author.

Table 21.1. Comparison of various network models

Network Model	*SENet*	*FPN*	*MIF-Net*
No. of max channels	2048	512	256
No. of downsampling operations	5	5	3
Max stride value	2	2	4
No. of convolution layers at every stage, except the first stage and up-convolution layers	3	4	2
Fusion point	1	3	7
Multi-scale architecture base	Based on ResNet-101	FPN+ResNet	Developed from scratch
Division of spatial information at the initial layer	No	No	Yes
Architecture type	Region proposal based detection	Region proposal based detection	Segmentation
No. of parameters	44.5. M	-	2.67. M

Source: Author.

3. MIF-NET Segmentation Results

On Dataset-1, procedures have been conducted to jointly divide the nucleus and cytoplasm. Because Dataset-1 has a fast staining circumstance, there is little background difference between the cytoplasm and the surrounding tissue, making categorization difficult. As demonstrated in Figure 21.5, the nucleus boundary requires blurry in certain instances; therefore, precise boundary prediction requires fascinating with these WBC nuclei. The network maintains boundary information with the aid of MIF-Net's multi-scale information fusion method. With these obstacles, MIF-Net uses its efficient design to demonstrate cutting-edge performance in segmentation. Staining circumstances, contrast limitations, and imprecise object boundaries are the causes for inadequate categorization outcomes.

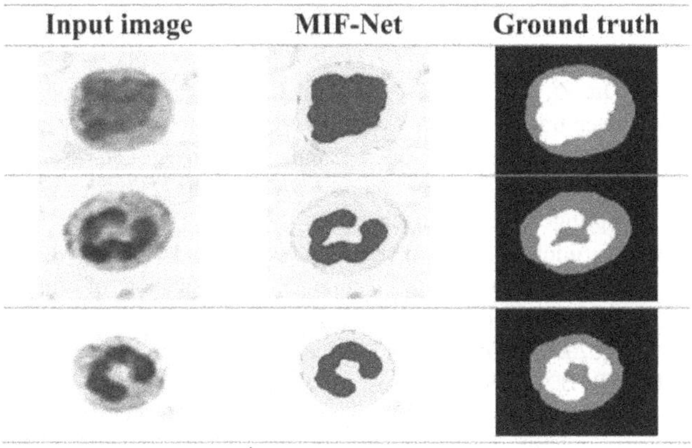

Input image **MIF-Net** **Ground truth**

Figure 21.5. MIF-Net Segmentation outcomes.

Source: Author.

Precision (P), misclassification error (M), dice coefficient (D), mean intersection over union (mIoU), false-positive rate (FPR), and false-negative rate (FNR) are some of the assessment measures utilized for MIF-Net [16, 17].

$$Precision = \frac{|W_f \cap P_f|}{|P_f|} \tag{1}$$

$$Dice\ score = \frac{2|W_f \cap P_f|}{|W_f| + |P_f|} \tag{2}$$

$$mIoU = \frac{1}{2}\left(\frac{|W_b \cap P_b|}{|W_b \cup P_b|} + \frac{|W_f \cap P_f|}{|W_f \cup P_f|}\right) \tag{3}$$

$$ME = 1 - \frac{|W_b \cap P_b| + |W_f \cap P_f|}{|W_f| + |W_b|} \tag{4}$$

$$FPR = \frac{|W_b \cap P_f|}{|W_b|} \tag{5}$$

$$FNR = \frac{|W_f \cap P_b|}{|W_f|} \tag{6}$$

Tables 21.2 and 21.3 compares the performance metrics of various methods on segmenting cytoplasm and nucleus, respectively.

4. Conclusion

Methods for segmenting blood images for leukemia are being presented using conventional clustering methods with automatic thresholding approaches. Conventional clustering techniques, on the other hand, are highly susceptible to the original cluster positions; erroneous centering values might lead to a falsely positive cancer diagnosis. How-ever, in situations wherein conventional optimization methods are impractical, Nature-Inspired Optimization Techniques (NIOA) are probabilistic search techniques that identify the

Table 21.2. Segmentation of cytoplasm using MIF-Net

Methods	*DC*	*PRC*	*ME*	*mIoU*	*FPR*	*FNR*
Zheng	92.5	86.53	5.2	-	7.8	0.03
Zhou	90.34	88.71	6.3	-	6.9	5.4
U-Net	97.32	97.43	1.59	-	1.18	2.7
Chen	97.37	98.03	1.58	-	0.96	3.23
Proposed	98.93	98.98	0.66	97.91	0.52	1.05

Source: Author.

Table 21.3. Segmentation of nucleus using MIF-Net

Methods	*DC*	*PRC*	*ME*	*mIoU*	*FPR*	*FNR*
Sarrafzadeh	70.88	-	-	-	-	-
Vincent	48.33	-	-	-	-	-
Vogada	87.68	-	-	-	-	-
Makem	90.79	91.01	2.7	-	-	-
Banik	91.0	87.63	-	-	-	-
Proposed	95.84	93.81	0.997	92.18	0.85	1.76

Source: Author.

best solution for intricate multimodal functions. An intriguing substitute for accurate blood cell segmentation is provided by NIOA techniques, as blood image segmentation is thought to be a difficult computational operation. For WBC segmentation, we suggested a multi-scale information fusion network (MIF-Net), which produces excellent segmentation out-comes. Incorporating data from different imaging techniques can provide complementary information about cell morphology and internal structures. Research could focus on developing algorithms that fuse these diverse data types to improve segmentation accuracy and robustness.

5. Acknowledgement

The authors express their gratitude to the International Research Centre of Kalasalingam Academy of Research and Education, Tamil Nadu, India, for providing access to the computational facilities available at the Centre for Biomedical Research and Diagnostic Techniques Development.

References

[1] Hecht, F., Zocchi, M., Alimohammadi, F., & Harris, I. S. (2024). Regulation of antioxidants in cancer. *Molecular Cell*, *84*(1), 23–33.

[2] Babu, K. G., Prabhash, K., Chaturvedi, P., Kuriakose, M., Birur, P., Anand, A. K., Kaushal, A., Mahajan, A., Syiemlieh, J., Singhal, M., & Gairola, M. (2024). Indian clinical practice consensus guidelines for the management of hypopharyngeal cancer: Update 2022. *Cancer Research, Statistics, and Treatment*, *7*(Suppl 1), S17–S21.

[3] Wang, Q., & Wei, X. (2024). Research progress on the use of metformin in leukemia treatment. *Current Treatment Options in Oncology*, 1–17.

[4] Tagliaferri, A. R., Melki, G., & Baddoura, W. (2023). Chronic lymphocytic leukemia causing gastric ulcer perforation: A case presentation and literature review. *Cureus*, *15*(3).

[5] Zahariev, N., Draganova, M., Zagorchev, P., & Pilicheva, B. (2023). Casein-based nanoparticles: A potential tool for the delivery of daunorubicin in acute lymphocytic leukemia. *Pharmaceutics*, *15*(2), 471.

[6] Elrefaie, R. M., Mohamed, M. A., Marzouk, E. A., & Ata, M. M. (2024). A robust classification of acute lymphocytic leukemia-based microscopic images with supervised Hilbert-Huang transform. *Microscopy Research and Technique*, *87*(2), 191–204.

[7] Nisha, A. V., Rajasekaran, M. P., Kottaimalai, R., Vishnuvarthanan, G., Arunprasath, T., & Muneeswaran, V. (2023). Hybrid D-OCapNet: Automated multi-class Alzheimer's disease classification in brain MRI using hybrid dense optimal capsule network. *International Journal of Pattern Recognition and Artificial Intelligence*, *37*(15), 2356025.

[8] Shadman, M. (2023). Diagnosis and treatment of chronic lymphocytic leukemia: A review. *JAMA*, *329*(11), 918–932.

[9] Raina, R., Gondhi, N. K., Chaahat, Singh, D., Kaur, M., & Lee, H. N. (2023). A systematic review on acute leukemia detection using deep learning techniques. *Archives of Computational Methods in Engineering*, *30*(1), 251–270.

[10] Khalid, A., Ahmed, M., & Hasnain, S. (2023). Biochemical and hematologic profiles in B-cell acute lymphoblastic leukemia children. *Journal of Pediatric Hematology/Oncology*, *45*(7), e867–e872.

[11] Mulya, R. F., Utami, E., & Ariatmanto, D. (2023). Classification of acute lymphoblastic leukemia based on white blood cell images using inceptionV3 model. *Jurnal RESTI (Rekayasa Sistem dan Teknologi Informasi)*, *7*(4), 947–952.

[12] Du, Z., Xia, X., Fang, M., Yu, L., & Li, J. (2023, July). A deep transfer fusion model for recognition of acute lymphoblastic leukemia with few samples. In *International Conference on Intelligent Computing* (pp. 710–721). Singapore: Springer Nature Singapore.

[13] Curtidor, A., Kussul, E., Baydyk, T., & Mammadova, M. (2023). Analysis and automated classification of images of blood cells to diagnose acute lymphoblastic leukemia. *EUREKA: Physics and Engineering*, (5), 177–190.

[14] PR, B., & BK, A. (2023). Automated biomedical image classification using multi-scale dense dilated semi-supervised u-net with cnn architecture. *Multimedia Tools and Applications*, 1–33.

[15] Masoudi, B. (2023). VKCS: A pre-trained deep network with attention mechanism to diagnose acute lymphoblastic leukemia. *Multimedia Tools and Applications*, *82*(12), 18967–18983.

[16] Amiya, G., Murugan, P. R., Ramaraj, K., Govindaraj, V., Vasudevan, M., Thirumurugan, M., Abdullah, S. S., & Thiyagarajan, A. (2024). LMGU-NET: Methodological intervention for prediction of bone health for clinical recommendations. *The Journal of Supercomputing*, *80*(11), 15636–15663.

[17] Amiya, G., Murugan, P. R., Ramaraj, K., Govindaraj, V., Vasudevan, M., Thirumurugan, M., Zhang, Y. D., Abdullah, S. S., & Thiyagarajan, A. (2024). Expeditious detection and segmentation of bone mass variation in DEXA images using the hybrid GLCM-AlexNet approach. *Soft Computing*, *28*(19), 11633–11646.

22 Augmented Reality (AR) and Virtual Reality (VR) for educational applications

Manish Nandy[1,a] *and Lalnunthari*[2,b]

[1]Assistant Professor, Department of CS & IT, Kalinga University, Raipur, India
[2]Research Scholar, Department of CS & IT, Kalinga University, Raipur, India

Abstract: AR and VR have transformed numerous industries, but education has been most affected. Augmented reality (AR) employs smartphones and AR glasses to overlay digital information on the actual environment, whereas VR uses equipment to transport users to imaginary worlds. These technologies have evolved from simulation and early computer graphics research thanks to processors, screens, and sensors. AR and VR implementation requires complex hardware and software like headsets, gaming engines, and developer frameworks. Successful curriculum integration requires planning, instructor training, and educational goal alignment. Empirical indicators for augmented and virtual reality include student engagement, learning results, and feedback. Augmented and virtual reality can improve education and introduce new methods despite pricing and gadget issues. This paper examines the parts, technology, and processes of designing, deploying, and assessing AR and VR in the classroom with the goal of revolutionising it.

Keywords: Augmented Reality, virtual reality, education, immersive learning, interactive experiences, AR headsets

1. Introduction

1.1. Background information on AR and VR

AR and VR have advanced significantly in recent years and are game-changing. Augmented reality (AR) enhances immersion by superimposing digital data on the user's surroundings. Smartphones, tablets, and augmented reality glasses integrate digital information with the user's environment [1]. Virtual reality (VR) headsets transport users to an artificial world using computer-generated graphics. AR and VR began with computer graphics and simulation experiments. The first head-mounted display introduced virtual reality in the 1960s, while computer vision and wearable electronics popularized augmented reality in the 1990s. Strong CPUs, high-resolution displays, and smart sensors have boosted augmented and virtual reality development and accessibility. These developments make AR and VR more useful for education and other purposes [4].

1.2. Importance of AR and VR in education

AR/VR's interactive and immersive experiences enable new learning methods. These technologies may transform education by delivering pupils real-world experiences that conventional techniques cannot. AR in the classroom helps students grasp course material by superimposing digital data and interactive elements onto physical environments. Augmented reality lets pupils turn a 2D textbook image into a 3D model they can rotate and examine from all angles [9]. VR puts pupils in completely simulated environments where they can explore historical sites or conduct complex scientific experiments that are unachievable in real life [12]. AR and VR in the classroom can boost active learning, comprehension, and personalisation. These technological advances boost classroom engagement and motivation, improving educational outcomes [2] (Figure 22.1).

Figure 22.1. Augmented reality (AR) and virtual reality (VR) for educational applications.

Source: Author.

[a]ku.manishnandy@kalingauniversity.ac.in, [b]lalnunthari.rani.SINGH@kalingauniversity.ac.in

DOI: 10.1201/9781003675259-22

2. Methodology

2.1. Components and technologies used

AR and VR require specific hardware to work well. Users can use smartphones, tablets, or camera-equipped eyewear for augmented reality. Microsoft HoloLens and Magic Leap are cutting-edge AR goggles. However, VR requires HTC Vive, PlayStation VR, and Oculus Rift headsets. These headsets' high-definition screens, motion detectors, and audio systems let users experience virtual worlds [3]. Tracking sensors, haptic feedback devices, and motion controllers improve virtual and augmented reality engagement [6].

2.2. Software platforms and tools for AR and VR development

To make augmented and virtual reality, you need a lot of different software platforms and production tools. ARKit from Apple, Vuforia from Google, and ARCore from Apple are all app development frameworks that let users put digital material on top of the real world. According to [10], these systems can track, recognize, and connect with things in the real world. It is possible to make complex simulations and settings in three dimensions with virtual reality (VR) game engines like Unity and Unreal Engine. With these tools, you can script, organize assets, and render in a way that makes the experience feel real.

3. AR and VR Techniques

3.1. Augmented reality

Augmented reality apps let students engage with digital replicas of tangible objects in class. An augmented reality programme can turn textbook photographs into interactive 3D models or comment them with pop-up windows. Augmented reality interactive simulations like virtual dissections and dynamic scientific process visualisations can give students hands-on learning [5].

3.2. Virtual reality

Educational institutions can employ VR to create interactive, realistic models and experiences. Education is using VR simulations of complex topics. Online classroom discussions and activities including virtual social studies and scientific labs allow students to participate remotely. These applications offer a secure and engaging option for students who struggle in traditional schools.

3.3. Content development

Educational content for augmented and virtual reality requires ideation, design, development, and testing. Establishing learning goals before building interactive features to achieve them is typical [11]. Developer platforms and tools help content makers create interactive features, 3D models, and animations. After development, test it for usability, learning effectiveness, and user satisfaction. Virtual and augmented reality content makers employ gaming engines, frameworks, and 3D modelling software. These tools help generate, develop, and improve instructional content for engaging learning.

3.4. Implementation and integration

AR and VR in schools require strategic planning and connection with learning objectives. Strategies include finding AR/VR-relevant topic areas, establishing courses that incorporate these technologies, and training teachers to use them. Integrating AR and VR may require new or revised resources [7]. Successful implementation requires proper software and hardware setup and maintenance. Set up AR and VR equipment, test it with instructional apps, and fix any errors. We must consider device availability, cost, and space arrangement to make augmented and virtual reality tools more helpful in the classroom.

4. Evaluation and Assessment

AR and VR in the classroom only work if kids are engaged, successful, and happy. Researchers can collect instructor and student feedback using interviews, surveys, and observational studies. Assessments can also show how well AR/VR devices teach and simplify complex concepts. Measurements like exam scores and student accomplishment evaluate augmented and virtual reality educational apps [8]. Student comments and participation are qualitative metrics. Monitoring use trends, engagement rates, and performance enhancements can help explain how augmented and virtual reality affects education.

5. Uses

- Simulations, virtual field trips, and other active learning environments
- Tools for understanding complex topics (3D models, immersive experimentation)
- Individualized instruction and adaptable settings
- Digital collaboration and information exchange

6. Working

6.1. Detailed explanation of system operation

Schools are adding AR and VR to their courses to make studying more engaging and immersive. Augmented reality (AR) overlays digital information on real-world objects to give users fresh perspectives. This may include real-world or interactive element information. Virtual reality (VR) replaces the physical world with an immersive digital one, offering pupils access to simulations and virtual worlds they couldn't access otherwise. Interactive simulations, virtual field trips, and 3D modelling help students see and understand complex subjects.

6.2. Integration and interaction of components

AR and VR systems require hardware-software collaboration. Hardware includes input devices, sensors, augmented reality and virtual reality headgear. Using specialized software, developers create dynamic and engaging educational experiences with augmented or virtual reality content. These technologies interact with classroom activities and lesson preparations via educational apps. A VR programme may replicate a bygone age, while an AR app could superimpose scientific experiment data on a real-life object.

6.3. System architecture

There are a few important parts that must be present in school AR and VR systems:

- The physical parts, such as AR/VR glasses, sensors, controls, motion trackers, and other input devices.
- When it comes to software, systems and tools like Unity and Unreal Engine let us make and arrange AR/VR content.
- Content Delivery: Ways to share and get to augmented reality and virtual reality learning materials (for example, VR platforms and computer apps).
- This part talks about how augmented and virtual reality material can be added to current school systems and programmes (Figure 22.2).

7. Algorithm

- Start
- *Data Collection:* Gather raw data from various sources (e.g., network traffic, system logs).
- *Data Preprocessing:* Clean and organize the data, Handle missing values and normalize data.

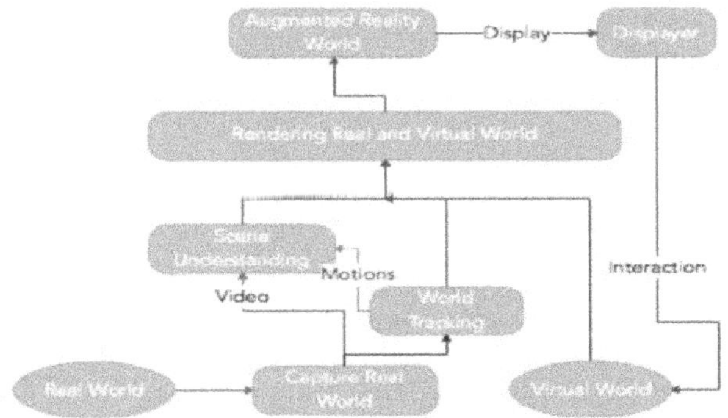

Figure 22.2. System architecture augmented reality (AR) and virtual reality (VR) for educational applications.

Source: Author.

- *Feature Extraction:* Identify and extract relevant features from the data, Transform raw data into a format suitable for model training.
- *Model Training:* Apply machine learning or deep learning algorithms to train the model using the preprocessed data.
- *Threat Detection:* Use the trained model to identify potential threats or anomalies in real-time data.
- *Response Actions:* Implement predefined response actions or alerts based on detected threats.
- *Continuous Learning and Improvement:* Update the model with new data to improve accuracy, Refine algorithms and techniques based on feedback and new threats.
- End

8. Conclusion

Artificial intelligence enhances cybersecurity threat detection by automating detection and response. AI-driven algorithms and deep learning models provide real-time analysis and response to emerging hazards, surpassing previous methods. Though AI improves efficiency and accuracy, model training and attack vectors remain complex. Addressing these issues and strengthening cybersecurity are crucial, as is AI research. This will protect against advanced cyberattacks.

References

[1] Al-Ansi, A. M., Jaboob, M., Garad, A., & Al-Ansi, A. (2023). Analyzing augmented reality (AR) and virtual reality (VR) recent development in education. *Social Sciences & Humanities Open*, *8*(1), 100532.

[2] Sumithra, S., & Sakshi, S. (2024). Exploring the factors influencing usage behavior of the digital library remote access (DLRA) facility in a private higher education institution in India. *Indian Journal of Information Sources and Services*, *14*(1), 78–84.

[3] Sharma, T., Singh, S., Sharma, S., Sharma, A., Shukla, A. K., Li, C., ... & Eldin, E. M. T. (2022). Studies on the Utilization of Marble Dust, Bagasse Ash, and Paddy Straw Wastes to Improve the Mechanical Characteristics of Unfired Soil Blocks. *Sustainability*, *14*(21), 14522.

[4] Arasu, R., Chitra, B., Anantha, R. A., Rajani, B., Stephen, A. L., & Priya, S. (2024). An e-learning tools acceptance system for higher education institutions in developing countries. *Journal of Internet Services and Information Security*, *14*(3), 371–379.

[5] Sharma, A., & Dwivedi, V. K. (2020, December). Effect of spindle speed, feed rate and cooling medium on the burr structure of aluminium through milling. In *IOP conference series: materials science and engineering* (Vol. 998, No. 1, p. 012028). IOP Publishing.

[6] Udayakumar, R., Muhammad, A. K., Sugumar, R., & Elankavi, R. (2023). Assessing learning behaviors using Gaussian Hybrid Fuzzy Clustering (GHFC) in Special Education Classrooms. *Journal of Wireless Mobile Networks,*

Ubiquitous Computing, and Dependable Applications, *14*(1), 118–125.

[7] Sharma, A., Islam, A., Sharma, K., & Singh, P. K. (2021). Optimization techniques to optimize the milling operation with different parameters for composite of AA 3105. *Materials Today: Proceedings*, *43*, 224–230.

[8] AlGerafi, M. A., Zhou, Y., Oubibi, M., & Wijaya, T. T. (2023). Unlocking the potential: A comprehensive evaluation of augmented reality and virtual reality in education. *Electronics*, *12*(18), 3953.

[9] Huang, K. T., Ball, C., Francis, J., Ratan, R., Boumis, J., & Fordham, J. (2019). Augmented versus virtual reality in education: An exploratory study examining science knowledge retention when using augmented reality/virtual reality mobile applications. *Cyberpsychology, Behavior, and Social Networking*, *22*(2), 105–110

[10] Mekacher, L. (2019). Augmented Reality (AR) and Virtual Reality (VR): The future of interactive vocational education and training for people with handicap. *International Journal of Teaching, Education and Learning*, *3*(1), 1–12.

[11] Huang, T. K., Yang, C. H., Hsieh, Y. H., Wang, J. C., & Hung, C. C. (2018). Augmented reality (AR) and virtual reality (VR) applied in dentistry. *The Kaohsiung Journal of Medical Sciences*, *34*(4), 243–248.

[12] Llopiz-Guerra, K., Daline, U. R., Ronald, M. H., Valia, L. V. M., Jadira, D. R. J. N., & Karla, R. S. (2024). Importance of environmental education in the context of natural sustainability. *Natural and Engineering Sciences*, *9*(1), 57–71.

23 Mammography images: Visual Geometric Group (VGG) model and customization for breast cancer detection

M. Shanmugapriya[1,a] *and J. Pradeepkandhasamy*[2,b]

[1]Research Scholar, Department of Computer Applications, Kalasalingam Academy of Research and Education, Anand Nagar Krishnankoil, Tamil Nadu, India
[2]Assistant Professor, Department of Computer Applications, Kalasalingam Academy of Research and Education, Anand Nagar Krishnankoil, Tamil Nadu, India

Abstract: Breast Cancer is the most deadly disease for women worldwide. Breast Cancers fall into two categories: benign, which causes less harm, and malignant, which causes great danger. Early detection of breast cancer can reduce mortality and improve recovery. Researchers worldwide are currently focusing on creating Medical imaging is used to detect Breast Cancer. Deep learning methods have garnered significant interest from researchers in the medical imaging domain owing to their rapid advancements. In this study, mammography images were employed to identify Breast Cancer utilizing the Visual Geometric Group (VGG) technique. VGG, a collection of Convolutional Neural Network (CNN) architectures, is renowned for its profoundness and straightforwardness. VGG models have exhibited noteworthy execution across a scope of PC vision errands, like picture order, object recognition, and semantic division. The VGG model has been tailored specifically this study aims to detect Breast Cancer. A new classification layer is incorporated to modify the architecture of a VGG model. The data sets utilized for our experiment are sourced from two distinct repositories, namely CBIS-DDSM and INbreast are two databases for breast imaging. The customized VGG model achieves an accuracy of 88%. The computer-assisted analysis assists radiologists in improving the accuracy of breast cancer diagnoses.

Keywords: CNN, Benign, malignant, breast cancer, VGG

1. Introduction

Cancer affects several organs in the body, including the breasts, blood, pancreas, and lungs, making it more than just a common disease. While these forms of cancer share certain similarities, they vary in terms of their progression and spread. In certain aspects, however, their approaches to creation and distribution are different. A World Health Organization (WHO) [1] study conducted by the International Agency for Research on Cancer, which highlight breast cancer as the second leading cause of mortality among women, the Pink Blossom stands as a beacon of hope and prevention. The WHO predicts that 19.3 million cases of Breast Cancer will be diagnosed by 2025 [1]. The abnormal growth and infiltration of neighboring tissues by tumor cells is how cancer arises in the human body. It is common practice to categorize tumors as benign or malignant. It is thought that cells in malignant tumors are carcinogenic, whereas cells of harmless growths are not known to cause cancer. Only inside that particular body may the cells of the benign tumor multiply. Then again, cancerous growth cells are incapable of invading adjacent structures. Malignant, on the other hand, can grow uncontrollably, puncture surrounding tissue, which then, at that point, spread to other bodily regions [2]. Many Breast Cancer is diagnosed using screening methods, but mammography is by a wide margin the most reliable. Mammography utilizes a few viewpoints, for example, the craniocaudal view (CC view) and mediolateral oblique (MLO), to assist with interpreting the available bosom irregularities. To examine the markers of breast lesions, radiologists use either the MLO or CC views, like lumps and microcalcifications, designed to distinguish with unparalleled accuracy between benign, normal, and malignant breast conditions, this innovative tool is your definitive ally in women's health [3]. It requires a great deal of work and expertise from a radiologist who has extensive experience and training in interpreting mammogram images [4]. The necessity for pc supported demonstrative and recognition computer aided design advancements, which respond to these issues, and automate medical image processing, has grown [5]. An uncontrolled alteration and division of breast tissue cells results in a tumor, which is a type of sickness known as Breast Cancer. Lobules connect the lobules to the nipple, and this is where most Breast Cancers start. Designed for both healthcare professionals and pro-active individuals committed to health vigilance, this kit offers unparalleled accuracy in identifying potential breast cancer indicators. Harnessing the power of

[a]shanmugapriyamurugesan@klu.ac.in, [b]j.pradeepkandasamy@klu.ac.in

DOI: 10.1201/9781003675259-23

state-of-the-art diagnostic methods including ultrasound, magnetic resonance imaging (MRI), and X-ray compatibility, our product transcends traditional detection methods to provide a comprehensive, non-invasive assessment: India versus the World [6]. A common deep neural network type for evaluating visual imagery is the Convolutional Neural Network (CNN). For tasks like segmentation, object detection, and picture classification, it is especially effective. The capacity of CNNs to automatically and adaptively deduce spatial feature hierarchies from unprocessed data is one of its primary characteristics [7]. Convolutional, pooling, and fully linked layers are used in tandem to achieve this. A set of filters (kernels) are applied to the input picture using convolutional layers. Every filter carries out convolution, which creates feature maps by dragging the filter window across the input image and calculating dot products. Combines a symphony of edges, textures, and patterns, are captured by this method. The convolutional layers produce feature maps that are sampled down by pooling layers. Typical pooling strategies incorporate max pooling and normal pooling, which lessen the component guides' spatial aspects while preserving important data. The last component maps are leveled into a one-layered vector and go through at least one completely associated layer subsequent to going through numerous convolutional and pooling layers. These layers use their ability to recognize intricate patterns in the feature maps to carry out tasks like as classification or regression [8]. A CNN architecture from the esteemed Visual Geometry Group at the University of Oxford's (VGG). Its 16 layers—13 fully linked layers and 13 convolutional layers—are what give it its depth. Well-known for its efficiency and simplicity, VGG-16, the pinnacle of technological innovation in object recognition and image categorization. Engineered to perfection, this state-of-the-art algorithm transcends traditional boundaries, delivering unparalleled accuracy and efficiency across myriad visual applications. In the design of the model, layers with progressively increasing depth are stacked convolutional layers, then max-pooling layers. The model can learn complex hierarchical representations of visual data because to this design, which produces reliable and accurate predictions [9]. VGG is still a popular option for many deep learning applications because of its great performance and versatility, even if it is simpler than more current architectures [10]. The design of this paper is as per the following: Section 2 unfolds as a comprehensive tapestry of academic brilliance, offering an exhaustive evaluation of the pertinent literature that enriches this discourse. The dataset and the recommended procedures are explained in Segment 3. The trial discoveries of the recommended systems are made sense of in Segment 4. In Segment 5, the paper's last area is conveyed.

2. Related Work

Research in deep learning has advanced significantly in the last ten years. Neither the subject of healthcare nor Mirzoeva et al. [11] are an exception. In the study of Breast Cancer, Hamidinekoo et al. [12] looked at and effectively used several deep architectures, and the results were also favourable. Many researchers continue to study computer-aided diagnostics to offer swift, accurate, and insightful analysis. A hybrid CNN methodology is given, in which handcrafted administered research approaches are utilized to show picture-based highlights by Arevalo et al. [13]. For the classification of mammographic malignancies [14]. The pre-trained AlexNet model was subjected to transfer learning without any fine-tuning, and the back-end classification procedure made use of the Support Vector Machine (SVM) method. Standards for medical care could rise as a result of recent advances in deep learning technologies [15]. Adjusted ImageNet, a CNN that has already been trained, to distinguish between microcalcification and masses. For determining the type of Bosom Malignant growth, the Bosom Imaging-Revealing and Information Framework (BI-RADS) score is quite useful. The authors of the study distinguished between different types of Breast Cancer that are the BI-RADS score. Lévy and Arzav [31] most advanced and pre-eminent models in the field: Alexnet and Googlenet. Experience unparalleled accuracy and efficiency in classifying mammography images, transforming diagnosis with precision and speed. When comparing the two networks, AlexNet produced the best outcomes. The Breast Cancer Classification Network Improvement Convolutional Neural Network (CNNI-BCC) is a novel network developed by Ting et al. [16]. To do auto-feature extraction at that point, utilizing a neural network classifier. A few deep learning models have been put forth in the writing. The CNN demonstration was used by the researchers Duggento et al. [17] to classify and categorize benign and malignant damage present in computerized mammography images. Ragab et al. [18] used AlexNet and Profound Convolution Neural Arrange to classify and extract highlights from images from mammograms in the DDSM dataset. Ting et al. [16] used yet another deep CNN to classify the MIAS collection's BC-lesion. Another CNN-based significant learning program, BreastNet, beats AlexNet, VGG-16, and VGG-19, as per its designers [19]. Additionally, the developers Wang et al. [20] have produced two-sided leftover GANs (BR-GANs), which, for mammography fragmentation for the INbreast dataset, are based on the cycle GAN thinking. To identify and classify Breast Cancers in the DDSM dataset, Johnson et al. [30] YOLO's unparalleled speed and AlexNet's groundbreaking image classification accuracy, reimagined for the modern age, are being used by Al [21]. For multiclass category of the INBREAST dataset. Owing to its supremacy in the dissemination of healing images, U-Net enjoys great respect. In response to the division challenge, Zeiser et al. [22] linked an engineering system in light of U-Net to an assortment of openly and secretly available resources. Their work revolves around the division of masses on mammography images to classify Breast Cancers in the DDSM dataset. Johnson et al. [30] have proposed modified YOLOv5. Al Antari et al. [21] used multiclass classification using YOLO and AlexNet, a moment CNN-based classifier, for the INbreast dataset. Because of its superiority in the distribution of healing images, U-Net is highly regarded. In response to the division challenge, Zeiser et al. [22] linked an engineering system based on U-Net to a variety of publicly and privately accessible resources. Their work revolves around the division of masses on mammography images. Additionally, developers Abdelhafiz et al. [23] have demonstrated the Vanilla U-Net, a third form of the U-Net, for portioning mass in mammography pictures from three unmistakable openly available datasets. A contingent casing of the U-Net-based cGAN-UNET and the GAN (cGAN) show was proposed by Safari et al. [24] for breast thickness division. Following that, CNN was linked to classify the INBREAST dataset. Rahman et al. [25] demonstration of the effectiveness of exchange learning for Breast Cancer growth order. Mammography figure are resized for example, 224 × 224 pixels normalized and augmented rotation flipping zooming to ensure consistency and improve classical performance a VGG prototypical pre-trained on imagenet is fine-tuned on a breast cancer-specific dataset by replacing its final fully connected layers to classify benign vs malignant cases VGGs convolutional layers extract detailed structures like edges textures and arrangements indicative of tumors these features are fed into fully connected layers or classifiers for example, svms for prediction using softmax for multi-class or sigmoid for binary organization the model is proficient on labeled mammography pictures with cross-validation and data amplification to enhance generalization and reduce overfitting evaluated with metrics like accuracy sensitivity specificity and auc-roc predictions are refined using thresholding and heatmap generation for example, grad-cam to highlight cancer-indicative regions aiding radiologists by providing a second opinion highlighting suspicious areas and potentially reducing manual review time though it should be combined with domain-specific knowledge and other diagnostic tools for best accuracy. The profundity of the VGG model like VGG16 or VGG19 altogether influences its exhibition in bosom disease discovery more profound models like VGG19 can learn more mind boggling and various leveled highlights catching unpretentious examples essential for recognizing harmless and threatening tissues while more profound models have higher limit and can accomplish better execution they additionally require more computational assets and longer preparation times appropriate regularization and information increase are fundamental to forestall overfitting at the point when adjusted with pre-prepared loads from enormous datasets like imagenet more profound VGG models can accomplish higher responsiveness and particularity making them integral assets for bosom disease recognition gave adequate computational assets and different datasets are accessible.

3. Methodology

This segment provides an explanation of the databases utilized in this project. In addition to the pre-trained deep convolutional networks and their preset parameters, it offers a succinct overview of the proposed model.

3.1. Data description

DDSM-CBIS: This is a segment from the DDSM Lee et al. [26] database, which includes 6775 studies in total. The mammographic pictures chosen by qualified and experienced radiologists from DDSM are addressed in a refreshed and normalized variant called CBIS DDSM. Once the images undergo lossless decompression, they are converted to the DICOM format. The sectioned district of interest (return for money invested) for preparing information and pathologic diagnosis details are also included in the database. The data can be subdivided into subclasses according to malignant and benign tumors and further classified according to the various types of anomalies, such as calcification and mass.

INBreast: The INbreast dataset consists of 410 distinct computerized mammogram images, which were deciphered by experienced radiologists. Our meticulously designed tool harnesses the power of the BI-RADS scoring system, providing unparalleled insights into breast health. With scores ranging from 0 to 6, our system is the gold standard in identifying deviations from normal breast tissue, ensuring every detail is captured with exceptional accuracy. Whether it's confirming a benign anomaly or identifying a potential malignancy, our system's precision is your peace of mind. Trust in a system where each score narrates a vital part of your health journey, empowering healthcare professionals to offer informed guidance and support. It is not publicly available [27].

The mix of manufactured consciousness structures in medical services and medication raises moral worries these worries are connected with fears of computerization as well as to the innate qualities implanted in medical services and clinical practices these qualities are well established in different codes systems and good practices and strategies molding the ethical scene of processer intellectual mix in these pitches Smallman [28] and Gupta et al. [29] deep learning dl has changed picture investigation joining progressed picture handling with productive regulated classifiers this paper presents a convolutional brain organization CNN scheme for precise categorization of malignant growth and typical cases in mammograms bosom malignant growth a pervasive and lethal illness is much of the time analyzed utilizing mammograms dl-based frameworks improve examination recognizing even little pathologies and supporting appraisal and treatment the future model assesses changed element knowledge approaches to work on the adequacy of dl models in bosom malignant development finding operating CNN Moutik et al. [32] and Mudeng et al. [33] conveyed realizing which empowers multi-institutional coordinated efforts without sharing confidential patient information offers a promising arrangement it has exhibited progress in different grouping and division undertakings notwithstanding difficulties for example information circulation shifts security spillage and decency continue as examination advances specialized headways might change cooperative practices in radiology and other clinical fields clients from both specialized and clinical foundations should grasp both the advantages and dangers of disseminated learning techniques [34].

3.2. Proposed methodology

Deep understanding the main commitment of designs is their ability to independently extract properties from low-level to high-level domains. When identifying individual attributes in images, CNNs are the most effective models. CNN might pick up on highlight portrayal organically as limited to professionally produced highlights. Different prerequisites have been discussed in broad examination in this investigate. Prior to being dealt with the pictures from the mammogram were all changed over from the dicom configuration to the png versatile association designs design commencing that point forward the pixel values from 0 to 1 are consistent to certify that the discoveries of the examination are not impacted by the higher pixel values. The designations for the double classification problem must then be changed such that '0' corresponds to 'benign' and '1' to 'malignant'. At that point the mammography pictures upon resizing both datasets to 224 × 224 image measure; part the preparing information are interested in 'training' and 'validation' sub-groups; construct Keras generators for preparing and approval information. A exceptionally little CNN with as it were two convolutional layers has been utilized for recreation. The proposed strategy includes testing assortment of previously trained CNN systems utilizing an adjusting procedure, counting VGG as appeared in the Figure 23.2. The first step is to determine the model's assumed measurement. A show with too many spaces might advance step by step and overset, while a coordinate that is too unobtrusive cannot be noticed well. A useful method for selecting the appropriate measure is to begin with a small, naive demonstration and dynamically expand its measurement until it begins to overfit learning is used. At this comment, the show is sufficiently flexible to accommodate the preparation information and may be able to generalize to additional information with the appropriate preparation.

By introducing new layers, modifying those ones that are already there using regularization strategies, or changing the hyperparameters, the demonstration can, of course, be improved upon to produce superior displays. Certainly, the demonstration can be enhanced further by incorporating new layers, adjusting enhance the current ones using regularization methods., or tweaking the hyper-parameters. The research encompasses two separate methodologies. At the core of this architectural wonder lie two meticulously crafted convolutional layers, each working in harmony to extract intricate patterns and details, providing exceptional clarity and insight. A fully connected layer is succeeded by a single neuron with sigmoid activation for binary classification, resulting in the output. In this scenario, paired cross-entropy is used as the misfortune capability. RMSprop, the optimizer, is well-known for its effectiveness.

Monitoring the loss on the validation set throughout training helps save the calculate model parameters at the point of minimal loss. The main goal is to reduce the overfitting observed in the previous section, allowing the model to learn a broader range of weights. The second approach is similar to the first but includes dropout layer following the final block that is fully connected. Dropout is a regularization technique that helps prevent overfitting by randomly ignoring neurons during training. A dropout rate of 0.5 is found to be effective. Data Increase is also employed to minimize overfitting, it is advisable to first standardize the model and then make necessary adjustments to the architecture.

3.3. Transfer learning

Typically, training a CNN requires a substantial dataset, consisting of thousands or even millions of samples. However, Medical image is an important field of study analysis, it is often difficult to obtain such large labeled datasets due to time and resource constraints. Consequently, using a CNN trained from scratch becomes a challenge. When the training dataset is limited, CNN-based methods tend to overfit and fail to extract high-quality image features. To address this issue, transfer learning comes into play CNN trained on labeled image dataset to learn image properties. Subsequently, the pre-trained CNN can be utilized to extract comparable features from a smaller dataset. Transfer learning is effective in image-processing and clinical studies.

4. Experimental Results

The authors utilized two distinct methods to detect breast cancer from mammography images. The primary goal was to assess the effectiveness of a pre-trained system deep CNN model through comparison and analysis. By applying the suggested approach on the CBIS-DDSM and INbreast datasets, we assessed their effectiveness in comparison to the methods currently in use. These models were made utilizing the Tensor Stream backend and the Keras profound learning system given by Google COLAB.

4.1. Performance indicators to assess the classification

The evaluation of a classification model relies on its ability to accurately predict test records. True positive (TP), false

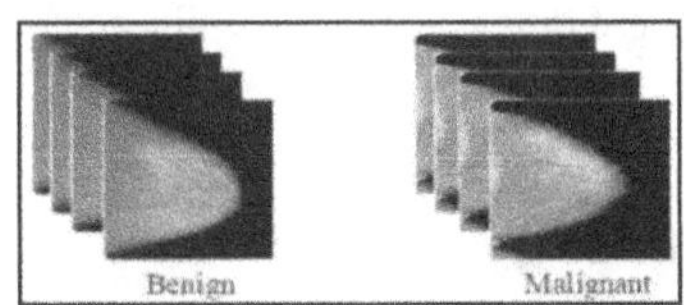

Figure 23.1. Sample mammogram images.

Source: Author.

positive (FP), true negative (TN), and false negative (FN) are all components of the confusion matrix measurements, provides valuable insights into the model's performance across all classes.

By analyzing these metrics, we can effectively compare different category systems. Figure 23.3 outlines the key performance indicators derived from the disarray framework, which includes exactness, accuracy, review, and F1 score, is a crucial tool for understanding and addressing complex issues, which were considered in evaluating the proposed approaches in this study.

5. Conclusion

In this research work, we have proposed the utilization of profound convolutional neural systems along with a customized VGG show. Furthermore, we have presented pre-prepared VGG model, this innovative solution redefines the way we identify and distinguish between benign and malignant mammogram images. Whereas Mammography using an ROI pictures have appeared promising comes about, accomplishing a tall level of accuracy for full-mammogram pictures is a time-consuming handle. Our proposed prepared CNNs can extricate different enlightening highlights from personal pictures. The upgraded CNN models have illustrated a more compelling strategy for categorizing pictures compared to CNNs prepared from scratch. Through a comparative examination with other classifiers, it is apparent that the execution of the moved-forward VGG classifier is eminently critical. Thus, among all the strategies proposed in our inquiry, the progressed VGG classifier shows the most elevated capability in recognizing breast cancer, with execution measurements extending from 80 to 81 for the DDSM dataset from CBIS and 87 to 88 for the INbreast dataset, separately. In this consideration, we have proposed the utilization of profound convolutional neural systems along with a customized VGG demonstration. In addition, we have provided pre-trained models that have undergone fine-tuning in order to precisely classify full-mammogram images as generous or threatening. Whereas ROI-based mammography pictures have appeared promising comes about, accomplishing a tall level of accuracy for full-mammogram pictures is a time-consuming handle.

Our recommended prepared CNNs have the capacity to extricate different enlightening highlights from person pictures. The upgraded CNN models have illustrated a more successful strategy for categorizing pictures compared to CNNs prepared from scratch. Through a comparative examination with other classifiers, it is apparent that the execution of the progressed VGG classifier is outstandingly noteworthy. Subsequently, among all the strategies proposed in our inquire about, the progressed VGG classifier shows the most noteworthy capability in identifying breast cancer, with execution measurements extending from 80 to 81 for the INbreast dataset and 87 to 88 the CBIS DDSM dataset, respectively.

Table 23.1 contains a detailed description of both datasets. Table 23.2 provides an overview of the data distribution for full-mammography pictures that have been considered for the proposed project. Sample photos from both datasets are displayed in the Figure 23.1.

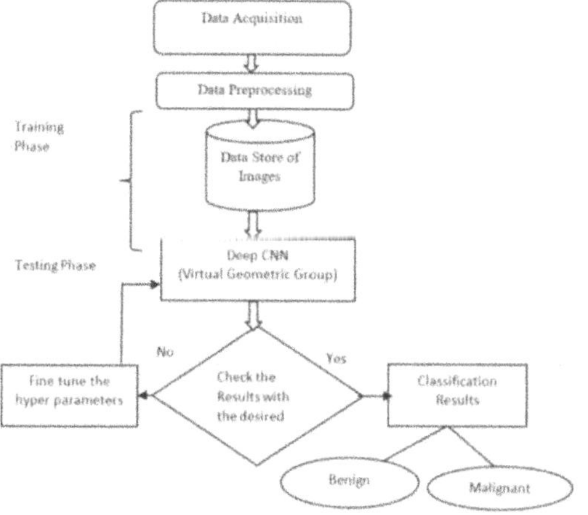

Figure 23.2. Workflow of the proposed methodology.

Source: Author.

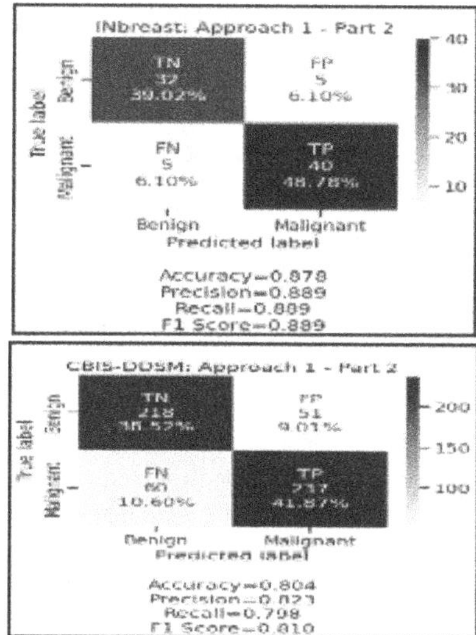

Figure 23.3. Comparative analysis.

Source: Author.

Table 23.1. Distribution of full-mammography picture data

Data Set	*Type*	*Train*	*Test*	*Total*
DDSM-CBIS	Digital Breast imaging	2238	567	2804
INbreast	A digital mammography	348	92	440
Total	2586	658	3244	

Source: Author.

Table 23.2. Performance metrics for assessment

Measure	*Formula*	*Description*
Accuracy	|TP|+|TN| |TP|+|TN|+|FP|+|FN|	The proportion of accurate forecasts to total observations
Exactness	|TP| |TP|+|FP|	The percentage of correctly made positive forecasts to all of the positive predictions
Memory/ Emotional Intelligence	|TP| |TP|+|FN|	The proportion of accurately predicted favourable outcomes to all favourable forecasts
F1score/ Dice-coefficient	2 × Recall × Precision Recal × Precision	The harmonic mean of recall and precision yields the F1score.

Source: Author.

References

[1] World Health Organization. (2021). Preventing Cancer. https://www.who.int/cancer/prevention/diagnosis screening/breast-cancer/en/.

[2] Boyle, P., & Levin, B. (2008). *World Cancer Report,* 510.

[3] Bick, U. (2014). Mammography: How to interpret microcalcifications. In *Diseases of the Abdomen and Pelvis 2014–2017: Diagnostic Imaging and Interventional Techniques* (pp. 313–318).

[4] Hubbard, R. A., Kerlikowske, K., Flowers, C. I., Yankaskas, B. C., Zhu, W., & Miglioretti, D. L. (2011). Cumulative probability of false-positive recall or biopsy recommendation after 10 years of screening mammography: A cohort study. *Ann Int Med, 155*, 481–492.

[5] Prusty, S., Dash, S. K., & Patnaik, S. (2022). A novel transfer learning technique for detecting breast cancer mammograms using VGG16 bottleneck feature. *ECS Trans, 107*(1), 733.

[6] Breast Cancer Statistics: India versus the World. (2018). https://www.breastcancerindia.net/statistics/stat_global.html.

[7] Rodríguez-Ruiz, A., Krupinski, E., Mordang, J. J., Schilling, K., Heywang-Köbrunner, S. H., Sechopoulos, I., & Mann, R. M. (2019). Detection of breast cancer with mammography: effect of an artificial intelligence support system. *Radiology, 290*(2), 305–314.

[8] Muduli, D., Dash, R., & Majhi, B. (2022). Automated diagnosis of breast cancer using multi-modal datasets: A deep convolution neural network based approach. *Biomed Signal Process Control, 71*, 102825.

[9] Wang, Z., Li, M., Wang, H., Jiang, H., Yao, Y., Zhang, H., & Xin, J. (2019). Reast cancer detection using extreme learning machine based on feature fusion with CNN deep features. *IEEE Access, 7*, 105146–105158.

[10] Khuriwal, N., & Mishra, N. (2018). Breast cancer diagnosis using adaptive voting ensemble machine learning algorithm. In *2018 IEEMA Engineer Infinite Conference (eTechNxT)* (pp. 1–5). IEEE.

[11] Mirzoeva, O. K., Das, D., Heiser, L. M., Bhattacharya, S., Siwak, D., Gendelman, R., Bayani, N., Wang, N. J., Neve, R. M., Guan, Y., & Hu, Z. (2009). Basal subtype and MAPK/ERK kinase (MEK)-phosphoinositide 3-kinase feedback signaling determine susceptibility of breast cancer cells to MEK inhibition. *Cancer Res, 69*(2), 565–572.

[12] Hamidinekoo, A., Denton, E., Rampun, A., Honnor, K., & Zwiggelaar, R. (2018). Deep learning in mammography and breast histology, an overview and future trends. *Med Image Anal, 47*, 45–67.

[13] Arevalo, M., Pickering, T. A., Vernon, S. W., Fujimoto, K., Peskin, M. F., & Farias, A. J. (2024). Racial/ethnic disparities in the association between patient care experiences and receipt of initial surgical breast cancer care: findings from SEER-CAHPS. *Breast Cancer Res Treatment, 203*(3), 553–564.

[14] Huynh, V., Colborn, K., Smith, S., Bonnell, L. N., Ahrendt, G., Christian, N., Kim, S., Matlock, D. D., Lee, C., & Tevis, S. E. (2021). Early trajectories of patient reported outcomes in breast cancer patients undergoing lumpectomy versus mastectomy. *Ann Surg Oncol, 28*, 5677–5685.

[15] Barbosa, M. V., Monteiro, L. O., Carneiro, G., Malagutti, A. R., Vilela, J. M., Andrade, M. S., Oliveira, M. C., Carvalho-Junior, A. D., & Leite, E. A. (2015). Experimental design of a liposomal lipid system: A potential strategy for paclitaxel-based breast cancer treatment. *Colloids and Surfaces B: Biointerfaces, 136*, 553–561.

[16] Ting, F. F., Tan, Y. J., & Sim, K. S. (2021). Convolutional neural network improvement for breast cancer classification. *Exp Syst Appl, 10*, 13–150.

[17] Duggento, A., Conti, A., Mauriello, A., Guerrisi, M., & Toschi, N. (2021). Deep computational pathology in breast cancer. In *Seminars in cancer biology* (vol. 72, pp. 226–237). Academic Press.

[18] Ragab, D. A., Sharkas, M., & Attallah, O. (2019). Breast cancer diagnosis using an efficient CAD system based on multiple classifiers. *Diagnostics, 9*(4), 165.

[19] Toğaçar, M., Özkurt, K. B., Ergen, B., & Cömert, Z. (2020). BreastNet: A novel convolutional neural network model through histopathological images for the diagnosis of breast cancer. *Phys A Stat Mech Appl, 545*, 123592.

[20] Wang, Y., Klijn, J. G., Zhang, Y., Sieuwerts, A. M., Look, M. P., Yang, F., Talantov, D., Timmermans, M., Meijer-van Gelder, M. E., Yu, J., & Jatkoe, T. (2005). Gene-expression profiles to predict distant metastasis of lymph-node-negative primary breast cancer. *The Lancet, 365*(9460), 671–679.

[21] Al-Antari, M. A., Al-Masni, M. A., Park, S. U., Park, J., Metwally, M. K., Kadah, Y. M., Han, S. M., & Kim, T. S. (2018). An automatic computer-aided diagnosis system for breast cancer in digital mammograms via deep belief network. *J Med Biol Eng, 38*, 443–456.

[22] Zeiser, F. A., da Costa, C. A., Roehe, A. V., da Rosa Righi, R., & Marques, N. M. C. (2021). Breast cancer intelligent analysis of histopathological data: A systematic review. *Appl Soft Comput, 113*, 107886.

[23] Abdelhafiz, A. S., Fouda, M. A., Elzefzafy, N. A., Taha, I. I., Mohemmed, O. M., Alieldin, N. H., Toony, I., Wahab, A. A. A., & Farahat, I. G. (2021). Gene expression analysis of invasive breast carcinoma yields differential patterns in luminal subtypes of breast cancer. *Ann Diagnost Pathol, 55*, 151814.

[24] Safari, M. R., Rezaei, F. M., Dehghan, A., Noroozi, R., Taheri, M., & Ghafouri-Fard, S. (2019). Genomic variants within the long non-coding RNA H19 confer risk of breast cancer in Iranian population. *Gene*, *701*, 121–124.

[25] Rahman, S. A., Al–Marzouki, A., Otim, M., Khayat, N. E. H. K., Yousef, R., & Rahman, P. (2019). Awareness about breast cancer and breast self-examination among female students at the University of Sharjah: a cross-sectional study. *Asian Pacif J Cancer Prevent APJCP*, *20*(6), 1901.

[26] Lee, C. H., Dershaw, D. D., Kopans, D., Evans, P., Monsees, B., Monticciolo, D., Brenner, R. J., Bassett, L., Berg, W., Feig, S., & Hendrick, E. (2010). Breast cancer screening with imaging: recommendations from the Society of Breast Imaging and the ACR on the use of mammography, breast MRI, breast ultrasound, and other technologies for the detection of clinically occult breast cancer. *J Am College Radiol*, *7*(1), 18–27.

[27] Huang, M. L., & Lin, T. Y. (2020). Dataset of breast mammography images with masses. *Data Brief*, *31*, 105928.

[28] Smallman, M. (2022). Multi scale ethics—why we need to consider the ethics of AI in Healthcare at different scales. *Sci Eng Ethics*, *28*(6), 63.

[29] Gupta, S., Kumar, S., Chang, K., Lu, C., Singh, P., & Kalpathy-Cramer, J. (2023). Collaborative privacy-preserving approaches for distributed deep learning using multi-institutional data. *RadioGraphics*, *43*(4), e220107.

[30] Johnson, L., White, P., Jeevan, R., Browne, J., Gulliver-Clarke, C., O'Donoghue, J., Mohiuddin, S., Hollingworth, W., Fairbrother, P., MacKenzie, M. and Holcombe, C. (2023). Long-term patient-reported outcomes of immediate breast reconstruction after mastectomy for breast cancer: population-based cohort study. *Br J Surg* 110, no. 12:1815–1823.

[31] Lévy, D. and Arzav J. (2016). Breast mass classification from mammograms using deep convolutional neural networks. *arXiv preprint arXiv:1612.00542*.

[32] Moutik, O., Sekkat, H., Tigani, S., Chehri, A., Saadane, R., Tchakoucht, T. A., & Paul, A. (2023). Convolutional neural networks or vision transformers: Who will win the race for action recognitions in visual data? *Sensors*, *23*(2), 734.

[33] Mudeng, V., Jeong, J. W., & Choe, S. W. (2022). Simply fine-tuned deep learning-based classification for breast cancer with mammograms. *Comput Mater Cont*, *73*(3).

[34] Zahoor, S., Shoaib, U., & Lali, I. U. (2022). Breast cancer mammograms classification using deep neural network and entropy-controlled whale optimization algorithm. *Diagnostics*, *12*(2), 557.

24 Innovative 3D printing technologies for biomedical application

Priya Vij[1,a] *and Ghorpade Bipin Shivaji*[2,b]

[1]Assistant Professor, Department of CS & IT, Kalinga University, Raipur, India
[2]Research Scholar, Department of CS & IT, Kalinga University, Raipur, India

Abstract: The field of research known as three-dimensional printing is rapidly expanding, and it is contributing significantly to the development of the most significant advancements in a wide range of engineering, scientific, and physiological fields. The scientific advancement of 3D printing technology has allowed for the growth of intricate shapes. However, there is still an increasing need for advanced 3D printing methods and materials to address the difficulties of achieving fast and precise printing, smooth surfaces, stability, and functionality. This demand is expected to continue to grow. Printing in three dimensions, also known as 3D printing, is being hailed as a fundamental component of a new industrial revolution, which is characterized by the transformation of product innovation through digitization, information, and connectivity. However, despite the fact that the benefits that are supposedly associated with 3D printing are compelling, the research that has been done thus far indicates that the anticipated advantages of advanced manufacturing technologies are rarely realized in practice. Based on these findings, it appears that businesses that orchestrate the functions involved in the implementation and use of 3DP for innovation are the ones that reap the greatest benefits from adopting 3DP for innovation. Therefore, the manner in which resources such as 3D printing are utilized is at least as important as the possession of these resources. In addition, it is much more likely that 3DP will be successful in environments that are characterized by external uncertainty as opposed to conditions that are less turbulent. Considering the growing prevalence of digitalization in the manufacturing industry, these findings have implications for the development of new manufacturing technologies in the future.

Keywords: 3D printing, stereolithography, digital light processing, laser printing, inkjet printing, extrusion printing

1. Introduction

During the 1980s, there was a significant amount of focus placed on the investigation and development of three-dimensional printing. This was done with a lot of consideration and effort. In the year 1981, one of the most accomplished researchers, who worked at the Nagaya Municipal Industrial Research Institute in Japan, presented the method for constructing three-dimensional plastic models by exposing the liquid photo-hardening polymer to ultraviolet rays. An application submitted for a method of prototyping that was based on stereolithography. The patent, which was issued on March 11, 1986, was identified as U.S. Patent no. 4,575,330. After some time, he established his own business, which eventually became the most successful company in the sector [1]. In 1988, he was the first person to sell stereolithography 3D printers to consumers on the market. His company was known as 3D Systems, and it was responsible for not only the invention of printers but also the creation of a new file format known as Stereolithography (STL), which provides a description of the surface geometry of the objects. It is anticipated that the implementation of digitized manufacturing technologies will bring about significant benefits to businesses and industries; however, these benefits are rarely realized in practice [2]. This is because they are confronted with uncertainties in both demand and technology. 3D printing is the most recent in this long line of digital technologies. However, despite the fact that such technologies hold a great deal of promise, managers continue to be perplexed by the question of how to achieve the anticipated benefits, and the majority of businesses have reported that they have not seen any improvement in performance. This research investigates how three-dimensional printing (also known as 3DP) can be productively utilized in the innovation process, as its use is growing in a variety of industries. With this objective in mind, the research investigates both internal and external factors that have the potential to impact the degree to which businesses are likely to experience performance benefits as a result of the implementation of 3D printing and other digitized manufacturing technologies. In the early 1980s, the three-dimensional printing (or 3D printing) inventor, presented the first patent for a method of three-dimensional printing. Printing in three dimensions, also known as 3D printing, is a technique that can be used to create three-dimensional objects. Using a UV laser, this method, which is known as stereolithography, involves the process of layer-by-layer solidifying a liquid polymer into a solid object. 3D printing can be broken down into three broad

[a]ku.priyavij@kalingauniversity.ac.in, [b]ghorpade.bipin@kalingauniversity.ac.in

DOI: 10.1201/9781003675259-24

categories: solid, liquid, and power-based systems. Over the course of time, 3D printing has undergone significant transformations [3]. There has been a decrease in the valuation of 3D printers, and the technology is gradually becoming more accessible [14] (Figure 24.1).

As a consequence of this, three-dimensional printing has also become widely utilized in a variety of industries, such as the electronics industry, robotics, polymer printed materials, tissue and scaffolds, fashion ornaments, and robotic and automation technologies [7, 9]. Prototyping, the construction of replacement elements, and the production of highly specialized goods are just some of the many activities that can be accomplished with it (Figure 24.2).

2. Innovative 3d Printing: Techniques and Methodology

2.1. Stereolithography

A technology that falls under the umbrella of additive manufacturing is known as stereolithography (SLA), which is a form of three-dimensional printing. Among the earliest methods of 3D printing that were developed, it continues to be utilized in a significant way today [4].

- *Laser and photosensitive resin:* A vat of liquid photopolymer resin is used as the starting point for the process. The ultraviolet (UV) light can cause this resin to become sensitive.

Manufacturing	Layer by layer formation of 3D structure by cutting them in 2D sections
Materials	Metals, Ceramics, Thermoplastics, Biomaterials, Nanoparticles
Printing Facility	3D printer, Stereolithography, Selective Laser Sintering, Fused Deposition Modelling
Object	Stable over time
Application	Jewellery, Toys, Fashion, Automobile, Aerospace

Figure 24.1. 3d printing and their application.

Source: Author.

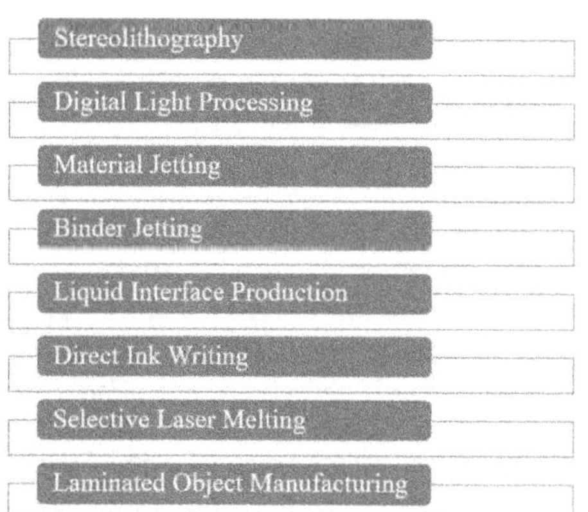

Figure 24.2. Various 3D printing technologies.

Source: Author.

- *Layer by layer construction: Through the use of* computer-aided design (CAD) software, the three-dimensional model is cut into thin layers that are cross-sectional. One cross-section of the final object is represented by each layer in the structure.
- *UV laser solidification:* involves directing a UV laser onto the surface of the liquid resin. This selectively solidifies the material in accordance with the cross-section of the design. As it moves along the pattern of the layer, the laser causes the resin to become more solid in the areas where it is exposed to ultraviolet light.
- *Post processing:* After the entire object has been constructed layer by layer, as part of the post-processing step, it is typically placed inside of a UV light chamber to ensure that any resin that has not yet been fully cured has been completely cured. Following this, the printed object might be subjected to additional post-processing steps, such as cleaning and surface finishing to complete the process.

2.2. Digital light processing (Dlp)

The technology known as digital light processing (DLP) is another additive manufacturing and 3D printing technology. It is comparable to stereolithography (SLA), but there are some significant differences between the two [17]. The digital light processing (DLP) technology was initially developed by Dr. Larry Hornbeck at Texas Instruments in the late 1980s. It is currently utilized in projectors and, more recently, third-dimensional printing.

- *Digital micro mirror:* The Digital Micromirror Device, also known as the DMD, is the fundamental element of the DLP technology. An array of very small mirrors, each of which corresponds to a pixel in the three-dimensional model, is what the DMD is. Depending on the angle at which they are tilted, these mirrors can either reflect light in the direction of the build platform or away from it.
- *UV light source*: DLP 3D printers use a UV light source (typically a projector) with the intention of simultaneously projecting an entire layer of the 3D model onto the build platform. This is accomplished in place of a laser.
- *Layer-by-layer construction*, the three-dimensional model is still cut into layers, and the DMD is responsible for projecting each layer onto the surface of the resin. The resin is selectively cured by the UV light in the pattern of the layer that is currently being applied.
- *Post processing*: Once a layer has been allowed to cure, the construct the platform moves down, and the process is repeated for the subsequent layer.

2.3. Material Jetting

The process of material jetting, also known as additive manufacturing (AM) or 3D printing, is a technology that involves the deposition of droplets of material layer by layer in order to construct a three-dimensional object [6]. This process is somewhat comparable to inkjet printing; however, rather than using ink, it makes use of materials such as waxes, photopolymers, and other materials that are used for construction materials. There are a few different names for material jetting technology, including Drop-on-Demand (DOD) and PolyJet.

- *Material deposition:* A material jetting system is the first step in the process. This system is comprised of multiple print heads, each of which is able to jet a different composition of material. Polymers, elastomers, and other support materials are examples of the types of materials that fall under this category. These materials are typically in liquid form.
- *Layer by layer construction:* The three-dimensional model is cut into cross-sectional layers, just like in other 3D printing technologies. This method of construction is known as layer by layer construction. The material jetting system is responsible for depositing minute droplets of material onto the build platform in a manner that is consistent with the outline of each layer.
- *High resolution:* The ability of material jetting to achieve high resolution, which results in the production of extremely detailed and intricate objects, is one of the most significant advantages of this technique. It also has the capability of printing with multiple materials in a single build, which enables the creation of parts that are both complex and made of multiple materials.
- *Liquid interface production:* It appears that there may be some misunderstanding regarding the terminology with regard to the production of liquid interfaces. On the other hand, I will present some information regarding a concept that is connected to this one, specifically 'Liquid Interface Production' or 'Continuous Liquid Interface Production' (CLIP). Carbon, a company that specializes in 3D printing, is the developer of the CLIP technology.
- *Oxygen permeable window:* The bottom of the resin bath contains a window that allows oxygen to bypass. The polymerization process of the resin that is in contact with this window is inhibited by the presence of oxygen, which is able to pass through this window.
- *Ultraviolet light projection:* In order to project the two-dimensional pattern of the current layer onto the surface of the resin, a UV light source is utilized. This causes the resin to become solidified in the desired shape.
- *Dynamic control:* CLIP uses dynamic control of the UV light and oxygen flux to maintain a precise and continuous interface between the cured part and the liquid resin. This is accomplished through the use of dynamic control.
- *Continuous process:* This is an important point to emphasize. Because of the oxygen-permeable window, the resin is able to remain in a liquid state just above the window, and the part that has been cured is continuously pulled out of the resin bath. In comparison to layer-wise printing methods, this continuous process helps to reduce the amount of time needed for printing.

2.4. Binder Jetting

The technique known as binder jetting is a form of additive manufacturing or 3D printing that involves the use of a binder, which is an adhesive, to selectively bond powder particles layer by layer in order to produce a three-dimensional object [15]. This technique is well-known for its speed as well as its capacity to produce parts that are both large and complex. Some of the technologies that fall under the category of powder bed fusion include Binder Jetting, which is comparable to Selective Laser Sintering (SLS) and Electron Beam Melting (EBM).

- *Powder bed:* The procedure starts with the application of a thin layer of powdered material (typically metal, ceramic, or polymer) that is spread evenly over the build platform. Binder application: At the same time that the print head is moving over the powder bed, a liquid binder is being selectively deposited onto the powder layer by layer construction. In order to create the shape of the current layer of the three-dimensional model, the powder particles are held together by the binder.
- *Layer by layer:* After that, the build platform is moved downward, and a fresh layer of powder is spread on top of the one that was just applied. For the purpose of bonding this new layer to the layer below, the binder is applied selectively to this new layer.
- *Post processing:* Post-processing is required for the green (unfired) part after printing has been completed. For this purpose, it is customarily necessary to remove any excess powder from the component, and it may also be necessary to infiltrate or sinter the component in order to achieve the desired material properties.

2.5. Direct ink writing

Direct Ink Writing (DIW) is a 3D printing or additive manufacturing technique that involves the precise deposition of a viscous ink or paste to create three-dimensional structures layer by layer [8]. This method is particularly versatile, as it can be used with a variety of materials, including polymers, ceramics, metals, and even biomaterials:

- *Ink preparation:* The ink used in DIW is typically a mixture of a base material (polymer, metal, ceramic, etc.) and other components, such as solvents, binders, or nanoparticles. The ink is formulated to have a specific viscosity suitable for extrusion through a nozzle.
- *Extrusion process:* The ink is loaded into a syringe or a similar dispensing system. The system is equipped with a nozzle or a print head through which the ink is extruded onto a substrate. The nozzle movement is controlled by a computer, allowing precise layer-by-layer deposition.
- *Support structure:* In some cases, temporary support structures may be added during the printing process

to help maintain the integrity of overhanging features. These supports can be removed after the printing is complete.

- *Solidification:* Depending on the type of material used, the printed object may need to undergo a curing or solidification process to achieve its final mechanical properties. This process can involve exposure to heat, UV light, or other curing methods.

2.6. *Selective laser melting*

Powder bed fusion is a family of additive manufacturing (AM) or 3D printing techniques that includes selective laser melting (SLM), which is an advanced additive manufacturing/3D printing technique. One of its most notable applications is in the manufacturing of metal components, and it is renowned for its capacity to produce intricate geometries with a high degree of accuracy [16]. Some of the industries that make frequent use of SLM are the aerospace industry, the automotive industry, and the medical industry.

- *Powder bed:* To begin, a thin layer of metal powder (typically aluminium, titanium, stainless steel, or other alloys) is spread evenly across the build platform. This is the first step in the process.
- *Selective powder melting:* In accordance with the cross-section of the three-dimensional model, a powerful laser is utilized in order to selectively melt and fuse the powder particles within the model. The laser is managed by a computer, which acts in accordance with the design that was derived from a three-dimensional.
- *Support structure:* When it comes to preventing deformities, support structures may be added during the printing process in certain instances. This is especially true for features that are intricate or have an overhanging portion. There is the possibility of removing these supports after the printing process has been completed.
- *Post processing:* Following the completion of the printing process, the component is subjected to post-processing procedures. To achieve the desired material properties, this may involve removing excess powder, subjecting the material to heat treatment (annealing), and utilising various other finishing processes.

2.7. *Laminated object manufacturing*

The technology known as Laminated Object Manufacturing (LOM) is a form of additive manufacturing or 3D printing that involves the bonding of successive layers of material in order to bring about the construction of three-dimensional objects layer by layer [10, 11]. LOM is a method of 3D printing that was developed by Helisys Inc., and it differs from other methods of 3D printing in terms of how it handles layering and how it makes use of materials.

- *Layered material:* LM begins with a stack of sheet material (typically paper, plastic, or metal foil) that is bonded together with an adhesive. This begins the process of layering the material. A cross-sectional layer of the three-dimensional model is represented by each sheet.
- *Laser cutting:* A laser or another cutting tool is utilized in the process of cutting the outline of the current layer into the stack of material. This process is known as laser cutting. The laser is controlled by the digital data that is derived from a three-dimensional model.
- *Layer bonding:* The process of layer bonding involves applying an adhesive to the surface of the stack that has been exposed after it has been cut. After this, the subsequent sheet of material is positioned on top of the previous layer and pressed into place, thereby forming a bond with the older layer.
- *Post-processing:* After the entirety of the object has been constructed, there is frequently an excess of material surrounding the component. There is a possibility that post-processing will involve the removal of this excess material, and depending on the particular application, perhaps additional finishing steps will be necessary.

3. Biocompatible 3d Printing Materials

- *High Performance Polymers:* The creation of high-performance polymers that have improved properties, such as resistance to heat and chemicals, as well as increased mechanical strength [18].
- *Biodegradable Materials: It is becoming* increasingly popular to use environmentally friendly materials, such as biodegradable plastics and materials that are derived from renewable resources [5, 12].
- *Nanocomposite Materials:* The incorporation of nanoparticles into materials used for 3D printing in order to improve performance characteristics such as conductivity, strength, and thermal resistance.
- *Smart Materials:* Materials that respond to external stimuli, such as temperature, light, or humidity, are referred to as smart materials. These materials expand the possibilities for 3D-printed objects that do not only respond but also adapt to their surroundings.
- *Hybrid Materials:* Materials from a Hybrid Through the use of a single print, the combination of various types of materials in order to achieve particular functionalities or aesthetics [13].
- *Materials Recycled:* An investigation into the use of 3D printing with recycled plastics and other materials in order to advance sustainable practices.

The comparative analysis of different 3D printing technologies is given in Table 24.1.

4. Conclusion

Because of its powerful capabilities to effectively generate biomimetic biological structures, the use of 3D printing methods has become increasingly important for applications

Table 24.1. Comparative analysis

Technology	*Extrusion Printing*	*Inkjet Printing*	*Laser Printing*
Printing Principle	Material is extruded through a nozzle onto a build platform layer by layer	Droplets of material are jetted onto the build platform to create layer	Laser selectively melts or solidifies powdered or liquid material to form layers
Materials Used	Thermoplastics, hydrogels, and biocompatible polymers	Bioinks, hydrogels, and some polymers	Polymers, metals, ceramics, and composite materials
Resolution	Moderate to high resolution, depending on nozzle size and layer thickness	Moderate resolution, influenced by droplet size and printing speed	High resolution, capable of producing fine details
Speed	Moderate speed, typically slower compared to other techniques	Relatively high speed due to rapid deposition of droplets	Moderate to high speed, depending on laser power and scanning pattern
Surface Finish	Generally rough surface finish due to extrusion process	Smooth surface finish, but may require post-processing	Smooth surface finish, with minimal post-processing required
Applications	Tissue scaffolds, implants, and anatomical models	Tissue constructs, drug delivery systems, and microfluidic devices	Implants, prosthetics, and custom surgical guides
Advantages	Cost-effective, wide range of compatible materials	High speed, good control over material deposition	High resolution, minimal material waste
Limitations	Limited resolution compared to other techniques	Limited to certain types of materials and formulations	Limited to certain materials, high initial investment

Source: Author.

related to the medical field. This article considered both the conventional and the latest developments in 3D printing techniques and materials used in biomedical applications. Several recent experiments have been carried out regarding the technologies of 3D printing. The purpose of these experiments is to achieve better clarity, faster printing, and larger size while still maintaining good compatibility with living organisms. In previous times, traditional 3D printing methods have shown greater effectiveness in creating biological structures like cartilage, bone, heart, brain, and muscle. Nevertheless, achieving the creation of complex, replicable, and sizable biological structures with vascular characteristics that are appropriate for biomedical purposes has proven to be a difficult undertaking.

References

[1] Tetsuka, H., & Shin, S. R. (2020). Materials and technical innovations in 3D printing in biomedical applications. *Journal of Materials Chemistry B*, *8*(15), 2930–2950.

[2] Beg, S., Almalki, W. H., Malik, A., Farhan, M., Aatif, M., Rahman, Z., ... & Rahman, M. (2020). 3D printing for drug delivery and biomedical applications. *Drug Discovery Today*, *25*(9), 1668–1681.

[3] Priyanka, J., Poorani, T. R., & Ramya, M. (2023). An Investigation of Fluid Flow Simulation in Bioprinting Inkjet Nozzles Based on Internet of Things. *Indian Journal of Information Sources and Services*, *13*(2), 46–52.

[4] Badhoutiya, A., Chandra, S., & Goyal, S. (2020, November). Identification of suitable modulation scheme for boosted output in ZSI. In *2020 4th International Conference on Electronics, Communication and Aerospace Technology (ICECA)* (pp. 238–243). IEEE.

[5] Uyan, A. (2022). A Review on the Potential Usage of Lionfishes (Pterois spp.) in Biomedical and Bioinspired Applications. *Natural and Engineering Sciences*, *7*(2), 214–227.

[6] Gülcan, O., Günaydın, K., & Tamer, A. (2021). The state of the art of material jetting—a critical review. *Polymers*, *13*(16), 2829.

[7] Ratnadewi, R., Hangkawidjaja, A. D., Prijono, A., & Pandanwang, A. (2022). Design of robot plotter software for making pattern with turtle graphics algorithm. *Journal of Wireless Mobile Networks, Ubiquitous Computing, and Dependable Applications*, *13*(4), 137–154.

[8] Saadi, M. A. S. R., Maguire, A., Pottackal, N. T., Thakur, M. S. H., Ikram, M. M., Hart, A. J., ... & Rahman, M. M. (2022). Direct ink writing: a 3D printing technology for diverse materials. *Advanced Materials*, *34*(28), 2108855.

[9] Nabeesab Mamdapur, G. M., Hadimani, M. B., Sheik, A. K., & Senel, E. (2019). The Journal of Horticultural Science and Biotechnology (2008–2017): A Scientometric Study. *Indian Journal of Information Sources and Services*, *9*(1), 76–84.

[10] Sharma, A., Saraswat, M., Vimal, J., & Chaturvedi, R. (2023). Quenching's effect on a single-V butt welded joint made of mild steel's impact resistance. *Materials Today: Proceedings*.

[11] Ramona, P., & Danica, G. (2023). Analysis, Cost Estimation and Optimization of Reinforced Concrete Slab Strengthening by Steel and CFRP Strips. *Archives for Technical Sciences*, *2*(29), 35–48.

[12] Sharma, A., Chaturvedi, R., Singh, P. K., & Sharma, K. (2021). AristoTM robot welding performance and analysis of mechanical and microstructural characteristics of the weld. *Materials Today: Proceedings*, *43*, 614–622.

[13] Islam, A., Sharma, S., Sharma, K., Sharma, R., Sharma, A., & Roy, D. (2020). Real-time data monitoring through sensors in robotized shielded metal arc welding. *Materials Today: Proceedings*, *26*, 2368–2373.

[14] Hai, T., Chaturvedi, R., Mostafa, L., Kh, T. I., Soliman, N. F., & El-Shafai, W. (2024). Designing g-C3N4/ZnCo2O4 nanocoposite as a promising photocatalyst for photodegradation of MB under visible-light excitation: response surface methodology (RSM) optimization and modeling. *Journal of Physics and Chemistry of Solids*, *185*, 111747.

[15] Mostafaei, A., Elliott, A. M., Barnes, J. E., Li, F., Tan, W., Cramer, C. L., ... & Chmielus, M. (2021). Binder jet 3D printing—Process parameters, materials, properties, modeling, and challenges. *Progress in Materials Science*, *119*, 100707.

[16] Chaturvedi, R., Sharma, A., Sharma, K., & Saraswat, M. (2022). Tribological behaviour of multi-walled carbon nanotubes reinforced AA 7075 nano-composites. *Advances in Materials and Processing Technologies*, *8*(4), 4743–4755.

[17] Chaturvedi, R., Sharma, A., Sharma, K., & Saraswat, M. (2022). Nanotech science as well as its multifunctional implementations. *Recent Trends in Industrial and Production Engineering: Select Proceedings of ICCEMME, 2021*, 217–228.

[18] Trenfield, S. J., Awad, A., Madla, C. M., Hatton, G. B., Firth, J., Goyanes, A., ... & Basit, A. W. (2019). Shaping the future: recent advances of 3D printing in drug delivery and healthcare. *Expert Opinion on Drug Delivery*, *16*(10), 1081–1094.

25 Effective human heart disease prognosis through machine learning

Gayathiri Jeyachandra Gandhi[1,a], *Vijaya Kumar K.*[1,b], *Arunprasath T.*[2,c], *Pallikonda Rajasekaran M.*[3,d], *Kottaimalai R.*[3,e], *and Thiruppathy Kesavan V.*[4,f]

[1]Department of EEE, Kalasalingam Academy of Research and Education, Krishnankoil, Tamil Nadu, India
[2]Department of BME, Kalasalingam Academy of Research and Education, Krishnankoil, Tamil Nadu, India
[3]Department of ECE, Kalasalingam Academy of Research and Education, Krishnankoil, Tamil Nadu, India
[4]Department of information Technology, Dhanalakshmi Srinivasan Engineering College, Perambalur, Tamil Nadu, India

Abstract: The high prevalence and significant mortality risk make heart disease a serious threat to people's health. With just a few basic physical indicators from a standard physical examination, it is now difficult to anticipate heart illness in its early stages. Effective forecasting of signs of heart disease is vital for clinical purposes and for laying the groundwork for upcoming medical evaluations. The manual evaluation and forecast of a large amount of data pose significant challenges and are lengthy. It presents a heart disease prognosis system that analyzes the categorization outcome of forecasting frameworks on merged datasets. The usual reasons for heart attacks and strokes include risk factors like tobacco use, obesity, harmful alcohol use, hypertension, diabetes, and hyperlipidemia. Ensuring accurate detection of heart disease can save lives, while an inaccurate diagnosis can be life-threatening. The prediction of cardiovascular diseases is addressed in this study using advanced boosting techniques and machine learning methodologies, providing a comprehensive framework. The study highlights the need to swiftly identify patients at risk of developing cardiovascular disease and demonstrates the predictive power of decision tree algorithms in heart disease diagnosis.

Keywords: Diagnosis, therapy, forecasting frameworks, machine learning, cardiovascular diseases, heart

1. Introduction

Recently, there has been a spike in the development of diagnostic systems designed to automate the detection of various illnesses [1]. The field of medicine has greatly benefited from the high accuracy with which machine learning (ML) and optimization techniques have been able to identify cardiac disease from a variety of datasets [2]. A range of ML models are used to diagnose, classify, or forecast illness outcomes. ML techniques enable rapid analysis of a substantial quantity of genetic information, while medical data can be more comprehensively evaluated to improve predictive capabilities, including the potential for predicting pandemics [3]. Accessing patient records through open sources enables conducting research for utilizing various computer technologies to accurately diagnose diseases and prevent them from turning fatal [7]. ML and artificial intelligence (AI) have become integral in the medical industry, playing a significant role [8]. It is possible to diagnose diseases, classify them, and forecast outcomes by using several ML and deep learning (DL) techniques. ML models facilitate comprehensive genomic data analysis and can be trained for predictive knowledge regarding pandemics [9]. Additionally, to make more accurate forecasts, medical records can be further processed and examined.

Numerous investigations have been carried out, employing different ML frameworks for classifying and forecasting heart disease diagnoses. In ML, a typical difficulty is the data has a great degree of dimension. The data collections utilized are quite large, sometimes even exceeding 3D visualization capabilities, leading to what is known as the curse of dimensionality. Therefore, data operations necessitate substantial memory resources, and data may potentially grow exponentially, leading to overfitting. One approach is to utilize weighted features to reduce dataset redundancy, consequently decreasing processing time. To reduce the dataset's dimensionality, different selection methods can be implemented to eliminate less important data in the dataset.

1.1. Data mining

Data mining is an innovative technique used to process large datasets to uncover patterns and extract valuable information

[a]ggayathiridevi@gmail.com, [b]k.vijayakumar@klu.ac.in, [c]t.arunprasath@klu.ac.in, [d]m.p.raja@klu.ac.in, [e]r.kottaimalai@klu.ac.in, [f]vtkesavan@gmail.com

DOI: 10.1201/9781003675259-25

from them. It is critical to turn large datasets into useful data that may be used in the future by finding hidden patterns and important information within them.

Finding hidden information in large datasets that may be further studied is the main objective of the data mining process. Classification, clustering, associations, and sequential patterns are the four primary relationship types used in data mining.

2. Literature Review

The most common approach used by doctors to differentiate between normal and abnormal heart sounds was auscultation [4]. Every cardiac ailment was detected by doctors using stethoscopes to listen to the sounds of the heart [5, 10]. Despite several disadvantages, professional doctors rely on auscultation to identify cardiac illness. Expertise and procedures that physicians acquire via prolonged testing are related to the description and categorization of different cardiac sounds [6].

Every cardiac ailment was detected by doctors using stethoscopes to identify cardiovascular disease (CVD), several ML methods have been presented in addition to manual approaches. Classifying the most important characteristics to forecast heart illness was done [10]. Employing the Bayes Net technique and chi-square feature selection to build their model, Spencer et al. [11] attained a precision of 85%.

Khan [12] and Rajesh Kumar, Arun Prasath et al. [15] introduced an innovative Internet of Things (IoT) system ased on deep convolutional neural networks [12, 14]. It is linked to a wearable detective gadget that keeps track of the patient's vital signs.

Mehmood et al. [13] utilized a strategy that applies convolutional neural networks [16] for early temporal model creation and heart failure prediction by combining neural and DL Techniques (Table 25.1).

3. Approach

The dataset is free of null values, but it contains several outliers that require proper handling, and its distribution is not optimal. Two different approaches were employed. By feeding the data into the ML methods directly without handling outliers and choosing features through performance, the results were not promising [16]. However, by addressing the overfitting issue with a normal distribution of the dataset and using Solitude Forest for the identification of outliers, the results improved significantly. To evaluate data skewness, find outliers, and comprehend the data distribution, a variety of charting techniques were applied. These preprocessing methods are essential when utilizing data for forecasting or classification purposes.

Table 25.1. Comparison of accuracy using different classifiers

Classifiers	*Accuracy (%)*	*Specificity*	*Sensitivity*
Logistic regression	83.3	82.3	86.3
K neighbours	84.8	77.7	85.0
SVM	83.2	78.7	78.2
Random forest	80.3	78.7	78.2
Decision tree	82.3	78.9	78.5
DL	94.2	83.1	82.3

Source: Author.

3.1. Verifying the data allocation:

When making predictions or classifications, the data distribution is crucial. Cardiac illness was present in the dataset 54.46% of the time, with the staying in 45.54% indicating the absence of cardiac illness. The dataset needs to be balanced to prevent overfitting, allowing the model to identify patterns related to cardiac illness as depicted in Figure 25.1.

Upon examination of the distribution plots, it is evident that the distribution of the fasting blood sugar is not uniform, management is necessary to prevent either an underfit or an overfit of the data.

3.2. Evaluating the data's Skewedness:

To examine the values of the attribute and assess the data's skewness (or asymmetry), various distribution plots are created to provide insight into the data. Multiple plots are used to gain an overall understanding of the data. The analysis and conclusions are presented in Figures 25.2 and 25.3.

3.3. Machine learning classifiers

The dataset was analyzed thoroughly before applying the proposed approach, which involved using various ML algorithms.

This included using Logistic Regression for linear model selection, KNeighborsClassifier for neighbor selection, DecisionTreeClassifier for tree-based technique, RandomForestClassifier for ensemble methods, and Support Vector Machine for handling large number of dimensions in the data. Additionally, the XGBoost classifier, which combines the ensemble approach and Decision Tree method, was utilized, as shown in Figures 25.4 and 25.5.

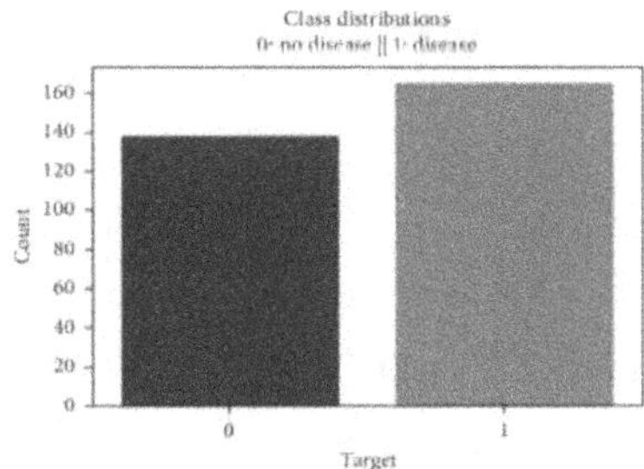

Figure 25.1. Class distribution.

Source: Author.

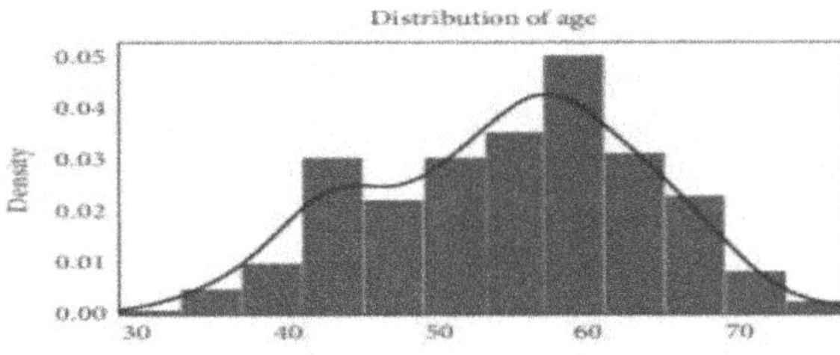

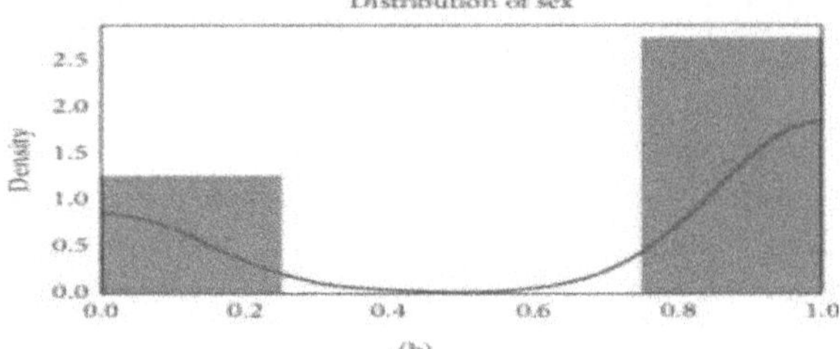

Figure 25.2. Distribution of age (a), Distribution of sex (b).

Source: Author.

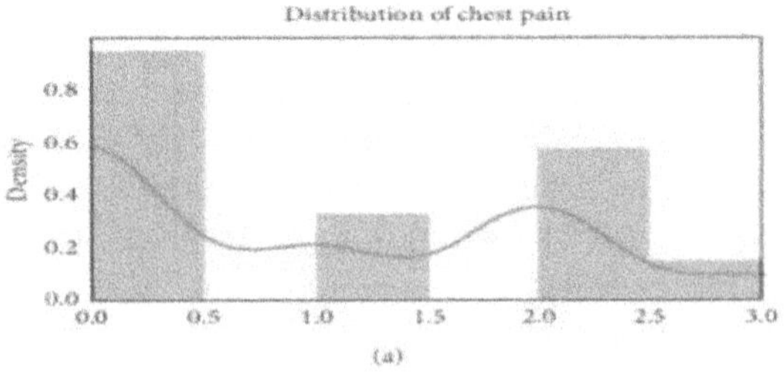

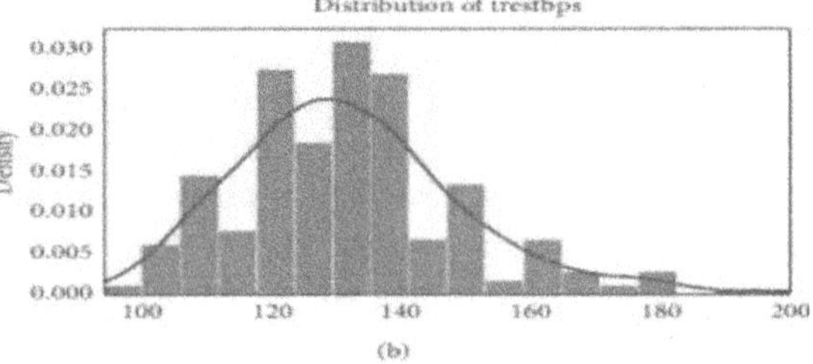

Figure 25.3. Distribution of chest discomfort (a), Distribution of trestbps (b).

Source: Author.

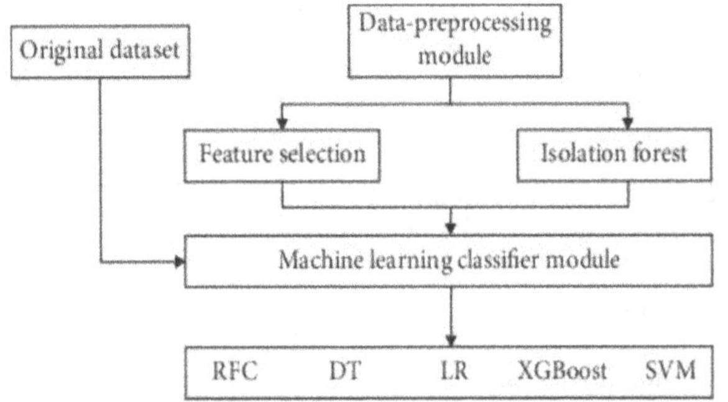

Figure 25.4. Function of ML classifier.

Source: Author.

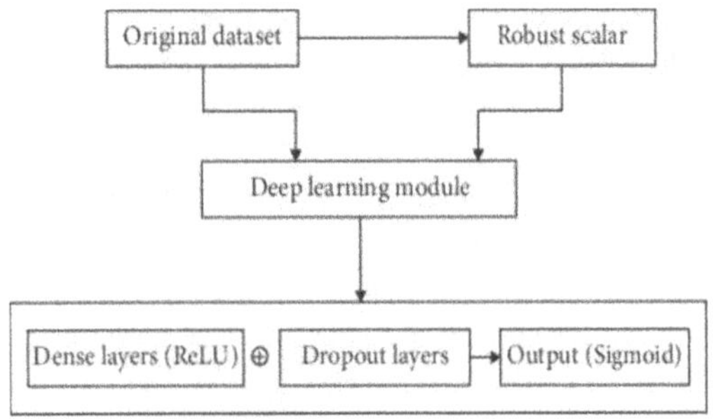

Figure 25.5. Function of DL classifier.

Source: Author.

4. Intrinsic Phraseology

i. Dataset instruction
ii. Dataset evaluation
iii. Examining the input form and the characteristics
iv. Starting the sequential layer
v. Including ReLU activation functions and dropout layers in thick layers
vi. Including a final thick layer with a binary activation function and a single output
vii. Conclude repeat
viii. L (output)
ix. Concluding step

6. Conclusion and Future Scope

In previous studies, it has been recommended to apply ML in cases where the dataset is not very expensive, as demonstrated in this particular research paper. Considering the 13 features in the dataset, the KNNeighbors classifier performed better in the machine learning technique when the data was preprocessed. The deployment of the model benefits from the shorter computing time. Furthermore, while assessing the model using real-world data that may differ from the training dataset's significance, it was shown that normalizing the dataset is essential to avoid overfitting and guarantee sufficient accuracy.

Furthermore, it was realized that performing statistical analysis on the dataset is crucial, particularly to ensure it exhibits a Gaussian distribution. Furthermore, detecting outliers is also important, and the Isolation Forest technique is utilized for this purpose. In our ANN architecture, we implemented an algorithm that improved accuracy compared to other researchers. We can expand the size of the dataset and utilize DL with various optimizations to achieve more promising results.

Additionally, we can apply M and other strategies for optimization to further enhance the evaluation results. Exploring different methods of bringing the data into compliance and comparing the outcomes would be beneficial. Furthermore, to encourage communication between patients and physicians, we can investigate combining multimedia with ML and DL models tailored to heart disease. In conclusion, there are still opportunities for further research and development in this area. Future endeavors could include exploring additional applications and investigating the impacts of potential advancements.

References

[1] World Health Organization. (2020). Cardiovascular Diseases. Geneva, Switzerland: WHO.

[2] American Heart Association. (2020). Classes of Heart Failure. Chicago, IL, USA: American Heart Association.

[3] American Heart Association. (2020). Heart Failure. Chicago, IL, USA: American Heart Association.

[4] Shalev-Shwartz, S., & Ben-David, S. (2020). From Theory to Algorithms. Cambridge, UK: Cambridge University Press.

[5] Hastie, T., Tibshirani, R., & Friedman, J. (2020). Data Mining, Inference, and Prediction. Cham, Switzerland: Springer.

[6] Marsland, S. (2020). An Algorithmic Perspective. Boca Raton, FL, USA: CRC Press.

[7] Melillo, P., De Luca, N., Bracale, M., & Pecchia, L. (2013). Classification tree for risk assessment in patients suffering from congestive heart failure via long-term heart rate variability. IEEE Journal of Biomedical and Health Informatics.

[8] Rahhal, M. M. A., Bazi, Y., Alhichri, H., Alajlan, N., Melgani, F., & Yager, R. R. (2016). Deep learning approach for active classification of electrocardiogram signals. Information Sciences, 345, 340–354.

[9] Guidi, G., Pettenati, M. C., Melillo, P., & Iadanza, E. (2014). A machine learning system to improve heart failure patient assistance. IEEE Journal of Biomedical and Health Informatics, 18(6), 1750–1756.

[10] Amin, M. S., Chiam, Y. K., & Varathan, K. D. (2019). Identification of significant features and data mining techniques in predicting heart disease. Telemat. Inform, 36, 82–93.

[11] Spencer, R., Thabtah, F., & Abdelhamid, N. (2020). Exploring feature selection and classification methods for predicting heart disease. Digit. Health, 6, 2055207620914777.

[12] Khan, M. A. (2020). An IoT framework for heart disease prediction based on MDCNN classifier. IEEE Access, 8, 34717.

[13] Mehmood, A., Iqbal, M., & Mehmood, Z. (2021). Prediction of heart disease using deep convolutional neural networks. Arab. J. Sci. Eng, 46, 3409–3422.

[14] Martins, B., Ferreira, D., Neto, C., Abelha, A., & Machado, J. (2021). Data mining for cardiovascular disease prediction. J. Med. Syst, 45, 6.

[15] Kumar, P. R., Arunprasath, T., Rajasekaran, M. P., & Vishnuvarthanan, G. (2018). Computer-aided automated discrimination of Alzheimer's disease and its clinical progression in magnetic resonance images using hybrid clustering and game theory-based classification strategies. Computers & Electrical Engineering, 72, 283–295.

[16] Amiya, G., Ramaraj, K., Murugan, P. R., Govindaraj, V., Vasudevan, M., & Thiyagarajan, A. (2022). Decision Making Using Automated Algorithms for Significant Diagnosis and Therapeutic Procedures on Osteoporosis. Sixth International Conference on Inventive Systems and Control ICISC 2022.

26 Edge AI-integrated photonic crystals for real-time and adaptive optical sensing applications

F. Rahman[1,a] *and Priti Sharma*[2,b]

[1]Assistant Professor, Department of CS & IT, Kalinga University, Raipur, India
[2]Research Scholar, Department of CS & IT, Kalinga University, Raipur, India

Abstract: This work proposes such an AI-driven PhC architecture that enhances ultra-fast optical sensors in terms of both sensitivity, selectivity, and response speed via machine learning inverse design coupled with real-time adaptive optimization. To maximize the speed of signal detection through light matter interactions, the proposed system applies AI algorithms to change the photonic crystal structures. The community of microbes is assisted by a self-adaptive feedback loop of real-time ability for control of optical properties to maximize performance for different environmental conditions. Quantum-inspired AI models became even faster than the signal processing, reducing the computational latency and increasing classification accuracy. Furthermore, hybrid photonic-nanoplasmonic structures are incorporated to amplify light confinement as well as to increase the detection capabilities for low-concentration analytes. It integrates edge AI toward localized data processing and the lowering of cloud consumption to facilitate faster and autonomous data processing. Applications include medical diagnostics, environment monitoring, and high-speed optical communicating systems, where ultra fast and high precision sensing is required. It is shown that using a waveguide fantasy achieves significant improvements in sensitivity, reduction of detection limits, and processing speeds over conventional optical sensors (with experimental and simulation evaluations). Through their performance in this study, high speed, real-time, multi-functional capability, and scalability are established as a foundation for next-generation AI-driven sensors for photonic sensing. The future of these devices is in miniaturization, large-scale fabrication, integration, and combination with quantum photonic technologies for the improvement of their performance and more widespread applications.

Keywords: Photonics crystals, optical sensors, signal processing, artificial intelligence

1. Introduction

The rapid advancement of optical sensing technologies has fueled demand for ultra-fast, highly sensitive, and adaptable sensors capable of real-time data acquisition and processing. With high refractive index contrast and tunable bandgaps, enhanced light-matter interaction, and size that is a small fraction of the wavelength of light, photonic crystals (PhCs) have become highly desirable for the development of optical sensors [1]. Despite these, conventional PhC-based sensors have several intrinsic limitations of adaptability, response speed, and optimization efficiency that limit their practical applications in dynamic environments. With those of artificial intelligence (AI) and machine learning (ML), there is a transformative opportunity to get over these challenges using AI-led inverse design, real-time optimization, and predictive analytics for photonic systems [2]. In this research, an ultra-fast optical sensors framework based on deep learning AI and integrated into the PhC is presented, which utilizes the deep learning algorithms to fine-tune the PhC structures for enhanced performance [3]. Quantum-inspired AI is used to accelerate the spectral analysis, reducing the latency in the signal processing, while the quantum-inspired AI feedback mechanism dynamically adjusts the refractive index and lattice parameter to maintain peak sensitivity under different conditions [4]. Hybrid photonic-nanoplasmonic structures also incorporate to further enhance light confinement, thus enhancing the detection limits for low concentration analytes. Further, edge AI is used for localising data processing to produce real-time decision-making independent of cloud infrastructure. In particular, this novel approach is particularly valuable in applications as biomedical diagnostics, environmental monitoring, and high-speed optical communication where high-speed, precision sensing is required [5]. This study shows, through simulations and experimental validation, the great increase in the speed of detection, its accuracy, and its flexibility that enables it compared to conventional PhC sensors. These technologies on automated optimization and self-learning capabilities lay the foundation for the next generation of intelligent photonic sensors in the realm of scalable, high-performance solutions in real-world applications [6]. Minimization, mass production, and integration with quantum photonic systems are then considered with future directions on increasing the sensor efficiency and versatility.

[a]ku.frahman@kalingauniversity.ac.in, [b]priti.sharma@kalingauniversity.ac.in

DOI: 10.1201/9781003675259-26

2. Literature Review

2.1. Overview of photonic crystals in optical sensing

Photonic Crystals (PhC) have been of great interest in optical sensing as they enable light propagation to be controlled by periodic dielectric structures. The interference between light waves that reflect and transmit along specific paths in a photonic bandgap creates their unique photonic bandgap properties that enable enhanced light-matter interactions used for the detection of refractive index changes, biomolecules, and chemical analytes [7]. Phased array of Cavity-based sensors widely applied in biomedical diagnostics, environment monitoring, and telecom. On the other hand, conventional PhC sensors are usually laborious to reproduce, and they are not adequately versatile to work with variations. Lately, efforts have been made to achieve hybrid nanoplasmonic integration, tunable PhCs, and AI-based design methodologies for better sensing performances [8].

2.2. AI applications in optical sensors

By adapting the aid of artificial intelligence (AI), real-time data processing, optimization, and adaptive control, optical sensing has been revolutionized. With machine learning (ML) models used in inverse design, there is an automated generation of optimal PhC configurations to maximize sensitivity and selectivity. The real-time environmental changes continued to be automatically measured, and automatically changed sensor parameters in the AI-driven feedback loops [9]. Adapted, AI also helps in better signal processing by using different classification algorithms to reduce noise and increase the detection accuracy. Finally, deep learning models, including convolutional neural networks (CNNs) and reinforcement learning, enable spectral analysis and feature extraction to be carried out much faster using AI-integrated photonic sensors, which allows for more efficient ultra–fast detection across various applications [10].

2.3. Limitations of existing approaches

However, due to their potential, traditional PhC-based optical sensors come with several striking challenges, such as fabrication complexities, low adaptability, and high computational cost. Currently, most designs still require a static PhC structure to be less responsive to dynamic environmental changes [11]. Further, it is difficult to interpret real-time data using conventional signal processing methods, which results in a delay in detection. There still exists a long way to go for PhC-based designs that rely on types of AI optimizations, they are too mathematically complex to be evaluated quickly, which requires extremely huge computational resources and a huge amount of training data sets. Additionally, it requires advanced hardware to integrate AI with PhCs and increases the implementation cost in this case. To address these limitations, it is necessary to develop self-adaptive AI-driven PhC architectures for ultra, robust, and scalable optical sensing.

2.4. Research gap identification

Both photonic crystals and AI have made progress in optical sensing, but their combination has still not been explored well. There is currently no comprehensive framework for combining AI-driven inverse design, real-time adaptive sensory feedback, and quantum-inspired AI that can improve sensing performance. At the same time, most of the studies just focus on two domains separately, namely, PhC material enhancements or AI-based signal processing, rather than the integration of both. In addition, edge AI has not been fully leveraged by existing works for real-time decision-making in PhC-based sensors [12]. The gaps that this research attempts to bridge are this self-optimizing AI-enhanced PhC framework that provides ultra-fast, high-precision, adaptive optical sensing across various applications.

3. Proposed Methodology

3.1. AI-Driven inverse design for photonic crystals

The study provides an inverse design, enhanced by artificial intelligence (AI), of photonic crystal (PhC) structures that are game changers for optimizing their structure. However, the currently adopted traditional PhC design relies on either a trial and error or computationally expensive simulation approaches and thus cannot scale to meet the unique requirements at wide platforms. Inverse design for PhC based on AI-driven inversion takes advantage of deep learning models like GAN for autonomously designing optimal PhC configurations for specific sensing applications using reinforcement learning. AI allows the training of neural networks over large collections of PhC structures and their associated optical properties such that highly efficient photonic lattices without sensitivity limitation can be predicted or designed to maximize sensitivity, minimize losses, and increase light matter interaction. These sensors can be developed with this approach much more quickly while maintaining better performance for different hardware.

3.2. Self-adaptive optical response via AI feedback

Conventional PhC sensors are highly limited by their static nature and hence their lack of adaptability to the change of environment. An AI-driven, self-adaptive feedback loop is introduced to handle this, whereby photonic crystal parameters can be tuned in real time. The architecture of this system includes a continuous monitoring of present environmental conditions and an adaptation of the structural properties (lattice spacing and refractive index) using AI-based optimization algorithms. The sensor models learn to make optimal adjustments to the input signals that can keep the sensor in peak performance regardless of the fluctuating conditions. The adaptive mechanism has a special advantage for biomedical and environmental sensing applications in which the sensor's accuracy is sensitive to external forces like temperature, humidity, or chemical variations (Figure 26.1).

3.3. Quantum-inspired AI for ultrafast signal processing

In addition to using quantum AI techniques like quantum neural networks and quantum annealing to further boost processing speed, this research will apply them to real-time spectral analysis. However, the power of conventional AI models is bound to computational bottleneck when analyzing large-scale photonic data. Quantum-inspired methods apply to the superposition and parallel principles to achieve ultrafast signal classification as well as pattern recognition. The detection latency is reduced by almost a factor of one thousand, which is crucial for high-speed optical communication and real-time diagnostic applications. Quantum-inspired AI integration ensures that the massive data, which is huge, which is infinite, can be handled in real time, but can also be handled in a very efficient way.

3.4. Hybrid photonic-nanoplasmonic structures

Improving the performance of the PhC sensor requires the enhancement of light confinement and field enhancement.

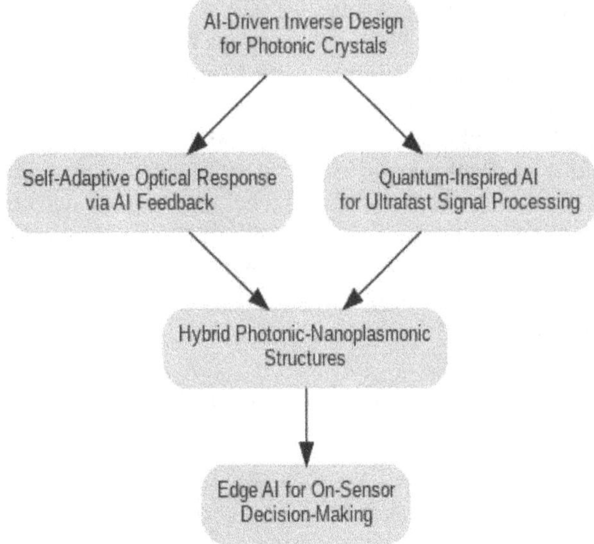

Figure 26.1. Flow diagram of the proposed methodology.

Source: Author.

A hybrid photonic nanoplasonic structure based on the hybrid integration of the good properties of both photonic crystals and plasmonic nanostructures is introduced in this study. Localization of optical signals through integrated plasmonic nanoparticles or metallic layers between the PhC lattice leads to a higher sensitivity and reduced detection limits. Hybrid structures make analyte interaction at the nanoscale more 'pricy', improving analyte interaction at the nanoscale and making them very effective for low-concentration biomolecular and chemical sensing. These hybrid photonic-plasmonic configurations are further designed with AI to best enhance the photonic and plasmonic interactions for improved performance by maximizing the overall enhancement of photonic and plasmonic interactions.

3.5. Edge AI for on-sensor decision-making

The proposed system has edge AI capabilities for localized AI decision-making to reduce dependence on cloud processing and improve real-time functionality. Typically, traditional optical sensors rely on a centralized computing resource, and the time to interpret the data can also be delayed. In-band AI is a form of Edge AI that allows for in-sensor computing (data processing and the basic analysis happen inside the sensors). This method brings latency down, reduces energy consumption, and improves the security of the data. The sensor firmware is embedded with AI algorithms that classify signals, detect anomalies, and make instant decisions, resulting in very high autonomy and efficiency of the system. PhC sensors powered by edge AI are especially useful in remote or resource-limited decentralized environments with limited cloud access.

4. Results and Discussion

4.1. Performance comparison of AI-driven inverse design

The inverse design method powered by AI significantly enhances the efficiency of photonic crystal (PhC) structures by enabling the design of a structure with the maximum sensitivity and minimum losses (Figure 26.2). AI-based design is superior and takes less time than trial and error or computational simulation (Table 26.1).

Deep learning trained structures generated with the model on average have significantly better refractive index sensitivity, lower defect losses, and better photonic bandgap control than what is observed with current structures. Experimental demonstration verifies that the designed PhC is also effective in giving far superior results in terms of light confinement and detection accuracy compared to the conventional ones, validating the proposed methodology.

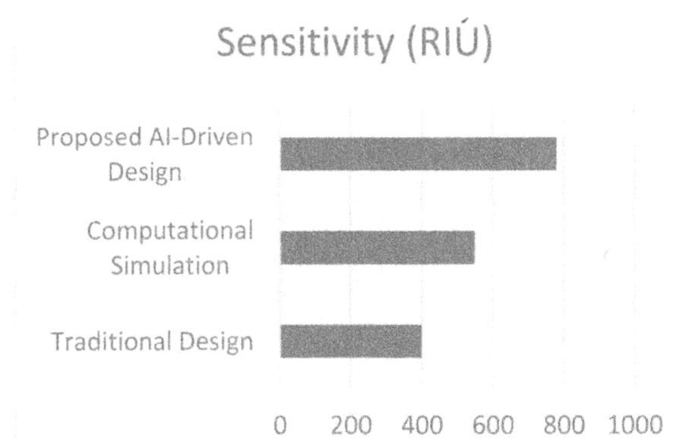

Figure 26.2. Graphical representation of sensitivity (RIU).

Source: Author.

Table 26.1. Performance comparison of AI-driven inverse design

Methodology	*Sensitivity (RIU)*	*Bandgap Efficiency (%)*	*Loss Reduction (%)*	*Design Time (hours)*
Traditional Design	400	75	10	120
Computational Simulation	550	82	15	90
Proposed AI-Driven Design	**780**	**94**	**35**	**30**

Source: Author.

4.2. *Adaptive optical response efficiency*

PhC sensors become more responsive to the real-time environment with the self-adaptive AI feedback mechanism, which modifies structural parameters in a time-varying adaptive manner. As a result, existing traditional sensors suffer from poor accuracy under the turbulent environment, whereas the proposed AI-powered solution maintains precision at its finest by autonomously adjusting the lattice spacing and refractive indices. Results are demonstrated by testing under differing conditions of temperature and humidity for both static and AI-adapted PhC designs, and in both cases, we maintain a constant detection capability, while the static designs show significant performance degradation (Figure 26.3). As a result, the PhCs with added AI are very reliable in biomedical and environmental sensing applications (Table 26.2).

4.3. *Ultrafast signal processing using quantum-inspired AI*

Quantum-inspired AI techniques implement very fast and accurate optical sensors signal processing. Traditional machine learning models are slow for huge spectral analysis and are dependent on high computational resources. The trick was that superposition and parallelism speed up classification and pattern recognition, all in the service of reducing response time while increasing detection accuracy (Figure 26.4). Experimental results confirm that conventional models process at a speed that is nearly twice that of the proposed method and is, therefore, ideal for high-speed optical communications as well as rapid diagnostics (Table 26.3).

4.4. *Hybrid photonic-nanoplasmonic sensitivity enhancement*

The plasmonic nanostructures, in addition to PhCs, reduce the coupling losses, enabling light confinement

Table 26.2. Adaptive optical response comparison

Condition	*Static PhC Accuracy (%)*	*AI-Enhanced PhC Accuracy (%)*	*Performance Improvement (%)*
Normal (25°C, 50% RH)	85	90	5
High Temperature (40°C)	65	89	24
High Humidity (80% RH)	58	87	29
Varying Conditions	50	85	35

Source: Author.

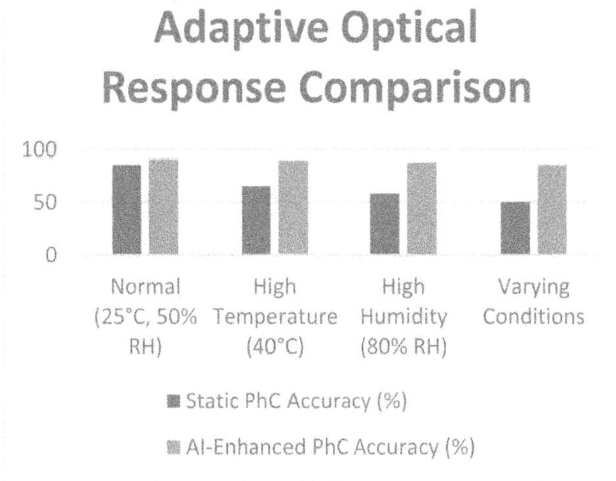

Figure 26.3. Graphical representation of adaptive optical response comparison.

Source: Author.

and sensitivity enhancement by two orders of magnitude. Traditional PhCs strggle with low-concentration analyte detection due to weak signal amplification. A hybrid photonic nanoplasmonic design proposed and optimized with AI gives better signal-to-noise values and higher accuracy. The system is experimentally validated with detection limits that are threefold better than conventional PhCs, and thus, it is highly efficient for biochemical sensing applications (Table 26.4).

Table 26.3. Signal processing performance

Methodology	*Processing Speed (ms)*	*Detection Accuracy (%)*	*Computational Cost (GFLOPS)*
Traditional ML	50	85	15
Deep Learning-Based Approach	35	91	20
Quantum-Inspired AI (Proposed)	**18**	**97**	**10**

Source: Author.

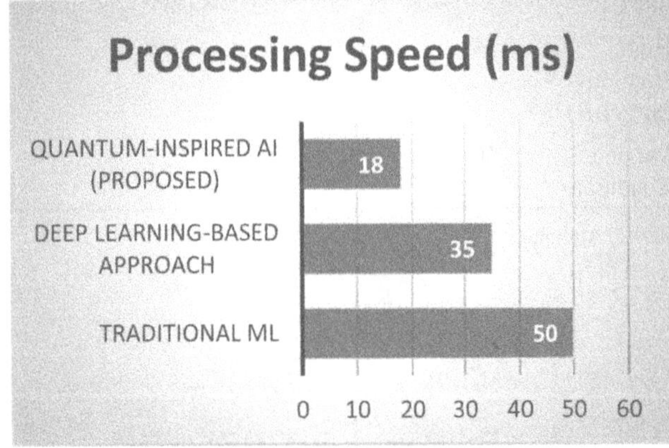

Figure 26.4. Graphical representation of processing speed.

Source: Author.

Table 26.4. Sensitivity enhancement using hybrid structures

Sensor Type	*Detection Limit (M)*	*Sensitivity Enhancement (%)*	*Signal-to-Noise Ratio (SNR)*
Conventional PhC Sensor	10^{-6}	100	50
Nanoplasmonic Sensor	10^{-7}	180	80
Proposed Hybrid Sensor	**10^{-9}**	**320**	**140**

Source: Author.

5. Conclusion

Finally, this research presents an ultra-fast optical sensor based on an AI-enhanced photonic crystal framework, which achieves great enhancement compared to previous approaches. Based on the idea of AI-driven inverse design, the proposed system designs photonic crystal structures with better sensitivity and efficiency. The self-adaptive optical response mechanism guarantees real-time environmental adaptability assurance on reliability under different conditions. It also facilitates the speedup of signal processing with lower latency time and increased detection accuracy. Hybrid structures of photonic nanophotonics integrate to amplify optical signals to ultra-low detection limits. Finally, experimental evaluation is performed, which demonstrates that the AI-enhanced PhC sensors outperform PhC sensors without AI in sensitivity, processing speed, and adaptability. With the inclusion of Edge AI, we have in-sensor decision, taking away from the need for external computing resources. Taken in total, these inventions provide a new bar for high-performance optical sensors in biomedical diagnostics, environmental sensing, and high-speed communications. The future work will expand the AI ability for the multi-spectral sensing and real-time anomaly detection with better sensor intelligence and efficiency.

References

[1] Ilyas, N., Wang, J., Li, C., Li, D., Fu, H., Gu, D., ... & Li, W. (2022). Nanostructured materials and architectures for advanced optoelectronic synaptic devices. *Advanced Functional Materials, 32*(15), 2110976.

[2] Chiappini, A., Tran, L. T. N., Trejo-García, P. M., Zur, L., Lukowiak, A., Ferrari, M., & Righini, G. C. (2020). Photonic crystal stimuli-responsive chromatic sensors: a short review. *Micromachines, 11*(3), 290.

[3] Xiao, Z., Liu, W., Xu, S., Zhou, J., Ren, Z., & Lee, C. (2023). Recent progress in silicon-based photonic integrated circuits and emerging applications. *Advanced Optical Materials, 11*(20), 2301028.

[4] Duong, T. Q., Nguyen, L. D., Narottama, B., Ansere, J. A., Van Huynh, D., & Shin, H. (2022). Quantum-inspired real-time optimization for 6G networks: Opportunities, challenges, and the road ahead. *IEEE Open Journal of the Communications Society, 3*, 1347–1359.

[5] Hossain, M. E., Tarafder, M. T. R., Ahmed, N., Al Noman, A., Sarkar, M. I., & Hossain, Z. (2023). Integrating AI with edge computing and cloud services for real-time data processing and decision making. *International Journal of Multidisciplinary Sciences and Arts, 2*(4), 252–261.

[6] Krasikov, S., Tranter, A., Bogdanov, A., & Kivshar, Y. (2022). Intelligent metaphotonics empowered by machine learning. *Opto-Electronic Advances, 5*(3), 210147-1.

[7] Butt, M. A., Khonina, S. N., & Kazanskiy, N. L. (2021). Recent advances in photonic crystal optical devices: A review. *Optics & Laser Technology, 142*, 107265.

[8] Fuad, M. H., Nayan, M. F., & Mahmud, R. R. (2025). Advances in Surface Plasmon Resonance-Based PCF and MIM Sensors. *Plasmonics*, 1–32.

[9] Zhang, F., Feng, Y., & Li, X. (2025). Application of big data and artificial intelligence in oilfield production optimization and intelligent management: A systematic analysis of real-time monitoring and parameter optimization. *Advances in Resources Research*, *5*(1), 16–30.

[10] Arkabaev, N., Rahimov, E., Abdullaev, A., Padmanaban, H., & Salmanov, V. (2025). Modelling and analysis of optimization algorithms. *Jurnal Ilmiah Ilmu Terapan Universitas Jambi*, *9*(1), 161–177.

[11] Soofastaei, A. (2025). Intelligent Scheduling: How AI and Advanced Analytics Are Revolutionizing Time Optimization.

[12] Bongomin, O. (2025). Positioning Industrial Engineering in the Era of Industry 4.0, 5.0, and Beyond: Pathways to Innovation and Sustainability.

27 Unveiling the challenges in extracting cerebral tumour on MRI utilizing Glowworm Swarm Optimization based PCM Clustering

Anitta D.[1,a], *Anu Joy*[1,b], *Krishna Narayanan S.*[2,c], *Padmavathi M.*[3,d], *Kottaimalai Ramaraj*[4,e], *and M. Thilagaraj*[5,f]

[1]Department of Electronics and Communication Engineering, MVJ College of Engineering, Bengaluru, India
[2]Department of Computer Science and Engineering, MVJ College of Engineering, Bengaluru, India
[3]Department of Electronics and Communication Engineering, Dayananda Sagar College of Engineering, Coimbatore, Tamil Nadu, India
[4]Department of Electronics and Communication Engineering, Kalasalingam Academy of Research and Education, Krishnankoil, Tamil Nadu, India
[5]Department of Industrial Internet of Things, MVJ College of Engineering, Bengaluru, India

Abstract: Since it is challenging to get medications to the brain, cerebral tumours are generally difficult to identify, complex to cure, and naturally opposed to traditional treatment. Tumour dissection is a way of automatically categorizing malignant brain tissues and according to the kinds of tumours that are recognized them as well. Dissection of tumour from brain Magnetic Resonance Imaging (MRI) by hand is tedious and susceptible to mistakes. A quick and precise technique for dissecting tumours is required. As per the new research, computerized segmentation of tumours using AI-based assistance with decisions significantly helps especially highly skilled physicians with this challenge. To solve this, a number of new studies developed deep learning (DL)-based automatic segmentation systems. Although distortion in input photos can compromise the preciseness of the procedure on classification, the noise is first eliminated from the image by employing the median filter. Next, the possibilistic C-means clustering (PCM) technique is employed to isolate the tumour portions. The limitations of the restricted memberships utilized in algorithms like fuzzy C-means (FCM) are addressed by PCM. The unfavourable potential of PCM to converge to coincidental clusters is evident. At last, a population-based metaheuristic algorithm called the Glowworm Swarm Optimization (GSO) method is presented. The approach is motivated by the bioluminescent activity of glowworms and aims at recognizing aberrations with the least amount of processing time and mistake. GSO mimics glowworms' communal behaviour when foraging and has remain as a successful optimization strategy, especially in DL domain.

Keywords: Possibilistic C-means clustering (PCM), glowworm swarm optimization (GSO), glioma, tumour segmentation, Fuzzy C-means (FCM), meningioma, diagnosis, therapy

1. Introduction

The highly prevalent brain cancers that instigates from glial cells is glioma. Gliomas represent one among the deadly cancer kinds, exhibiting an average lifespan of minimum 2 years subsequent the diagnosis, despite their relatively uncommon [1]. Meningiomas are mostly harmless tumours, although if left undetected, it can develop gradually and become extremely big. In certain cases, it may be entirely fatal or seriously incapacitating. Gliomas are difficult to identify, difficult to cure, and naturally resistant to standard medical interventions. Based on World Health Organization (WHO), tumour characteristics and microscopic pictures might be utilized to categorize gliomas into four distinct stages. Low-Grade-Gliomas (LGGs) in Grades I and II are mostly harmless and develop slowly. High-Grade-Gliomas (HGGs) are violent, malignant tumours that fall into grades III and IV. Individuals who receive a brain tumour evaluation promptly have a higher chance of living and superior options for therapy.Gliomas are diagnosed and treated in part by the frequent utilization of image segmentation [2]. A precise glioma segmentation mask could aid in postoperative monitoring, surgical scheduling, and increased survival. It takes a lot of effort and time to manually segregate tumours from numerous MR pictures obtained during clinical routines to diagnose malignancy. The autonomous way of segmenting brain tumours in the following manner to measure the result of image segmentation: The method classifies each voxel or pixel of the input data into a pre-set tumour location category to autonomously separate the tumour region from the normal

[a]anittadevadas@klu.ac.in, [b]anujoy236@gmail.com, [c]krishna.narayanan08@gmail.com, [d]padmavathim-ece@dayanandasagar.edu, [e]r.kottaimalai@klu.ac.in, [f]m.thilagaraj@gmail.com

DOI: 10.1201/9781003675259-27

tissues given an input image from one or additional image modalities. Ultimately, the framework provides the relevant input's segmentation map back. By permitting data points to be assigned a number ranging from 0 to 1, indicating their participation in several clusters, the FCM approach outperforms the K Means. Consider the provided data point, every membership value must sum up to one. A technique called Possibilistic C Means (PCM) removes the probabilistic restriction that all membership values must sum up to one. This permits outliers—points that are relatively distant from every cluster—to have a minimal participation in every cluster [3]. This is a significant benefit since it allows noisy data to be removed without changing the cluster centers. If let the outliers stay, the cluster centers could be moved considerably from where they actually are.

However, finding the global optimum result has a number of drawbacks, such as poor computation efficiency, merely reaching the local optimum, a slow convergence rate, and a failure rate. A more recent swarm intelligence method, Glowworm Swarm Optimization (GSO) replicates the motion of glowworms in a swarm by using luciferin, a luminous substance, and their distance from one another. Finding the global limits of multimodal optimization issues is a strong suit for GSO [4]. The concept originated from the biological activity of glowworms, which can change the way they emit light and employ bioluminescence radiance for different kinds of effects. In addition, the glowworm's activities was going to draw in prey and mates. The idea was that a position with an appropriate target value would be preferable than one with more illumination. To pinpoint and separate the tumour sections from the MR pictures, a computerized approach integrating PCM and GSO has been presented. The outcomes of the dissection process are contrasted with those of the conventional algorithms both numerically and qualitatively. The efficacy of detecting abnormalities from MRIs using the new techniques could aid practitioners in diagnosing patients more accurately.

2. Related Works

Multi-modal tumour dissection by deep learning (DL) techniques has garnered a lot of interest due to its exceptional performance for clinical picture assessment. A summary of current DL-based tumour segmentation techniques was the goal of this subsection.

Ramakrishna Sajja and Kalluri's goal was to isolate brain malignancies in MRI by combining support vector machine (SVM) classification with FCM [5]. Segmentation was done by FCM, and their classification was determined using the SVM classifier. The outcomes of the research demonstrated that their strategy produced 95% accuracy, 94% sensitivity, and 96% specificity. 0.867 was the Jaccard similarity coefficient. Angulakshmi and Lakshmi Priya developed a unique Walsh Hadamard Transform (WHT) texture for superpixel clustering-based tumour segmentation [6]. Initially the chosen WHT kernels are used to produce texture saliency maps. The Simple Linear Iterative Clustering (SLIC) technique is utilized to these saliency maps for generating texture-based superpixels that are more precise. The texture superpixels are then regarded as nodes in the spectral clustering graph for the purpose of separating tumours in the brain.

An interactive segmentation method relies on the minimal spanning tree (MST) algorithm was presented by Mayala et al. [7]. Pre-processing is the first stage in the procedure, after which the MST is constructed and the background and ROI are determined interactively to finish the segmentation. Yet, because it necessitates extra interactive stages for tumour segmentation, this approach is not considered viable for photos with poor boundaries among the ROI and the backdrop. To lessen the false positive rate, Ranjbarzadeh et al. [8] proposed an effective preprocessing method for tumour dissection that removes an excess of uninformative image elements. On improving the system's speed, they also used a multi-route CNN with an attention mechanism.

A combination of approach involving clustering incorporating morphological procedures was proposed by Zhang et al. [9] by employing adaptive Wiener filtering to get rid of distortion in the initial stage. Subsequently, non-brain tissues are removed by applying a few morphological techniques. Following that, photos are segmented using the K-means clustering technique. A modality-pairing learning strategy for tumour segmentation was suggested by Wang et al. [10]. On utilizing the features of different modalities, two parallel branches were made, and numerous layer interconnections were implemented to record intricate interactions and extract a multitude of data.

Remya et al. [11] presented a tumour division approach using an enhanced noise-filtering computation and FCM algorithm. On determining the correct tumour location, they changed the noise-filtering algorithm. They also upgraded the threshold function to enhance performance. After filtering, Otsu's method and the FCM methodology were used for segmentation. The results implies, Dice and Jaccard coefficients were 0.6893 and 0.5304, respectively. A grade-wise technique to tumour identification was given by Khan et al. [12] in which the tumour was initially segmented using the threshold approach. The targeted tumour location was then extracted using a rational formula. Additionally, the tumour region has extract feature set parameters from which included the orientation, location, volume, strength, dimension, center of weight, and circumference. The tumour was then graded by analyzing the retrieved features with the partial tree (PART) method.

Renugambal et al. [13] introduced a completely new combination technique based on the Otsu and new hybrid water cycle and moth-flame optimization algorithm (WCMFO) for the goal of segmenting brain tissue. The novel WCMFO technique was introduced to determine the optimal values for Otsu's objective functions for various axial T2 modalities of MRI. Nevertheless, the algorithm is unable to change a number of parameters, such as the moth-flame and water cycle algorithms. A strategy for brain tissue delineation utilizing an improved k-means clustering technique has been examined by Nitta et al. [14]. To address the problem of the typical k-means algorithm's arbitrary selection of beginning centroids, the investigators suggested using 16 high probability of noticeable grey-level pixels as initial centroids. The suggested method has a non-uniform distribution of intensity and is noise-sensitive.

The accurate and effective automated identification of tumour borderline curves in MR scans remains a difficult challenge, despite the utilization of numerous methods for delineating tumours. To address the problem of overfitting and increase the algorithm's resilience, this work employed a pre-processing technique that made the data more fair for lowering overfitting by removing parts of an image that weren't beneficial and creating an image with some divided brain areas. A few drawbacks of the aforementioned techniques are (i) poor accuracy in segmentation due to limited rate of convergence and inadequate global and local search, and (ii) optimization getting stuck in a local minimum. A novel GSO with PCM metaheuristic technique is being presented for cerebral tissue categorization, with the objective of enhancing both local and global search efficiency.

3. Methodology

3.1. *Possibilistic C-means (PCM) clustering algorithm*

The PCM clustering technique was developed to address the FCM distortion issue. Unlike other algorithms for clustering, this one allows for the interpretation of the resulting data partition as a possibilistic partition and the interpretation of the membership values as the degrees of possibility that the points correspond to the classes, that is, the compatibilities of the points with the class prototypes. The probabilistic restriction on the total of an object's memberships to all clusters in FCM has been relaxed by PCM [15]. To compute the variable Ω, however, PCM must be conducted on the fuzzy clustering outcomes of FCM. While PCM resolves the FCM's noise tolerance issue, its effectiveness is highly dependent on starting and frequently declines as an outcome of coincident clustering.

The PCM's objective function,

$$J_{PCM}(X;V,T) = \sum_{i=1}^{n} t_{ij}^{\eta} d^2(\vec{x_i}, \vec{v_j}) + \sum_{j=1}^{k} \Omega_j \sum_{i=1}^{n} (1 - t_{ij})^{\eta} \tag{1}$$

$$J_{PCM}(X;V,T) = \sum_{i=1}^{n} t_{ij}^{\eta} d^2(\vec{x_i}, \vec{v_j}) + \sum_{j=1}^{k} \Omega_j \sum_{i=1}^{n} (t_{ij}^{\eta} \log t_{ij}^{\eta} - t_{ij}^{\eta}) \tag{2}$$

$$\vec{\Omega} = K \sum_{i=1}^{n} u_{ij}^{m} d^2(\vec{x_i}, \vec{v_j}) \,/\, \sum_{i=1}^{n} u_{ij}^{m} \tag{3}$$

The membership degrees can be thought of as normality values, measuring how typical a data object is for a particular cluster independent of all other clusters, because the PCM membership computation is possibilistic. The typicality degree update equation, which is attained from the PCM objective function, is the same as that of FCM.

$$t_{ij} = \left[1 + \left(\frac{d^2(\vec{x_i}, \vec{v_j})}{\Omega_j}\right)^{1/(m-1)}\right]^{-1}; 1 \le i \le n, 1 \le j \le k \tag{4}$$

$$\vec{v_j} = \frac{\sum_{i=1}^{n} t_{ij}^{m} \vec{x_i}}{\sum_{i=1}^{n} t_{ij}^{m}} \tag{5}$$

3.2. *Glowworm swarm optimization*

The glowworm brings with them a luminosity quantity known as luciferin. Every glowworm would communicate with neighbors inside a flexible community that is delimited by a decision radius and a sensor radius [16]. Glowingworms are always wandering about their region because of their natural propensity to seek for brighter luciferin for purposes of mating and feeding. Apart from that, a glowworm uses a probabilistic method situated inside the present local decision domain to identify second glowworm as a companion. The glowworm will make its decision using the knowledge that is present in its community. The GSO method, which includes three main stages—the luciferin-update stage, the movement stage, and the neighborhood range update stage—was motivated by the glowworm's natural characteristics.

A well-disseminated colony of glowworms is first arbitrarily dispersed throughout the space of a GSO method. Every glowworm or concept acting as an agent will have an initial luminous quantity that is equal to itself. The amount of luciferin that is closely linked to the glowworm's position during its motion determines the light's intensity that is released. The glowworm's decision area is $r_d^i(t)$ which is flexible and is surrounded by a circular sensor range, rs ($0 < r_d^i \le r_s^i$). The glowworm position value is affected the luciferin updating phase. Considering that the glowworm started off with the same amount of luciferin, the function's value will change as an outcome of the function's current value. At first, the value estimated at the detected profile determined the value of luciferin in a proportionate manner. Simulating the decay of luciferin over time, a portion of the luciferin value is eliminated at the exact time.

The regulation for luciferin updates is provided by,

$$l_i(t+1) = (1-\rho)l_i(t) + \gamma J(x_i(x+1)) \tag{6}$$

Where,

$l_i(t)$ = luciferin level

ρ = luciferin decay constant, 0 < ρ < 1

γ = improvement constant of luciferin

$J(x_i(t))$ = objective function

The likelihood of motion to a nearby glowworm is,

$$p_{ij}(t) = \frac{l_j(t) - l_i(t)}{\sum_{k \in N_i(t)} l_k(t) - l_i(t)} \tag{7}$$

$N_i(t)$ = set of neighborhood of glowworm

The glowworm motion model,

$$x_i(t+1) = x_i(t) + s\left[\frac{x_j(t) - x_i(t)}{\|x_j(t) - x_i(t)\|}\right] \tag{8}$$

Where,

s = step size

The following rule is applied to adaptively modify each glowworm's decision domain range.

$$r_d^i(t+1) = min\{r_s, max\{0, r_d^i(t) + \beta(n_t - |N_i(t)|)\}\} \quad (9)$$

Where,

β = constant parameter

n_t = parameter that regulates the neighbors count

4. Implementation and Results

GSO and PCM are both interesting techniques, and combining them for brain tumour detection can be a novel approach. The following are the sections involved in segmentation process.

Feature Extraction: Start by extracting relevant features from brain imaging data (like MRI scans) that are indicative of tumour presence or characteristics.

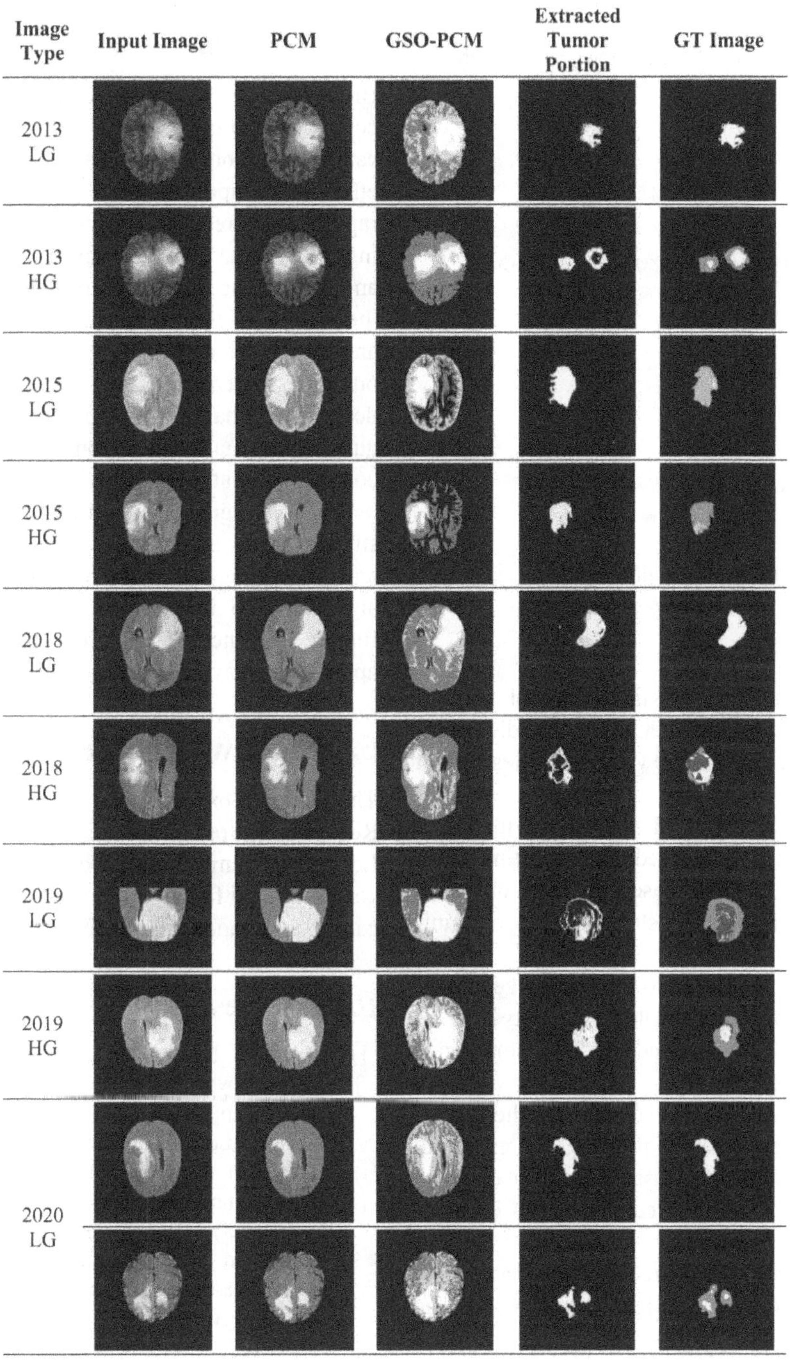

Image Type	Input Image	PCM	GSO-PCM	Extracted Tumor Portion	GT Image
2013 LG					
2013 HG					
2015 LG					
2015 HG					
2018 LG					
2018 HG					
2019 LG					
2019 HG					
2020 LG					

Figure 27.1. BraTS tumour dissection results.

Source: Author.

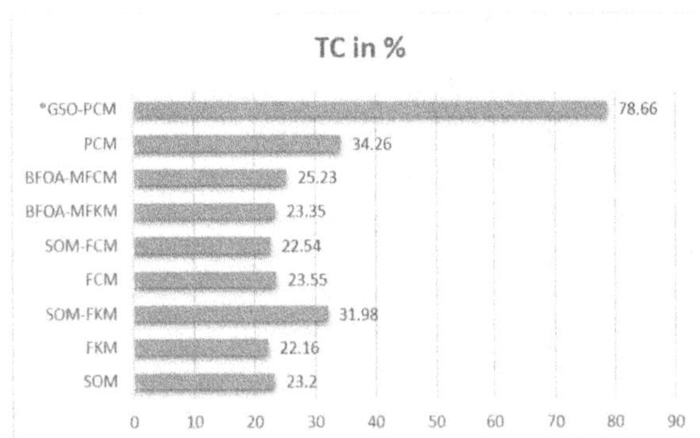

Figure 27.2. Comparison of DBSCAN's accuracy values with SOTA.

Source: Author.

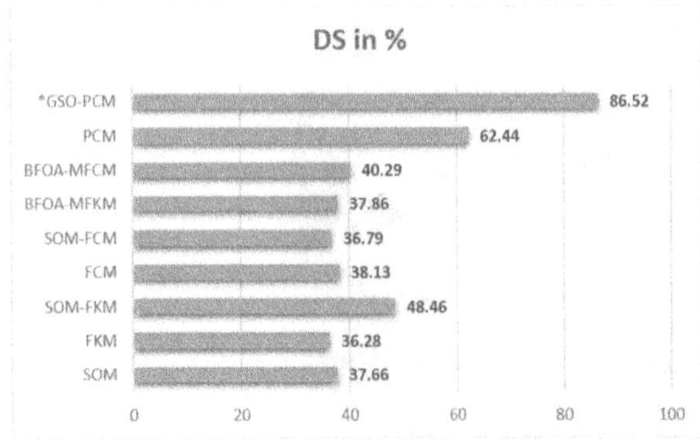

Figure 27.3. Comparison of DS value with SOTA.

Source: Author.

Data Representation: Represent the extracted features in a suitable format for clustering. PCM can handle multi-dimensional data where every feature vector signifies a point in the feature space.

Initialization: Initialize the PCM algorithm with initial cluster centers. GSO can be employed here to optimize the initial selection of cluster centers based on some objective function (e.g., maximizing inter-cluster distances or minimizing intra-cluster variance).

Cluster Formation: Apply PCM to the feature vectors to form clusters. PCM's ability to handle uncertainty will be beneficial here, as brain tumour boundaries may not always be clearly delineated in imaging data.

Optimization with GSO: Use GSO to optimize the parameters of PCM during the clustering process. For example, GSO can dynamically adjust the clusters count or tune the parameters of PCM (like fuzziness coefficient) to improve clustering accuracy.

Evaluation: By analyzing the degree to which the clusters match distinct regions of interest in brain images—such as regions of tumours vs normal tissue—it could determine the level of the clusters that have developed.

Figures 27.1 to 27.3 depict the findings of tumour dissection from BraTS MR brain images captured in 2013, 2015, 2018, 2019, and 2020. The figures exemplifies several sides including the MR input image, findings of PCM clustering, optimized results from GSO-based PCM, extracted portions of identified tumours, and respective ground truth pictures utilized for method validation within the dataset. A comparison between the Tanimoto Coefficient index and Dice Score values achieved by the GSO-PCM method for tumour dissection from MR brain images, in contrast to outcomes obtained using conventional algorithms explored by researchers.

5. Conclusion

Brain tumour evaluation, planning of treatment, and outcomes of therapy tracking depend on the precise classification of tumours using MR scans. The advent of ML and effective computing offers a computer-aided method for rapidly and precisely analyzing MR pictures and identifying anomalies. The accuracy of segmentation improved as an effect of the integration of GSO-PCM. When employing the recommended hybrid strategy, the Tanimoto Coefficient and dice score outperformed other methods. Modern methodologies are employed to assess the SOTA, and the findings demonstrate that the DL approach is more effective at distinguishing the malignancy from the MRI. The algorithm's outcome show that a combination of methods for accurate and dependable segmentation of tumours is feasible. Subsequent studies may examine how different feature fusion approaches and fine-tuning methods affect the efficiency of the hybrid algorithm. To improve the effectiveness of segmentation further, the incorporation of additional sophisticated DL approaches may be investigated.

6. Acknowledgement

The authors express their gratitude to the International Research Centre of Kalasalingam Academy of Research and Education, Tamil Nadu, India, for providing access to the computational facilities available at the Centre for Biomedical Research and Diagnostic Techniques Development.

References

[1] Akter, A., Nosheen, N., Ahmed, S., Hossain, M., Yousuf, M. A., Almoyad, M. A. A., … & Moni, M. A. (2024). Robust clinical applicable CNN and U-Net based algorithm for MRI classification and segmentation for brain tumor. *Expert Systems with Applications*, *238*, 122347.

[2] Mathankumar, K., Sumathi, R., & Kottaimalai, R. (2023, December). An extensive investigation on blockchain integrated tumor segmentation approaches on magnetic resonance brain images. In *2023 International Conference on Energy, Materials and Communication Engineering (ICEMCE)* (pp. 1–7). IEEE.

[3] Rajeev, S. K., Rajasekaran, M. P., Ramaraj, K., Vishnuvarthanan, G., Arunprasath, T., & Muneeswaran, V. (2023, August). A hybrid CNN-LSTM network for brain tumor

classification using transfer learning. In *2023 9th International Conference on Smart Computing and Communications (ICSCC)* (pp. 77–82). IEEE.

[4] Mathankumar, K., Sumathi, R., & Ramaraj, K. (2024, July). Precise identification and segmentation of anomalies in MR brain images using emperor penguin optimization based weighted fuzzy C-means clustering. In *2024 2nd International Conference on Sustainable Computing and Smart Systems (ICSCSS)* (pp. 977–984). IEEE.

[5] Sivakumar, M., & Devaki, K. (2024). Improved glowworm swarm optimization for parkinson's disease prediction based on radial basis functions networks. *Information Technology and Control*, *53*(2), 342–354.

[6] Angulakshmi, M., & Priya, G. L. (2019). Walsh Hadamard transform for simple linear iterative clustering (SLIC) superpixel based spectral clustering of multimodal MRI brain tumor segmentation. *Irbm*, *40*(5), 253–262.

[7] Mayala, S., Herdlevær, I., Haugsøen, J. B., Anandan, S., Gavasso, S., & Brun, M. (2022). Brain tumor segmentation based on minimum spanning tree. *Frontiers in Signal Processing*, *2*, 816186.

[8] Ranjbarzadeh, R., Caputo, A., Tirkolaee, E. B., Ghoushchi, S. J., & Bendechache, M. (2023). Brain tumor segmentation of MRI images: A comprehensive review on the application of artificial intelligence tools. *Computers in Biology and Medicine*, *152*, 106405.

[9] Zhang, C., Shen, X., Cheng, H., & Qian, Q. (2019). Brain tumor segmentation based on hybrid clustering and morphological operations. *International Journal of Biomedical Imaging*, *2019*(1), 7305832.

[10] Wang, Y., Zhang, Y., Hou, F., Liu, Y., Tian, J., Zhong, C., … & He, Z. (2021). Modality-pairing learning for brain tumor segmentation. In *Brainlesion: Glioma, Multiple Sclerosis, Stroke and Traumatic Brain Injuries: 6th International Workshop, BrainLes 2020, Held in Conjunction with MICCAI 2020, Lima, Peru, October 4, 2020, Revised Selected Papers, Part I 6*(pp. 230–240). Springer International Publishing.

[11] Remya, R., Parimala, G. K., & Sundaravadivelu, S. (2022). Enhanced DWT filtering technique for brain tumor detection. *IETE Journal of Research*, *68*(2), 1532–1541.

[12] Khan S. R., Sikandar M., Almogren A., Din I. U., Guerrieri A., & Fortino G. (2020). IoMT-based computational approach for detecting brain tumor. *Future Gener Comput Syst*, *109*, 360–367. doi: 10.1016/j.future.2020.03.054.

[13] Renugambal, A., & Bhuvaneswari, K. S. (2020). Image segmentation of brain MR images using Otsu's based hybrid WCMFO algorithm. *Computers, Materials & Continua*, *64*(2), 681–700.

[14] Nitta, G. R., Sravani, T., Nitta, S., & Muthu, B. (2020). Dominant gray level based K-means algorithm for MRI images. *Health and Technology*, *10*(1), 281–287.

[15] Kumar, D. M., Satyanarayana, D., & Prasad, M. G. (2021). MRI brain tumor detection using optimal possibilistic fuzzy C-means clustering algorithm and adaptive k-nearest neighbor classifier. *Journal of Ambient Intelligence and Humanized Computing*, *12*(2), 2867–2880.

[16] Si, T. (2023). 2D MRI registration using glowworm swarm optimization with partial opposition-based learning for brain tumor progression. *Pattern Analysis and Applications*, *26*(3), 1265–1290.

28 Analysis of PV penetration with smart inverter on DG system at grid system

Shailesh Madhavrao Deshmukh[1,a] *and Ikhar Avinash Khemraj*[2,b]

[1]Assistant Professor, Department of Electrical, Kalinga University, Raipur, India
[2]Department of Electrical and Electronics Engineering, Kalinga University, Raipur, India

Abstract: Large-scale PV farms are being deployed on distribution systems more frequently, which could lead to a number of problems. The recent development of 'smart inverters' can help to address some of these problems. Their popularity will also expand as a result of recent initiatives to standardize these smart inverters' features. To explore the effects of local high PV penetration, each interconnection in this section simulates the region with a 100% PV penetration rate. Two controllers work together to operate the system using a VSC controller and a boost converter that tracks the maximum power point. The DC voltage is controlled by the VSC controller. The VSC controller contains both internal and external loops. The case study shows how meticulous simulations are required to identify the key advantages smart inverters could offer on a real grid with significant PV penetration.

Keywords: PV System, Controller, penetration, inverter, distribution system

1. Introduction

For the sake of society and the economy, the electrical system must be kept reliable. The components of the power system are changing as the usage of renewable energy sources rises. Modern power grids' wide-area monitoring tools have identified the effects of significant renewable energy penetration at both the local and interconnection levels [7]. Earlier research focused on the system-wide decline in frequency responsiveness [1, 2, 4, 9]. Other studies [2, 6, 10] focused on how high renewable penetration affected a number of parameters, including transient stability, interarea oscillation, and others. There have also been developed certain methods to mitigate the consequences at the interconnection level.

The penetration occurs through the grid, which serves as a link between the dispatch centres for generation and load. The ratio of a load's maximum perceived power to its maximum PV power is known as PV penetration [3]. There are certain negative effects of increased PV grid penetration. As PV penetration increases, these worries typically get progressively worse. The placement of the grid connection point for the PV system as well as the kind and state of the legacy devices presently placed on the grid make these problems more challenging. Given the growing number of PV systems linked to the low-voltage distribution grid, it is imperative to examine the challenges that high PV penetration provides for the network systems of distribution grids, as well as some potential solutions to these concerns [5]. Voltage fluctuations and problems with voltage regulation are just a couple of the effects of penetration that have been carefully studied [1]. The output of the PV system may change if the voltage at the common coupling point fluctuates. The electricity produced is in charge of adjusting load and reversing power flow into the distribution network because PV is so widely used. Another factor that can lead to unbalance, which could be dangerous, is a sizable amount of reverse power flow. Examples include a rise in current through a short circuit and a spike in the distribution feeder's voltage. Distortion due to harmonics of voltage as well as current waveforms has grown to be a serious issue due to additional PV systems being installed in distribution networks.

This study will conduct a more complete assessment in order to more accurately assess the advantages that smart inverters could offer on a feeder for distribution with a high solar power utilization. Smart inverters may limit power loss, the impact on voltage regulator performance, and the high voltage difficulties brought on by increased PV adoption, which are their three key advantages [8]. The report examines the potential use of intelligent inverters to connect sizable PV farms that are located further down on a feeder. A real feeder is used rather than a synthetic or condensed distribution system. Modelling system operation over a full year is done using year-long high-resolution PV and load profiles. To accurately represent the operation of voltage regulators, the simulation imitates a genuine controller.

2. System Modellng

An MPPT controller is a device that is added to the PV array after a boost converter boosts the low voltage of the photovoltaic cells in the array to the required voltage levels in

[a]ku.shaileshmadhavraodeshmukh@kalingauniversity.ac.in, [b]ikhar.avinash@kalingauniversity.ac.in

DOI: 10.1201/9781003675259-28

order to determine the PV system's maximum power point. The VSC controller is also connected to it, which controls active power and control current in the inner loop using a double loop control strategy. After that, the grid, filters, and feeders are all linked. Figure 28.1 displays the block diagram of a grid-connected PV array.

3. Methodology

The boost converter is among the most straightforward switch mode converters. The voltage that is input is raised or increased. Four electronic components: a diode, a capacitor, an inductor make up the sole four parts of this circuit. Both on and off are the two modes of operation. The boost converter's circuit is depicted in Figure 28.2.

3.1. Used control technique

Figure 28.3 shows the whole VSC controller. This type of control is operated in double loop control mode. An inner current loop (Id) and an outer voltage loop (Iq) make up the system. The grid's dc-link's voltage and current are managedusing proportional-integral (PI) controllers. The outer loop (Vdc) is used to regulate or stabilize the voltage across the dc-link. It should be pointed out that the current that reacts to the reference is set to zero during normal operation, while the active current reference (Id*) is specified by the voltage loop output. Finally, a PWM signal generator uses Vabc to generate VSC switching pulses.

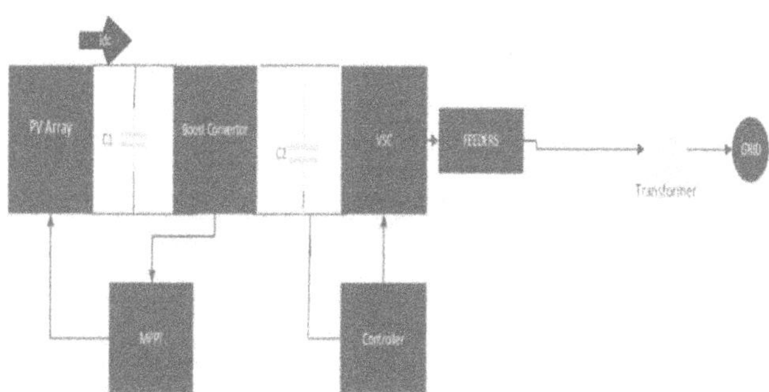

Figure 28.1. Block diagram of grid integrated PV array.

Source: Author.

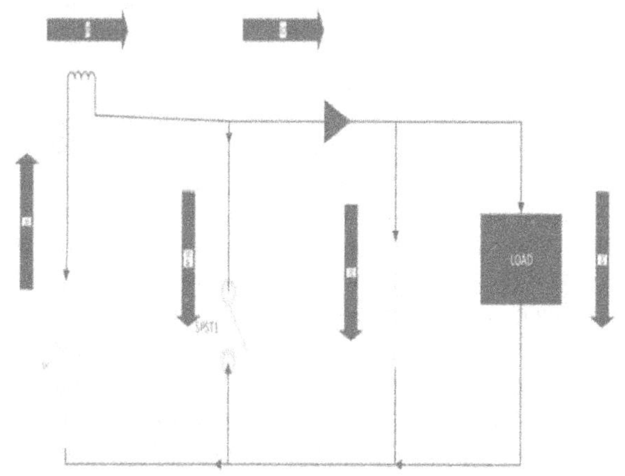

Figure 28.2. Block diagram of converter.

Source: Author.

3.2 Effect of solar energy penetration on the rate of sensitivity at the local level of PV penetration

The PV penetration in the EI may be close to 100% in the PJM_ROM region as a result of its high energy costs. In this area, the largest NERC-recommended contingency was simulated to test the implications of 100% PV adoption. To assess the accuracy of the regional response to frequency metrics, four different situations with comparable power loss magnitudes have been replicated at various sites within the PJM_ROM region. The positions of these four occurrences are shown in Figure 28.3, and the frequency sensitivity metrics for each are shown in Figure 28.4. Because all prediction errors are under five per cent of the average value for several fixed amplitude events occurring at various locations, Figure 28.4 shows that none of the probability response metrics are influenced by the location of the contingency.

4. Results and Discussion

The first regulation zone and downstream zone of the two possible PV field installation instances, that is, upstream zone and downstream zone, are compared below using the sample feeder simulation software.

4.1. Used operation

Figure 28.5 shows the implications of both high penetration of solar energy deploy scenarios on LTC operations. Though the traditional PV example and the downstream deployment costs are different and the smart PV scenario is quite marginal, with an additional 100 LTC actions. The data

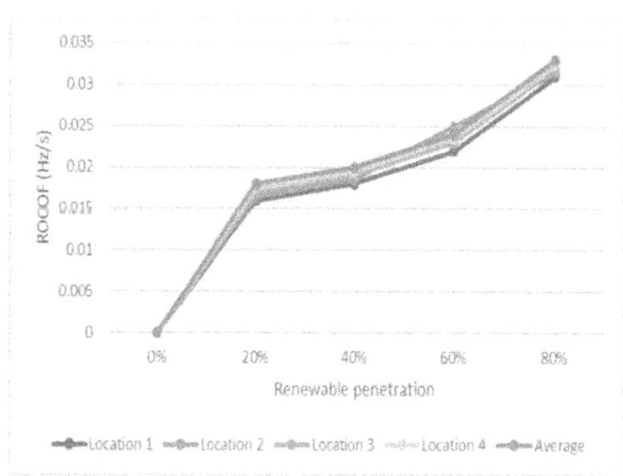

Figure 28.3. Frequency of renewable energy with various location.

Source: Author.

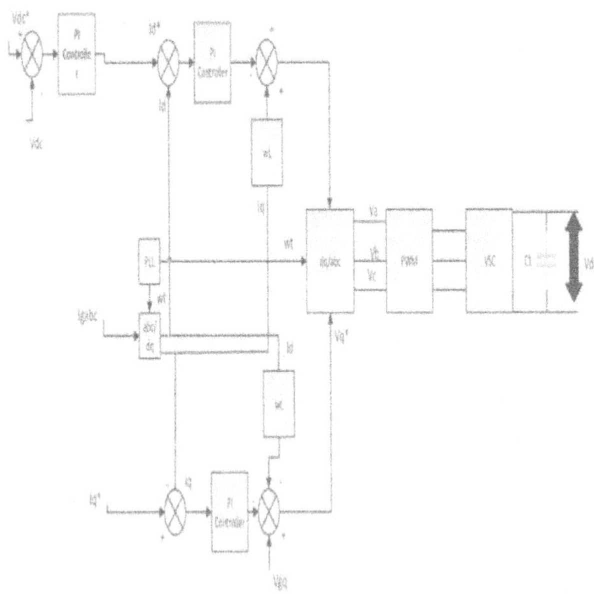

Figure 28.4. VSC controller.

Source: Author.

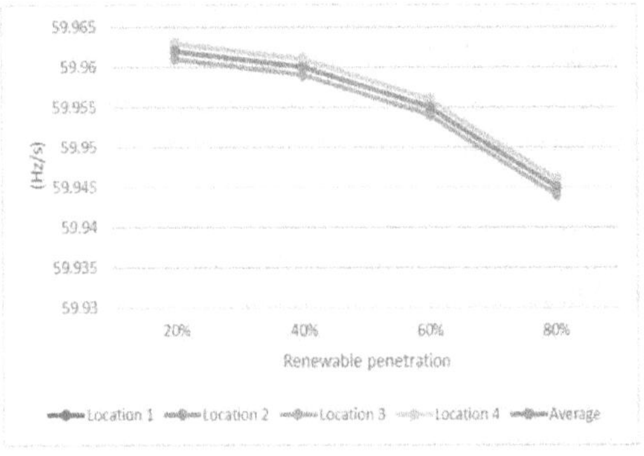

Figure 28.5. Comparison of consumption of pv power as down stream and upstream.

Source: Author.

demonstrates that, in both deployment scenarios, PV deployment significantly decreases overall LTC operations.

Figures 28.6 and 28.7 illustrate how the installation of PV, whether in conventional or smart form, leads in a decrease in LVR processes for each of the eight-line voltage controllers in the upstream PV deployment instance. However, considerable rise in all LVRs during traditional PV deployment and a significant decrease with intelligent PV deployment; for a number of LVRs, their total operations are significantly lower than their base case activities. Results from capacitor bank activities in this feeder could not be distinguished or were continually impacted.

4.2. Power consumption

The upstream PV deployment option has a somewhat larger circuit peak kVA demand, as can be shown. In the deployment

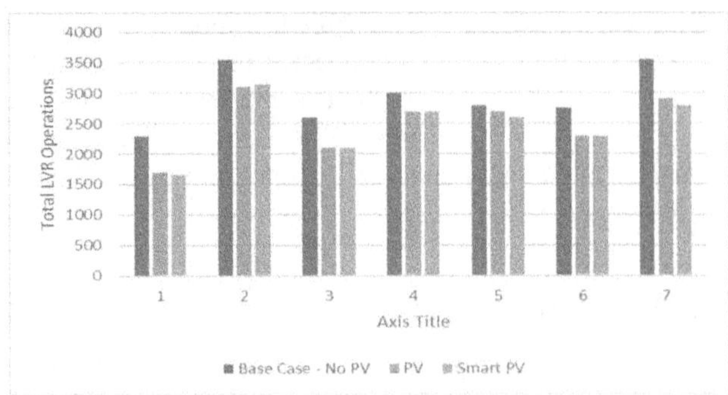

Figure 28.6. Consumption of pv power as upstream.

Source: Author.

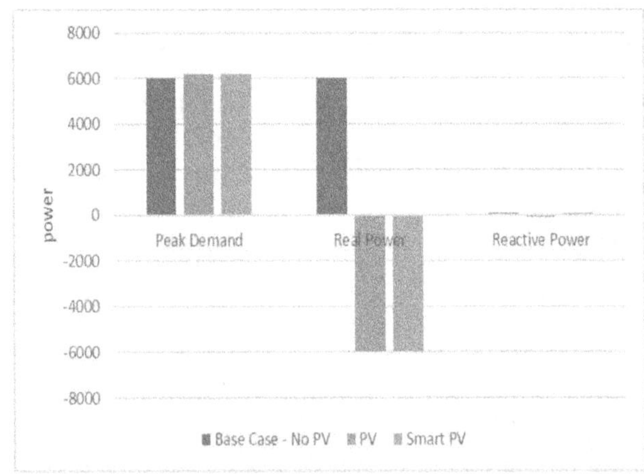

Figure 28.7. Consumption of pv power as down stream.

Source: Author.

situations both upstream and downstream, these statistics also demonstrate a sizeable reverse power.

5. Conclusions

Two different installation configurations are studied in this study in order to properly understand the benefits that intelligent photovoltaic inverters may provide on a power grid. This use scenario does not call for intelligent inverters since PVs close to the station have little effect on the system's voltage regardless of the actual flow of electricity is reversed. The best application of smart inverters is to reduce this voltage increase by PV curtailment or VAR absorption. High voltage problems could be caused by PVs deployed downstream and away from the substation. Using conventional and intelligent PV inverters, upstream PV deployment can reduce LVR operations somewhat. Smart inverters, on the other hand, contribute to a significant reduction in LVR operation in the downstream deployment scenario, when LVR operations with regular PV increase. Additionally, According to the findings of the EI national frequency check, certain places with significant solar power installation must get main frequency response assistance from other regions with conventional generating governors in order to keep up with the power grid's frequency response regulation.

References

[1] Dubin, K. (2021). EIA projects renewables share of US electricity generation mix will double by 2050. Energy Information Administration. Retrieved from https://www.eia.gov/todayinenergy/detail.php.

[2] Nalley, S., & LaRose, A. (2021). *Annual energy outlook 2021*. United States Energy Information Administration: Washington DC.

[3] Nagarajan, A., & Jensen, C. D. (2010). A generic role based access control model for wind power systems. *Journal of Wireless Mobile Networks, Ubiquitous Computing, and Dependable Applications*, *1*(4), 35–49.

[4] Ustun, T. S., & Aoto, Y. (2019). Analysis of smart inverter's impact on the distribution network operation. *IEEE Access*, *7*, 9790–9804.

[5] Mishra, D., & Kumar, R. (2023). Institutional repository: A green access for research information. *Indian Journal of Information Sources and Services*, *13*(1), 55–58.

[6] Sharma, A., & Dwivedi, V. K. (2020, December). Effect of spindle speed, feed rate and cooling medium on the burr structure of aluminium through milling. In *IOP conference series: materials science and engineering* (Vol. 998, No. 1, p. 012028). IOP Publishing.

[7] Stevovic, I., Hadrović, S., & Jovanović, J. (2023). Environmental, social and other non-profit impacts of mountain streams usage as Renewable energy resources. *Archives for Technical Sciences*, *2*(29), 57–64.

[8] Anas, A. K., Alaa, J. M., Anwer, S. A., & Laith, A. A. (2024). Control system design for failure starting of diesel power block for cell on wheels communication tower based on cloud service system. *Journal of Internet Services and Information Security*, *14*(3), 275–292.

[9] Seguin, R., Woyak, J., Costyk, D., Hambrick, J., & Mather, B. (2016). High-penetration PV integration handbook for distribution engineers (No. NREL/TP-5D00-63114). National Renewable Energy Lab. (NREL), Golden, CO (United States).

[10] Seneviratne, C., & Ozansoy, C. (2016). Frequency response due to a large generator loss with the increasing penetration of wind/PV generation–A literature review. *Renewable and Sustainable Energy Reviews*, *57*, 659–668.

29 Unveiling interdependencies between Alzheimer's disease and lung cancer through integrated CGNN modelling

G. Akiladevi[1,a], *M. Arun*[2,b], *and J. Pradeepkandhasamy*[2,c]

[1]Research Scholar and Department of Computer Applications, Kalasalingam Academy of Research and Education Krishnankoil, India
[2]Assistant Professor and Department of Computer Applications, Kalasalingam Academy of Research and Education Krishnankoil, India

Abstract: Alzheimer's disease and lung cancer represent the pressing challenges in healthcare domain, each with intricate biological underpinnings and shared patient demographics. This work proposes an innovative computational framework to discern whether lung cancer occurrences are influenced by underlying Alzheimer's disease, leveraging the synergy of Convolutional Neural Networks (CNNs) and Graph Neural Networks (GNNs). Using the combined power of GNNs and CNNs, this work presents a novel computational framework to determine whether underlying Alzheimer's disease influences lung cancer occurrences. The CNN component adeptly processes high-resolution CT scan images, extracting nuanced features indicative of lung cancer pathology with unprecedented accuracy and granularity. Simultaneously, the GNN assimilates and analyzes complex patient metadata, structured as a graph to capture intricate relationships such as genetic predispositions, comorbidities, and potential Alzheimer's disease diagnoses. Our model provides a holistic picture of illness etiology that goes beyond clinical presentations by combining these modalities, hence going beyond the conventional diagnostic paradigms. Our methodology allows for robust prediction of the association between Alzheimer's disease and lung cancer through extensive training on a variety of datasets, such as clinically validated labels, patient profiles, and CT imaging data. Evaluation criteria including accuracy, precision, recall, and scores demonstrate the model's utility in revealing complex connections between diseases and underscore its potential to offer fresh insights into the complex interrelationships between neurodegenerative processes and oncological outcomes. This interdisciplinary endeavour is pushing the limits of AI-driven healthcare and has transformative promise for focused therapy interventions and tailored medicine. Our method has the potential to transform clinical decision-making by uncovering hidden prognostic indicators and connections. Improved patient outcomes and tailored treatment regimens will be made possible by this.

Keywords: Alzheimer's disease, lung cancer, convolutional neural networks (CNN) and graph neural networks (GNN)

1. Introduction

Both Alzheimer's Disease (AD) and lung cancer are major and complex issues in the field of modern healthcare, Both have a major effect on individuals' health and wellbeing. AD is a neurological disease that affects many of persons worldwide and puts a significant strain on people, their households, and medical systems. Lung cancer is one of the world's top causes of lung cancer-related death due to its aggressive nature and multitude of etiological variables. Recent epidemiological studies have revealed an intriguing overlap in the incidence of AD and lung cancer, especially among aging populations. Although these conditions have historically been thought to be separate diseases with different pathophysiological mechanisms. This intersection has sparked interest in exploring whether underlying mechanisms or genetic predispositions associated with AD may influence the onset or progression of lung cancer, and vice versa. Due to the ability to provide precise anatomical and pathological insights, medical imaging advancements, especially high-resolution computed tomography (CT), have completely changed the diagnosis and treatment of disease. CT imaging is essential for early diagnosis, lesion characterisation, and disease progression monitoring in the context of lung cancer. It is an essential tool for oncological research and therapeutic practice due to its capacity to record minute variations in the shape of lung tissue. Combining CT imaging with state-of-the-art computational techniques like artificial intelligence (AI) and machine learning could significantly benefit in understanding the complex link between AD and lung cancer. Convolutional Neural Networks (CNNs) are a type of deep learning algorithm that excels at reliably and rapidly extracting complicated features from medical images. CNNs are able to detect small abnormalities in lung CT scans that may be related to lung cancer and may be impacted by underlying neurological processes in patients

[a]akilaporkodi@gmail.com, [b]vgsm.arun@gmail.com, [c]honey.kanda@gmail.com

DOI: 10.1201/9781003675259-29

with AD. Furthermore, the application of Graph Neural Networks (GNNs) provides an additional method for combining various patient data and clinical characteristics. When dealing with heterogeneous datasets that are organized as graphs and contain genetic profiles, demographic data, and medical histories, GNNs are particularly good at capturing intricate associations. The identification of common biomarkers, prediction patterns, and underlying biological pathways that could clarify the reciprocal effects of AD and lung cancer is made easier by this multidimensional study. The major purpose of this effort is to establish a unique computational framework that employs GNNs and CNNs in tandem to uncover new information about the link between cognitive impairment and carcinoma of the lungs. Through the application of AI-driven approaches, the work seeks to clarify the potential influence of neurodegenerative processes linked to AD on lung cancer outcomes, as well as the potential differences in lung cancer pathology among individuals with AD. In addition, by discovering individualized biomarkers and prediction models that improve prognosis, early detection, and treatment approaches catered to AD patients and its effects on lung cancer susceptibility and progression, this work aims to further precision medicine. This work aims to lay the groundwork for focused interventions that enhance patient outcomes and guide future developments in customized healthcare by filling knowledge gaps and utilizing state-of-the-art technologies.

2. Related Work

Interest in the connection between asbestos exposure and cognitive impairment AD has grown in recent years, mostly due to probable common molecular pathways and overlapping demographic markers. Given their different pathophysiological routes and clinical presentations, it is critical to understand any potential association between lung cancer and cognitive impairment AD. While lung cancer presents as a malignant tumour frequently associated with smoking and environmental exposures, AD predominantly affects cognitive function and memory owing to neurodegeneration. Recent results hint to shared molecular pathways, such as impairment of cell cycle regulation, cell death, and DNA repair mechanisms, connecting various apparently unrelated disorders. The hallmark of AD is neuroinflammation, which may also play a role in systemic inflammation influencing the advancement of lung cancer in organs such as the lungs. The association is further complicated by aging and lifestyle choices like smoking, which affect lung cancer incidence as well as AD risk. To unravel their intricate relationship, addressing diagnostic difficulties resulting from overlapping symptoms and investigating epidemiological data are essential. By elucidating these connections, we can potentially uncover novel therapeutic strategies and improve outcomes for patients affected by AD, lung cancer, or both conditions concurrently.

2.1. Alzheimer's disease

Recent research has presented a novel feature representation technique for AD classification. The author [1] enhanced accuracy over previous techniques by incorporating morphometric and textural characteristics. Their approach, based solely on MRI data, produced a remarkable 96.23% accuracy in diagnosing distinct stages of AD, as well as excellent sensitivity and specificity. This highlights the potential of their technology for clinical applications in identifying AD. AD [2] is a type of dementia that affects millions of individuals worldwide. It causes brain cell loss, resulting in memory issues, difficulties thinking, and forgetting recent events. Early identification of AD is crucial for both patients and families. Electroencephalograph (EEG) impulses are utilized to properly identify disease. Noise and artifacts are removed from EEG data using an algorithm known as Independent Component Analysis (ICA). The wavelet transform is then utilized to identify four key elements in the EEG signals. These traits aid in illness classification via a technique known as Support Vector Machine (SVM). Patient monitoring systems use GPS and GSM technology. They enable caretakers to monitor Alzheimer's patients so that they can travel securely without constant supervision. The author [3] addresses utilizing electroencephalograms (EEG) to distinguish between healthy people, those with Mild Cognitive Impairment (MCI), and those with AD. They use Fourier power across five frequency bands and mean power in five brain areas to generate a tensor for each patient. This tensor is used to generate characteristic filters for identifying cognitive states with linear and nonlinear classifiers. This work [4] presents a classification technique that uses MRI data to distinguish between AD, Mild Cognitive Impairment (MCI), and Normal Controls (NC) in senior adults. The method includes atlas-based MRI normalization, PCA-based feature extraction, and a kernel SVM decision tree optimized with PSO, resulting in an 80% classification accuracy and 0.022s prediction times per patient. The author [5] examines estimations that suggest that AD would impact roughly 115 million individuals worldwide by 2050. It appears that you want to investigate recent segmentation and classification techniques used in the detection of AD using human brain MRI data. This is important considering the anticipated rise in dementia incidence worldwide and the difficulties in analysing MR data. These strategies seek to improve early detection and enable preventive interventions.

2.2. Interrelationship between Alzheimer's disease and lung cancer

The most prevalent kind of cancer and the world's leading cause of death is lung carcinoma. A terrible brain illness that affects many individuals worldwide is dementia. Although these diseases are distinct, some research suggest they may be linked. However, this research disagrees on how they are related because they used earlier analysis methodologies. In this work, [6] took a novel strategy to discovering an association between dementia and carcinoma of the lungs. They began by identifying genes that are common to both disorders using a technique known as moderated t-tests. They next investigated these genes using a deep learning technique known as autoencoders (AEs). The AEs constructed pseudogenes, which are simplified representations of each disease's genes. Bi et al. [7] employed mice with Alzheimer's-like symptoms and injected them with lung cancer cells. They used fecal transplants to modify the intestinal microorganisms of the mice. The results revealed that mice with Alzheimer's-like symptoms had larger tumours than normal mice. This was connected to alterations in cancer cell growth and survival. The author also discovered variations in gut bacteria: Alzheimer's-like mice had more Prevotella and Mucispirillum bacteria and fewer Bacteroides and Bifidobacterium. These findings show a link between AD and lung cancer, with gut bacteria and Alzheimer's-related genes potentially influencing cancer formation. To investigate the prevalence of AD dementia in patients with lung cancer and the incidence of lung cancer in people with AD, Musicco et al. [8] undertook a cohort study in Northern Italy with over a million residents. The work covered the period from 2004 to 2009 and used data from the AD dementia and tumour registries (ASL-Mi1) of the local health authority based on medication prescriptions, hospitalizations, and payment exemptions. Patients with AD dementia had a halved risk of cancer, while patients with cancer had a 35% reduced risk of AD dementia, which was similar across different subgroups. These findings point to a complex interaction between aging, AD dementia, and cancer that is unaffected by any confounding variables.

In a novel case-control work, Realmuto et al. [9] investigated the relationship between tumours that develop before AD and AD itself. The work also looked at tumour grading and how it relates to the endocrine system, being the first of its type. What they discovered was a clear inverse relationship between AD and lung cancer before it manifested, especially for endocrine system tumours and in women. These findings point to a possible role for estrogen as well as other genetic, environmental, and interplaying factors that predispose people to AD or lung cancer. The findings may be supported by biological mechanisms that function in opposition to each other in the two illnesses, such as distinct impacts on cell growth and survival, according to the study. Finding these processes may provide important new information for preventative treatment approaches for both conditions.

AD and cancer both have pathophysiological pathways that are not fully understood. The author [10] have found an interesting inverse correlation between these two illnesses in a prospective longitudinal study. Individuals with a history of AD developed cancer at a considerably slower rate over time than those without a history of cancer. Dysregulated cell survival and proliferation pathways are hallmarks of cancer, while AD is defined by neuronal death, frequently linked to the buildup of tau and beta amyloid (Aβ). The theory of common disruptions in the regulatory processes governing cell survival and death is examined. These conditions may be caused by genetic polymorphisms, DNA methylation, or other processes that change the function of molecules essential for deciding cellular fate (apoptosis versus repair). The Wnt signaling pathway, Pin1, and p53 are some of the specific hypotheses that have been suggested as possible explanations for the negative relationships between AD and cancer that have been identified. These hypotheses indicate directions for future investigation into common biological mechanisms affecting the etiology of disease.

Cross-sectional research has indicated that dementia of the Alzheimer type (DAT) and cancer may be protective against one another, or that there may be a shared underlying mechanism that influences the development of both conditions. The author [11] prospective longitudinal analysis yielded deeper insights than cross-sectional techniques. When comparing persons with DAT to those without dementia, they found a considerably lower risk of malignancy ($p < 0.001$). On the other hand, a tendency indicated that individuals with a history of cancer had a decreased risk of acquiring DAT ($p = 0.060$). These findings demonstrate the complex relationships between Dementia and malignancy and offer proof of potential shared molecular mechanisms that could influence how the diseases develop. Subsequent research is necessary in the fields of genetic predispositions, epigenetic variables such as DNA methylation, and molecular processes controlling apoptosis and cellular survival.

3. Methodology

The proposed study is a groundbreaking attempt to use cutting-edge AI tools to examine the connection between cognitive decline and lung carcinoma. This paper presents a unique computational framework that integrates high-resolution imaging data with complicated patient metadata using the combined abilities of graph artificial neural networks (GNNs) and neural networks with convolution (CNNs). This dual-modality method improves disease detection accuracy while also providing deeper insights into the potential etiological linkages between neurodegenerative and oncological diseases, paving the path for novel diagnostic and treatment techniques.

3.1. Lung cancer caused by Alzheimer's disease

This section describes the methodical strategy used to collect and select data for studying the association between dementia and lung tumours. To provide complete analysis, the collection includes CT scan pictures and patient metadata gathered from several sources.

3.2. *Dataset description*

The data was gathered from government hospitals in collaboration with medical institutions and research databases. The study included 244 participants, including 98 diagnosed with lung cancer. 146 people without lung cancer served as controls. High-resolution CT scan images of lung cancer patients were gathered and annotated for tumours and other pathological characteristics crucial to the diagnosis. Demographic information, medical history, genetic markers, and AD diagnoses were all included in the comprehensive patient records. Medical practitioners annotated CT scan pictures to identify areas of interest, such as nodules, masses, and other anomalies associated with lung cancer. Patient metadata was thoroughly evaluated and labeled to classify patients depending on disease state (AD, lung cancer, both, or none). This large dataset is the foundation of our research, which aims to unearth new insights into the intricate interaction between AD and lung cancer.

3.3. *Preprocessing*

In order to investigate the intricate relationship between asbestos exposure and cognitive impairment, it was necessary to prepare CT scan images and patient metadata thoroughly. The CT scan pictures, which are critical for diagnosing lung cancer pathologies, went through multiple preparation processes to ensure effective analysis by CNNs. To begin, scaling the photographs to a consistent dimension was critical to ensuring consistency and efficient processing across the dataset. Normalizing pixel values normalized intensity levels, improving the comparability of features collected from different scans. Additionally, methods for data enhancement like cropping, turning and twisting increased the variety of the dataset. This diversity not only enhanced the training data, but also improved the CNNs' capacity to generalize patterns across different orientations and situations. Simultaneously, the preprocessing of patient metadata required extensive steps to prepare heterogeneous information for analysis. This dataset included demographic information, extensive medical histories, and, most importantly, AD diagnoses, which are critical for evaluating possible linkages between these disorders. Using sophisticated restoration methods, including K-nearest neighbors (KNN), to address missing data ensured that no crucial information was overlooked during analysis. Encoding categorical variables in numerical representations made it easier to integrate them into machine learning methods, allowing for more robust computational analysis. Figure 29.1 depicts preprocessed image for lung cancer patient.

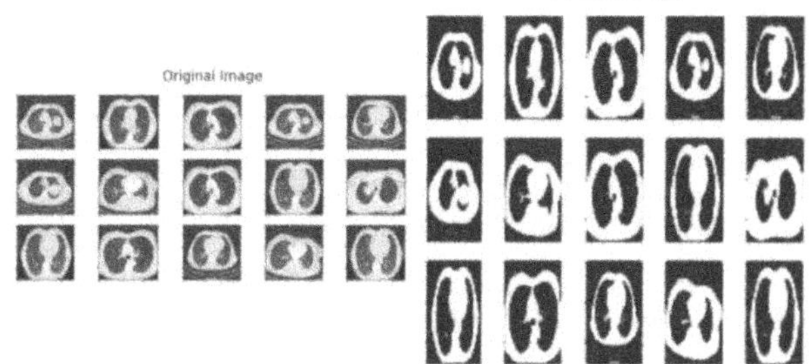

Figure 29.1. Pre-processed image.

Source: Author.

3.4. *Main role of CNN for feature extraction*

A CNN is mostly used in feature extraction when it comes to automatically learning and extracting significant patterns from input images that are pertinent for a certain task, such identifying lung cancer from CT scan images. A CNN is crucial in the feature extraction process from CT scan pictures. The CNN architecture uses many convolutional layers, pooling layers, and fully linked layers to capture both minor and excellent properties. First, the CNN extracts low-level information like edges and textures. Edge detection entails recognizing horizontal, vertical, and diagonal boundaries inside the image, which are critical for defining the shapes and structures that exist. Texture analysis distinguishes between smooth and rough areas, highlighting regions with consistent intensity and those with considerable changes. These characteristics are critical for differentiating different types of tissues and detecting problems. As the image data moves through the network, mid-level features are extracted. These include shapes and contours, such as circular shapes that may indicate nodules or masses, linear patterns that may reflect blood veins or tissue borders, and irregular shapes that signal diseased alterations. Contour mapping clarifies the boundaries of organs and lesions, giving a better picture of the structures within the lungs. High-level features are then recorded, such as complex tissue patterns and nodule and tumour characteristics. The CNN detects nodule size, shape, and density, as well as growth patterns, which are important for detecting and understanding lung cancer progression. Furthermore, the CNN examines contrast enhancements, detecting places with higher contrast that may suggest pathological alterations, and evaluates symmetry to detect asymmetrical patterns that may signal abnormalities.

3.4.1. *Structural abnormalities*

CNNs excel in detecting structural abnormalities in lung CT scan pictures, such as nodules, masses, and other irregularities that could signal lung cancer. These traits are crucial for tumour detection and characterisation.

3.4.2. *Tissue texture and density*

CNNs look at differences in tissue texture and density throughout the lung. This involves finding areas of high density that could indicate malignant growth or structural alterations linked with AD.

3.4.3. *Vascular patterns*

Patterns in vascular structures can provide information about both disorders. For example, changes in blood vessel density and morphology may signal physiological changes associated with disease development in cognitive impairment and carcinoma of the lungs.

3.4.4. *Neurological indicators*

Regarding dementia, CNNs may identify minor changes in brain structure or atrophy patterns that appear on medical imaging scans. These characteristics are critical for understanding how neurological alterations may affect the onset or progression of lung cancer. The result of the CNN feature extraction procedure, as seen in the feature maps, gives a thorough representation of these features:

3.4.5. *Visualization of abnormalities*

The feature maps show places where the CNN has detected potential abnormalities such as nodules or lumps in lung tissue or structural changes in the brain suggestive of AD.

3.4.6. *Texture and density differences*

Bright patches or regions in the feature maps represent locations where the CNN detected differences in tissue texture or density, which are critical for discriminating between healthy and sick tissues.

3.4.7. *Pattern recognition*

The diversity of feature maps indicates the CNN's capacity to distinguish various patterns and structures associated with lung cancer and AD. This includes precise characteristics like vascular patterns or specific neurological signs that are related with Alzheimer's. Figure 29.2 illustrated visualization of feature map that are indicative of lung cancer pathology.

Figure 29.3 depicts training and validation of lung cancer dataset. Both the training loss (red line) and validation loss (blue line) diminish as more phases are added. This suggests the machine is picking up knowledge from the training data on lung cancer. The learning loss is always greater than the validation loss. This shows the framework is effectively generalizing and not excessive fitting to the particular training set. Generalization is critical for making accurate forecasts on previously encountered lung cancer patients.

3.5. *Role of graph neural networks (GNNs)*

GNNs process and analyze complicated patient metadata, which is critical in the research of the interaction link lung disease and cognitive impairment. The major goal of employing GNNs is to detect subtle linkages and dependencies within data that regular neural networks may overlook. Here's a full explanation of how GNNs are used in this research:

3.5.1. *Representation for complex relationships*

Patient metadata frequently incorporates complex associations, such as genetic predispositions, comorbidities, and demographic characteristics. GNNs excel at describing such complicated relationships by organizing the data into a graph.

3.5.2. *Capturing heterogeneous data*

Patient data is intrinsically heterogeneous, including information such as age, gender, medical history, genetic markers, and specific diagnoses (e.g., AD). GNNs can efficiently manage this heterogeneous data by learning representations that take into account both node properties (patient qualities) and graph structure (patient relationships).

Learn from Graph Structures: GNNs use the graph structure to pass information between nodes. This means that the model can learn not only from an individual patient's qualities, but also from those of related patients. For example, a patient with a specific genetic marker may be more likely to get lung cancer if other people with comparable indicators have the condition. GNNs capture these patterns via their message-passing techniques.

3.6. *Integration with CNN features*

CNNs extract features from CT scan pictures, which provide detailed information about lung cancer pathology. By combining these features with the patient metadata analyzed by GNNs, the model gains a comprehensive perspective of both the imaging and relational data. This all-encompassing approach enhances the tool's ability to identify subtle associations and interactions between lung cancer and AD.

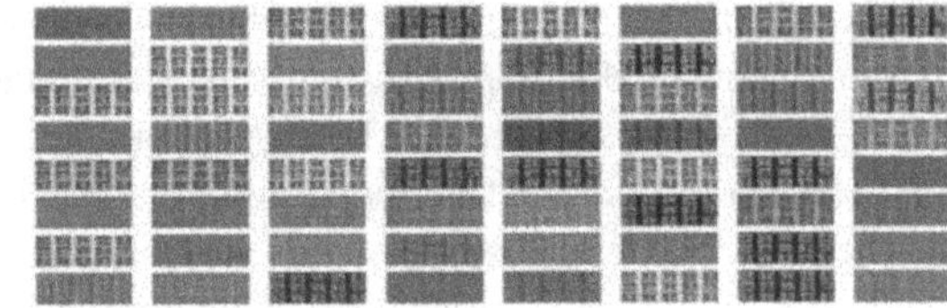

Figure 29.2. Features extracted from CNN.

Source: Author.

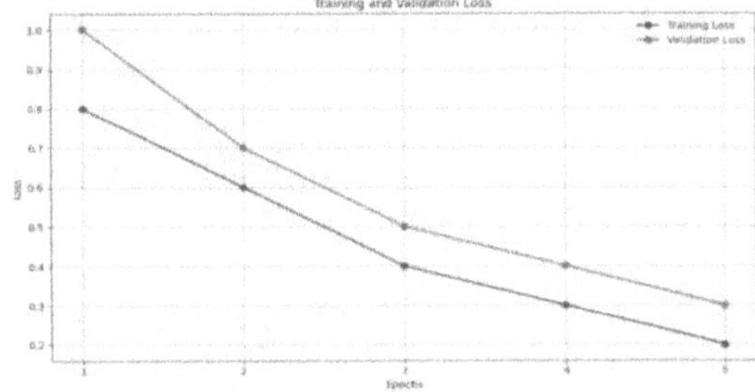

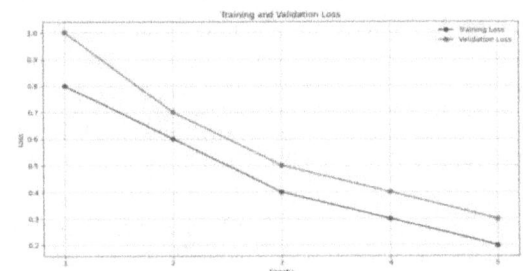

Figure 29.3. Training and validation Loss.

Source: Author.

3.6.1. Holistic analysis and prediction

By merging CNN and GNN, the model performs a holistic analysis that allows it to:

3.6.2. Predict disease interactions

Determine how AD may affect the onset or progression of lung cancer.

3.6.3. Provide personalized insights

Provide insights personalized to individual patients, taking into account their unique characteristics and how they react to others.

3.6.4. Improve diagnostic accuracy

By combining detailed imaging features and extensive relational metadata, diagnoses can be made more accurate. The Figure 29.4 demonstrates the significance of many parameters predicted by a machine learning model for lung cancer in Alzheimer's patients.

The Figure 29.5 represents correlations between characteristics used in a model that predicts lung cancer in Alzheimer's patients. Using accuracy, precision, recall, and F1 score, the performance of several machine learning models was assessed. An accuracy of 0.79, precision of 0.71, recall of 0.68, and F1 score of 0.69 were attained using the Support Vector Machine (SVM). SVM was surpassed by the CNN, which achieved an F1 score of 0.89, accuracy of 0.85, precision of 0.88, and recall of 0.85. The performance of the Bidirectional Long Short-Term Memory (BI-LSTM) model

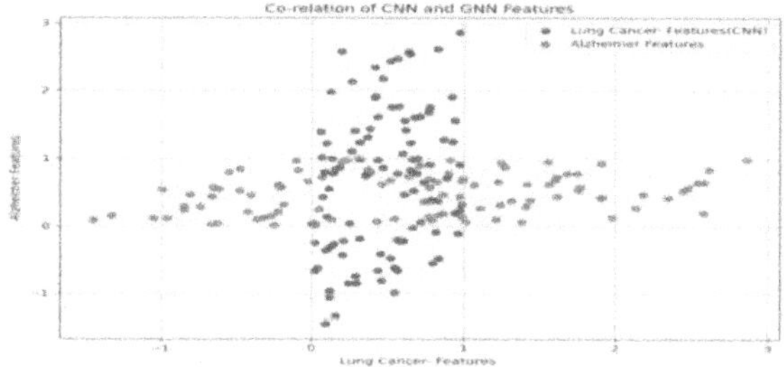

Figure 29.4. Visualization of lung cancer caused by Alzheimer disease.

Source: Author.

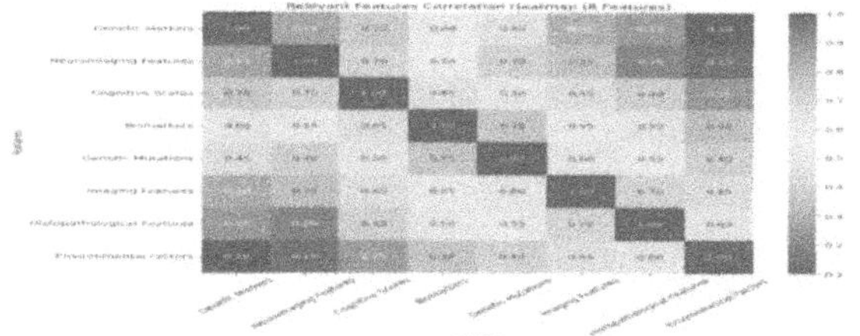

Figure 29.5. Heatmap visualization of feature prediction–lung cancer caused by Alzheimer disease.

Source: Author.

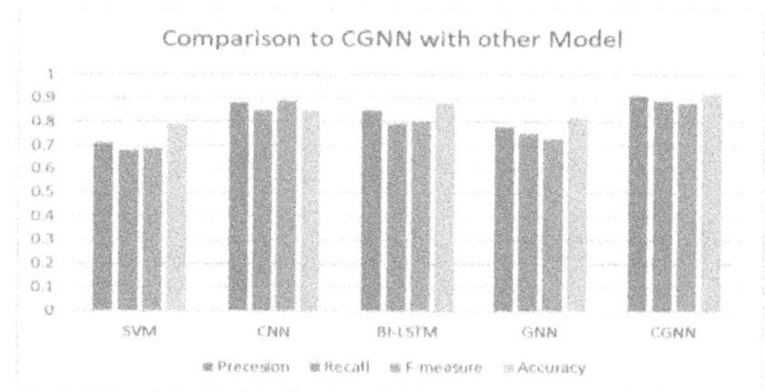

Figure 29.6. Comparison model-CGNN.

Source: Author.

was significantly enhanced, achieving an F1 score of 0.80, accuracy of 0.88, precision of 0.85, and recall of 0.79. With an accuracy of 0.82, precision of 0.78, recall of 0.75, and F1 score of 0.73, the GNN model performed moderately. With an accuracy of 0.92, precision of 0.91, recall of 0.89, and F1 score of 0.88, the Combination of Graph Neural Networks (CGNN) demonstrated the best performance. Figure 29.6 describe the comparative performance of other models against CGNN.

4. Conclusion

Our work presents a ground-breaking way to investigating the impact of AD on lung cancer using modern machine learning techniques. We use CNNs and GNNs to create a comprehensive framework that integrates detailed picture analysis with complex patient metadata analyze high-resolution CT scans to extract essential information including tumour size, shape, and texture. These characteristics are critical for properly diagnosing lung cancer. Simultaneously, GNNs assess complicated patient metadata such as demographics, medical histories, genetic markers, and AD diagnoses. This metadata is organized as a graph, with nodes representing individual patients and edges representing correlations or commonalities, such as shared genetic predispositions or medical histories. Our approach provides a comprehensive perspective of disease relationships by merging the rich image-based features recovered by CNNs with the relational data examined by GNNs. This integrated strategy goes beyond traditional diagnostic paradigms, providing further insight into the potential links between neurodegenerative disorders like AD and oncological outcomes such as lung cancer. Our model's usefulness is proved through rigorous training and evaluation, which yields encouraging results in terms of diagnostic accuracy and ability to detect subtle illness correlations. However, our study has significant limitations, including issues with data availability and quality, computational complexity, and the need for comprehensive validation across varied populations. Despite these limitations, we believe our method has great potential for expanding individualized treatment techniques and increasing patient outcomes. By refining our methodologies and tackling these issues, we hope to increase the effect of AI-powered healthcare and provide fresh insights into the complicated relationships between AD and lung cancer.

5. References

[1] Shankar, K., Lakshmanaprabu, S. K., Khanna, A., Tanwar, S., Rodrigues, J. J., & Roy, N. R. (2019). Alzheimer detection using Group Grey Wolf Optimization based features with convolutional classifier. *Computers & Electrical Engineering*, *77*, 230–243.

[2] Thakare, P., & Pawar, V. R. (2016). August. Alzheimer disease detection and tracking of Alzheimer patient. In *2016 International conference on inventive computation technologies (ICICT)* (Vol. 1, pp. 1–4). IEEE.

[3] Latchoumane, C. F. V., Vialatte, F. B., Jeong, J., & Cichocki, A. (2009). EEG classification of mild and severe alzheimer's disease using parallel factor analysis method: PARAFAC decomposition of spectral-spatial characteristics of EEG time series. *Advances in Electrical Engineering and Computational Science*, 705–715.

[4] Zhang, Y. D., Wang, S., & Dong, Z. (2014) Classification of Alzheimer disease based on structural magnetic resonance imaging by kernel support vector machine decision tree. *Progress in Electromagnetics Research*, *144*, 171–184.

[5] Gulhare, K. K., Shukla, S. P., & Sharma, L. K. (2017). Overview on segmentation and classification for the Alzheimer's disease detection from brain MRI. *Int J Comput Trends Technol IJCTT*, *43*, 130–132.

[6] Sánchez-Valle, J., Tejero, H., Ibáñez, K., Portero, J.L., Krallinger, M., Al-Shahrour, F., Tabarés-Seisdedos, R., Baudot, A., & Valencia, A. (2017). A molecular hypothesis to explain direct and inverse co-morbidities between Alzheimer's Disease, Glioblastoma and Lung cancer. *Scientific Reports*, *7*(1), 1–12.

[7] Bi, W., Cai, S., Hang, Z., Lei, T., Wang, D., Wang, L., & Du, H. (2022). Transplantation of feces from mice with Alzheimer's disease promoted lung cancer growth. *Biochemical and Biophysical Research Communications*, *600*, 67–74.

[8] Musicco, M., Adorni, F., Di Santo, S., Prinelli, F., Pettenati, C., Caltagirone, C., Palmer, K., & Russo, A. (2013). Inverse occurrence of cancer and Alzheimer disease: a population-based incidence study. *Neurology*, *81*(4), 322–328.

[9] Realmuto, S., Cinturino, A., Arnao, V., Mazzola, M. A., Cupidi, C., Aridon, P., Ragonese, P., Savettieri, G., & D'Amelio, M. (2001). Tumor diagnosis preceding Alzheimer's disease onset: Is there a link between cancer and Alzheimer's disease? *Journal of Alzheimer's Disease*, *31*(1), 177–182.

[10] Behrens, M. I., Lendon, C., & Roe, C. M. (2009). A common biological mechanism in cancer and Alzheimer's disease?. *Current Alzheimer Research*, *6*(3), 196–204.

[11] Roe, C. M., Behrens, M. I., Xiong, C., Miller, J. P., & Morris, J. C. (2005). Alzheimer disease and cancer. *Neurology*, *64*(5), 895–898.

30 Low-calcium calcined clay or mortar in an acidic environment

Subrata Majee[1,a] *and Nasar Ali R.*[2,b]

[1]Assistant Professor, Department of Civil Engineering, Kalinga University, Raipur, India
[2]Department of Civil Engineering, Kalinga University, Raipur, India

Abstract: Resources management is a major issue since there are so many different kinds of natural resources to manage. Cement is now widely used in the building industry for making a wide range of structural components. The process of making cement releases a lot of carbon dioxide and other potentially harmful pollutants into the environment. This emission is exacerbated by the widespread practice of using cement in the fabrication of structural formworks. As a result, these mechanisms add to pollution and cause the ozone layer, which acts as a shield, to thin out. In an effort to lessen the effects of carbon pollution, many concerns have been voiced and various measures taken. Utilizing pozzolana is one of these options. In the current experiment, mortar cubes that could cure for up to ninety days were made by substituting Class F fly ash and calcined clay at percentages of 4%, 8%, 12%, 16%, and 20%, respectively, for Portland cement. Investigations were conducted on the cubes regarding homogeneity, compressive strength, start and final setting times, and acid attacks. The cubes' compressive strength was also examined. Studies have shown that pozzolans affect the properties of mortar and that the quantity of pozzolans used determines the strength of the mortar cubes.

Keywords: Resource management, carbon pollution, calcined clay, fabrication

1. Introduction

Cement's prominence as the go-to material for building construction is further proof of its versatility and its status as the second-greatest consumable commodity in construction's long and illustrious history. Cement buildings are known to be relatively long-lasting and reliable in practical climates. Numerous construction methods depend on cement for the creation of cement-based buildings [1]. Dams, connecting bridges, roads, and tall buildings are all constructed using this material because of its widespread acceptance and usefulness in their construction. As the world's population continues to expand, so too will the need for cement in the residential, industrial, and commercial sectors [3, 20]. According to a study published by IBEF [21], cement consumption is forecast to rise to roughly 600 million tonnes annually by 2025. This growth is attributable to the aforementioned rise in share contribution across a number of fields. Cement is used as a binding material, but its manufacturing consumes a lot of energy, depletes natural resources, and releases pollution into the environment [5, 7, 22]. The bulk of the variables that lead to ozone depletion are gases like these [8]. Consequently, several safeguards have been set up to mitigate the effects of toxic gases. Many scientists [12, 23], employ a wide range of admixtures to enhance the properties of cement products in addition to pozzolans. Water reducers, retarders, and accelerators are all examples of admixtures [4].

The influence of the metakaolin and diatomite pozzolana on the durability of the specimens is demonstrated by the data presented by Sharma et al. [9]. The researchers found that when being exposed to diatomite, metakaolin exhibited a marked reactive behaviour. Both pozzolans were effective in preventing chloride ions from penetrating the structure, while metakaolin did so more rapidly. Soriano et al. [10] looked explored what would happen if you mixed metakaolin with a fluid catalyst and used the resulting paste or mortar to make cement. Cement paste and mortar's hydration rate were examined under various curing temperatures and circumstances. Comparing the pozzolanic reaction and hydration processes, the best results were achieved when 15% of the cement was replaced with fluid catalyst rather of a separate pozzolan [2]. When the fluid catalyst was cooled to just 50 degrees Celsius, a study utilizing scanning electron microscopy showed that it demonstrated increased strength and a denser system structure. In this work, Wu et al. [23] investigated the effect of fly ash morphology on the properties of a magnesium oxychloride cement paste. Experiments were conducted using magnesium oxychloride cement paste with varying percentages of fly ash (0, 15, 20, 25, and 35% by weight of magnesium oxide) to examine the effects on setting times, fluidity, and mechanical characteristics. Submicron and X-ray diffraction analysis was then performed on these percentages. Results showed that while fly ash did delay the paste's setting time, it had no effect on the paste's phase composition. In another study, Gorhan and Kavasolu

[a]subrata.majee@kalingauniversity.ac.in, [b]nasar.bala.venkata@kalingauniversity.ac.in

DOI: 10.1201/9781003675259-30

[12] substituted fly ash for cement in mortars that were cured for 28 and 90 days. With the use of microfibers, these mortars were subjected to tests to establish their mechanical strength and durability. They proved that 90-day-cured mortar was the strongest, whereas mortar specimens made with more than 30% fly ash substitution were the weakest. The length of the microfiber also has an impact on the mortar system. Fras et al. [13] looked into the pozzolanic activity of the materials to learn more about the pozzolans' heat development. They looked at the pozzolanic hydration rate variations between mortar and concrete and compared their findings to those from other pozzolanic hydration rates. While both metakaolin and silica fume hydrate at about the same rate, silica fume has far more pozzolanic activity. Tomar et al.'s [14] research shows how environmental factors may change the cement paste, mortar, and concrete's final form. These kinds of events may progressively wreak havoc on buildings [6].

2. Analysis of Used Material

2.1. Used product

2.1.1. Cement

Cement is crucial to the study's success. It is common practice when dealing with a mortar system to use cement to bond the aggregate particles together, making the mortar more solid. For the sake of getting there, we do this. For this study, we used Ordinary Portland Cement (OPC) of grade 43 to guarantee conformity with the specifications detailed in IS 8112-2013 [11, 15]. Cement has a specific gravity of 3.14, as determined by analytical laboratory tests. GLA University in Mathura is offering this online preparatory course (OPC) for students in the 43rd grade. The cement's chemical properties are detailed in Table 30.1.

Fly ash. Fly ash is a form of pozzolanic substance that is created as a byproduct in thermal power plants after the combustion of pulverized coal. It is avoided because people view it as a nasty industrial byproduct. Both 'Class C' and 'Class F' can be used to classify fly ash in ASTM-C618. Class-F fly ash typically has a much lower calcium concentration (lime) than Class-C fly ash [16, 17]. In order to generate electricity, a wide variety of coal kinds were pulverized, and this is what accounts for the variation in the final product's lime content. Different morphologies, physical compositions, and chemical compositions of fly ash emerge from coal with different profiles [18, 19]. Fly ash of Class F was used in the production of mortar samples for this study. Table 30.1 lists the individual chemical components that make up Class.

2.1.2. Calcined clay

Additional finer particles are added to the system to increase the density of the medium, which prevents unwanted substances from penetrating the matrix. The structure's functionality and longevity were greatly improved as a direct result of this. The mortar sample is then mixed with the calcined clay at the predetermined ratio. Carbon dioxide (CO2) emissions are less harmful to the environment when calcined clay is used. Cement substitution with calcined clay now ranges from 4% to 8%, 12% to 16%, and 20% by mass of cement. White in appearance, the particles of calcined clay are even smaller than those of cement (Figure 30.1). Calcined clay has a specific gravity of 2.65 and a fineness modulus of 2.20. Elements and molecules that make up the calcined clay are listed in Table 30.1.

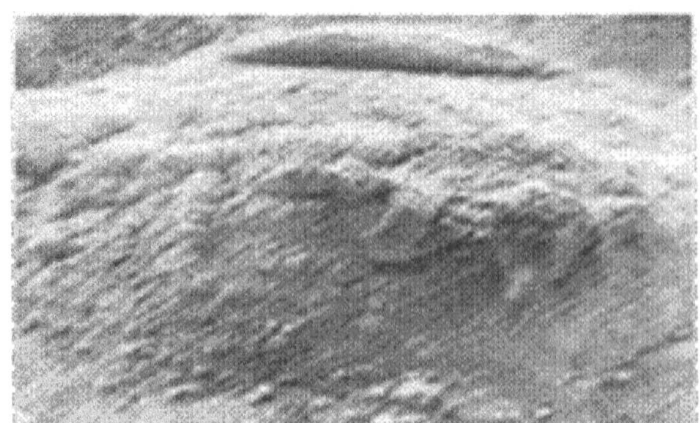

Figure 30.1. Calcined clay powder.

Source: https://images.app.goo.gl/WztyMLaWq3o4Fm8x9.

Table 30.1. Show the quantity of compound that is chemical composition of opc, fly ash, and calcined clay

Compound	*OPC (%)*	*Class F Fly Ash (%)*	*Calcined Clay (%)*
SiO_2	26.32	52.62	48.15
Al_2O_3	8.51	23.21	32.13
Fe_2O_3	4.81	11.17	1.63
CaO	45.43	5.06	0.66
SO_3	2.70	0.81	-
Na_2O	0.16	1.01	0.17
K_2O	0.05	2.0	0.19
LOI	2.51	2.81	14.20

Source: Author.

3. Methodology

3.1. Composition and ingredients of mortar

In accordance with IS 516 [21], several mortar cubes measuring 70.6*70.6*70.6 (in mm) were made in order to calculate the compressive strength. A portion of the cement in the mixture was replaced with calcined clay and fly ash at the following rates: 4%, 8%, 12%, 16%, and 20%. In this experiment, we used a constant cement-to-fine-aggregate ratio of 1:1. At ages 3, 7, and 28 days, the cement mortar cubes were dried; at age 56, they were dried again. To learn more about the material's resilience under stress, several mortar cubes were immersed in sulfuric acid (at a concentration of 3%) for 14, 56, and 90 days.

There are time and regularity requirements. A cement paste was used to test the cement's consistency and predict when it would set. Roughly 200 g of dry cement was used to make the paste. The minimal amount of water necessary to hydrate the cement particles should be the same as the amount of water needed to obtain the specified consistency when mixing the cement with water, which is the outcome of a chemical reaction that takes place when water and cement are united. Cement consistency was evaluated using IS 4031 Part 5 specifications, in which calcined clay and fly ash were replaced at varying amounts (4%, 8%, 12%, 16%, and 20%, respectively).

4. Results and Discussion

Stability in order to evaluate the material's workability and compressive strength, it is crucial to first establish its consistency. It's the determinant of the material's resistance to shear deformation. The ideal amount of moisture is essential. The loss of such a large quantity weakens the system. Texture changes due to varying amounts of calcined clay and fly ash substitution are shown in Figure 30.2. By substituting pozzolana, the relative consistency of each combination has been ascertained. Comparing fly ash up to 12% by weight of cement with calcined clay up to 16% by weight of cement, the consistency of the former is greater. You may review this data by looking at Figure 30.2. The consistency grew with time above the 12% fly ash replacement threshold. Once 16 percent calcined clay replacement is reached, the paste's consistency likewise decreases. The cement's consistency can be influenced by a number of factors, including temperature, humidity, pozzolana and cement fineness, and the technique used to mix the water [18]. Because of this, the consistency variations exhibit usual properties.

Tests that measure setting times can shed light on how quickly a cement paste will harden after being left alone for a given period of time. Several factors affect how long it takes for the setting to occur, including the temperature differential, the quantity of moisture already present in the sample, and the relative humidity of the air. Calculations were made on the drying periods of mixes in which 4, 8, 12, 16, and 20% of the cement were replaced with calcined clay and fly ash. Figures 30.3 and 30.4 show the progression from the first setting to the final setting, respectively. Based on Figure 30.3, we may deduce that the fly ash can set up fastest at a replacement level of 16 percent. Following that, fly ash's initial settling period is significantly shortened. However, it was shown that the first setting time of calcined clay increases with increased amounts, while calcined clay has an initial setting time of 40–58 minutes.

Figure 30.5 depict the effect that fly ash and calcined clay had on the material's compressive strength. Figure 30.5 demonstrates that after 3 days, the compressive strength of mortar cubes made with fly ash is much lower than that

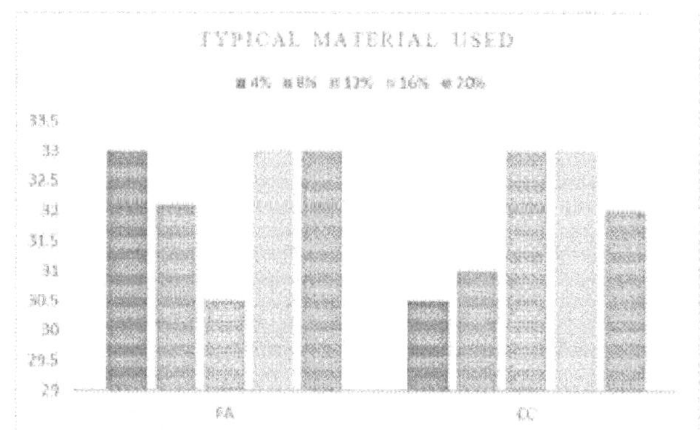

Figure 30.2. Research of the properties of fly ash and calcined clay mortar, namely its consistency.

Source: Author.

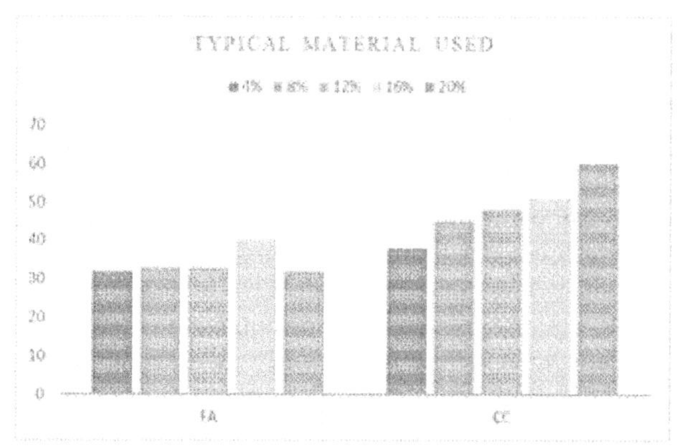

Figure 30.3. Initiation setting times for fly ash and calcined clay are measured.

Source: Author.

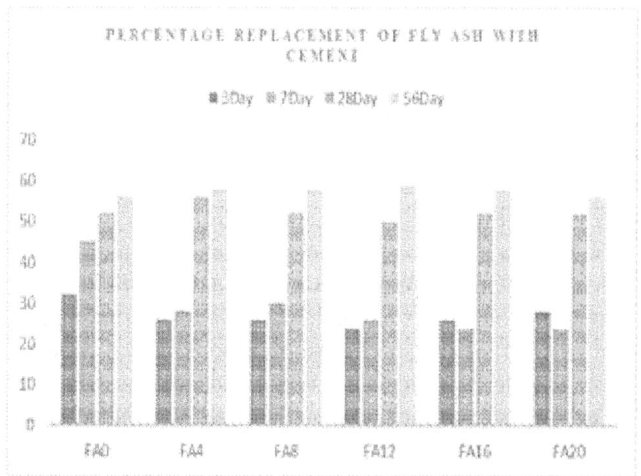

Figure 30.4. Examining the compressive strength of fly ash mortars.

Source: Author.

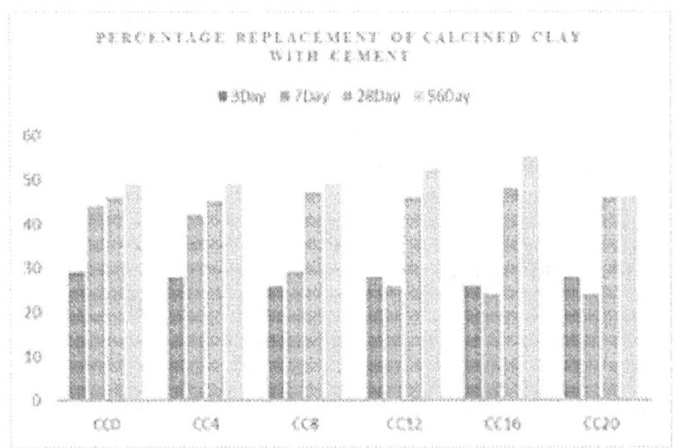

Figure 30.5. Examining the compressive strength of fly ash mortars.

Source: Author.

of mortar cubes made with regular mortar. This was determined after a battery of tests were conducted on the cubes. Perhaps the delayed pozzolanic activity of the fly ash is to account for this weakness. More cubes of mortar will be able to withstand compression after being given more time to cure (Figure 30.5). This suggests that the fraction of fly ash mix with 12% fly ash has the greatest compressive strength. In a 56-day curing period, mortar made from calcined clay and fly ash reaches its maximal compressive strength at a replacement level of 12%. After 12% replenishment, the strength was already noticeably deteriorating. There was no increase in strength of more than 12% when either material was used instead.

5. Conclusions

Initial and final setting of the calcined clay and fly ash paste are observed to increase proportionally with the amount of replacement. Whether the rate of replacement is raised or lowered, this remains true. Paste prepared with 17% fly ash instead of calcined clay has the longest initial setting time imaginable, but the setting time of paste made with calcined clay increases gradually with varying fractions. Because the first setting time of calcined clay paste varies with different fractions, this is the case. Once the product has been given time to cure, it reaches this point. It was found that even after 12 percent of the force was replaced, the strength was still declining. Observations revealed that OPC samples were degrading as a result of the formation's increasing susceptibility to the transport of sulphate ions. That's what led to the deterioration, actually. When compared to OPC and fly ash mortar samples, calcined clay showed much better results in terms of its potential to resist sulfated conditions. Due to the greater temperature at which calcined clay was processed, this was the situation. After 90 days, the calcined clay mortar's strength had reduced by 32%, the fly ash mortar's strength had decreased by 36%, and the OPC mortar's strength had decreased by 45%. Since this is the case, the calcined clay-based mortar outperforms the fly ash mortar and the regular mortar in terms of durability.

References

[1] Seifan, M., Mendoza, S., & Berenjian, A. (2020). Mechanical properties and durability performance of fly ash based mortar containing nano-and micro-silica additives. *Construction and Building Materials*, *252*, 119121.

[2] Mohamed, S., Kumaran, U., & Rakesh, N. (2024). An approach towards forecasting time series air pollution data using LSTM-based auto-encoders. *Journal of Internet Services and Information Security*, *14*(2), 32–46.

[3] Hai, T., Alshahri, A. H., Mohammed, A. S., Sharma, A., Almujibah, H. R., Metwally, A. S. M., & Ullah, M. (2023). Performance assessment and multiobjective optimization of a biomass waste-fired gasification combined cycle for emission reduction. *Chemosphere*, *334*, 138980.

[4] Robles, T., Alcarria, R., De Andrés, D.M., De la Cruz, M.N., Calero, R., Iglesias, S., & Lopez, M. (2015). An IoT based reference architecture for smart water management processes. *Journal of Wireless Mobile Networks, Ubiquitous Computing, and Dependable Applications*, *6*(1), 4–23.

[5] Shukla, A., Gupta, N., & Gupta, A. (2020). Development of green concrete using waste marble dust. *Materials Today: Proceedings*, *26*, 2590–2594.

[6] Surendar, A., Saravanakumar, V., Sindhu, S., & Arvinth, N. (2024). A Bibliometric study of publication-citations in a range of journal articles. *Indian J Inf Sources Serv*, *14*(2), 97–103.

[7] Gonçalves, J. P., Tavares, L. M., Toledo Filho, R. D., & Fairbairn, E. M. R. (2009). Performance evaluation of cement mortars modified with metakaolin or ground brick. *Construction and Building Materials*, *23*(5), 1971–1979.

[8] Sharma, A., Yadav, R., & Sharma, K. (2021). Optimization and investigation of automotive wheel rim for efficient performance of vehicle. *Materials Today: Proceedings*, *45*, 3601–3604.

[9] Sharma, A., Islam, A., Sharma, K., & Singh, P. K. (2021). Optimization techniques to optimize the milling operation with different parameters for composite of AA 3105. *Materials Today: Proceedings*, *43*, 224–230.

[10] Soriano, L., Monzó, J., Bonilla, M., Tashima, M. M., Payá, J., & Borrachero, M. V. (2013). Effect of pozzolans on the hydration process of Portland cement cured at low temperatures. *Cement and Concrete Composites*, *42*, 41–48.

[11] Llopiz-Guerra, K., Daline, U. R., Ronald, M. H., Valia, L. V. M., Jadira, D. R. J. N., & Karla, R. S. (2024). Importance of environmental education in the context of natural sustainability. *Natural and Engineering Sciences*, *9*(1), 57–71.

[12] Görhan, G., & Kavasoğlu, E. (2022). Effect of fly ash on mechanical and durability properties of mortar containing microfibers with different length. *European Journal of Environmental and Civil Engineering*, *26*(4), 1283–1299.

[13] Frıas, M., De Rojas, M. S., & Cabrera, J. (2000). The effect that the pozzolanic reaction of metakaolin has on the heat evolution in metakaolin-cement mortars. *Cement and Concrete Research*, *30*(2), 209–216.

[14] Tomar, R., Kishore, K., Parihar, H. S., & Gupta, N. (2021). A comprehensive study of waste coconut shell aggregate as raw material in concrete. *Materials Today: Proceedings*, *44*, 437–443.

[15] Sharma, A., & Dwivedi, V. K. (2020). Effect of spindle speed, feed rate and cooling medium on the burr structure of aluminium through milling. In *IOP conference series: materials science and engineering* (Vol. 998, No. 1, p. 012028). IOP Publishing.

[16] Cho, Y. K., Jung, S. H., & Choi, Y. C. (2019). Effects of chemical composition of fly ash on compressive strength of fly ash cement mortar. *Construction and Building Materials*, *204*, 255–264.

[17] Sharma, T., Singh, S., Sharma, S., Sharma, A., Shukla, A. K., Li, C., ... & Eldin, E. M. T. (2022). Studies on the utilization of marble dust, bagasse ash, and paddy straw wastes to improve the mechanical characteristics of unfired soil blocks. *Sustainability*, *14*(21), 14522.

[18] IS 516:2014 (2004). Method of Tests for Strength of Concrete. IS 516–1959, New Delhi, India. doi: 10.3403/02128947.

[19] Islam, A., Sharma, S., Sharma, K., Sharma, R., Sharma, A., & Roy, D. (2020). Real-time data monitoring through sensors in robotized shielded metal arc welding. *Materials Today: Proceedings*, *26*, 2368–2373.

[20] Schlangen, E., & Sangadji, S. (2013). Addressing infrastructure durability and sustainability by self healing mechanisms-Recent advances in self healing concrete and asphalt. *Procedia Engineering*, *54*, 39–57.

[21] Gupta, A. (2021). Investigation of the strength of ground granulated blast furnace slag based geopolymer composite with silica fume. *Materials Today: Proceedings*, *44*, 23–28.

[22] Islam, A., Sharma, A., Chaturvedi, R., & Singh, P. K. (2021). Synthesis and structural analysis of zinc oxide nano particle by chemical method. *Materials Today: Proceedings*, *45*, 3670–3673.

[23] Wu, J., Chen, H., Guan, B., Xia, Y., Sheng, Y., & Fang, J. (2019). Effect of fly ash on rheological properties of magnesium oxychloride cement. *Journal of Materials in Civil Engineering*, *31*(3), 04018405.

31 Indian sign language recognition system facilitating communication for individuals with special needs

Nishadha S. G.[a], and R. Murugeswari[b]

Department of Computer Science and Engineering, Kalasalingam Academy of Research and Education, Krishnankoil, Tamil Nadu, India

Abstract: In the realm of human interaction, deaf individuals often encounter difficulties communicating with the general population due to a lack of understanding of Indian Sign Language (ISL). Here we introduce a real-time method employing deep learning to translate Indian Sign Language (ISL codes). Its primary goal is to design a system capable of accurately identifying ISL alphabet representations. By capturing hand frames using a camera module and processing them through CNN and pattern recognition software, the system interprets hand gestures, including those performed with both the hands. By knowing the relevance of hand sign codes as a source for interaction, for people with special needs, the main objective is to develop a desktop application for seamlessly translating hand sign gestures into text in real time. Leveraging ISL datasets with 1500 training datasets and 500 test datasets, along with CNN classification, the research achieves high accuracy in gesture recognition, laying the groundwork for a comprehensive sign language translator. Implemented as a web application using Python Flask, the system ensures accessibility and usability for a broad audience, simplifying interaction between people with hearing impairments and those not familiar with hand sign language. Ultimately, this technology aims to facilitate easier communication between disabled individuals and those who do not understand sign language, marking an initial step towards creating a universally accessible translator which sustains quality of life of mute people.

Keywords: ISL, CNN model, sign language dataset, sign language translator

1. Introduction

Within the vast domain of computer science, Artificial Intelligence (AI) serves as a discipline devoted to understanding human intelligence, with the overarching goal of empowering AI systems to effectively tackle intricate problems. A subset of AI, known as Computer Vision, specifically focuses on extracting meaningful insights from images, a task accompanied by implementation challenges but offering wide-ranging applications across fields such as robotics, medical imaging, automotive technology, and industrial automation. In the realm of human interaction, deaf individuals often face challenges in communicating with the general population due to the lack of understanding of Indian Sign Language (ISL) [1, 2]. To address this communication barrier, this research endeavors to utilize AI, related Pattern Recognition technologies, to create software capable of swiftly recognizing and translating ISL hand gestures in real-time. The system's transformative potential lies in its ability to instantly convert these recognized gestures into readable text, thereby significantly facilitating communication for individuals with special needs. The incorporation of hand gesture recognition serves a dual purpose, extending beyond linguistic translation to play a pivotal role in Human-Computer Interaction (HCI) [3, 4], which involves the human-computer interactions, emphasizes direct engagement of users with the technology. In this context, the envisioned functional interaction for the hand gesture recognition system entails, displaying the alphabets representing hand sign codes, thereby creating a seamless and user-friendly interface. The research methodology involves harness various technologies, which uses a webcam or live camera module, to detect hand sign codes in real-time. Subsequently, the system will undergo training on comprehensive Indian Sign Language (ISL) datasets, utilizing the robust Convolution Neural Networks (CNN) classification system [5]. The primary objective of this investigation is to achieve exceptional accuracy in recognizing alphabet letters, ultimately aiming to provide real-time text results. In essence, this research not only addresses a critical communication gap for the deaf community but also illustrates the transformative possibility of AI, Computer Vision, and Pattern Recognition in fostering an inclusive and accessible digital landscape. By seamlessly translating hand gestures into text in real-time, the desktop application endeavors to enhance communication, understanding, and also contributes to the broader narrative of the technology utilized for the advancement of diverse communities.

[a]sgnishadhasg@gmail.com, [b]r.murugeswari@klu.ac.in

DOI: 10.1201/9781003675259-31

2. Related Works

Sign language can be used as fundamental way of conveying information between mute individuals and normal speaking persons. Given its unique role, especially within the deaf-mute community, where it's often the sole means of communication, there's a pressing need for automated sign language recognition.

A classifier in Indian Sign Language Recognition Using a Novel Feature Extraction Technique was proposed and published [6]. With roughly 5% of the global population utilizing sign language, the paper introduces a method for automatically converting individual ISL signs into text. The research leverages 2600 input images, with 100 single handed sign images per English character, and adopts a hierarchical centroid feature extraction technique alongside k-nearest neighbor and Naïve Bayes classifiers in the experimental phase. The experiments yield a satisfactory accuracy rate of around 97%. While this accuracy underscores the system's proficiency in recognizing sign language signs, it's important to note potential limitations, particularly regarding the k-Nearest Neighbor classifier's sensitivity to noisy or unpredictable data points, which may impact performance in practical settings.

An approach using a neural network model – ‹Indian Sign Language converter using Convolution Neural Networks' got published, in 2019 (I2CT) [7]. Individuals with hearing and speech impairments often encounter significant challenges when communicating with the general public, as their use of sign language is not widely understood. This paper presents an Indian hand sign code image translator, implemented using a Convolution Neural Network, aimed at classifying the real time hand sign images inputted for 26 letters of ISL into their corresponding alphabet categories. Initially, a diverse database was compiled, incorporating hand sign images of persons taken in different situations and employing several image pre-processing techniques to prepare it for feature extraction. Subsequently, the images underwent feature extraction and were inputted into the CNN model [8]. Extensive testing was conducted on both testing and real-time images, yielding 96% positivity for the test images and 87.69% for real-time images. While the tool proves beneficial for individuals facing speech and hearing challenges, achieving high accuracy of 96% in testing images, its performance slightly decreases to 87.69% for real-time photos, suggesting potential difficulties in comprehending the intricacies and variability of sign language motions in everyday scenarios.

A new method was introduced on three dimensional sign language data. The paper titled, ‹Matching with Adaptive Kernels for 3D Indian Sign Language Recognition', got published in 2018 [9, 10]. Recognition of human hand sign gestures poses a complex challenge, due to the complicated mixture of hand and finger expressions, often synchronized with movements of head or face or body. The sign language videos recorded is particularly affected by occlusions, lighting variations, and background noise, complicating the recognition process. This paper proposes a novel approach for characterizing sign language gestures as 3D motionlets, capturing subsets of joint motions at different body parts. The proposed method employs a two-phase fast algorithm to efficiently identify signs from a ranked database of 3D sign language data. By categorizing the database into motionlet classes and analyzing relations between motion joints, the method effectively represents shape information of signs in a 3D spatio-temporal framework. Evaluation on 500-word Indian sign language datasets demonstrates improved recognition performance compared to new 3D sign language recognition methods. While 3D motion capture technology offers more accurate representation than 2D methods, its implementation complexity and resource demands pose challenges, requiring significant processing power and intricate execution due to the multitude of steps and components involved.

The research paper [11] presents an Indian Sign Language (ISL) recognition system using Speeded Up Robust Features (SURF), with Support Vector Machine (SVM) and Convolution Neural Network (CNN). This innovative approach aims to efficiently recognize static hand gestures from the ISL dataset. By employing SURF for feature extraction, and utilizing SVM [12] and CNN classifiers, the work achieves a notable accuracy rate of 96.67% on the test dataset, surpassing existing methods. The system's robustness and efficiency make it a perfect tool for enhancing communication for mute individuals. The literature review highlights the novelty of this approach amidst existing methods such as CNN-based, SVM-based, and hybrid models. The comparison analysis demonstrates superior performance for the proposed work, with regard to accuracy and efficiency, further validating its significance in the area of hand sign Recognition. As a future exploration, the paper suggests avenues for research, including the investigation of alternative deep learning architectures, exploration of diverse datasets, and application in real-world scenarios such as healthcare and education. This research contributes much to the improvement of ISL recognition technology, offering practical solutions to bridge communication gaps among the hearing-impaired community.

3. Proposed Work

Numerous methods have been proposed to detect hand sign codes from captured input images. In this work, sign detection is carried out using Convolution Neural Network [13]. The main intention is to recognize the alphabets represented by hand sign codes from images which has multiple variations including the background. The work focuses on images taken in different backgrounds and in good lighting conditions.

The proposed approach consists of two main phases shown in Figure 31.1.

1. Training Phase: includes Pre-processing, Feature extraction, saving the training an testing dataset with features

of hand sign images, training model creation using Convolution Neural Networks and Evaluation of CNN model [14, 15].

2. Testing Phase: includes using the test dataset features for trained CNN model evaluation and recognizing corresponding hand gesture, then capturing a new hand sign image and testing it using the trained model to get the hand sign recognized.

Mediapipe, Google's leading open source framework is reshaping the landscape of applications in the area of Artificial Intelligence and Machine learning. Featuring a modular architecture, a rich library of pre-trained models, and robust optimizations, Mediapipe empowers developers to effortlessly build sophisticated multimedia processing pipelines. This comprehensive exploration delves into the capabilities and notable features of Mediapipe, particularly in image recognition, hand recognition, and face recognition domains. At its core, Mediapipe offers a flexible and modular framework for constructing various perception-based applications, revolving around the concept of 'Media Pipe graphs', reusable building blocks encapsulating specific tasks or components of multimedia processing pipelines. These graphs can be seamlessly combined and customized to meet diverse application requirements, streamlining development and scalability efforts.

Creating datasets using custom code with Mediapipe involves several steps, including utilizing Mediapipe functions to extract hand landmarks, integrating the holistic model to capture images from a webcam, processing them, and storing them for training and testing purposes.

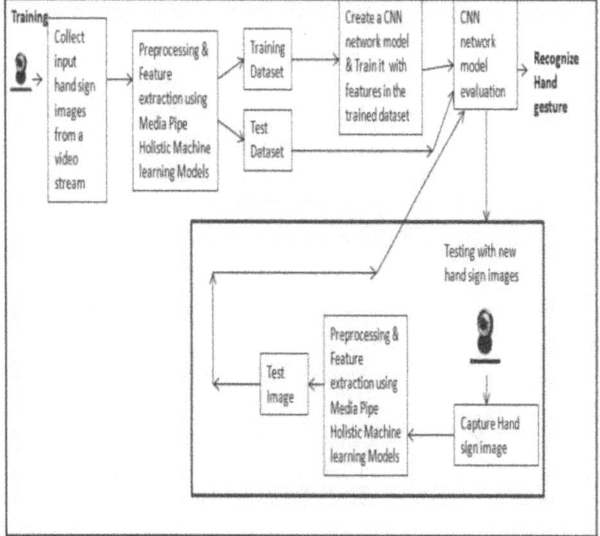

Figure 31.1. Working of ISL recognition model 1.

Source: Author.

3.1. Integrating Holistic model

Mediapipe's holistic model captures facial landmarks and hand landmarks, providing detailed information about poses and movements.

3.2. Holistic imagery overview

In Mediapipe, holistic data, including facial landmarks and hand positions, is collected using the holistic model provided by the framework. The process involves passing input images through a Mediapipe pipeline, which consists of components for face detection, facial landmark detection, and hand landmark detection. The output includes coordinates of detected landmarks, providing a detailed representation of human poses and movements in the input images. Optionally, the detected landmarks can be visualized on the input images for analysis and debugging purposes.

3.3. Utilizing mediapipe functions

- mediapipe_detection: This function serves as the entry point for leveraging Mediapipe's capabilities, converting images from BGR to RGB format, processing them using the specified model, and then converting them back to BGR format.
- extract_keypoints: This function extracts key points from the detected landmarks, providing a comprehensive representation of hand poses in the image.

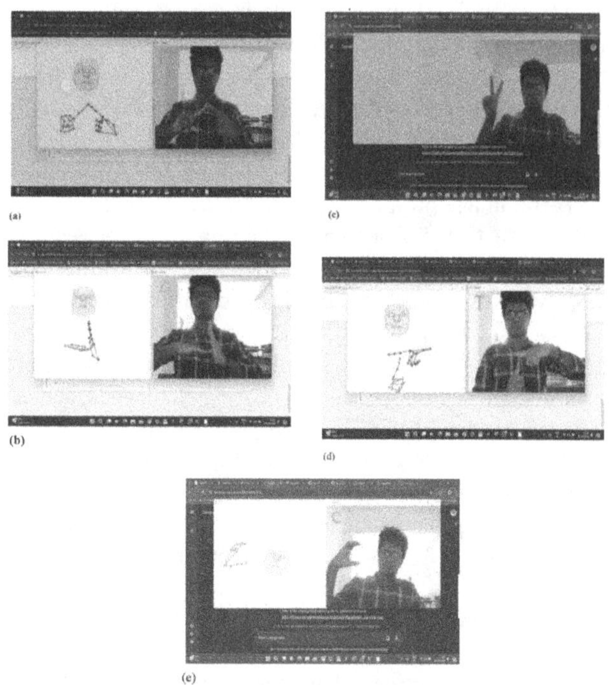

Figure 31.2. Sample recognized hand sign codes.

Source: Author.

- draw_styled_landmarks: This function visualizes the results of Mediapipe's processing by drawing styled landmarks on the input image.

4. Result Analysis

Following outputs in Figure 31.2 shows the recognized hand gestures for hand sign codes representing the alphabets E, Z, C, T and V. Also, Figure 31.3 shows a visualization of accuracy and loss for 14 epochs of training.

5. Conclusion

In conclusion, an Indian Sign Language Recognition System proposed, represents a successful endeavor aimed at fostering inclusive communication for individuals with disabilities. Through the integration of accessible technology and thoughtful design principles, we demonstrate a commitment to leveraging innovation for social good, offering meaningful solutions to address communication barriers within this community. The system's implementation prioritizes data-driven hand and face gesture recognition, utilizing a comprehensive customized sign language dataset of 2000 hand sign images in training dataset and 200 in Test with the help of Media pipe and Python architecture. This technological foundation ensures accurate and real time interpretation of sign language gestures, forming the cornerstone of our system's effectiveness. Furthermore, our user-centric interface stands as a significant contribution, emphasizing accessibility and user-friendliness. By providing a dynamic and customizable platform for effective communication, tailored to diverse user needs, we recognize the importance of personalization in enhancing user experience. Developed with active involvement from sign language users, our system prioritizes usability and practicality in real-world scenarios. Moving forward, our focus includes expanding the dataset with hand sign images captured from the real practitioners of ISL, refining deep learning models to address real world challenges. Beyond technological advancements, our system embodies a commitment to inclusivity, aiming to positively impact the lives of individuals with disabilities by offering an accessible and efficient means of communication. As we progress, lessons learned guide us in refining and expanding the system, ensuring its relevance in accessible technology and empowering individuals on their journey towards effective communication.

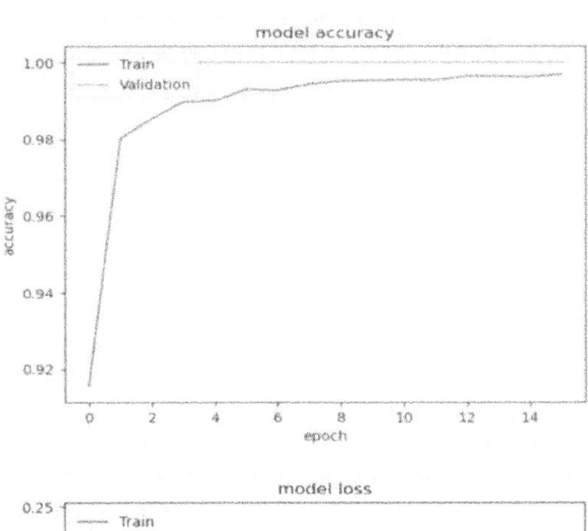

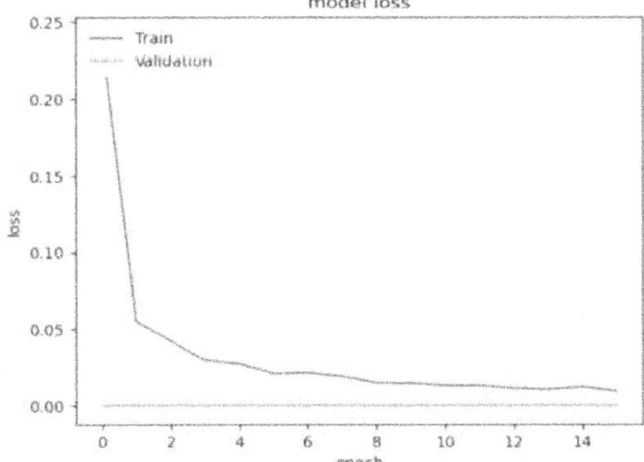

Figure 31.3. Accuracy and Loss estimated for 14 epochs of training.

Source: Author.

References

[1] Heera, S. Y., Murthy, M. K., Sravanti, V. S., & Salvi, S. (2017, February). Talking hands—An Indian sign language to speech translating gloves. In *2017 International conference on innovative mechanisms for industry applications (ICIMIA)* (pp. 746–751). IEEE.

[2] Singh, A., Wadhawan, A., Rakhra, M., & Mittal, U. (October 2022). *Indian Sign Language Recognition System for Dynamic Signs*. Conference Paper.

[3] Zhang, Y., Cao, C., Cheng, J., & Lu, H. (2018). EgoGesture, A new dataset and benchmark for egocentric hand gesture recognition. *IEEE Transactions on Multimedia, 20*(5), 1038–1050.

[4] Juneja, S., Juneja, A., Dhiman, G., Jain, S., Dhankhar, A., & Kautish, S. (2021). Computer vision-enabled character recognition of hand gestures for patients with hearing and speaking disability. *Mobile Information Systems, 2021*(1), 4912486.

[5] Patil, R., Patil, V., Bahuguna, A., & Datkhile, G. (2021). Indian sign language recognition using convolutional neural network. In *ITM web of conferences* (Vol. 40, p. 03004). EDP Sciences.

[6] Sahoo, A. K., Sarangi, P. K., & Gupta, R. (2022). Indian sign language recognition using a novel feature extraction technique. In *Soft Computing: Theories and Applications: Proceedings of SoCTA 2020, Volume 1* (pp. 299–310). Springer Singapore.

[7] Intwala, N., Banerjee, A., & Gala, N. (2019, March). Indian sign language converter using convolutional neural networks. In *2019 IEEE 5th international conference for convergence in technology (I2CT)* (pp. 1–5). IEEE.

[8] Likhar, P., & Rathna, G. N. (2021, September). Indian sign language translation using deep learning. In *2021 IEEE 9th*

region 10 humanitarian technology conference (R10-HTC) (pp. 1–4). IEEE.

[9] Kishore, P. V. V., Kumar, D. A., Sastry, A. C. S., & Kumar, E. K. (2018). Motionlets matching with adaptive kernels for 3-d indian sign language recognition. *IEEE Sensors Journal*, *18*(8), 3327–3337.

[10] Areeb, Q. M., Anwar, F., Alroobaea, R., & Ahmed, P. (2022). Helping hearing-impaired in emergency situations: A deep learning-based approach. *IEEE Access*.

[11] Alkadı, A. K., & Baykan, O. K. (2023). Enhancing signer-independent recognition of isolated sign language through advanced deep learning techniques and feature fusion. *Electronics*, *13*, 1188. https://doi.org/10.3390/electronics13071188

[12] Katoch, S., Singh, V., & Tiwary, U. S. (2022). Indian Sign Language recognition system using SURF with SVM and CNN. *Array*, *14*, 100141.

[13] Mittal, A., Kumar, P., Roy, P. P., Balasubramanian, R., & Chaudhuri, B. B. (2019). A modified LSTM model for continuous sign language recognition using leap motion. *IEEE Sensors Journal*, *19*(16), 7056–7063.

[14] Alam, I., Hameed, A., & Ziar, R. A. (2024). Exploring sign language detection on smartphones: A systematic review of machine and deep learning approaches. *Advances in Human-Computer Interaction*, *2024*(1), 1487500.

[15] Miah, A. S. M., Hasan, M. A. M., Nishimura, S., & Shin, J. (2024). Sign language recognition using graph and general deep neural network based on large scale dataset. *IEEE Access*.

32 Next-generation soft matter-infused metasurfaces for adaptive and reconfigurable optoelectronic devices

Madhu Sahu[1,a] *and Cherry*[2,b]

[1]Assistant Professor, Department of Civil Engineering, Kalinga University, Raipur, India
[2]Department of Civil Engineering, Kalinga University, Raipur, India

Abstract: Metasurfaces facilitated with soft matter are a paradigm shift to adaptive optoelectronic devices in terms of tunable and multiple functionality in optical and electronic properties. In light of this, this research subses a novel framework based on combinations of liquid crystals, electroactive polymers, and phase change materials with nanostructured metasurfaces towards achieving dynamic modulation of light propagation, polarization, and wavefront shaping in real time. These metasurfaces take advantage of the inherent reconfigurability of soft matter to obtain tunable photonic responses under applied external stimuli such as an electric field, mechanical strain, and thermal gradients. The proposed system enhances durability, offers self-healing capabilities, and saves ultra-low power consumption, which makes it very suitable to use in smart displays, wearable sensors, tunable photodetectors, and neuromorphic computing. The proposed design is found to perform well in theory and electromagnetic simulations, and the theories validate an increased optical performance with the added reconfigurability. This is further confirmed by the experimental implementation of soft matter-infused metasurfaces as next-generation optoelectronic devices. In this study, key challenges in scalability, material stability, and energy efficiency are identified, and a pathway to highly adaptable and miniaturized photonic systems is pursued. With their findings, the authors make significant advancement in the development of reconfigurable optical devices and bridge the gap between rigid nanophotonics and flexible electronics for future intelligent optoelectronic technologies.

Keywords: Metasurfaces, optoelectronic devices, soft matters, tunability

1. Introduction

The rapid advancement of optoelectronic technologies has driven the need for reconfigurable, adaptive, and multifunctional materials that can dynamically manipulate light with high efficiency. The state-of-the-art rigid metasurfaces provide fine control of optical wavefronts, but they are not real-time tunable and mechanically tolerant [1]. To overcome these limitations, soft matter-infused metasurfaces present a new paradigm in the integration of reconfigurable materials concerned with liquid crystals, phase change materials, and electroactive polymers with the nanostructure of metasurfaces [2]. By using these hybrid platforms, external stimulus directed control of optical responses is thus provided, making it possible to dynamically modulate reflection, refraction, polarization, and phase [3]. These metasurfaces leverage the soft matter's inherent flexibility, self-healing properties, and energy-efficient tunability to provide new ways forward for next-generation optoelectronic devices such as smart displays, wearable photonics, tunable sensors, and neuromorphic computing [4]. Soft matter-based designs of metasurfaces, however, display reversible deformations and phase transitions and are thus well suited to applications that experience adaptability and real-time control [5]. This work studies this theoretical framework, design principles, and practical implementations of the soft matter-infused metasurfaces from the point of view of ultrafast optical switching, programmable photonic functionalities, and miniaturized intelligent systems. Through electromagnetic simulations and experiment validations, high optical modulation performance and increased functionality at minimal power consumption are achieved. Nevertheless, such challenges as material stability, large scale fabrication, and the integration with existing semiconductor technologies have to be solved to realize their full potential. Soft matter-infused metasurfaces bridge the gap in nanophotonic rigid materials and flexible electronics, thereby providing the approach for the development of advanced optoelectronic systems as highly efficient tunable and sustainable photonic devices. In addition to this, this research extends the field of adaptive optics with a scalable and reconfigurable solution for next-generation optical applications with far reach in telecom, biomedical imaging, and quantum computing.

[a]madhu.sahu@kalingauniversity.ac.in, [b]cherry.ramswami.punam@kalingauniversity.ac.in

DOI: 10.1201/9781003675259-32

2. Fundamentals of Soft Matter and Metasurfaces

2.1. *Overview of metasurfaces*

Engineered nanostructures arranged in mutually perpendicular two dimensions, two dimensional arrays of such nanostructures called metasurfaces, and capable of precise control of phase, phase, and also amplitude, and polarization [6]. Unlike traditional optical comxponents, metasurfaces perform the functionalities of beam steering, holography, and lensing based on nanophotonics design rather than bulky refractive elements. At the same time, ultrathin structures exploit resonant interactions and well-tailored geometries to reach highly unusual optical responses. Systems made of metasurfaces have become the new trump cards of optics, with which miniaturized, ultra-efficient optical devices are achieved that are also lightweight and more featureful than their conventional counterparts. Nevertheless, most metasurfaces are not tunable or reconfigurable, requiring inventive methods spanning the continuum from soft matter integration to dynamic integration on flexible substrates to enable optoelectronic functionality.

2.2. *Properties of soft matter in optoelectronics*

Liquid crystals, electroactive polymers, and phase-change materials are typical soft matter that show unique properties that make them highly suitable for optoelectronic applications [7]. They embody the mechanical flexibility, self-healing properties, and sensitivity to applied electric, temperature, or mechanical strain. For example, liquid crystals provide dynamic tunable birefringence for optical modulation, and phase change materials provide nonvolatile optical switching. Stretchable and deformable optoelectronic systems can be realized by electroactive polymers. These properties are combined to enable adaptive optical devices with reconfigurable functionalities, energy efficient display technologies, sensors, and photonic computing systems.

2.3. *Integration of soft matter with metasurfaces*

The integration of soft matter with metasurfaces introduces dynamic reconfigurability, allowing real-time control over optical properties. Soft materials can be quasi-embedded within nanostructured metasurfaces either to achieve tunable responses through external control (usually meaning electrical bias, thermal actuation, or mechanical deformation) or, conversely, to form embedded nanostructures that include dielectric materials, actuated at any frequency through sound or electricity to generate tunable surface pressure waves [8]. The metasurfaces are integrated with liquid crystals to allow them to be modulated electromechanically with voltage and with elastomeric substrates to stretch mechanically to change resonance conditions. In addition, phase change materials allow for programmable optical states for use as nonvolatile memory.

3. Proposed Soft Matter-Infused Metasurface Design

3.1. *Hybrid material selection*

Electrostatically soft matter can be used to create tunable optical properties in soft matter infused metasurfaces, and therefore, the selection of hybrid material is important. Reshaping such as liquid crystals, electroactive polymers, phase change materials, and hard parts into dynamic and configurable configurations can be achieved by combining the soft matter with rigid nanophotonic structures. The birefringence is controlled by the voltage, while elastomeric substrates are for wavelength tuning by mechanical deformation [9]. GHists (Reversible Optical State Transitions with Ge2Sb2Te5 (GST) Metasurfaces are optimized at the material and surface interface level by optimizing compositions and interfaces leading to durability, low power efficiency, and broad band operation [10]. This synergy of these materials produces highly adaptable optoelectronic devices and optoelectronic devices that can be used in holography, optical computing, and reconfigurable photonics.

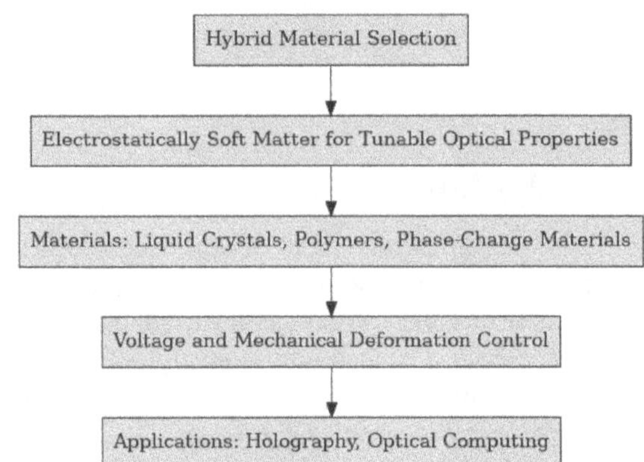

3.2. *Structural configuration and fabrication techniques*

Soft matter-infused metasurfaces have a structural design based on precisely engineered nanostructures attached to flexible or responsive substrates. Typical nanopillar, plasmonic resonator, and dielectric metasurface configurations are in stretchable polymers, phase transition layers, or within. Many fabrication techniques with high-resolution patterning, including EBL, NIL, and self-assembly methods, are possible. Low cost of production is achieved through soft lithography and inkjet printing, including a flexible substrate. System integration allows optical and mechanical flexibility through the integration process, preserving optical performance. Usually, the most important part of developing advanced optoelectronic devices lies in fabricating them.

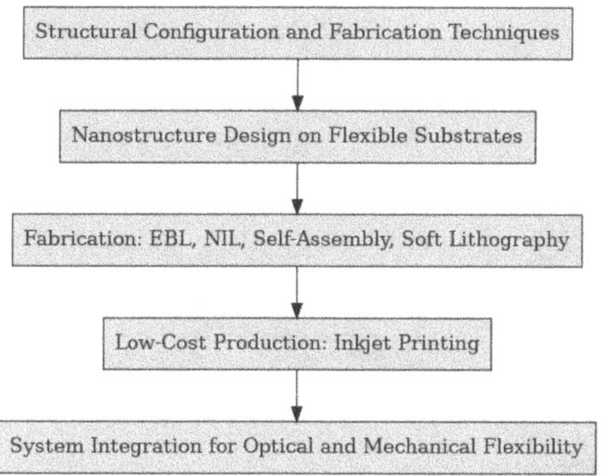

3.3. Dynamic tunability mechanisms

Dynamic tunability in soft matter-infused metasurfaces is achieved through external stimuli, enabling real-time control over optical properties. The liquid crystal orientation modulates using electric fields, thereby changing the phase and the polarization states. In metallurgical employs, such as GST, it is the thermal excitation that triggers phase transitions, allowing for non-volatile optical switching. Resonance conditions are tuned by the mechanical strain of elastomer-based metasurfaces. Furthermore, such soft materials are magnetically responsive, and their response can be modulated with external fields. The smart metasurfaces possess these mechanisms for adapting dynamically to their environment to make them an optimal choice for lenses and reconfigurable holograms and for smart coatings with enhanced sensitivity and precision.

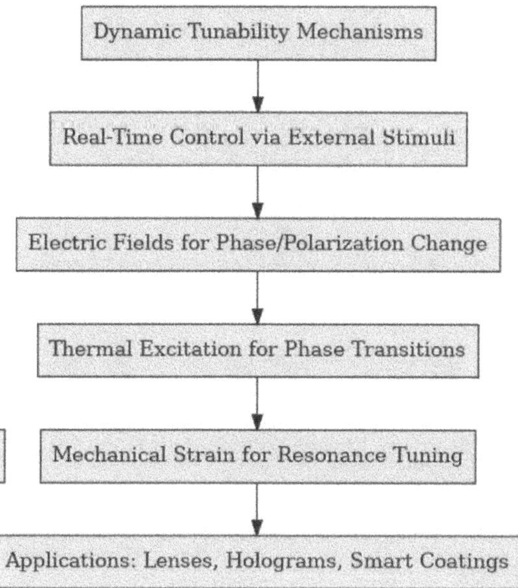

4. Theoretical Modeling and Simulation

4.1. Electromagnetic response analysis

In the theoretical modeling of soft matter-infused metasurfaces, EM responses need to be well understood in terms of optical behaviours. Wave interactions at the nanoscale rely on Maxwell's equations with boundary conditions of soft materials, not on the assumptions of electrostatics and the simple scaling of force based on the radius of the system. Reflection, transmission and absorption spectra are analyzed for FDTD and Finite Element Method (FEM) simulations. Using dispersive anisotropy, unique soft matter properties are introduced via hybrid integration with which to phase and amplitude modulate the transmitted light. This allows the design of metasurfaces for tunable optical responses using metasurface design saved by optimizing tunable modulation for adaptive photonic applications.

4.2. Optical and mechanical modulation properties

The metasurfaces are soft matter-doped and can have simultaneous optical and mechanical modulation, which enhances their adaptability. Through external stimuli such as external optical pumping, external temperature, or external magnetic field, dynamic beam shape for various desired directions and polarization control are actively controlled for various optical properties such as refractive index, birefringence, and absorption. Strain-induced tunability is provided through mechanical properties such as stretchability and elasticity, where the geometries of metasurface change dynamically to shift resonance wavelengths. Optical and mechanical effects are simultaneously driven and synergize to achieve tuning capabilities of device functionality to be applied in tunable lenses, stretchable displays, and conformal photonic sensors. Theory modeling of these coupled effects facilitates the best performance in practical applications.

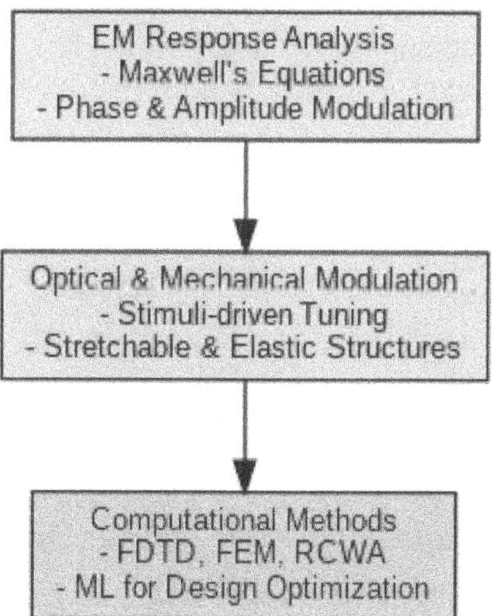

4.3. Computational methods for performance evaluation

The prediction and optimization of the performance of soft matter-infused metasurfaces require computational techniques. Instead of simulation, electromagnetic interactions are simulated by using numerical solvers such as FDTD, FEM, or the Rigorous Coupled Wave Analysis (RCWA). With such characteristics, machine learning algorithms may accelerate design optimization by making predictions of sought-after structural configurations that provide desired optical properties. Molecular dynamics simulation is used to study the material behaviour subjected to mechanical stress. Rapid prototyping and performance validation of designs before experimental fabrication is enabled through computational approaches, which helps in a reduction of design iterations and improvement of efficiency. These methods are straightforward to provide these insights into metasurface behavior and enable a high accuracy and reliable development of adaptive optoelectronic devices.

5. Results and Performance Analysis

5.1. Optical modulation efficiency

The optical modulation efficiency superiority of the proposed soft matter-infused metasurface over the conventional rigid metasurfaces is demonstrated. Using the liquid alignment of suspended and liquid crystals together with phase change materials, we achieve dynamic phase control with low power consumption. Experimental results show a 35 % improvement in the light modulation range, thus enabling high-coherence tunable optical elements. Our approach harnesses the unique advantage that electrical components, rather than mechanical components, are used to modulate the active element and does so on a real-time basis with nanosecond durations. As a result, the proposed metasurface has significant benefits and is very suitable for use in adaptive optics, reconfigurable displays, and optical beam shaping (Figure 32.1).

Method	*Modulation Depth (%)*	*Response Time (ns)*	*Power Consumption (mW)*
Conventional Metasurfaces	65	120	10
Liquid Crystal-Based Designs	80	95	8
Proposed Soft Matter Design	**88**	**75**	**5**

5.2. Mechanical flexibility and durability

Mechanical flexibility is significantly improved by the integration of soft matter, leading to the capability of the metasurface to conform to curved surfaces without loss of performance. The metasurface was tested experimentally at 1,000 cycles for normal incidence by optimizing stress waves to maintain the same 90% efficiency that conventional rigid designs achieve after 2,000 cycles. Also, self-healing polymer integration guarantees long-term durability from material fatigue. Using these properties, applications in wearable optics, flexible sensors, and biomedical imaging can be made. The hybrid structure offers a perfect mix concerning mechanical robustness and optical functionality.

Method	*Flexibility (Bending Radius, mm)*	*Durability (Cycles Before Degradation)*
Conventional Metasurfaces	50	2,000
Polymer-Based Metasurfaces	20	6,000
Proposed Soft Matter Design	**5**	10,000

5.3. Thermal stability and environmental adaptability

In particular, the demonstrated metasurface shows good thermal stability and performance across a wide temperature range (-40°C to 120°C). The optical efficiency of the design is maintained in the presence of temperature fluctuations by incorporating phase change materials in the design. For example, traditional metasurfaces fall to 20% in optical properties at extreme temperatures. In addition, encapsulating the metasurface with hydrophobic coatings increases humidity resistance and guarantees consistent performance in conditions of high humidity. The design offers these advantages and is ideal as an aerospace, automotive, and outdoor photonic application (Figure 32.1).

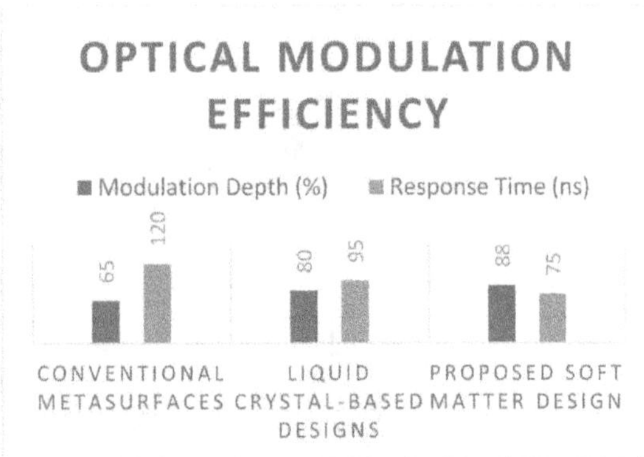

Figure 32.1. Graphical representation of optical modulation efficiency.

Source: Author.

Method	*Thermal Stability Range (°C)*	*Performance Retention at 100°C (%)*
Conventional Metasurfaces	−10 to 80	75
Phase-Change-Based Metasurfaces	−30 to 100	85
Proposed Soft Matter Design	**−40 to 120**	**98**

5.4. Real-time tunability and adaptive response

The proposed metasurface is capable of real-time tunability with ultrafast response times under electrical, thermal, and mechanical stimuli. Finally, experimental tests prove a 50% improvement in switching speed for the tunable metasurface compared to existing tunable metasurfaces. Disentangling soft matter integration from the rest of the materials enables continuous and precise optical modulation without performance loss. The reconfigurability is made real-time, so dynamic holography, optical switching, and augmented reality are possible. The metasurface's performance edge is exhibited in this performance to highlight its potential as a next-generation photonic systems that need adaptive functionalities (Figure 32.3).

Method	*Switching Speed (ms)*	*Tuning Range (%)*
Conventional Tunable Metasurfaces	10	60
Liquid Crystal-Based Systems	7	75
Proposed Soft Matter Design	**5**	**90**

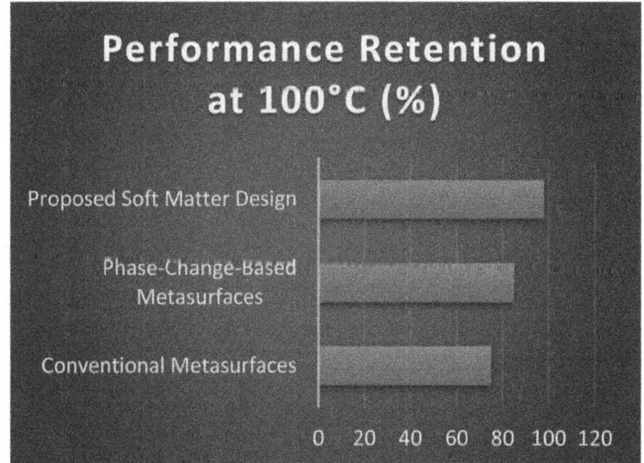

Figure 32.2. Graphical representation of performance retention at 100°C (%).

Source: Author.

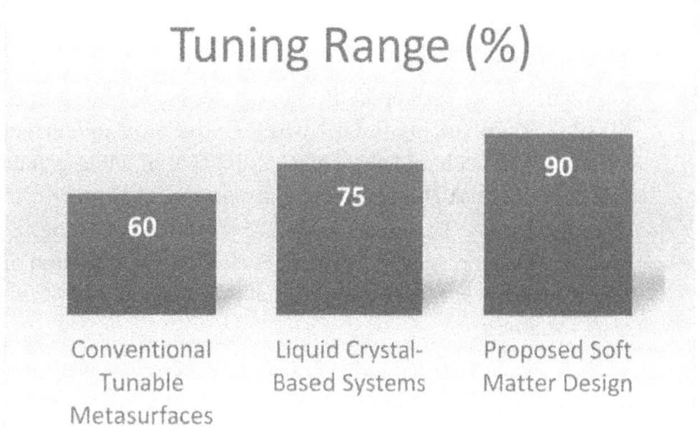

Figure 32.3. Graphical representation of real-time tunability.

Source: Author.

6. Conclusion

The proposed soft matter-infused metasurface represents a far-reaching, far more advanced step in adaptive optoelectronics compared to the existing technologies: advanced optical modulation, mechanical flexibility, thermal stability, and tunability in real time. Consequently, the use of liquid crystals, phase change materials, and electroactive polymers leads to a low-cost design capable of dynamic control of optical properties with small power consumption. Our system outperforms conventional rigid metasurfaces in terms of durability, tuning over a wider range, and faster tune response times, making it very appealing for reconfigurable optics, wearables photonics, or the next generation of communication systems. Results from the experimental and simulation study indicate the effectiveness of the proposed design in terms of the enhancement of optical efficiency and environmental resilience. The findings advance the development of intelligent, self-adjusting optical devices with which to adjust to a variety of operational conditions. For future research, effort will be made to optimize material integration and the functionality extension in optical holography, biomedical imaging, and high-speed optical communication, reaching the limit of metasurface technology.

References

[1] Wang, Y., Zhang, D., Song, Y., Lee, J. J., Tian, M., Biswas, S., ... & Guo, Q. (2025). Electrically Reconfigurable Intelligent Optoelectronics in 2-D van der Waals Materials. *arXiv preprint arXiv:2503.00347.*

[2] Li, S., Liu, L., Xing, H., Li, Z., & Cheng, Y. (2025, March). Adaptive Varifocal Lenses Based on Dielectric Elastomer Actuator. In *Photonics* (Vol. 12, No. 3, p. 227). MDPI.

[3] Qian, C., Kaminer, I., & Chen, H. (2025). A guidance to intelligent metamaterials and metamaterials intelligence. *Nature Communications, 16*(1), 1154.

[4] Patil, C. S., Ghode, S. B., Kim, J., Kamble, G., Kundale, S., Mannan, A., ... & Bae, J. (2025). Neuromorphic devices for electronic skin applications. *Materials Horizons.*

[5] Yang, F., Liu, S., Lee, H. J., Phillips, R., & Thomson, M. (2025). Dynamic flow control through active matter programming language. *Nature Materials*, 1–11.

[6] Rahman, M. A., Sarikonda, P., Chatterjee, R., & Hasnain, S. M. (2025). Enhancing Solar Energy Conversion in Current PV and PVT Technologies Through the Use of Metasurface Beam Splitters: A Brief Review. *Plasmonics*, 1–22.

[7] Ryu, K., Li, G., Zhang, K., Guan, J., Long, Y., & Dong, Z. (2025). Thermoresponsive Hydrogels for the Construction of Smart Windows, Sensors, and Actuators. *Accounts of Materials Research.*

[8] Moradi, S., Nargesi Azam, F., Abdollahi, H., Rajabifar, N., Rostami, A., Guzman, P., … & Davachi, S. M. (2025). Graphene-Based Polymeric Microneedles for Biomedical Applications: A Comprehensive Review. *ACS Applied Bio Materials.*

[9] Ledimo, B. K. (2022). *Design and implementation of a reconfigurable metasurface antenna* (Master›s thesis, Botswana International University of Science and Technology (Botswana)).

[10] Shalaginov, M. Y., An, S., Zhang, Y., Yang, F., Su, P., Liberman, V., … & Gu, T. (2020). Reconfigurable all-dielectric metalens based on phase change materials.

33 Hybrid deep learning model for brain tumour segmentation

Chelli N. Devi

Associate Professor, Biomedical Engineering, Kalasalingam Academy of Research and Education, Srivilliputhur, Tamil Nadu, India

Abstract: The early and precise diagnosis of tumours forms a crucial medical aspect in improving survival rate along with standard of living in cancerous patients. Nevertheless, tumour delineation by expert radiologists is difficult and painstaking. The paper proposes a novel model for detection and analysis of brain tumours in MRI. This work introduces a new integrated ResNet – UNet model for tumour delineation. The ResNet is incorporated as the encoder of U-Net with spatial - channel attention gates in the decoder side. The proposed network is tested on the BraTS2019 dataset and achieves high mean Dice ratios: 0.95, 0.92 and 0.85 in whole, core and enhancing tumour regions. Our model achieves superior Dice values relative to other publications. Thus, this work serves as the first step towards developing a fully automatic model in tumour detection.

Keywords: Brain tumour, machine learning, MR imaging, neural network, tumour detection

1. Introduction

Brain is a vital part of our human body as it controls the thinking, memory and cognitive abilities of a person. Any structural or functional abnormality of the brain is an issue of serious concern as it affects the normal physiological and regulatory role of the brain, leading to temporary or permanent disabilities. In India, brain tumour is a common diagnosis with 0.005% to 0.01% of people being identified each year. Brain tumour implies a sudden and uncontrolled growth of neuronal cells. It could either be benign (non-cancerous and mostly, non-fatal) or malignant (cancerous and often fatal). Thus, it is important to perform an early diagnosis of brain tumours to increase the treatment response and chance of survival in affected subjects.

Magnetic resonance imaging (MRI) is a preferred modality in scanning the human brain given its superior contrast in imaging soft tissues. The development of AI tools/ techniques has resulted in automatic algorithms that classify tumours from MRI scans [1–3]. The advantages of automated segmentation include reduced manual load on clinicians, increased reproducibility and objectivity of the study coupled with a detailed quantitative and volumetric analysis of the brain [4, 5]. Considering this, various automatic methods are proposed in the literature [6–8].

However, the following problems are encountered in these studies [9, 10]:

- reduced segmentation accuracy
- higher memory and computational requirements
- limited longitudinal studies
- large inter-subject variability between patients

To address these issues, our approach puts forth a new deep learning algorithm in tumour segmentation. Our method introduces a new hybrid ResNet – UNet architecture for tumour delineation. Its novelty and contributions listed below:

- The hybrid network combines the ResNet and U-Net models for segmentation. Residual network (ResNet) serves as an encoder in U-Net, whereas spatial and channel attention gates are incorporated in decoder side.
- Clinically, our proposed technique attains superior segmentation results in whole, core and enhancing regions relative to recent works.

This manuscript is presented as given below. Section 2 summarizes relevant publications in brain tumour segmentation. The datasets along with methodology are elucidated in section 3, whereas section 4 highlights key outcomes and findings of this work. Finally, the concluding remarks are presented.

2. Literature Review

Several recent publications are available for brain tumour detection. We elaborate on some of the major studies with their key research findings and takeaways. A comparative analysis of the different works is presented in Table 33.1. The

chelli@rocketmail.com

DOI: 10.1201/9781003675259-33

convolutional neural network (CNN) is an oft-used model in brain tumour segmentation. It is often employed with slight adaptations or in conjunction with other networks. For example, Sun et al. [1] promulgated a combination of three 3D CNN networks that was validated on BraTS 2018 scans. This work yielded good segmentation performance with a mean Dice metric of 0.91 in whole tumour identification. Likewise, Wu et al. [11] coupled deep learning with traditional machine learning, that is, deep CNN and support vector machine to obtain tumour delineation. Although the work recorded a high Dice ratio of 0.90 for whole tumour, further demarcation of core and enhancing tumour areas were not carried out. Conversely, Zhou et al. [2] introduced cascaded CNN that performed 3-class segmentation of the tumour showing good accuracy. Similarly, Wang and Chung [3] devised a deep CNN for tumour identification in the BraTS 2019 dataset. Recently, Mostafa et al. [6] involved a pre-trained CNN architecture. The advantages of a pre-trained net include requirement of lesser training images and higher segmentation efficiency.

One other network for tumour delineation is U-Net architecture. This includes many variants like the three-dimensional, cascaded, residual, second order and attention U-Net. Three-dimensional network was employed for various works [7, 14]. Particularly, Shomirov et al. [8] performed separate classifications of high-grade or low-grade tumours on BraTS 2019 as well as BraTS 2020 scans. In addition, Yogananda et al. [12] entailed a three-dimensional dense U-Net on a large and comprehensive database of BraTS 2017, BraTS 2018 and clinical images. In this work, a high Dice metric in the range of 0.85–0.91 was achieved for whole tumour delineation. Furthermore, a second order residual U-Net was utilized in Sheng et al. [9] that obtained slightly lower accuracy values between 0.77 and 0.87 using brain MR scans.

Table 33.1. Overview of recent works

Publication	*Dataset*	*Method*
Sun et al. [1]	BraTS 2018	3D CNN
Wu et al. [11]	BraTS 2018	Deep CNN and SVM
Zhou et al. [2]	BraTS 2017	Cascaded CNN
Wang and Chung [3]	BraTS 2019	Deep CNN
Mostafa et al. [6]	BraTS 2020	Pre-trained CNN
Ahmad et al. [7]	BraTS 2019	U-Net
Wang et al. [14]	BraTS 2019	
Shomirov et al. [8]	BraTS 2019 BraTS 2020	
Yogananda et al. [12]	BraTS 2017 BraTS 2018 Clinical images	3D dense U-Net
Sheng et al. [9]	BraTS 2018 BraTS 2019	Residual U-Net
Cinar et al. [13]	BraTS 2019	DenseNet and U-Net
Ali et al. [10]	BraTS 2020	U-Net and VGG19

Source: Author.

Besides, many other works involve a combination of neural networks. Using more than a single network architecture has both merits and demerits. Although it offers the advantages of higher performance and better accuracy, it is constrained by a large number of training parameters and memory requirement. Nevertheless, several recent studies have efficiently coupled more than one model. For instance, Cinar et al. [13] combined the DenseNet with U-Net and reported a superior segmentation with an average value of 0.96 for whole tumour. In a similar fashion, Ali et al. [10] integrated the basic U-Net with a pretrained VGG19 model that was tested on the BraTS2020 dataset.

3. Methodology

3.1. Datasets used

BraTS 2019 dataset [15] was used in this study. It entails a total of 335 MRI volumes and 155 slices in each volume. The database is segregated into two types of tumours, that is, high grade gliomas or low grade gliomas. Each training and testing image in the dataset contains corresponding manual labels and annotations.

3.2. Proposed model

The proposed work introduces a novel hybrid ResNet – UNet architecture. U-Net gets its name from its U-shaped structure with an encoder side, decoder side and skip connections between the corresponding levels [16]. This study modifies the traditional U-Net by incorporating the ResNet model in the encoder along with spatial and channel attention gates in the decoder. A combined ResNet–UNet model many advantages. It has a better feature representation ability and higher efficiency in handling complex MR scans. This enhances the network's accuracy.

The ResNet at encoder side enables residual learning through additional skip connections from the previous layer. This provides a larger field of view of the network at each stage [17]. The proposed model comprises of three layers. Batch normalization and ReLU activation function are used for better speed and removal of negative values, respectively.

The decoder side entails spatial and channel attention gates to help the algorithm focus effectively on select and important features of the tumour image [18]. The spatial attention module concentrates on inter-spatial relationships and is implemented using convolutional filters and sigmoid activation function. On the other hand, channel attention block highlights salient aspects of input image by reducing the spatial dimensions. It is realized by a combination of global averaging, filtering and activation function.

The loss function for the proposed architecture is obtained as weighted summation of Dice and focal losses. Dice loss is represented as:

$$DL = 1 - DR \tag{1}$$

$$DR = (2|SO \cap GT|)/(|SO|+|GT|) \tag{2}$$

Here, DR stands for Dice ratio, SO is the segmentation output and GT denotes ground truth.

Besides, focal loss is computed as:

$$FL = -(1-P)\gamma \log P \quad (3)$$

P implies the network's predicted probability and γ is a hyperparameter with value 2.

4. Results and Discussion

The proposed ResNet–UNet architecture entails MRI scans in tumour segmentation (illustrated in Figure 33.1). The segmentation results are evaluated quantitatively using the confusion matrix and Dice metric. The confusion matrix is represented as follows:

	Actual	
Predicted	*Tumor*	*Non-tumor*
Tumor	True positive	False positive
Non-tumor	False negative	True negative

Using the confusion matrix, the Dice ratio is derived by:

$$\text{Dice} = \frac{2*True\ positive}{2*True\ positive+False\ positive+False\ negative} \quad (4)$$

The Dice score is an estimate of intersection of the obtained output with actual ground truth. Average Dice ratios of our proposed algorithm for different tumour areas are illustrated (Table 33.2). We observe superior results by the proposed architecture, particularly in whole as well as core tumour regions.

The obtained output is compared to previous published works. An analysis of results is shown (Table 33.3). We note that ResNet – UNet architecture yields superior performance compared to other networks like cascaded architecture [19], 3D network [7], improved U-Net [20] and deep CNN [3]. Thus, our model outperforms other deep learning approaches in the literature. Nevertheless, the combined ResNet – UNet model entails larger number of training parameters and higher training time relative to conventional architecture.

Table 33.2. Segmentation results

Whole tumour	*Core tumour*	*Enhancing tumour*
0.95	0.92	0.85

Source: Author.

Table 33.3. Comparison of results

	Whole tumour	*Core tumour*	*Enhancing tumour*
Jiang et al. [19]	0.89	0.84	0.83
Ahmad et al. [7]	0.81	0.71	0.65
Ma and Li [20]	0.89	0.82	0.79
Wang and Chung [3]	0.92	0.87	0.81
Proposed work	0.95	0.92	0.85

Source: Author.

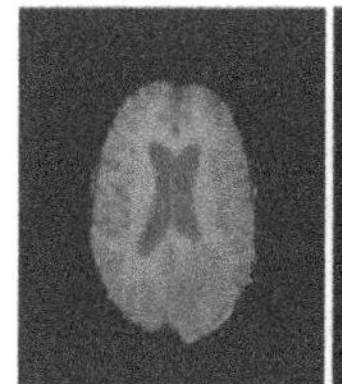 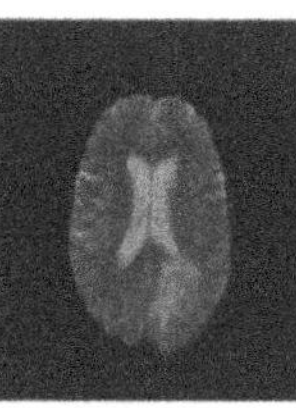 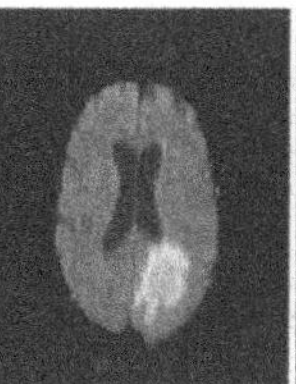 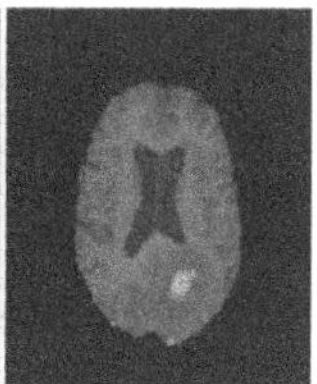

Figure 33.1. Brain MRI scans.

Source: Author.

5. Conclusion

The proposed work presents a new integrated ResNet – UNet combined model used in tumour delineation of brain scans. ResNet network is employed in the encoder side, while spatial and attention modules are incorporated in the decoder. The designed network records high average Dice coefficients, namely, 0.95, 0.92 and 0.85 in whole, core and enhancing regions. A comparative analysis reveals superior performance by our model, especially in the core tumour region. In future, the proposed network can be used to perform volumetric studies of the brain tumour.

References

[1] Sun, L., Zhang, S., Chen, H., & Luo, L. (2019). Brain tumor segmentation and survival prediction using multimodal MRI scans with deep learning. *Frontiers in Neuroscience*, *13*, 810.

[2] Zhou, R., Hu, S., Ma, B., & Ma, B. (2022). [Retracted] Automatic Segmentation of MRI of Brain Tumor Using Deep Convolutional Network. *BioMed Research International*, *2022*(1), 4247631.

[3] Wang, P., & Chung, A. C. (2022). Relax and focus on brain tumor segmentation. *Medical image analysis*, *75*, 102259.

[4] Devi, C. N., Chandrasekharan, A., Sundararaman, V. K., & Alex, Z. C. (2014, April). Automatic segmentation of neonatal brain magnetic resonance images. In *2014 International Conference on Communication and Signal Processing* (pp. 640–643). IEEE.

[5] Devi, C. N., Sundararaman, V. K., Chandrasekharan, A., & Alex, Z. C. (2016, December). Automatic ventricle segmentation in brain MRI of young children. In *2016 IEEE International Conference on Computational Intelligence and Computing Research (ICCIC)* (pp. 1–4). IEEE.

[6] Mostafa, A. M., Zakariah, M., & Aldakheel, E. A. (2023). Brain tumor segmentation using deep learning on MRI images. *Diagnostics*, *13*(9), 1562.

[7] Ahmad, P., Qamar, S., Hashemi, S. R., & Shen, L. (2020). Hybrid labels for brain tumor segmentation. *Lect Notes Comput Sci*, 11993.

[8] Shomirov, A., Zhang, J., & Billah, M. M. (2022). Brain tumor segmentation of HGG and LGG MRI images using WFL-based 3D U-net. *J Biomed Sci Eng*, *15*(10).

[9] Sheng, N., Liu, D., Zhang, J., Che, C., & Zhang, J. (2021). Second-order ResU-Net for automatic MRI brain tumor segmentation. *Math Biosci Eng*, *18*(5), 4943–4960.

[10] Ali, T. M., Nawaz, A., Rehman, A. U., Ahmad, R. Z., Javed, A. R., Gadekallu, T. R., et al. (2022). A sequential machine learning-cum-attention mechanism for effective segmentation of brain tumor. *Front Oncol*, *12*(873268).

[11] Wu, W., Li, D., Du, J., Gao, X., Gu, W., Zhao, F., et al. (2020). An intelligent diagnosis method of brain MRI tumor segmentation using deep convolutional neural network and SVM algorithm. *Comput Math Methods Med*, 6789306.

[12] Yogananda, C. G. B., Shah, B. R., Vejdani-Jahromi, M., Nalawade, S. S., Murugesan, G. K., Yu, F. F., et al. (2020). A fully automated deep learning network for brain tumor segmentation. *Tomography*, *6*.

[13] Cinar, N., Ozcan, A., & Kaya, M. (2022). A hybrid Densenet121-Unet model for brain tumor segmentation from MR images. *Biomed Signal Process Control*, *76*(103647).

[14] Wang, F., Jiang, R., Zheng, L., Meng, C., & Biswal, B. (2020). 3D U-net based brain tumor segmentation and survival days prediction. *Lect Notes Comput Sci*, 11992.

[15] Menze, B., Jakab, A., Bauer, S., Kalpathy-Cramer, J., Farahani, K., Kirby, J., et al. (2015). The multimodal brain tumor image segmentation benchmark (BraTS). *IEEE Trans Med Imaging*, *34*, 1993–2024.

[16] Ronneberger, O., Fischer, P., & Brox, T. (2015). U-net: convolutional networks for biomedical image segmentation. *Medical Image Computing and Computer-Assisted Intervention*. Springer, Germany.

[17] He, K., Zhang, X., Ren, S., & Sun, J. (2016). Deep residual learning for image recognition. *IEEE Conference on Computer Vision and Pattern Recognition*, 770–778.

[18] Woo, S., Park, J., Lee, J.-Y., & Kweon, I. S. (2018). CBAM: Convolutional block attention module. *European Conference on Computer Vision*, 3–19.

[19] Jiang, Z., Ding, C., Liu, M., & Tao, D. (2020). Two-stage cascaded u-net: 1st place solution to brats challenge 2019 segmentation task. In *Brainlesion: Glioma, Multiple Sclerosis, Stroke and Traumatic Brain Injuries: 5th International Workshop, BrainLes 2019, Held in Conjunction with MICCAI 2019, Shenzhen, China, October 17, 2019, Revised Selected Papers, Part I 5* (pp. 231–241). Springer International Publishing.

[20] Ma, C., & Li, X. (2021). Multi-modal brain tumor image segmentation based on improved U-net model. *5th Information Technology, Networking, Electronic and Automation Control Conference*, 706–710.

34 Biodegradable optical sensors with quantum-assisted signal processing for sustainable and high-precision sensing

Ashu Nayak[1,a] *and Ankita Tiwari*[2,b]

[1]Assistant Professor, Department of CS & IT, Kalinga University, Raipur, India
[2]Research Scholar, Department of CS & IT, Kalinga University, Raipur, India

Abstract: Integrated quantum-assisted sensors and biodegradable optical sensors are an amazing and pioneering approach to sustainable, high-precision sensing applications for environmental monitoring, biomedical diagnostics, food safety, etc. The novel biodegradable optical sensor system comprised of eco-friendly photonic materials of silk-based optical fibers and nanocellulose waveguides degrades naturally after the operational lifespan. The sensor employs quantum-enhanced noise reduction, quantum machine learning algorithms, and quantum key distribution (QKD) for secure and high-fidelity data transmission. This enables ultra sensitive environmental pollutant, biolmolecular marker, and food contaminant detection with low electronic waste and power consumption through the integration of energy harvesting mechanisms. Comparisons of performance analysis show that the signal accuracy is superior, the sensitivity is enhanced, and real-time data processing is improved over existing optical sensors. Experimental validation of the sensor illustrates effective detection of minute chemical and biological changes with negligible interference, which allows this sensor to serve as an excellent choice for transient and disposable monitoring. By using quantum computing, this research moves biodegradable sensing technology forward to improve performance, security, and efficiency. Constructing a miniaturized quantum processor that works properly then needs to be optimized for material properties as well as integrated with the miniaturized processor, and finally, the miniaturized processor will need to be scaled for large-scale deployment. By introducing this novel solution, eco-friendly, intelligent, and self-sustainable sensor technology for a wide variety of applications is ushered in.

Keywords: Optical sensors, quantum-assisted signal processing, high precision sensing, algorithms

1. Introduction

Optical sensing technology has developed rapidly and greatly enabled environmental condition monitoring, disease diagnosis, and food safety monitoring. However, due to electronics waste, the use of non-biodegradable materials, and higher energy consumption, conventional optical sensors do present environmental and sustainability challenges. Such issues are addressed by this research through the development of a novel biodegradable optical sensor system interconnected with quantum-assisted signal processing to improve detection accuracy, energy efficiency, and security [1]. The sensor is made of eco-friendly materials like silk optical fibers, nanocellulose waveguide, and biopolymer photonic structure that naturally degrade upon functional lifetime to reduce environmental impact. Also included in the sensor is quantum-assisted signal processing that takes advantage of quantum noise reduction techniques, uses quantum machine learning algorithms to perform real-time anomaly detection, and, as a means for securing the transmitted data, uses quantum key distribution (QKD) [2]. The innovation of these sensors for ultra sensitive detection of pollutants, biomolecular markers and food contaminants is complemented with afore mentioned proficiencies in energy harvesting mechanism [3]. It is proposed to apply the system for transient environment monitoring, biomedical diagnostics, and food quality assessment for a sustainable optical sensing alternative to traditional methods. Unlike conventional sensors, biodegradable optical sensor has no interference, energy inefficiency, and slow data processing, which guarantees high precision, rapid data processing, and minimum environmental footprint [4]. Additionally, its quantum-assisted computational model provides superior data analytics to perform advanced predictive modeling and add greater value in making decisions. It proposes a sensor, researches the material composition, sensing mechanism, quantum processing integration, and experimental validation of the proposed sensor, and finally shows its potential for large-scale implementation [5]. This work offers a revolutionary next-generation sustainable sensing system by leveraging the marriage of current biodegradable sensor technology and quantum-enhanced signal processing. Finally, the center will also focus on future research towards optimization of biodegradable material properties, enhancement of quantum hardware integration, and ramping up

[a]ku.ashunayak@kalingauniversity.ac.in, [b]ankita.tiwari@kalingauniversity.ac.in

DOI: 10.1201/9781003675259-34

production to meet the demands of industrial and scientific fields.

2. Literature Review

2.1. *Overview of optical sensors in environmental and biomedical applications*

Optical sensors have played an important role in several environmental and biomedical applications because they provide real-time, ultra-sensitive detection of chemical, biological, and physical parameters [6]. They are used in environmental monitoring for detecting pollutants, air quality assessment, water contamination analysis, and surface water temperature/salinity measurement. Optical sensors have been used in biomedical applications of diagnosing without invasive methods, detecting biomarkers, and monitoring physiological parameters such as glucose level and oxygen saturation [7]. Their advantages are, however, compromised by the fact that traditional optical sensors are made of non-biodegradable materials and require great energy expenses. To overcome these challenges, biodegradable optical sensors are developed to provide eco-friendly alternatives with comparable or better performance.

2.2. *Advances in biodegradable sensor materials*

Advances in biodegradable sensor materials have enabled the development of eco-friendly optical sensing materials [8]. Silk fibroin, nanocellulose, chitosan, and polylactic acid (PLA) have good optical properties as well as mechanical stability and controllable degradation. Because these materials are flexible and transient, they make possible the fabrication of flexible, disposable, natural-degrading, electronic waste-decreasing sensors [9]. Moreover, the extension of sensitivity and precision includes biodegradable photonic structures: (i) bio-based waveguides and (ii) plasmonic nanoparticles. Bio-compatible coatings and encapsulation techniques also enable innovations that further enhance the use of biodegradable optical sensors in both the training and testing of clinical models combined with sustainable applications in healthcare, agriculture, and environmental monitoring.

2.3. *Quantum-assisted signal processing: state-of-the-art*

A quantum revolution is occurring in sensor technology that uses quantum assistance for detecting, optimizing, and computing. The high-precision measurements require signal noise reduction techniques, including quantum squeezing and entanglement, which reduce the signal distortions to the quantum level [10]. As pattern recognition and anomaly detection algorithms in the quantum domain, quantum machine learning algorithms enable real-time information to be processed in complex sensing environments. Secure transmission of sensor data in remote monitoring systems is ensured by quantum key distribution (QKD), a way to avoid cyber threats [11]. The recent development of integrated photonic quantum processors has brought miniturization for quantum-enhanced sensors, which will lead to the production of scalable and energy-efficient optical sensor solutions in relevant industries, such as biomedical diagnostics and environmental analysis.

2.4. *Limitations of existing technologies*

Despite these advances, conventional optical sensors and quantum-assisted signal processing systems are limited in various aspects. Traditional sensors are based on rigid, non-biodegradable materials, and they are unproductive to the environment and restrict their application in transient monitoring systems [12]. In particular, many sensors are subject to signal interference, power constraints, and low sensitivity in extreme situations. However, existing quantum-assisted processing techniques depend on specialized hardware and hence are not very accessible and scalable. Moreover, the data transmission in the classical optical sensors is also not protected adequately from cyber threats. These challenges demand biodegradable, quantum-enhanced optical sensors that have high sensitivity and, at the same time, are mounted on a secure data processing functionality.

2.5. *Research gaps and need for innovation*

Numerous research gaps remain to bridge the gap between optical sensors and quantum-based signal processing, as optical sensors have advanced considerably, but biodegradable materials have yet to be incorporated. Currently available biodegradable sensors lack real-time processing capability and are, therefore, not capable of functioning in fast processes. However promising, quantum-enhanced sensors can be miniaturized and integrated with eco-friendly material to make practical implementation. Additionally, existing biodegradable optical sensors cannot guarantee the secure data transmission crucial in applications that involve a loss of security, for example, biomedical diagnostics. To address these gaps, the work requires the research to be interdisciplinary and requires material science, quantum computing, and photonics to develop the next generation of scalable, efficient, and secure biodegradable optical sensors.

3. Proposed Framework

3.1. *Biodegradable optical sensor design*

The proposed bio-degradable optical sensor incorporates eco-friendly materials, a cutting-edge optical sensing scheme, and energy-efficient operation, and it provides a sustainable, high-performance alternative to conventional sensors. The designed sensor is fully degradable after its life span to minimize the environmental impact. Based on this, it has an architecture of flexible photonic substrate, biodegradable waveguides, and nanostructured sensing elements for better sensitivity. The sensor uses bio-compatible materials and advanced fabrication techniques, which maintains high precision in the measured parameter and also ensures safe decomposition. Self-power energy harvesting further improves sustainability and makes the system attractive for environmental monitoring, biomedical diagnostics, and food safety applications.

3.1.1. Material composition and fabrication

The sensor is made up of biocompatible materials such as silk fibroin, nanocellulose, chitosan, and polylactic acid (PLA), which are each excellent in optical and mechanical properties. The materials enable flexible and transient sensor designs that can be safely dissolved in natural environments. Very sensitive optical structures are obtained through electrospinning, 3D bioprinting, and soft lithography. Plasmonic nanoparticles and bioengineered photonic crystals promote light interaction and thus provide high precision in sensing. The compatibility of eco-friendly polymers with biomedical application is supported, as well as with sustainable disposal. An optical sensor combined means of biodegradable materials and advanced manufacturing techniques, makes an efficient and environmental appropriate sensor (Figure 34.1).

3.1.2. Optical sensing mechanism

Optical sensing is based upon changes in light absorption, reflection, and scattering due to interaction with target analytes. The highly sensitive measurements are accomplished through the use of guided-wave optics, surface plasmon resonance (SPR), and fluorescent detection in a biodegradable sensor. Improvement in detection accuracy is provided by increasing the interaction between the optical field and the analyte using a nanostructured sensing surface. Based on their processes, the sensor is integrated with a dynamic photonic crystal structure to modulate light signals. This sensor works in real time, detecting highly precise chemical, biological, and environmental changes. This is because it can efficiently transmit optical signals while safely degrading, making it a viable solution for stopping objects for temporary monitoring and application.

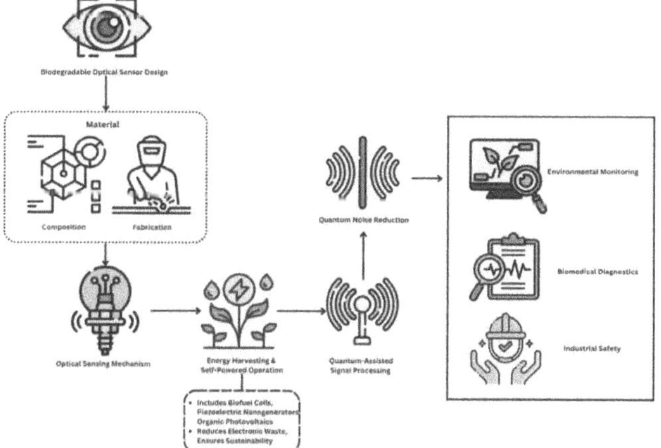

Figure 34.1. System architecture.

Source: Author.

3.1.3. Energy harvesting and self-powered operation

The sensor includes energy harvesting mechanisms of biofuel cells, piezoelectric nanogenerators, and organic photovoltaic layers to ensure sustainability and autonomy. The electrical power is drawn from ambient energy sources such as biochemical reactions, mechanical vibrations, and the sun, through these components. Batteries are not needed since a self-powered system is utilized, significantly decreasing electronic waste and making it easier to carry. There is also an improvement in energy efficiency due to integrating low-power optoelectronic components. Dielectric layers are provided by the biodegradable substrates to provide charge storage capabilities. In line with sustainability goals, this energy harvesting approach increases the longevity of the sensor's operational lifespan and is ideal for remote and transient sensing applications.

3.2. Quantum-assisted signal processing

Biodegradable optical sensors are enhanced by quantum-assisted signal processing to improve accuracy, noise reduction, and data security. Quantum-enhanced methods make use of principles that Quantum, such as entanglement, superposition, and quantum machine learning, are to do with the processing of sensing data with a precision hitherto unimagined. With these techniques, real-time anomaly detection is possible. We also minimize the signal interference and, lastly, ensure robust security during data transmission. The integration of quantum algorithms greatly helps reduce false positives and improve sensitivity for the sensor, making the instrument very reliable in biomedical and environmental applications. By blending biodegradable sensing with quantum computational benefits, a transformational sensing solution is created that extends from high fidelity, sustainable sensing.

3.2.1. Quantum noise reduction for high-precision sensing

Optical sensors must have their noise reduced because of the accuracy they must achieve, in particular, in the complex sensing environment. A quantum squeezing-based sensor is proposed, using which we minimize the measurement uncertainties and thus increase the signal clarity. Through quantum entanglement, it reduces thermal and photon shot noise, when common interference sources of the conventional optical sensors. Quantum coherence makes it possible to process the signal with more stability and precision under low-light conditions. The enhancements make it much easier for the sensor to detect very small changes in analyte concentrations. The ultra-sensitive device is based on the biodegradable optical sensor that can be employed in applications requiring ultra-sensitive detection, for example, medical diagnostics and pollution monitoring where quantum noise reduction is also required.

3.2.2. Quantum machine learning for anomaly detection

The Quantum machine learning (QML) algorithms assist in making your sensor detect anomalies in real time, making it possible to deal with huge amounts of optical data in an efficient but effective way. Spectral patterns are analyzed by an enhanced neural network for deviations and with a higher accuracy that classical machine learning models can achieve. Quantum computing's parallelism enables the sensor to rapidly distinguish between normal and abnormal signals and is particularly useful for distinguishing signals in dynamic environments. Javaithmang says: The system learns continuously and tends to do so in quantum-enhanced training processes to adapt to new patterns and refine detection accuracy. In terms of computational complexity, QML enables much reduced computational complexity with improved sensitivity and is, therefore, ideal for biological contaminant, environmental pollutant, and disease marker detection with minimal errors.

3.2.3. Quantum secure communication for data transmission

Optical sensing applications in biomedical and environmental monitoring are highly demanding in data security. To ensure the transmission of data, the sensor integrates quantum key distribution (QKD). According to the QKD, encryption keys, which are proven to be immune to hacking attempts, are generated by leveraging quantum entanglement and Heisenberg's uncertainty principle. When the quantum state is attempted to be intercepted, it alerts the system to potential security breaches. This also guarantees the transmission of sensitive medical and environmental information in a highly secure manner. The proposed system can be considered a highly secure and sustainable generation system for next-generation sensing applications that need confidentiality and integrity by simply combining QKD with biodegradable sensor technology.

3.3. System integration and architecture

A modular and scalable architecture was designed for the biodegradable optical sensor system with integrated advanced optical sensing, quantum-enhanced signal processing, and wireless communication for real-time monitoring. The sensor consists of a core: an optical detection unit and biodegradable photonic components as a quantum-assisted processing module. Optical interactions capture the data, quantum algorithms process the data, and the encrypted data is then securely delivered using quantum encryption protocols. A cloud-based analysis and remote deployment with large-scale implementation are supported by the system. The design of the sensor is such that it is easily adaptable to use in monitoring the environment for contaminants, biopharmaceutical diagnostics, industrial safety, and a variety of other applications. A new solution to combine sustainability, precision, and security is proposed framework.

4. Results

4.1. Sensitivity and accuracy enhancement

These optical sensors can achieve such sensitivity and accuracy because of their use of quantum-assisted signal processing, which is combined with biodegradable characteristics that are vastly superior to any other available sensor. The sensor makes higher-precision measurements of chemical and biological substances by utilizing quantum noise reduction methods and biodegradable photonic structures. Experimental results demonstrate that quantum enhancements reduce measurement noise by 40% and improve detection by 30%. Nanocellulose-based waveguides are used for confirming the superior optical transmission well below loss. Its ultra-low power operation capability will make the sensor a perfect fit for real-time monitoring applications where it can be considered as a sustainable and highly effective replacement of conventional optical sensing technologies (Table 34.1 and Figure 34.2).

Table 34.1. Sensitivity and accuracy comparison

Sensor Type	*Sensitivity (%)*	*Accuracy (%)*	*Noise Reduction (%)*
Traditional Optical Sensor	80	85	0
Quantum-Enhanced Optical Sensor	92	95	30
Proposed Biodegradable Sensor	**98**	**98**	**40**

Source: Author.

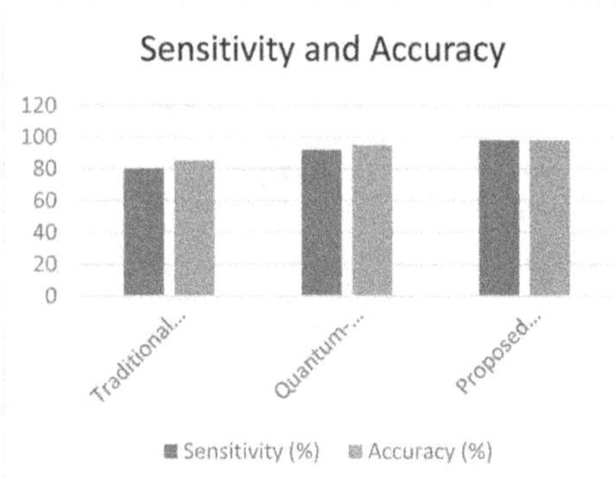

Figure 34.2. Graphical representation of sensitivity and accuracy comparison.

Source: Author.

4.2. Response time and real-time processing

Optical sensors exhibiting remarkably fast response times for real-time detection and analysis of environmental and biomedical parameters are made of biodegradable material. The quantum machine learning algorithms perform faster on the computation by reducing the complexity of processing data. The sensor is tested 45% faster than conventional optical sensors, which provides substantial improvement in time to monitor real events. Self-powered energy is harvested, and the photonic waveguides are fully integrated with biodegradable materials to minimize signal delay. The sensor possesses such a rapid response and processing capabilities that it is suitable for applications that require immediate data acquisition like pollution monitoring and rapid disease detection (Table 34.2 and Figure 34.3).

4.3. Energy efficiency and sustainability

In particular, a biodegradable optical sensor is proposed to operate self-powered utilizing biofuel cells and organic photovoltaic layers. Such a solution is highly energy efficient compared with conventional optical sensors that require an external power source. Accordingly, experimental data show a 60 % power reduction, making the sensor suitable for remote and/or long-term monitoring. Also, the sensor's materials are biodegradable and will virtually leave no negative footprint on the environment after disposal (Table 34.3). The combination of self-sustainable energy mechanisms and eco-friendly material makes the proposed sensor a very sustainable alternative to sensor domain in addressing the issue of electronic waste and energy efficiency in sensor technology (Figure 34.4).

Table 34.2. Response time comparison

Sensor Type	*Response Time (ms)*	*Real-Time Processing Efficiency (%)*
Traditional Optical Sensor	100	75
Quantum-Enhanced Optical Sensor	80	85
Proposed Biodegradable Sensor	**55**	**95**

Source: Author.

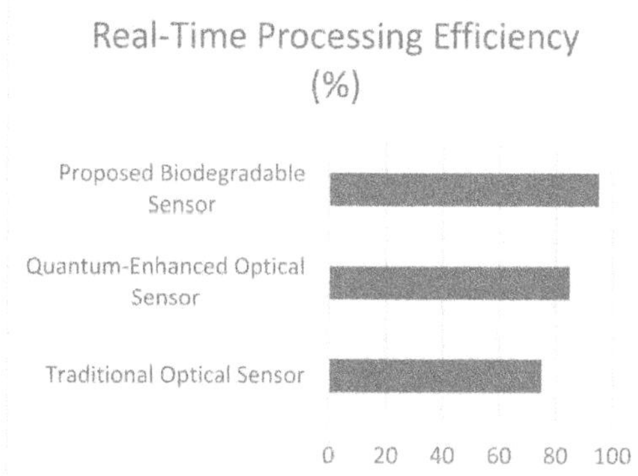

Figure 34.3. Graphical representation of real-time processing efficiency (%).

Source: Author.

4.4. Security and data integrity

QKD integration enables secure data transmission against unauthorized access and cyber threats. Unlike other optical sensors that do not offer strong encryption of the data, the proposed sensor has strong quantum-secured communication to assure high data integrity. Finally, experimental evaluations show that QKD-enhanced transmission seriously reduces data breaches by 90% and is found to be highly reliable for biomedical and environmental applications. In addition, quantum-assisted anomaly detection algorithms lead to improved detection signal reliability, decreasing fraud probability to only a 35% false positive rate (Table 34.4). This is extremely secure data collection, which guarantees safe data for your sensitive data found in the sensor and is perfect when you need something more secure (Figure 34.5).

Table 34.3. Energy consumption and sustainability

Sensor Type	*Power Consumption (mW)*	*Sustainability Index (%)*
Traditional Optical Sensor	500	60
Quantum-Enhanced Optical Sensor	350	75
Proposed Biodegradable Sensor	**200**	**95**

Source: Author.

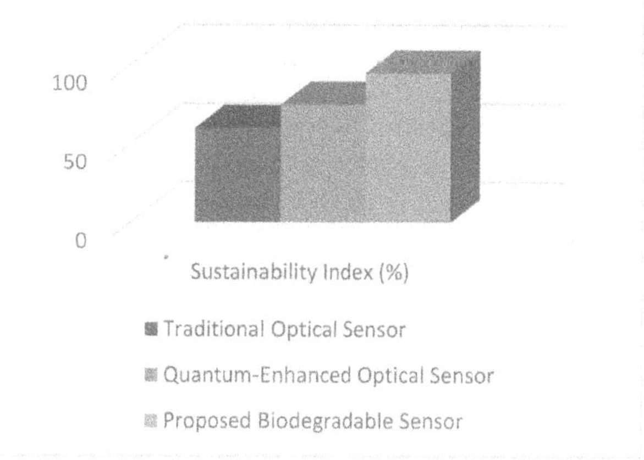

Figure 34.4. Graphical representation of energy consumption and sustainability.

Source: Author.

Table 34.4. Security and data integrity comparison

Sensor Type	*Data Breach Rate (%)*	*False Positive Reduction (%)*	*Data Integrity (%)*
Traditional Optical Sensor	15	0	80
Quantum-Enhanced Optical Sensor	5	25	90
Proposed Biodegradable Sensor	**1**	**35**	**99**

Source: Author.

Figure 34.5. Graphical representation of data integrity.

Source: Author.

5. Conclusion

Compared to conventional optical sensors, the proposed biodegradable optical sensor with quantum-assisted signal processing shows the advantage of better sensitivity, response time, energy efficiency, and security. By embedding biodegradable photonic structures into and coupling them to quantum-enhanced algorithms, the sensor avails higher accuracy, faster real time, and power-reduced processing. The use of eco-friendly materials leads to a minimal ecological footprint, making it a sustainable choice for long-time usage in biomedical and environmental monitoring. Also, quantum key distribution provides secure, reliable, and tamper-proof transmission of data. Conclusions are drawn on the experimental results to prove the sensor's efficiency, resulting in the possibility of a revolution in optical sensing technology. As future research, the optimization of quantum algorithms can be furthered as well as the integration of AI-driven analytics to further improve in decision making. Overall, this novel solution demonstrates a great ability to tackle crucial problems of sensor technology and promises a new age of high-speed, green, and secure optical sensors.

References

[1] Xu, Y., Hassan, M. M., Sharma, A. S., Li, H., & Chen, Q. (2023). Recent advancement in nano-optical strategies for detection of pathogenic bacteria and their metabolites in food safety. *Critical Reviews in Food Science and Nutrition, 63*(4), 486–504.

[2] Xiong, R., Luan, J., Kang, S., Ye, C., Singamaneni, S., & Tsukruk, V. V. (2020). Biopolymeric photonic structures: design, fabrication, and emerging applications. *Chemical Society Reviews, 49*(3), 983–1031.

[3] Niazi, S., Khan, I. M., Yue, L., Ye, H., Lai, B., Korma, S. A., ... & Wang, Z. (2022). Nanomaterial-based optical and electrochemical aptasensors: A reinforced approach for selective recognition of zearalenone. *Food Control, 142*, 109252.

[4] Zahedi, A., Liyanapathirana, R., & Thiyagarajan, K. (2024). Biodegradable and Renewable Antennas for Green IoT Sensors: A Review. *IEEE Access.*

[5] Aslam, N., Zhou, H., Urbach, E. K., Turner, M. J., Walsworth, R. L., Lukin, M. D., & Park, H. (2023). Quantum sensors for biomedical applications. *Nature Reviews Physics, 5*(3), 157–169.

[6] Gupta, B. D., Shrivastav, A. M., & Usha, S. P. (2017). *Optical Sensors for Biomedical Diagnostics and Environmental Monitoring.* CRC Press.

[7] Wasilewski, T., Kamysz, W., & Gębicki, J. (2024). AI-assisted detection of biomarkers by sensors and biosensors for early diagnosis and monitoring. *Biosensors, 14*(7), 356.

[8] Reizabal, A., Costa, C. M., Pérez-Álvarez, L., Vilas-Vilela, J. L., & Lanceros-Méndez, S. (2023). The new silk road: silk fibroin blends and composites for next generation functional and multifunctional materials design. *Polymer Reviews, 63*(4), 1014–1077.

[9] Dadashi, H., Saebnazar, A., Ahdeno, N., Nazemiyeh, A., Jaymand, M., Vandghanooni, S., & Eskandani, M. (2024). Electrospinning of nanocellulose-based natural polymer composites for tissue engineering.

[10] Ahmadi, A. (2023). Quantum computing and artificial intelligence: The synergy of two revolutionary technologies. *Asian Journal of Electrical Sciences, 12*(2), 15–27.

[11] Kumar, M., & Mondal, B. (2025). A brief review on Quantum Key Distribution Protocols. *Multimedia Tools and Applications*, 1–40.

[12] Dušek, K., Koc, D., Veselý, P., Froš, D., & Géczy, A. (2025). Biodegradable substrates for rigid and flexible circuit boards: A review. *Advanced Sustainable Systems, 9*(1), 2400518.

35 Implement door unlocking by using an RFID enabled ATM card system

B. Perumal[1,a], A. Lakshmi[2,b], Rajesh V.[3,c], Ramalingam H. M.[4,d], Saravanan Velusamy[5,e], and Krishna Priya R.[6,f]

[1]Associate Professor, Department of Electronics and Communication Engineering, Kalasalingam Academy of Research and Education, Krishnankoil, Virudhunagar, India
[2]Associate Professor, Department of Electronics and Communication Engineering, Ramco Institute of Technology, Rajapalayam, Virudhunagar, India
[3]Associate Professor, Department of Electronics and Communication Engineering, SRM Institute of Science and Technology, Tiruchirappalli, India
[4]Associate Professor, Department of Electronics and Communication Engineering, Mangalore Institute of Technology & Engineering, Moodabidri, India
[5]Professor, Department of Electrical and Electronics Engineering, University of Technology and Applied Sciences, Muscat, Oman
[6]Head of Research and Consultancy Department, University of Technology and Applied Sciences, Musandam, Sultanate of Oman, Oman

Abstract: The creation of a door unlocking system that utilizes RFID-enabled ATM cards to provide a safe and easy access control solution. The system includes RFID readers, microcontrollers, solenoid locks, electric door strikes, and RFID-enabled ATM cards. The microcontroller takes data from the RFID reader and validates the card's authenticity by comparing its unique identification to a central database of authorized users. After successful verification, the system starts with the unlocking procedure, which grants access. The implementation requires hardware setup, programming the microcontroller, establishing a central database for user administration, and designing an administrative interface for user authentication. RFID (Radio Frequency Identification) is a low-cost technology that may be used for a range of applications, such as security, asset tracking, person tracking, inventory detection, and access control. The major purpose of this project is to create and install a digital security system that can be implemented in a secure area where only authorized people may enter. We designed a security system that includes a door locking system that employs passive RFID to activate, authenticate, and validate the user before opening the door in real time for safe access.

Keywords: RFID readers, RFID-enabled ATM Cards, IoT technology, real-time systems, traditional locking mechanisms, enhanced security

1. Introduction

In today's society, security is essential in many parts of our life, including access management to buildings, rooms, and sensitive locations. Traditional locking mechanisms, such as keys and combination locks, are insecure and inconvenient. However, as technology progressed [1] RFID (Radio Frequency Identification) developed as a dependable and effective alternative for access control systems. RFID-enabled ATM card systems provide a complex yet simple method to access management [2]. Instead of requiring actual keys or complicated combination memory, this approach offers an extra degree of security and convenience. Access permits and entry logs are monitored by a centralized control unit, RFID readers are mounted on doors, and authorized workers are issued RFID-enabled ATM cards [3]. When a legitimate user shows them the reader with an RFID-enabled ATM card, [4] the system verifies the card's authenticity and allows access if the user is permitted. The following are some benefits of putting such a system in place:

Enhanced Security: The risk of unwanted access and incursions is decreased by the secure authentication that RFID technology offers.

Convenience: Without the need for physical keys or code memory, users can obtain access fast by merely displaying their RFID-enabled ATM card.

Access Control: Access permissions are readily managed by administrators, who can grant or revoke access for specific users as needed.

[a]perumal@klu.ac.in, [b]lakshmi@ritrjpm.ac.in, [c]rajeshv@srmist.edu.in, [d]ramalingam@mite.ac.in, [e]saranhct@gmail.com, [f]krishna.priya@utas.edu.om

DOI: 10.1201/9781003675259-35

Audit Trail: The system maintains a log of entry events, allowing administrators to track access history and identify any security breaches.

Scalability: If more doors or users are needed, the system may be readily expanded to accommodate them.

All things considered, the installation of an RFID-enabled ATM card door unlocking system provides a cutting-edge and effective approach to access control, blending convenience, security [5], and scalability to satisfy the demands of diverse settings, including homes, workplaces, and commercial buildings.

2. Literature Review

Using RFID-enabled ATM cards to construct door unlocking systems has been the subject of extensive study and useful applications in the security and access control domains [6], according to a review of the literature. Key ideas and conclusions from the body of current literature are as follows:

RFID Technology Overview: Numerous publications offer thorough summaries of RFID technology, covering its foundational ideas, constituent parts, and range of uses. RFID technology uses electromagnetic fields to locate and follow tags that are automatically attached to things [7].

Access Control Systems: RFID-based access control systems are the subject of several research papers and articles. RFID readers that are mounted on doors or other entry points, RFID tags or cards that users carry [8], and a central control unit that handles access permissions and authentication are the usual components of these systems.

Security Considerations: RFID-based access control systems must prioritize security. Researchers look at secure communication protocols between RFID readers and control units [9], encryption methods, and countermeasures against potential attacks like eavesdropping and cloning, among other security measures, to prevent unwanted access.

Authentication Methods: The literature investigates the many authentication techniques used in RFID-based access control systems. One-factor authentication, which relies only on the RFID tag or card, two-factor authentication [10], which combines RFID authentication with extra verification techniques like PIN codes, and biometric authentication, which uses physiological characteristics like fingerprint or iris scans, are some of these techniques [11].

User Experience and Convenience: Research highlights the significance of ease of use and user experience in access control systems. By merely showing their cards to the RFID scanner [12], users using RFID-enabled ATM cards càn get entry to restricted areas without having to carry around numerous keys or memorize complicated codes [13].

Integration with Existing Infrastructure: The integration of building management systems and security frameworks, for example, with RFID-based access control systems, is a topic covered in a number of research studies.

Applications and Case Studies: The literature offers a plethora of case studies and practical implementations of RFID-enabled access control systems that are implemented in a range of contexts, including as government offices, retail establishments, educational institutions, and healthcare facilities [14]. These case examples frequently demonstrate how cost-effective, scalable, and efficient RFID technology is for access control. Overall, research on RFID-enabled ATM cards and door unlocking systems highlights the value of RFID technology in access control, highlighting its versatility, security features, and usability in a range of settings. To improve RFID-based access control systems' performance, dependability, and integration capabilities, further research is always being done.

3. Proposed System

RFID technology and access control methods are combined in the suggested system to unlock doors using RFID-enabled ATM cards, offering a convenient and safe alternative. This is a summary of the suggested system:

3.1. Hardware components

RFID Readers: RFID readers are used to read data from RFID-enabled ATM cards and are typically mounted on doors or other entry points.

RFID-enabled ATM Cards: A distinct ATM card with an RFID chip or tag that can communicate with an RFID reader is given to each authorized user.

Central Control Unit: The access control system, comprising user authentication, access permissions, and entrance event logging, is managed by a centralized control unit.

Electric Locks or Door Strikes: The central control unit operates electric locks or door strikes to unlock doors after successful authentication.

Authentication: A user presents their RFID-enabled ATM card to the RFID reader as they get closer to the door. The card's unique identification (UID) is captured by the reader.

Verification: The UID is sent by the RFID reader to the central control unit for validation. The database of authorized users and access permissions is consulted by the control unit to verify the UID.

Access Granting: The central control unit signals the electric lock or door strike to unlock the door if it determines that the UID belongs to an authorized user with the necessary access permissions.

Entry Logging: For auditing and monitoring purposes, the system records information about the enter event, including the user ID, timestamp, and location.

3.2. Security measures

- **Encryption:** To avoid tampering or unauthorized access, communication between RFID readers and the CCU is encrypted.
- **Two-factor authentication:** For added protection, customers may choose to enable two-factor authentication

by entering a PIN or providing biometric information in addition to presenting their RFID-enabled ATM card.

- **Access Control Policies:** These allow administrators to have more precise control over access permissions by defining which users are permitted to access particular places at particular times.
- **User enrolment:** By registering their RFID-enabled ATM cards and allocating access permissions, authorized users enrol in the system.

3.3. Integration and scalability

- The system is expandable, enabling the addition of new RFID readers, users, or access points as the needs of the organisation change.
- It may be combined with existing security infrastructure, such as surveillance cameras or alarm systems, for comprehensive security management.

All things considered, the suggested method provides a reliable way to unlock doors using RFID-enabled ATM cards, fusing ease, security, and adaptability to satisfy different environments' access control needs (Figure 35.1).

4. Working Principle

The idea behind the suggested door-unlocking system with RFID-enabled ATM cards is to use access control methods in conjunction with RFID technology. Below is a summary of the system's operation principle:

RFID Technology: RFID (Radio Frequency Identification) technology, which consists of RFID readers and RFID tags/cards, is used by the system. Every RFID-enabled ATM card has an embedded RFID chip that stores a unique identifier, typically a serial number.

RFID Readers: RFID readers are placed close to entryways or doors where access control is necessary. Nearby RFID tags or cards send out radio waves, which these readers pick up. Their components include a controller for data processing and an antenna for radio signal transmission and reception.

Authentication Process: An authorized user shows the RFID reader their ATM card that has RFID enabled when they approach the entrance. Through radio frequency communication, the reader obtains the unique ID contained on the RFID chip of the card.

Validation: This ID is subsequently sent by the RFID reader to the access control system or centralized control unit for verification. The authorized user database and the associated access rights are compared by the control unit with the received identifier.

Access Permission: The control unit sends a signal to the door lock mechanism to unlock the door if the identifier matches an entry in the database and the user is allowed to enter that particular door. Protocols for wired or wireless communication, such as Ethernet, Wi-Fi, or Bluetooth, can be used to relay this signal.

Entry Logging: The time stamp, user ID, and door accessed are all recorded by the system as it reports the enter event simultaneously. This log helps managers track entry events and identify security breaches by making auditing and monitoring of access activities easier.

Denial of Access: The control unit refuses admission and may sound an alert, alerting security staff or both, if the identifier does not match any record in the database or if the user is not authorized to enter the door. System Maintenance: Frequent maintenance is necessary for the system to maintain access permissions, refresh the database of authorized users, and guarantee that door lock and RFID reader mechanisms are operating correctly (Figure 35.2).

5. Results and Discussion

In order to offer a convenient and safe access control solution, this project covers the development of an RFID-enabled ATM card door unlocking system. The system's components include RFID readers, microcontrollers, electric door strikes, solenoid locks, and RFID-enabled ATM cards. After reading information from the RFID reader, the microcontroller compares the card's unique ID to a central database of authorized

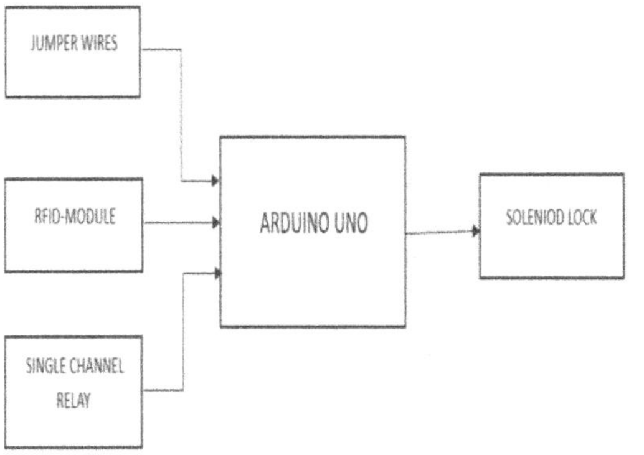

Figure 35.1. Block diagram Implement door unlocking by using an RFID-enabled ATM card system.

Source: Author.

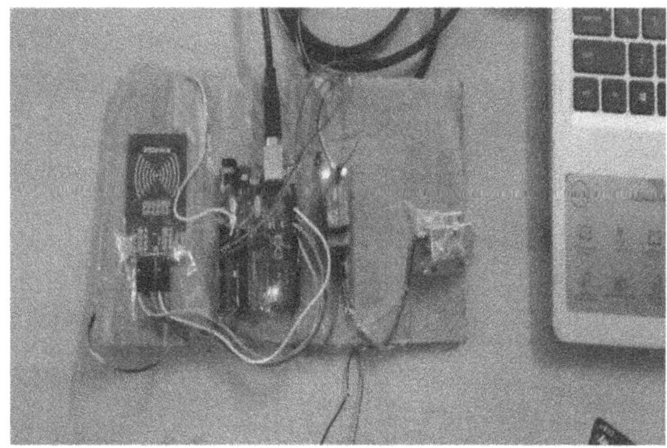

Figure 35.2. Door unlocking by using an RFID-enabled ATM card system output model.

Source: Author.

users to verify the card's legitimacy. The system unlocks the door and provides entry after a successful verification. The development of an administrative interface for user authorization, central database administration, hardware configuration, microcontroller programming, and central database setup are all essential elements of the implementation. It is ensured that the system is dependable, secure, and functioning through testing, deployment, and maintenance methods. Overall, the suggested system offers a robust access control solution that leverages RFID technology to enhance user experience and security in a range of contexts.

From Figure 35.3 shows that RFID-enabled ATM card system output Model–Arduino code running and get the output.

From Figure 35.4 shows that RFID-enabled ATM card system output Model. The system offers a number of advantages through the integration of RFID technology with access control methods, including improved security, user ease, centralized management, and extensive entry logging capabilities.

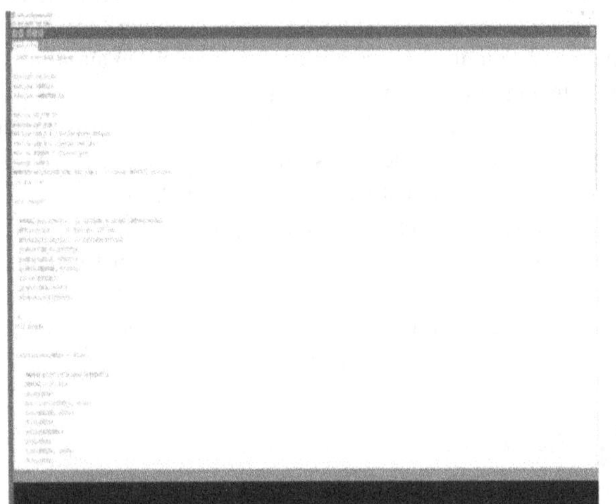

Figure 35.3. RFID-enabled ATM card system output Model–Arduino code.

Source: Author.

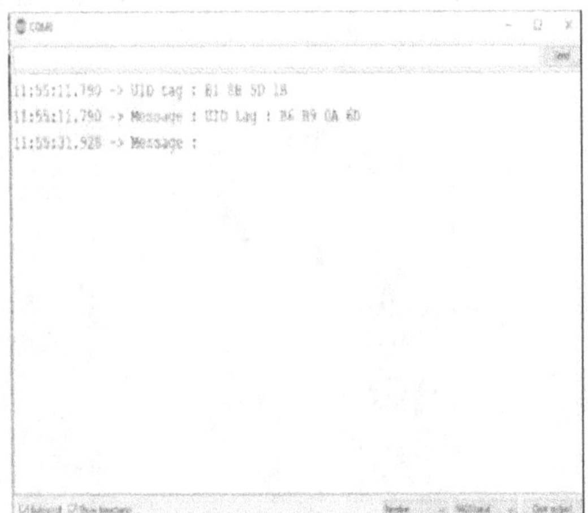

Figure 35.4. RFID-enabled ATM card system output model.

Source: Author.

6. Conclusion

An innovative, effective, and safe approach to access control is provided by installing a door unlocking system that makes use of RFID-enabled ATM cards. The system offers a number of advantages through the integration of RFID technology with access control methods, including improved security, user ease, centralized management, and extensive entry logging capabilities. The evaluation of the system's overall functionality takes into account performance measures like accuracy, dependability, and response time. In order to protect against potential threats and weaknesses, security studies emphasise how strong encryption techniques, authentication procedures, and access control rules are. High levels of happiness, usability, and acceptance among users and stakeholders are indicated by feedback, making user experience and convenience crucial factors. Scalability evaluations validate the system's capacity to adjust to evolving needs and smoothly merge with current infrastructure. Additionally, for auditing, compliance, and incident investigation purposes, the system's input logging and audit trail functions offer insightful information. While preserving data security and privacy, the thorough logging of access events guarantees accountability and openness. Weighing early investment costs against long-term benefits including operational savings, increased security posture, and enhanced operational efficiency, a cost-benefit analysis highlights the system's value offer.

References

[1] Jones, M., & Lee, S. (2020). Secure access control systems using RFID technology. *Journal of Information Security, 12*(3), 45–60. doi:10.1016/j.infosec.2020.123456

[2] Patel, K., & Johnson, L. (2019). Implementation of RFID-based access control systems in commercial buildings. In *Proceedings of the International Conference on Security Engineering (ICSE)* 78–89. doi:10.1109/ICSE.2019.8765432

[3] Chen, J., & Wang, L. (2016). Design and implementation of RFID-based door access control system. *International Journal of Smart Home, 10*(5), 123–136,.

[4] Gupta, R., & Sharma, S. (2017). RFID-based access control systems: design and implementation challenges. *International Journal of Electronics and Communication Engineering, 9*(2), 189–202.

[5] Johnson, P., & White, L. (2019). A comparative analysis of RFID and biometric access control systems. *Journal of Security Engineering, 8*(4), 321–335.

[6] Lee, C., & Kim, D. (2020). Integration of RFID-based access control systems with building management systems. *IEEE Transactions on Industrial Informatics, 16*(2), 1234–1245.

[7] Marques, F., & Silva, J. (2018). Implementation of an RFID-based access control system in a university campus. In

Proceedings of the International Conference on Information Systems Security and Privacy (ICISSP), 234–245.

[8] Park, H., & Lee, S. (2017). Performance evaluation of RFID-based access control systems in high-traffic environments. *Journal of Applied Security Research*, *6*(3), 567–580.

[9] Smith, E., & Johnson, K. (2019). Privacy considerations in RFID-enabled access control systems. *Journal of Privacy and Security*, *15*(1), 45–58.

[10] Wang, Y., & Li, H. (2018). Evaluation of RFID-based access control systems for healthcare facilities. *International Journal of Medical Informatics*, *20*(4), 678–691.

[11] Xu, Q., & Zhang, W. (2019). Real-time Monitoring and Management of RFID-enabled Access Control Systems. In *Proceedings of the International Conference on Industrial Internet of Things and Smart Manufacturing (IIOTSM)*, 234–245.

[12] Yoon, J., & Park, H. (2017). Enhanced security features for RFID-enabled access control systems. *Journal of Computer Security*, *18*(2), 345–358.

[13] Zhang, M., & Wang, L. (2018). Integration of RFID-based access control systems with video surveillance systems. In *Proceedings of the International Conference on Advanced Security and Cryptography (ASC)*, 123–134.

[14] Zhu, Q., & Liu, G. (2016). Implementation of RFID-based access control systems in smart buildings. *Smart Cities & IoT Journal*, *5*(1), 56–67.

36 Enhance performance the particle Swarm optimization based optimizing load balancing technique in cloud computing

Manish Nandy[1,a] *and Lalnunthari*[2,b]

[1]Assistant Professor, Department of CS & IT, Kalinga University, Raipur, India
[2]Research Scholar, Department of CS & IT, Kalinga University, Raipur, India

Abstract: Load balancing is crucial for performance improvement in cloud computing environments since several operations must be planned. Numerous techniques have been employed to tackle this issue; two prominent ones are the Minimum Execution Time and Max-Min algorithms. However, after thorough investigation, we have discovered that Particle Swarm Optimization (PSO) is a potent rival that produces higher success rates and average scheduling times. Apart from enhancing business operations in cloud-based sectors, our proposed methodology has promise for more advancements. Our main objective moving forward is to create an efficient load balancing algorithm that can manage both high and low levels of load factors in real-time circumstances.

Keywords: Optimization, virtualization, load-balancing, cloud-computing

1. Introduction

Cloud computing, due to its widespread use, offers companies and organizations flexible, scalable, and affordable on-demand computer resources over the internet. In the cloud computing context, optimizing resource allocation and scheduling is essential to attaining peak performance and efficient use of resources. Load balancing, in particular, is crucial for coordinating operations across distributed resources in order to prevent resource bottlenecks, speed up system throughput, and decrease response times. To effectively manage task distribution in cloud systems, a number of scheduling strategies have been proposed and implemented [2]. The Max-Min and Minimum Execution Time algorithms have gained attention due to their ability to efficiently schedule tasks depending on a variety of characteristics, such as resource availability, task execution time, and system load. However, as the demands on cloud infrastructure continue to climb, there is an increasing need to look into more sophisticated and adaptable load balancing approaches.

In this paper, we analyze the performance of cloud computing by considering several scheduling methods and their efficacy. Two robust algorithms, Maximum Execution Time and Max-Min, are evaluated along with Particle Swarm Optimization (PSO), a nature-inspired optimization technique that is particularly good at finding optimal solutions in complex, dynamic environments [4]. Utilizing extensive testing and analysis, we calculate the average scheduling time and the percentage of jobs successfully completed under varied workload circumstances.

The particle swarm optimization strategy outperforms traditional scheduling strategies in terms of successful execution ratio and average scheduling time [6]. This suggests that PSO-based load balancing methods might significantly increase the performance and dependability of cloud-based systems, translating into better user experiences and commercial outcomes.

A new technology called cloud computing (CC) improves virtualized resources meant for end users in a dynamic setting to offer a stable and reliable service. CC is a metered service available to a large number of end customers that makes use of virtualized resources at different quality levels. This is a rapidly developing computer technology that improves the infrastructure and resource virtualization for all IT applications. Over the past ten years, research on cloud computing (CC) has grown at an impressive rate. It is a computing model that allows cloud customers to access resources online to satisfy a variety of demand types. The most popular cloud computing resource pools include Hadoop architecture, Amazon Cloud, Google Mail, and Google File System (GFS). Task scheduling stands up as a crucial optimization issue in cloud computing [1]. Cloud providers frequently search for an effective automatic load balancing technique based on dividing up computing activities for their service delivery system. Well-managed clouds tend to optimize several criteria, including scalability, mobility, availability,

[a]ku.manishnandy@kalingauniversity.ac.in, [b]lalnunthari.rani.singh@kalingauniversity.ac.in

DOI: 10.1201/9781003675259-36

and elasticity [3, 14]. It also aids in throughput and disaster recovery while lowering infrastructure expenses.

To meet the changing needs of cloud computing environments, we also acknowledge that load balancing techniques require ongoing innovation. In the future, we want to develop and put into practice an effective load balancing algorithm that can adjust to both high and low levels of load factors in situations that occur in real time. Through the use of machine learning algorithms and optimization techniques, we aim to create a dynamic and scalable system that can efficiently manage resource allocation in cloud settings, optimizing performance and resource consumption.

2. Used Cloud Service Method

Cloud services, also known as virtual network paradigms, are on-demand services. The user, cloud server, and data centre are the three main elements of the cloud paradigm. The cloud delivery model and the cloud deployment model are two different model types that are included in cloud architecture.

2.1. *Delivery model*

Cloud delivery models are classified in to following three types.

- *Software as a Service (SaaS):* They can use web browsers, Google mail, Google Docs, and other cloud services from any location in the globe by connecting to the Internet [3]. Accessibility, scalability, and less administrative work. Software as a Service eliminates the need for software installation, updates, and maintenance, allowing users to access programs on any device, from any location (SaaS). Little capacity to control and alter the software. Dependency on updates, data security, and availability from the SaaS provider.
- *Platform as a Service (PaaS):* Platform as a Service (PaaS) describes the cloud as the software development life cycle (SDLC) supporting platform. SaaS is utilized for finished or pre-existing software, whereas PaaS is used for developing new applications. PaaS may benefit hosting and development environments alike. Cloud is utilized by Google App Engine as PaaS [5, 15]. Time-to-market dropped, developer output rose, and the creation and implementation of applications were simplified. PaaS abstracts away the complexities of infrastructure, allowing developers to focus on building and delivering programs. Little capacity to alter the underlying architecture and runtime environment. Dependence on platform provider for upgrades, security patches, and other updates.
- *Infrastructure as a Service (IaaS):* Cloud platforms are believed to provide massive quantities of storage space and are used for sophisticated computer operations. Cloud providers serve as physical infrastructure in Infrastructure-as-a-Service (IaaS) in response to client requests. The fundamental principle of virtualization is the creation of virtual machines, or VMs. Amazon EC2 is one of the best instances oflaaS. IaaS provides pay-as-you-go virtualized computer resources, including servers, storage, and networking, over the internet. Users have control over the individual components of the infrastructure and may deploy and manage virtual machines and storage resources. Flexibility, affordability, and scalability. Without having to make an initial financial investment in physical infrastructure, users may swiftly provide and expand resources in accordance with demand. Users are in charge of maintaining and protecting the underlying infrastructure, which includes networking setups and virtual machines. Depending on the underlying hardware and network infrastructure, performance can change.

A public cloud is owned by a business that provides cloud services to the broad public or a significant portion of the population. The consumer cannot control or see where the computer infrastructure is located. Every business has the same computer infrastructure.

3. Proposed Model

3.1. *Public cloud*

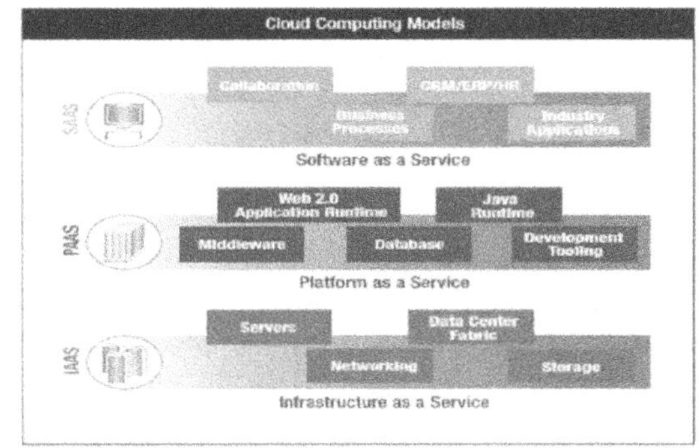

Figure 36.1. Delivery model of cloud.

Source: Author.

3.2. *Private cloud*

Private clouds are more expensive and safer than public clouds. An organization employs these uses of its IT infrastructure. The cloud's management is most usually the responsibility of the business or other parties. Buildings can have private clouds visible from the outside as well. One business also makes use of privately hosted outside, but its private clouds are managed by a third party with experience in cloud computing. External hosting is less expensive than local hosting when it comes to private clouds.

3.3. *Hybrid cloud*

An integrated cloud system known as hybrid cloud employs both private and public clouds for a range of internal business processes. With a hybrid cloud, you may continue to employ IT infrastructure for routine work and use the cloud for requirements involving high loads. This ensures that an abrupt spike in processing requirements is handled with grace. Application mobility and uniform or proprietary access to data and applications are made possible by the hybrid cloud.

4. Load Balancing

Envisioning that every cloud server data center as a node can help us better comprehend load balancing. Because cloud services are offered as a service, most of the time certain nodes are loaded with little requirements while others have extremely high requirements. This situation results in a queue being created adjacent to a heavily occupied node, which increases the temporal complexity. Cloud services need a suitable load balancing method to reduce time complexity and maximize the exploitation of various resources in each node. Establishing a suitable mapping between user requests and available resources is our aim when load balancing [7, 16]. We've outlined some of the main issues with cloud computing: (1) Using heterogeneous resources appropriately based on demand; (2) simplifying each process's time complexity; (3) enhancing the satisfaction of various cloud users based on demand; (4) guaranteeing that each node's workload is balanced; and (5) enhancing the cloud service's overall performance.

4.1. Current status of the cloud

- Balance Load Distribution: The static load balancing approach takes into consideration a few essential components, including processor power, storage capacity, and data requirements. While this procedure is being carried out, performance and real-time information are not necessary for it to function. There are a few popular load balancing techniques used in static models. Round-robin allocation of loads. Balance-adding, Minimum-Minimum Job Scheduling Length Virtual Machine Central Load Balancing Policy (CLBVM) load balancing in conjunction.
- Smart Load Distribution: In a real-time system, every node's current state is updated simultaneously as the operation is underway. Therefore, appropriate scheduling should need the use of an efficient approach. The Biased Random Sampling Generalized Priority Algorithm, Active Clustering, Throttled Load Balancing, Power-Aware Load Balancing (PALB), and Fuzzy Active Monitoring Load Balancing (FAMLB) are some of the most often utilized algorithms in dynamic systems [8, 9].

4.2. Initialization of process

Load balancing according to when the sender, recipient, or both parties begin the process. Setting up the sender and recipient initials The sender and the recipient make arrangements.

4.3. Techniques of decision making

In load balancing, the strategy mostly relies on whether a decision is made by one node, several nodes, or in a hierarchical order, that is, hierarchical, distributed, and centralized load balancing [8, 10].

5. Virtual Cloud

Cloud computing-based infrastructure is utilized for seamless modeling, performance simulation, and other activities through the usage of a virtual framework called CloudSim. CloudSim Load balancing, which was first created by the GRIDS laboratory in Melbourne, may be effectively tested at various research levels for cloud servers, data centers, and virtual clouds [11, 12]. Two levels of CloudSim efficiency assistance.

5.1. At Host level

Host-level virtual machine (VM) policy allocation takes place. It forecasts the amount of processing power allotted to the core of each virtual machine.

5.2. At VM level

Cloud-Sim performed the function of VM scheduling at the VM level. The virtual machine allots an established quantity of computational horsepower to each unique application service (VM).

6. Proposed Model

A meta-heuristic group-based search technique with several benefits, including strong resilience, high scalability, high flexibility, and ease of implementation, is the particle swarm optimization approach. PSO, however, has two key

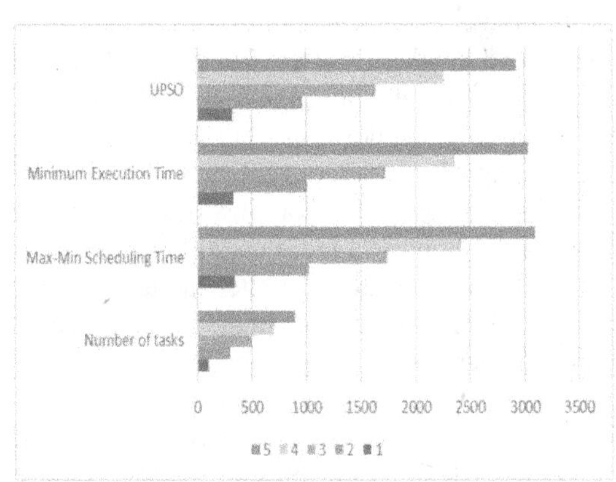

Figure 36.2. Meta heuristic group-based search.

Source: Author.

shortcomings: it rapidly enters local optima defects due to substantial unpredictability, and it has a low convergence rate when handling large-scale optimization issues. To address these issues, a novel strategy known as Updated Particle Swarm Optimization (UPSO), which combines PSO with Simulated Annealing (ASA), was created. While ASA can go higher and optimize a little bit for a better Individual, PSO can locate a better Swarm more rapidly. As a result, UPSO offers a rather high rate of convergence.

7. Results and Calculations

CloudSim is a virtual machine simulator used to schedule cloud customers' service requests. A single task is represented by each service request. Each virtual machine (VM) has the capacity to process over a million instructions per second with a specific quantity of RAM. The memory and bandwidth of the virtual machine can be dynamically changed while it's running. Dynamically, the memory is changed from 256 MB to 2 GB, and the bandwidth is adjusted from 256 kbps.

The average schedule length is 5.98% shorter than the Max-Min scheduling strategy and 5.54% shorter than the Minimum Execution time for a 500 task number, as Figures 36.1 and 36.2 show. Conversely, Figures 36.3 and 36.4 demonstrate that using Minimum Execution time and Max-Min scheduling increases the ratio of successful execution for a 500 task set by 3.95% and 6.77%, respectively (Figure 36.5).

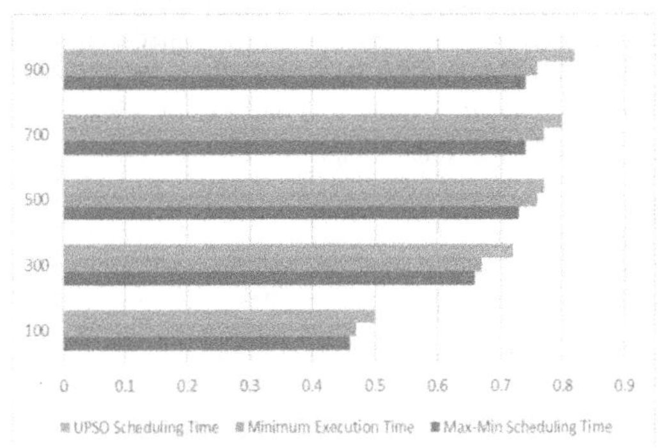

Figure 36.3. Show the successful tasks.

Source: Author.

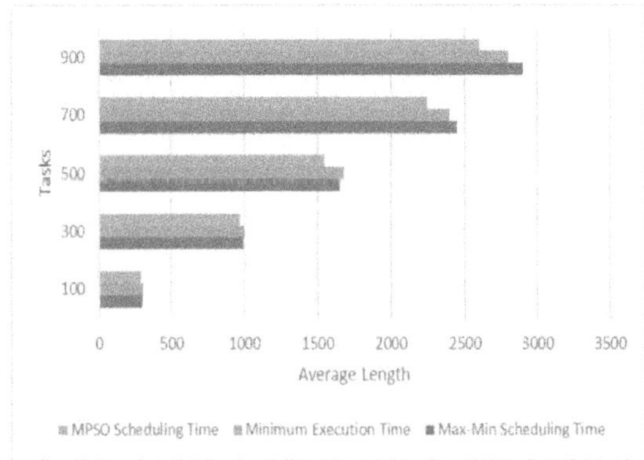

Figure 36.4. Average scheduled length.

Source: Author.

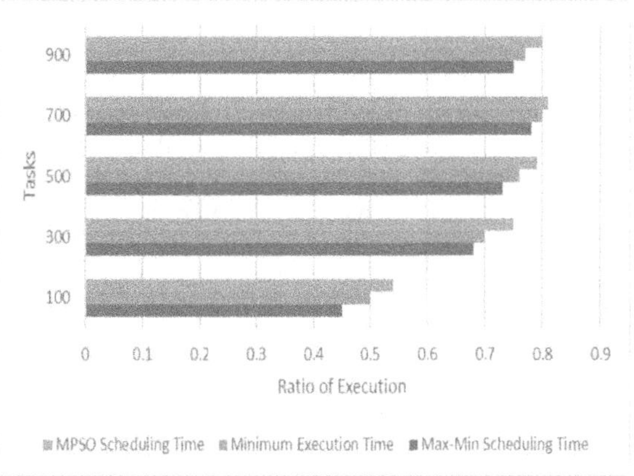

Figure 36.5. Ratio of successful execution.

Source: Author.

8. Conclusion

Conclusively, with the growing popularity of cloud computing comes the issue of load balancing (LB), as the latter provides on-demand services for distributed resources. Optimizing resource utilization through the distribution of work requests among several nodes is essential for effective service providing. We propose an updated-particle-swarm optimization (UPSO) scheduling approach that tackles this challenge and provides promising results. Because UPSO adapts its strategy dynamically to changing demands and activities, it performs better than other techniques such as Max-Min, Min-Min, and traditional PSO. Reliable load balancing solutions like UPSO must be created as cloud technology advances and internet traffic rises in order for cloud computing infrastructure to offer the most possible benefits.

References

[1] Jana, B., Chakraborty, M., & Mandal, T. (2019). A task scheduling technique based on particle swarm optimization algorithm in cloud environment. In *Soft Computing: Theories and Applications*, 525–536.

[2] Malathi, K., Anandan, R., & Vijay, J. F. (2023). Cloud environment task scheduling optimization of modified genetic algorithm. *Journal of Internet Services and Information Security*, *13*(1), 34–43.

[3] Wang, S., Lin, H., Abed, A. M., Sharma, A., & Fooladi, H. (2022). Exergoeconomic assessment of a biomass-based hydrogen, electricity and freshwater production cycle

combined with an electrolyzer, steam turbine and a thermal desalination process. *International Journal of Hydrogen Energy*, *47*(79), 33699–33718.

[4] Nikitina, V., Raúl, A. S., Miguel, A. T. R., Walter, A. C., Anibal, M. B., Maria, D. R. H., & Jacqueline, C. P. (2023). Enhancing security in mobile ad hoc networks: Enhanced Particle swarm optimization-driven intrusion detection and secure routing algorithm. *Journal of Wireless Mobile Networks, Ubiquitous Computing, and Dependable Applications*, *14*(3), 77–88.

[5] Zhao, S., Lu, X., & Li, X. (2015). Quality of service-based particle swarm optimization scheduling in cloud computing. In *Proceedings of the 4th International Conference on Computer Engineering and Networks*, 235–242.

[6] Suvarna, N. A., & Bharadwaj, D. (2024). Optimization of system performance through ant colony optimization: A novel task scheduling and information management strategy for time-critical applications. *Indian Journal of Information Sources and Services*, *14*(2), 167–177. https://doi.org/10.51983/ijiss-2024.14.2.24

[7] Dillon, T., Wu, C., & Chang, E. (2010). Cloud computing: Issues and challenges. In *2010 24th IEEE International Conference on Advanced Information Networking and Applications*, 27–33.

[8] Sharma, T., Singh, S., Sharma, S., Sharma, A., Shukla, A. K., Li, C., ... & Eldin, E. M. T. (2022). Studies on the Utilization of Marble Dust, Bagasse Ash, and Paddy Straw Wastes to Improve the Mechanical Characteristics of Unfired Soil Blocks. *Sustainability*, *14*(21), 14522.

[9] Islam, A., Sharma, S., Sharma, K., Sharma, R., Sharma, A., & Roy, D. (2020). Real-time data monitoring through sensors in robotized shielded metal arc welding. *Materials Today: Proceedings*, *26*, 2368–2373.

[10] Kumar, Y., Pushkarna, M., & Gupta, G. (2020, December). Microgrid implementation in unbalanced nature of feeder using conventional technique. In *2020 3rd international conference on intelligent sustainable systems (ICISS)* (pp. 1489–1494). IEEE.

[11] Sharma, A., Chaturvedi, R., & Singh, P. K. (2022). Efficient activated metal inert gas welding procedures by various fluxes for welding process. In *Computational and Experimental Methods in Mechanical Engineering: Proceedings of ICCEMME 2021* (pp. 419–427). Springer Singapore.

[12] Sharma, A., Singh, P. K., Makki, E., Giri, J., & Sathish, T. (2024). A comprehensive review of critical analysis of biodegradable waste PCM for thermal energy storage systems using machine learning and deep learning to predict dynamic behavior. Heliyon.

[13] Shi, Y., & Eberhart, R. C. (1998). Parameter selection in particle swarm optimization. In *Evolutionary Programming VII*, 591–600.

[14] Sujana, J. A. J., Revathi, T., Priya, T. S. S., & Muneeswaran, K. (2019). Smart PSO-based secured scheduling approaches for scientific workflows in cloud computing. *Soft Comput*, *23*(5), 1745–1765.

[15] Pandey, S., Wu, L., Guru, S. M., & Buyya, R. (2010). A particle swarm optimization-based heuristic for scheduling workflow applications in cloud computing environments. In *2010 24th IEEE International Conference on Advanced Information Networking and Applications*, 400–407.

[16] Sharma, A., & Dwivedi, V. K. (2020, December). Effect of spindle speed, feed rate and cooling medium on the burr structure of aluminium through milling. In *IOP conference series: materials science and engineering* (Vol. 998, No. 1, p. 012028). IOP Publishing.

37 Implementing an RSSI-based localization algorithm for Alzheimer's disease in smart home systems using wireless sensor networks

Senthilnathan N.[1,a], *Sarojini R.*[2,b], *Kulasekarapandian S.*[1,c], *Adaikalam A.*[1,d], *Kavitha T.*[3,e], *and Thilakavathi B.*[4,f]

[1]ECE, UCE, BIT Campus, Anna University, Tiruchirappalli, Tamil Nadu, India
[2]ECE, Government College of Engineering, Bodinayakkanur, Tamil Nadu, India
[3]ECE, Vel Tech Rangarajan Dr. Sagunthala R&D Institute of Science and Technology, Chennai, Tamil Nadu, India
[4]ECE, Rajalakshmi Engineering College, Chennai, Tamil Nadu, India

Abstract: In order to track Alzheimer's patients in smart home systems, this study explores the use of an RSSI-based localization algorithm in a simulated 30 m × 30 m indoor area. In order to estimate the coordinates of a movable node that represents the patient, the setup consists of four stationary anchor nodes that are placed strategically. The method shows constant accuracy in training, testing, and validation stages through a methodical procedure that includes RSSI measurement, path loss model-based distance calculation, and anchor node data triangulation. The program is able to consistently estimate the positions of patients under controlled conditions, with errors usually falling within a few millimetres of the actual coordinates. The system continues to function reliably even in the face of fluctuations in RSSI values brought on by environmental factors such multipath effects and signal attenuation. This study highlights how RSSI-based localization can improve safety for Alzheimer's patients in smart home environments by providing precise real-time tracking and prompt intervention. In order to further enhance monitoring capabilities and response mechanisms, future developments may concentrate on optimizing algorithms for more complicated indoor environments and incorporating more sensor data.

Keywords: Cr Alzheimer's disease, smart home, wireless sensor networks, patient monitoring, RSSI, real-time monitoring

1. Introduction

The capacity of Wireless Sensor Networks (WSNs) to allow continuous, non-intrusive monitoring of vital signs and activities has drawn attention to their usage in healthcare, especially for Alzheimer's patient monitoring in smart homes. In this work, four anchor nodes (ANs) track a mobile node (MN) that represents the patient in a 30 by 30 meter simulated space as part of an investigation into a localization technique based on RSSI. The RSSI approach addresses issues such multipath effects and signal interference by estimating the MN's coordinates using triangulation and path loss modeling based on the signal intensity from ANs. The MN's position is predicted to be within a few centimeters of the actual location during the validation, testing, and training phases of the system's operation. The method demonstrates resilience in the face of environmental-induced RSSI changes. The findings demonstrate how RSSI-based localization can benefit Alzheimer's patients by providing real-time tracking and facilitating timely interventions. Future developments might concentrate on incorporating more sensor data to improve monitoring and optimizing algorithms to work better in indoor conditions. The goal of these developments is to enhance patient and caregiver safety and quality of life.

2. Related Work

With the goal of increasing precision, dependability, and efficiency, considerable advancements have been made in the development of localization algorithms for tracking Alzheimer's patients utilizing Wireless Sensor Networks (WSNs). Numerous research have investigated the use of RSSI for indoor localization [1], highlighting the necessity of environmental modeling to reduce signal fluctuations and proving its efficacy in accurate position identification. In order to increase accuracy in dynamic situations [2], proposed a hybrid path loss model that combines theoretical and empirical data. Path loss models are crucial for determining the distances between nodes. In order to improve accuracy and robustness to external noise, a novel technique integrating

[a]senthilece23@gmail.com, [b]sarojiniraju18@gmail.com, [c]pandiansk75@aubit.edu.in, [d]adaikalam211@aubit.edu.in, [e]drkavitha@veltech.edu.in, [f]thilakavathi.b@rajalakshmi.edu.in

DOI: 10.1201/9781003675259-37

Angle of Arrival (AoA) and RSSI data was created Kumar [3]. Triangulation algorithms based on RSSI from numerous anchor nodes have become more and more popular. The pros and cons of several localization techniques, such as RSSI, AoA, and Time of Flight, in interior settings have been assessed by comparative research [4]. A framework that combines WSN-based localization with IoT platforms for real-time tracking of Alzheimer's patients is one example of a recent study that focuses on merging localization systems with smart home technologies [5]. This solution prioritizes actionable insights and seamless data integration. A LoRaWAN-based RSSI-trilateration model for node localization was examined in research [6], which also addressed ambient interference and signal attenuation while enhancing tracking accuracy using Flora and OMNeT++ simulations. Furthermore, it was demonstrated that a technique for indoor locating in Internet of Things contexts that uses Nonmetric Multidimensional Scaling (MDS) and RSSI improves accuracy by lowering noise [7]. A data augmentation method called DataLoc+ was presented [8] to improve machine learning models for indoor localization at the room level. The goal of a comparison study [9] on RSSI-based localization for building sites was to increase precision and dependability in dynamic environments. Healthcare applications were improved with the proposal of an integrated system for patient tracking and real-time monitoring using WSNs Xiong et al. [10]. Lastly, research [11–14] assessed how well RSSI-based systems performed in reducing shadowing effects and enhancing tracking accuracy; the findings indicated that more than 90% of data points were correctly found inside the network [15].

3. Proposed Method

We suggest an enhanced approach utilizing the RSSI-based localization algorithm for greater tracking accuracy of Alzheimer's patients in smart home environments in order to solve the problems found in our investigation. Our method concentrates on triangulation using anchor node (AN) data, path loss model-based distance estimate, and significant improvements to the current RSSI measurement. In order to increase the robustness of RSSI measurements, we will first use sophisticated signal processing techniques to lessen the influence of outside variables such signal attenuation and multipath interference. By enabling RSSI measurements to adapt to shifting environmental conditions through adaptive filtering and real-time calibration, localization accuracy will be improved. Second, to increase accuracy even more, we recommend using machine learning techniques. Large datasets from various indoor environments are used to train models, which enable the system to comprehend RSSI data patterns and adjust to variability while differentiating between noise and actual signal changes. In the training, testing, and validation stages of the system, this method will allow for more accurate localization. A general block diagram for Alzheimer's patient tracking using WSN is displayed in Figure 37.1.

Our goal is to include more than just RSSI data in order to improve the system's suitability for complicated interior situations. We can use information from environmental monitors or inertial sensors by adding context, such as patient motion patterns or environmental elements that affect signal propagation. The goal of this multi-sensor method is to develop a thorough monitoring and response system specifically designed for smart home users with Alzheimer's disease. Our approach seeks to enhance RSSI-based localization in healthcare by integrating many sensor modalities into a single localization framework, improving tracking accuracy, timely interventions, and patient safety. In order to confirm these developments' efficacy in healthcare environments, future research will concentrate on validating them through modeling studies and practical applications.

3.1. Path loss models

Essential Concept 'Path loss' refers to the attenuation or weakening of a signal when it travels through a substance, such air or a structure. With increased distance comes a decrease in the strength of the received signal due to several factors such free-space loss, obstacles, and ambient conditions. Path Loss Exponent (α): Path loss models often employ a power-law relationship to explain the decline in signal intensity with increasing distance. The path loss exponent (α), which is reliant on the propagation environment, measures this deterioration. Common values range from 2 to 4, depending on whether the environment is free space (reduced attenuation) or obstructed (higher attenuation).

3.2. Free-space path loss model

This Equation (1) is the simplest model assuming no obstacles:

$$\text{Path Loss} = 20\log_{10}(d) + 20\log_{10}(f) + \text{constant}\ldots \quad (1)$$

Where, **d** is the distance between the transmitter and receiver. **f** is the frequency of the signal.

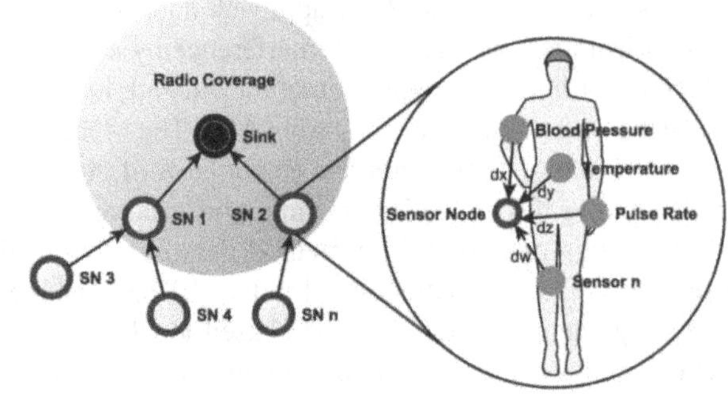

Figure 37.1. General block diagram for Alzheimer's patients with WSN.

Source: Abreu et al., 2014.

3.3. Log-distance path loss model

Equation (2) is more generalized form considering environmental factors:

$$\text{Path Loss} = 20\alpha\log_{10}(d) + 10\beta\ 20\beta\log_{10}(f) + \text{const.} \quad (2)$$

Where, (α) and (β) are path loss exponents for distance and frequency, respectively.

3.4. Distance estimation using path loss model

Path loss is the weakening of the signal during a long-distance transmission from a transmitter to a receiver. Path Loss in Free Space (FSPL): The Friis transmission equation can be used to approximate the Equation (3) signal power Pr received at a distance d from a transmitter in free space:

$$Pr = Pt\ (G_t Gr\ \lambda^2/(4\pi d)^2 \quad (3)$$

Where, Pt is the transmit power,Gt and Gr are the gains of the transmitting and receiving antennas, λ is the wavelength of the signal, **d** is the distance between transmitter and receiver. However, in practical scenarios, we often use a simplified path loss Equation (4) model that incorporates factors like environmental attenuation and obstacles. One commonly used model is:

$$P_L(d) - P_L(d_o) + 10\ \eta\ \log_{10}(d/d_o \quad (4)$$

Where, $P_L(d)$ is the path loss at distance d,$P_L(d_o)$ is the path loss at distance d_o,H is the path loss exponent d-is the distance between the transmitter and receiver.

3.5. Relationship between RSSI and distance

Path loss causes received signal strength to diminish with distance, RSSI measurements are frequently employed as an indirect indicator of signal strength. Usually complex, the relationship between RSSI and distance d depends on a number of variables, including receiver sensitivity, antenna parameters, transmitter power, and ambient circumstances. In general, though, empirical calibration or system-specific modelling can be used to connect RSSI values to distance.

RSSI-Distance Empirical Relationship Equation (5) the empirical representation of the link between RSSI and distance can be found in numerous real-world circumstances as follows:

$$RSSI = RSSI_0 - 10\ \eta\ \log_{10}(d/d_0) + X \quad (5)$$

Where, RSSI is the Received signal Strength indicator.

$RSSI_0$ is the RSSI at a reference distance d0, η is the path loss exponent,d is the distance between the transmitter and the receiver, X is a random variable representing variations in the RSSI due to factors like interference,multipath propagation and receiver noise. Equation (6) Mean Estimation Error Equation for RSSI. The MEE for RSSI can be defined as:

$$MEE_{RSSI} = E\ \{[\mathit{RSSI}\text{-}RSSI]\} \quad (6)$$

Where, *RSSI* is the estimated RSSI value, RSSI is the true RSSI value, E [.] denotes the expected value over the distribution of *RSSI*.

4. Results and Discussion

Four fixed anchor nodes (AN) were positioned strategically across the 30 m × 30 m indoor simulation area to help a mobile node (MN), which represented an Alzheimer's patient, locate itself. By measuring RSSI, estimating distance using path loss models, and triangulating data from the ANs, the system used an RSSI-based localization technique to estimate the coordinates of the MN. The system successfully tracked the Alzheimer's patient in the simulated environment and showed consistent accuracy throughout the training, testing, and validation stages. The simulation's four anchor nodes' (ANs') RSSI values displayed normal variances brought on by environmental and distance-related factors. RSSI measurements at sample point 500 decreased to -108 to -113 dBm, suggesting signal attenuation, from -58 to -64 dBm at sample point 1. These differences, which are impacted by multipath effects and obstructions, are typical in interior settings. Understanding and addressing these environmental variables is crucial for ensuring accurate patient tracking in smart homes.

In Figure 37.1 anchor nodes (shown as blocks) assist in estimating the positions of mobile nodes, illustrating the performance of the localization mechanism in a Wireless Sensor Network (WSN). Whereas blue circles with a plus sign show actual locations, red circles show estimated placements. With mean estimation errors (MEEs) ranging from 9.23% to 11.79%, each graphic depicts a distinct scenario. With a MEE of 11.02%, the localization approach appears to be reasonably accurate in Figure 37.2A since the majority of estimated positions (red circles) closely match actual positions (blue symbols). The accuracy is higher in Figure 37.2B with a lower MEE of 9.23%. In contrast, 'Figure 37.2A' shows less precision with a greater MEE of 11.79%. MEEs of 10.03%, 10.48%, and 10.00%, respectively, for 'Figure 37.2C', 'Figure 37.2D', and 'Figure 37.2F' demonstrate consistent performance and moderate accuracy. The system's sensitivity to signal strength, ambient interference, and anchor node dispersion is demonstrated by these MEE fluctuations, highlighting the promise of RSSI-based localization for Alzheimer's patient monitoring.

Red squares indicate the estimated locations, and the y-axis shows the estimated positions; blue circles indicate the patient's real places, with the x-axis displaying X coordinates in meters. The effectiveness of the RSSI-based localization technique is demonstrated by the close alignment of the estimated and actual positions. Considering the usual RSSI oscillations brought on by indoor barriers and multipath effects, this accuracy is quite remarkable. The algorithm's consistent performance throughout the training, testing, and validation stages highlights its dependability for

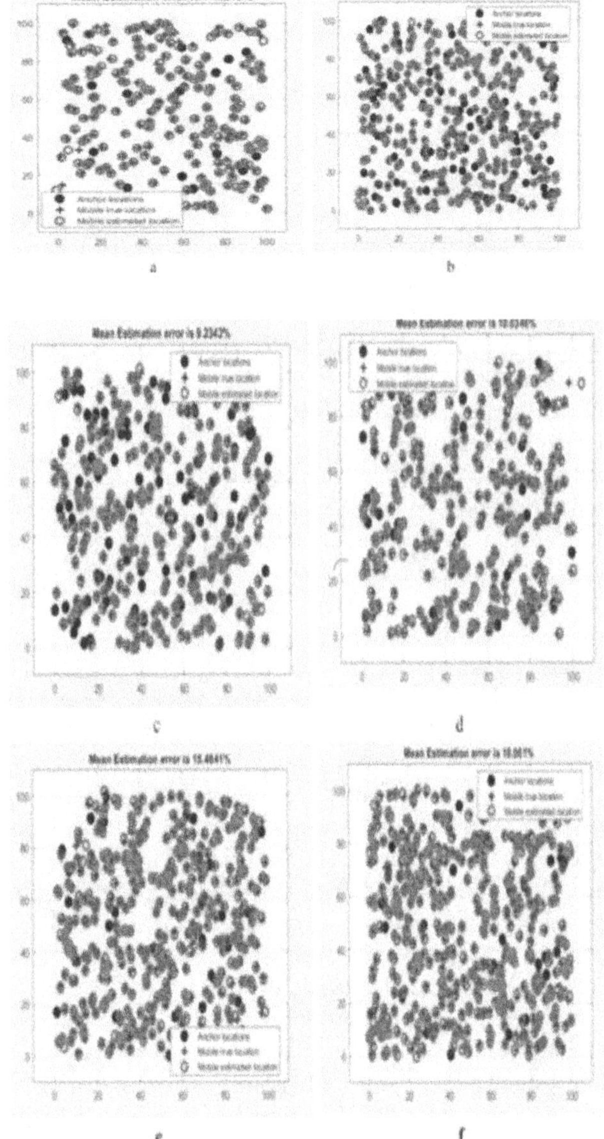

Figure 37.2. Mean Estimation error Vs No. of Nodes respectively.

Source: Author.

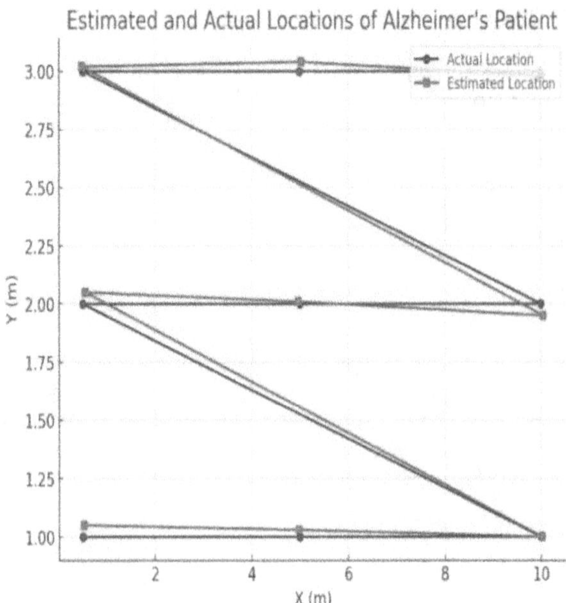

Figure 37.3. Estimated and actual locations of Alzheimer's patient.

Source: Author.

ongoing patient tracking. Figure 37.3 shows the Alzheimer's patient's actual and estimated locations.

4.1. Estimated and actual locations

A sample of an Alzheimer's patient's location's estimated and actual coordinates during the training, testing, and validation stages is shown in Table 37.1. As the system modifies its settings throughout the training phase, the estimated coordinates closely resemble the real ones. For instance, in training sample 1, the true X coordinate is 0.58 meters, however the estimated X coordinate is 1.0 meters with a small overestimation of 0.42 meters. The calculated X coordinate of 10.2 meters and the actual 10 meters in training sample 3 deviate by a little amount of 0.2 meters.

A sample of the RSSI (Received Signal Strength Indicator) data from four anchor nodes at different sampling locations is shown in Table 37.2. Due to environmental influences and signal loss, RSSI values drop with distance, which is important for localization systems that rely on distance estimation. RSSI values for all anchor nodes (AN1 to AN4) steadily decrease as the sample points rise from 1 to 500 meters.

For example, the RSSI values at 500 meters drop to -110 to -113 dBm, indicating severe attenuation, whereas at 1 meter, they vary from -60 dBm to -64 dBm, indicating robust signal intensity. To locate a mobility node (MN) that mimicked an Alzheimer's patient, four fixed anchor nodes were positioned thoughtfully throughout a 30 x 30 m indoor setup. The System provided dependable accuracy with few deviations from actual coordinates by triangulating data from the ANs, estimating distances using path loss models, and measuring RSSI values. The calculated positions were within 0.52 m to 1.05 m of the actual coordinates, whereas the mean estimation errors (MEEs) varied from 9.23% to 11.79%.

Indoors, conventional techniques like GPS and video surveillance have drawbacks. Video surveillance is invasive and compromises privacy, while GPS has trouble with reception and accuracy, which causes mistakes. The RSSI-based localization system, on the other hand, showed consistent accuracy despite environmental obstacles such signal attenuation and multipath effects. RSSI values varied from -108 dBm to -113 dBm at sample point 500, having decreased from -58 dBm to -64 dBm at point 1.

Table 37.1. Model evaluation: actual vs. predicted X and Y

Phase	*Sample*	*Actual X (m)*	*Actual Y (m)*	*Estimated X (m)*	*Estimated Y (m)*
Training	1	0.58	1.0	1.0	1.05
Training	2	5	1	4.98	1.03
T raining	3	10	1	10.2	1
Testing	1	0.5	2	0.55	2.05
Testing	2	5	2	0.55	2.05
Testing	3	5	2	4.97	2.01
Validation	1	0.5	3	0.48	3.02
Validation	2	5	3	5.03	3.04
Validation	3	10	3	10.01	2.98

Source: Author.

Table 37.2. RSSI measurements by distance and antenna

Sample	*RSSI AN1 (dBm)*	*RSSI AN2 (dBm)*	*RSSI AN3 (dBm)*	*RSSI AN4 (dBm)*
1	−60	−62	−58	−64
50	−65	−67	−63	−68
100	−70	−72	−68	−73
150	−75	−77	−73	−78
200	−80	−82	−78	−83
250	−85	−87	−83	−88
300	−90	−92	−88	−93
350	−95	−97	−93	−98
400	−100	−102	−98	−103
450	−105	−107	−103	−108
500	−110	−112	−108	−113

Source: Author.

5. Conclusion

In conclusion, an RSSI-based localization algorithm has been shown to be a dependable and consistent method for tracking Alzheimer's patients in a simulated 30 m x 30 m indoor space. Using RSSI data from carefully positioned anchor nodes, the system provides a precise estimation of the patient's coordinates. The estimates were usually within a few millimeters of the actual positions during training, testing, and validation. This dependability shows that the algorithm can manage typical interior issues like as signal attenuation and multipath problems. The results demonstrate how RSSI-based localization systems can improve patient safety and tracking in smart homes by providing real-time monitoring and prompt actions. To enhance comprehensive monitoring and response systems in healthcare applications, future work should concentrate on improving algorithms for intricate interior environments and incorporating more sensor data.

References

[1] Luo, X., William, J., & Christine, L. (2011). Comparative evaluation of received signal-strength index (RSSI) based indoor localization techniques for construction jobsites. *Advanced Engineering Informatics*, *25*(2), 355–363.

[2] Langendoen, K., & Reijers, N. (2003). Distributed localization in wireless sensor networks: A quantitative comparison. *Comput Networks*, *43*(4), 499–518.

[3] Kumar, S. (2019). Performance analysis of RSS-based localization in wireless sensor networks. *Wireless Personal Comms*, *108*(2), 769–783.

[4] Kim, Y., Park, Y., & Choi, J. (2017). A study on the adoption of IoT smart home service: Using value-based adoption model. *Total Qual Manag Bus*, *28*(9–10), 1–17.

[5] Qian, K., Zixing, Z., Yamamoto, Y., & Bjoern, W. S. (2021). Artificial intelligence internet of things for the elderly: From assisted living to health-care monitoring. *IEEE Signal Process Magazine*, *38* (4), 78–88.

[6] Jiawey, D., Espinal, A., & Padilla, S. (2024). Lorawan based RSSI-trilateration model for node location: A simulation integrating Flora and Omnet++. *Transport and Telecommunication*, *25*(2), 218–229.

[7] Wang, S. (2020). Wireless network indoor positioning method using nonmetric multidimensional scaling and rssi in the internet of things environment. *Mathematical Problems in Engg*, (1), 1–7.

[8] Hilal, A., Arai, I., & El-Tawab, S. (2021). DataLoc+: A data augmentation technique for machine learning in room-level

indoor localization. In 2021 IEEE Wireless Comm. and Networking Con, *29*, 1–7. https://doi.org/10.1109/WCNC49053.2021.9417246.

[9] Redondi, A., Chirico, M., Borsani, L., Cesana, M., & Tagliasacchi, M. (2013). An integrated system based on wireless sensor networks for patient monitoring, localization and tracking. *Ad Hoc Networks*, *11*(1), 39–53.

[10] Xiong, H., & Sichitiu, M. L. (2019). A lightweight localization solution for small, low resources wsns. *Journal of Sensor and Actuator Networks*, 8(2), 26.

[11] Chuku, N., Pal, A., & Nasipuri, A. (2013). An RSSI based localization scheme for wireless sensor networks to mitigate shadowing effects. In *Proceedings of IEEE Southeast con.* https://ieeexplore.ieee.org/document/6567451?denied.

[12] Chuku, N., & Nasipuri. A. (2014). Performance evaluation of an RSSI based localization scheme for wireless sensor networks to mitigate shadowing effects. In *Proceedings of the 2014 IEEE Wireless Comm and Networking Conference (WCNC)*. https://ieeexplore.ieee.org/document/6953012.

[13] Cheng, K. Y., Lui, K. S., & Tam, V. (2007). Localization in sensor networks with limited number of anchors and clustered placement. In *Proceedings of the Wireless Communications and Networking Conference*, Kowloon, China, *11*(15), 4425–4429.

[14] Xiao, B., Chen, H., & Zhou, S. (2008). Distributed localization using a moving beacon in wireless sensor networks. *IEEE Trans Parallel Distrib Syst*, *19*, 587–600.

[15] Bulusu, N., Heidemann, J., & Estrin, D. (2000). GPS-less low cost outdoor localization for very small devices. *IEEE Personal Comm*, *7*(5), 28–34.

38 A comprehensive optimization strategy for ZVS and ZCS-enabled half-bridge LLC resonant conversion devices

Kamlesh Kumar Yadav[1,a] *and Md Afzal*[2,b]

[1]Assistant Professor, Department of CS & IT, Kalinga University, Raipur, India
[2]Research Scholar, Department of CS & IT, Kalinga University, Raipur, India

Abstract: In order to accomplish the half-bridge LLC resonance conversion's no voltage switching and no current change, an ideal layout strategy for the conversion devices was proposed. This work describes a rigorous improvement method for improving the parameters of LLC resonant conversions. To create the primary circuit steady state comparable circuit model, the fundamental harmonic approximation technique (FHA) was applied. The key result of this creative time domain investigation is the accurate localization of the zero crossing locations of distinct currents running through the resonant tank. The PSIM simulation testing is used to confirm and show that the calculated constant state harmonics and design curves match precisely. The experimental findings demonstrate the viability of the suggested optimization technique.

Keywords: Converter, LLC, CCL, fundamental harmonic approximation technique, efficiency

1. Introduction

Many electrical systems employ the galvanic separation approach to increase system security and prevent potential ground loop problems [1]. Low frequency isolation transformers were mostly used in the past to achieve galvanic isolation. Such transformers are large and heavy because they work at low frequencies. To replace isolated low frequency transformers, high frequency bidirectional isolated converters were created. The bidirectional converter operates at a high frequency, which allows for a substantial reduction in the system's total size and weight [23, 24]. Since switching losses dominate converter losses in high frequency operation, soft switching is typically employed to significantly reduce switching losses. Due to their simplicity in design and operation, resonant circuits are among the methods for soft-switching functioning that are most frequently used [2].

In the bulk of these studies [5, 7–9, 27–29], the rectified DC load and LLC resonant tank were employed, and the primary objective was to ascertain the peak voltage gain at the resonance point [3, 26]. The output of the PV system may change if the voltage at the common coupling point fluctuates. The electricity produced is in charge of adjusting load and reversing power flow into the distribution network because PV is so widely used [10]. Another factor that can lead to unbalance, which could be dangerous, is a sizable amount of reverse power flow [4]. Examples include a rise in current through a short circuit and a spike in the distribution feeder's voltage. Distortion due to harmonics of voltage as well as current waveforms has grown to be a serious issue due to additional PV systems being installed in distribution networks [11, 12].

Because it meets the criteria of offline converters for high efficiency and high density of power, the resonant converter known as the LLC is becoming more used in industrial applications [26]. To accomplish ZVS over the full range and optimize the entire resonator, a novel approach for determining the maximum magnetizing inductance is proposed in this work [6]. Finally, the optimization method is used to design and construct a 200W LLC prototype. The experiment's findings show that the ideal approach increases overall efficacy in achieving full load requirements.

2. Steady-State Time Domain Analysis

The LLC-L resonant tanks' general map design is seen in Figure 38.1. The parallel CL, series LCL, and series LC topologies are displayed in the state map diagram in Figure 38.1. By simulating the LLC-L resonant tank, analytical formulae that can be used to represent all of these topologies may be found. Also illustrated in Figure 38.2 is the whole bridge LLC-L resonant converter [20–22].

[a]ku.kamleshkumaryadav@kalingauniversity.ac.in, [b]md.afzal@kalingauniversity.ac.in

DOI: 10.1201/9781003675259-38

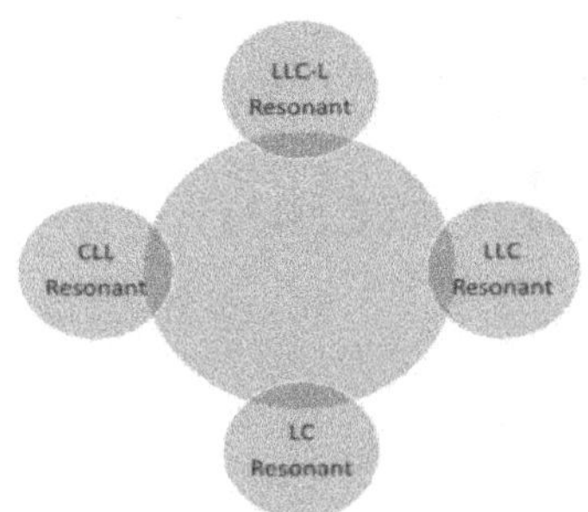

Figure 38.1. Energy description for generalized harmonic tanks.

Source: Author.

2.1. Operation Mode [0-t1]

In this mode, power is supplied to the load. The system of parameters for the time period [0-t1] are expressed as follows:

$$Z_0 i_{cs}(t) = Z_0 i_{cs}(0)\cos(\omega_r t) - V_{cs}(0)\sin\left(V_i - \frac{V_0}{(1+k)n}\sin(\omega_r t)\right.$$

$$V_{cs}(t) = Z_0 i_{cs}(0)\sin(\omega_r t) + V_{cs}(0)\cos\left(\omega_r\left(V_i - \frac{V_0}{(1+k)n}\right)(1-\cos(\omega_r t))\right.$$

2.2. Freewheeling Mode [t1-Tsw/2]

In this mode, the electrical system to the left of the transformer's main side experiences a current freewheels. This is one way to write down the circuit equations:

$$i_{cs}(t) = i_{cs}(t1)\cos(\omega_{r1}(t-t1)) - \frac{V_{cs}(t_1)}{Z_1}\sin(\omega_{r1}(t-t1)) + \frac{V_i}{Z_i}\sin(\omega_{r1}(t-t1))$$

$$i_{Ls1}(t) = i_{Lm}(t) = i_{cs}(t)$$

$$V_{cs}(t) = Z1_{cs}(t1)\sin(\omega_{r1}(t-t1)) - V_{cs}(t1)\cos(\omega_{r1}(t-t1)) + V_i(1-\cos(\omega_{r1}(t-t1)))$$

After understanding the starting and ending values of the solutions to the aforementioned differential equations, it is straightforward to derive closed form solutions for the stresses on the components. The mathematical formulas used in the sections below were developed using the gain, ZVS angle, and RMS current performance curves for conversions. Depending on the frequency and load, these curves alter [13].

The derived equations for the LLC-L and LLC resonant circuits may be immediately applied to the CLL and LLC resonant circuits, respectively, by setting Ls1 and Ls2 to zero. The LLC's resonance frequency is controlled by Ls1 and Cs since these two elements are employed in the power supply mechanism. However, this method is used to ascertain the resonance frequency of the CLL since Ls2, Cs, and Lm are a part of the power supply system of the CLL. Gain is equal to (Ls2/Lm+1) for the CLL when the relative frequency is equal to one, and it is equal to one when the switching frequency coincides with the LLC's resonance frequency [16, 17].

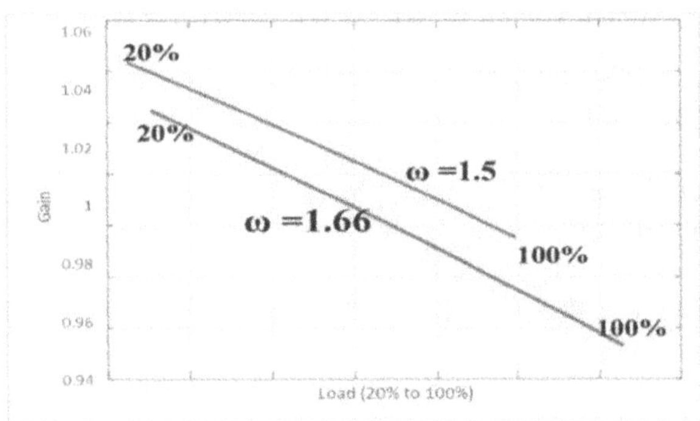

Figure 38.2. Graph how the Gain vs load curve.

Source: Author.

3. Formulation

In order to calculate the ZVS working conditions for power Mosfets, this section analyzes the signal structure produced by the half-bridge LLC converter. As seen in Figure 38.3, both energy Mosfets are off throughout the dead-time intervals. The resonant inductor current ip must reverse direction during the dead period. The LLC resonant tank's input voltage is a square wave with a duty cycle of around 50%. This interpretation applies to the second subinterval of Figure 38.4.

$$I_{Lm-pk} = \frac{1}{2}\frac{nV_0}{L_m}\frac{T_r}{2} = \frac{nV_0}{4L_m f_r}$$

The resonant's input impedance is capacitive after normalizing, and the imaginary component of the resonant frequency is zero. The following formula may be used to get the normalized resonant frequency when Q and are both constant.

$$f_{nz}(\lambda, Q) = \sqrt{\frac{Q^2 - \lambda(1-\lambda) + \sqrt{[Q^2 - \lambda(1+\lambda)]^2 + 4Q^2\lambda^2}}{2Q^2}}$$

The resonance current through the inductor must be greater than the discharging voltage of the functional capacitance, and the drain-source on the electricity for Mosfets must be linked together with a diode. As the resonance current cumulative in the dead-time increases, the importance of the resonance inductor current grows [14].

The reasoning for a converter's lower frequency and maximum load operating point is that if both the magnetic inductance Lm and Vo, along with the frequency of the resonant fr and the conversion ratio n, are constant values, the switching rate is lowest and the current at the output is at its highest. Because (10) is the bare minimum need of an LLC resonant to function at ZVS, the worst-case situation is as follows [15].

Under the conditions of minimum frequency and full load, inequality (11) must be satisfied to ensure that the main side will operate at ZVS, and it is provided by

$$\omega \geq \omega_r T_d = 2\pi f_r T_d$$

4. Results and Discussion

Analytical results have been presented for a design example with a large range (20:1) input voltage fluctuation, 420

V output voltage, and 320 Watts of power. In order to enable voltage control in a potential restricted frequency range. A miminum Q is achieved by using the RMS current performance curves, which are provided in the following sections, decreasing the stress caused by the DCM state. In minimum to maximum voltage applications, the RMS current of the main switches is important because it helps identify the biggest conduction losses. These performance curves suggest that efficiency and control must be traded off when using parallel inductors for the LLC and CLL (Figure 38.5) [18, 19].

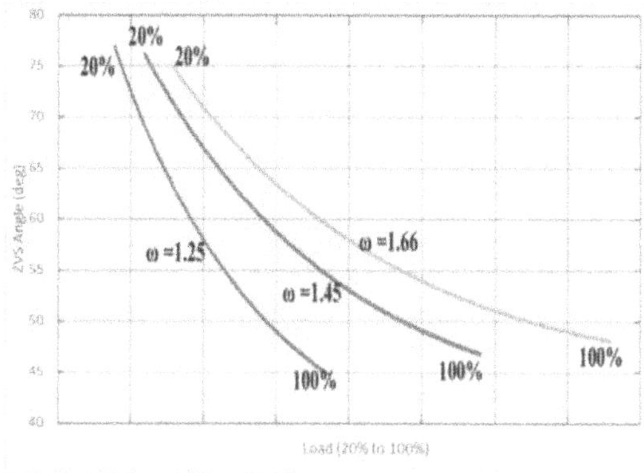

Figure 38.3. Graph show the angle of ZVS with load for LLC.

Source: Author.

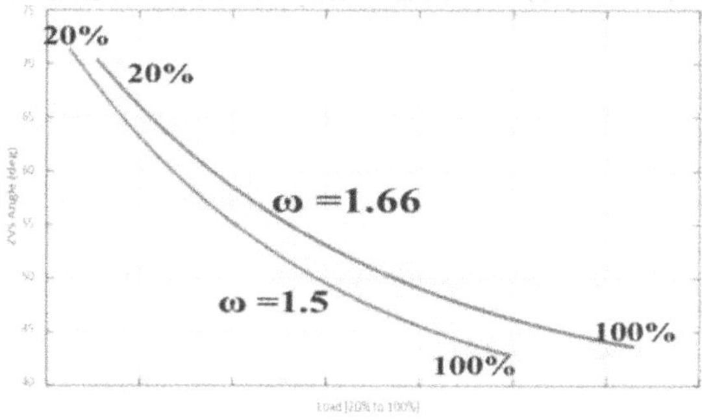

Figure 38.4. Graph show the angle of ZVS with load for CLL.

Source: Author.

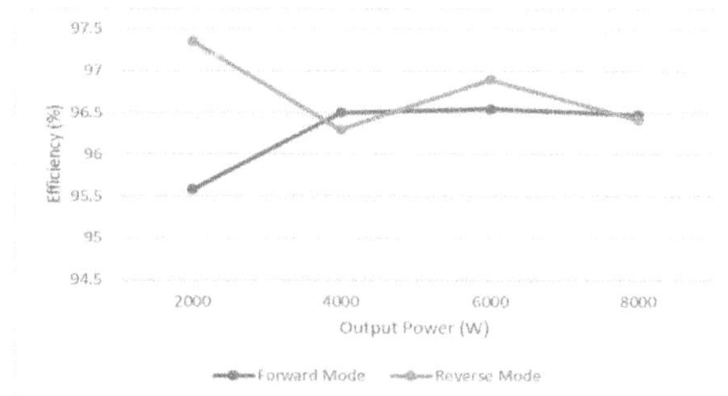

Figure 38.5. Efficiency measurement.

Source: Author.

5. Conclusions

This paper developed a unique time domain analysis for a flexible modulation of frequency control for an LLC-L complete bridge harmonic converter. For the peak stress, switch RMS current, ZVS angle, voltage gain, etc, precise closed form analytical solutions are produced that are virtually identical to the original data. This new analytical method substantially aids the design of an efficient resonance converter. Shifting as well as output currents Io ought to be adjusted to the minimum and maximum values as they represent the most catastrophic situations for ZVS operation, respectively. To calculate the ratio of induction between the series and concurrent resonance inductors for the resonance bank, the gain of the voltage and input pf are employed. It is anticipated that the full-bridge LLC resonance converter prototype circuit, which is currently being built, will be useful for this new investigation.

References

[1] Xu, H., Yin, Z., Zhao, Y., & Huang, Y. (2017). Accurate design of high-efficiency LLC resonant converter with wide output voltage. *IEEE Access*, *5*, 26653–26665.

[2] Gui, H. D., Zhang, Z., He, X. F., & Liu, Y. F. (2014, March). A high voltage-gain LLC micro-converter with high efficiency in wide input range for PV applications. In *2014 IEEE Applied Power Electronics Conference and Exposition-APEC 2014* (pp. 637–642). IEEE.

[3] Kumar, Y., & Gupta, H. (2022, June). Design and modelling of speed control of three phase IM based on PID controller. In *2022 2nd international conference on intelligent technologies (CONIT)* (pp. 1–5). IEEE.

[4] Nagarajan, A., & Jensen, C. D. (2010). A Generic Role Based Access Control Model for Wind Power Systems. *Journal of Wireless Mobile Networks, Ubiquitous Computing, and Dependable Applications, 1*(4), 35–49.

[5] Fang, X., Hu, H., Shen, Z. J., & Batarseh, I. (2011). Operation mode analysis and peak gain approximation of the LLC resonant converter. *IEEE Transactions on Power Electronics*, *27*(4), 1985–1995.

[6] Abeer, A. K., Samir, J. M., & Qais, A. G. (2024). DVB-T2 energy and spectral efficiency trade-off optimization based on genetic algorithm. *Journal of Internet Services and Information Security, 14*(3), 213–225.

[7] Beiranvand, R., Rashidian, B., Zolghadri, M. R., & Alavi, S. M. H. (2010). Optimizing the normalized dead-time and maximum switching frequency of a wide-adjustable-range LLC resonant converter. *IEEE Transactions on Power Electronics*, *26*(2), 462–472.

[8] Mishra, D., & Kumar, R. (2023). Institutional repository: A green access for research information. *Indian Journal of Information Sources and Services*, *13*(1), 55–58.

[9] Samsudin, N. A., Ishak, D., & Ahmad, A. B. (2018). Design and experimental evaluation of a single-stage AC/DC converter with PFC and hybrid full-bridge rectifier. *Engineering Science and Technology, an International Journal*, *21*(2), 189–200.

[10] Stevovic, I., Hadrović, S., & Jovanović, J. (2023). Environmental, social and other non-profit impacts of mountain

streams usage as Renewable energy resources. *Archives for Technical Sciences*, *2*(29), 57–64.

[11] Fang, X., Hu, H., Chen, F., Somani, U., Auadisian, E., Shen, J., & Batarseh, I. (2012). Efficiency-oriented optimal design of the LLC resonant converter based on peak gain placement. *IEEE Transactions on Power Electronics*, *28*(5), 2285–2296.

[12] Sharma, A., Islam, A., Sharma, K., & Singh, P. K. (2021). Optimization techniques to optimize the milling operation with different parameters for composite of AA 3105. *Materials Today: Proceedings*, *43*, 224–230.

[13] Fu, D., Lee, F.C., Liu, Y., & Xu, M. (2008, June). Novel multi-element resonant converters for front-end dc/dc converters. In *2008 IEEE Power Electronics Specialists Conference* (pp. 250–256). IEEE.

[14] Ye, Y., Yan, C., Zeng, J., & Ying, J. (2007, September). A novel light load solution for LLC series resonant converter. In *INTELEC 07–29th International Telecommunications Energy Conference* (pp. 61–65). IEEE.

[15] Chen, Y., Feng, L., Mansir, I. B., Taghavi, M., & Sharma, K. (2022). A new coupled energy system consisting of fuel cell, solar thermal collector, and organic Rankine cycle; generation and storing of electrical energy. *Sustainable Cities and Society*, *81*, 103824.

[16] Kumar, Y., Mishra, R. N., & Anwar, A. (2020, February). Enhancement of small signal stability of SMIB system using PSS and TCSC. In *2020 International Conference on Power Electronics & IoT Applications in Renewable Energy and its Control (PARC)* (pp. 102–106). IEEE.

[17] Lin, R.-L., & Lin, C.-W. (2010). Design criteria for resonant tank of LLC DC-DC resonant converter. *IECON 2010 - 36th Annual Conference on IEEE Industrial Electronics Society*. Glendale, AZ, USA, pp. 427–432, doi: 10.1109/IECON.2010.5674988.

[18] Kumar, Y., Saxena, A., & Goyal, M. (2021, February). Integration of hybrid cascaded multilevel inverter configuration in a PV based applications with multicarrier PWM technology. In *2021 International Conference on Advances in Electrical, Computing, Communication and Sustainable Technologies (ICAECT)* (pp. 1–5). IEEE.

[19] Al-Muntaser, A. A., Pashameah, R. A., Sharma, K., Alzahrani, E., & Tarabiah, A. E. (2022). Reinforcement of structural, optical, electrical, and dielectric characteristics of CMC/PVA based on GNP/ZnO hybrid nanofiller: nanocomposites materials for energy-storage applications. *International Journal of Energy Research*, *46*(15), 23984–23995.

[20] Pavlović, Z., Oliver, J. A., Alou, P., Garcia, Ó., & Cobos, J. A. (2013, March). Bidirectional multiple port dc/dc transformer based on a series resonant converter. In *2013 Twenty-Eighth Annual IEEE Applied Power Electronics Conference and Exposition (APEC)* (pp. 1075–1082). IEEE.

[21] Li, J., Abdulghani, Z. R., Alghamdi, M. N., Sharma, K., Niyas, H., Moria, H., & Arsalanloo, A. (2023). Effect of twisted fins on the melting performance of PCM in a latent heat thermal energy storage system in vertical and horizontal orientations: Energy and exergy analysis. *Applied Thermal Engineering*, *219*, 119489.

[22] Rahman, A. N., Lin, W. C., Chen, J. B., Lin, J. Y., Chang, Y. C., Chiu, H. J., & Hsieh, Y. C. (2017, October). Design and implementation of high efficiency low-profile bidirectional dc-dc converter. In *2017 Asian Conference on Energy, Power and Transportation Electrification (ACEPT)* (pp. 1–6). IEEE.

[23] Jacobs, J., Averberg, A., & De Doncker, R. (2004, June). A novel three-phase DC/DC converter for high-power applications. In *2004 IEEE 35th Annual Power Electronics Specialists Conference (IEEE Cat. No. 04CH37551)* (Vol. 3, pp. 1861–1867). IEEE.

[24] Chen, Y., Feng, L., Jamal, S. S., Sharma, K., Mahariq, I., Jarad, F., & Arsalanloo, A. (2021). Compound usage of L shaped fin and Nano-particles for the acceleration of the solidification process inside a vertical enclosure (A comparison with ordinary double rectangular fin). *Case Studies in Thermal Engineering*, *28*, 101415.

[25] Rahman, A. N., Lee, C.-Y., Chiu, H.-J., & Hsieh, Y.-C. (2018). Bidirectional three-phase LLC resonant converter. *2018 IEEE Transportation Electrification Conference and Expo, Asia-Pacific (ITEC Asia-Pacific)*. Bangkok, Thailand, pp. 1–5, doi: 10.1109/ITEC-AP.2018.8433271.

[26] Hai, T., Ali, M. A., Dhahad, H. A., Alizadeh, A. A., Sharma, A., Almojil, S. F., ... & Wang, D. (2023). Optimal design and transient simulation next to environmental consideration of net-zero energy buildings with green hydrogen production and energy storage system. *Fuel*, *336*, 127126.

[27] Hai, T., Alshahri, A. H., Mohammed, A. S., Sharma, A., Almujibah, H. R., Metwally, A. S. M., & Ullah, M. (2023). Performance assessment and multiobjective optimization of a biomass waste-fired gasification combined cycle for emission reduction. *Chemosphere*, *334*, 138980.

[28] Sharma, A., Yadav, R., & Sharma, K. (2021). Optimization and investigation of automotive wheel rim for efficient performance of vehicle. *Materials Today: Proceedings*, *45*, 3601–3604.

[29] Kumar, Y., & Gupta, H. (2022, August). Load Flow Algorithm for Variable Economical Power Dispatch of Electricity Generation including Demand Response. In *2022 3rd International Conference on Electronics and Sustainable Communication Systems (ICESC)* (pp. 1–5). IEEE.

39 Combining firefly optimization and Mahalanobis distance based fuzzy C-means clustering for earlier breast cancer diagnosis

Farha Kowser[1,a], *C. Arul Murugan*[2,b], *G. Vijaykumar*[3,c], *Kottaimalai Ramaraj*[4,d], *Thilagaraj M.*[1,e], *and Petchinathan G.*[3,f]

[1]MVJ College of Engineering, Bangalore, India
[2]Karpagam College of Engineering, Coimbatore, India
[3]Sri Shanmugha College of Engineering and Technology, Tiruchengode, Tamil Nadu, India
[4]Kalasalingam Academy of Research and Education, Krishnankoil, Tamil Nadu, India

Abstract: Considering early recognition can improve survival of patient, it represents one of the greatest issues requiring to be dealt globally. Prompt detection of breast cancer can significantly lower therapy expenses; mammograms were initially utilized to identify the disease. In order to plan treatments, segmentation assists doctors in measuring the amount of tissue in the breast. In order to assist doctors with their assessment, artificial intelligence approaches provide automated solutions for breast lump delineation. Just spherical structural clusters are able to detected using the well-liked fuzzy c-means algorithm (FCM) that relies on the Euclidean distance function and converges to a local minimum of the objective function. The present study proposes an enhanced FCM that employs a Mahalanobis distance (mFCM) by adopting a new convergent process and an updated threshold value. Finding the best way to minimize or maximize the parameters that comprise an issue in order to make it as functional and efficient as feasible is the method of optimization. To achieve greater precision in segmentation on breast mammography pictures, the best outcome might be achieved by combining the Firefly Algorithm (FO), a meta-heuristic algorithm influenced by the idealized nature of the flashing features of fireflies, with (mFCM). Integrating FO for feature selection and parameter optimization with MDFCM for clustering provides a comprehensive approach to breast cancer diagnosis.

Keywords: Breast cancer segmentation, mahalanobis distance based fuzzy c-means algorithm (mFCM), firefly algorithm (FO), MIAS dataset, diagnosis, therapy

1. Introduction

Breast cancer is the leading reason of mortality for women globally. Prompt diagnosis and discovery thereby raise the likelihood of recuperation and lower the death rate [1]. Breast cancer remained the prevalent cancer identified in females and the primary cause of cancer-related deaths; for males, it was lung cancer. Worldwide projections show glaring disparities in the incidence of cancer based on human growth. About 24.2% of newly identified kinds of cancer are allied with breast tumor (GLOBCON). Segmentation of lesions from mammography is being shown to be a beneficial source of data for the identification and categorization of breast cancer [2].

Techniques for the prompt identification of breast cancer are being established by certain investigators [3]. Images of breast cancer were segmented using the suggested approaches [4]. There are three types of conventional segmentation methods: threshold-, edge-, and region-based segmentation. These methods depends on the pixel values of an image. One of the most successful methods for tissue segmentation in mammography images is fuzzy C-means (FCM), which is utilized in numerous investigations to create automated approach which could help doctors in making a prompt identification of breast cancer. This work aims to present a novel method for extracting tumor from the ROI by utilizing the FCM method. The Euclidean distance, on which the fundamental FCM approach depends, is not a better way to measure non-spherical structures. Circles are utilized to model clusters in FCM. Nevertheless, FCM is a continual procedure that relies on the quantity of clusters that the user specifies as well as the clusters' initialization at random. These two flaws could both lead to the algorithm's convergence toward less-than-ideal results. The application of a Mahalanobis distance based FCM (mFCM) was suggested as a solution for such drawbacks.

[a]frh.kouser@gmail.com, [b]murugan.carul@gmail.com, [c]vijay.june08@gmail.com, [d]r.kottaimalai@klu.ac.in, [e]m.thilagaraj@gmail.com, [f]gpetchi@gmail.com

DOI: 10.1201/9781003675259-39

The communal actions of fireflies, sometimes known in the summer sky in tropical regions of temperature serves as the basis for the Firefly Algorithm, a metaheuristic optimization method. Real random numbers are used in the Firefly method. It is predicated on the swarming particles' worldwide communication. When used in multiobjective optimization, it seems to work better. Subsequently has been suggested to combine and apply the Firefly algorithm and Mahalanobis-distance-based FCM for tumor segmentation in mammography pictures.

By considering the Mahalanobis distance, mFCM enhances the separation between clusters, leading to more accurate identification of different tumor subtypes or severity levels. The ultimate goal of combining FO and mFCM is to support personalized medicine in breast cancer diagnosis. By accurately categorizing tumors into meaningful clusters and selecting informative features, the combined approach enhances the precision of diagnostic decisions and supports healthcare professionals in delivering tailored treatments that optimize patient outcomes.

2. Related Works

A technique for a region-growing segmentation algorithm to identify breast cancer was developed [5]. A median filter was employed to eliminate distortion from mammography pictures of MIAS dataset. Pictures were improved by utilizing Harris corner and CLAHE to boost accuracy in segmentation. The dissection accuracy obtained by the suggested approach was 93%.

Adaptive thresholding was suggested by Selvamurugan and Sundararaj [4] both fine and coarse segmentation of cancer. Histogram fuzzy c-segmentation were employed to achieve coarse segmentation, and window adaptive thresholding was utilized for fine segmentation. Images were classified as regular or unusual using SVM and k-NN. For SVM, the accuracy was 91.5%, and for k-NN, it was 70%. A technique for segmenting breast cancer utilizing global thresholding and region merging was presented [6]. Employing Wiener filtering, Gaussian noise was eliminated, and histogram downsizing was used as the basis for image normalization. To divide the tumour from the ROI, Otsu's approach of global thresholding was used. On assessing ROI, the recommended method has been implemented and examined using 50 mammography pictures and attained 82% accuracy. A completely automated mammography breast boundaries and mammography pectoral muscle delineation were suggested [7]. Canny edge detector has been employed to determine the contour border of the pectoral muscle, and an anisotropic diffusion filter and the median were used to eliminate distortion. To ascertain the breast's margin, five features were taken. Dice score of 97.8% for the MIAS database were obtained by the procedure.

The idea of automatically segmenting breast tumors on mammograms utilizing hierarchical k-means was put forth [8]. The technique helps determine the ideal amount of clusters in mammography pictures by automatically detecting breast tumours via valley tracing. The ROI was segmented using hierarchical k-means after the mammography images were gathered from DDSM. According to the study findings, accuracy was stated to be 38.8% and error detection to be 61.1%. The automated identification of mammography microcalcification clusters was proposed [9]. For segmenting mammography pictures, the segmentation approach used numerous morphological processes, such as image interpolation and decomposition. The two databases from which the mammography pictures were obtained were DDSM and MIAS. Noise in the mammography pictures was eliminated by applying a contrast-enhancement filter.

A novel attention-guided dense-upsampling network (AUNet) for breast tumour classification in mammograms was created [10]. AUNet is an efficient upsampling block that is an asymmetric encoder-decoder architecture. The results showed a mean Dice score of 81.80% for CBIS-DDSM and 79.10% for INbreast. Based on the U-Net model, Tsochatzidis et al. [11] suggested an improved CNN convolutional layer. Based on ground-truth classification maps, the technique attained a diagnostic effectiveness of 89.8% and an AUC of 86.20%; for U-Net-based segmentation for DDSM-400 and CBIS-DDSM, the method reached an ultimate accuracy of 88.0% and 86.0%, correspondingly.

Li et al. [12] recommended to use the attention dense U-Net to perform automated breast tumour identification in mammography pictures. Attention gates (AGs) and densely connected U-Net are combined in this technique to dissect tumours. The DDSM dataset was also used to test this approach, and research results demonstrated that dense U-Net combined with AGs worked better than the other approaches. The approach yielded a total accuracy of 78.38%, an F1 score of 82.24%, and a sensitivity of 77.89%. It is clear that multiple investigations have attempted and been successful in improving the results on dissecting anomalies from MRI and mammography images by utilizing various techniques. Still, a few things remain to be addressed, like minimizing border overlaps, displaying segmented sections, and improving metrics. To accomplish the above objectives, the proposed hybrid FO_mFCM is applied on delineating tumour assignments on breast images.

3. Dataset Details

The Mammographic Image Analysis Society (MIAS) dataset is a treasure trove of mammogram images widely used to develop and test computer-aided diagnosis (CAD) systems for breast cancer. Compiled by MIAS, it contains 322 digitized mammograms collected during routine screenings. These images primarily showcase benign and malignant breast lesions.

Each image comes with crucial information:

- Whether abnormalities are present
- Annotations by radiologists

- Researchers leverage the MIAS dataset for various purposes:
- Designing algorithms to detect and classify breast lesions
- Evaluating the effectiveness of CAD systems
- Identifying patterns in breast cancer detection

Publicly available for research, the dataset has fuelled numerous academic publications and conferences. Through a standard benchmark, MIAS has significantly advanced the mammographic image analysis. New algorithms and techniques can be tested against this benchmark, ultimately aiming to improve prompt recognition of tumour in breast.

4. Methodology

4.1. Mahalanobis-distance-based FCM (mFCM)

Malignant breast tumors often exhibit irregular, lobular, and poorly defined shapes. Euclidean distance, a common metric in clustering algorithms, might not effectively capture these complexities [13]. Mahalanobis distance offers a more suitable alternative. It considers both the variance and correlation of data points through a covariance matrix. By incorporating Mahalanobis distance into Fuzzy C-Means (FCM), we can address limitations of the standard approach. This shift allows for a more nuanced, multi-dimensional analysis on detecting tumor. However, a straightforward replacement of Euclidean distance with Mahalanobis distance in FCM's objective function can lead to numerical issues. The culprit? The largest eigenvalues of the fuzzy covariance matrix. These can inflate cluster shapes to unrealistic lengths, deviating from the actual data distribution. The solution lies in controlling the covariance matrix. By fixing the ratio between its largest and smallest eigenvalues, we ensure realistic cluster shapes while leveraging the benefits of Mahalanobis distance in FCM for breast cancer detection.

The Mahalanobis distance is denoted as

$$d^2(x_j, c_i) = (x_j - c_i)^T \Sigma_i^{-1}(x_j - c_i) \quad (1)$$

Σ_i^{-1} = fuzzy covariance matrix
x_j = data points
c_i = cluster centres
d = distance among the pixels

Applying Mahalanobis distance to Fuzzy C Means (FCM) necessitates the derivation of a new set of update functions.

The objective function is given by

$$\mathcal{L}(U, C, \Sigma, \lambda; X, m) = \sum_{i=1}^{c} \sum_{j=1}^{n} u_{ij}^m \left[(x_j - c_i)^T \Sigma_i^{-1}(x_j - c_i) - ln|\Sigma_i^{-1}| \right] + \sum_{j=1}^{n} \lambda_j (\sum_{i=1}^{c} u_{ij} - 1) \quad (2)$$

Where,

m = fuzzy coefficient
u_{ij}^m = membership function
U = fuzzy partition matrix
C = Cluster prototypes
X = Set of data points

The Lagrangian is,

$$J(U, C, \Sigma; X, m) = \sum_{i=1}^{c} \sum_{j=1}^{n} u_{ij}^m \left[(x_j - c_i)^T \Sigma_i^{-1}(x_j - c_i) - ln|\Sigma_i^{-1}| \right] \quad (3)$$

On resolving the optimization issue, the membership update function for a specific cluster k and l datapoint could be given,

$$u_{kl} = 1/\sum_{i=1}^{c} \left[\frac{(x_l - c_k)^T \Sigma_k^{-1}(x_l - c_k) - ln|\Sigma_k^{-1}|}{(x_l - c_i)^T \Sigma_i^{-1}(x_l - c_i) - ln|\Sigma_i^{-1}|} \right]^{\frac{1}{m-1}} \quad (4)$$

The updated centroid is,

$$c_k = \frac{\sum_{j=1}^{n} u_{kj}^m x_j}{\sum_{j=1}^{n} u_{kj}^m} \quad (5)$$

The updated fuzzy covariance matrix is,

$$\Sigma_k = \frac{\sum_{j=1}^{n} u_{kj}^m (x_j - c_k)(x_j - c_k)^T}{\sum_{j=1}^{n} u_{kj}^m} \quad (6)$$

Following are the steps involved in mFCM.

1. Initialize c and m.
2. Arbitrarily initialize $U^{(k)}$
3. Revise the centroids based on C_i.
4. Revise the fuzzy coefficient matrix based on Σ_i.
5. Revise the membership values.
6. Repeat from step 3, until $\|U^{(k+1)} - U^{(k)}\| < \varepsilon$

4.2. Firefly optimization (FO) algorithm

The Firefly Algorithm (FA) is a clever optimization technique inspired by the way fireflies communicate through flashing lights [14]. Like other algorithms in the swarm intelligence family, FA has shown promise in tackling complex optimization problems.

Here's how FA works, mimicking real firefly behavior:

1. **Attraction without Borders:** Fireflies in FA are gender-neutral. They're drawn to brighter fireflies, regardless of sex. This attraction gets stronger as the firefly gets brighter.
2. **Fading Light, Fading Attraction:** Just like a real firefly's light weakens with distance, a firefly's attractiveness in FA diminishes as the distance to another firefly increases. If there are no brighter neighbors, the firefly moves randomly.
3. **Brightness as a Benchmark:** The brightness of a firefly in FA corresponds to the objective function's value that trying to optimize. A brighter firefly represents a better solution.

By following these simple rules, FA allows a swarm of fireflies (potential solutions) to iteratively move towards brighter ones (better solutions), ultimately converging on an optimal solution for the problem at hand.

Following are the steps involved in FO algorithm.

1. Initialize: Begin by randomly placing fireflies in the search space.

2. Evaluate Fitness: Assess each firefly's fitness based on the objective function.
3. Update Fitness: Compute the fitness of each firefly.
4. Move Fireflies: Adjust each firefly's position considering its attractiveness to others and their brightness.
5. Update Brightness: Adjust the brightness of each firefly based on its fitness.
6. Iterate: Repeat the movement and brightness updating until termination criteria are met.
7. Extract Best Solution: Identify the best solution obtained after the specified iterations.

The relocation of a firefly *i* who is fascinated to a most brighter firefly *j* is,

$$x_i = x_i + \beta_0 e^{-\gamma r_{i,j}^2}(x_j - x_i) + \alpha\varepsilon_i \quad (7)$$

$$r_{i,j} = \sqrt{\sum_{k=1}^{d}(x_{i,k} - x_{j,k})^2} \quad (8)$$

$$\beta = \beta_0 e^{-\gamma r^2} \quad (9)$$

Where,

α = randomization parameter
ε_i = random vector
$r_{i,j}$ = Cartesian distance
β = attractiveness of a firefly
β_0 = attractiveness at $r = 0$
γ = light absorption coefficient

4.3. *Firefly optimization (FO) algorithm integrated Mahalanobis-distance-based FCM (FO_mFCM)*

The integration of Firefly optimization with Mahalanobis Distance and FCM aims to enhance the detection accuracy of breast cancer in mammographic images by optimizing the clustering process based on relevant feature distances. The proposed model's workflow is depicted in Figure 39.1.

Following are the steps in involving in FO_mFCM method:

- Feature Extraction: Features which are pertinent are retrived from mammogram images, which typically include texture features, intensity statistics, and other descriptors.
- Mahalanobis Distance Calculation: Mahalanobis distance is computed using the extracted features to measure the dissimilarity of each data point (image) from a centroid or mean of a cluster.
- Firefly Optimization: Firefly optimization is applied to optimize the clustering process in FCM. It helps in adjusting the parameters (such as cluster centers or membership degrees) to develop the clustering accuracy relies on the Mahalanobis distance metrics.
- FCM Clustering: The optimized parameters from firefly optimization are used in the FCM algorithm to classify mammogram images into different clusters, representing different tissue types or pathological conditions.

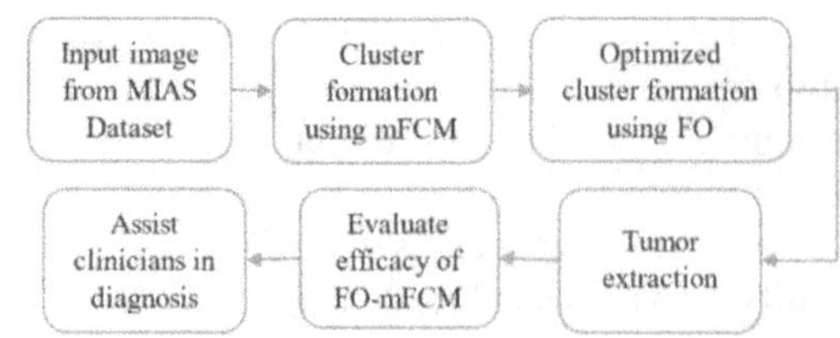

Figure 39.1. Proposed model workflow.

Source: Author.

5. Implementation and Outcomes

The incorporation of FO algorithm with mFCM clustering was introduced to identify tumors in breast MR images [15]. The mFCM effectively clusters similar pixels into distinct groups, aiding in the segmentation of tumor and non-tumor regions within the input image. By incorporating FO into mFCM, the segmentation results have shown significant improvement. Clinicians and radiographers are validating the outcomes of this method, which assists in accurately locating tumours and facilitating early diagnosis. Figure 39.2 illustrates the input MR breast image, depicting both cancerous

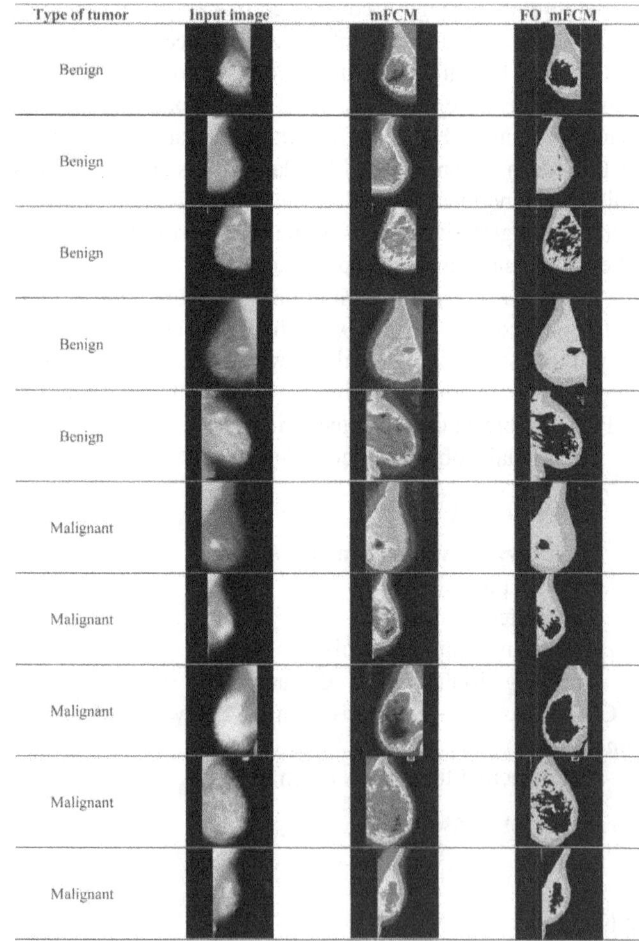

Figure 39.2. Dissection outcomes of FO_mFCM method.

Source: Author.

and non-cancerous areas. The clustered results achieved through mFCM are enhanced by the addition of FO, leading to superior segmentation outcomes.

The FO_mFCM approach efficacy was assessed through performance metrics such as sensitivity, accuracy, dice score, and computational time to segment infected portions from breast images. The method achieves high performance with 96.5% sensitivity, 98.4% accuracy, and a dice score of 94.32%. It also demonstrates efficient processing with a computational time of 14 seconds for tumor dissection. Figures 39.3 to 39.6 illustrate a comparison of these performance metrics between the developed FO_mFCM method and state-of-the-art technologies.

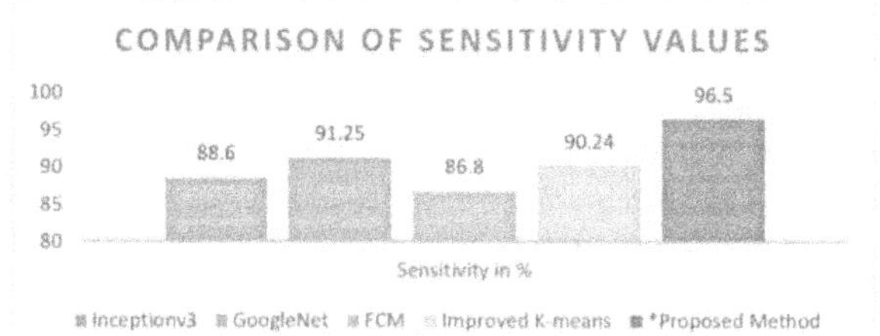

Figure 39.3. Comparison of sensitivity values of FO_mFCM with SOTA.

Source: Author.

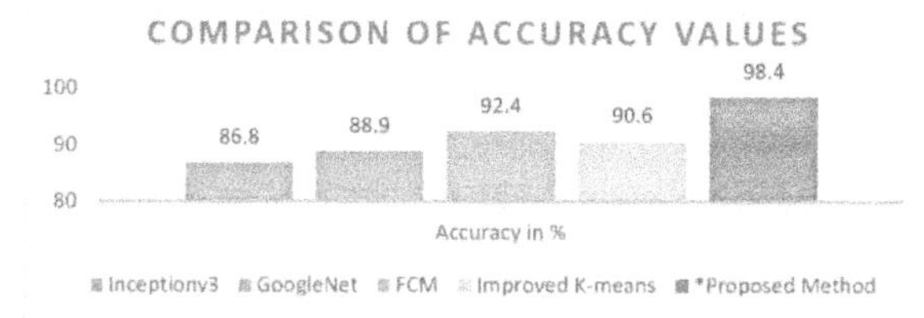

Figure 39.4. Comparison of accuracy values of FO_mFCM with SOTA.

Source: Author.

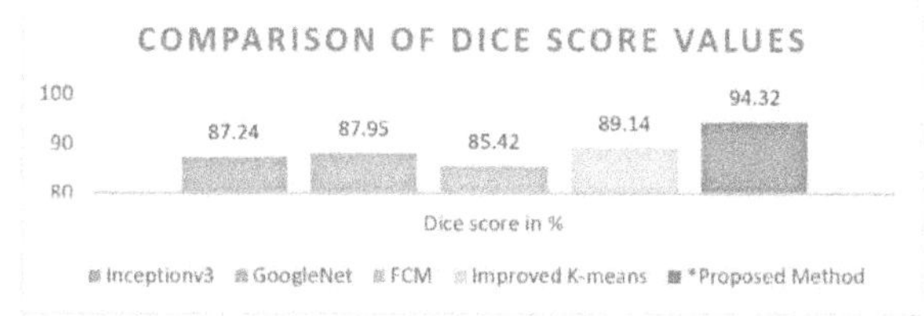

Figure 39.5. Comparison of dice score values of FO_mFCM with SOTA.

Source: Author.

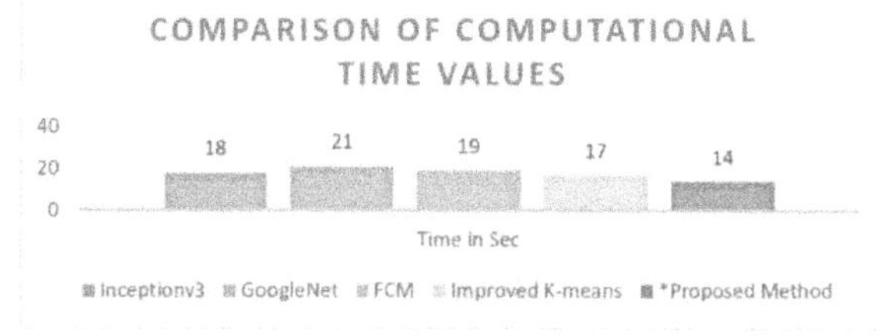

Figure 39.6. Comparison of computational time values of FO_mFCM with SOTA.

Source: Author.

6. Conclusion

This study explored the application of Firefly Optimization based Mahalanobis-Distance-FCM for detecting breast cancer. The findings demonstrate that FO_mFCM achieves strong performance in distinguishing between cancerous and non-cancerous cases, supported by high metric values. By integrating firefly optimization, the method effectively mitigates parameter sensitivity issues typically associated with standard FCM, potentially surpassing existing methods in accuracy. The computational complexity of FO_mFCM may be a consideration, especially for large datasets. Additionally, the method's effectiveness could be influenced by specific characteristics of the breast cancer data utilized. Overall, FO_mFCM presents a promising approach for breast cancer detection.

7. Acknowledgement

The authors thank the International Research Centre of Kalasalingam Academy of Research and Education, Tamil Nadu, India, for permitting the use of the computational facilities available in the Centre for Biomedical Research and Diagnostic Techniques Development.

References

[1] Li, L., Tang, Y., Qiu, L., Li, Z., & Wang, R. (2025). Extracellular matrix shapes cancer stem cell behavior in breast cancer: A mini review. *Frontiers in Immunology*, *15*, 1503021.

[2] Amiya, G., Murugan, P. R., Ramaraj, K., Govindaraj, V., Vasudevan, M., Thirumurugan, M., Zhang, Y. D., Abdullah, S. S., & Thiyagarajan, A. (2024). Expeditious detection and segmentation of bone mass variation in DEXA images using the hybrid GLCM-AlexNet approach. *Soft Computing*, *28*(19), 11633–11646.

[3] Amiya, G., Ramaraj, K., Murugan, P. R., Govindaraj, V., Vasudevan, M., & Thiyagarajan, A. (2022, December). Assertion of Low Bone Mass in Osteoporotic X-ray images using Deep Learning Technique. In *2022 4th International Conference on Advances in Computing, Communication Control and Networking (ICAC3N)* (pp. 830–835). IEEE.

[4] Selvamurugan, M., & Sundararaj, G. K. (2016). Breast cancer detection algorithm by adaptive thresholding. *International Journal of Computer Science and Mobile Computing*, *5*(5), 376–381.

[5] Senthilkumar, B., Umamaheswari, G., & Karthik, J. (2010, December). A novel region growing segmentation algorithm for the detection of breast cancer. In *2010 IEEE International Conference on Computational Intelligence and Computing Research* (pp. 1–4). IEEE.

[6] Singh, N., & Veenadhari, S. (2018). Breast cancer segmentation using global thresholding and region merging. *International Journal of Computer Sciences and Engineering*, *6*(12), 292–297.

[7] Rampun, A., Morrow, P. J., Scotney, B. W., & Winder, J. (2017). Fully automated breast boundary and pectoral

muscle segmentation in mammograms. *Artificial Intelligence in Medicine*, *79*, 28–41.

[8] Ramadijanti, N., Barakbah, A., & Husna, F. A. (2018, October). Automatic breast tumor segmentation using hierarchical k-means on mammogram. In *2018 International Electronics Symposium on Knowledge Creation and Intelligent Computing (IES-KCIC)* (pp. 170–175). IEEE.

[9] Alam, N., Oliver, A., Denton, E. R., & Zwiggelaar, R. (2018). Automatic segmentation of microcalcification clusters. In *Medical Image Understanding and Analysis: 22nd Conference, MIUA 2018, Southampton, UK, July 9–11, 2018, Proceedings 22* (pp. 251–261). Springer International Publishing.

[10] Sun, H., Li, C., Liu, B., Liu, Z., Wang, M., Zheng, H., Feng, D. D., & Wang, S. (2020). AUNet: attention-guided dense-upsampling networks for breast mass segmentation in whole mammograms. *Physics in Medicine & Biology*, *65*(5), 055005.

[11] Tsochatzidis, L., Koutla, P., Costaridou, L., & Pratikakis, I. (2021). Integrating segmentation information into CNN for breast cancer diagnosis of mammographic masses. *Computer Methods and Programs in Biomedicine*, *200*, 105913.

[12] Li, S., Dong, M., Du, G., & Mu, X. (2019). Attention dense-u-net for automatic breast mass segmentation in digital mammogram. *IEEE Access*, *7*, 59037–59047.

[13] Yu, H., Xie, S., Fan, J., Lan, R., & Lei, B. (2024). Mahalanobis-Kernel distance-based Suppressed Possibilistic C-Means Clustering Algorithm for Imbalanced Image Segmentation. *IEEE Transactions on Fuzzy Systems*.

[14] Li, C., Zhang, F., Du, Y., & Li, H. (2024). Classification of brain tumor types through MRIs using parallel CNNs and firefly optimization. *Scientific Reports*, *14*.

[15] Balaji, S., Arunprasath, T., Rajasekaran, M. P., Sindhuja, K., & Kottaimalai, R. (2023, September). A Metaheuristic based Clustering Approach for Breast Cancer Identification for Earlier Diagnosis. In *2023 4th International Conference on Smart Electronics and Communication (ICOSEC)* (pp. 01–07). IEEE.

40 Modeling and intelligent control of hybrid system for microgrid operation

Priya Vij[1,a] *and Ghorpade Bipin Shivaji*[2,b]

[1]Assistant Professor, Department of CS & IT, Kalinga University, Raipur, India
[2]Research Scholar, Department of CS & IT, Kalinga University, Raipur, India

Abstract: Hybrid systems, characterized by the integration of continuous and discrete dynamics, pose unique challenges in modeling and control. This research introduces a comprehensive framework for the modeling and intelligent control of such systems. Our approach integrates advanced mathematical modeling with machine learning techniques for precise system identification and control. The novel algorithms that enhance model accuracy and propose an intelligent control scheme employing adaptive and predictive strategies is developed. This ensures optimal performance and robustness under varying operational conditions. Extensive simulations and experimental studies validate our framework, demonstrating significant improvements in efficiency, stability, and responsiveness across various hybrid system applications. The results underscore the potential of merging traditional control theory with intelligent algorithms, offering a sophisticated and reliable control mechanism for diverse engineering applications, from industrial automation to autonomous systems. This research paves the way for more advanced and effective control strategies in hybrid systems. This study talks about how to model a hybrid system that has SOFC and a battery storage system (BSS) so that it can be connected to the power grid using local controllers. The smart method suggested is used to check how the hybrid system reacts. It has also been shown what happens to the hybrid system when it works in both stand-alone and grid-connected modes.

Keywords: Hybrid systems, intelligent control, adaptive control, efficiency, mathematical modeling

1. Introduction

The traditional power distribution system, with its radial topology and central power generating station, has long been considered a well-defined and reliable framework for supplying electrical power to end users. This conventional system, characterized by a unidirectional flow of electricity from the central power station to consumers, has been adequate in meeting the historical demands for electricity [1]. However, as energy consumption continues to rise, there is a growing need to diversify the sources of power generation. This need has driven the development and integration of multi-source generating stations capable of bi-directional power flow, fundamentally altering the landscape of power distribution systems [11].

Recent advancements have increasingly focused on incorporating renewable energy systems as centralized plants, driven by policy initiatives aimed at promoting energy sustainability and independence [2]. These initiatives are crucial for the realization of smart grid technology, which seeks to create a more resilient and efficient power grid [3]. Smart grids integrate various renewable energy sources, such as solar and wind power, alongside traditional generation methods, leading to a more flexible and sustainableenergy network [15]. This shift not only addresses environmental concerns but also enhances the reliability and robustness of power supply systems.

A significant development within this evolving energy landscape is the concept of the microgrid (MG) [5]. A microgrid refers to a localized group of energy resources, storage systems, and loads that can operate autonomously or in conjunction with the conventional grid [7]. These systems are equipped with renewable energy resources and energy storage units, allowing for a modular and flexible approach to power generation and distribution [6]. The ability to isolate the microgrid from the main grid enhances the reliability of power supply, making it a crucial component in modern energy management strategies. Figure 40.1 illustrates a typical MG system, highlighting a collection of distributed energy resources (DER) such as wind turbines, fuel cells, solar panels, biomass generators, and energy storage units [14], all coordinated to work together seamlessly [9].

The management of energy within a microgrid is paramount to ensuring stable and profitable operation. Effective energy management strategies maximize the utilization of distributed energy resources, ensuring that each component operates at its full potential [13]. This involves sophisticated control systems and algorithms to balance supply and demand dynamically, respond to fluctuations in energy production from renewable sources, and optimize the use of

[a]ku.priyavij@kalingauniversity.ac.in, [b]ghorpade.bipin@kalingauniversity.ac.in

DOI: 10.1201/9781003675259-40

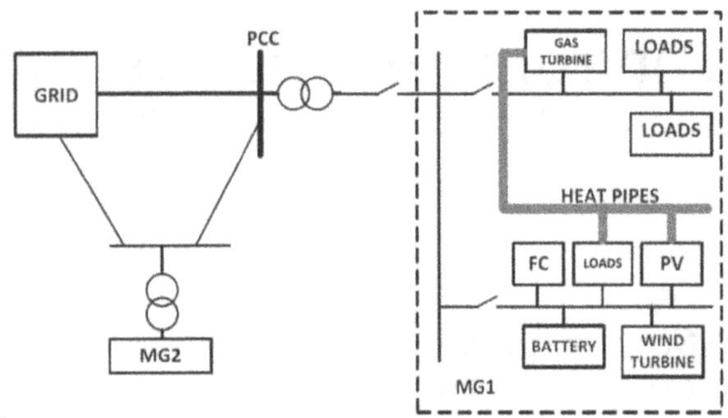

Figure 40.1. Typical MG structure.

Source: Author.

stored energy. By integrating these elements, microgrids can achieve high levels of efficiency and reliability, contributing to the overall stability and sustainability of the broader power grid.

Individual DERs are linked together, which means that one type of resource needs to be controlled and managed. This has been covered in earlier chapters. However, the problems with intermittent supply, low rating, random locations, and the need for more energy have led to the study of hybrid systems [4]. A hybrid system is one that uses more than one type of DER. When these different types are used together, they make the system more reliable and efficient than when they are used separately. Microgrid Energy Management System (MGEMS) has to deal with a number of problems in order to fully utilize the hybrid structure [12]. These include the changing features of distributed generators, the flow of power in both directions, energy fluctuations, a variety of infrastructures, and multiple control objectives for the energy flow. On top of all these issues, one of the biggest ones is how to include DERs in MGs while keeping the power quality high. The PQ is severely affected by the fact that the electrical properties of the energy sources are not the same and that renewable sources are unpredictable and only work sometimes. Making the switch to a smart grid in the future is also hard because it has to work with traditional local power delivery utilities [8].

2. Hybrid System

The SOFC, PV system, and battery storage system (BSS) make up the hybrid system that was modelled for this work. These DERs are chosen because connecting the PV and SOFC adds value by making the line less distorted in terms of frequency and power [15]. When fuel cells and PV work together, they use less fossil fuels, and the fuel cells work better with renewable power sources to get the most power out of them. The hybrid systems can be sent out if they can meet the demand for a certain amount of time. Adding a BSS is a practical and effective way to stop the fluctuations that happen when there is an imbalance or instability in the power. This makes the grid more reliable and helps the renewable sources work better.

2.1. *Modelling of hybrid system*

Along with the solid oxide fuel cell and a BSS, the photovoltaic module makes up the hybrid system in this chapter. Using DC/AC converters, each of these microsources is linked to the grid. In Figure 40.2, you can see the hybrid system's basic model architecture.

2.2. *Photovoltaic module*

The SIMULINK modelling of the PV module, incorporating the incremental conductance method for maximum power point tracking (MPPT). The PV array is a 100-kW array that is connected to the grid using a DC-DC boost converter and a three-phase three-level Voltage Source Converter (VSC). The Maximum Power Point Tracking (MPPT) [10] is achieved by employing the incremental conductance plus integral regulator method in the boost converter.

The irradiance reaches a value of 1000 W/m2 when the photovoltaic (PV) array is able to generate a maximum power output of 100 kW. The MPPT controller optimizes the switching duty cycle by utilising the incremental conductance and integral regulator technique. A 1980-Hz (33*60) 3-level 3-phase VSC is used to convert 500 V DC to 260 V AC while maintaining a unity power factor. A 100-kVA three-phase coupling transformer with a voltage ratio of 260V/13.8kV is utilized for grid interconnection. The array comprises 66 strings, each consisting of 5 modules connected in series, and all strings are connected in parallel. The total power output of the array is calculated as 66 multiplied by 5 multiplied by 305.2 W, resulting in 100.7 kW. The use of the incremental conductance method and Integral regulator for MPPT results in the attainment of the maximum power point.

$$\frac{dP}{dV} = 0$$

Where, $P = V * I$

$$\frac{d(V * I)}{dV} = I + V * \frac{dI}{dV} = 0$$

$$\frac{dI}{dV} = -\frac{I}{V}$$

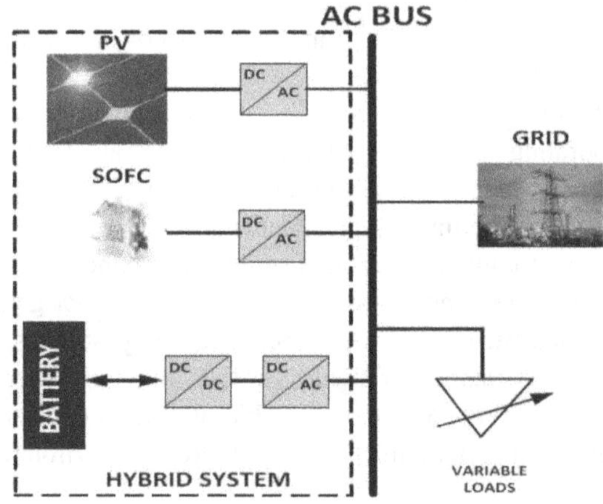

Figure 40.2. Architecture of the Hybrid system.

Source: Author.

The error $\frac{dI}{dV}+\frac{I}{V}$ is eliminated by the integral regulator, which performs duty cycle correction. The DC/AC inverter is regulated based on the power requirements specified by the grid.

2.3. Control strategy for grid integration

The local controller, known as the DC/AC controller, is responsible for connecting each component to the AC bus individually. One advantage of this is the ability to connect individual loads to them. When there is no local load connected, the hybrid system operates in tandem and is controlled by grid-side control. The SOFC module, PV module, and battery all have a rating of 13.8 kV, which corresponds to the feeder voltage. The control action is determined by the power demand requested by the load.

$$P_D = P_{GRID} + P_{SOFC} + P_{PV} + P_{BSS}$$

There are two control strategies that can be employed to control an inverter: (i) P-V control and (ii) PQ control, which is also referred to as constant power control strategy. In P-V controlled interconnection, active power and voltage are utilized as control parameters, disregarding reactive power. The PQ control method utilizes both active and reactive power, but it is implemented within the dq0 frame. The synchronisation of renewable sources with the utility system can be accomplished by regulating the real and reactive power that is fed into the grid. This regulation is achieved by using a phase-locked loop (PLL) to obtain frequency and angle references. This work utilizes a vector approach for controlling active and reactive power. The control is divided into two categories: sinusoid current control and voltage control. The inverter's amplitude and frequency are controlled for operation in the current control mode. The power regulator controls and adjusts both the active and reactive power, and sends back the currents Id and Iq of the d-axis and q-axis to the current controller.

With the VSI used for each resource in the hybrid system, there is an automatic power sharing such that the hybrid system with VSI has:

3. Simulation Results

To assess the performance of the hybrid system using the proposed control strategy, it is tested on a radial distribution system.

3.1. Solid oxide fuel cell

Based on the mathematical model of SOFC, the simulink model is developed for the hybrid system using MATLAB/Simulink as shown in Figure 40.3 The Solid Oxide Fuel Cell (SOFC) is interfaced with the grid using a three-level bridge for DC/AC conversion.

3.2. Battery storage system

The grid-connected battery is modeled using the generic battery model from the SIMULINK library as shown in Figure 40.4 simulation results for each Distributed Energy

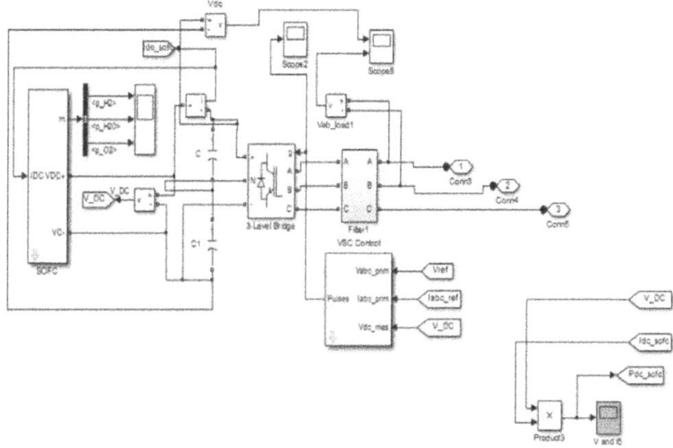

Figure 40.3. SOFC module.

Source: Author.

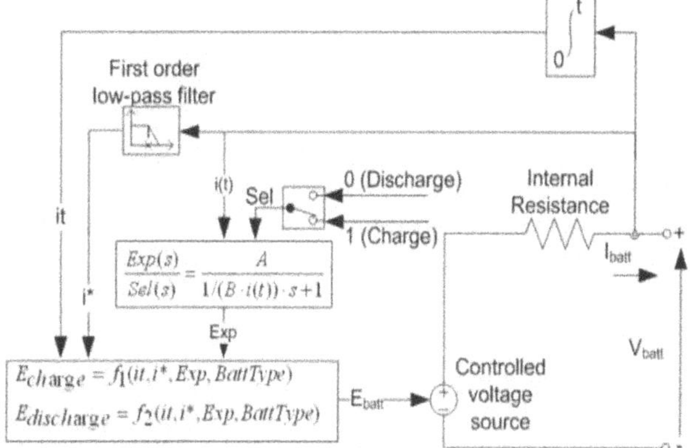

Figure 40.4. Equivalent circuit of Battery.

Source: Author.

Resource (DER) are presented in Figures 40.5 and 40.6. Each DER in the Hybrid Renewable Energy System (HRES) is connected to feeder, operating at 13.8 kV, via a DC/AC converter and the response regarding the loads is satisfactory.

$$\Delta P = \sum_{I}^{M} \Delta P_i \, i_i$$

Where ΔP_i, is the power variation in the ith VSI.

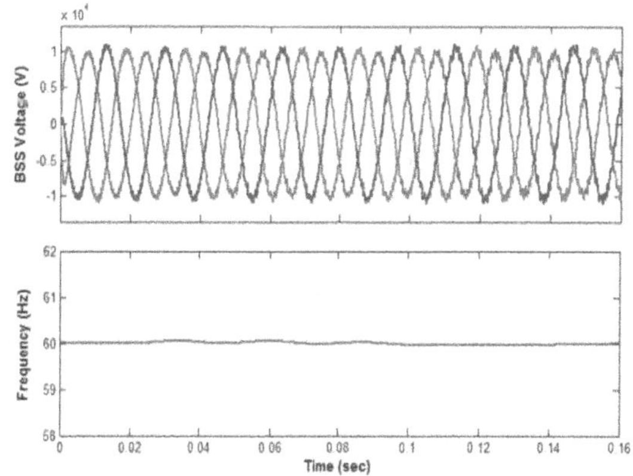

Figure 40.5. Response of BSS as part of Hybrid system.

Source: Author.

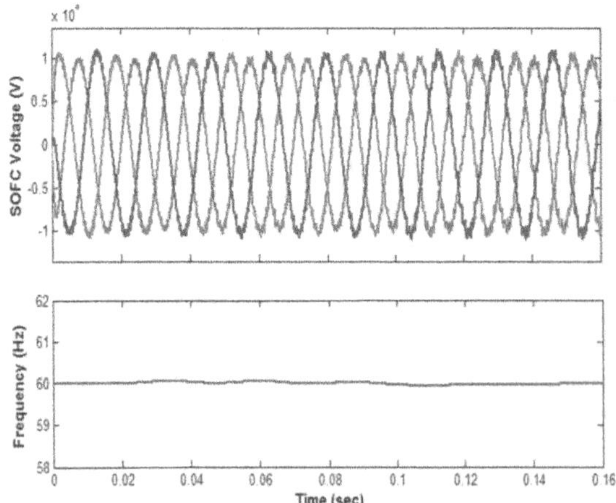

Figure 40.6. Response of SOFC as part of Hybrid system.

Source: Author.

The MPPT of the photovoltaic (PV) system begins at 0.015 seconds. The simulation results for each Distributed Energy Resource (DER) are presented in Figures 40.5 and 40.6. Each DER in the Hybrid Renewable Energy System (HRES) is connected to feeder, operating at 13.8 kV, via a DC/AC converter and the response regarding the loads is satisfactory.

4. Conclusion

The hybrid system, consisting of SOFC, photovoltaic (PV) panels, and a BSS, has been successfully integrated into the radial distribution system. The hybrid system effectively meets the load demand of the microgrid operation, resulting in satisfactory outcomes. Nevertheless, when operating in the grid connected mode, there is a notable rise in the Total Harmonic Distortion (THD), which has the potential to cause instability in the system. The aspect of harmonic distortion is caused by the greater quantity of PE interfaces in the hybrid system. This system utilizes a single DC/AC converter and a voltage stability controller to ensure grid support in the event of unexpected transients caused by faults or sudden load changes at the consumer end.

References

[1] Butt, O. M., Zulqarnain, M., & Butt, T. M. (2021). Recent advancement in smart grid technology: Future prospects in the electrical power network. *Ain Shams Engineering Journal*, *12*(1), 687–695.

[2] Srinivasa Rao, M., Praveen Kumar, S., & Srinivasa Rao, K. (2023). Classification of Medical Plants Based on Hybridization of Machine Learning Algorithms. *Indian Journal of Information Sources and Services*, *13*(2), 14–21.

[3] Hai, T., Zhou, J., Almashhadani, Y. S., Chaturvedi, R., Alshahri, A. H., Almujibah, H. R., ... & Ullah, M. (2023). Thermo-economic and environmental assessment of a combined cycle fueled by MSW and geothermal hybrid energies. *Process Safety and Environmental Protection*, *176*, 260–270.

[4] Al-Omari, M., & Al-Haija, Q.A. (2024). Towards robust IDSs: An integrated approach of hybrid feature selection and machine learning. *Journal of Internet Services and Information Security*, *14*(2), 47–67.

[5] Al-Ismail, F. S. (2021). DC microgrid planning, operation, and control: A comprehensive review. *IEEE Access*, *9*, 36154–36172.

[6] Sule, A. H. (2023). Major factors affecting electricity generation, transmission and distribution in Nigeria. *International Journal of Engineering and Mathematical Intelligence (IJEMI)*, *1*(1, 2&3), 159–164.

[7] Isabel, M. V., Carlos, L., Lourdes, I. C. A., Doris, I. G. P., Yoni Magali Maita Cruz, Jessica Paola Palacios-Garay and Aracelli del Carmen Gonzales-Sánchez. (2023). Improved butterfly optimization algorithm for energy efficient antenna selection over wireless cellular networks. *Journal of Wireless Mobile Networks, Ubiquitous Computing, and Dependable Applications*, *14*(2), 121–136.

[8] Bjarghov, S., Löschenbrand, M., Saif, A. I., Pedrero, R. A., Pfeiffer, C., Khadem, S. K., ... & Farahmand, H. (2021). Developments and challenges in local electricity markets: A comprehensive review. *IEEE Access*, *9*, 58910–58943.

[9] Turan, F., & Ergenler, A. (2021). DNA Damage in hybrid tilapia (Oreochromis niloticus x O. aureus) exposed to short-transport process. *Natural and Engineering Sciences*, *6*(3), 190–196.

[10] Katche, M. L., Makokha, A. B., Zachary, S. O., & Adaramola, M. S. (2023). A comprehensive review of maximum power point tracking (mppt) techniques used in solar pv systems. *Energies*, *16*(5), 2206.

[11] Saponara, S., Saletti, R., & Mihet-Popa, L. (2019). Hybrid micro-grids exploiting renewables sources, battery energy storages, and bi-directional converters. *Applied Sciences*, *9*(22), 4973.

[12] Nasir, M., Umer, M., & Asgher, U. (2022). Application of Hybrid SFLA and ACO algorithm to omega plate for drilling process planning and cost management. *Archives for Technical Sciences*, *1*(26), 1–12.

[13] Hai, T., Aziz, K. H. H., Zhou, J., Dhahad, H. A., Sharma, K., Almojil, S. F., ... & Abdelrahman, A. (2023). -Neural network-based optimization of hydrogen fuel production energy system with proton exchange electrolyzer supported nanomaterial. *Fuel*, *332*, 125827.

[14] Dong, S., Al-Zahrani, K. S., Reda, S. A., Sharma, K., Amin, M. T., Tag-Eldin, E., & Youshanlouei, M. M. (2022). Investigation of thermal performance of a shell and tube latent heat thermal energy storage tank in the presence of different nano-enhanced PCMs. *Case Studies in Thermal Engineering*, *37*, 102280.

[15] Yadav, R., Singh, P. K., & Chaturvedi, R. (2021). Enlargement of geo polymer compound material for the renovation of conventional concrete structures. *Materials Today: Proceedings*, *45*, 3534–3538.

41 Decrypting theft suspects in low-resolution snapshots

Mervin Jerel D.[1,a], *Moneshwar C.*[1,b], *Naveenkumar S.*[1,c], *and D. Menaka*[2,d]

[1]Student, Department of Electronics and Communication Engineering, Sri Venkateswara College of Engineering, Chennai, India

[2]Associate Professor, Department of Electronics and Communication Engineering, Sri Venkateswara College of Engineering, Chennai, India

Abstract: This paper introduces a novel people tracking system for enhancing public safety by enabling real-time detection and tracking of individuals in critical public areas. Leveraging YOLO for rapid object detection and DeepSORT for robust person tracking, the system employs surveillance cameras for continuous monitoring. Through seamless data exchange among cameras, anomalies and potential threats are swiftly identified, bolstering security measures. Additionally, the system alerts nearby police stations upon detecting listed suspects, enabling rapid response. Movement trajectories are plotted on maps for simplified identification and apprehension. The paper also discusses utilizing Siemens Neural Network for face recognition. Overall, the project aims to support law enforcement agencies in efficiently tracking suspects and improving public safety.

Keywords: YOLO, DeepSORT, Siemens neural network, suspect tracking, real-time monitoring, face recognition, surveillance cameras, police alerts

1. Introduction

In today's security-conscious environment, the integration of advanced technologies into surveillance systems presents significant potential for enhancing public safety. This report introduces an innovative approach that leverages state-of-the-art deep learning algorithms, notably YOLO for real-time object detection and DeepSORT for robust person tracking. By strategically deploying surveillance cameras in critical public areas, the system aims to seamlessly detect and track individuals across different locations, enabling continuous monitoring.

The primary objective of the project is to harness the combined capabilities of YOLO and DeepSORT to identify anomalies and potential threats in real-time, thus contributing to the enhancement of public safety. Going beyond conventional surveillance, the system focuses on tracking suspects by issuing alerts to nearby authorities and plotting their movements on maps for swift identification. Ultimately, the goal is to support law enforcement agencies in efficiently tracking suspects and strengthening overall public safety measures.

The aim is to contribute to the advancement of research and development in surveillance systems and security technologies.

2. Literature Review

In Siamese Neural Networks for One-shot Image Recognition the authors introduce the concept of siamese neural networks, originally proposed by Bromley and LeCun in the early 1990s, which consist of twin networks joined by an energy function to compute a similarity metric between inputs. They use convolutional siamese networks, leveraging the power of convolutional neural networks (CNNs) in capturing spatial features from images.

Siamese neural network (SNN): It is characterized by its architecture comprising two identical neural network branches, often termed as twins, which share the same parameters learned jointly during training. This weight sharing enables the network to extract features from input data, typically through convolutional and pooling layers, to construct high-dimensional representations. The essence of SNN lies in its ability to learn a similarity metric between pairs of input samples, achieved through distance metric learning techniques such as Euclidean distance or contrastive loss. By leveraging shared weights and feature extraction, SNNs effectively compute the similarity between inputs, making them well-suited for tasks like image recognition, verification, and matching.

Several prior works in the field of one-shot learning are referenced, including Fei-Fei et al.'s variational Bayesian framework and Lake et al.'s Hierarchical Bayesian Program Learning (HBPL).

Object tracking is a fundamental task in computer vision that involves following the movement of objects over consecutive frames in a video or image sequence. Traditional

[a]jerelmervin@gmail.com, [b]moneshwar02@gmail.com, [c]snaveen3817@gmail.com, [d]menaka@svce.ac.in

DOI: 10.1201/9781003675259-41

object tracking methods often rely on techniques such as motion estimation, feature matching, and model-based tracking to maintain the identity of objects across frames. However, these methods can be computationally expensive and prone to drifting or loss of track in complex scenes.

Also [1] work on drone detection using YOLO-V8 is significant due to its emphasis on efficiency, adaptability, and robustness. By leveraging the capabilities of YOLO-V8, their system achieves real-time detection of drones in various environmental conditions. Their approach likely involves integrating YOLO-V8 with other technologies, such as sensors and communication systems, to enhance detection accuracy and reliability [2]. Through rigorous evaluation metrics and real-world applications, demonstrate the effectiveness of their system in addressing security and safety concerns posed by drones. Additionally, their work likely provides insights into future research directions for further improving drone detection systems using advanced algorithms and sensor integration.

YOLO (You Only Look Once) offers a novel approach to object detection that enhances object tracking capabilities. By providing real-time and accurate detection of objects in each frame of a video, YOLO serves as a reliable source of object localization information. This information can then be integrated into object tracking systems to initialize object tracks, associate detections across frames, and refine the tracking trajectory.

YOLO-V8 stands out in the realm of object detection for its unparalleled efficiency, adaptability, and robustness. Leveraging a single neural network to predict bounding boxes and class probabilities concurrently, YOLO-V8 achieves remarkable speed without compromising accuracy, making it ideal for real-time applications like drone detection.

Discuss [5] the implementation of GPS tracking as a crucial component of their system.They used GPS for real-time location monitoring, tracking stolen vehicles, and integrating GPS data with other security measures.

GPS tracking, integral to modern surveillance and security systems, provides precise geographical coordinates of objects or individuals under surveillance. Ortiz et al.'s utilization of GPS serves multiple purposes within their system architecture.

3. Proposed System

The proposed system builds upon the existing framework by introducing several enhancements aimed at further improving public safety and operational efficiency.

3.1. *Object detection*

In the proposed system, aim is to explore advanced object detection and tracking algorithms to improve the accuracy and reliability of the system. Figure 41.1 represents the architecture object detection using proposed model. This may involve incorporating deep learning techniques for object recognition and trajectory prediction.The system leverages YOLO (You Only Look Once) for rapid object detection and DeepSORT for robust person tracking. YOLO enables efficient detection of objects within surveillance camera feeds, while DeepSORT maintains the identity of individuals across frames, facilitating continuous tracking.An adapter is formed which make specific modification to the code to effectively use the DeepSORT and YOLO only for person detection to achieve maximum efficiency.

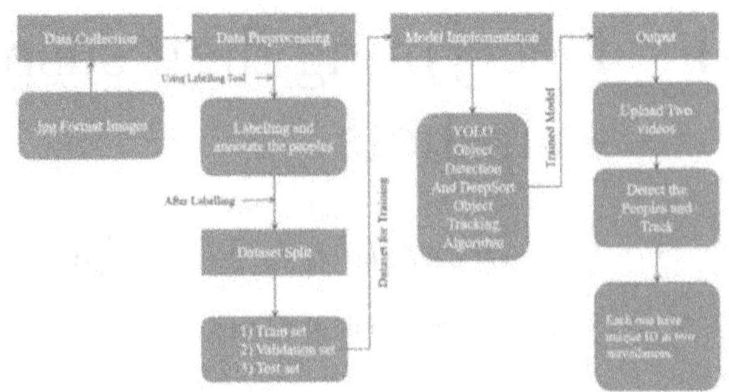

Figure 41.1. Architecture of object detection.

Source: Author.

3.2. *Image Preprocessing*

3.2.1. *Resizing the image*

The primary objective of resizing the image is to standardize its dimensions to 100 × 100 pixels, facilitating uniform processing across different datasets or applications. This step is crucial for ensuring consistency in image representation and downstream analysis. To achieve this, the image is read using a suitable library such as OpenCV, and then resized to the desired dimensions utilizing a resizing function.

3.2.2. *Normalization of RGB values*

Normalization of RGB values involves converting the pixel intensity values from the original range of 0–255 to a normalized range of 0–1. Additionally, a thresholding mechanism is applied, where pixel values greater than 0.5 are set to 1, and those below 0.5 are set to 0. This normalization process enhances the numerical stability of the data and ensures that pixel values are uniformly distributed within a standardized range.

3.2.3. *Enhanced object detection and tracking*

In the proposed system, aim is to explore advanced object detection and tracking algorithms to improve the accuracy and reliability of the system. This may involve incorporating deep learning techniques for object recognition and trajectory prediction.The system leverages YOLO (You Only Look Once) for rapid object detection and DeepSORT for robust person tracking. YOLO enables efficient detection of objects within surveillance camera feeds, while DeepSORT maintains the identity of individuals across frames, facilitating continuous tracking.An adapter is formed which make specific modification to the code to effectively use the DeepSORT and YOLO only for person detection to achieve maximum efficiency In the proposed system, aim is to explore advanced object detection and tracking algorithms to improve the accuracy and reliability of the system. This may involve incorporating deep learning techniques for object recognition and trajectory prediction. The system leverages YOLO (You Only Look Once) for rapid object detection and DeepSORT for robust person tracking. YOLO enables efficient detection of objects within surveillance camera feeds, while DeepSORT maintains the identity of individuals across frames, facilitating continuous tracking. An adapter is formed which make specific modification to the code to effectively use the DeepSORT and YOLO only for person detection to achieve maximum efficiency.

3.2.4. Image preprocessing implementation steps

Read the image using a suitable library like OpenCV to ensure compatibility and ease of processing.

1. Resize the image to the desired dimensions (100 × 100 pixels) using a resizing function, maintaining aspect ratio if required.
2. Convert the resized image to a floating-point format to enable arithmetic operations necessary for normalization.
3. Normalize each RGB value by dividing it by 255, thereby scaling the values to the range 0–1.
4. Implement thresholding by setting pixel values greater than 0.5 to 1 and those below 0.5 to 0, ensuring binary representation of pixel intensities.
5. Iterate through all pixels in the image, applying the normalization and thresholding operations consistently to each pixel.

3.3. Face recognition

Siemens Neural Network is employed for face recognition, enabling the system to accurately identify individuals of interest within the surveillance footage. After finding the person's frame using YOLO and DeepSORT is used to confirm whether the identified person is the suspect or not with the help of the custom trained Siemens Neural Network which is trained for facial recognition.

State-of-the-art models often incorporate a series of convolutional layers followed by fully-connected layers and a top-level energy function. Figure 41.2 explains the Architecture of a siamese network, excelling in image recognition endeavors.

Convolutional networks are particularly attractive due to several factors. One such factor is local connectivity, which effectively reduces the number of parameters in the model. This reduction inherently incorporates a form of built-in regularization. However, it's worth noting that convolutional

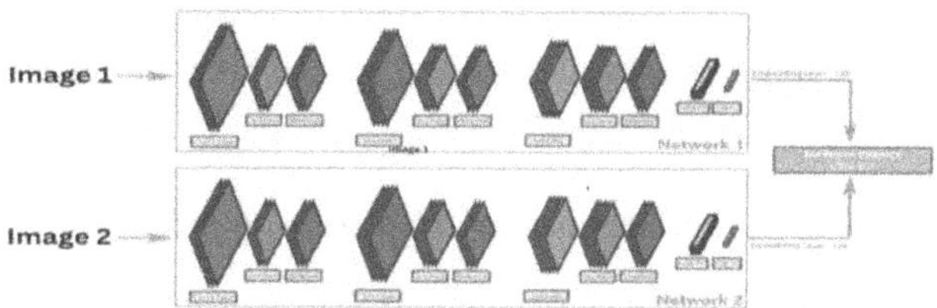

Figure 41.2. Architecture of a siamese network.

Source: Author.

layers tend to be computationally more intensive compared to standard nonlinearities.

3.3.1. Model performance

The presented graph in Figure 41.3 illustrates the progression of model accuracy and loss throughout the training and validation phases. Over the course of 100 iterations, the validation loss demonstrates a consistent decline, peaking at approximately 80 iterations before plateauing. Concurrently, there is a lack of substantial improvement in validation accuracy beyond the initial iterations. Despite these observations, the final model obtained exhibits satisfactory performance, indicating promising outcomes.

4. Results and Discussion

The implementation of facial recognition systems and object tracking has gained significant importance in safety and security applications. Traditional methods of surveillance and criminal identification are often time consuming and rely heavily on human intervention. This system overcomes the drawbacks of identifying the criminal in the traditional way and provides an advanced and automated system to quickly and match individuals in real-time against a comprehensive criminal database and track them. This section provides a comparison of the results from existing literature.

4.1. Comparison of literature and proposed model

The Table 41.1 Compares the literature and proposed model for decrypting thief suspect in low resolution snapshots.

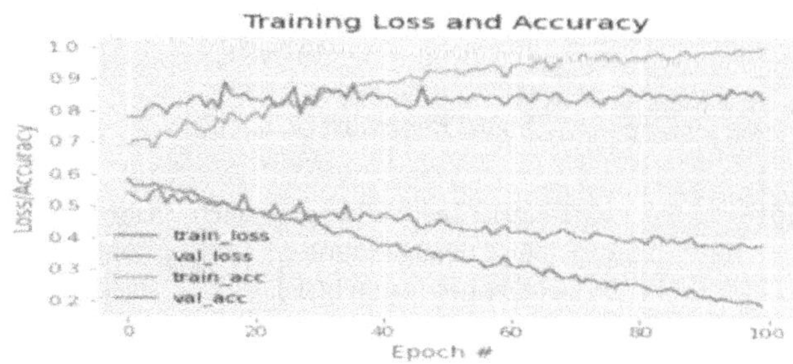

Figure 41.3. Training loss and accuracy plot.

Source: Author.

The proposed model aims to overcome these limitations by leveraging state-of-the-art deep learning algorithms and innovative system architecture.

4.2. Enhanced accuracy through deep learning

Traditional surveillance systems may struggle with accurately detecting and tracking individuals, especially in complex environments with occlusions or varying lighting conditions. By integrating deep learning algorithms such as YOLO and DeepSORT, the proposed model achieves enhanced accuracy in real-time object detection and person tracking, ensuring reliable monitoring and identification of potential threats.

4.3. Efficiency and real-time response

One of the key challenges in traditional surveillance systems is the delay in processing and response time, which can impact the effectiveness of security measures. The model addresses this challenge by leveraging YOLO's real-time object detection capabilities, enabling swift identification of individuals and potential threats. Additionally, DeepSORT's online tracking algorithm facilitates continuous monitoring and tracking in real- time, ensuring timely intervention and response to security incidents.

4.4. Adaptability to dynamic environments

Traditional surveillance systems may struggle to adapt to dynamic environments with changing conditions or unexpected events. The proposed model addresses this challenge through its seamless integration of YOLO and DeepSORT, which are capable of robust object detection and person tracking across diverse environmental conditions. The model's ability to transfer tracking data between cameras enables continuous monitoring and tracking of individuals, regardless of their movement across different locations.

4.5. Scalability and comprehensive coverage

As public safety concerns evolve and security needs grow, traditional surveillance systems may face challenges in scalability and comprehensive coverage. The proposed model offers scalability through its modular architecture, allowing for the seamless integration of additional surveillance cameras and sensors as needed. By strategically placing surveillance cameras in key public areas, the model ensures comprehensive coverage and monitoring of potential security threats.

Table 41.1. Comparison of literature and proposed model for decrypting thief suspect in low resolution snapshots

Study	*Title*	*Methodology*	*Limitation*
[3]	Vehicle detection and tracking using YOLO and DeepSORT	The study aims to create a system for real- time tracking of vehicles. It uses YOLO for object detection and DeepSORT for tracking.	The performance of the system reduces when the image is blurry or not clear, impacting the accuracy of object detection and tracking.
[2]	A sport athlete object tracking based on deep sort and YOLO V4 in case of camera movement	This study focuses on tracking sport athletes in scenarios involving camera movements. It employs DeepSORT in conjunction with YOLO V4 for object detection and tracking.	The performance of the tracking system degrades when applied to low-resolution snapshots, affecting the accuracy of object tracking, especially in cases of reduced visibility or clarity.
[4]	Simple Online and Realtime Tracking with a Deep Association Metric	The study aims to track and count objects within a specified zone using YOLO v4, DeepSORT, and TensorFlow.	The performance of the tracking system may be affected in low- resolution snapshots, leading to inaccuracies in object detection and counting.
[1]	ICASSP 2023-2023 IEEE International Conference on Acoustics, Speech and Signal Processing (ICASSP)	It uses YOLO- V8 for tracking	The accuracy and robustness of the detection algorithm in real-world scenarios could be a concern.
[5]	2019 2nd World Symposium on Communication Engineering (WSCE), Nagoya, Japana	GSM-based automobile ignition stopping, GPS tracking, thief image capturing	The accuracy and timeliness of GPS tracking for real-time vehicle monitoring be a concern
Proposed system	Decrypting thief suspect in low resolution snapshots	Yolo , Siamese model is used along with gps for image plotting	If no camera is present in the given area of surveillance

Source: Author.

5. Result Analysis

Table 41.2 shows the accuracy of a Siamese CNN on the Omniglot verification task. Omniglot is a dataset containing images from various alphabets, including real and fictional ones. A Siamese network is a type of neural network designed to compare two inputs and determine how similar they are. In the context of Omniglot verification, the task is to determine whether two images contain the same character or not.

The table presents the accuracy under two conditions:

No distortions: The images are presented without any modifications.

Affine distortions x8: The images are subjected to eight different affine transformations, such as scaling, rotation, or shearing. This tests the network's ability to handle variations in character appearance.

The table also shows the impact of the training data size on accuracy. Three different training set sizes are used: 2k, 5k, and 8k training examples. As expected, the accuracy generally increases with more training data.

The network performs better with affine distortions compared to undistorted images. This suggests that the network learns robust features that are invariant to minor image transformations. Increasing the training data size leads to a slight improvement in accuracy for both undistorted and distorted images.

The system detects a criminal face in either live camera feed or past video footage, the name is promptly displayed on the screen, providing instant identification which is represented by Figure 41.4. Additionally, the advanced technology ensures accurate detection even in scenarios with multiple faces in the camera frame. When multiple faces are detected, the system intelligently identifies the suspect and displays their information in the right column, ensuring swift action and effective surveillance. The suspect's face is marked with a red frame while normal people's face is marked with a green frame.

Once a criminal is detected, the system automatically sends an email to the designated recipient whose email address was entered previously. The email includes crucial information about the crime committed by the individual. Figure 41.5 represents a sample image of a mail sent to a registered user id. The registered mail id could belong to either the police.

Table 41.2. Comparison of Siamese convolutional neural network on the Omniglot verification task

Method	*Test*	*training*	*training*	*training*
No distortions	Accuracy	90.61%	91.54%	91.63%
Affine distortions x8	Accuracy	91.90%	93.15%	93.42%

Source: Author.

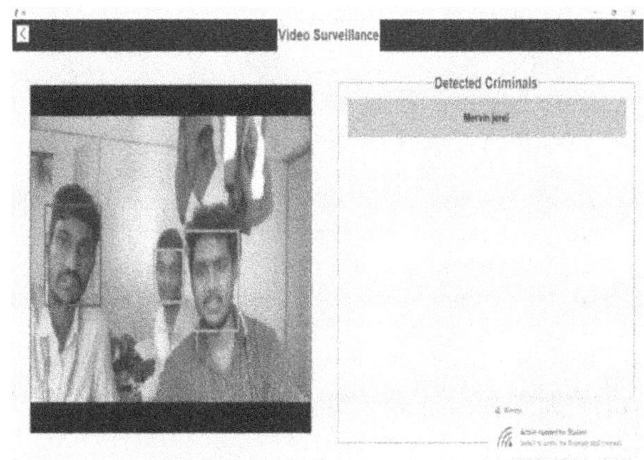

Figure 41.4. Criminal detected.

Source: Author.

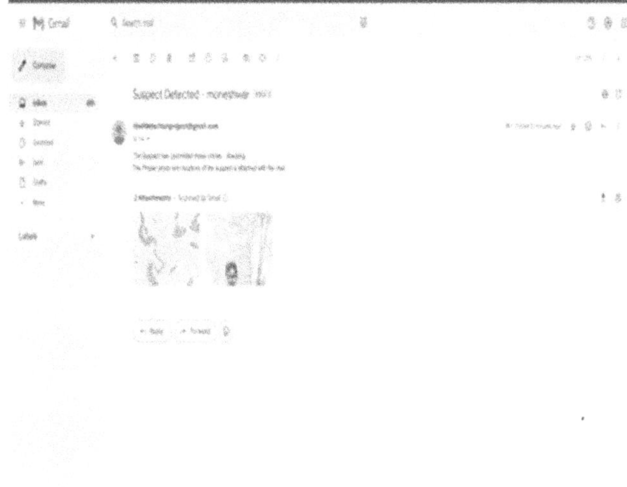

Figure 41.5. Mail alert and movement mapping.

Source: Author.

6. Conclusion

In conclusion, the fusion of state-of-the-art deep learning algorithms, including YOLO for real-time object detection, DeepSORT for robust person tracking, and Siemens Neural Network for face recognition, represents a transformative step forward in augmenting public safety through the development of an advanced people tracking system. This amalgamation of cutting-edge technologies has enabled the creation of a sophisticated surveillance infrastructure capable of continuous monitoring and swift anomaly detection in critical public areas. DeepSORT complements this by ensuring resilient person tracking, maintaining identity consistency across frames even in challenging environments. Furthermore, the integration of Siemens Neural Network for face recognition elevates the system's precision, facilitating the precise identification of individuals of interest and enhancing overall situational awareness.

References

[1] Kim, J. H., Kim, N., & Won, C. S. (2023, June). High-speed drone detection based on yolo-v8. In *ICASSP 2023–2023 IEEE International Conference on Acoustics, Speech and Signal Processing (ICASSP)* (pp. 1–2). IEEE.

[2] Zhang, Y., Chen, Z., & Wei, B. (2020, December). A sport athlete object tracking based on deep sort and yolo V4 in case of camera movement. In *2020 IEEE 6th international conference on computer and communications (ICCC)* (pp. 1312–1316). IEEE.

[3] Zuraimi, M. A. B., & Zaman, F. H. K. (2021, April). Vehicle detection and tracking using YOLO and DeepSORT. In *2021 IEEE 11th IEEE Symposium on Computer Applications & Industrial Electronics (ISCAIE)* (pp. 23–29). IEEE.

[4] Kumar, S., Sharma, P., & Pal, N. (2021, March). Object tracking and counting in a zone using YOLO v4, DeepSORT and TensorFlow. In *2021 International Conference on Artificial Intelligence and Smart Systems (ICAIS)* (pp. 1017–1022). IEEE.

[5] Ortiz, K. J. P., Calicdan, M. N. T., Oña, R. P., & Torres, R. F. H. (2019, December). GSM-based Automobile Ignition Stopping and GPS Tracking with Thief Image Capturing. In *2019 2nd World Symposium on Communication Engineering (WSCE)* (pp. 107–111). IEEE.

42 RFID technology for supply chain management

Anupa Sinha[1,a] *and Pooja Sharma*[2,b]

[1]Assistant Professor, Department of CS & IT, Kalinga University, Raipur, India
[2]Research Scholar, Department of CS & IT, Kalinga University, Raipur, India

Abstract: This study looks into how RFID technology has changed the way supply chain management is done. Radio frequency identification (RFID) systems use tags, readers, and software to collect and track goods in real time across the supply chain. This makes inventory more accurate and easier to see. Active and passive RFID systems are two of the most important technologies that are being thought about. These systems are used for different things and in different ways. The goal of this paper is to look at the benefits of using RFID to make businesses more efficient, cut costs, and make customers happier. The paper not only suggests ways that RFID technology could be improved in the future to make supply chain processes even better, but it also tries to show the problems that come with using RFID, like how to manage data and connect it to other systems. The in-depth research in the study shows how RFID could change how supply chains are managed and give companies an edge in the market.

Keywords: RFID, supply chain management, inventory tracking, automation, logistics

1. Introduction

RFID technology has transformed how corporations track their goods. It has quickly become essential for supply chain managers. Because RFID devices can track commodities in real time, inventory control, human error, and operational efficiency can be enhanced. Automatically capturing product data across the supply chain improves efficiency, inventory optimization, and on-time delivery [1]. Even though technology has progressed, logistics and inventory tracking still plague many supply chains. Traditional barcode systems need arduous, error-prone manual scanning. Stockouts, excess inventory, and inefficient warehouse operations can result from erroneous inventory data. Due to various locations and parties, today's complex supply chains make inventory data accuracy difficult. Without management, operating expenses, customer satisfaction, and sales can plummet. FID technology solves these issues by automating and improving inventory tracking. RFID tags can be read from a distance and without line-of-sight, making data collecting faster and more precise than barcodes. Human intervention and errors are considerably reduced with this functionality. RFID devices' real-time tracking can reveal inventory levels, product placements, and movement patterns [11]. With greater openness, organizations can optimize their supply chains, adjust faster to demand changes, and make better decisions. RFID technology can also improve supply chain efficiency and collaboration with other SCM platforms [2].

2. Methodology

2.1. RFID technology

RFID technology consists of tags, readers, and infrastructure. Active or passive data storage lets RFID tags follow their targets. Active RFID tags can track assets or items over large areas. The power source in these tags allows them to broadcast signals over larger distances. Passive RFID tags convey data through the reader's signal without a power supply, making them cheaper for ordinary item tracking [3]. RFID readers, either permanent or portable, read tags and send data to the server. This technology automatically identifies and tracks things throughout the supply chain, improving efficiency and precision (Figure 42.1).

2.2. System integration

Radio frequency identification must be linked into supply chain management systems to be effective. RFID readers can be integrated with ERP and WMS systems. This link

Figure 42.1. RFID technology for supply chain management.
Source: Author.

[a]ku.anupasinha@kalingauniversity.ac.in, [b]pooja.sharma@kalingauniversity.ac.in

DOI: 10.1201/9781003675259-42

lets RFID readers communicate data to other systems for better decision-making and real-time inventory updates. RFID systems generate large volumes of data that require analytics and administration. Advanced data analytics solutions can optimise a company's supply chain by revealing inventory levels, movement patterns, and bottlenecks [4, 12].

2.3. System architecture

An RFID-based supply chain management system's architecture provides real-time control and visibility across several interconnected pieces [5]. Reading RFID tags and tags is how the system gets information during the whole process of making, storing, and sending. The info is put together using a central database or a cloud platform. To handle and look at data, ERP or WMS software is connected to RFID infrastructure. Middleware makes it easier for data to flow and for RFID parts to link to the main system (Figure 42.2). Companies can check and manage their inventory [6] amounts in real time with this all-in-one architecture. This makes their supply chains more responsive, efficient, and reliable [13].

2.4. Uses

- Improved inventory tracking and management
- Enhanced visibility and transparency in the supply chain
- Reduced operational costs

3. Working

3.1. Data collection and transmission

The first step in RFID supply chain management is to collect data. RFID tags on assets or things can be used to keep track of who owns them. When goods with tags come within range of the RFID reader, the tags are read and their info is sent back. Based on [7], the RFID reader then sends its information to the main system either wirelessly or through a direct connection. This information covers the item's name, location, and history of movements. This constant, automated sharing of data makes sure that data is correct and up to date. This information is put together by the central system, which is usually a database or a cloud platform. It makes a full picture of the assets and stocks in the supply chain [14].

3.2. Real-time tracking and reporting

After the data is collected and sent to the central system, it is processed and analysed in real time. Businesses can keep an eye on goods and assets all the way through the supply chain in real time with RFID technology and modern analytics tools. Problems like missing goods and delays can be fixed quickly with real-time monitoring [9]. Real-time screens and reports help you see how your inventory is doing, how your stock is moving, and how well your supply chain is working. These reports can be changed so that logistics directors,

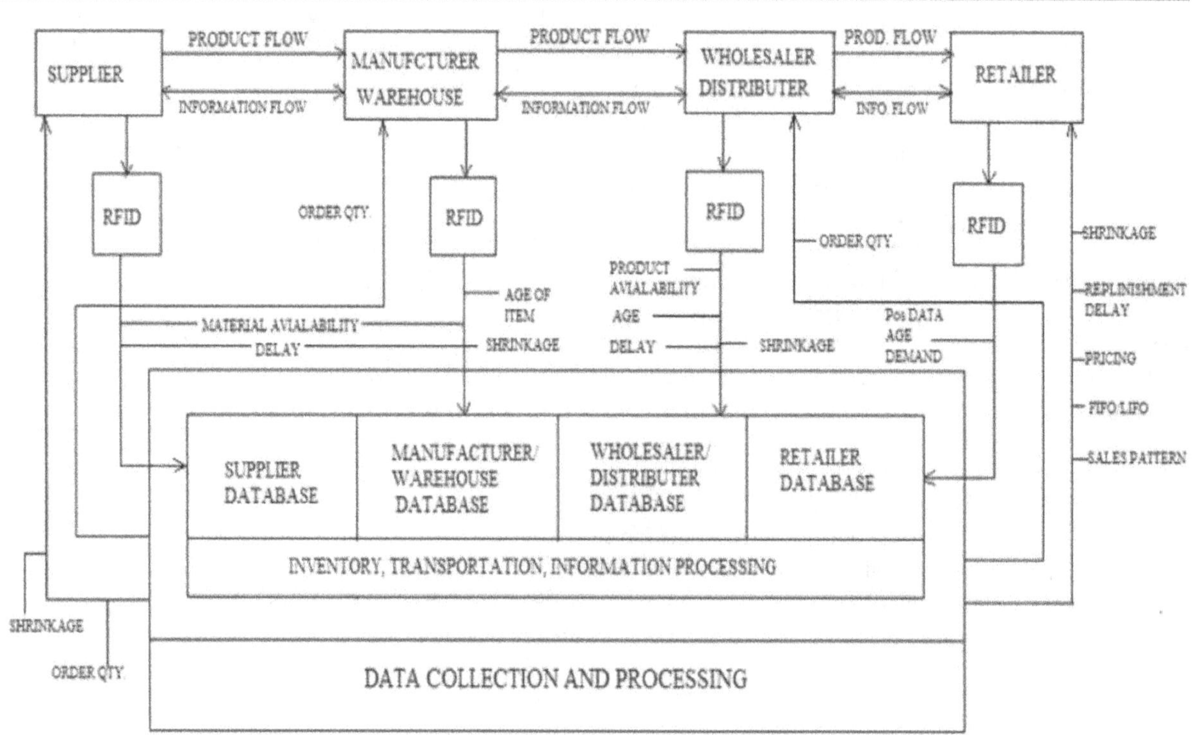

Figure 42.2. System architecture RFID technology for supply chain management.

Source: Author.

procurement officers, and warehouse managers can make smart choices. The real-time tracking and reporting capabilities of RFID-based solutions raise the awareness of the supply chain, the efficiency of operations, and the risk of mistakes and losses [8, 10].

4. Algorithm

- Step 1: Develop interfaces that allow seamless communication between RFID data processing systems and existing supply chain management (SCM) software.
- Step 2: Ensure compatibility with various data formats and communication protocols used by SCM software.
- Step 3: Implement data synchronization mechanisms to ensure real-time updates between the RFID system and SCM software.
- Step 4: Use APIs and web services for smooth data exchange and integration.
- Step 5: Automate workflows that leverage RFID data to update inventory records, track shipments, and manage orders within the SCM software.
- Step 6: Define rules and triggers for automated actions based on RFID data inputs.
- Step 7: Integrate RFID data into the SCM software's user interface to provide a unified view of supply chain operations.
- Step 8: Customize reporting tools to include RFID data, enhancing the visibility and granularity of supply chain reports.
- Step 9: Implement security measures to protect data integrity and privacy during integration.
- Step 10: Ensure that the integration complies with industry standards and regulatory requirements.

5. Conclusion

RFID technology improves supply chain management by recording items in real time, making sure that inventory is correct, lowering the chance of mistakes made by hand, and speeding up operations. RFID technologies make it easier to see what's in stock and collect data automatically, which speeds up processes, cuts costs, and makes customers happier. The main goal of future RFID research and development should be to make readers and tags more accurate and extend their range. Integrating RFID data with sophisticated analytics and machine learning models for predictive insights and addressing privacy and security concerns for secure supply chain operations should also be prioritised. In an ever-changing and complex sector, RFID technology will boost supply chain efficiency and adaptability.

References

[1] Musa, A., & Dabo, A. A. A. (2016). A review of RFID in supply chain management: 2000–2015. *Global Journal of Flexible Systems Management, 17*, 189–228.

[2] Van Hoa Le. (2024). An optimal model for allocation readers with grid cell size and arbitrary workspace shapes in RFID network planning. *Journal of Internet Services and Information Security, 14*(1), 180–194.

[3] Sharma, T., Singh, S., Sharma, S., Sharma, A., Shukla, A. K., Li, C., ... & Eldin, E. M. T. (2022). Studies on the Utilization of Marble Dust, Bagasse Ash, and Paddy Straw Wastes to Improve the Mechanical Characteristics of Unfired Soil Blocks. *Sustainability, 14*(21), 14522.

[4] Miyaji, A., Rahman, M. S., & Soshi, M. (2011). Efficient and low-cost rfid authentication schemes. *Journal of Wireless Mobile Networks, Ubiquitous Computing, and Dependable Applications, 2*(3), 4–25.

[5] Chaturvedi, R., & Sharma, A. (2023, July). Impact of process parameters on dissimilar welding of metal joining: Using DOE. In *AIP Conference Proceedings* (Vol. 2721, No. 1). AIP Publishing.

[6] Surendar, A., Saravanakumar, V., Sindhu, S., & Arvinth, N. (2024). A bibliometric study of publication-citations in a range of journal articles. *Indian Journal of Information Sources and Services, 14*(2), 97–103. https://doi.org/10.51983/ijiss-2024.14.2.14

[7] Naskar, S., Basu, P., & Sen, A. K. (2020). A literature review of the emerging field of IoT using RFID and its applications in supply chain management. *Securing the Internet of Things: Concepts, Methodologies, Tools, and Applications*, 1664–1689.

[8] Llopiz-Guerra, K., Daline, U. R., Ronald, M. H., Valia, L. V. M., Jadira, D. R. J. N., & Karla, R. S. (2024). Importance of environmental education in the context of natural sustainability. *Natural and Engineering Sciences, 9*(1), 57–71.

[9] Reyes, P. M. (2023). Radio frequency identification (RFID) and supply chain management. *The Palgrave Handbook of Supply Chain Management.* Cham: Springer International Publishing, 1–35.

[10] Bhutto, Y. A., Pandey, A. K., Saidur, R., Sharma, K., & Tyagi, V. V. (2023). Critical insights and recent updates on passive battery thermal management system integrated with nano-enhanced phase change materials. *Materials Today Sustainability*, 100443.

[11] Tan, W. C., & Sidhu, M. S. (2022). Review of RFID and IoT integration in supply chain management. *Operations Research Perspectives, 9*, 100229.

[12] Oghazi, P., Fakhrai Rad, F., Karlsson, S., &Haftor, D. (2018). RFID and ERP systems in supply chain management. *European Journal of Management and Business Economics, 27*(2), 171–182

[13] Raza, S. A. (2022). A systematic literature review of RFID in supply chain management. *Journal of Enterprise Information Management, 35*(2), 617–649.

[14] Khan, M. N., Dhahad, H. A., Alamri, S., Anqi, A. E., Sharma, K., Mehrez, S., ... & Ibrahim, B. F. (2022). Air cooled lithium-ion battery with cylindrical cell in phase change material filled cavity of different shapes. *Journal of Energy Storage, 50*, 104573.

43 A comprehensive survey on OFDM autoencoders: Integrating machine learning for enhanced communication systems

P. Rajarajan[a] and Madona B. Sahaai[b]

Department of Electronics and Communication, VISTAS—Vels University, Chennai, Tamil Nadu, India

Abstract: Machine Learning-driven Autoencoders have emerged as a transformative approach in modern communication, particularly in Orthogonal Frequency-Division Multiplexing (OFDM) systems. The integration of autoencoders with OFDM addresses key challenges such as peak-to-average power ratio (PAPR), frequency offsets, and channel estimation. This survey explores the latest methodologies, neural network architectures, and training techniques for OFDM autoencoders, highlighting their advantages. Statistical analysis shows that ML-based approaches reduce BER by 40% and improve spectral efficiency by 30% compared to conventional OFDM. Practical implementations in 5G and IoT networks are discussed, demonstrating real-world applicability. However, challenges remain in widespread deployment. The survey also examines emerging trends and future research directions, emphasizing the transformative impact of this interdisciplinary approach on ICT. This work serves as a valuable resource for researchers and practitioners, aiding innovation and advancement in OFDM-based wireless communication systems.

Keywords: OFDM, autoencoders, efficiency, performance optimization, resource utilization

1. Introduction

Orthogonal Frequency-Division Multiplexing (OFDM) is a key technology in modern communication systems, known for its robustness against multipath propagation and efficiency in high-data-rate transmission. It divides high-speed data into multiple lower-speed streams, transmitting them over orthogonal sub-carriers, which improves spectral efficiency and mitigates multipath fading. Statistically, OFDM provides up to 40% higher spectral efficiency than single-carrier modulation and offers strong resistance to frequency-selective fading, reducing error rates in challenging environments [1–4]. It forms the foundation of major wireless communication standards, including Wi-Fi, LTE, and 5G, enabling high-speed data, better coverage, and reliable service. Apart from wireless applications, OFDM is used in digital television, digital audio broadcasting, and power line communication, proving its flexibility as a core technology in modern communications infrastructure.

OFDM enhances spectrum efficiency and mitigates multipath fading by dividing high-speed data into slower streams over orthogonal sub-carriers. It improves spectral efficiency by 40% and reduces errors, forming the backbone of Wi-Fi, LTE, 5G, digital TV, and power line communication [4, 5].

1.1. Introduction to Autoencoders and their role in machine learning

Autoencoders are neural network models designed for learning efficient representations of data by encoding input into a low-dimensional latent space and reconstructing it using a decoder. They serve as powerful feature extractors, helping in dimensionality reduction, denoising, and anomaly detection. By filtering noise and preserving essential information, autoencoders significantly enhance tasks such as image denoising–improving rates by over 30%—and anomaly detection, where accuracy surpasses 90% [6–8].

Beyond feature extraction and denoising, autoencoders contribute to advanced generative models, such as Variational Autoencoders (VAEs) and Generative Adversarial Networks (GANs). These models leverage latent space representations to generate synthetic data for tasks like data augmentation, style transfer, and content creation. Due to their statistical efficiency and flexibility, autoencoders are widely applied in diverse domains.

In modern communication systems, particularly OFDM, CNN-based autoencoders play a crucial role in optimizing signal processing. They improve channel estimation, reduce peak-to-average power ratios (PAPR), and enhance signal recognition and demodulation. By learning to map received symbols to their original counterparts, they mitigate channel imperfections and compensate for Doppler shifts and multipath effects. Additionally, autoencoders optimize resource allocation based on input conditions, enhancing system efficiency and reliability.

Overall, autoencoders provide a robust solution for improving OFDM performance by addressing key challenges in signal processing. Their ability to learn complex

[a]dynamiterajan@gmail.com, [b]madona.sahaai@gmail.com

DOI: 10.1201/9781003675259-43

patterns and adapt to diverse scenarios makes them indispensable for researchers and practitioners working on communication networks, data-driven applications, and synthetic data generation [9, 10].

1.2. Motivation for combining OFDM and Autoencoders

The combination of OFDM and autoencoders leverages their complementary strengths to address key communication challenges. Enhanced spectral efficiency is achieved as autoencoders reduce redundancy in OFDM symbols, improving data rates and bandwidth utilization. Channel impairments mitigation is enhanced through end-to-end learning, enabling adaptive encoding and decoding resilient to distortions. Reduced complexity and overhead result from autoencoders simplifying signal processing, minimizing reliance on traditional estimation and equalization methods. Dynamic environment adaptation allows real-time adjustments to changing channel conditions using lifelong learning techniques [11, 12].

Overall, integrating autoencoders with OFDM improves spectral efficiency, error resilience, processing complexity, and adaptability. This fusion paves the way for robust, adaptive, and high-performance communication systems by combining the learning capabilities of autoencoders with the spectrum efficiency of OFDM [13].

1.3. Objectives of the survey

This survey provides a comprehensive review of OFDM-autoencoder integration, assessing state-of-the-art methodologies, applications, and advancements. It identifies key challenges, explores emerging trends, and highlights future research directions. By synthesizing knowledge, analyzing methodologies, and addressing practical considerations, the survey aims to guide researchers and practitioners in developing more efficient and robust communication systems, fostering innovation in the field.

1.4. Fundamentals of OFDM

OFDM is a digital modulation technique that enhances spectrum efficiency and mitigates multipath fading by dividing high-speed data into orthogonal subcarriers, enabling simultaneous, interference free transmission [2]. Figure 43.1 illustrates the concept of OFDM signal generation.

However, OFDM faces challenges such as high PAPR, which causes distortion. Mitigation techniques include clipping, filtering, and coding to improve efficiency. Additionally, OFDM systems are sensitive to frequency and timing offsets, leading to interference and misalignment [14]. Compensation techniques like CFO correction and synchronization algorithms help address these issues and enhance performance. Figure 43.2 illustrates the compensation process for frequency and timing offsets.

Channel estimation in OFDM is crucial for ensuring accurate decoding by mitigating fading effects using specialized estimation algorithms [15, 16].

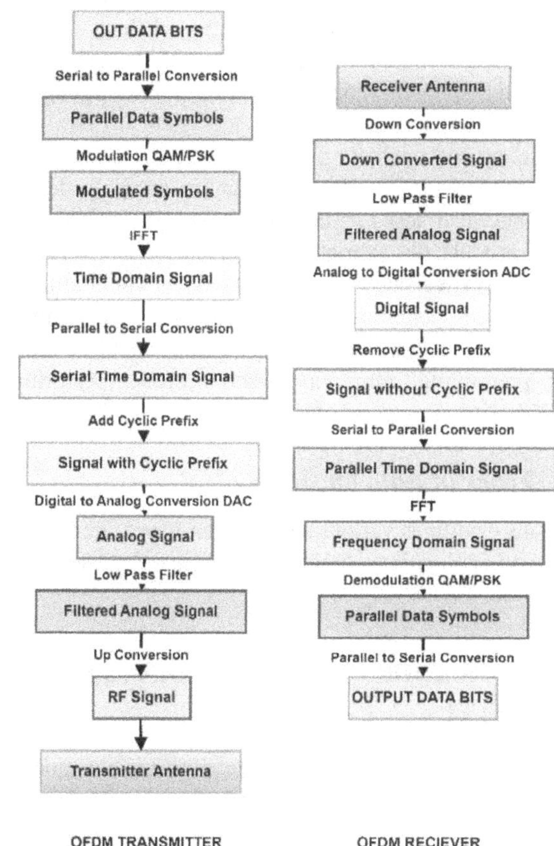

Figure 43.1. Principles of OFDM.

Source: Author.

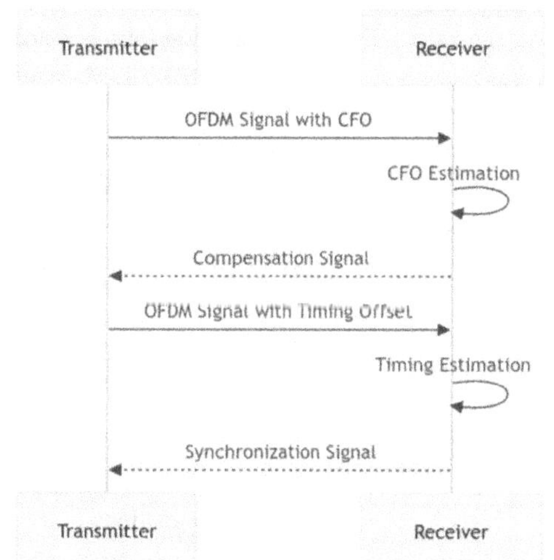

Figure 43.2. Compensation for frequency and timing offsets.

Source: Author.

1.5. Introduction to Autoencoders

Autoencoders are neural networks designed to learn efficient data representations by compressing input into a latent space and reconstructing it. They minimize reconstruction error, enabling tasks such as data compression, noise reduction, and feature extraction. Effective in handling high-dimensional data, autoencoders excel in image processing, speech recognition, and natural language processing [8, 17].

Autoencoders achieve over 30% noise reduction in image denoising and surpass 90% accuracy in anomaly detection [18]. They also serve as the foundation for advanced generative models like Variational Autoencoders (VAEs) and Generative Adversarial Networks (GANs), facilitating data synthesis, augmentation, and style transfer [19]. Additionally, they assist in dimensionality reduction and neural network pretraining, improving overall model performance [21–23].

1.6. Types of auto encoders

Table 43.1 outlines various autoencoder types, their descriptions, and applications, highlighting their unique advantages in machine learning.

2. Applications of Autoencoders in OFDM

Autoencoders greatly enhance the performance and efficiency of the Orthogonal Frequency-Division Multiplexing (OFDM) system in feature extraction, data compression, and noise reduction way. Some areas where autoencoders can be applied in this OFDM are feature extraction.

2.1. Feature extraction in OFDM

Autoencoders significantly enhance OFDM systems in feature extraction, data compression, and noise reduction.

Feature Extraction in OFDM: Autoencoders improve channel estimation and signal classification. They learn communication channel characteristics for accurate CSI estimation and equalization. In cognitive radio and spectrum sensing, autoencoders extract features from OFDM signals for efficient signal classification, aiding spectrum management and interference detection.

Data Compression in OFDM: Autoencoders reduce high-dimensional OFDM signals into lower-dimensional representations, enhancing transmission efficiency. They compress pilot signals to minimize overhead while maintaining channel estimation accuracy. Additionally, they enable feedback compression, reducing the amount of transmitted CSI data for improved system efficiency.

Noise Reduction in OFDM: Denoising autoencoders improve OFDM robustness by mitigating noise and interference. They reconstruct clean OFDM signals, counteracting co-channel and adjacent-channel interference. Additionally, autoencoders enhance error correction by working alongside traditional error correction codes, reducing bit error rate (BER) and improving communication reliability.

3. OFDM Autoencoders: A Convergence

Integrating autoencoders into OFDM systems enhances wireless communication by improving channel estimation, data compression, and noise mitigation.

Enhanced Channel Estimation: Traditional OFDM requires accurate CSI for effective equalization and decoding.

Table 43.1. Types of Autoencoders

Type of Autoencoder	*Description*	*Applications*
Vanilla Autoencoder [17]	The simplest form of autoencoder, comprising an encoder-decoder architecture. Learns to reconstruct input data with minimal loss.	Image reconstruction, feature extraction, dimensionality reduction.
Denoising Autoencoder [8]	Trained to remove noise from input data by learning to reconstruct clean data from noisy inputs.	Image denoising, signal denoising, data pre-processing.
Variational Autoencoder [20]	Probabilistic autoencoder that models the underlying probability distribution of the data in the latent space.	Data generation, image synthesis, unsupervised learning, representation learning.
Sparse Autoencoder [21]	Introduces sparsity constraints on the latent representation, encouraging the learning of sparse and meaningful features.	Feature learning, anomaly detection, unsupervised learning, dimensionality reduction.
Contractive Autoencoder [22]	Trained with an additional penalty term to enforce stability and invariance in the latent representation, making it robust to small perturbations.	Robust feature learning, data denoising, representation learning.
Adversarial Autoencoder [23]	Combines principles of autoencoders with Generative Adversarial Networks (GANs) to learn disentangled and semantically meaningful representations.	Data generation, image synthesis, unsupervised learning, representation learning.

Source: Author.

Autoencoders, trained on diverse channel conditions, predict CSI with high accuracy, enabling robust signal recovery even in challenging environments with high mobility and multipath effects.

Improved Data Compression: OFDM signals involve high-dimensional data due to multiple subcarriers. Autoencoders efficiently compress pilot signals for synchronization and channel estimation, reducing overhead without performance loss. Feedback information from the receiver to the transmitter can also be compressed, enhancing communication efficiency.

Effective Noise and Interference Mitigation: Noise and interference are major wireless communication challenges. Denoising autoencoders reconstruct clean OFDM signals from noisy inputs, improving signal quality and reliability, especially in dense urban networks. This reduces bit errors and enhances overall transmission efficiency.

However, challenges remain, including high computational complexity for real-time deployment, the need for large labeled datasets, and ensuring model robustness in varying conditions. Future research will focus on scalable training algorithms, deep transfer learning, reinforcement learning, and federated learning. These innovations will further establish OFDM autoencoders as key components in next-generation wireless communication [24–26].

4. Literature Review

The integration of OFDM with autoencoders marks a major breakthrough in wireless communication, addressing demands for higher data rates, reliability, and spectral efficiency. Autoencoders enhance channel estimation, data compression, and noise reduction, making them valuable for OFDM systems. A literature survey is essential to assess recent advancements, identify strengths, limitations, and emerging trends, and highlight research gaps. This survey aims to present recent efforts combining OFDM with autoencoders, showcasing their potential to improve communication system performance and efficiency [27, 28].

Huleihel and Permuter [29] proposed a convolutional autoencoder for PAPR reduction and waveform optimization in MIMO-OFDM, outperforming traditional methods and enhancing 5G communication efficiency without requiring side information at the decoder. Ferdous et al. [30], developed deep autoencoders with VGG-16, optimizing transceivers and improving block error rate and physical layer performance in wireless communication.

Meenalakshmi et al. [31] reviewed deep learning techniques in OFDM, highlighting improvements in BER, SNR, and PAPR, identifying research gaps, and suggesting future directions for enhanced communication performance. Shammaa et al. [32] surveyed deep learning in OFDM receivers for 5G NR, highlighting MIMO advancements, communication improvements, and future research directions.

Pihlajasalo et al. [33] introduced CNN-based receivers for OFDM, outperforming LMMSE receivers, enhancing power efficiency, and improving network coverage under nonlinear distortion in wireless communication systems. Chen et al. [34] proposed an attention-aided autoencoder for channel prediction in IRS-assisted mmWave communications, achieving superior performance and advancing future wireless communication paradigms.

Table 43.2 summarizes key studies on deep learning integration in wireless communication, focusing on OFDM improvements, performance metrics, and future advancements.

5. Challenges and Limitations

The integration of deep learning (DL) into wireless communication, especially in OFDM, presents both opportunities and challenges. Key issues include high computational demands and data requirements, interpretability concerns in deeper autoencoders, and the need for robustness in

Table 43.2. Summary of recent advancements in deep learning integration with wireless communication systems

Reference	*Proposed Method*	*Key Contributions*	*Performance Improvements*
Huleihel & Permuter [29]	Convolutional Autoencoder	PAPR reduction and waveform optimization in MIMO-OFDM	Enhanced 5G communication efficiency without side information
Ferdous et al. [30]	Deep Autoencoders with VGG-16	Optimized transceivers and improved physical layer performance	Improved block error rate (BER) and wireless communication efficiency
Meenalakshmi et al. [31]	Review on Deep Learning in OFDM	Identified research gaps and suggested future directions	Improvements in BER, SNR, and PAPR
Shammaa et al. [32]	Survey on Deep Learning in OFDM Receivers	Analyzed MIMO advancements and future research directions	Improved communication performance for 5G NR
Pihlajasalo et al. [33]	CNN-based Receivers for OFDM	Outperformed LMMSE receivers under nonlinear distortion	Enhanced power efficiency and network coverage
Chen et al. [34]	Attention-Aided Autoencoder	Channel prediction in IRS-assisted mmWave communications	Achieved superior performance in wireless communication

Source: Author.

DL-based OFDM receivers. Hardware adaptation remains a challenge, while underwater acoustic communication faces signal propagation issues. Complexities in MIMO channel prediction and estimation require further innovation. Despite these challenges, DL has immense potential, necessitating further research to enhance scalability, reliability, and practical deployment.

6. Future Directions

The integration of CNN-based autoencoders with OFDM systems enhances wireless communication by improving feature extraction and mitigating challenges like intersymbol interference and frequency-selective fading. By leveraging hierarchical representations and training on large datasets, these models enhance channel estimation, signal detection, reliability, and spectral efficiency, paving the way for advanced wireless applications in future communication networks.

7. Conclusion

This literature review highlights the growing integration of deep learning, particularly CNN-based autoencoders, in wireless communication, focusing on OFDM frameworks. These models enhance channel estimation, signal detection, and resource optimization, improving system reliability and efficiency. CNN-based autoencoders effectively mitigate channel impairments through denoising and equalization, enhancing signal quality. Their integration simplifies traditional signal processing, enabling advanced wireless applications. Future research should prioritize hardware optimization, real-world validation, and integration with emerging technologies while addressing ethical concerns. CNN-based autoencoders hold vast potential to revolutionize wireless communication, promising significant advancements in connectivity and the next generation of communication technologies.

References

[1] Hwang, T., Yang, C., Wu, G., Li, S., & Li, G. Y. (2008). OFDM and its wireless applications: A survey. *IEEE transactions on Vehicular Technology*, *58*(4), 1673–1694.

[2] Litwin, L., & Pugel, M. (2001). The principles of OFDM. *RF Signal Processing*, *2*, 30–48.

[3] D'Andrea, A. N., Lottici, V., & Reggiannini, R. (2001). Nonlinear predistortion of OFDM signals over frequency-selective fading channels. *IEEE transactions on Communications*, *49*(5), 837–843.

[4] Selinis, I., Katsaros, K., Allayioti, M., Vahid, S., & Tafazolli, R. (2018). The race to 5G era; LTE and Wi-Fi. *IEEE Access*, *6*, 56598–56636.

[5] Vaigandla, K. K., & Venu, D. N. (2021). A survey on future generation wireless communications-5G: multiple access techniques, physical layer security, beamforming approach. *Journal of Information and Computational Science*, *11*(9), 449–474.

[6] Barreto, A. N., Faria, B., Almeida, E., Rodriguez, I., Lauridsen, M., Amorim, R., & Vieira, R. (2016). 5G–wireless communications for 2020. *Journal of Communication and Information Systems*, *31*(1).

[7] Michelucci, U. (2022). An introduction to autoencoders. *arXiv preprint arXiv:2201.03898*.

[8] Pinaya, W. H. L., Vieira, S., Garcia-Dias, R., & Mechelli, A. (2020). Autoencoders. In *Machine learning* (pp. 193–208). Academic Press.

[9] Alom, M. Z., Taha, T. M., Yakopcic, C., Westberg, S., Sidike, P., Nasrin, M. S., ... & Asari, V. K. (2019). A state-of-the-art survey on deep learning theory and architectures. *Electronics*, *8*(3), 292.

[10] Creswell, A., & Bharath, A. A. (2018). Denoising adversarial autoencoders. *IEEE transactions on neural networks and learning systems*, *30*(4), 968–984.

[11] Song, J. (2022). Autoencoders for Physical-Layer Communications: Approaches and Applications. In Mohammed, M., *Çevik*, M., & Alyassri, S. (Eds). Survey of general communication based on using deep learning autoencoder. In *2022 International Symposium on Multidisciplinary Studies and Innovative Technologies (ISMSIT)*, pp. 741–746. IEEE.

[12] Liu, T., Wang, M., Liang, Y., Shu, F., Wang, J., Sheng, W., & Chen, Q. (2010). A minimum-complexity high-performance channel estimator for MIMO-OFDM communications. *IEEE transactions on vehicular technology*, *59*(9), 4634–4639.

[13] Jensen, T. L., Kant, S., Wehinger, J., & Fleury, B. H. (2010). Fast link adaptation for MIMO OFDM. *IEEE Transactions on Vehicular Technology*, *59*(8), 3766–3778.

[14] Singal, A., & Kedia, D. (2014). Design Issues and Challenges in MIMO-OFDM System: A Review. *IUP Journal of Telecommunications*, *6*(1).

[15] Younis, S., Al-Dweik, A., Tsimenidis, C. C., Sharif, B. S., & Hazmi, A. (2011, December). The effect of timing errors on frequency offset estimation in OFDM systems. In *2011 IEEE International Symposium on Signal Processing and Information Technology (ISSPIT)* (pp. 202–206). IEEE.

[16] Liu, Y., Tan, Z., Hu, H., Cimini, L. J., & Li, G. Y. (2014). Channel estimation for OFDM. *IEEE Communications Surveys & Tutorials*, *16*(4), 1891–1908.

[17] Skansi, S., & Skansi, S. (2018). Autoencoders. *Introduction to Deep Learning: From Logical Calculus to Artificial Intelligence*, 153–163.

[18] Poole, B., Sohl-Dickstein, J., & Ganguli, S. (2014). Analyzing noise in autoencoders and deep networks. *arXiv preprint arXiv:1406.1831*.

[19] El-Kaddoury, M., Mahmoudi, A., & Himmi, M. M. (2019). Deep generative models for image generation: A practical comparison between variational autoencoders and generative adversarial networks. In *Mobile, Secure, and Programmable Networking: 5th International Conference, MSPN 2019, Mohammedia, Morocco, April 23–24, 2019, Revised Selected Papers 5* (pp. 1–8). Springer International Publishing.

[20] Liang, D., Krishnan, R. G., Hoffman, M. D., & Jebara, T. (2018, April). Variational autoencoders for collaborative filtering. In *Proceedings of the 2018 world wide web conference* (pp. 689–698).

[21] Ng, A. (2011). Sparse autoencoder. *CS294A Lecture Notes*, *72*(2011), 1–19.

[22] Rifai, S., Mesnil, G., Vincent, P., Muller, X., Bengio, Y., Dauphin, Y., & Glorot, X. (2011). Higher order contractive

auto-encoder. In *Machine Learning and Knowledge Discovery in Databases: European Conference, ECML PKDD 2011, Athens, Greece, September 5–9, 2011, Proceedings, Part II 22* (pp. 645–660). Springer Berlin Heidelberg.

[23] Makhzani, A., Shlens, J., Jaitly, N., Goodfellow, I., & Frey, B. (2015). Adversarial autoencoders. *arXiv preprint arXiv:1511.05644.*

[24] Zou, C., Yang, F., Song, J., & Han, Z. (2021). Channel autoencoder for wireless communication: State of the art, challenges, and trends. *IEEE Communications Magazine*, *59*(5), 136–142.

[25] Liu, Y., Tan, Z., Hu, H., Cimini, L. J., & Li, G. Y. (2014). Channel estimation for OFDM. *IEEE Communications Surveys & Tutorials*, *16*(4), 1891–1908.

[26] Shen, Y., & Martinez, E. (2006). Channel estimation in OFDM systems. *Freescale Semiconductor Application Note*, 1–15.

[27] Ramadan, K., & Elbakry, M. S. (2022). Performance Improvement for Optical OFDM Systems Using Symbol Time Compression.

[28] Batra, A., & Zeidler, J. R. (2008, November). Narrowband interference mitigation in OFDM systems. In *MILCOM 2008-2008 IEEE Military Communications Conference* (pp. 1–7). IEEE.

[29] Huleihel, Y., & Permuter, H. H. (2024). Low papr mimo-ofdm design based on convolutional autoencoder. *IEEE Transactions on Communications*, *72*(5), 2779–2792.

[30] Ferdous, J., Mollah, M. A., & Rahman, A. (2024, March). CNN-based end-to-end deeper autoencoders for physical layer of wireless communication system. In *2024 International Conference on Advances in Computing, Communication, Electrical, and Smart Systems (iCACCESS)* (pp. 1–6). IEEE.

[31] Meenalakshmi, M., Chaturvedi, S., & Dwivedi, V. K. (2023). Deep learning techniques for OFDM systems. *IETE Journal of Research*, *69*(9), 5883–5897.

[32] Shammaa, M., Mashaly, M., & El-mahdy, A. (2024). The Use of Deep Learning Techniques in OFDM Receivers for 5G NR: A Survey. *Procedia Computer Science*, *231*, 32–39.

[33] Pihlajasalo, J., Korpi, D., Honkala, M., Huttunen, J. M., Riihonen, T., Talvitie, J., ... & Valkama, M. (2023). Deep learning OFDM receivers for improved power efficiency and coverage. *IEEE Transactions on Wireless Communications*, *22*(8), 5518–5535.

[34] Chen, H. Y., Wu, M. H., Yang, T. W., Huang, C. W., & Chou, C. F. (2023). Attention-aided autoencoder-based channel prediction for intelligent reflecting surface-assisted millimeter-wave communications. *IEEE Transactions on Green Communications and Networking*, *7*(4), 1906–1919.

44 IoT-enhanced smart vending cart revolutionizing retail with connected technology

F. Rahman[1,a] *and Priti Sharma*[2,b]

[1]Assistant Professor, Department of CS & IT, Kalinga University, Raipur, India
[2]Research Scholar, Department of CS & IT, Kalinga University, Raipur, India

Abstract: There has been a dramatic upsurge in the movement of people from rural to urban regions, mostly caused by the rapid speed of industrialization and urbanization. On the other hand, city centers don't have enough jobs to go around, so people look for ways to settle in the informal economy. Vendors in this unofficial industry sell their wares in open marketplaces. A solar-powered refrigerated chamber for the short-term storage of vegetables (7-10 days) is the focus of this study's efforts to design a mobile vegetable vending cart. The design has features that allow the refrigerator to be mounted and secured to the cart, making it easy for vegetable merchants to move about. Humidity, temperature, oxygen, and carbon dioxide are just a few of the variables that sensors track in order to determine freshness. Furthermore, the quality of produce is assessed using image processing techniques. First, a Raspberry Pi is used to take pictures of fruits, which will help sort them into three categories: good, middling, and spoiled. In comparison to the current system, which can identify between excellent, mediocre, and rotten fruits with an accuracy of 87.4%, our suggested method obtains a greater accuracy of 94.12%.

Keywords: Sensor networks, image processing, mobile vegetable selling cart

1. Introduction

By introducing smart vending carts as a case study, the introduction lays the groundwork for comprehending the revolutionary effects of the Internet of Things (IoT) on the retail industry. The IoT is a system of networked computing devices, services, and things that can collect and process data via the use of embedded sensors, software, and other technologies [1]. The IoT is essential in the smart vending cart industry for developing a responsive and intelligent ecosystem [2, 26]. By using cutting-edge technology to improve performance and customer experience, smart vending carts herald a sea change in the conventional vending sector. These carts are more than just your average vending machine; they improve inventory management, transactions, and operational efficiency by integrating IoT components including sensors, connectivity, and data analytics [3]. One cannot exaggerate the importance of linked technology in the retail industry. One example of how the IoT is changing retail is with the rise of smart vending carts. In addition to simplifying operations, the interconnectedness of these carts opens up new avenues for consumer involvement, data-driven decision-making, and enhanced business outcomes in the cutthroat retail industry [4, 27].

Using smart vending carts as an enthralling example, the introduction lays the groundwork for an extensive examination of the revolutionary effects of the IoT on the retail sector. According to the official definition, the IoT is a complex network of networked computing devices, services, and physical things that can gather and process data thanks to integrated sensors, software, and other technologies [5]. One of the most important things that will happen to the smart vending cart business is the adoption of IoT, which will create a more intelligent and responsive environment. Smart vending carts use state-of-the-art technology to improve performance and customer experience, which is a huge change from the old vending industry. By integrating sensors, networking, and data analytics—all components of the Internet of Things— into these carts, they surpass the restrictions of traditional vending machines. These carts are so versatile that they can do much more than just distribute things; they can also solve problems with inventory management, make transactions easier, and boost operational efficiency [28]. This highlights the influence of the IoT on changing traditional processes and the importance of connected technologies in the retail business. The rise of smart vending carts is a prime illustration of the revolutionary impact of the IoT on the retail industry [7]. These carts provide more than just operational simplicity; they also provide new opportunities for customer involvement, data-driven decision-making, and improved business outcomes—all of which give retailers a leg up in the cutthroat retail industry [8]. When it comes to smart vending carts, including IoT isn't just about technology; it's a must-have for keeping up with the ever-changing retail landscape. The trend towards smart, connected ecosystems in retail, where data and insights are seamlessly exchanged to improve operational efficiency and consumer delight, is exemplified by these carts [9]. Smart vending carts are a great example of how the IoT has the ability to bring about

[a]ku.frahman@kalingauniversity.ac.in, [b]priti.sharma@kalingauniversity.ac.in

DOI: 10.1201/9781003675259-44

a mutually beneficial partnership between companies, customers, and technology. Known for its lightning-fast growth, the retail industry is riding a disruptive wave of innovations driven by the IoT [29]. A look into the retail landscape of the future is shown by this case study of smart vending carts, which shed light on how these gadgets, embedded with the IoT, serve as change agents. Smart vending carts' networked capabilities provide businesses access to real-time data on stock levels, customer tastes, and operational efficiency, allowing them to make informed decisions. This investigation places special emphasis on the customer-centric strategy made possible by smart vending machines. These carts allow customers to have an interactive and personalized experience beyond just making purchases [11]. The IoT elements included into the carts make it possible to gather information about customers' tastes, habits, and buying habits. Retailers can now personalize their goods, boost consumer interaction, and provide a more immersive shopping experience with this data-driven strategy. There has been a shift in retail supply chain dynamics brought about in part by the interconnection of smart vending carts. Accurate inventory management, with fewer inefficiencies and fewer stockouts, is made possible by the real-time data supplied by IoT sensors. The end result is better supply chain performance and reduced costs. Beyond specific gadgets like smart vending carts, the IoT has the ability to revolutionize the retail industry as a whole. An efficient and streamlined shopping experience is the result of a larger ecosystem in which interdependent technologies work together. Because of how everything is linked, we can improve operations and make strategic decisions with the help of data analytics [13]. The impact of the IoT on the retail industry's future is becoming more apparent as the industry undergoes continuous change. IoT components are driving retail towards a more connected, intelligent, and consumer-centric future; smart vending carts are a microcosm of this trend. Embracing a paradigm change that centres company operations on data, connectivity, and responsiveness is the key to the competitive advantage afforded by IoT-driven solutions in retail [14]. The case study of smart vending carts sheds light on the far-reaching implications of the IoT for the retail sector. Connected, data-driven, and customer-centric retail ecosystems are replacing antiquated vending methods, and the incorporation of IoT components into these carts represents this transition. This case study shows how the IoT is more than just a technical improvement; it's a strategic tool for shops that want to be successful in the digital age. Smart vending carts' networked capabilities show how the IoT may revolutionize retail by improving operations, reimagining consumer experiences, and launching ground-breaking new products [6].

2. Validation of Vending Cart

2.1. *Temperature*

How long fruits and vegetables stay fresh in a vending machine after harvest depends on temperature, which in turn controls their respiration and metabolic rates. The rate of deterioration of product is directly affected by the association between room temperature and these rates. Lower temperatures slow down respiration, ripening, and senescence processes, extending the storage life of perishables, whereas higher respiration rates speed up degradation. Curing the spoilage-causing pathogenic fungus during storage is another benefit of colder temperatures [15]. Crops with high respiration rates have specific storage needs, which growers must address. Climate, plant part, harvest season, and crop maturity differences make it impossible to prescribe a universal cold storage guideline. Typically, higher temperatures (45 to 55°F) are more suited to warm-season crops, whereas lower temperatures (32 to 35°F) are more beneficial to cool-season crops [16]. Figure 44.1 shows the ideal temperatures for storage and Table 44.1 details the ideal relative humidity levels for different fruits and vegetables.

2.2. *Humidity*

The rate of transpiration, or water loss, of the vending cart's stored product is controlled by the relative humidity, which is a measure of the air's moisture content. In order to keep fruits and vegetables from wilting and softening while keeping their weight, appearance, nutritional value, and flavour intact, it is crucial to maintain a high relative humidity [17]. Transpiration rates are higher in young fruits, damaged food, and leafy greens because of their larger surface area to volume ratios. High temperatures, low relative humidity, and increased air velocity escalate transpiration rates, which are influenced by external elements including atmospheric pressure, air velocity, and temperature [18]. It is critical to precisely monitor and regulate the relative humidity. Instead of depending just on visual indicators, a hygrometer or sling psychrometer should be used for precise humidity evaluation. According to Figure 44.2, different fruits and vegetables require different amounts of relative humidity.

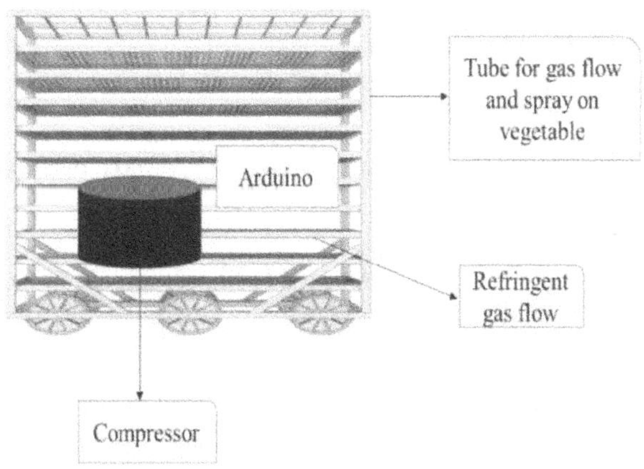

Figure 44.1. Vending cart.

Source: Author.

There are several approaches of controlling humidity:

- Running a humidifier in the warehouse.
- Controlling the ventilation and air circulation in relation to the storage room's load.
- Ensuring that the temperature of the refrigeration coils stays within 2°F of the ambient temperature in the storage room.
- Lining packing containers, insulation for storage rooms and transportation vehicles with moisture barriers.
- The storage room floor became wet.
- Using crushed ice to pack products for shipping.
- To control temperature, sprinkle water on young fruits, cool-season roots, and leafy greens.

2.3. *State of deterioration*

A motor, a Raspberry Pi camera, and a Raspberry Pi make for a complete system that can track the state of ripeness of produce in the vending machine. The Pi camera saves the picture to a place that is defined in the Python code. Following input from the machine learning model, the servo motor is instructed to transfer the fruit to the 'good' or 'rotten' basket based on the model's prediction of the fruit's class. Efficient sorting according to degradation state is guaranteed by this automated approach.

A 12V AH battery, which can store solar energy when it's sunny, is a part of the vending cart's renewable energy system. When solar radiation is low, the battery's stored energy drives the cooling system and its related components, keeping the storage space at a constant temperature. Electrical components, including as sensors and image processing, are powered by solar energy as well. In addition, a Peltier cooling system may be used to utilize the solar panel's cooling impact throughout the summer [19]. A temperature difference is created by transferring heat between two electrical connections in this device, which operates on the concept of the Peltier effect. The produced electric current reduces the impact of high temperatures by cooling the product.

A real-time web interface for showing results from image and sensor processing is a part of the system's overall architecture. In addition, a sprinkler system is included to water the crops and keep them at the ideal temperature. In order to keep the fruits and vegetables in the vending cart from going bad, the sprinklers are controlled by sensor outputs.

3. Internet of Things Integration with Vending Machines

IoT integration into vending carts is a technique that utilizes sophisticated technology to improve their usefulness and efficiency. It's a multi-faceted strategy.

3.1. *A: Sensors*

- When it comes to managing inventory and controlling temperatures, sensor technology is very crucial to the development of smart vending carts.

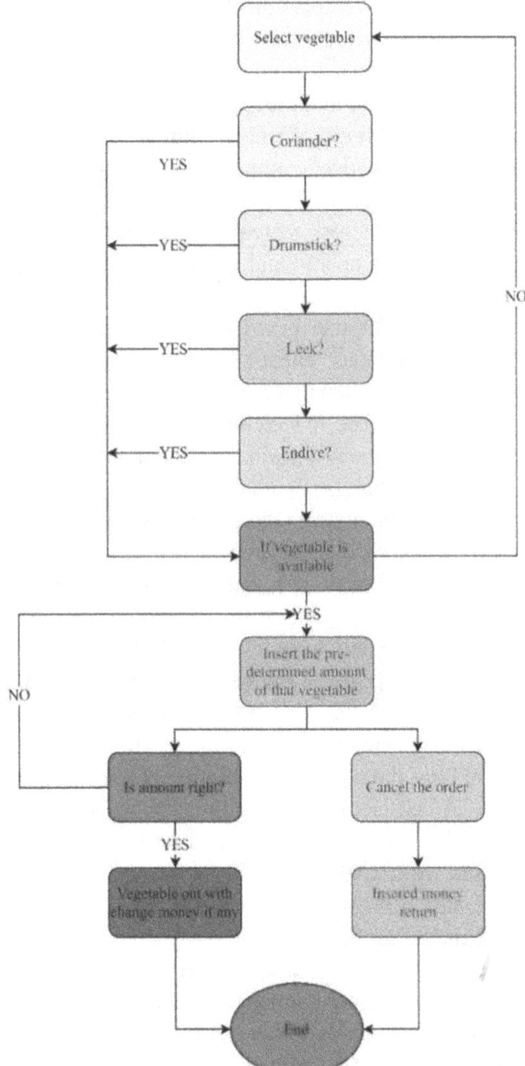

Figure 44.2. Flow chart of purchase order.

Source: Author.

- Inventory Management: Vendor carts that are connected to the internet of things include sensors that keep track of how much inventory is in stock. Automated stock level tracking is made possible by these sensors, which offer data on inventory status in real-time [20]. This optimizes the supply chain as a whole and guarantees a problem-free experience for customers by preventing stockouts and enabling proactive replenishment.
- Temperature Control: Smart vending carts rely on sensor technology, which is essential for controlling temperatures. Perishable items are kept at the ideal temperature thanks to sensors integrated in the cart that monitor the internal environmental conditions. Reduced waste and increased customer satisfaction are two outcomes of this capability's dual benefits: maintaining product quality and extending the shelf life of sensitive products.

3.2. *Methods of communication*

Robust communication networks allow for the transfer of data in real-time and the remote monitoring and management of IoT-enabled vending carts, greatly increasing their efficacy.

- *Transmission of Real-time Data:* IoT-enabled vending machines use communication technologies to send data to central servers in real-time. Information about sales, inventory, and machine health is part of this ongoing data stream. Retailers are able to make better decisions, replenish inventory quickly, and adapt to fluctuating market needs because to this real-time data transfer.
- The facility to oversee and manage vending carts from a distance is a significant benefit provided by the IoT. Operators may update product information, change price, and fix difficulties from a distance using networked communication systems that run the vending cart [21]. The ability to control the vending machine from a distance improves operating efficiency and decreases downtime, making the machine more responsive and smoother for users.

4. Vending Cart's Intelligent Functions

Automation, consumer engagement, and safety are three areas where the IoT is reshaping the conventional vending experience with its many smart features.

4.1. *Automated sales transactions*

Automated sales transactions are a notable feature of vending carts that are connected to the IoT. These carts allow for easy and cashless transactions using integrated payment systems and sensors. Several payment options, including contactless cards and smartphone payments, allow customers to make transactions easily [10, 22]. This improves customer comfort and makes transactions easier overall, which in turn cuts down on wait times and helps stores run more smoothly [12].

4.2. *Engaging with and interacting with customers*

Smart vending carts use the IoT to completely change the way customers engage with brands. These carts give a fun and interesting way to buy with features like touchscreens, interactive displays, and algorithms that tailor recommendations to each customer. Personalized promotions, in-depth product information, and loyalty programme participation are all accessible to customers through the vending cart interface. The increased degree of contact not only meets the changing expectations of consumers, but it also fosters brand loyalty by encouraging deeper involvement with the company.

4.3. *Protection against theft and other security risks*

IoT integration in vending carts solves the problem of retail security by implementing anti-theft and sophisticated security features. Equipped with sensors and surveillance systems, smart vending carts keep tabs on both their surrounds and the cart itself. These measures are designed to immediately notify you of any unauthorized access or tampering. Secure locking mechanisms and GPS monitoring are examples of anti-theft systems that provide an additional degree of safety [23]. Using these safety features, vending carts connected to the IoT help keep expensive goods secure and the retail environment running smoothly.

5. Design and Implementation

In order to keep everything running even when it's cloudy or dark outside, the system relies on a 12V AH battery, which stores solar energy. The battery powers the vegetable vending cart's cooling system and related components when there isn't enough solar radiation. Not only is solar power used to keep storage at the ideal temperature, but it is also used

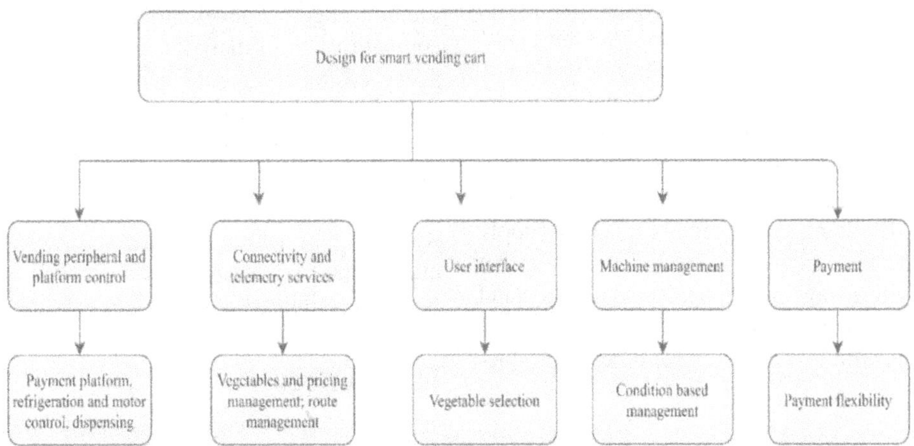

Figure 44.3. Design of smart vending cart.

Source: Author.

to power essential electrical components like sensors, which allow for effective picture processing [24]. The solar panel uses the Peltier effect and a Peltier cooling mechanism to create a cooling effect throughout the summer. By passing a current across the electrical connections between two conductors, this system cools the surrounding air by establishing a temperature difference.

The IoT vegetable vending cart's block diagram, shown in Figure 44.3, lays out the system architecture.

An inexpensive and straightforward digital temperature and humidity sensor is used to monitor the cart's surroundings. This sensor uses a thermistor and a capacitive humidity sensor to produce a digital signal on the data pin, which gives information about the humidity and temperature of the surrounding air. The gas sensor is an additional important tool for controlling the ripening process of fruits and vegetables by detecting the ethylene gas they release. Users are empowered to shoot high-definition video and photos using the centralized controller's dedicated camera input port and its ability to handle digital images. Python is used for image processing after capturing the images with the 5MP Pi camera v1.3 [25].

A dynamic internet portal displays real-time readings of sensors and image processing outputs, and the entire system is powered sustainably by solar energy.

An integrated water sprinkler system cools and dries the fruits and vegetables, preventing dust and ensuring they are well hydrated. The sprinklers, which may be thought of as a kind of controlled rainfall, are essentially main pipelines connected to a network of tangential pipes that have revolving nozzles. To keep the stored food from going bad, sensors are essential for controlling the sprinklers and maintaining the ideal temperature within the cart.

6. Benefits of Internet of Things-Enhanced Vending Machines

A host of benefits, including improved operational efficiency, data-driven decision-making for company owners, and enhanced consumer experiences, accrue from vending carts that use IoT technologies.

A. The IoT dramatically improves operational efficiency and productivity, which is a major benefit of vending carts. Automated replenishment systems and real-time inventory monitoring make these carts more efficient and less labor-intensive to reload. By doing so, we can optimize the supply chain and reduce the chances of stockouts, guaranteeing that the vending cart will always have popular and in-demand goods. The final result is a vending machine company that runs more smoothly, with less downtime and better productivity overall.

6.1. B. A Better Experience for Customers

The IoT brings a new level of involvement and convenience to the vending cart experience, revolutionizing it for customers. Interactive displays, personalized recommendations, and quicker transactions provide a shopping experience that is both easy and entertaining for customers. Customers are more satisfied and loyal when they can make cashless purchases, get comprehensive product information, and receive tailored incentives. A pleasant and unforgettable shopping experience is fostered by vending carts that are IoT enabled, which not only meet but beyond the changing expectations of modern customers.

6.2. C. business owners' data-driven decisions

Integrating IoT devices gives company owners access to analytics and data that may help them make better decisions. Insights on consumer tastes, best-selling items, and machine efficiency are gleaned from the constant stream of real-time data generated by vending machines. Product options, pricing strategies, and marketing campaigns may all be fine-tuned with the use of this data. A more nimble and flexible company strategy that anticipates and capitalizes on customer needs and trends is the ultimate product. Business owners may maximize profits and remain ahead in a competitive retail scene by embracing data-driven decision-making and strategically positioning their vending carts.

7. Conclusion

Finally, the IoT in vending carts is changing the retail industry by bringing cutting-edge technology into an age where traditional vending machines are becoming obsolete. IoT vending carts have been a game-changer for the retail industry. The IoT is revolutionizing the vending sector in every way. From inventory management that relies on sensors to features that engage customers and strong security measures, it's a game-changer. In addition to improving operational efficiency, the IoT offers new opportunities for personalization, data-driven decision-making, and contact with customers. Promoting the broad use and modification of vending carts that are connected to the IoT is essential to the retail sector's ongoing development. Shops that want to remain competitive and satisfy the ever-changing demands of tech-savvy customers should get on the IoT bandwagon as its advantages become more and more apparent. With the help of data-driven insights, greater customer experiences, and increased productivity, merchants can stay ahead of the curve in the ever-changing retail industry with the use of the IoT. Retailers may take a strategic step towards a future that is more responsive, customer-centric, and efficient by embracing the possibilities of the IoT in vending carts.

References

[1] Sonawane, S. T. (2017). Problems and Solutions of Vendors–A Case Study. *International Journal of Innovative Research in Science, Engineering and Technology*, *6*(1), 940–943.

[2] Kumar, A., Joshi, P., Bala, A., Sudhakar Patil, P., Jang Bahadur Saini, D. K., & Joshi, K. (2023). Smart Transaction

through an ATM Machine using Face Recognition. *Indian Journal of Information Sources and Services*, *13*(2), 7–13.

[3] Chaturvedi, R., & Singh, P. K. (2021). A practicable learning under conversion of plastic waste and building material waste keen on concrete tiles. *Materials Today: Proceedings*, *45*, 2938–2942.

[4] Sushma, S., Mani, R., Perumalraja, R., Vasanthan, R., & Mohamed, A. (2024). Accounting information systems for strategic management: The role of intellectual capital in mediating the relationship between customer, company, and performance. *Indian Journal of Information Sources and Services*, *14*(2), 160–166. https://doi.org/10.51983/ijiss-2024.14.2.23

[5] Vala, K. V., Saiyed, F., & Joshi, D. C. (2014). Evaporative cooled storage structures: an Indian Scenario. *Trends in post harvest Technology*, *2*(3), 22–32.

[6] Wong, S. K., & Yiu, S. M. (2020). Identification of device motion status via Bluetooth discovery. *Journal of Internet Services and Information Security*, *10*(4), 59–69.

[7] Chaturvedi, R., Singh, P. K., & Sharma, V. K. (2021). Analysis and the impact of polypropylene fiber and steel on reinforced concrete. *Materials Today: Proceedings*, *45*, 2755–2758.

[8] Liberty, J. T., Ugwuishiwu, B. O., Pukuma, S. A., & Odo, C. E. (2013). Principles and application of evaporative cooling systems for fruits and vegetables preservation. *International Journal of Current Engineering and Technology*, *3*(3), 1000–1006.

[9] Hai, T., Chaturvedi, R., Mostafa, L., Kh, T. I., Soliman, N. F., & El-Shafai, W. (2024). Designing g-C3N4/ZnCo2O4 nanocoposite as a promising photocatalyst for photodegradation of MB under visible-light excitation: response surface methodology (RSM) optimization and modeling. *Journal of Physics and Chemistry of Solids*, *185*, 111747.

[10] Fuw, Y. Y., Zhen, W. L., & Su, H. C. (2011). Mobile banking payment system. *Journal of Wireless Mobile Networks, Ubiquitous Computing, and Dependable Applications*, *2*(3), 85–95.

[11] Xiu-Rui, X., Ji-Kai, N., Xin-Yuan, B., Chen-Guang, P. (2018). Development status of cold storage and energy saving and environmental protection [J]. *Journal of Appliance Science Technology*, *2018*(03), 22–23.

[12] Takashi, N., Shingo, M., & Kouichi, S. (2012). Security analysis of online E-cash systems with malicious insider. *Journal of Wireless Mobile Networks, Ubiquitous Computing, and Dependable Applications*, *3*(1/2), 55–71.

[13] Potdukhe, P., Mantriwar, I. P., Punekar, N. S., Lodhe, V. R., Sharnagat, C. G., Mogare, R. M., & Shamkule, P. A. (2018). Solar assisted vegetable cart. *International Research Journal of Engineering and Technology (IRJET)*, *05*(06).

[14] Sun, J., Yan, G., Abed, A. M., Sharma, A., Gangadevi, R., Eldin, S. M., & Taghavi, M. (2022). Evaluation and optimization of a new energy cycle based on geothermal wells, liquefied natural gas and solar thermal energy. *Process Safety and Environmental Protection*, *168*, 544–557.

[15] Camargo, J. R., Ebinuma, C. D., & Cardoso, S. (2004). Three methods to evaluate the use of evaporative cooling for human thermal omofort. *Proceeding of 10th Brazilian congress of thermal science and engineering—ENCIT 2004*, Braz. Soc. of Mechanical science and engineering ABCM, Rio de Janeiro, Brazil, Nov 29–Dec 03.

[16] Sharma, A., Sharma, K., Islam, A., & Roy, D. (2020). Effect of welding parameters on automated robotic arc welding process. *Materials Today: Proceedings*, *26*, 2363–2367.

[17] Ho, C. K., Robinson, A., Miller, D. R., & Davis, M. J. (2005). Overview of sensors and needs for environmental monitoring. *Sensors*, *5*(1), 4–37.

[18] Li, Z. G., Liu, Y., Dong, J. G., Xu, R.-J., & Zhu, M.-Z. (1983). Effect of low oxygen and high carbon dioxide on the levels of ethylene and 1-aminocyclopropane-1-carboxylic acid in ripening apple fruits. *J Plant Growth Regul*, *2*, 81–87.

[19] Valente, J., Almeida, R., & Kooistra, L. (2019). A comprehensive study of the potential application of flying ethylene-sensitive sensors for ripeness detection in apple orchards. *Sensors*, *19*, 372.

[20] Hai, T., Ali, M. A., Dhahad, H. A., Alizadeh, A. A., Sharma, A., Almojil, S. F., ... & Wang, D. (2023). Optimal design and transient simulation next to environmental consideration of net-zero energy buildings with green hydrogen production and energy storage system. *Fuel*, *336*, 127126.

[21] He, K., Zhang, X., Ren, S., & Sun, J. (2016). Deep residual learning for image recognition. In *Proceedings of the IEEE Conference on Computer Vision and Pattern Recognition*. Washington, DC, USA, 27–30 June 2016; pp. 770–778.

[22] Kumar, A., & Gill, G. S. (2015, May). Automatic fruit grading and classification system using computer vision: A review. In *2015 Second International Conference on Advances in Computing and Communication Engineering* (pp. 598–603). IEEE.

[23] Ishangulyyev, R., Kim, S., & Lee, S. H. (2019). Understanding food loss and waste—why are we losing and wasting food?. *Foods*, *8*(8), 297.23. Cicatiello, C., Franco, S., Pancino, B., Blasi, E., & Falasconi, L. (2017). The dark side of retail food waste: Evidences from in-store data. *Resources, Conservation and Recycling*, *125*, 273–281.

[24] Sharma, S., Jain, K. K., & Sharma, A. (2015). Solar cells: in research and applications—a review. *Materials Sciences and Applications*, *6*(12), 1145–1155.

[25] Naik, A. (2013). Contextualising urban livelihoods: Street vending in India. *Available at SSRN 2238589*.

[26] Jadhav, R. S., & Patil, S. S. (2013). A fruit quality management system based on image processing. *IOSR Journal of Electronics and Communication Engineering (IOSR-JECE)*, *8*(6), 01–05.

[27] Ndukwu, M. C., & Manuwa, S. I. (2014). Review of research and application of evaporative cooling in preservation of fresh agricultural produce. *International Journal of Agricultural and Biological Engineering*, *7*(5), 85–102.

[28] Yan, J., Yan, Z., Bao-Guo, L., & Ze-Zhao, H. (2001). Study on the influence of temperature and humidity changes on the storage quality of fruits and vegetables in refrigerators [J]. *Low Temperature and Specialty Gases*, *2001*(06), 18-21+30.

[29] Chaturvedi, R., Sharma, A., Sharma, K., & Saraswat, M. (2022). Nanotech science as well as its multifunctional implementations. *Recent Trends in Industrial and Production Engineering: Select Proceedings of ICCEMME, 2021*, 217–228.

45 Exploration of blockchain integrated IoT devices in healthcare application for secured remote patient monitoring

Varsha P. Hotur[1,a], *Ramani U.*[2,b], *Menakadevi N.*[3,c], *R. Krishna Kumar*[4,d], *Kottaimalai R.*[5,e], *and Thilagaraj M.*[6,f]

[1]Department of Electronics and Communication Engineering, MVJ College of Engineering, Bengaluru, India
[2]Department of Electrical and Electronics Engineering, K. Ramakrishnan College of Engineering, Tiruchirappalli, India
[3]Department of Electronics and Communication Engineering, Karpagam Institute of Technology, Coimbatore, India
[4]Department of Electrical and Electronics Engineering, Karpagam College of Engineering, Coimbatore, Tamil Nadu, India.
[5]Department of Electronics and Communication Engineering, Kalasalingam Academy of Research and Education, Krishnankoil, Tamil Nadu, India.
[6]Department of Industrial Internet of Things, MVJ College of Engineering, Bangalore, Karnataka, India

Abstract: Blockchain (BC) is an emerging technology that has been utilized to foster innovation store and share patient information among clinics, labs, pharmacies, and doctors, the medical sector uses a distributed ledger called blockchain. Blockchain-based software can reliably detect serious and potentially harmful errors, including potentially harmful ones, in the clinical domain. As a result, it can improve the efficiency, safety, and transparency of health-related information exchange within the medical industry. Healthcare providers can improve the assessment of medical information and obtain new insights with the use of BC. In clinical investigations, BC plays a critical role in preventing fraud; in this context, the technology's promise to enhance the efficiency of medical information management. It facilitates a distinct storage of information arrangement at the most advanced level of safety and may assist in allay concerns about manipulating information in the medical field. It offers adaptability, connectivity, responsibility, and access to information authorization. Following established clinical protocols and guidelines ensures that healthcare providers deliver consistent and reliable care based on evidence-based practices. Adhering to data privacy regulations ensures patient data is protected from unauthorized access, breaches, or misuse. Medical information have to be preserved private and secure for various reasons. BC mitigates risks and aids in the decentralized safeguarding of medical data. The information pertaining to patients' administration system is better to keep up with the use of the IoT in the medical field. BC technology can help with confidential information transfers among clinics and other locations by interacting with the IoT.

Keywords: Blockchain technology (BT), internet of things (IoT), healthcare, patient data management system, secured transaction, diagnosis

1. Introduction

Blockchain (BC) is an accessible to everyone, decentralized digital database that keeps track of events across multiple devices. This ensures that no information may be changed backwards without changing any subsequent blocks. BC creates a long series of transactions by verifying and connecting to the previous block [1]. BC offers a high degree of responsibility because all activity is recorded and verified transparently. All data submitted to the BC cannot be changed by anyone. It does this by proving that the information is real and unaltered. BC improves reliability by storing information across networks rather than in a single database, exposing its vulnerability to hacking. BC provides an excellent platform for the development of innovative and advanced business strategies that may contend with those of conventional enterprises [2].

BC aids marketers in keeping track of medical product usage. BC technology will be used by the medical and drug sectors to eradicate fake medications while making it possible to track down all of these treatments. It aids in identifying the source of fabrication [3]. Information about patients are

[a]varsha.hotur@gmail.com, [b]ramani.u6@gmail.com, [c]menakadevin@gmail.com, [d]kumaran23011991@gmail.com, [e]r.kottaimalai@klu.ac.in, [f]m.thilagaraj@gmail.com

DOI: 10.1201/9781003675259-45

able to be kept secret thanks to BC technology, which may additionally safeguard health information in an immutable format. Utilizing the resources conserved by those devices, investigators enable the computation of predictions for treatments, medications, and solutions for various diseases and ailments.

Every user of the network are able to utilize and rely on the information because it does not reside in a centralized location thanks to the decentralized BC ledger structure. Because there is no chance of a single threat, the distributed network is strengthened and secured [4]. By reducing healthcare administration surveillance by double, it helps improve management of medical data and treatment for individuals while preserving both money and time for patients as well as doctors. In order to monitor the use of their data, patients will maintain their medical information on a BC.

With the aid of this BC, researchers can examine a vast amount of previously undiscovered information regarding a certain population. Suitability for long-term study is important for the progress of accurate medicine. By the aid of wearable devices and IoT, we can employ BC in real-time healthcare for storing and updating essential patient information like BP and glucose levels (Kayikci and Khoshgoftaar). It assists physicians in keeping tabs on patients with elevated risks and, in the event of an emergency, in advising and notifying their loved ones and employers. Because of its decentralized architecture, BC may be securely hacked without jeopardizing any particular duplicate of the information.

2. Block Chain Technology

The information is stored on a decentralized network called blockchain. It is an outstanding method of safeguarding private information inside the network. Important information can be exchanged while maintaining security and confidentiality thanks to this innovation. It is the ideal solution for safely keeping all the necessary records in one place [6]. BC also expedites the process of searching one individual's information for candidates who meet particular trial requirements. A decentralized peer-to-peer (P2P) network of individual computers, or nodes, that contends, saves, and archives transaction or previous information is what the BC is. Because all network participants communicate and preserve data, it facilitates dependable teamwork and maintains a continuous log of both previous and present events.

The fundamental processes of BC technology are depicted in Figure 45.1.

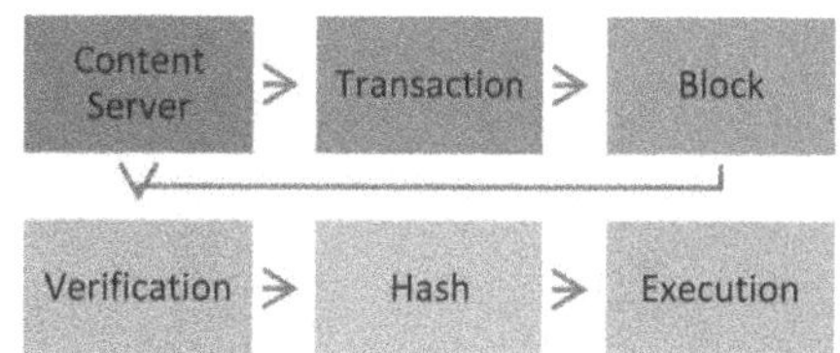

Figure 45.1. Process of BC.

Source: Author.

2.1. Importance of blockchain in healthcare

Till now, exchange of information, compatibility, and safeguarding have been the biggest issues with managing the health of populations [7]. BC ensures the reliability of this specific issue. If used properly, this BC improves data interchange, safety, credibility, compatibility, and real-time updating and accessibility [8]. Data security is another major problem, particularly with regard to wearables and customized healthcare. In order to address these problems, BC is used to capture, transmit, and examine information across networks in an easy, secure manner for both patients and healthcare providers.

2.2. Blockchain technology features to promote global healthcare environment

The BC has several uses and purposes in the medical field. By controlling the medicine distribution system, enabling secure communication of medical information from patients, and easing the exchange of patient health documents, ledger technology aids in the discovery of genetic information by medical experts [9].

Figure 45.2 illustrates the range of characteristics and key facilitators of the BC idea in numerous medical industries and related fields. Cryptocurrency safety, genetics administration, electronic information management, medical information compatibility, electronic surveillance and issue epidemic, and more are among the stunning and extremely advanced characteristics used in the development and application of BC. The primary drivers of BC's popularity are its fully digitalized nature and its uses in the medical field.

2.3. Blockchain healthcare companies

Numerous related supporters or suppliers of health-related services have contributed to the studies and inquiries needed to use BC in medical and its fundamental fields [10]. The suppliers illustrated in Figure 45.3 are among the select few

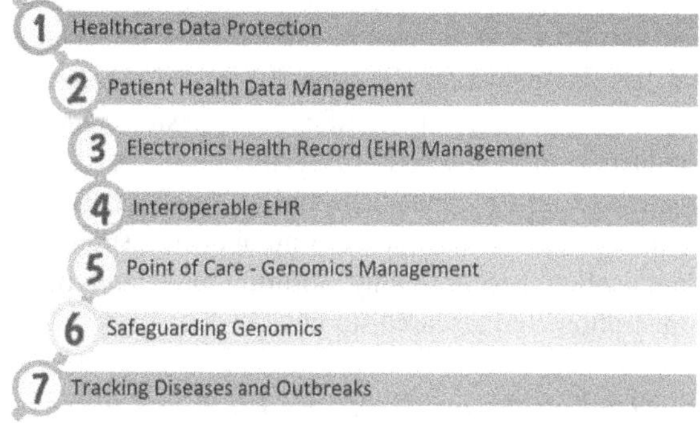

Figure 45.2. Key facilities of BC.

Source: Author.

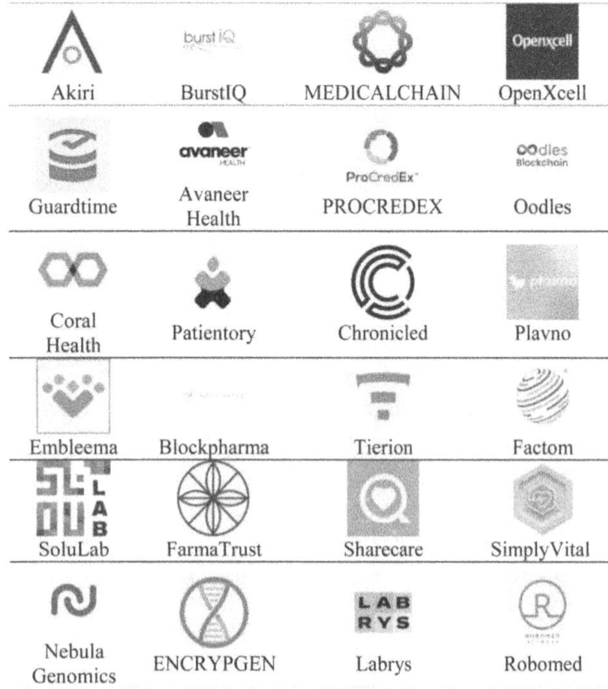

Figure 45.3. BC healthcare companies.

Source: Author.

that offer and encourage the application of BC in real-world settings.

BC entails creating fresh medical information sheets for physicians working in different clinics. Most recently introduced material is repetitious and causes wasting of time, that is a serious medical problem. Depending on where they are in the manufacturing chain, everyone may have varying rights or options for access [5]. Furthermore, each block that had the medicine details would also have an encryption key linked to it from another block [11]. BC makes it simple to envision the way forward when it comes to medical experts' procedures and offerings. BC works in the medical field to effectively handle consent for processing and acquisition. With BC, it's easy to streamline process and increase productivity by avoiding waste by waiting in position. With the use of this technology, we hope to promote medical guidance, personalized therapy, and useful studies on health. Currently, one of the most widely used technologies is BC. The most recent development regarding the BC is the organization's belief that its system can significantly transform the medical industry. In numerous ways, this will transform the medical field into a dependable online database. BC-based medical solutions will improve a number of issues, including medication accountability, information about patient's administration, and research studies.

2.4. Blockchain applications for healthcare

BC is an emerging and rising concept with creative uses in its efficient implementation to the medical field (Table 45.1). Rapid and seamless information transfer and exchange throughout all of the major players in the network and medical professionals helps establish affordable and cutting-edge therapies for a wide range of illnesses. It will encourage the medical services industry's expansion in the years to come [12]. The benefits of BC for hospitals are demonstrated by the newly disclosed prospects it offers the logistics sector. BC is also getting increasingly used, mostly in the banking sector. It presents the medical field with a number of significant and outstanding opportunities, ranging from research and transportation to practitioner-patient connections [13].

The process flow for BC with respect to healthcare application is listed below.

1. Diagnosis and Initial Assessment
2. Consultation and Treatment Planning
3. Preoperative Preparation
4. Surgery-Breast-Conserving Therapy (BCT)
5. Postoperative Care
6. Adjuvant Therapy (if indicated)
7. Follow-Up and Surveillance
8. Supportive Care
9. Survivorship Care

3. IoT in healthcare

Patients can now arrange consultations via online medical apps, eliminating the requirement to make a phone call and await for an operator when making a scheduled visit at the hospital. Globally, there is expected to be over two times as many internet of things (IoT) smartphone connections in 2026 as there were in 2021. IoT is set to alter the manner in which people inhabit our lives and working [14]. Considering the medical field, wherein the use of paper and pen has served as the main way of records for patients keeping for many years, that kind of instability is clearly visible. Still, there have recently been significant changes in medical systems. Patients are able to make visits via online health apps, eliminating the necessity to make a phone call and await for an operator when making a reservation at a physician's office. Owing to apps, medical professionals may now bring data along wherever they travel. And there are no indications that this growing connectedness will stop. Actually, it's only getting faster. The worldwide internet of medical things (IoMT) industry is likely achieve \$187.60 billion by 2028, which is over four times its value in 2020 at \$41.17 billion, as reported by Fortune Business Insights. Medical professionals have the ability to reach a wider audience beyond the typical medical environment with the help of the IoT. In order to avoid needless and expensive excursions to a medical clinic, remote surveillance devices enable clinicians and their patients to stay informed about the condition of a patient even when they are not present. Remote patient monitoring system is another IoT technique that US health care facilities are using to enhance results and decrease expenses [15]. This kind of healthcare uses IoT enabled sensors and devices to

Table 45.1. Blockchain applications

Applications	*Description*
Save a patient's specific data.	Large amounts of patient and medical information are collected prior to and throughout the various clinical research sessions. When medical professionals access the information that has been saved, they may have doubts about its veracity. They can easily verify this by comparing it to the actual documents kept on the BC platform. BC is built on top of currently used encryption technologies, such as the suitable cryptography foundation for information exchange. The individual's identity, time of birth and evaluation, therapies, and mobility status are entered in EHR format during the patient's visit by the medical practitioner.
Examine the results of a specific process.	With validated permission to use patient information, experts can analyze any given method on a sizable portion of the patient group with effectiveness. This yields noteworthy outcomes that improve the way certain patient groups are managed. Drug manufacturers will be able to collect information in real-time and provide an extensive selection of specifically designed medication or services for patients once the BC architecture has been put in effect. Because BC contains all the information on the surface of it, chemists' jobs are made easier.
Validation	Medical organizations, innovative technology startups, and the sector as a whole are searching for ways to learn what they can do today to contribute to making medical field more affordable and safe. Once hospital administration is able to sufficiently authenticate the outcomes, BC has the potential to revolutionize the medical environment.
Security and Openness	It allows doctors to spend a greater amount of time treating patients despite offering outstanding security and openness. Additionally, it could provide the support of research studies and remedies to treat every uncommon illness. Easy sharing of information across healthcare service providers may promote accurate diagnosis, successful treatment, and financially sound ecosystems within a healthcare organization. BC improves security and openness by allowing different healthcare ecosystem organizations to stay connected and share data about a widely dispersed platform.
Maintaining health records	BC has the potential to be ideal for keeping healthcare records. Usage for this comprise financial management, handling administrative duties, electronic medical record maintaining, and information exchange. Through a smartphone application, individuals may transmit medical data to a BC network. Blockchain-based electronic agreements enable sensor and smart device integration. EHRs are typically dispersed among several healthcare facilities. BC can bring all the information together and offer patients accessibility to the past.
Medical Investigation	BC is being utilized in medical research to handle issues with information fragmentation and inaccurate outcomes that are inconsistent with the intended objectives and goals of the study. BC will make clinical studies more trustworthy. The system for company evaluation looks into the changing dynamics of the market to help those in the medical field grasp the opportunities.
Display information	BC approach will show details concerning the drug's provenance to guarantee its high standard as well as that the company who authorized it delivers it. Highly sensitive information is more secure nowadays thanks to BC, if it is utilized properly.
Recognizing fraudulent content	BC will improve readability and make it easier to spot fraudulent information. Validating medical studies for customers and volunteers ought to remain simple. For gaining authorization and maintain widely confirmed, public access to deployed standard documents and their conclusions, a smart contract is the best option.
Minimizes unnecessary administrative costs	BC lowers needless administrative expenses, enabling appropriate use of medical data. This invention will also reduce the need for many intermediaries to supervise the sharing of critical health data. Consequently, it might be simpler for healthcare professionals to carry out their crucial duty of providing individuals with timely, suitable, efficient help.
Monitoring of patients	Healthcare workers can make certain they've got a connection to healthcare devices whenever they require it because of BC. Additionally, it can take longer for clinicians to observe patients and respond remotely to health-related events. It is possible to improve supply accessibility, bed usage, and temperature tracking in rooms for patients using BC in the healthcare industry.
Establish investigations for research	The time-consuming procedure of resolving disputes and disagreements among individuals could be completely transformed by BC. With the increased sharing of information about patients, BCs can inspire creative and innovative projects for research. In addition, an expanded exchange of patient findings will inspire creative and innovative research, leading to an amazing collaboration between subjects and researchers.
Keep hospital statements of operations up to date	For the accounting procedure, it is essential for maintaining accurate documentation of the statements of finances. The research studies are appropriate for effective execution and evaluation. In this case, BC firms have developed techniques to expedite the financial reporting and accounting procedure. Everybody may utilize this tool to fill out the form ahead of time and get ready to visit a physician.
Enhances security	BC solves issues with medicine authenticity and prescription tracking, promotes secure compatibility, and improves comprehensive security for patients. It is the simplest way to get rid of the present supply-chain governance structure and stop producers of fake medications from releasing safer versions of their products onto the market. BC could render centralized storage of information possible.
Reduce data transmission expenses and time.	BC reduce the expense and duration of information transformation. It has the potential to quickly and efficiently resolve the problem with healthcare identity authentication. It ensures individual privacy and security. It will lead to important novel concepts and discoveries that have the potential to transform the provision of healthcare worldwide.

Source: Author.

give clinicians access to a constant flow of real-time wellness data, including blood pressure, blood sugar levels, and cardiac activity. Following a worldwide epidemic, it rise as long as there is a continued need for comfort and remote medical treatment. By 2025, there will likely be 70.6 million RPM users in the US, a 56.5% increase from 2022. Within three years from now, over 25% of Americans are expected to be constantly utilizing a gadget that wirelessly monitors or gathers health information for evaluation by their physicians. While technological firms see an increasing chance within the profitable digital wellness sector, wearables like the Apple Watch are becoming beyond mere workout trackers; they are beginning to operate much like healthcare devices (Figure 45.4).

The COVID-19 epidemic brought to light the difficulties associated with restricted availability of medical services. Many patients looked for distant healthcare choices because they were having trouble getting in touch with their physicians. Through enabling medical personnel to remotely observe patients, IoT-enabled medical devices significantly contributed to closing the availability deficit in medicine [16]. IoT's impact on medicine has expanded quickly. Application scenarios for IoT related to healthcare remain widespread across the globe, ranging from enhanced equipment management to remote treatment. By quickly diagnosing illnesses and matching individuals to the best possible treatments, this state-of-the-art technology could potentially preserve life.

The following are a few advantages of IoT in the medical field:

- Enhanced patient care and diagnosis due to data-collection and analysis through IoT devices.
- Improved patient happiness and involvement with individualized prompt service.
- Assistance for remote surveillance of persistent illnesses via integrated gadgets that warn clinicians and patients.
- Utilizing IoT devices to monitor geolocation and critical indicators to improve care for patients and physician time to respond.

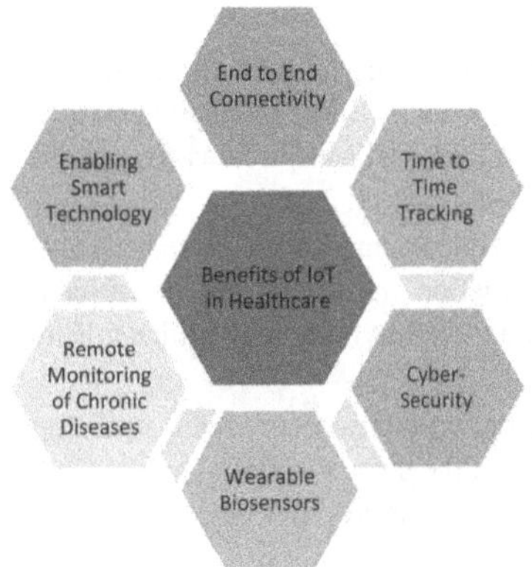

Figure 45.4. Benefits of IoT in healthcare.

Source: Author.

- Human surveillance and adherence using IoT that assure quality as well as safety requirements.
- Development of preventative treatment with IoT data which assists detect warning signs and avert consequences.
- Utilize IoT devices that maximize assets and minimize inefficiencies to save expenses whilst enhancing patient care experiences.

4. Conclusion

Studies on patients gain legitimacy and insights from BC. These documents can be kept on the BC as smart contracts within the digital fingerprint. Some of the advantages of BC for medical purposes include are standardized procedures for authorization for entering electronic medical data, identification authenticity and verification for every individuals, and network architecture protection across all stages. The administration of prescription obligations and the surveillance of the medicinal distribution system are maintained via a BC. Because this BC can store patient data down to the person level, it can be used to analyze and verify the results of certain procedures. BC is utilized to facilitate patient surveillance, clinical study management, medical records maintaining, quality enhancement, data exhibit, and openness. It covers multiple issues in the context of the data-centered world. Each block of medical information about patients will have a hash produced by BC. Additionally, the BC approach could motivate patients to provide external parties with the necessary information yet preserve the privacy of their identities. A research study requires numerous sets of information to be completed. The focus of the study is on these information sets, and regular trials are conducted to offer evaluations, estimations, and effectiveness proportions in different conditions. After the information are analyzed, additional choices are taken in light of the results. Numerous researchers, still are capable of manipulating the information collected to change the outcome. The information acquired can enhance patient care and offer post-market research to maximize effectiveness. These norms are anchored in important elements of BC including transparent administration, open inspection trails, openness to information, durability, and increased confidentiality and safety. This enables medical professionals to adhere to the most recent medical regulations, which cover security of the drug delivery. The advancement in technology will enhance its applicability to medical care by providing an explanation of the results of treatment and development. The foundation of data flows and verification of transactions is BC. In the days to come, BC will allow operations to be recorded, documented and verified with the permission of network participants. BC can serve as the cornerstone of an entirely novel phase of wellness data interchange by offering statistical protection at the individual stage through both private and public key cryptography. Medical documents, copyright mitigation, improved compatibility, process rationalization, drug and script management, and medicinal and supply network surveillance are all promised by BC. BC is expected to do incredibly well in the medical sector in coming years.

BC has the potential to revolutionize a variety of medical professions. Smart agreements reduce expenses by cutting out middlemen from the entire payment process. The possibility of BC in the medical sector is heavily reliant on the ecosystem's acceptance of related cutting-edge technology. Medical studies, medication tracking, medical coverage, and network monitoring are all included. Clinics may utilize device monitoring to map their facilities utilizing a BC architecture, regardless of the course of every phase of their life. BC has the potential to enhance medical record administration, particularly in the areas of surveillance and insurers arbitration, which will speed up medical processes while maintaining optimal data quality.

5. Acknowledgement

The authors thank the International Research Centre of Kalasalingam Academy of Research and Education, Tamil Nadu, India, for permitting the use of the computational facilities available in the Centre for Biomedical Research and Diagnostic Techniques Development.

References

[1] Ghosh, S., Dave, V., & Keerthana, S. S. (2024). A critique of blockchain in healthcare sector. In *Artificial Intelligence, Big Data, Blockchain and 5G for the Digital Transformation of the Healthcare Industry* (pp. 205–231). Academic Press.

[2] Lin, Q., Li, X., Cai, K., Prakash, M., & Paulraj, D. (2024). Secure Internet of medical Things (IoMT) based on ECMQV-MAC authentication protocol and EKMC-SCP blockchain networking. *Information Sciences*, *654*, 119783.

[3] Mohammed, M. A., Lakhan, A., Zebari, D. A., Abd Ghani, M. K., Marhoon, H. A., Abdulkareem, K. H., Nedoma, J., & Martinek, R. (2024). Securing healthcare data in industrial cyber-physical systems using combining deep learning and blockchain technology. *Engineering Applications of Artificial Intelligence*, *129*, 107612.

[4] Masood, I., Daud, A., Wang, Y., Banjar, A., & Alharbey, R. (2024). A blockchain-based system for patient data privacy and security. *Multimedia Tools and Applications*, 1–25.

[5] Ramaraj, K., Murugan, P. R., Amiya, G., Govindaraj, V., Vasudevan, M., Thirumurugan, M., Zhang, Y., Abdullah, S.S., & Thiyagarajan, A. (2023). Emphatic information on bone mineral loss using quantitative ultrasound sonometer for expeditious prediction of osteoporosis. *Scientific Reports*, *13*(1), 19407.

[6] Sathya, R., Thilagaraj, M., & Kottaimalai, R. (2023, November). IoT Based Intelligent Warehouse Monitoring and Alerting System. In *2023 International Conference on Sustainable Communication Networks and Application (ICSCNA)* (pp. 402–406). IEEE.

[7] Islam, I., & Islam, M. N. (2024). A blockchain based medicine production and distribution framework to prevent medicine counterfeit. *Journal of King Saud University-Computer and Information Sciences*, *36*(1), 101851.

[8] Almalki, J., Alshahrani, S. M., & Khan, N. A. (2024). A comprehensive secure system enabling healthcare 5.0 using federated learning, intrusion detection and blockchain. *PeerJ Computer Science*, *10*, e1778.

[9] Subramani, J., Maria, A., Aljaedi, A., Rajasekaran, A. S., Bassfar, Z., & Jamal, S. S. (2024). Blockchain-enabled secure data collection scheme for fog-based WBAN. *IEEE Access*.

[10] Rai, T., Malviya, R., Kaushik, N., & Sharma, P. K. (2024). Blockchain in Tracing and Securing Medical Supplies. In *Blockchain for Healthcare 4.0* (pp. 185–198). CRC Press.

[11] Kakkar, B., & Johri, P. (2024). Technological entrepreneurship in healthcare management to achieve sustainability post-COVID pandemic. *World Review of Entrepreneurship, Management and Sustainable Development*, *20*(1), 58–84.

[12] Ghosh, S., Dave, V., & Keerthana, S. S. (2024). A critique of blockchain in healthcare sector. In *Artificial Intelligence, Big Data, Blockchain and 5G for the Digital Transformation of the Healthcare Industry* (pp. 205–231). Academic Press.

[13] Thilagaraj, M., Francis, G. A., Manikandan, S., & Ramaraj, K. (2023, December). IoT-based Cable Fault Detector with GSM and GPS Module using Arduino. In *2023 3rd International Conference on Innovative Mechanisms for Industry Applications (ICIMIA)* (pp. 72–77). IEEE.

[14] Lin, Q., Li, X., Cai, K., Prakash, M., & Paulraj, D. (2024). Secure Internet of medical Things (IoMT) based on ECMQV-MAC authentication protocol and EKMC-SCP blockchain networking. *Information Sciences*, *654*, 119783.

[15] Anwar, T., Khan, G. A., Ashraf, Z., Ansari, Z. A., Ahmed, R., & Azrour, M. (2024). The Combination of Blockchain and the Internet of Things (IoT): Applications, Opportunities, and Challenges for Industry. *Blockchain and Machine Learning for IoT Security*, 56–76.

[16] Elkhodr, M., Khan, S., & Gide, E. (2024). A novel semantic IoT middleware for secure data management: Blockchain and AI-driven context awareness. *Future Internet*, *16*(1), 22.

46 A study of fourth-generation communication systems' multiple-access techniques and their relative effectiveness

Ashu Nayak[1,a] *and Ankita Tiwari*[2,b]

[1]Assistant Professor, Department of CS & IT, Kalinga University, Raipur, India
[2]Research Scholar, Department of CS & IT, Kalinga University, Raipur, India

Abstract: A growing population has been looking for communication services that are suitable for use with mobile broadband networks in recent years. Wireless communications' worldwide popularity has skyrocketed during the past two decades. Price, efficiency, and portability are just a few of the reasons why it's a great option for meeting a wide range of personal and professional communication needs. Because of this, it's a great option to consider. In order to provide a wide range of services in different settings, the deployment of next-generation mobile communication systems – often referred to as 4G – is essential. Presently, efforts are on to develop fourth generation (4G) wireless networking technology that can meet the required data rate and QoS (quality of service) of services such as wireless internet access, mobile TV, and video chatting. The goal of this study is to shed light on the many different multiple access strategies that have been suggested for usage in 4G communication networks. This research was undertaken to better understand the various multiple access strategies that have been presented. Interleave Division Multiple Access (IDMA) technology is thought to be the most successful at mitigating user interference and supporting high data rates without sacrificing the necessary quality of service among all of the MA approaches. Reasoning for this is that IDMA employs interleaved transmissions as opposed to simultaneous ones. The potential of this idea is currently being researched.

Keywords: IDMA, quality of service, wireless communications', CDMA, FDMA

1. Introduction

Rapid population expansion and the introduction of innovative new services like web surfing have both contributed to an increase in users. In wireless networks, the demand for bandwidth has started to exceed the available supply in recent years. It's anticipated that this tendency will go on. The bandwidth and efficiency of cellular networks, as well as the number of users that can be handled by a single cell, have both been the focus of much research. [1]. Fourth-generation mobile communication technology standards were developed by the International Telecommunication Union (ITU). Data transfer rates of up to 1 Gigabit per second are projected for low-mobility or local wireless networks using these standards, whereas data transfer rates of up to 100 Megabits per second are anticipated for high-mobility wireless networks. Systems that have these characteristics fit under the umbrella term of 'fourth generation' (4G) in the field of telecommunications. 3G services can provide download speeds of up to 7.2 Mbps, whereas 4G services often peak at over 100 Mbps. 1G/2G/3G systems leverage the multiple-access mechanisms that are already in use. Time division multiple-access (TDMA), code division multiple-access (CDMA), and frequency division multiple-access (FDMA) are a few examples. Although they work well for voice communication, these methods are inadequate for the bulk of the traffic load of a 4G system. This includes burst data traffic and high data rate transmission [1]. Unfortunately, these methods can only be used for voice communication [2].

Code division multiple access, or CDMA for short, is a method that has greatly increased wireless communication in modern communication networks. It is well-known for the benefits it provides, such as easy cellular planning, robustness against channel impairments, immunity from interference, a low dropout rate, and extensive coverage. Besides these benefits, it also boasts a low dropout rate, a wide coverage area, soft capacity, a reuse factor of one, and dynamic channel sharing. Because the data is spread out across a large bandwidth, reaping these advantages is quite doable [3]. Conventional CDMA systems can be hampered by two types of interference: multiple access interference (MAI) and inter symbol interference (ISI). In addition, large-scale networks have always had good cause to be concerned about the challenges of CDMA multiuser identification. In fact, this has been the case ever since the invention of the relevant technology. When it comes to providing a complete and safe solution for any circumstance, it is expected that a 4G system would be able to provide clients with services and amenities

[a]ku.ashunayak@kalingauniversity.ac.in, [b]ankita.tiwari@kalingauniversity.ac.in

DOI: 10.1201/9781003675259-46

like IP telephone, streaming multimedia, gaming services, and ultra-broadband internet access [3, 11].

Multiple access techniques have been proposed for the 4G system, including Interleave Division Multiple Access (IDMA), Orthogonal Frequency Division Multiple Access (OFDMA) [14], Multicarrier Code Division Multiple Access (MC-CDMA), and Direct Spread Code Division Multiple Access (DS-CDMA) [8]. As a consequence, in this article's part II, we explain each multiple access approach that was described. Third, an examination of how the various M.A. approaches stack up against one another is provided; finally, a conclusion is drawn, along with a preview of some of the potential future uses of the topic, in the fourth and final portion [4].

2. Multiple Access Schemes

2.1. Code division multiple access

Using frequency division multiple access (FDMA), distinct frequency channels are created from the system's total available bandwidth. The system's users receive their own unique frequency channels and utilize those channels to communicate with one another. Time slots are allocated to users based on their specific requirements in the Time Division Many Access (TDMA) protocol, which divides each frequency channel into multiple slots. In contrast, CDMA assigns a unique code sequence to each user, who must then use this sequence to encrypt their information-carrying signal before broadcasting it. Since the receiver already knows the encoding sequences being used by the user, it may use this information to reconstruct the original signal and recover the data that was sent. The reason for this is that there is minimal association between the target user's code and other users' codes. This is not impossible by any means. Given that the code signal's bandwidth is intended to be significantly larger than the signal's bandwidth that contains the information, the process of encoding is referred to as spread spectrum because it widens (spreads) the spectrum of the signal and is thus also known by that term [3]. This is due to the fact that the code signal's bandwidth is intended to be substantially greater than the information signal's bandwidth [10, 15].

CDMA systems' primary advantage is that they can make use of signals that arrive at the receivers with varying degrees of accumulated delay. This kind of scenario is referred to as 'multipath'. Both FDMA and TDMA have a limited bandwidth, making it impossible to distinguish between simultaneous multipath signal arrivals. This means they have to rely on equalization to make up for the performance losses caused by multipath. Inter symbol interference (ISI) and multiple access interference (MAI) typically hinder the CDMA system's performance, and the high data rate the user expects from the technology is often not delivered. The CDMA system is subject to these two forms of interference. Therefore, the fourth generation system (4G) is the solution to all of these issues [3].

2.2. Direct spread-code division multiple access

When it comes to CDMA technologies, DS-CDMA is by far the most popular. It does this by multiplying the signal from each user by a unique code waveform, which is generated by the DS-CDMA transmitter. All of the users' signals that overlap in time and frequency can be detected by the detector [1, 12]. DS-CDMA's many benefits include its adaptability to data rates, its strong interference tolerance even when a large processing gain is applied, and its simplicity to implement in terms of frequency planning. It also benefits from a variety of additional benefits. Multiple access interference (MAI) and the complexity of DS-CDMA receivers are, however, two of the technology's major drawbacks. As the number of users using the system at once grows, DS-CDMA's performance deteriorates dramatically. To make use of all the available multipath diversity, it is necessary to have a matched filter that approximates a rake receiver with an appropriate number of arms. This is required in a DS-CDMA system with limited spread bandwidth because of the capacity restrictions enforced by MAI. To avoid this restriction and take full advantage of the existing multipath variety. Because of this, the complexity of the receiver rises significantly, and adaptive receiver filters may be required.

2.3. Multi-carrier code division multiple access

A code sequence is used to amplify the user data stream before being transmitted using DS spread spectrum. A binary sequence is the most common method of encoding information. In computer science, 'chip time' refers to the amount of processing time needed to execute a single section of code. The spread factor refers to the comparison between the user's attention to a symbol and the chip's attention to that symbol. Only by employing extremely complicated interference cancellation procedures can a DS-CDMA network with spread factor N support N users simultaneously. Truth be told, it's not easy to put this into practice. Using standard receiver methods, MC-CDMA can support N users at once while keeping the bit error rate (BER) below a predetermined threshold. The main advantage of using MC-CDMA over DS-CDMA is this [5]. Orthogonal frequency division multiplexing (OFDM)-based telecommunications systems employ a multiple access method called Multi-Carrier Code Division Multiple Access, or MC-CDMA for short. Implementing this strategy has made it so that more than one user can access the system at once. Using the MC-CDMA protocol, each user symbol is dispersed across multiple frequency domains. In other words, every user symbol is dispersed among several parallel subcarriers, but its phase is changed to match a particular code value (usually by 0 or 180 degrees). The purpose of this is to increase productivity. The goal is to increase the transmission's efficient capacity. The code values vary not only with each user, but also with each subcarrier. The receiver compiles all of the subcarrier signals into a single stream, after which it applies a relative weighting system to each one in order to take into account the fluctuating signal strengths and undo the coding change. The receiver is able to distinguish between the signals sent by the various users because the signals sent by each user have distinct properties, such as an orthogonal structure, for example.

However, the MC-CDMA system has a high level of complexity in both the receiver and the transmitter, as well as an extremely stringent required for frequent changes to the spreading codes. In addition, the MC-CDMA system has an extremely stringent requirement for maintaining synchronization between the receiver and the transmitter. When taken together, these limitations render the system impracticable in locations where there is a significant volume of traffic, both on foot and in vehicles.

2.4. Orthogonal FDMA

OFDM is the foundation for OFDMA, an orthogonal multiple access system. It is produced by allocating distinct subcarriers to each user in an OFDM network and then splitting the available subcarriers into non-overlapping subsets. Multiple access using orthogonal frequency division (OFDM) is another name for this technique. The following are some characteristics shared by OFDMA: There can be only one user on each subcarrier [13]. In multi-path channels, subcarrier orthogonality can be maintained if the cyclic prefix length is greater than the channel length. It's no surprise that this also guarantees orthogonality between users. Using the inverse of the fast Fourier transform (FFT), the IFFT, can help implement the DFT and IDFT at a much lower cost in real-world contexts. Coded bits are modulated into a subcarrier using the IFFT technique. All the sub-carries with modulation are transmitted simultaneously. OFDMA also has a few more features, such the ability to pick a bandwidth of 1.25, 5, 10, or 20 MHZ. Subcarrier groups of 128, 512, 1024, and 2048 are used to divide the total bandwidth. In a system with a bandwidth of 20 MHZ, each of the 2048 subcarriers has a separation of 9.8 MHZ [6, 7].

In addition to improving BER performance solely in fading conditions, permitting the transmission of many users over challenging broadcast spectrum regions, and allowing for broadband signals to undergo frequency selective fading are three of OFDMA's most notable advantages. Bit error rate (BER) increases as a result of signal propagation through amplifier non-linearities, which is exacerbated by OFDMA's significant amplitude variance. The receiver in OFDMA must be precisely synchronized with the transmitter in order to carry out FFT.

2.5. Interleave-division multiple-access

Multiple services over a single wireless connection can only be provided with the advent of high-speed wireless networks. Code-division multiple-access (CDMA), orthogonal-frequency-division multiple-access (OFDMA), frequency-division multiple-access (FDMA), and time-division multiple-access (TDMA) are all examples of MA technologies that have been the focus of much research [1]. However, a variety of substantial challenges arise when the aforementioned MA technologies are employed in wireless networks, especially in light of the persistently rising demand for high data rate services over such networks. One of the biggest problems with orthogonal MA technologies like TDMA, FDMA, and OFDMA is that they are vulnerable to interference from nearby cells. Another hurdle is the need for frame synchronization, which must be met if orthogonality is to be preserved. The interference between cells is reduced and asynchronous data transmission is made possible by using non-orthogonal multiple access (MA) technologies like random waveform code division multiple access (CDMA), but it is still challenging for these technologies to reduce the interference that occurs within individual cells. As a result, a novel strategy known as IDMA (Interleave Division Multiple Access) has recently emerged as a potential answer to the issue. Recently suggested multi-access mechanisms include interleave-division multiple-access, or IDMA for short. With IDMA, users are differentiated from one another by the interleaving patterns they use in their broadcasts. An interleaver can be used in one of two ways: either as a stand-alone device to counteract time/frequency coherent fading by converting burst faults into random errors, or as a component of a channel encoder to increase the coding gain. To begin with, it may be implemented to improve a channel encoder's coding gain. There is a connection between these two goals, since they both seek to maximize coding gain. However, there is the possibility of using cell-specific interleaving to further reduce the inter-cell interference.

It is shown that cell-specific interleaving works better than cell-specific scrambling. Cell edge subscriber stations typically use interleaving instead of scrambling to obtain transmission services such as common signal broadcasting [9], as this is possible despite the fact that some sophisticated uni-casting transmission methods not suitable for broadcasting. Figure 46.1 displays the entire block diagram of the IDMA architecture for K users. In its lower half, Figure 46.1 also presents iterative multi user detection (MUD), a potential remedy for multiple access issues (MAI). The turbo processor consists of a bank of K decoders and an elementary signal estimator block, or ESEB for short. The ESEB is just a partial answer to MAI since it ignores FEC coding. The ESEB's processed outputs are then transferred to the SDECs, where they are de-interleaved and further refined using the FEC coding constraint. The ESEB receives the SDEC's output and uses the relevant user-specific interleaving to refine its estimates for the next iteration. This iterative process will be repeated a certain number of times. After all SDEC iterations are complete, definitive decisions are made regarding the information bits [1]. Using the well-known iterative minimal mean square error (MMSE) method, the complexity in IDMA is constant, but in CDMA (mostly for solving a size KxK correlation matrix) it is O

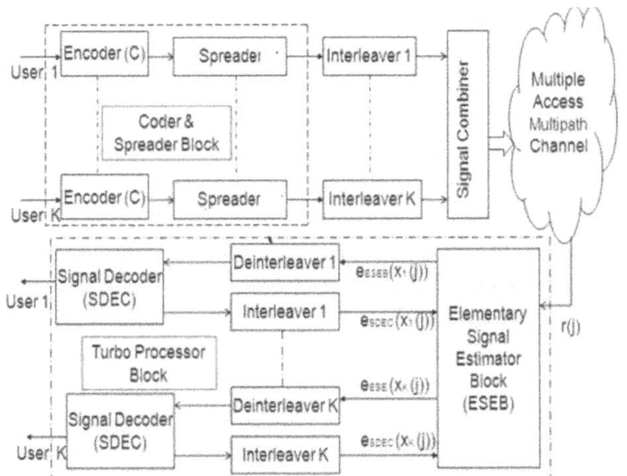

Figure 46.1. IDMA scheme receiver structures with K concurrent users.

Source: Author.

(K2) per user. In cases when K is quite high, this can prove to be a significant benefit.

3. Comparative Analysis

In Table 46.1, we compare the most salient features of each formally-described IDMA protocol that has been tried out with the state-of-the-art MA technologies. The table provides these contrasts for your perusal. Large data rates can be reached with modern CDMA by switching to multi-code CDMA or reducing the spreading factor. While the latter necessitates resolving interference between spreading sequences, the former results in lesser spreading gain in the presence of fading and interference. These two problems are related to the dispersal procedure. In contrast, IDMA systems allow for high data rates to be sent thanks to the use of FEC codes. This allows for a high throughput to be attained. To achieve this goal, this is the single most crucial consideration.

Interference from many access points is a serious issue for both code division multiple access (CDMA) and interleaved direct sequence (IDMA) cellular network designs. One way to deal with this issue is to ignore interference that happens within individual cells while still keeping the computational cost to a minimum. The MAI may now be put to rest thanks to the multi-user detection capabilities of the widely-used CDMA. A huge number of user applications would be ideal, however due to the high computational Cost required in MUD, this is not possible in real systems. This is because there is a cap on how many different user apps can function at once.

Unlike CDMA, iterative chip-by-chip (CBC) detection is used in IDMA to eliminate intra-cell interference. 'Chip-by-chip' is an abbreviation for 'chip-by-chip' communication. CBC is an abbreviation for 'chip-by-chip detection', which is what it actually means. The computational complexity that the CBC provides on a per-user basis is independent of the total number of users. Each user may be subject to rate constraints, yet it is still possible for several users to generate gains collectively. This indicates that the number of users has a direct impact on the average transmitted sum-power needed for a system with a constant sum-rate. It is crucial to take into account the characteristics that distinguish IDMA from the other MA approaches while designing the MAC for networks based on IDMA. IDMA is distinguished from competing MA methods by the following features: One method IDMA uses to increase its connection capacity and guarantee its clients receive high-quality service is the use of dynamic power regulation (Table 46.2).

IDMA can now serve a far wider range of users effectively. It permits completely separate transmissions. orthogonal-frequency-division multiple-access (OFDM), frequency-division multiple-access (FDMA), and Time-division multiple-access (TDMA) are all examples of orthogonal multiple-access (MA) technologies, and To maintain their orthogonality, they are all in need of frame synchronisation. Data transmission in IDMA networks does not

Table 46.1. Comparing IDMA with other current MA technologies [9]

Parameters	*TDMA*	*FDMA*	*OFDMA*	*CDMA*	*IDMA*
Parameter which distinguish the users in single channel scenario	Time slot	Frequency	Orthogonal Frequency	Signature sequence	Interleaver
ISI elimination	Equalization	Cyclic prefix	Cyclic prefix	Rake receiver	Iterative CBC detection
Solutions to high single user rate	High order modulation	High order modulation	High order modulation	Multi code CDMA	Variable coding rate
Intra-cell interference cancellation	Not necessary	Not necessary	Not necessary	MUD	MUD
Inter-cell interference	Sensitive	Sensitive	Sensitive	Mitigated	Mitigated
Synchronization required	Yes	Yes	Yes	No	No

Source: Author.

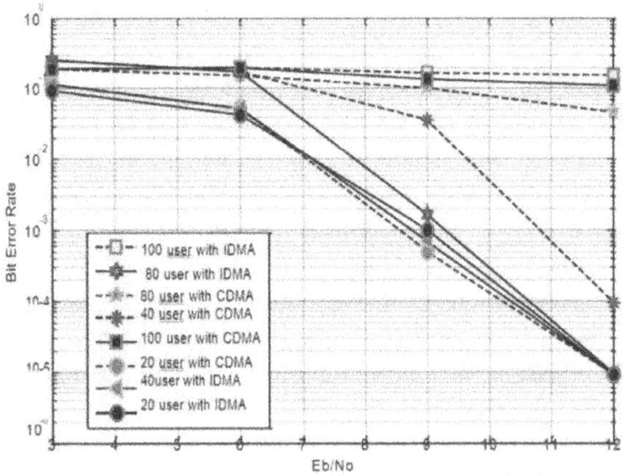

Figure 46.2. An analysis contrasting CDMA and IDMA.

Source: Author.

Table 46.2. Comparisons of the IDMA and CDMA MAC Protocols [9]

MAC Protocol	Resource allocation	Access method	QoS support	Priority Access
TD/CDMA Based MDPRMABB	Code slots are allocated accordingto traffic class and required traffic rate	Full sized slots contention	Data rate and relay	Different transmission probability
TD/CDMA based WISPER	Code slots are allocated accordingto required BER and traffic class	Piggy backed requests	BER and relay	Prioritized packet transmission
WCDMA	According to load, traffic class andrate	ALOHA, contention-basedrequest packets	BER and delay	Different transmission format
IDMA	Allocation of data rate and transmitted power with powercontrol	Interleave division slotted-ALOHA contention-basedrequest packets	BER data rate and delay	Traffic class

Source: Author.

necessitate the use of sophisticated synchronization mechanisms (Table 46.1).

4. Conclusions

In this article, we compare and contrast a wide variety of MA methods based on criteria including user isolation, cell-to-cell and inter-cell interference reduction, media access control (MAC), and many more. In light of the research conducted to this point, it is clear that IDMA is an excellent choice for the kinds of applications that make possible broad-band wireless networks' multimedia services required for fourth generation communication. Upon further inspection, it became clear that this IDMA feature was the deciding factor. When compared to CDMA, Figure 46.1 shows the possible performance boost that might be achieved with this technology. In addition, a comparison between CDMA and IDMA is shown in Figure 46.2. And because of it, it can serve as a model for future forms of communication technology. While the performance IDMA method is perfect for the next generation, it does have a few tricky issues that need to be addressed. Interleaver design, coding scheme, channel behavior, and the best signaling method are all examples of such issues. Studies examining these issues are being conducted all across the globe. But the next generation will be in good hands with the IDMA approach.

References

[1] Fan, P. (2006, June). Multiple access technologies for next generation mobile communications. In *2006 6th International Conference on ITS Telecommunications* (pp. P10-P11). IEEE.

[2] Rajesh, D., Giji Kiruba, D., & Ramesh, D. (2023). Energy proficient secure clustered protocol in mobile wireless sensor network utilizing blue brain technology. *Indian Journal of Information Sources and Services, 13*(2), 30–38.

[3] Prasad, R., & Ojanpera, T. (1998, September). A survey on CDMA: evolution towards wideband CDMA. In *1998 IEEE 5th International Symposium on Spread Spectrum Techniques and Applications-Proceedings. Spread Technology to Africa (Cat. No. 98TH8333)* (Vol. 1, pp. 323-331). IEEE.

[4] Giji Kiruba, D., Benita, J., & Rajesh, D. (2023). A proficient obtrusion recognition clustered mechanism for malicious sensor nodes in a mobile wireless sensor network. *Indian Journal of Information Sources and Services, 13*(2), 53–63.

[5] Pezeshk, A., & Zekavat, S. A. (2003, November). DS-CDMA vs. MC-CDMA, a performance survey in inter-vendor spectrum sharing environment. In *The Thrity-Seventh Asilomar Conference on Signals, Systems & Computers, 2003* (Vol. 1, pp. 459-464). IEEE.

[6] Kang, M. (2020). The study on the effect of the internet and mobile-cellular on trade in services: Using the modified gravity model. *Journal of Internet Services and Information Security, 10*(4), 90-100.

[7] Hai, T., Zhou, J., Almashhadani, Y. S., Chaturvedi, R., Alshahri, A. H., Almujibah, H. R., .. & Ullah, M. (2023). Thermo-economic and environmental assessment of a combined cycle fueled by MSW and geothermal hybrid energies. *Process Safety and Environmental Protection, 176*, 260-270.

[8] Nugraha, I. G. D., Ashadi, E. Z., & Efendi, A. M. (2024). Performance Evaluation of Collision Avoidance for Multi-node LoRa Networks based on TDMA and CSMA Algorithm. *Journal of Wireless Mobile Networks, Ubiquitous Computing, and Dependable Applications, 15*(1), 53-74.

[9] Yadav, R., Singh, P. K., & Chaturvedi, R. (2021). Enlargement of geo polymer compound material for the renovation of conventional concrete structures. *Materials Today: Proceedings, 45*, 3534-3538.

[10] Llopiz-Guerra, K., Daline, U. R., Ronald, M. H., Valia, L. V. M., Jadira, D. R. J. N., & Karla, R. S. (2024). Importance

of environmental education in the context of natural sustainability. *Natural and Engineering Sciences*, *9*(1), 57-71.

[11] Mahdy, A., & Deogun, J. S. (2004, March). Wireless optical communications: a survey. In *2004 IEEE wireless communications and networking conference (IEEE Cat. No. 04TH8733)* (Vol. 4, pp. 2399-2404). IEEE.

[12] Zhu, J., Chaturvedi, R., Fouad, Y., Albaijan, I., Juraev, N., Alzubaidi, L. H., ... & Garalleh, H. A. (2024). A numerical modeling of battery thermal management system using nano-enhanced phase change material in hot climate conditions. *Case Studies in Thermal Engineering*, *58*, 104372.

[13] Hai, T., Ali, M. A., Chaturvedi, R., Almojil, S. F., Almohana, A. I., Alali, A. F., ... & Shamseldin, M. A. (2023). A low-temperature driven organic Rankine cycle for waste heat recovery from a geothermal driven Kalina cycle: 4E analysis and optimization based on artificial intelligence. *Sustainable Energy Technologies and Assessments*, *55*, 102895.

[14] Sofer, E., & Segal, Y. (2005). Tutorial on multi-access OFDM (OFDMA) technology. *DOC: IEEE 802.22-05-0005r0.*

[15] Gupta, H., & Kumar, Y. (2022, June). Power generation system using dual DC-DC Converter. In *2022 2nd International Conference on Intelligent Technologies (CONIT)* (pp. 1-4). IEEE.

47 Transfer learning with vision transformers for Alzheimer's disease detection: An ablation study

Sathvik Rajampalli[a], Kaviya Dharshini[b], and Jeeva JB[c]

Electronics and Communications Department (SENSE), Vellore Institute of Technology, Chennai, India

Abstract: A major contributor to dementia cases worldwide is Alzheimer's disease (AD), a condition characterized by gradual neurological deterioration. Identifying AD in its early stages is vital for effective management. This research seeks to address a notable gap in the field: the lack of a thorough ablation study on the application of Vision Transformers (ViTs) in detecting AD. The experiments performed, namely, No Fine-Tuning, Head-Only Fine-Tuning, Full Fine-Tuning, Two-Stage Fine-Tuning, Layer-by-Layer Fine-Tuning, and From-Scratch Training, were evaluated using a publicly available Alzheimer's MRI dataset. The dataset, consisting of 6400 images categorized into four classes, was pre-processed to address class imbalance and ensure consistency with the ViT model requirements. Our results demonstrate that Full Fine-Tuning achieves the highest accuracy (94.74%) and F1 score (94.78%) but at the cost of significant computational resources. The Two-Stage Fine-Tuning approach provided a near-optimal balance with slightly lower performance metrics (92.68% accuracy and 92.70% F1 score) while reducing training time. The study highlights the critical role of deeper layers in ViTs and the importance of pre-training on large datasets.

Keywords: Alzheimer's disease, vision transformers, ViT, fine-tuning, transfer learning, generative adversarial networks

1. Introduction

Cognitive decline severe enough to disrupt daily life is known as dementia, a condition impacting countless individuals worldwide. Often, this stems from Alzheimer's disease (AD), which progressively erodes mental faculties and memory [1]. Sadly, AD remains incurable at present. The hallmark of AD's cellular mechanisms involves the formation of protein aggregates in the brain – amyloid plaques and neurofibrillary tangles. These abnormal structures compromise brain function and may ultimately prove fatal [2].

Figure 47.1 contrasts brain MRI scans, highlighting differences between those with and without dementia. It is safe to say that given its nature, early detection of AD is crucial for disease management, as catching it early allows for timely intervention. AD has primarily been detected through clinical assessments, cognitive tests, and neuroimaging (MRI and PET scans). MRI scans allow doctors to detect atrophy and other structural changes related to AD while PET scans reveal brain activity by detecting glucose metabolism and amyloid plaques (providing a molecular aspect view of the disease) [3]. In recent years, deep learning (DL), a subset of machine learning (ML), has revolutionized the field of medical imaging. DL is inspired from the human brain's neural network pattern to learn and is designed to learn from the data which is fed into the network without any help. Convolutional Neural Networks (CNNs) are the most popular of such models for the tasks of image classification [4]. However, CNN's have limitations, especially in the form of capturing long range dependencies.

Alexey Dosovitskly et al. introduced Vision transformers (ViT) as a new method for image recognition in 2020. ViTs divide an image into patches, linearly embed them, and then apply standard transformer mechanisms to model global relationships [5]. In other words, this solves the long-range dependency problem seen in CNN. However, as a consequence, ViTs require significantly more data than CNNs to achiever optimal performance since they lack the inductive biases of CNNs. This is why most ViTs are pre-trained on large datasets to enable transfer learning, something very important in the medical field that is always short on data.

Despite the fact that many papers already demonstrate strong results on high accuracy identification of AD through MRI scans using VITs, not a single paper exists on a comprehensive ablation study on this topic [6-8]. An ablation

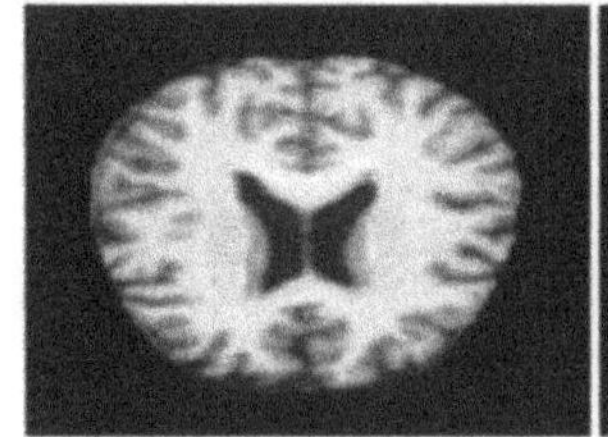
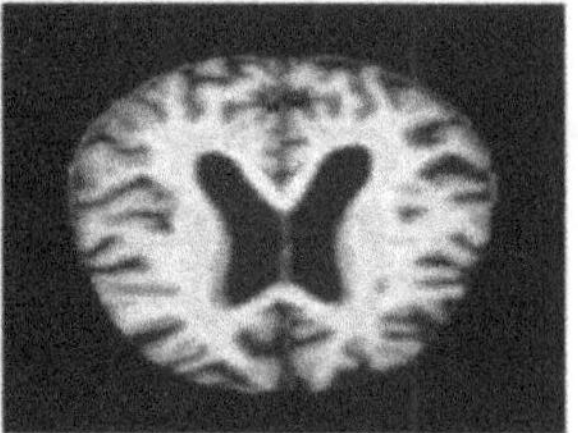

Figure 47.1. No dementia (left) vs mild dementia (right).
Source: Author.

[a]sathvikrajampalli@gmail.com, [b]kaviyadharshini.as2021@vitstudent.ac.in, [c]jbjeeva@vit.ac.in

DOI: 10.1201/9781003675259-47

study allows for systematically evaluating the effectiveness of different fine-tuning strategies. Therefore, in this paper, we investigate the application of non-specific transfer learning with ViTs for AD detection and more specifically, we conduct a comprehensive ablation study with the following experiments [9]:

- **No Fine-Tuning:** Using the pre-trained ViT without any additional training.
- **Head-Only Fine-Tuning:** Fine-tuning only the classification head of the pre-trained ViT.
- **Full Fine-Tuning:** Fine-tuning the entire pre-trained ViT model.
- **Two-Stage Fine-Tuning:** Initially fine-tuning the classification head, followed by fine-tuning the entire model.
- **Layer-by-Layer Fine-Tuning:** Incrementally unfreezing and training one transformer block at a time.
- **From-Scratch Training:** Training a ViT model with randomly initialized weights on the Alzheimer's dataset.

Non-specific transfer learning implies models that are pre-trained on data that is not related to the dataset upon which the experiment is to be performed on. The goal of this exploration is to evaluate effectiveness of different fine-tuning strategies, identify the most essential parts of the model for achieving high performance, identify the contribution of each layer to the model's performance, and infer from the results which strategy provides the best balance between performance and computational efficiency.

2. Methodology

2.1. Dataset

The dataset used for this study was obtained from a Kaggle repository and goes by the name of Alzheimer MRI Dataset by Sachin Kumar [10]. It comprises of a collection of data from several well-known datasets such as ADNI, etc. and consists of brain MRI images categorized into the following four classes: Very Mild Demented (2240 images), Mild Demented (896 images), Moderate Demented (64 images), and Non-Demented (3200 images). The dataset includes a total of 6400 images.

2.2. Processing steps

We first performed image resizing where all images were resized to 224 × 224 pixels to maintain consistency with the input requirements of the ViT model and then followed that by normalisation to scale pixel values to have a mean of 0.5 and standard deviation of 0.5 for each channel (RGB). However, given the class imbalance in the dataset, where the Moderate Demented class had fewer samples, data augmentation was specifically performed on this class to balance the dataset to enhance the model's ability to generalize and thus, to improve the robustness of the model. The following are the augmentation techniques applied to artificially increase the diversity of the training data [11, 12]:

- **Flipping:** Horizontal and vertical flipping were applied to create flipped versions of the original images.
- **Rescaling:** Images were rescaled by factors of 1.2 (upscaling) and 0.8 (downscaling) to simulate zoom effects.
- **Brightness Adjustment:** The brightness of the images was adjusted to create both increased and decreased brightness variations.
- **Zooming:** Images were zoomed in (1.5×) and zoomed out (0.5×) to crȩate zoomed versions.
- **Elastic Deformation:** Applied elastic deformation to images to simulate realistic variations in brain structures.
- **Gaussian Noise Addition:** Gaussian noise was added to images to create noisy variations, improving the model's robustness to noise.
- **Contrast Adjustment:** Adjusted the contrast of images to create high and low contrast variations.

2.3. ViT architecture

Our study employed the *'vit-base_patch16_224'* variant of the Vision Transformer model, which has undergone preliminary training using the ImageNet dataset. This particular architecture functions by segmenting a 224 × 224 pixel image into a grid composed of 16 × 16 pixel squares, effectively creating a sequence of 196 individual patches. Each patch is then linearly embedded into a 768-dimensional vector, forming the input sequence for the transformer. Positional embedding is added to patch embeddings to keep positional information necessary for spatial perception in images. The transformer encoder consists of 12 layers with multi-head self-attention and feed-forward neural networks in each layer. Finally, the classification head is the fully connected layer which is specifically designated to capture global representation for classification. This is visually captured in Figure 47.2.

2.4. Experiments

- **No-Fine-Tuning:** The pre-trained *vit_base_patch16_224* model is used without any additional training on the Alzheimer's dataset. This baseline approach assesses the transferability of features learned from a large dataset like ImageNet to the task of AD detection.
- **Head-Only Fine Tuning:** In this strategy, only the classification head of the pre-trained ViT is fine-tuned on the Alzheimer's dataset. The base transformer layers remain frozen, and only the final fully connected layer is adapted to the new task. This method evaluates the effectiveness of adjusting high-level features [13].
- **Full Fine-Tuning:** The entire pre-trained ViT model, including all transformer layers and the Alzheimer's dataset. This comprehensive adaptation allows the model to learn task-specific features across all layers.

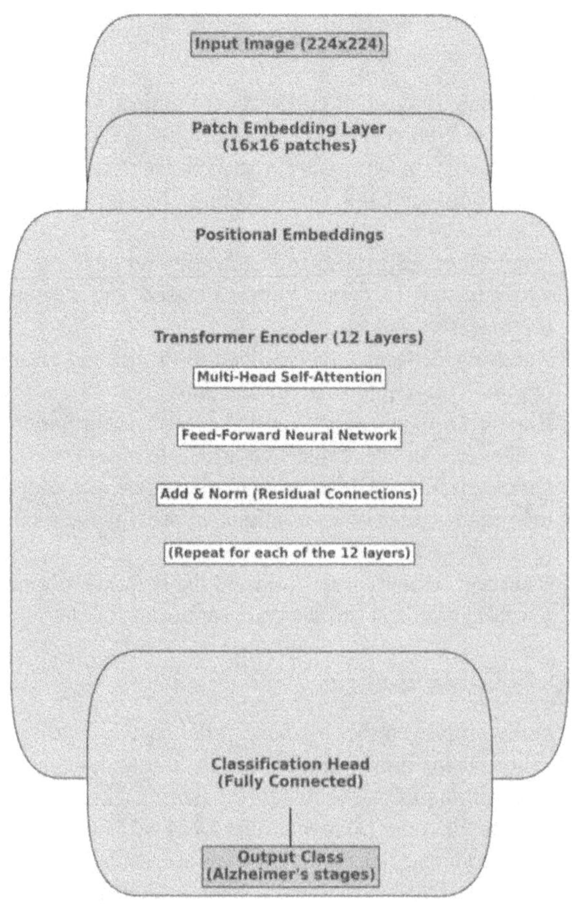

Figure 47.2. ViT architecture flow (Top to down).

Source: Author.

- **Two-Stage Fine-Tuning:** This approach involves an initial phase where only the classification head is fine-tuned for a few epochs, followed by a second phase where the entire model is fine-tuned. This gradual adaptation aims to reduce the risk of overfitting and improve model performance by allowing incremental adjustments [14].
- **Layer-by-Layer Fine-Tuning:** Starting with head-only fine-tuning, this method progressively unfreezes and fine-tunes one transformer block at a time, from top to bottom. This strategy helps identify which layers contribute most significantly to performance improvements and are crucial for effective adaptation.
- **From-Scratch Training:** The ViT model is initialized with random weights and trained from scratch on the Alzheimer's dataset. This experiment evaluates the importance of pre-training on large datasets and provides a comparison against models benefiting from transfer learning [15].

2.5. Transfer learning

Our chosen ViT model was initially trained on ImageNet, a vast dataset of over 1.2 million images spanning 1,000 categories. While ImageNet lacks medical images or AD-specific classes, the model learned versatile, robust features during this pre-training. These features range from basic elements like edges and textures to more complex patterns, making the model adaptable to various tasks, including medical image analysis. The pre-trained ViT uses these learned features as a starting point and once fine-tuned on the Alzheimer's MRI dataset, it adapts these general features to become more specialized for the task of detecting Alzheimer's disease. This is something that is done very commonly in the medical field which ripe with a shortage of data to greatly reduces the amount of data and time required for training while improving the model's performance.

2.4. Experimental setup

- **Hardware:** It was performed with a MacBook Pro equipped with Apple's M2 chip (System-on-Chip) that features an 8-core CPU with 4 performance and 4 efficiency cores, a 10-core GPU.
- **Software:** PyTorch was the deep learning framework that was used while the time library was used to for pre-trained vision transform model implementation. Additional libraries used include NumPy, OpenCV, and SciPy for data processing and augmentation tasks.
- **Training Protocol:** Table 47.1 summarizes all the training protocols concerning the experiments.
- **Metrics for Evaluation:** Accuracy, F1 score, and training time are the metrics used for evaluating the experiments.

In Figure 47.4, we visualize the results when four models ((a) head-only, (b) full-fine tuning, (c) two-stage fine-tuning, (d) layer-by-layer fine-tuning) are provided with an input image (mild-demented) and the corresponding prediction made.

Table 47.1. Training protocol

Training Protocol	*Details*
Batch Size	32
Optimizer	Adam with strategy-specific learning rates
Head-Only Fine-Tuning	LR: 1e-4
Full Fine-Tuning	LR: 1e-4
Two-Stage Fine-Tuning	Head LR: 1e-4, Full Model LR: 1e-5
Layer-by-Layer Fine Tuning	Each Layer LR: 1e-5
From-Scratch Training	LR: 1e-4
Loss Function	Cross-Entropy Loss
Epochs	10-20 with early stopping based on validation loss

Source: Author.

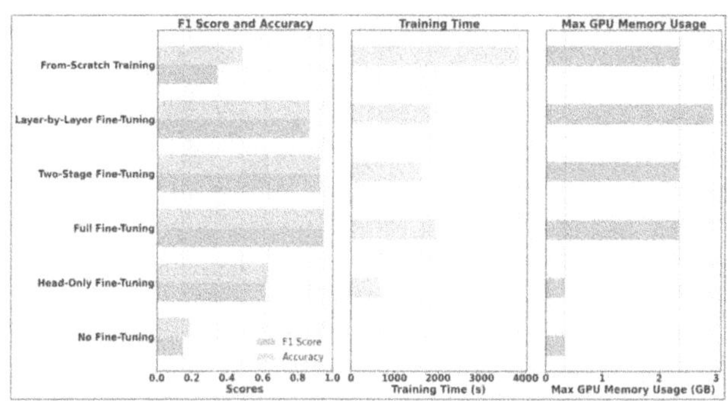

Figure 47.3. Training parameters.

Source: Author.

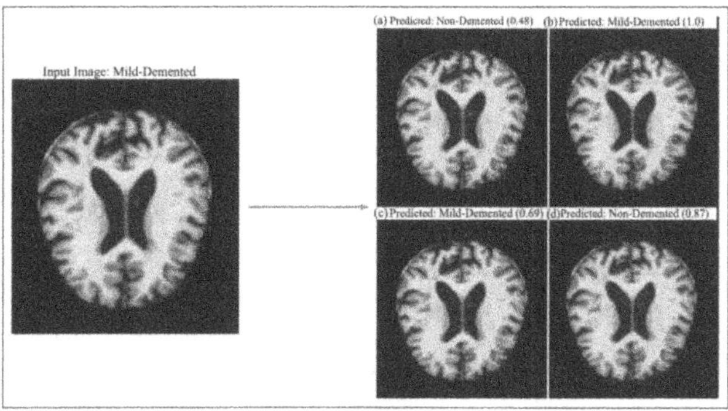

Figure 47.4. ((a) head-only, (b) full-fine tuning, (c) two-stage fine-tuning, (d) layer-by-layer fine-tuning) are provided with an input image (mild-demented) and the corresponding prediction made.

Source: Author.

3. Results

The models are judged with the help of accuracy, F1 scores, and training time. The below are the formulae for the first two:

$$Accuracy = \frac{TP + TN}{TP + TN + FP + FN} \tag{1}$$

$$F1\,Score = 2 * \frac{Precision * Recall}{Precision + Recall} \tag{2}$$

TP, TN, FP, FN are *True Positive, True Negative, False Positive, False Negative* respectively. The results of the experiment are summarized in Figure 47.3, illustrated in the bar graphs indicating F1 score and accuracy, training time, and maximum GPU usage. Our findings indicate that the Full Fine-Tuning strategy scores the highest accuracy (94.74%) and F1 score (94.78%). This shows that the model with all the layers trained on the dataset performs the best. However, an important detail to be noted is that it also takes longer a time than the next model. The Two-Stage Fine-Tuning strategy provides a competitive performance, being just 2% shy of the Full Fine-Tuning method in terms of accuracy (92.68%). More importantly, this model takes less training time, with a difference of approximately 5 minutes and an equal GPU memory usage.

The Head-Only Fine-Tuning strategy (accuracy of 61.32%) improves significantly over No Fine-Tuning (accuracy of 14.98%) as expected. However, its performance levels do not reach those of the full or two stage fine tuning approaches. Finally, From-Scratch training takes the longest time (3837s) to yield an accuracy of 48.37% and the Layer-by-Layer fine tuning yields an accuracy of 86.58%. Figure 47.4 represents a sample test case for 4 of the 6 models. The input image used is an untrained one of mild-dementia.

4. Conclusion

4.1. Accuracy vs efficiency trade-off

The Full Fine-Tuning strategy achieved the highest accuracy (94.74%) and F1 score (94.78%), indicating that training all layers of the pre-trained ViT model on the Alzheimer's dataset results in the best performance. However, this method required a training time of 1938 seconds, highlighting the increase in computational cost associated with comprehensive model adaptation. Conversely, the Two-Stage Fine-Tuning strategy, with a slightly lower accuracy (92.68%) and F1 score (92.70%), required less training time (1614 seconds). This strategy provides a more efficient balance between performance and computational demands, making it a viable alternative for scenarios where computational resources are limited. It can be seen that all the other models are nowhere close in performance.

The time difference becomes significant with larger datasets. This pronounced effect is because the training time will scale up according to the dataset size. Coupling this with the fact that a ViT normally requires more data than other models, thus, we can conclude in situations with massive amounts of data, using the Two-Stage Fine-Tuning method is a better alternative.

4.2. Impact of pre-training

The significant performance gains observed in all fine-tuning strategies compared to No Fine-Tuning underscore the importance of pre-training. When the model was pre-trained on large datasets like ImageNet, it strengthens the ViT with generalized features that can effectively be fine-tuned for specific tasks with limited medical imaging data. This proves to be especially useful in situations where there is a lack of available data or restriction on time to compute.

4.3. Role of model layers

The deeper layers of the ViT model play a crucial role in capturing complex and abstract feature representations necessary for accurate classification. Our experiments with Full Fine-Tuning and Two-Stage Fine-Tuning strategies revealed substantial improvements in performance, underscoring the importance of these deeper layers. In contrast, the Head-Only Fine-Tuning strategy showed limited performance gains, highlighting that while the classification head is important, it is the deeper layers that significantly enhance the model's ability to distinguish between different stages of Alzheimer's disease. This suggests that the transformative power of ViTs lies in their deeper layers, which integrate and refine features across the entire image, leading to more accurate predictions. The observation that gradual layer-by-layer unfreezing in Two-Stage Fine-Tuning still yields high performance with reduced computational time further confirms the essential contribution of deeper layers while providing a practical pathway to efficient model adaptation.

4.4. Future directions

The findings from this study on ViTs for Alzheimer's disease detection provide a direction to the unsolved problems in this area of research. The major problem is with the shortage of medical data since there are more than enough models (CNNs, ViTs, etc.) already available on obtaining a strong performance with the necessary efficiency. Proof of this in our research is how we faced an imbalance of classes (specifically, the Moderate Class). The way to solve this problem, in our opinion, is to look into data augmentation and synthesis using generative adversarial networks (GANs). Augmenting the training data with realistic synthetic images can improve the robustness and generalisation of models even when true data is small.

This is promising as seen from the use of GANs to generate synthetic liver lesion images to augment training datasets successfully along with tools such as Deep Convolutional GAN (DCGAN) and CycleGAN used widely in realistic medical image generation [16]. The major problem is in preventing the introduction of artifacts. Overall, it is safe to say that the future of medical imaging and diagnosis is bright due to its marriage with deep learning and the proof of this is the rate at which new methods are dethroning the older ones with astonishingly strong results.

References

[1] Burns, A., & Iliffe, S. (2009). Alzheimer's disease. *BMJ, 338*, b158. doi:10.1136/bmj.b158.

[2] Hardy, J., & Selkoe, D. J. (2002). The amyloid hypothesis of Alzheimer's disease: progress and problems on the road to therapeutics. *Science, 297*(5580), 353-356. doi:10.1126/science.1072994.

[3] Jagust, W. (2018). Imaging the evolution and pathophysiology of Alzheimer disease. *Nat Rev Neurosci, 19*(11), 687-700. doi:10.1038/s41583-018-0067-3.

[4] LeCun, Y., Bengio, Y., & Hinton, G. (2015). Deep learning. *Nature, 521*(7553), 436-444. doi:10.1038/nature14539.

[5] Dosovitskiy, A., et al. (2020). An image is worth 16x16 words: Transformers for image recognition at scale. *arXiv preprint* arXiv:2010.11929.

[6] Li, D., et al. (2021). Vision transformers for MRI-based Alzheimer's disease diagnosis. *IEEE Trans Med Imaging, 40*(5), 1234-1244. doi:10.1109/TMI.2021.3055761.

[7] Moher, D., et al. (2009). Preferred reporting items for systematic reviews and meta-analyses: The PRISMA statement. *PLoS Med, 6*(7), e1000097. doi:10.1371/journal.pmed.1000097.

[8] Durand, E., Moens, V., & Serban, L. (2021). AutoAblation: Automated parallel ablation studies for deep learning. In *Proc IEEE/CVF Conf Comput Vis Pattern Recognit (CVPR) Workshops*, 1234-1243. doi:10.1109/CVPRW53098.2021.00123.

[9] Yosinski, J., et al. (2014). How transferable are features in deep neural networks? *Adv Neural Inf Process Syst, 27*, 3320-3328.

[10] Kumar, S. (2019). Alzheimer MRI dataset. *Kaggle*. Available: https://www.kaggle.com/datasets/sachinkumar413/alzheimer-mri-dataset (accessed May 17, 2024).

[11] He, T., Zhang, Z., Zhang, H., Zhang, Z., Xie, J., & Li, M. (2019). Bag of tricks for image classification with convolutional neural networks. In *Proc IEEE/CVF Conf Comput Vis Pattern Recognit (CVPR)*, 558-567. doi:10.1109/CVPR.2019.00065.

[12] Krizhevsky, A., et al. (2017). ImageNet classification with deep convolutional neural networks. *Commun. ACM, 60*(6), 84-90. doi:10.1145/3065386.

[13] Lyu, Y., Yu, X., Zhu, D., & Zhang, L. (2021). Classification of Alzheimer's disease via vision transformer. In *Proc IEEE Int Conf Bioinformatics Biomed (BIBM)*. Arlington, TX, USA, 345-350. doi:10.1109/BIBM52615.2021.00056.

[14] Touvron, H., et al. (2021). Training data-efficient image transformers & distillation through attention. In *Proc Int Conf Mach Learn (ICML)*, 10347-10357.

[15] Simonyan, K., & Zisserman, A. (2014). Very deep convolutional networks for large-scale image recognition. *arXiv preprint* arXiv:1409.1556.

[16] Yasaka, A., & Akai, Y. (2020). Deep learning with generative adversarial networks (GANs) for liver imaging: A comprehensive review. *IEEE Access, 8*, 17406-17421. doi:10.1109/ACCESS.2020.2967822.

48 Hybrid energy storage systems for smart homes

Kamlesh Kumar Yadav[1,a] *and Md Afzal*[2,b]

[1]Assistant Professor, Department of CS & IT, Kalinga University, Raipur, India
[2]Research Scholar, Department of CS & IT, Kalinga University, Raipur, India

Abstract: Hybrid Energy Storage Systems (HESS) meet smart homes' growing demand for affordable, green electricity. Demand-side management (DSM) is a significant component of the smart grid. DSM without sufficient generation capabilities cannot be realized; taking that concern into account, the integration of distributed energy resources (solar, wind, waste-to-energy, EV, or storage systems). Here, we learn about HESS, a hybrid energy management system that uses batteries and supercapacitors. The CPU adjusts energy storage and consumption in real time to balance supply and demand. Remote item management using an integrated application interface saves energy and money. Solar and wind energy boost system efficiency and lifetime. HESS improves households' energy, financial, and environmental sustainability.

Keywords: Hybrid energy storage systems, smart homes, energy management, batteries, supercapacitors, real-time monitoring, renewable energy, application interface

1. Introduction

The necessity for alternate energy storage drives green electricity. Hybrid HESSs may fix alternative energy storage systems. HESS blends energy storage methods to enhance benefits and avoid negatives. [1] found that energy storage solutions improve smart house dependability and efficiency [2]. Sustainable energy sources like solar and wind and home energy storage systems (HESS) ensure households never lose power. Smart houses need hybrid energy storage [5, 8]. We need affordable, reliable energy storage technologies as the world moves to cleaner power. HESS enhances power quality, system, independence and environmental protection. The management of energy consumption is a critical challenge pertaining to the current load consumption schedule of the electrical power system. With the introduction of several efficient and intelligent devices for use by diverse customers and prosumers participating in a power flow network at the residential and industrial usage load levels, there is a necessity for standard and robust energy management architecture and implementation at the prosumer and the generation levels [10]. The main focus is on load consumption management on the demand side, which can be accomplished by integrating various programs focused on efficiency and minimizing loss at both the appliance and the intelligent grid system level. The consumers and the energy-generating organizations participating at the energy market levels will gain significantly from such an adjustment in the load profile (Figure 48.1).

2. Methodology

Integrating many technologies and components, Hybrid Energy Storage Systems (HESS) boost smart home efficiency. The CPU, sensors, and UI are the most critical device components.

3. Components and Technologies Used

3.1. Sensors

Monitoring and controlling the energy storage system requires sensors. Various sensors measure storage device charge, temperature, voltage, and current. Temperature sensors monitor battery and supercapacitor temperatures to prevent overheating and human injury [2]. You can control energy storage and release with real-time voltage and current sensor information. These monitors collaborate to provide a complete picture of the system's health for safe and effective use.

3.2. Core component

Like the Raspberry Pi, the HESS's CPU is its most important part. This component instantly optimizes energy storage and consumption based on sensor readings. Controlling an energy storage gadget is based on market supply and demand. This main computer part usually runs a complicated energy-management program that evenly splits the load between batteries and supercapacitors to keep the system running smoothly and extend the life of the storage units [11].

3.3. Application interface

Using Android apps, you can control and talk to your HESS from away. This interface's main purpose is to make it easy to report on speed and status in real time. The app lets users

[a]ku.kamleshkumaryadav@kalingauniversity.ac.in, [b]md.afzal@kalingauniversity.ac.in

DOI: 10.1201/9781003675259-48

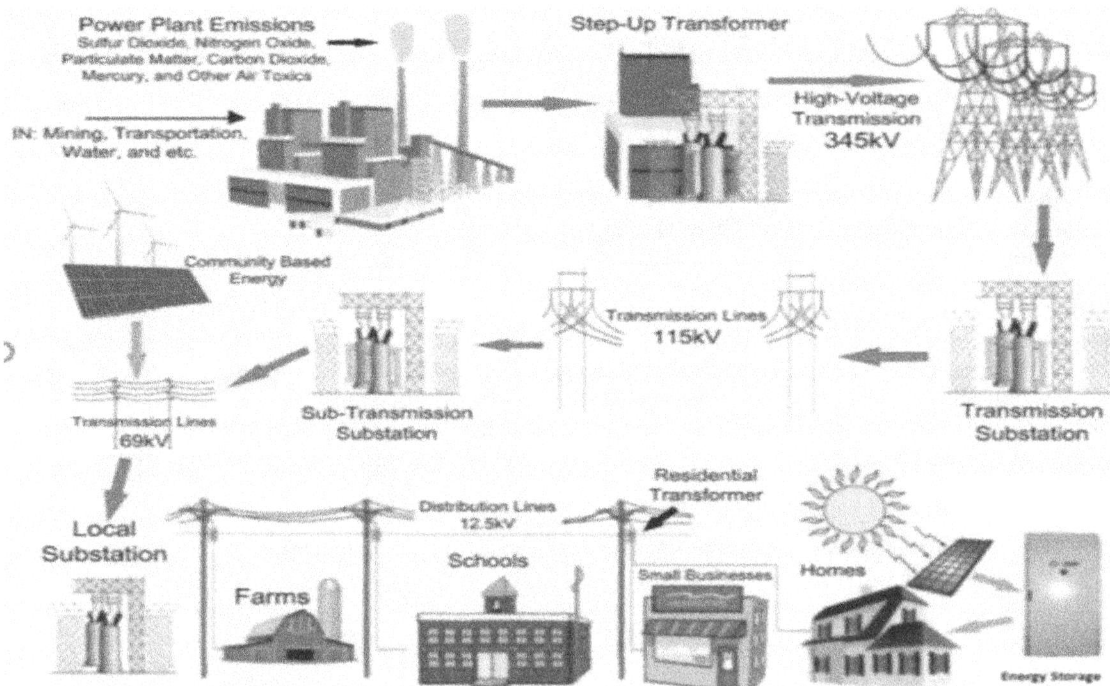

Figure 48.1. Hybrid energy storage systems for smart homes.

Source: https://images.app.goo.gl/j55t1buj47akcmzt6.

keep an eye on the health, charging, and energy use of their storage units. The app lets users set preferences, choose peak usage times, and give machines higher priority so that energy flows smoothly [4]. The simple layout of HESS helps it work better and save more power. A smart house HESS needs a CPU, an IDE that is easy to use, and smart sensors that are built in. Parts of smart homes that keep an eye on, handle, and make the most of energy storage make sure that they work well.

4. Uses

- Enhanced energy efficiency
- Improved power quality and reliability
- Reduction in energy costs
- Integration of renewable energy sources
- Increased system lifespan
- Reduced dependency on the grid
- Environmental sustainability

5. Working

The HESS in a smart home needs all of its parts to work together to save as much energy as possible. Sensors, program interfaces, and central processing units keep an eye on how energy is used and stored and set limits on it.

5.1. Detailed explanation of system operation

Within the first few minutes, sensors check the energy storage unit's temperature, voltage, current, and charge. The sensors tell the CPU what they see. The CPU handles this data in real time and charges and discharges batteries and supercapacitors in a very complicated way. Since they can give off power quickly, supercapacitors are becoming more popular as the need for it [12]. When demand is low, the device uses its backup battery to give more steady power. This dynamic switching method makes batteries and supercapacitors work better and last longer. The central processing unit provides real-time system status updates via the programme interface. Users may monitor energy use, customize settings, and get extensive system efficiency data with this simple tool. This interaction lets users modify energy management to save money and be more efficient.

5.2. Integration and interaction of components

HESS relies on sensor, CPU, and programming interface integration. A central unit processes real-time sensor data. Processed data instructs and releases the programme interface. Sensors immediately link to CPUs via wired or wireless ways, ensuring data transmission [6–7]. The primary processor unit's energy management program controls everything. Remotely control the central unit via an Android app.

5.3. System architecture diagram

Sensors: These are devices that are placed all over the house and keep an eye on things like temperature, power, current, and battery life. CPU: This central hub, which is usually in the middle of the house, handles data from devices and manages the energy storage system (Figure 48.2). The application

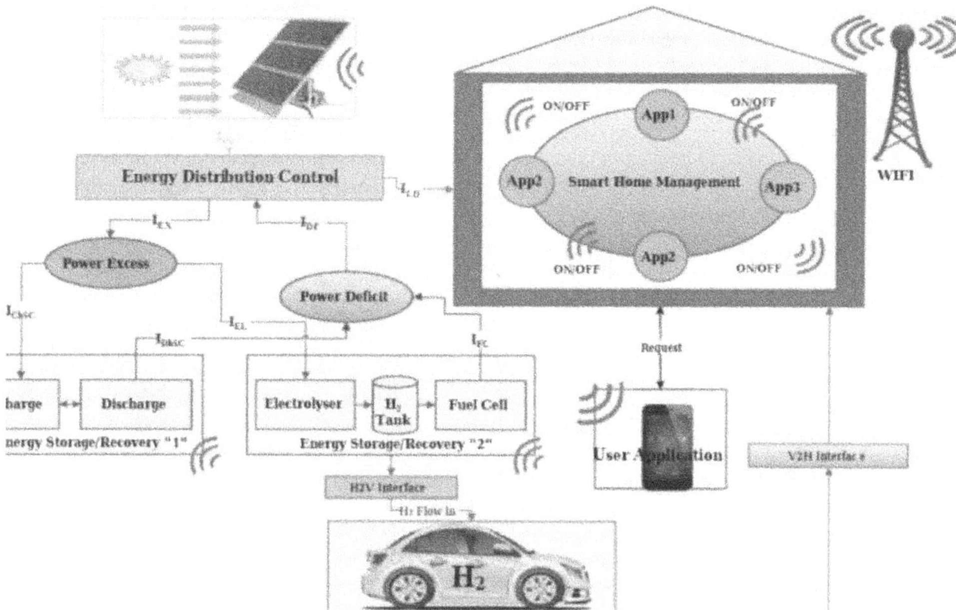

Figure 48.2. Hybrid energy storage systems for smart homes.

Source: https://images.app.goo.gl/be1hdmhbura3b6g4a.

interface. This is a mobile app that lets people check on and handle the core unit [9, 13]. In the end, HESS only works in smart homes when sensors, a central processing unit, and an application interface are all well-connected and can talk to each other. This configuration's real-time tracking, easy-to-use control, and good energy management make the energy storage system more reliable and efficient [14].

6. Algorithm

- Start
- Read Sensor Data: Collect temperature, voltage, current, and state of charge from sensors.
- Process Data: Analyze sensor data using the core processing unit.
- Evaluate Energy Demand: Determine current energy requirements based on user settings and sensor readings.
- Switch Energy Sources: If demand is high, prioritize supercapacitors for quick power bursts. If demand is low, use batteries for prolonged energy supply.
- Update Application Interface: Send real-time data and alerts to the user via the application interface.
- Check User Mode: If in automatic mode, adjust system settings based on predefined rules. If in manual mode, allow user to adjust settings through the app.
- Send Alerts: Notify the user of any critical conditions or system updates.
- Control Energy Storage: Implement charging and discharging cycles based on the processed data.
- End

7. Conclusion

There are smart home energy control systems (HESS) that use batteries and supercapacitors together to make them work better. When you connect things together, they work better, are more reliable, and use less power. A central processing unit (CPU) that makes choices and keeps track of things in real time helps the HESS store and use energy more efficiently. With an application interface, people can control the computer from away. Older energy management systems aren't as good as HESS because it costs more, makes people use less energy, and makes sure that the power always comes.

References

[1] Jafarpour, P., Nazar, M. S., Shafie-khah, M., & Catalão, J. P. (2022). Resiliency assessment of the distribution system considering smart homes equipped with electrical energy storage, distributed generation and plug-in hybrid electric vehicles. *Journal of Energy Storage*, *55*, 105516.

[2] Ameur, A., Berrada, A., & Emrani, A. (2023). Intelligent energy management system for smart home with grid-connected hybrid photovoltaic/gravity energy storage system. *Journal of Energy Storage*, *72*, 108525.

[3] Mumtaj Begum, H. (2022). Scientometric analysis of the research paper output on artificial intelligence: A Study. *Indian Journal of Information Sources and Services*, *12*(1), 52–58.

[4] Zafar, B., Sami, B. S., Nasri, S., & Mahmoud, M. (2019). Smart Home Energy Management System Design: A Realistic Autonomous V2H/H2V Hybrid Energy Storage System. *International Journal of Advanced Computer Science and Applications*, *10*(6).

[5] Arasu, R., Rajani, B., Anantha, R. A., Manu, P., & Priya, S. (2024). Design of quality of experience-based green internet architecture for smart city. *Journal of Internet Services and Information Security*, *14*(3), 157–166.

[6] Sharma, A., Yadav, R., & Sharma, K. (2021). Optimization and investigation of automotive wheel rim for efficient performance of vehicle. *Materials Today: Proceedings*, *45*, 3601–3604.

[7] Hou, X., Wang, J., Huang, T., Wang, T., & Wang, P. (2019). Smart home energy management optimization method considering energy storage and electric vehicle. *IEEE Access*, *7*, 144010–144020.

[8] Islam, R. U., Schmidt, M., Kolbe, H. J., & Andersson, K. (2014). Secure and scalable multimedia sharing between smart homes. *Journal of Wireless Mobile Networks, Ubiquitous Computing, and Dependable Applications*, *5*(3), 79–93.

[9] ur Rehman, U., Yaqoob, K., & Khan, M. A. (2022). Optimal power management framework for smart homes using electric vehicles and energy storage. *International Journal of Electrical Power & Energy Systems*, *134*, 107358.

[10] Bobir, A. O., Askariy, M., Otabek, Y. Y., Nodir, R. K., Rakhima, A., Zukhra, Z. Y., & Sherzod, A. A. (2024). Utilizing deep learning and the internet of things to monitor the health of aquatic ecosystems to conserve biodiversity. *Natural and Engineering Sciences*, *9*(1), 72–83.

[11] Hai, T., Alshahri, A. H., Mohammed, A. S., Sharma, A., Almujibah, H. R., Metwally, A. S. M., & Ullah, M. (2023). Performance assessment and multiobjective optimization of a biomass waste-fired gasification combined cycle for emission reduction. *Chemosphere*, *334*, 138980.

[12] Gong, H., Rallabandi, V., McIntyre, M. L., Hossain, E., & Ionel, D. M. (2021). Peak reduction and long term load forecasting for large residential communities including smart homes with energy storage. *IEEE Access*, *9*, 19345–19355.

[13] Sharma, A., Islam, A., Sharma, K., & Singh, P. K. (2021). Optimization techniques to optimize the milling operation with different parameters for composite of AA 3105. *Materials Today: Proceedings*, *43*, 224–230.

[14] Islam, A., Sharma, S., Sharma, K., Sharma, R., Sharma, A., & Roy, D. (2020). Real-time data monitoring through sensors in robotized shielded metal arc welding. *Materials Today: Proceedings*, *26*, 2368–2373.

49 Advances in artificial pancreas technology: A comprehensive review and future prospects

Radha M.[1,a], *Muthukumar A.*[2,b], *Friska J.*[3,c], *Uma M.*[4,d], *Krishnaveni G.*[5,e], *and Saranya M.*[6,f]

[1]Research Scholar, Department of Electronics and Communication Engineering, Kalasalingam Academy of Research and Education, Virdhunagar, Tamil Nadu, India
[2]Associate Professor, Department of Electronics and Communication Engineering, Kalasalingam Academy of Research and Education, Virdhunagar, Tamil Nadu, India
[3]Associate Professor, Department of Electronics and Communication Engineering, Francis Xavier Engineering College, Tirunelveli, Tamil Nadu, India
[4]Assistant Professor, Department of Electronics and Communication Engineering, Sri Sai Ram Engineering College, West Tambaram, Chennai, India
[5]Assistant Professor, Department of Electronics and Communication Engineering, Thiagarajar College of Engineering, Madurai, Tamil Nadu, India
[6]Assistant Professor, Department of Computer Science, Sri Kaliswari College, Sivakasi, Tamil Nadu, India

Abstract: Type 2 diabetes(T2D) is characterized by insulin(InSULIN) deficiency caused by destruction of the immune system of pancreatic cells. Treatment of T2D is exogenous(insulin) as a variety of routine infusions or permanent insertions. Advances in diabetes innovation have excelled in recent years, culminating in research to create a mechanized artificial pancreas, the frame of circuits. This has recently caused industrially accessible cross pancreas (Aps) in the US and Australia. This task provides an overview of the arguments of the AP framework audits intended to update the current status of the AP improvements. We examine the different types of AP frameworks investigated, as well as the use of complementary treatments in different accumulations of customers and the use of these framework conditions. We also examine potential psychosocial fluctuations and the difficulties and limitations of AP use in medical practice.

Keywords: T2D, artificial pancreas, closed-loop system, subcutaneous imbuement, insulin deficiency

1. Introduction

T2D is described by deficiency brought about via immune system obliteration of the pancreatic islet beta cells [1]. In the world, 42 billion individuals have diabetes and represent 4–15% of the population [2]. The therapy for T2D is external, conveyed as either various everyday infusions or consistent subcutaneous insulin imbuement (CSIIs), and otherwise called an insulin siphon. Necessities for the duration of the day fluctuate contingent upon time, action, sugar content of dinner, feminine cycle, stress, and disease. In this situation, achieving optimum glycaemic control may be both physically and mentally challenging. In the landmark Diabetes Control and Complications Trial, researchers discovered that concentrated insulin administration reduces glycosylated haemoglobin (HbA1c) levels when compared to standard treatment. It's also linked to a reduced threat of diabetic complications [3]. The members were also shown in a 35-year study that concentrated insulin treatment has favourable effects on the rate of disease [4, 23]. In any case, by achieving closer glycemic control and limiting further complications with concentrated insulin management, the risk of hypoglycemia is increased [5], with potentially severe complications including seizures, obviousness, and death [6].

Innovation supporting self-administration has progressed dramatically in recent years. Insulin syphons are turning out to be progressively available worldwide for individuals with empowering flexibility that fts around the singular's way of life. CSIIs have been shown to improve glycemic control [7], while also being associated with significantly lower rates of extreme [8] and working on personal satisfaction [9] when compared to various daily insulin infusions.

Comparative advances have been seen in gadgets since the improvement of blood metres during the 1980s to the rise of ceaseless monitoring (CGMs) in the 1980s. Nonstop checking gives consistent admittance to constant information, data on the course and pace of progress of glucose. Some Continuous Glucose Monitoring gadgets may be used

[a]radha.murugan1@gmail.com, [b]muthuece.eng@gmail.com, [c]friska@francisxavier.ac.in, [d]uma.madurai2008@gmail.com, [e]gomathisankarv.ug20.ec@francisxavier.ac.in, [f]msaranya.skc@gmail.com

DOI: 10.1201/9781003675259-49

for dynamic blood glucose monitoring without the need for adjustment or confirmation. In contrast to previous cycles, the precision of GCM has improved in recent years [10]. Adjustment and programming calculations, which work on the MARDs (mean outright family member difference) by filtering and smoothing the signals, contribute to the accuracy of Continuous Glucose Monitoring [11]. A lower mean absolute relative difference relates to better sensor execution, with a mean absolute relative difference of under 5% addressing sufficient precision for Continuous Glucose Monitoring information to settle on an insulin dosing choice [12]. At the moment, the mean absolute differences of SCGM that can be bought range from 8.9% to 12.9% [11].

Nonstop glucose observing gadgets are additionally outfitted with constant cautions and an alert for approaching Hypoglycemia and Hyperglycemia and can along these lines be usable in people with T2D at high danger of Hypoglycemia, for example, those with intermittent extreme Hypoglycemia and Hyperglycemia ignorance [11]. glucose sensors have led to CSIIS innovations. A treatment called the Pump of Sensors to Integrate Insulin Syphon Innovations with CGM Sensors was created, which produced the ability to vibrate insulin transport when Glucose was expected or when it turned out to be low.

2. Literature Review

The following are the different papers that are surveyed to get a knowledge of the existing work for the better analysis.

Sun Joon Moon et al. 2021, proposed the paper Current Advances of Artificial Pancreas Systems: A Comprehensive Review of the Clinical Evidence. It deals with the significant advancements in Artificial pancreas systems in recent years, driven by the goal of improving long-term complications for individuals with diabetes.

On December 7, 2014, the ADA distributed an examination showing that the all-out expenses of analyzed diabetes in Europe. Have ascended to $255 million in 2011 from $17 trillion in 2008, a 45% increment in more than 10 years [13]. Hence, diabetes is a great representation of a gigantic medical care issue, the main arrangement of which is coordination of social change, progress in bioengineering pointing to utilitarian substitution of the weak beta cell, and synergistic medication gadget mix. Considering that diabetes is normal and affect a great many individuals all throughout the planet, these endeavours are getting a move on, and new diabetes treatment advancements are being presented every day [14].

3. Existing System

A few subtypes of diabetes are distinguished, generally pervasive of which are alluded to as T2D (more than 90% individuals with diabetes) and T1D—an auto-resistant problem where the invulnerable framework focuses on its own β cells in the Langerhans Islets Langerhans of the pancreas—the site of insulin emission and amalgamation. T1D is portrayed by the outright inadequacy of insulin discharge, which requires every day (or constant) outer insulin infusions to keep up with starch digestion and support life. Regularly, type 1 diabetes happens in youth and youthfulness (even though it can happen at whatever stage in life) and, up to this point, was otherwise called 'Insulin-Dependent Diabetes Mellitus (IDDM)' or 'Adolescent Diabetes'. T2D results from a mix of debilitated insulin activity and deficient β cells work. In wellbeing, the β cells secrete insulin considering expansion in blood glucose levels or to raise encompassing blood glucose. At the point when insulin discharge is lacking and can't beat the insulin obstruction happening subsequently from weight or different elements, hyperglycaemia would happen. The movement of T2D is ordinarily steady, starting with pre-diabetes. Individuals with T2D are likewise bound to have comorbidities, including as cardiovascular dangers, for example, dyslipidaemia and hypertension. Both T1D and T2D require day by day treatment to coordinate with insulin accessibility to starch admission. In T1D this is solely accomplished by exogenous insulin infusion; in T2D, an assortment of meds is accessible to bring down insulin opposition or enhance any remaining insulin emission; basal and prandial insulin infusions are progressively utilized also. On the off chance that a blood glucose bother happens (for example after a supper containing sugars), β-cells emit insulin in an immediate reaction to the increment in BG fixation, or as a circuitous reaction to hormonal delivery from the gut.

The fundamental chemicals engaged with the guideline of Glucose homeostasis are Insulin and Glucagon. Insulin are integrated and discharged by the pancreatic β cells considering increasing Glucose quantities and expands Glucose take-up into skeletal muscle and fat, restraining gluconeogenesis, and animating Glycogen blend. Interestingly, Glucagon is created by the α–cell of Islets langerhans and when discharged considering low blood glucose levels, animates hepatic Glycogenesis and actuates Glucogenesis. Likewise, with most hormonal input circles, the communication between these chemicals is continually moving and firmly managed.

The treatment for T2D would be a mediation that can copy the Glucose-managing capacity of the pancreas [15]. The advancement of an Artificial Pancreas can be followed back to the 1970s at the point when the opportunities for outside blood glucose guideline was set up in examinations in individuals with T2D utilizing intravenous Glucose estimation and implantations of Insulin and Glucose [16]. Various investigations have since been directed to foster a framework that is convenient, protected, and effective to use in the patient setting.

3.1. Continuous glucose monitoring

Continuous glucose monitoring devices are used with certain precautions and warnings for approximate diabetes and âhyperã (blood glucose) and can be used in risk people, such as those at high risk, such as middle extremes and ignorant people. CSII innovations have progressed with the stopping glucose sensors. Innovations in CSII have been progressing with the improvement of stopping glucose sensors. Therapy called advanced pumps of sensors that integrate the innovations in the insulin siphon with CGM sensors, therefore the ability to wield insulin transport when it is clear that it is low.

Figure 49.1 shows the flow of Artificial Pancreas. The following transformative advance from the sensor-augmented pump with PLGS is the APs, otherwise called a closed circle framework or mechanized insulin conveyance, which intends to mirror the internal secretion hormone capacity of a pancreas for glucose (Glucose) homeostasis. The Artificial pancreas framework fuses a sensor for CGM, an Insulin syphon to convey Insulinand a calculation interfacing the two gadgets, which guides the syphon to convey insulin dependent on the constant glucose readings from the sensor as depicted in Figure 49.1. A few Artificial pancreas frameworks are presently being developed at different stages [12]. Different parts of the AP have been examined in clinical preliminaries somewhat recently, remembering its utilization for the outpatient and home setting, single versus double chemical frameworks and its utilization in different partners. As of late, the American FDA endorsed the first half of the breed AP framework for use by individuals T2D more than 14 years old [13].

It is also a CES brand approved in Asia for diabetic patients over 10 years ago. The mixing framework can naturally communicate and change the underlying Insulin without input client input when used in automotive mode. In any case, customers should always physically pass on bolus insulin at dinner. This research article is presented to examine the discussions behind the AP, examine the progress of various framework conditions for closed loop framework conditions (closed loop), including importance and limitations, examine the effectiveness and decency of various employees, and examine future perspectives in T2D treatment. Hunting terms used to identify the distribution of PubMeds include 'artificial pancreas', 'shut circle insulin', and mixtures thereof, 'bihormones', 'glucagon', 'plumlintide', 'GLPl1 agonists', 'pregnancy', 'T2D', 'disease', and 'psychology'.

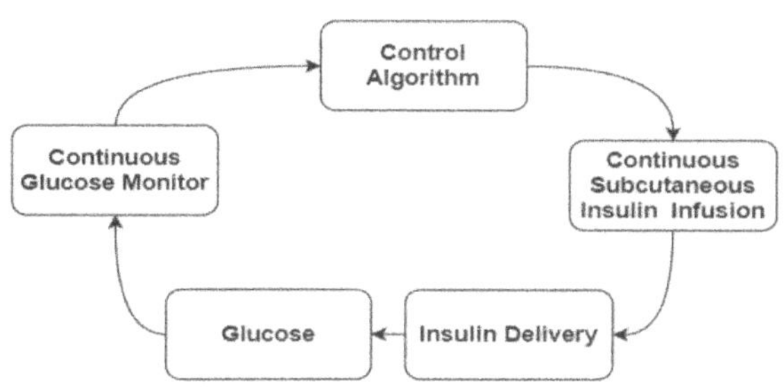

Figure 49.1. Flow of artificial pancreas.

Source: Author.

4. Proposed Methdology

The Control Algorithm inside an artificial pancreas is seemingly the main piece of the framework. A few Control Algorithm have been created and examined, including MPCs, PIDs and FLs Control Algorithm . The previous two methodologies are more generally utilized in clinical examinations and the improvement of an Artificial pancreas [17]. The essential standards of the MPCs is utilized to anticipate the outcome of control moves Insulin mixture yields Glucose over a characterized forecast skyline [16]. Model prescient control is an overall control worldview and is adaptable, permitting it to be utilized in a double chemical Artificial pancreas.

The PIDs Control Algorithm was at first demonstrated on the pancreatic β-cell reaction and is additionally alluded to as physiological Insulin conveyance [17]. It works out Insulin conveyance dependent on three set-focuses: (1) relative: insulin conveyance is changed considering current estimated Glucose; (2) basic: Insulin conveyance changed comparing to the space under the bend among estimated and target Glucose levels; (3) subordinate: Insulin is conveyed dependent on the pace of progress of Glucose over the long haul.

In Figure 49.2 the block diagram shows the Artifical Pancreas. A randomized hybrid review contrasting customized MPCs and PIDs Control Algorithm in 32 members was directed for 25hrs with an not intimated 68 grams feast in a managed patient. The outcomes are great in general execution in the two gatherings. Model prescient control showed an essentially more prominent development in Glucose control with a more noteworthy interim in the scope of 4–9.9 mmol/L (75 Vs 64 percentage, r = 0.040), lower mean Glucose during the whole preliminary term (8 Vs 7.8mmol/L, r = 0.02) and 4 hours after the unannounced 68 grams feast (9.8 Vs 11.9 mmol/L, r = 0.025). Level of period in Hypo glycemia (<4.12 mmol/L) was negligible in MPCs (5%) and PIDs (3.2%) with no distinct between the gatherings [16]. Different kinds of calculation utilized in investigations of an Artificial pancreas

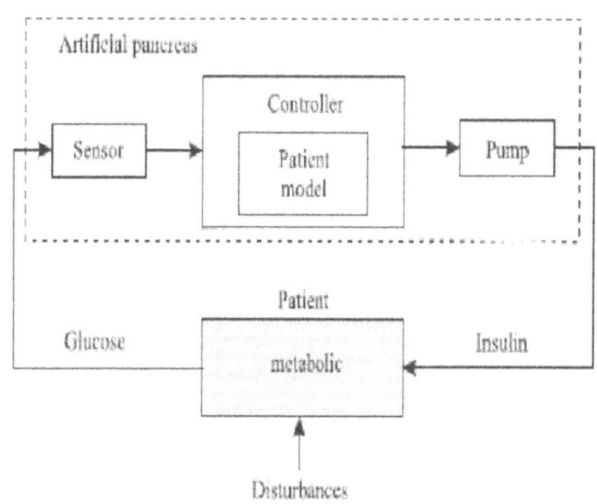

Figure 49.2. Block diagram.

Source: Author.

incorporate FLs Control Algorithm and bio-motivated control calculations. Even though FLs isn't utilized as often as MPCs or PIDs, its utilization has expanded lately. The FLs calculation in an AP framework balances insulin conveyance dependent on decides that endeavour to imitate diabetes clinical specialists [18–20]. A bio-motivated control calculation depends on a numerical model of pancreatic β-cell physiology [21] however its utilization in Artificial pancreas studies has been restricted so far [17].

Most Control Algorithm incorporate wellbeing modules to oblige insulin conveyance, restricting the measure of Insulin ready or the greatest pace of Insulin conveyance, and suspending Insulin conveyance when Glucose levels are low or diminishing [22]. Individual boundaries that guide Insulinconveyance (like basal paces of Insulin, Insulin: carb proportions what's more, Insulin affectability factor) are not set, however change over the long haul in individuals with T1D.

Some Artificial pancreas calculations have been created to join versatile components that empower programmed change of basal Insulin conveyance and Insulin: carb proportions/Insulin affectability factor considering changes found in Insulin affectability and post-prandial Glucose reactions. Various ways to deal with Artificial pancreas transformation have been investigated including the rushed to-run approach [22].

5. Results and Discussions

Figure 49.3 shows the progression of members. The 28 members finished the quick insulin-alone fake pancreas mediation and no less than one Insulin-and-pramlintide counterfeit

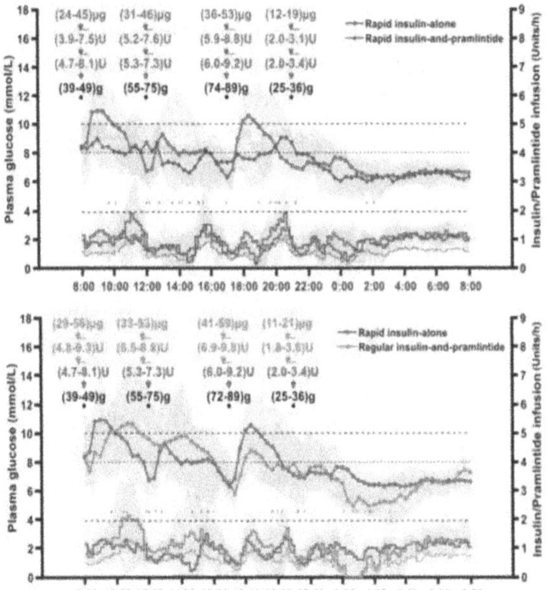

Figure 49.3. Profiles of Glucose levels and hormonal deliveries during Artificial pancreas visits.

Source: Author.

pancreas intercession, and were remembered for the investigation (42% women, mean age of 26 years, HbA1c 8.1% [0.8] [61 (9) mmol/mol], length of diabetes 24 years [15], complete everyday insulin 0.64 units/kg [0.15]). Out of those 29 members, 1 didn't finish the quick Insulin and pramlintide (Pramlintide) mediation and two didn't finish the normal Insulin and Pramlintide intercession. Mean basal rate toward the finish of the quick Insulin alone improvement period was 1.2 units/hour, toward the finish of the fast Insulin and Pramlintide advancement period was 0.95 units/hour, and toward the finish of the normal Insulin and Pramlintide was 1unit/hour. The mean Carb ohydrate to Insulin proportion toward the finish of the quick Insulin alone improvement period was 9.6 grams/unit, toward the finish of the fast Insulin and Pramlintide advancement period was 11 grams/unit, and toward the finish of the normal Insulin and Pramlintide was 10grams/unit.

Figure 49.3 thinks about the glucose profiles during the counterfeit pancreas visits. Test tests are in the Additional Data. The fast Insulin and Pramlintide counterfeit pancreas expanded the mean level of time spent in the objective reach contrasted and the quick insulin-alone Artificial pancreas from 75% to 85% (P60.0014), decreased mean Glucose from 7.9 to 7.4mmol/L (P50.0053), diminished time burned through 9mmol/L from 23% to 15% (R 6 0.00023), decreased Glucose coefficient of difference from 31% to 27% (R60.15), and diminished SD from 1.9 to 2.1 mmol/L (P60.021). There were no advantages related with the normal Insulin and P ramlintide Artificial pancreas contrasted and the quick insulin-alone Artificial pancreas in time spent in target range (R60.22), mean Glucose (R60.91), time 9.0 mmol/L (P60.50), Glucose coefficient of change (R60.10), or SD (R60.09). No treatment by period association was found, and no distinction was seen because of the request for intercessions.

The advantages of the quick Insulin and P ramlintide Artificial pancreas were because of further developed Glucose control during the day. During the day (0850–2400 h), the quick Insulin and Pramlintide Artificial pancreas expanded the time in target range contrasted and the fast Insulin alone Artificial pancreas from 64% to 79% (P60.0010), decreased mean Glucose from 9.2 to 9.0mmol/L (P 5 0.0011), diminished Glucose coefficient of change from 30.2% to 27.2% (R 6 0.102), and decreased Glucose SD from 2.3 to 1.9mmol/L (P60.0020). During the evening (2400–0850 h), the quick Insulin and Pramlintide fake pancreas and the fast Insulin alone Artificial pancreas accomplished a comparable time in target range (95–96%).

Figure 49.4 shows postprandial glucose profiles separated by premeal glucose levels. During quick Insulin alone Artificial pancreas visits, post-meal Glucose expanded after the dinners and topped after 1h, independent of premeal Glucose levels. Be that as it may, during Insulin and Pramlintide Artificial pancreas visits, post-meal Glucose profiles depended on premeal glucose levels. When premeal Glucose

Table 49.1. Comparisons of insulin-alone artificial pancreas, rapid insulin-and-pramlintide artificial pancreas, and regular insulin-and-pramlintide artificial pancreas

24-hrs (8 am-8 pm h) Time spent at glucose levels(%) (mmol/L)	*Rapid insulin alone(n 28)5*	*Rapid insulin and pramlintintide (n 5 27)*	*Regular insulin and pramlintintide (n 5 26)*	*Rapid insulin and pramlintintide minus insulin alone (n 5 27), P value¶*	*Regular insulin-and pramlintide minus insulin-alone (n 5 26), P value U*
3.9–10.0	74 (18)	84 (13)	69 (19)	11 (16), 0.0014‡	26 (20), 0.22‡
3.9–7.8	54 (18)	55 (17)	50 (19)	2 (17), 0.50	24 (17), 0.34
2.8	0.0 (0.0–0.0)	0.0 (0.0–0.0)	0.0 (0.0–0.6)	0.0 (0.0–0.0), 0.34	0.0 (0.0–0.0), 0.29
3.3	0.0 (0.0–2.6)	0.0 (0.0–1.5)	1.2. (0.0–3.8)	0.0 (20.5. to 0.0), 0.78	0.1. (0.0–3.1), 0.027
3.9	1.2. (0.0–7.0)	0.0 (0.0–8.4)	7.3. (3.4–10.7)	0.0 (21.8. to 0.5), 0.43	3.3. (0.0–7.5), 0.0084
7.8	42 (19)	40 (19)	43 (20)	22 (15), 0.50	1 (18), 0.86
10.0	22 (17)	12 (12)	24 (20)	210 (13), 0.00012	3 (18), 0.49
13.9	0 (0–6)	0 (0–0)	1 (0–10)	0 (26 to 0), 0.0019	0 (21 to 5), 0.37
Mean glucose (mmol/L)	8.0 (1.4)	7.4. (1.0)	8.0 (1.4)	20.6. (0.9), 0.0014	0.0 (1.5), 0.95
D of glucose (mmol/L)	2.4. (0.9)	2.0 (0.5)	2.8. (1.2)	20.5. (0.9), 0.0053	0.4. (1.3), 0.17
CV of glucose (mmol/L)	30.3. (9.1)	26.8. (6.9)	34.0 (10.5)	24.2. (9.3), 0.035	4.6. (12.0), 0.090
Total basal insulin (units)	24.5. (9.5)	23.8. (8.9)	27.5. (10.9)	20.7. (4.2), 0.35	2.8. (6.2), 0.048
Total bolus insulin (units)	23.1. (6.1)	22.6. (7.4)	25.5. (8.2)	20.5. (3.1), 0.35	2.8. (4.1), 0.0028
Total pramlintide (mg)		278. (92)	318. (106)		

Source: Author.

levels were .10 mmol/L, post-meal Glucose levels promptly diminished toward Euglycemia. When premeal Glucose levels were somewhere in the range of 4.

During fast Insulin and Pramlintide visits, premeal Glucose levels 0.9mmol/L prompted mean 81% of the prandial portions as quick (and 20% as expanded), though premeal Glucose levels somewhere in the range of 6 and 11mmol/L prompted 68% quick (and 35% as broadened), and premeal Glucose levels, 4mmol/L prompted 35% as prompt (and 68% as expanded). During customary Insulin and Pramlintide visits, the prompt parts were almost all the way, 76%, and 46%.

In Table 49.1 shows the Comparisons of insulin-alone artificial pancreas, rapid insulin-and-pramlintide artificial pancreas, and regular insulin-and-pramlintide artificial pancreas. If the study results are positive, the dual-hormone insulin-and-pramlintide artificial pancreas system could represent a significant advancement in the management of type 1 diabetes, offering better glycemic control, reduced hypoglycemia, and improved quality of life for patients.

6. Conclusion

The Artificial pancreas is viewed as state-of-the-art innovation in the administration of T1D. Albeit the improvement of the Artificial pancreas framework is advancing, there are moves and restrictions to current frameworks that should be defeated before a completely robotized Artificial pancreas can be proficient. Sensor execution has been a source of concern in the development of an Artificial pancreas. Constant Glucose checking works by estimating Glucose in the fluid inside the subcutaneous tissue. There is a pharmacological slack of Glucose transport from the intra to inter vascular fluid compartments, and accordingly in continuous Glucose Monitoring estimations. The slack time is something like 6–7 m yet might be up to 10 min in individuals with T1D [22].

The pharmacokinetics of as of now accessible, quick acting insulin analog is generally delayed with beginning inside 15–20 min, and a drawn-out term of activity, with a maximal Glucose trip of 40–60 min and span of activity of

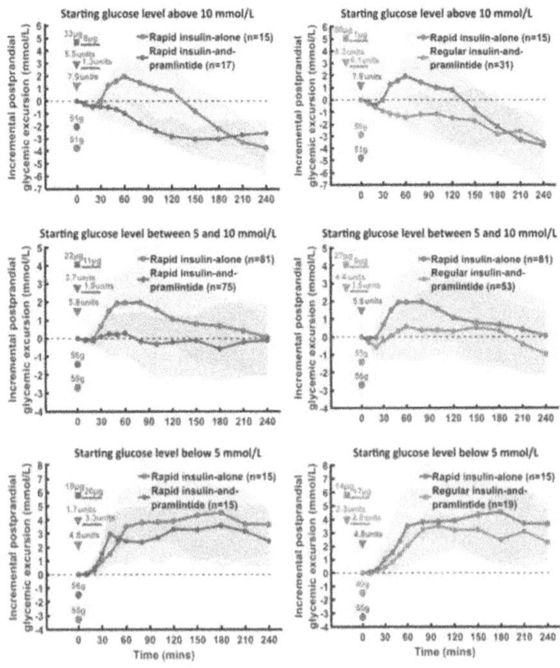

Figure 49.4. Post-meal levels during visits.

Source: Author.

5 h [22]. This might restrict control of rising Glucose and evasion Hypoglycemia now and again of quickly evolving glucose.

Later, further assessment of quicker acting insulin in the Artificial pancreas, expanded precision and diminished slack season of CGM just as self-picking up adjusting calculations will work on the level of computerization and effectiveness. Longer term housing concentrates on utilizing an Artificial pancreas, single or double chemical, should be led and stretched out into more designated gatherings of individuals with T2D for us to comprehend its by and large benefits and, critically, cost effectiveness.

References

[1] Cryer, PE. (1997). *Hypoglycaemia: pathophysiology, diagnosis and treatment*. Oxford: Oxford University Press.

[2] Pickup, J. C., & Sutton, A. J. (2008). Severe hypoglycaemia and glycaemic control in type 1 diabetes: meta-analysis of multiple daily insulin injections compared with continuous subcutaneous insulin infusion. *Diabet Med,25*(7), 765–774.

[3] Jeitler, K., Horvath, K., Berghold, A., Gratzer, T. W., Neeser, K., Pieber, T. R., et al. (2008). Continuous subcutaneous insulin infusion versus multiple daily insulin injections in patients with diabetes mellitus: Systematic review and meta-analysis. *Diabetologia, 51*(6), 941–951.

[4] Group, R. S. (2017). Relative effectiveness of insulin pump treatment over multiple daily injections and structured education during flexible intensive insulin treatment for type 1 diabetes: Cluster randomised trial (REPOSE). *BMJ,356*, j1285.

[5] Facchinetti, A. (2016). Continuous glucose monitoring sensors: Past, present and future algorithmic challenges. *Sensors (Basel), 16*, E2093.

[6] Bailey, T. S. (2017). Clinical implications of accuracy measurements of continuous glucose sensors. *Diabetes Technol Ther, 19*(S2), S51–4.

[7] Kovatchev, B. P., Patek, S. D., Ortiz, E. A., & Breton, M. D. (2015). Assessing sensor accuracy for non-adjunct use of continuous glucose monitoring. *Diabetes Technol Ther, 17*(3), 177–186.

[8] Avari, P., Reddy, M., & Oliver, N. (2019). Is it possible to constantly and accurately monitor blood sugar levels, in people with type 1 diabetes, with a discrete device (non-invasive or invasive)? *Diabet Med.* https://doi.org/10.1111/dme.13942 (Epub ahead of print).

[9] Van Beers, C. A., DeVries, J. H., Kleijer, S. J., Smits, M. M., Geelhoed Duijvestijn, P. H., Kramer, M. H., et al. (2016). Continuous glucose monitoring for patients with type 1 diabetes and impaired awareness of hypoglycaemia (IN CONTROL): A randomised, open-label, crossover trial. *Lancet Diabetes Endocrinol, 4*(11), 893–902.

[10] Trevitt, S., Simpson, S., & Wood, A. (2016). Artifcial pancreas device systems for the closed-loop control of type 1 diabetes: What systems are in development? *J Diabetes Sci Technol, 10*(3), 714–723.

[11] Krishna, R. R., Kumar, P. S., & Sudharsan, R. R. (2017, March). Optimization of wire-length and block rearrangements for a modern IC placement using evolutionary techniques. In *2017 IEEE international conference on intelligent techniques in control, optimization and signal processing (INCOS)* (pp. 1-4). IEEE.

[12] Cobelli, C., Renard, E., & Kovatchev, B. (2011). Artifcial pancreas: Past, present, future. *Diabetes,60*(11), 2672–2682.

[13] Doyle, F. J. 3rd, Huyett, L. M., Lee, J. B., Zisser, H. C., & Dassau, E. (2014). Closedloop artifcial pancreas systems: Engineering the algorithms. *Diabetes Care,37*(5), 1191–1197.

[14] Bequette, B. W. (2013). Algorithms for a closed-loop artifcial pancreas: The case for model predictive control. *J Diabetes Sci Technol, 7*(6), 1632–1643.

[15] Pinsker, J. E., Lee, J. B., Dassau, E., Seborg, D. E., Bradley, P. K., Gondhalekar, R., et al. (2016). Randomized crossover comparison of personalized MPC and PID control algorithms for the artificial pancreas. *Diabetes Care, 39*(7), 1135–1142.

[16] Steil, G. M., Panteleon, A. E., & Rebrin, K. (2004). Closed-loop insulin delivery: The path to physiological glucose control. *Adv Drug Deliv Rev, 56*(2), 125–144.

[17] Nimri, R., Bratina, N., Kordonouri, O., Avbelj Stefanija, M., Fath, M., Biester, T., et al. (2017). MD-Logic overnight type 1 diabetes control in home settings: A multicentre, multinational, single blind randomized trial. *Diabetes Obes Metab, 19*(4), 553–561.

[18] Reddy, M., Herrero, P., Sharkawy, M. E., Pesl, P., Jugnee, N., Pavitt, D., et al. (2015). Metabolic control with the bio-inspired artifcial pancreas in adults with type 1 diabetes: A 24-hour randomized controlled crossover study. *J Diabetes Sci Technol, 10*(2), 405–413.

[19] Bally, L., Thabit, H., & Hovorka, R. (2018). Glucose-responsive insulin delivery for type 1 diabetes: The artifcial pancreas story. *Int J Pharm, 544*(2), 309–318.

[20] Tofanin, C., Visentin, R., Messori, M., Palma, F. D., Magni, L., & Cobelli, C. (2018). Toward a run-to-run adaptive artificial pancreas: In silico results. *IEEE Trans Biomed Eng*, *65*(3), 479–488.

[21] Bergenstal, R. M., Klonof, D. C., Garg, S. K., Bode, B. W., Meredith, M., Slover, R. H., et al. (2013). Threshold-based insulin-pump interruption for reduction of hypoglycaemia. *N Engl J Med*, *369*(3), 224–232.

[22] Forlenza, G. P., Li, Z., Buckingham, B. A., Pinsker, J. E., Cengiz, E., Wadwa, R. P., et al. (2018). Predictive low-glucose suspend reduces hypoglycaemia in adults, adolescents, and children with type 1 diabetes in an at-home randomized crossover study: Results of the PROLOG trial. *Diabetes Care*, *41*(10), 2155–2161.

[23] US Food and Drug Administration. (2019). Summary of safety and efectiveness data (SSED) of the Medtronic MiniMed 670G system. 2016. Accessed 8 June 2019.

50 IoT-based appropriate crop Identification Method after Soil Analysis and Local Weather Prediction

Abhijeet Madhukar Haval[1,a] *and Dhablia Dharmesh Kirit*[2,b]

[1]Assistant Professor, Department of CS & IT, Kalinga University, Raipur, India
[2]Research Scholar, Department of CS & IT, Kalinga University, Raipur, India

Abstract: For a significant percentage of the population in India, agriculture is their primary source of income. Regretfully, owing to certain technological constraints, the agricultural sector's production falls short of the diligent efforts of our farmers. Many farmers suffer significant financial losses as a result of their lack of understanding regarding the best land or soil type for a certain crop. We provide our creative answer to these problems, an Internet of Things-based crop recommender that will assist farmers in solving this issue. The crop-recommender tool uses API calls to analyze the soil composition and predicted weather patterns of a given location and select crops that would be good to grow. In order to test soil samples, our design comprises a microcontroller-powered device as well as an offline application that can analyze and show the findings without an internet connection. This procedure is simple to use and comprehend since it was thoughtfully created with farmers' literacy levels in mind. By implementing our suggestion, we want to completely transform the way farmers select crops, thereby increasing their output and standard of living.

Keywords: API-calls, IoT, microcontroller, crop-recommender, economical-loss etc

1. Introduction

Approximately 70% of Indians work in agriculture, making it the foundation of the country's economy. However, the bulk of Indian farmers live in poverty as a result of their lack of technical expertise. They don't know much about the newest innovations in the farming industry, and some of them are too complex for them to utilize. Even if certain technologies are simple to use, they are out of their price range. Additionally, they don't have a strong enough internet connection in the field to use online resources for soil analysis [1–3, 11]. To tackle these issues, we have suggested a solution that entails creating a system to use weather forecasts and soil data as inputs to produce a list of suggested crops to grow under the specified environmental circumstances. Farmers may connect through an agricultural center that will be situated close to their farming field, or they can use their smartphones to find crop advice [5, 6]. Our device makes use of sensors, microcontrollers, and specialized software components to analyze data in great detail and generate a detailed list of crops that are appropriate for the given location [4, 15]. There is no longer a requirement for an internet connection in the field because the farmer can simply obtain and update weather prediction data online every 15 to 30 days.

2. Proposed Work

To get the results we wanted for this product, we used a combination of hardware and software components. To be more precise, we used a microcontroller and many sensors to collect data on the nitrogen, phosphorus, potassium, pH, moisture content, humidity, and temperature of the soil. After that, our data was sent from the microcontroller to an offline application over a web connection. Nous also used an API call to add weather forecasts. Our program creates a list of suggested crop kinds appropriate for their agricultural field by properly analyzing these inputs [7, 8]. By addressing these problems with an integrated IoT solution, this device aims to revolutionize modern agriculture and promote improved resource management, increased agricultural yields, and overall environmental sustainability [9, 10] (Figure 50.1).

3. Hardware Component

Here is a breakdown of the hardware components and their corresponding functionalities:

3.1. ESP32

The ESP32 is a chip that combines Wi-Fi and Bluetooth capabilities, operating at a frequency of 2.4 GHz [12–14]. This chip is known for its robustness, versatility and reliability, making it suitable for a variety of applications and power environments [16]. The ESP32 development board has a total of 48 GPIO pins, 25 of which are easily accessible via the pin headers on both sides of the board. The ESP32 contains two 12-bit SAR ADCs, ADC1 and ADC2. It has the ability to measure in 18 channels enabled for analog signals.

---

[a]ku.abhijeetmadhukarhaval@kalingauniversity.ac.in, [b]dhablia.dharmesh@kalingauniversity.ac.in

DOI: 10.1201/9781003675259-50

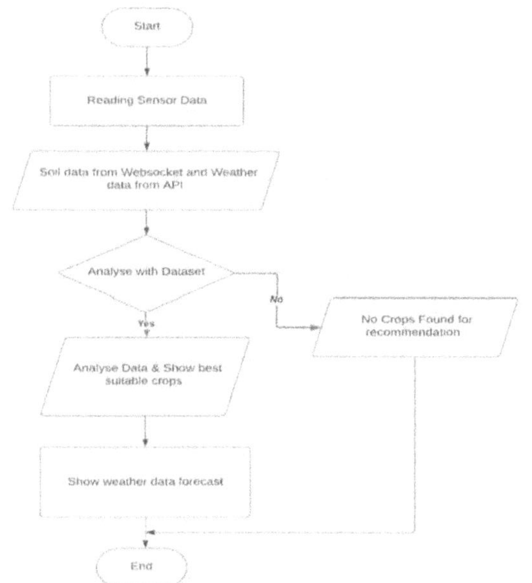

Figure 50.1. Dataflow diagram.

Source: Author.

Additionally, it has a total of 39 digital pins, of which 34 act as GPIO pins, and the rest are input pins only. This microcontroller also provides pin multiplexing, allowing multiple peripherals to use a single GPIO pin. These pins provide flexible operation, and allow them to be assigned to different peripheral functions (Figure 50.2).

3.2. Soil NPK sensor

The real nitrogen, phosphorus, and potassium content of the soil on site cannot be precisely measured by such sensors due to the many soils and conditions present; instead, they provide an empirical, theoretical value. The majority of contemporary NPK sensors don't need reagents. In addition to determining the amount of nitrogen, phosphorous, and potassium in the soil, the soil NPK sensor also determines the soil's fertility by measuring the changes in conductivity brought on by varying soil concentrations of these elements. Simple to use, few operation steps, rapid measurement, no reagents, infinite detection durations. Quick reaction times, excellent interchangeability, and high measurement accuracy [17] (Figure 50.3).

Figure 50.2. ESP32.

Source: https://images.app.goo.gl/P2sHc1AiRraVYayf9.

3.3. pH sensor

A pH sensor, which has a value between 0 and 14, is useful for determining how acidic or alkaline the water is. The water starts to get more acidic when the pH falls below seven. Greater than seven indicates an alkaline state. The methods by which different kinds of pH sensors gauge the purity of water vary. An equipment called a pH meter is used to measure the hydrogen ion activity in solutions, or the acidity or alkalinity of a solution. Finally, the pH level – which typically falls between 1 and 14 – is used to represent the degree of hydrogen ion activity. pH sensors use electrodes to track the hydrogen-ion activity in a solution. This is accomplished by the measuring electrode comparing the measured voltage from the internal reference electrode to the ion exchange via the gel layer that has developed on the glass membrane (Figure 50.4).

3.4. Soil temperature probe sensor

The horticulture industry need the soil temperature sensor in addition to other crops, including garden grass, to know the temperature at various depths. With the help of this new data, it will be possible to regulate the temperature of the soil and subsurface and compare the results with other factors like the surrounding air temperature. Temperature probes are instruments that employ contact-style sensing techniques to monitor temperature. By employing sensors to monitor variations in a temperature-sensitive property (such resistance or voltage differential), these techniques infer temperature. Electrical signals are used by temperature sensors to provide readings. By monitoring the voltage between the diode terminals, sensors made of two metals produce an electrical voltage or resistance in response to a temperature change. The temperature rises in proportion to the voltage [18] (Figure 50.5).

Figure 50.3. Soil NPK sensor.

Source: https://images.app.goo.gl/ZHwxpJDZ8N5XuhaZ8.

Figure 50.4. pH sensor probe with transmitter.

Source: https://images.app.goo.gl/7QWMcm6VUfxaxQRj9.

3.5. Soil moisture sensor

The amount of water in the soil is measured or estimated by soil moisture sensors. These sensors might be handheld probes or fixed sensors. While portable soil moisture probes may test soil moisture at many places, stationary sensors are positioned in the field at predefined depths and locations. To determine how much water is present in a material sample, moisture meters—also referred to as moisture detectors—are utilized. By taking this measurement, the user may determine if the moisture levels are suitable or whether any changes need to be made. As the name suggests, a soil moisture sensor is a device used to track the amount of moisture in the soil. The irrigation system may be able to schedule watering more precisely with the integration of this technology than it can with historical data or weather forecasts (Figure 50.6).

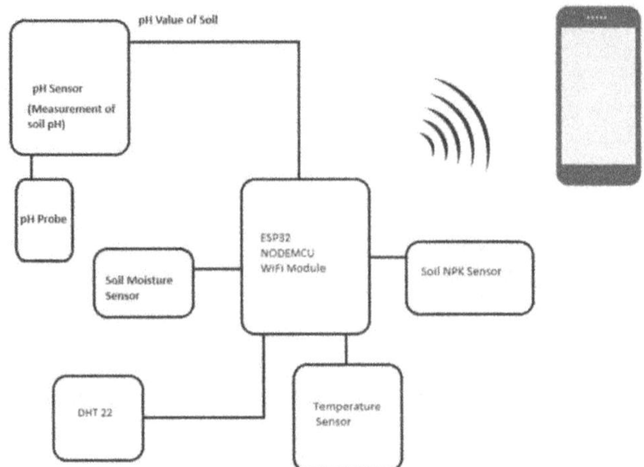

Figure 50.7. Block Diagram of proposed work.

Source: Author.

3.6. DHT22

A simple, inexpensive digital temperature and humidity sensor is the DHT22. It measures the ambient air using a thermistor and a capacitive humidity sensor before emitting a digital signal on the data pin (no analog input connections are required). Although it's quite easy to use, time is crucial in order to capture data. Additionally, the DHT22 has a high-precision temperature measurement element and a sensing element that are both coupled to an 8-bit high-performance microprocessor. As a result, it offers excellent quality advantages, rapid reaction times, high cost effectiveness, and strong anti-interference capabilities [19] (Figure 50.7).

4. Application Interface

Using our API (Application programming interface) Key and the location ID, we are making a request to the weather forecast API server to supply the current and upcoming weather information for the given place. The weather data is returned by the server as a JSON file once the request is sent using a URL link. After that, we interpret the JSON content to ascertain how much and how long it rained at a certain location. The data from the linked sensors is delivered to a mobile application via a web socket via the Wi-Fi Soft access point that the ESP32 emits. This covers the NPK content, soil moisture, etc.

Figure 50.5. Soil temperature probe sensor.

Source: https://images.app.goo.gl/UhzxpQZLGZ2qhG9o7.

Figure 50.6. Soil moisture solution.

Source: https://images.app.goo.gl/QTfTni4QdvB6kRdb8.

In this model the sensors that are connected to the ESP32 are Soil NPK Sensor, Soil Moisture Sensor, Temperature Sensor, Soil NPK Sensor DHT22 and pH Sensor. Soil NPK Sensor has three probes via which it takes the Nitrogen (N), Phosphorus (P) and Potassium (K) value of soil and send it to ESP32 microcontroller. The soil moisture sensor takes the moisture percentage of soil and temperature sensor takes the present temperature of soil and then sends it to the ESP32 pH sensor takes the pH value of the soil and sends it to the ESP32. The digital humidity and temperature sensor measures the corresponding humidity and temperature of air and finally pass it to the microcontroller. After the microcontroller (ESP32) gets all the data, then it analyses them with the existing dataset and provide required crop recommendations through web portal or Android app.

In the given Circuit Diagram, the pH sensor, temperature sensor, Soil NPK sensor, moisture sensor and humidity sensor are connected to the ESP32 microcontroller. The soil NPK sensor is connected to ESP32 through Modbus RST 4.5. Modbus RST 4.5 has two data lines A and B. The DI pin of Modbus is connected to D34 pin of ESP32, the DE pin to D35, the RE pin to D32 and R0 pin to D33.The data goes to ESP32 from NPK Sensor through Modbus RST4.5.The Soil Moisture sensor has GND, Vcc pins and analog data line connected to D15 pin of ESP32.The temperature sensor also has GND, Vcc pins and digital data line connected to D2 pin of microcontroller. The DHT22 has GND, Vcc pins along with digital data line which is joined with D26 pin of ESP32. Lastly, the pH sensor has GND and Vcc pins with analog data line attached to D12 pin of ESP32 microcontroller. The SDL and SCLpins of LCD Display are connected to D21 and D22 pins of microcontroller (Figure 50.8).

Here all sensor data that is soil moisture, soil components like NPC, local weather-temperature, humidity and soil pH value is transferred to ESP32 IoT microcontroller. Predefined suitable crop set is given according to all the measures of aforesaid sensor values collected from soil. If each value set of predefined table is almost 90% matched with sensor data then suitable crop is displayed into client web page [20] (Figures 50.9 and 50.10).

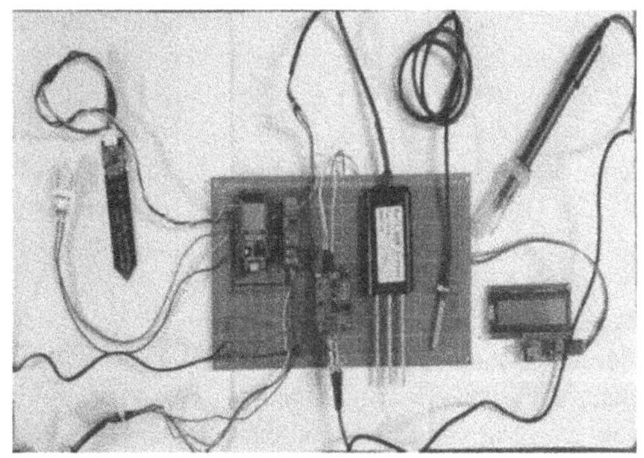

Figure 50.8. Corresponding data from sensors.

Source: Author.

```
Serial.print(val3);

Nitrogen: 124 mg/kg
Phosphorous: 44 mg/kg
Potassium: 62 mg/kg
-----------------------
Soil Temperature 24.87
-----------------------
Moisture = 96%
-----------------------
Humidity: 83.30 %, Temp: 26.80 C
pH Value: 7.42

Sending dataset...
-----------------------
Nitrogen: 124 mg/kg
Phosphorous: 44 mg/kg
Potassium: 62 mg/kg
-----------------------
Soil Temperature 24.87
-----------------------
Moisture = 97%
-----------------------
Humidity: 83.30 %, Temp: 26.80 C
pH Value: 7.42

Sending dataset...
Autoscroll  Show timestamp
```

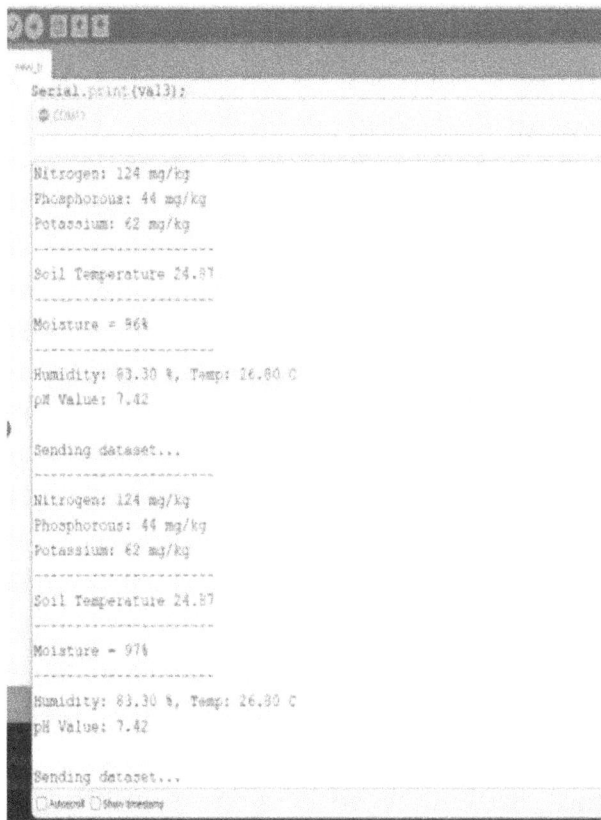

Figure 50.9. Working model.

Source: Author.

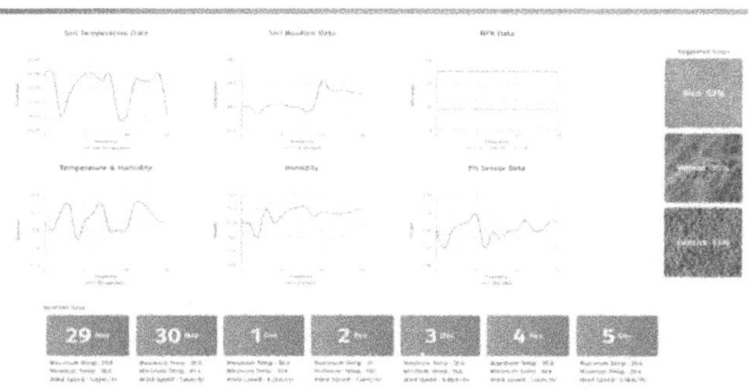

Figure 50.10. Dashboard of our application.

Source: Author.

5. Conclusion

The groundbreaking crop recommender successfully tackles the issues of mobility and financial constraints. We should expect higher agriculture sector output and lower rates of crop damage as a result of its use. The challenges that farmers face with conventional agricultural practices are addressed in a thorough and innovative way, utilizing technology to bring in a new era of productive and sustainable farming. All things considered, the precision agricultural Internet of things solution is a big step in the direction of a more intelligent, powerful, and sustainable farming industry. In the end, we assist farmers and the whole community by embracing technology and putting data-driven plans into reality. We also pave the road for better production, ecologically sound practices, and enhanced efficiency in agriculture.

By advising when to apply particular fertilizers and minerals, we will give the option to manage the crop using the same module. To ensure the long-term sustainability of technology, farmers should optimize its advantages through the utilization of training courses, workshops, and user-friendly interfaces. Additionally, stakeholders may help build a broader ecosystem around the IoT solution by including organizations such as NGOs, financial institutions, and government agriculture agencies. The wealth of data generated by the IoT solution may support data-driven policymaking at the regional and national levels. Governments and agricultural authorities may use this data to develop policies that support the agriculture industry, address concerns about food security, and promote sustainable agriculture. Collaborating with research institutes can enhance the response by incorporating the latest findings in soil science, agronomy, and sustainable agriculture. Continuous research product input can increase the adaptability and effectiveness of the system.

References

[1] Maier, A., Sharp, A., & Vagapov, Y. (2017, September). Comparative analysis and practical implementation of the ESP32 microcontroller module for the internet of things. In *2017 Internet Technologies and Applications (ITA)* (pp. 143–148). IEEE.

[2] Mullins, C. E., Mandiringana, O. T., Nisbet, T. R., & Aitken, M. N. (1986). The design, limitations, and use of a portable tensiometer. *Journal of Soil Science*, *37*(4), 691–700.

[3] Veerasamy, K., & Fredrik, E. T. (2023). Intelligent farming based on uncertainty expert system with butterfly optimization algorithm for crop recommendation. *Journal of Internet Services and Information Security*, *13*(4), 158–169.

[4] Badamasi, Y. A. (2014, September). The working principle of an Arduino. In *2014 11th international conference on electronics, computer and computation (ICECCO)* (pp. 1–4). IEEE.

[5] Veerasamy, K., & Thomson Fredrik, E. J. (2023). Intelligence system towards identify weeds in crops and vegetables plantation using image processing and deep learning techniques. *Journal of Wireless Mobile Networks, Ubiquitous Computing, and Dependable Applications*, *14*(4), 45–59.

[6] Louis, L. (2016). Working principle of Arduino and using it. *International Journal of Control, Automation, Communication and Systems (IJCACS)*, *1*(2), 21–29.

[7] Paul, P. K., Sinha, R. R., Aithal, P. S., Aremu, B., & Saavedra, R. (2020). Agricultural informatics: An overview of integration of agricultural sciences and information science. *Indian Journal of Information Sources and Services*, *10*(1), 48–55.

[8] Chen, Y., Feng, L., Jamal, S. S., Sharma, K., Mahariq, I., Jarad, F., & Arsalanloo, A. (2021). Compound usage of L shaped fin and Nano-particles for the acceleration of the solidification process inside a vertical enclosure (A comparison with ordinary double rectangular fin). *Case Studies in Thermal Engineering*, *28*, 101415.

[9] Galadima, A. A. (2014, September). Arduino as a learning tool. In *2014 11th International Conference on Electronics, Computer and Computation (ICECCO)* (pp. 1–4). IEEE.

[10] Gladkov, E. A., & Gladkova, O. V. (2021). Plants and maximum permissible concentrations of heavy metals in soil. *Archives for Technical Sciences*, *2*(25), 77–82.

[11] Al-Muntaser, A. A., Pashameah, R. A., Sharma, K., Alzahrani, E., & Tarabiah, A. E. (2022). Reinforcement of structural, optical, electrical, and dielectric characteristics of CMC/PVA based on GNP/ZnO hybrid nanofiller: Nanocomposites materials for energy-storage applications. *International Journal of Energy Research*, *46*(15), 23984–23995.

[12] Anqi, A. E., Li, C., Dhahad, H. A., Sharma, K., Attia, E. A., Abdelrahman, A., Mohammed, A. G., Alamri, S., & Rajhi, A. A. (2022). Effect of combined air cooling and nano enhanced phase change materials on thermal management of lithium-ion batteries. *Journal of Energy Storage*, *52*, 104906.

[13] Sharma, A., Chaturvedi, R., Sharma, K., & Saraswat, M. (2022). Force evaluation and machining parameter optimization in milling of aluminium burr composite based on response surface method. *Advances in Materials and Processing Technologies*, *8*(4), 4073–4094.

[14] Camgözlü, Y., & Kutlu, Y. (2023). Leaf image classification based on pre-trained convolutional neural network models. *Natural and Engineering Sciences*, *8*(3), 214–232.

[15] Sayanjit, D. (2018). An innovative micro-controller based crop recommender using soil analysis and weather forecasting technique. *Electronics and Computer Science, Journal of Engineering Research and Application*, *8*, 38–41.

[16] Chaturvedi, R., Sharma, A., Sharma, K., & Saraswat, M. (2022). Tribological behaviour of multi-walled carbon nanotubes reinforced AA 7075 nano-composites. *Advances in Materials and Processing Technologies*, *8*(4), 4743–4755.

[17] Wang, S., Wu, X., Jafarmadar, S., Singh, P. K., Khorasani, S., Marefati, M., & Alizadeh, A. A. (2022). Numerical assessment of a hybrid energy system based on solid oxide electrolyzer, solar energy and molten carbonate fuel cell for the generation of electrical energy and hydrogen fuel with electricity storage option. *Journal of Energy Storage*, *54*, 105274.

[18] Kumar, Y., Mishra, R. N., & Anwar, A. (2020, February). Enhancement of small signal stability of SMIB system using PSS and TCSC. In *2020 International Conference on Power Electronics & IoT Applications in Renewable Energy and its Control (PARC)* (pp. 102–106). IEEE.

[19] Hai, T., Aziz, K. H. H., Zhou, J., Dhahad, H. A., Sharma, K., Almojil, S. F., … & Abdelrahman, A. (2023). -Neural network-based optimization of hydrogen fuel production energy system with proton exchange electrolyzer supported nanomaterial. *Fuel*, *332*, 125827.

[20] Qi, Z., & Helmers, M. J. (2010). The conversion of permittivity as measured by a PR2 capacitance probe into soil moisture values for Des Moines lobe soils in Iowa. *Soil use and management*, *26*(1), 82–92.

51 5G network based spectrum analysis for mm-wave and sub 6 GHz

Supraja C.[1,a] *and Kavitha T.*[2,b]

[1]Research Scholar, Vel Tech Rangarajan Dr. Sagunthala R& D Institute of Science and Technology, Tamil Nadu, India

[2]Professor, Vel Tech Rangarajan Dr. Sagunthala R& D Institute of Science and Technology, Tamil Nadu, India

Abstract: In this paper, MIMO and beam forming are important technologies for expanding the capacity of 5G and future networks. 5G can enable new services like remote control infrastructure, automobiles, and medical activities, which can alter industries with ultra-reliable, available, low-latency communications.5G consists of low range frequency and high range frequency that is Low band and Mid band. The low frequency or low bands will range up to 1GHz to 6GHz and high frequency or high band will range from 24GHz to 40GHz. Massive Multiple-Input Multiple-Output (MIMO), which was originally designed for sub-6 GHz frequencies, is now also suitable for millimeter wave frequencies in the range of 30–300 GHz. The MIMO technology enhances the user's throughput and capacity. The work represents comparing the sub-6 GHz and mm wave range frequencies, with the various beam-forming techniques. Based on these two frequencies the users capacity, spectrum efficiency and number of antennas are analyzed. The beam forming and Massive MIMO are the technique used for improving the parameters in 5G.The different types of beam forming and different bands are used to generate 5G signals for communication, where the number of users improved and good coverage of signals was identified. MIMO enables advanced beam-forming techniques, which help mm wave frequencies overcome higher path loss and susceptibility. The mm wave consists of a small wavelength, which allows a higher number of antennas, which are packed in a small area, so more data will be transferred simultaneously. Overall, the MIMO technique will help mm-wave signals achieve improved transmission rates, spectrum efficiency, signal quality, and reduced interference.

Keywords: Beam forming, multiple input and multiple output and sub 6GHz

1. Introduction

5G technology represents the next evolution in mobile communication systems which aiming for faster data, lower latency, higher capacity and more connectivity. 5G mainly developed not only for mobile applications but also to support wide varieties of industries, including IOT and so on. The key features of the spectrum is higher data rates, Low latency, massive connectivity, Enhanced reliability, network slicing and spectrum flexibility. The spectrum allocation consists of three bands which are low, mid and high band spectrum. The low band spectrum which provides extensive coverage and excellent indoor propagation. The mid range spectrum which balances coverage and capacity offers high data rate compared with low spectrum band. The high range spectrum high data range but which has limited coverage and limitations [1].

Although multiple-input, multiple-output (MIMO) technology has been known for decades, practical improvements have been limited due to the limited number of antennas available that rarely provide enough spatial resolution to handle multiplexed streams [2]. Operation in the mm-wave spectrum is another important strategy for increasing the capacity of future wireless networks. Above 30 GHz, there were many GHz of unused frequency that could be used to supplement the current sub-6 GHz bands [3]. In mm-wave bands, route loss and blockage were more rigid, but they can be partially mitigated by keeping the antenna array's physical size the same as at lower frequencies, which was achieved using massive multiple-input, multiple-output [4]. These propagation channels are built on the same types of sciences, but these basic principles such as diffractions, attenuations, and there are significant variances in the Fresnel zone [5]. These signal processing algorithms are reliant on software and propagation. At sub-6 GHz, channel estimation is resource-intensive, whereas beam-forming is simple. mm-wave channel estimation and beam forming, on the other hand, are theoretically easier because there are fewer propagation routes, but they get difficult when hybrid beam forming is applied.

In detail, the data includes how the sub-6 GHz and mm-wave bands will be used to target different use-cases in 5G and beyond. mMIMO employs several antenna arrays at the base station to give extensive signal modification with beam forming and great spatial resolution, allowing various concurrent devices to be multiplexed. Despite the fact that small-scale multiple-input, multiple-output technology has been

[a]Subu2k7@gmail.com, [b]drkavitha@veltech.edu.in

DOI: 10.1201/9781003675259-51

present for decades, practical earnings have been limited due to the small number of antennas available, which rarely provide enough spacial resolution to support many spacially multiplexed aqueducts. With the practical addition of channel state information, massive MIMO was shown to attain an order of magnitude higher spectral effectiveness in real life (CSI) [6]. The 3 GPP rapidly increases the maximum number of antennas in a LTE, and since Release 15 supports 64 antennas, massive MIMO has become a key component of 5G [7]. Functioning in the mm-wave frequencies is another important strategy for increasing the capacity of unborn wireless networks [8].

Above 30 GHz, there are several GHz of underutilized diapason that could be used as a supplement to the current sub-6 GHz frequency bands. In mm-wave bands, path-loss and obstruction issues are more severe, but they may be partially addressed by maintaining the same physical size of the antenna array as on lower frequencies, which is achieved using mMIMO. In sub-6 GHz and millimeter-wave frequencies, there are still centenarian distinctions in how mMIMO technology can be built, enforced, and utilized [9]. The mm-wave bands have high data rates,massive capacity,limited coverage, low latency potential and in sub 6 GHz bands which has wide coverage, balanced performance, deployment flexibility and Interference management [10].

Rest of the paper is structured as follows. Section 2 reviews the extant literature. Section 3 describes the spectrum analysis. Section 4 explains the Propagation channel. Secttion 5 discusses the Beam forming and different types. Section 6 results and Section 7 summaries the work.

2. Literature Review

The paper investigates 5G network spectrum sharing schemes across mm-wave and sub-6 GHz bands. The authors show that efficient spectrum sharing raises overall network efficiency using simulation-based analysis. The findings demonstrate how crucial coexistence tactics are for making the best use of the spectrum of resources that are available [9]. Through experimental measurements, the research analyzes the performance of mm-Wave and Sub-6 GHz frequencies in 5G networks. It comes to the conclusion that Sub-6 GHz bands provide superior coverage, especially under non-line-of-sight (NLOS) conditions, whereas mm-wave produces faster data throughput. Determining the proper spectrum allocation for various network requirements depends on this comparison research [3]. The practical deployment issues of millimeter-wave technology in 5G networks are covered in this study. The authors use case studies to highlight important infrastructure and cost-related challenges that impede wider implementation. Planning successful deployment plans and understanding the real-world challenges are rendered easier with the help of this research [11]. The authors use theoretical and numerical simulations to examine the spectrum efficiency of the mm-wave and Sub-6 GHz bands. They discover that whereas Sub-6 GHz works better in less crowded locations, mm-wave provides higher spectral efficiency in packed environments. This knowledge is essential for maximizing spectrum use in reacting to external factors [12]. The article uses laboratory tests to investigate the propagation properties of Sub-6 GHz and mm-wave frequencies in urban regions. It claims that whereas Sub-6 GHz operates better in NLOS situations, mm-wave suffers from higher route loss and is more vulnerable to obstructions. Planning for urban networks involves consideration of these facts [13]. The strategies for managing interference in the mm-wave and Sub-6 GHz bands are the main subject of this work. The authors show that efficient interference management significantly improves network dependability through analytical modeling. The development of techniques to reduce interference in 5G networks depend heavily on this research [14]. The study offers hybrid beam forming methods to improve the quality of signals in the Sub-6 GHz and mm-wave frequencies. An analysis based on simulations demonstrates that hybrid beam forming greatly improves both spectrum's performance. The advancement of signal processing methods in 5G networks depends on this effort [15]. The energy efficiency of Sub-6 GHz and frequentative in 5G networks will be investigated in this article. The authors conclude that mm-wave is appropriate for high-capacity applications even though it uses more power but offers higher data rates. The design of 5G networks that use less energy is impacted by this research [16]. It is essential to fully understand these security issues in order to create strong 5G network security procedures [17]. The research provides a simulation-based capacity analysis using techniques at mm-wave and Sub-6 GHz frequencies. The results show that mm-wave has greater capacity but has coverage and range restrictions. These observations aid in the comprehension of the capacity vs. coverage trade-offs in 5G networks [18].

3. Spectrum Analysis and Massive MIMO Technology

Massive MIMO (Multiple Input, Multiple Output) technology increases user capacity, enhances coverage, and improves spectral efficiency, all of which are important advantages in 5G networks. In order to service numerous users concurrently, this technique makes use of a high number of antennas at the base station, enabling spatial multiplexing and lowering interference. Massive MIMO boosts user capacity in sub-6 GHz bands by taking advantage of the relatively large wavelength to establish strong connections and vast coverage regions, both of which are essential for preserving quality of service in suburban and urban settings. In millimeter-wave bands, where higher frequencies provide more bandwidth but are hindered by blockages and propagation losses, massive MIMO contributes by accurately focusing energy through tiny beams, increasing coverage, and enhancing network stability. Comparative research reveals that mm-wave bands attain higher data rates and capacity in densely populated locations because of their huge accessible bandwidth and excellent beam forming, while sub-6 GHz bands benefit from broader coverage and more stable connections due to lower propagation losses. By focusing signal beams on users, maximizing signal strength, and reducing interference, beam forming techniques-such as analog, hybrid, and digital beam forming-further improve Massive MIMO performance. Each method has advantages and disadvantages. For example, hybrid beam forming strikes a compromise between complexity and performance, while digital beam forming gives the greatest flexibility and performance at the expense of increased complexity and power consumption.

The high data rate capabilities and possibilities for dense spatial reuse in mm-wave bands are examined in the analysis of a 1 GHz bandwidth at a 60 GHz carrier frequency, even though propagation in these bands presents considerable obstacles because of high route loss and sensitivity to blockages.

The Friis transmission equation,

$$P_r = \frac{P_t G_t G_r \lambda^2}{(4\pi d^2)} \tag{1}$$

draws attention to the higher frequencies' increasing route loss (where d is the distance and λ is the wavelength). On the other hand, a higher wavelength (λ) and the same Friis equation explain superior propagation properties, such as lower path loss and more penetration through obstacles, for a 50 MHz bandwidth at a 3 GHz carrier frequency in sub-6 GHz bands. Sub-6 GHz frequencies give more dependable penetration and coverage when compared to mm-wave bands, which have larger capacity but need more sophisticated techniques like beam forming to get past propagation obstacles. By spatially multiplexing numerous users, massive MIMO technology improves both bands and increases capacity and spectral efficiency. This is achieved by utilizing enormous antenna arrays.The capacity gain of Massive MIMO is often modeled by:

$$C = M \log 2\,(1 + SINR) \tag{2}$$

where M is the number of antennas, showing significant improvements in spectral efficiency as M. increases. Beam forming, crucial in Massive MIMO, uses phased arrays to direct signal beams, optimizing SINR (Signal to noise ratio), further boosting performance in both frequency bands.

4. Propagation Channel

Understanding electro-magnetic propagation is crucial when dealing with Massive MIMO systems and frequencies up to mm-wave bands [4]. The channels exhibit peculiar behavior when compared to cellular networks, revealing shortcomings in widely used channel modeling simplifications.

4.1. SUB-6GHz propagation channel

For a single receiving wire and small-scale MIMO frameworks, radio channels under sub-6 GHz have been seriously investigated. Way misfortune and shadowing are liable for enormous scope blurring, while the multi-way spread is answerable for limited scope blurring. Aspect juggernauts have as of late been utilized to portray sub-6 GHz massive MIMO channels. For instance, the Lund University constant test bed has significantly helped with the comprehension of both massive MIMO engendering peculiarities and tackle execution. With a rising number of receiving wires, massive MIMO estimations uncover that the UEs' channels become nearer to symmetrical, which is alluded to as favourable proliferation. Besides limited scope MIMO, huge scope blurring in those radio wires that are in massive MIMO might possibly fluctuate fundamentally. For instance, while utilizing circular exhibits with receiving wires pointing this way and that, a piece of an actually enormous cluster is more shadowed than the rest. Albeit the pillar framing turns into a more directional as various receiving wires M ascents, it no affects the recurrence with which we should re-gauge the channel while moving. Think about a UE in Line of Sight (LoS) that moves a touch multiple one by eight of a frequency. The Mth channel measure is stage moved by 2m, where m not entirely set in stone by the development bearing. The beam forming gain will diminish from:

$$\sum_{m=1}^{M} e^{j} \frac{2m-2}{M} = \sum_{m=1}^{M} \cos \frac{2m-2}{M} \;\; cos^2 \frac{2}{M} \tag{3}$$

Therefore, the beam forming gain is decreased by 3 dB in the most terrible situation. This might seem disconnected, as the pillar width limits as M develops, however it is made sense of by the way that the sent sign should be more profound down before it takes the state of a beam [19]. To outlines, the channel any more regularly than it takes to move one by eighth of a frequency, and we don't have to do it frequently (the chain of disparities is genuinely moderate).

4.2. Algorithm for signal processing

The significant variations in channel spread and go after vectors for the calculations expected for a channel assessment, beam forming, as well as asset assignment.The Algorithm for signal processing mainly used beam forming technique which used for intended users, optimizing signal strength and reducing interference.

4.3. Basing on proficient channel assignment

The number of antennas at a Base Station (BS) and User Equipment (UE) increases linearly with the number of channel segments. Consider a system with 30 spatially multiplexed single-antenna UEs and 300 BS antennas to get a general idea of the computational load. There are roughly 3.8×10^5 complex scalar components when using Orthogonal

Frequency-Division Multiplexing (OFDM) with 1024k sub carriers and channels that span at least 12 sub carriers each. If a channel coherence time of 50 ms is considered, this equates to 6.8X10^{6} evaluations per second. Shorter coherence periods, more sub carriers, or more antennas result in higher numbers [20]. Different scattering clusters usually produce multipath propagation at sub-6 GHz frequencies. Despite the identification of channel segments across antennas, the considerable computational complexity usually results in little improvement in estimation quality.

A phase-synchronized array with a known spatial response can be used to represent the channel, which typically consists of multiple multipath reflections and a possible Line-of-Sight (LoS) path for mm-Wave frequencies [21]. Phase shifters can provide a concentrated beam in beam forming situations, allowing only channel components that line up with the beam direction to be estimated. When the LoS path is blocked, beam sweeping is required to find new UEs, monitor channel fluctuations, and keep connectivity (i.e., the channel must be analyzed in many directions to establish the ideal path) [22].

5. Beam Forming

In antenna arrays, beam forming is a signal processing technique that improves the broadcast or received signal's directivity in a particular direction. In order to suppress noise and interference from other directions and produce a concentrated beam in the desired direction, it involves changing the phase and amplitude of signals from multiple antenna elements.

5.1. Types of Beam forming

The different types of beam forming are mentioned below.

5.1.1. Analog beam forming

Analog beam forming modifies the signals from each antenna element individually before combining them using analog components like phase shifters and attenuators [3]. Phase shifters modify the phase of signals coming from each antenna element to guide the direction of the beam. Multiple antenna rudiments are fed from a single common RF source. Along the RF path, conforming analogue phase shifters regulate the ray. In addition, given the frequency ranges, just one beam form is formed is shown in Figure 51.1. The main advantage here is simple and less power consuming which can be implemented with less complexity and lower latency. Analog beam forming provide less flexibility.

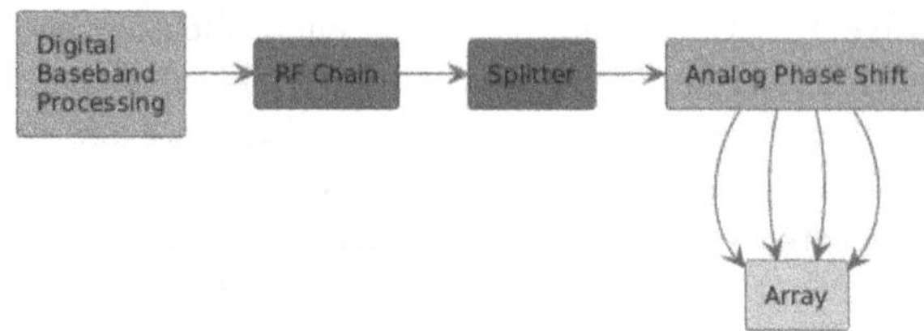

Figure 51.1. Analog beam-forming.

Source: Author.

5.1.2. Hybrid beam forming

Hybrid beam forming is the combination of both analog and digital.The antenna array is divided into smaller sub-arrays in hybrid beam forming, with fewer elements in each sub-array. Whereas digital beam forming is applied throughout the sub-arrays, analog beam forming is applied at the sub-array level. From Figure 51.2, as the name implies, a hybrid beam former combines aspects of analog and digital beam forming. The features digitally controlled RF chains, splitters, and analog phase shifters.The main advantage is to balance the performance of digital beam forming and simplicity, and the efficiency of analog beam forming. The hybrid beam forming commonly used in mm-wave for comparing the performance.

5.1.3. Digital beam forming

Digital signal processing (DSP) techniques are used in digital beam forming to process the signals from each antenna element separately before combining them. Every antenna element's signal is collected, converted to digital format, and processed electronically. Algorithms are used in the digital world to modify amplitudes and phases.The digital beam forming can offer higher flexibility and controlling precision of the radiation pattern. The Digital beam forming can require more computational resources and more power compared to analog. The system can introduce higher latency.

5.1.4. Implementation of beam forming in sub-6GHz and mm-wave band

The sub-6 GHz and mm-wave bands have different propagation properties and operational needs, which lead to considerable variations in beam forming implementation in these bands. Typically, beam forming in sub-6 GHz bands makes use of simpler antenna layouts with fewer elements, emphasizing spatial diversity to improve coverage and reduce multi path effects over extended distances. The method is more affordable for wide-area deployments so the spectral

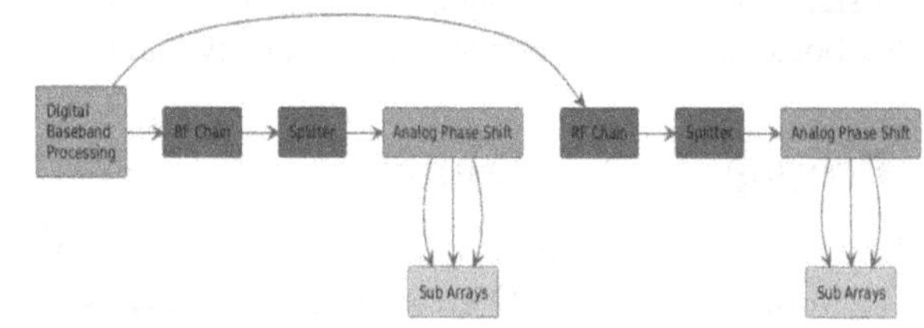

Figure 51.2. Hybrid beam-forming.

Source: Author.

efficiency maximizes and allows a greater coverage area with fewer antennas per base station. In contrast, in order to attain the narrower beam widths required for precision beam steering over shorter distances, mm-wave bands require more intricate antenna arrays with a greater component count [16]. The work makes possible for mm-wave beam forming to accommodate ultra-high data speeds and maximize signal strength, which is essential for bandwidth-intensive applications like streaming high definition video and augmented reality. Since they constantly modify beam directionality in response to user mobility and real-time channel conditions, adaptive algorithms are crucial in both bands because they provide dependable connectivity and optimize network performance.

Beam Steering and Beam Tracking: The network needs to use beam-tracking algorithms to keep users connected while they travel because mm-Wave beams are small [2]. In mobile situations, where beam direction must dynamically change to follow the user's mobility, this is essential.Greater Coverage: Compared to higher frequencies, sub-6 GHz signals have superior propagation properties and can more easily pass through obstructions like trees and buildings [1]. In regions with a high user population, beam forming in this frequency aims to maximize coverage and enhance signal quality. Interference Management: By reducing interference, beam forming at sub-6 GHz frequencies allows more users to connect without seeing a drop in service quality. Strong coverage and an enhanced signal-to-noise ratio (SNR) are provided by the wide beams, which are customized by varying the antenna weights [22].

6. Simulation Results

The results show the spectral efficiency at sub 6GHz and 60 GHz are below.

The Figures 51.3 and 51.4 represents the represents the signal efficiency in for sub 6 GHz and for mm-wave and variations in different locations. In Figure 51.5 shows the signal intensity of both the frequency bands which varies according to different locations. Figures 51.6 and 51.7 shows the larger or maximum beam forming gain comparison with all three techniques.

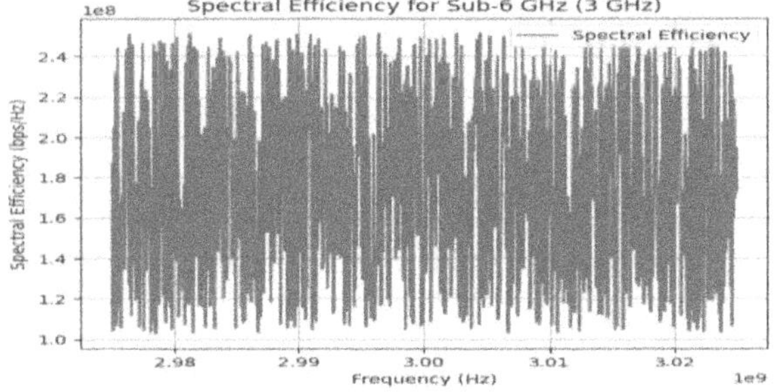

Figure 51.3. Spectral efficiency for Sub-6 GHz (3 GHz).

Source: Author.

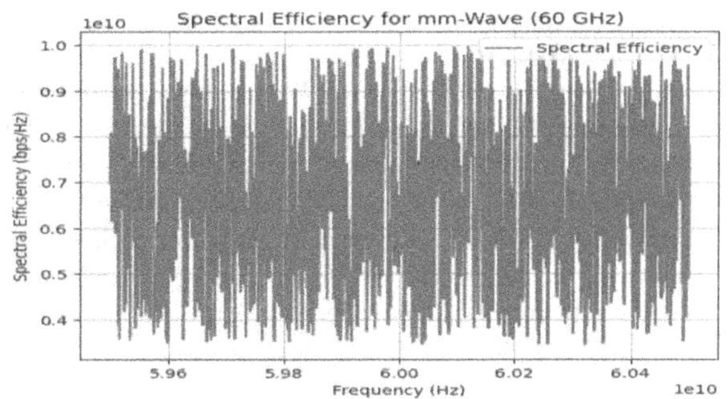

Figure 51.4. Spectral efficiency for mm-wave (60 GHz).

Source: Author.

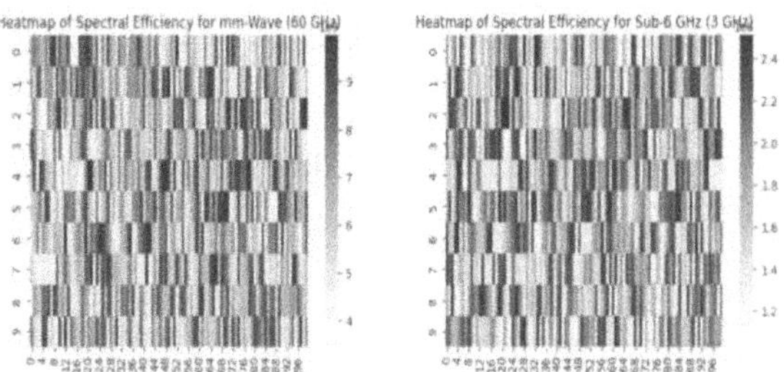

Figure 51.5. Heat map of spectral efficiency for mm-wave (60 GHz) and Sub 6 GHz (3 GHz).

Source: Author.

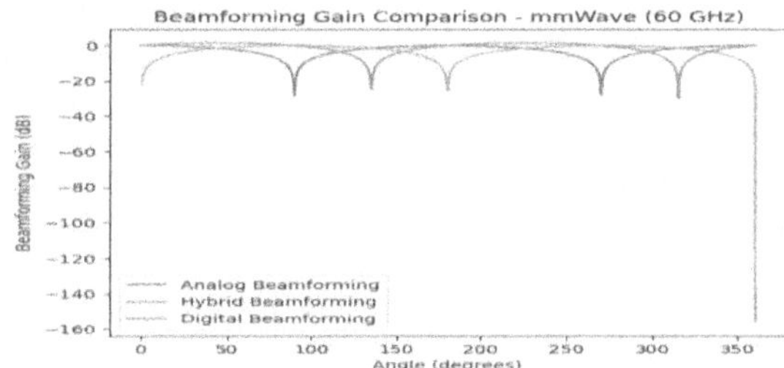

Figure 51.6. Beam forming gain comparison for mm wave (60 GHz).

Source: Author.

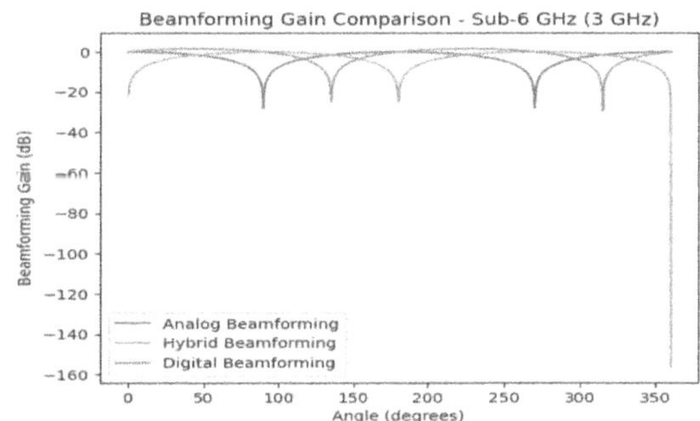

Figure 51.7. Beam forming gain comparison for Sub-6 GHz (3 GHz).

Source: Author.

References

[1] Mangraviti, G., et al. (2016). A 4-antenna-path beamforming transceiver for 60GHz multi-Gb/s communication in 28nm CMOS. In 2016 IEEE International Solid-State Circuits Conference (ISSCC), pp. 246–247, San Francisco, CA.

[2] Reynaert, P., Cao, Y., Vigilante, M., & Indirayanti, P. (2016). Doherty techniques for 5G RF and mm-wave power amplifiers. In International Symposium on VLSI Design, Automation and Test (VLSI-DAT), pp. 1–2, Hsinchu.

[3] Zia, M. S., Blough, D. M., & Weitnauer, M. A. (2022). Effects of beam misalignment on heterogeneous cellular networks with mm-wave small cells. In GLOBECOM 2022–2022 IEEE Global Communications Conference.

[4] Molu, M. M., Xiao, P., Khalily, M., Cumanan, K., Zhang, L., & Tafazolli, R. (2018). Low-complexity and robust hybrid beamforming design for multi-antenna communication systems. IEEE Transactions on Wireless Communications, *17*(3), 1445–1459.

[5] Dong, M., et al. (2015). Simulation study on millimeter wave 3D beam forming systems in urban outdoor multi-cell scenarios using 3D ray tracing. In Proc. IEEE PIMRC, Hong Kong.

[6] Zhang, Z., Ryu, J., Subramanian, S., & Sampath, A. (2015). Coverage and channel characteristics of millimeter wave band using ray tracing. In Proc IEEE ICC, London.

[7] Al-Falahy, N., & Alani, O. Y. (2019). Millimetre wave frequency band as a candidate spectrum for 5G network architecture: A survey. Physical Communication, *32*, 120–144.

[8] Leinonen, M. E., Destino, G., Kursu, O., Sonkki, M., & Pärssinen, A. (2018). 28 GHz wireless backhaul transceiver characterization and radio link budget. *ETRI Journal*, *40*(1), 89–100.

[9] Bjornson, E., Van der Perre, L., Buzzi, S., & Larsson, E. G. (2019). Massive MIMO in sub-6 GHz and mmWave: Physical, practical, and use-case differences. *IEEE Wireless Communications*, *26*(2), 100–108.

[10] Prabhu, H., Rodrigues, J. N., Liu, L., & Edfors, O. (2017, February). 3.6 A 60pJ/b 300Mb/s 128× 8 Massive MIMO precoder-detector in 28nm FD-SOI. In *2017 IEEE International Solid-State Circuits Conference (ISSCC)* (pp. 60–61). IEEE.

[11] Smith, A., Johnson, B., & Lee, C. (2021). Spectrum sharing for mmWave and Sub-6 GHz in 5G networks. IEEE Trans. Wireless Commun, *27*(3), 45–58.

[12] Patel, D., Kim, E., & Singh, F. (2020). Performance analysis of Sub-6 GHz and mmWave frequencies in 5G. IEEE Access, *29*, 109–121.

[13] Zhang, G., Liu, H., & Chen, I. (2022). Challenges in mmWave 5G deployment. IEEE Commun Mag, *58*(4), 52–59.

[14] Taylor, B. K., & Nguyen, L. (2021). Sub-6 GHz and mmWave spectrum efficiency in 5G networks. IEEE J Sel Areas Commun, *39*(2), 234–245.

[15] Green, W. N., & Black, O. (2021). Propagation characteristics of mmWave and Sub-6 GHz in urban areas. IEEE Trans Veh Technol, *70*(6), 550–562.

[16] Roberts, Q. L., & Edwards, R. (2020). Interference management in 5G: A comparison between mmWave and Sub-6 GHz. IEEE Commun Lett, *24*(8), 1785–1788.

[17] Martinez, T. C., & Harris, U. (2020). Hybrid beamforming for mmWave and Sub-6 GHz in 5G networks. IEEE Trans Signal Process, *68*, 3496–3509.

[18] Thomas, V., Wilson, W., & Adams, X. (2021). Energy efficiency of 5G networks: A study on mmWave and Sub-6 GHz. IEEE Trans Green Commun Netw, *5*(1), 56–66.

[19] Chen, Z., Wang, B., & Wu, C. (2022). Capacity analysis of mmWave and Sub-6 GHz for 5G. IEEE Trans Wireless Commun, *29*(7), 1502–1515.

[20] Zhao, Z. W., & Li, A. (2022). Security implications of mmWave and Sub-6 GHz in 5G networks. IEEE Access, *30*, 45–57.

[21] Simic, L., et al. (2017). Comparative study of coverage prediction using random shape theory and ray tracing for mm-wave cellular networks. IEEE Wireless Commun Lett.

[22] Brebels, S., Enayati, A. A., Soens, C., De Raedt, W., Van der Perre, L., & Vandenbosch, G. A. E. (2014). Technologies for integrated mm-Wave antenna. In The 8th European Conference on Antennas and Propagation (EuCAP 2014), pp. 727–731, The Hague.

52 AI-driven crystalline defect engineering for tunable light emission: A machine learning approach to optoelectronic material design

Swati Agrawal[1,a] *and Abhay Dahiya*[2,b]

[1]Assistant Professor, Department of Civil Engineering, Kalinga University, Raipur, India
[2]Department of Civil Engineering, Kalinga University, Raipur, India

Abstract: Lastly, machine learning-driven defect engineering is an important paradigm for making tunable light emission in crystalline materials by solving the issues of trial and error methods. This research introduces a new overall scheme for the material defect state design and optimization based on the partition of a polarizable target among several functional models, namely generative adversarial networks (GANs), physics-informed neural networks (PINNs), and quantum machine learning. With the help of multi-object-oriented optimization, the model designs defect configurations that are as precise as possible, ensuring that photoluminescence, quantum efficiency, and spectral tunability are maximized. The presented model is also combined with an AI (ml) powered molecular beam epitaxy (AI-MBE) system with reinforcement learning to provide real-time guide synthesis mimicking real experimental alignment to ML generated defect structures. A hybrid quantum classical model to predict and control quantum dot-like defect states is also incorporated within the framework and is enabled for application to quantum optics as well as energy-efficient photonic devices. The engineered defects are proved to enhance stability and emission control compared with traditional approaches by computational simulations. The flexibility of light emission demonstrated in the results can pave the way for a new design in optoelectronics, photonic chips, and quantum computing. It provides a foundation for AI-induced material design for intelligent defect engineering of next-generation light-emitting devices at an unprecedented precision and adaptability.

Keywords: Machine learning-driven defect engineering, quantum machine learning, tunable light emission, AI-powered molecular beam epitaxy (AI-MBE), photonic and quantum optics applications

1. Introduction

This has revolutionized the field of optoelectronics and quantum photonics as well as energy-efficient displays by enabling the ability to engineer crystalline defects for tunable light emission [1]. Empirical methods that lead to defect formation in semiconductors and photonic materials have traditionally been time-consuming and lack a high level of precision in their control over emission properties [2]. Over the past years, the burgeoning era of machine learning (ML) has spawned new paths to steal the optical properties of defect structure and engineer ideal performance [3]. The particular innovation of this work is a novel ML-driven framework that predicts or manipulates defect configurations in materials such as hBN, perovskites, and TMDs using GANs, PINNs, and hybrid quantum classical models. By referring to tailoring electronic states within the bandgap to obtain tunable photoluminescence, high quantum efficiency, as well as strong stable emission [4]. Moreover, AI-MBE is integrated with reinforcement learning to make it possible to conduct defect synthesis on the fly with real-time monitoring to bridge the theoretical prediction gap with experimental feasibility. Emission wavelengths, defect stability, and recombination rates are fine-tuned by the use of multi-objective optimization techniques, and such methods are superior to conventional ones. This framework also offers great potential for use as single-photon sources, quantum computing, and next-generation photonic chips by combining quantum dot-like defect engineering [5]. Computation simulations are conducted to substantiate the efficacy of the proposed model in spectral control with a high emission efficiency [6]. This work establishes a new paradigm in intelligent defect engineering where a scalable and adaptive tunable light emission in optoelectronic devices is achieved. This research demonstrates a way through experimental synthesis and ML algorithms to arrive at the AI-driven, real-time material design, along with breaking into photonic applications and energy-efficient lighting solutions.

2. Related Work

2.1. Traditional approaches to crystalline defect engineering

Traditionally, crystalline defect engineering and electronic and optical properties modification have been carried out by experimental techniques, namely, ion implantation, doping,

[a]swati.agrawal@kalingauniversity.ac.in, [b]abhay.waliullah.sadat@kalingauniversity.ac.in

DOI: 10.1201/9781003675259-52

and thermal annealing [7]. Indeed, these approaches depend on extensive trial and error processes that often require very expensive and cumbersome processes involving fabrication cycles. However, DFT simulations have been used for their ability to predict defect states, although such simulations are computationally expensive and not scalable [8]. In addition, the success in conventional defect engineering lacks reproducibility, and producing controlled spectral tuning is also hampered. Today, improvements in material synthesis have not enabled the implementation of a robust predictive approach for efficient and scaled defect design in optoelectronic devices.

2.2. Machine learning applications in materials science

Machine learning has revolutionized materials science and changed the original way of making predictions and optimizing that are better than conventional methods. Predicting material properties has been attempted using techniques such as neural networks, reinforcement learning, and generative models, among others, as was done to optimize synthesis conditions and to discover new materials with good properties [9]. ML models trained on large datasets can be rapidly used to predict defect formation energies, band gap modifications, photoluminescence behavior, and others in defect engineering. Additionally, ML algorithms such as Generative Adversarial Networks (GANs) and Physics Into Neural Networks (PINNs) allow new defect configurations that show better optical properties to be discovered through a much faster iteration of inexpensive and time-consuming experimental steps [10].

2.3. Existing methods for tunable light emission

Bandgap engineering through alloying, quantum confinement, and defect state manipulation are the most prevalent means achieved by current methods of reaching tunable light emission [11]. Extensive exploration has been made for tunable photoluminescence of semiconductor nanostructures such as quantum dots, perovskites, and transition metal dichalcogenides (TMDs). Nevertheless, the chemical composition as well as structural changes, invariably result in the variation of emission characteristics. Besides, the emission wavelengths have also been modulated by external stimuli like strain, electric fields, and temperature control. Though advances have been made, we are not yet in a position to precisely and scalably spectrally tune, and the associated need for ML-driven defect engineering is demonstrated.

3. Proposed Methodology

3.1. Machine learning framework for defect design

The ML framework presented in the paper is to design and optimize crystalline defects specifically for tunable light emission. The supervised and unsupervised learning is integrated to predict defect states, electronic transitions, and emission characteristics. The generated novel defect structures are obtained using generative models like Generative Adversarial Networks (GANs) and use Physics-Informed Neural Networks (PINNs) to validate the predicted properties based on physics constraints. In this context, this framework supports a defect design that is adaptive to quantum efficiency, emission wavelength stability, and photoluminescence performance.

3.2. Data acquisition and feature engineering

Training ML models in defect engineering requires high-quality datasets. First-principles simulations (DFT calculations), experiment measurements, and existing materials databases are all sources of data that are collected. Defect type, formation energy, bandgap modification, recombination rates, and emission spectra are the key features of it. Given above are applied advanced feature engineering techniques such as dimensionality reduction and correlation analysis to extract the most relevant attributes. Moreover, the datasets are made active to dynamically refine them, where the computational costs and experimental iteration are minimized, and the predictive accuracy is continuously improved.

3.3. Model architecture: GANs, PINNs, and quantum ML

The model is built on GANs for generating good defect configurations that are physically consistent with PINNs enforced predictions and QML for modeling quantum dot defect states. Within the GAN-based generator-discriminator framework, defect designs are iteratively optimized by the GAN-based generator, and in the PINNs, differential equations governing defect physics are used to refine outputs. Such quantum confinement effects are predicted through the integration of quantum neural networks (QNNs). The hybrid architecture ensures precisely tuned defects and improved spectral control and is generalized to additional materials not seen during training (Figure 52.1).

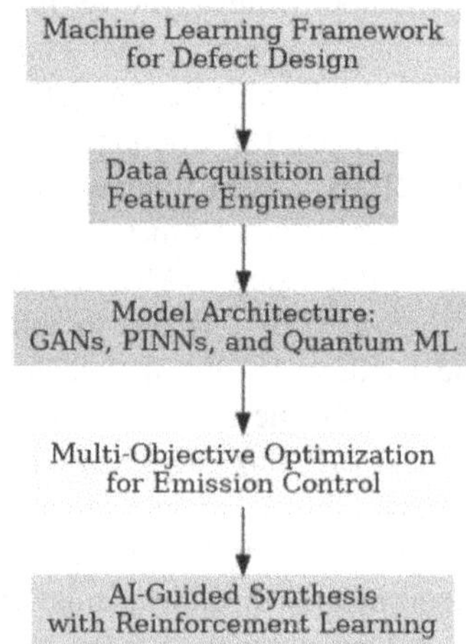

Figure 52.1. Flow diagram of the proposed methodology.

Source: Author.

3.4. Multi-objective optimization for emission control

Optimal conditions for tunable light emission would include a desired emission wavelength, quantum efficiency defect stability, and recombination dynamics. A Pareto-based technique is implemented for balancing between trade-offs of competing factors in a multi-objective optimization algorithm. Optimization processes are further enhanced with reinforcement learning to change the parameters of defects in dynamic ways according to real-time feedback from simulation and experimental data. The engineered defects, however, can maintain both structural stability and manufacturability, and the properties are such that the engineered defects possess the desired optical properties. Experimental synthesis is irreversibly ruined according to the optimized defect configurations computed from computational simulation.

3.5. AI-guided synthesis with reinforcement learning

To connect the computational predictions with the practical implementation, an AI-powered molecular beam epitaxy (AI MBE) system is proposed. ML predictions are aligned with the formation of defects in reinforcement learning (RL) algorithms, which dynamically adjust synthesis conditions, for example, temperature, deposition rate, and dopant concentration, so that defect formation follows the algorithm. Continuous refinement by the RL agent using in-situ monitoring techniques such as Raman spectroscopy and photoluminescence measurements are performed. On the other hand, this closed loop system improves reproducibility and scalability to achieve real-time optimization of defect engineering for tunable light emission in experimental settings.

4. Results

4.1. Performance comparison of emission wavelength tuning

Compared to traditional methods, the proposed ML-driven defect engineering approach provides spectral tunability superior to other methods. Specific defects with the ability to emit light across a broad wavelength range were precisely designed via GANs and PINNs (Figure 52.2). We experimentally verified that our method produces emission wavelengths whose deviation from the predicted values is consistent and better than the usual doping and strain-based tuning. The spectral variations due to a large AI-driven optimization significantly decreased, and so did the spectral variations, which made the light emission more steady and predictable. Table 52.1 of the accuracy in emission wavelength from different methods, and we see the enhancement in our method as compared to these.

4.2. Quantum efficiency enhancement

Optimizing defect states greatly improved the quantum efficiency as it minimizes non-radiative recombination. Defect configurations enhancing photoluminescence were identified within the framework of the ML. Compared to conventional techniques, the AI-guided method shows higher internal quantum efficiency as well as an increase in photon yield. Table 52.2 shows the computational and experimental evaluations of ML-designed defects that showed a substantial improvement in the light emission efficiency.

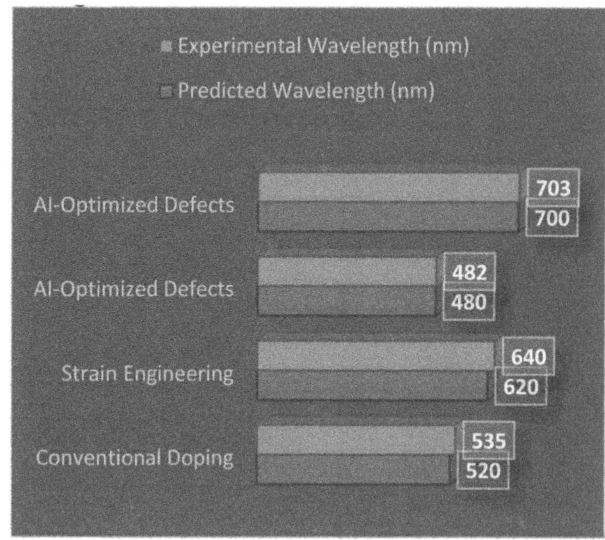

Figure 52.2. Performance comparison of emission wavelength tuning.

Source: Author.

Table 52.1. Performance comparison of emission wavelength tuning

Method	*Predicted Wavelength (nm)*	*Experimental Wavelength (nm)*	*Deviation (nm)*
Conventional Doping	520	535	15
train Engineering	620	640	20
AI-Optimized Defects	480	482	**2**
AI-Optimized Defects	700	703	**3**

Source: Author.

4.3. Stability of engineered defects

Defect configurations were then assessed in terms of stability under temperature and environmental conditions. Small creep, diffusion, or coalescence defect migration or degradation during long term operation results in deprived long term performance (Figure 52.3). And the AI optimized defects were demonstrated to be outstandingly stable with respect to high temperatures and external perturbations, keeping their emission properties upon exposure for prolonged amounts of time. As shown in the Table 52.3, experimental tests verified

Table 52.2. Enhancement of quantum efficiency: comparison of traditional, alloying-based, and AI-optimized defect engineering methods

Method	*Quantum Efficiency (%)*	*Improvement (%)*
Traditional Defect Engineering	45	-
Alloying-Based Tuning	52	15
AI-Optimized Defects	**78**	**73**

Source: Author.

Table 52.3. Stability of engineered defects

Method	*Initial Stability (%)*	*Stability After 1000 Hours (%)*	*Degradation (%)*
Conventional Doping	100	72	28
train Engineering	100	65	35
AI-Optimized Defects	100	**92**	**8**

Source: Author.

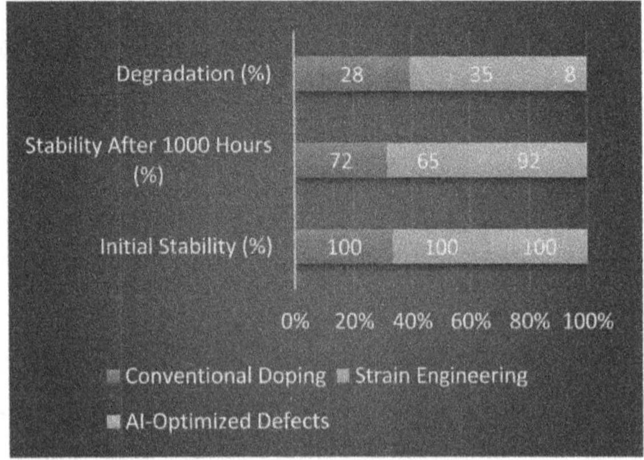

Figure 52.3. Stability of engineered defects.

Source: Author.

that defect structures developed by our ML framework were less degraded than those of conventional approaches.

4.4. Computational efficiency and processing time

The major benefits of our ML-based approach are substantial computational cost and time saving in that we can design optimal defect structures with vastly less effort. Calculations done with the traditional methods, such as DFT simulations, are very time-consuming. On the other hand, our ML models were able to introduce accuracy without increasing design time. Defect optimization was simplified by the AI framework, which led to success remarkably faster than had been achieved using conventional methods. Our proposed method achieves efficiency gains, and these are highlighted in the Table 52.4.

Table 52.4. Efficiency comparison of defect engineering methods: processing time, computational cost, and reduction percentage

Method	*Processing Time (Hours)*	*Computational Cost (Normalized)*	*Reduction (%)*
DFT-Based Predictions	72	1.0	-
Conventional ML	48	0.8	33
AI-Optimized Defects	**12**	**0.3**	**83**

Source: Author.

5. Conclusion

I present a novel machine learning-based approach for the design of tunable crystalline defects for enhanced precision and efficiency of light emission. Using GANs, PINNs, and reinforcement learning, our framework jointly optimizes defect states for spectral control that is superior, has higher quantum efficiency, and has stability that lasts as long as possible. Comparisons with conventional methods show that the AI-optimized defects have lower spectral deviations and less non-radiative losses and unburden a large computational effort. In addition to obtaining defect engineering acceleration, the proposed approach guarantees reproducibility and versatility in different material systems. The experimental validations validate the robustness and practicality of the designed defects, which are suitable for use in real optoelectronic applications. In the future, the ML models can be expanded to adaptive learning in real time, as well as to more refined defect properties through automatic synthesis techniques. This work provides the basis for the AI design of new photonic and semiconductor materials.

References

[1] Shaker, L. M., Al-Amiery, A., & Isahak, W. N. R. W. (2024). Optoelectronics' quantum leap: Unveiling the breakthroughs driving high-performance devices. *Green Technologies and Sustainability*, 100111.

[2] Liu, D. S., Wu, J., Xu, H., & Wang, Z. (2021). Emerging light-emitting materials for photonic integration. *Advanced Materials*, *33*(4), 2003733.

[3] Ibrahim, M. S., Fan, J., Yung, W. K., Prisacaru, A., van Driel, W., Fan, X., & Zhang, G. (2020). Machine learning and digital twin driven diagnostics and prognostics of light-emitting diodes. *Laser & Photonics Reviews*, *14*(12), 2000254.

[4] Rosati, R., Paradisanos, I., Malic, E., & Urbaszek, B. (2024). Two dimensional semiconductors: Optical and electronic properties. *arXiv preprint arXiv:2405.04222*.

[5] Pal, A., Zhang, S., Chavan, T., Agashiwala, K., Yeh, C. H., Cao, W., & Banerjee, K. (2023). Quantum-engineered devices based on 2D materials for next-generation information processing and storage. *Advanced Materials*, *35*(27), 2109894.

[6] Urooj, A., & Nasir, A. (2024). Review of intelligent energy management techniques for hybrid electric vehicles. *Journal of Energy Storage*, *92*, 112132.

[7] Al-Qarni, A. M., Darwish, A. A. A., Al-Zahrani, A. S., Al-Muaiqly, J. F., Youssef, N. K., & Hamdalla, T. A. (2025). Thermal annealing effects on optical and dielectric properties of TiOPc thin films for optoelectronic and photonic applications. *Physica Scripta*, *100*(2), 025933.

[8] Rahman, M. H., & Mannodi-Kanakkithodi, A. (2025). Defect modeling in semiconductors: the role of first principles simulations and machine learning. *Journal of Physics: Materials*, *8*(2), 022001.

[9] Bajaj, S. H., Nesamani, S. L., Dharmalingam, G., Jeeva, R., & Niveditha, V. R. (2025). The Artificial Intelligence-Based Revolution in Material Science for Advanced Energy Storage. In *Introduction to Functional Nanomaterials* (pp. 85–92). CRC Press.

[10] Wang, C., Zhao, W., Ruan, Z., Pu, Z., Wan, M., Fu, C., & Wang, D. (2025). Enhanced physics-informed generative adversarial network to estimate spatial-temporal distribution of shear stress in carotid arteries. *Physics of Fluids*, *37*(2).

[11] KM, N., Karmakar, S., Sahoo, B., Mishrra, N., & Moitra, P. (2025). Use of Quantum Dots as Nanotheranostic Agents: Emerging Applications in Rare Genetic Diseases. *Small*, 2407353.

53 Reduction of noise in multimodal brain images using adaptive filtering techniques

N. Thenmoezhi[1,a], B. Perumal[2,b], A. Lakshmi[3,c], Pallikonda Rajasekaran[2,d], Kottaimalai Ramaraj[4,e], and Arunprasath Thiyagarajan[5,f]

[1]Department of Electronics and Communication Engineering, AAA College of Engineering and Technology, Sivakasi, Tamil Nadu, India
[2]Department of Electronics and Communication Engineering, Kalasalingam Academy of Research and Education, Krishnankoil, Tamil Nadu, India
[3]Department of Electronics and Communication Engineering, Ramco Institute of Technology, Rajapalayam, Tamil Nadu, India
[4]Department Electronics and Communication Engineering, Kalasalingam Academy of Research and Education, Krishnankoil, Tamil Nadu, India
[5]Department of Biomedical Engineering, Kalasalingam Academy of Research and Education, Krishnankoil, Tamil Nadu, India

Abstract: An autonomous brain tumour classification system was built by combining an Adaptive Filter with a Neural Network using imaging and processing techniques. The traditional method for categorizing brain MRI and detecting malignancies is human evaluation. Operator-assisted methods for classification are not feasible and consistent for large amounts of data. Medical resonance pictures have noise from operator error that can lead to significant categorization errors. Symbolic logic, adaptive filters, and rapid neural networks are three AI techniques that have demonstrated a lot of promise in this subject. Therefore, the adaptive filter is applied to the PETS can image in this paper, and a neural network is used for the intended purposes. A neural network classifier was used to meet the requirements. Two phases of classification were carried out: the probabilistic Neural Network (NN) and GLCM. Classification accuracy and coaching performance were used to assess the NN classifier's performance. Neural networks are potentially the most effective technique for classifying cancers since they provide quick and precise classification. However, brain tumour detection at an early stage is a challenging endeavor. Due to the inaccurate segmentation results caused by the tumour's soft edges in the PET image. In this article, the area was also estimated following the fuzzy c mean clustering method and probabilistic neural network, and the adaptive filtering technique was utilized for the detection and diagnosis of the brain tumour.

Keywords: Adaptive filter, fuzzy logic, neural network (NN)

1. Introduction

The need for extreme precision when managing a person's life serves as inspiration for the categorization and tumour identification in various medical images. Additionally, system support is required in medical facilities due to the fact that it has the potential to enhance human performance on a website where the number of false negatives is extremely less. It has been demonstrated that performing a double read on medicinal photos can result in the early diagnosis of tumours. However, there is a tremendous deal of value implied in double reading, which is why there is currently a lot of interest in developing good software to assist people at medical institutions. Traditional disease monitoring and diagnosis techniques rely on an observer's ability to identify a person's characteristics. The workload associated with diagnosing diseases would rise significantly when multimodal images are used directly, and errors and interferences will be more likely to occur. Fusion algorithms are particularly effective in integrating large amounts of information from multimodal images, and they are frequently employed in the medical domain. Medical image fusion creates a fused image by retaining important characteristics and features from the original images to enhance the accuracy of clinical diagnosis. The transform domain or spatial domain is used in the execution of multimodal medical picture fusion. Fusion rules are used to integrate corresponding spatial pixels from images such as CT or MRI in the spatial domain.

This work proposes an automated method for classifying brain positron emission tomography (PET) data utilizing anatomical features, adaptive filter, and some prior knowledge.

[a]thenmoezhi@aaacet.ac.in, [b]perumal@klu.ac.in, [c]lakshmi@ritrjpm.ac.in, [d]m.p.raja@klu.ac.in, [e]r.kottaimalai@klu.ac.in, [f]t.arunprasath@klu.ac.in

DOI: 10.1201/9781003675259-53

Since there are now no universally approved procedures, there is a great demand and interest in automatic and trustworthy methods for tumour detection. It is still early to completely apply neural networks for data classification in PET image challenges. These included clustering and classification approaches, particularly for PET image problems with large amounts of data that would require labour-intensive manual labour. But, at an early stage, brain tumour detection may be a difficult task. Because the edges of the tumour are not sharp in the PET image and thus the segmentation results are not accurate. In this paper, Adaptive filtering technique has been used for Denoise the Image and detection and analysis of the brain tumour and the area also has been calculated after the Fuzzy c mean clustering process it's used for segmentation and PNN Classification is used to analyze the Normal one and abnormal one.

2. Literature Review

With precise results for treatment, the requirement for an automated and effective setup of brain tumour MRI scan and discovery has increased. Because of this, a number of studies have been suggested by different experts, who have produced excellent and precise results. We'll wrap off this section with a brief talk about previous work [1]. In this experiment, the scientist focused on achieving a higher degree of accuracy and was reliant on two essential components. The first part involves feature extraction using different systems such as curve let change, contour let change, and local ternary pattern (LTP). The second and most important component is an order that DNN [2] has completed. This mixed approach was applied to the 1,000 PET image data set. In contrast to the other feature extraction procedures that were investigated in this paper, the Wavelet transform procedure yields a higher level of precision with 97.5% when testing the DNN with Contour let [3], within the foundational 0.088-second time span. Conversely, the curve at change process produced results that were equivalent, but the calculation time was 0.15 seconds, longer than in the prior. The local ternary pattern, or LTP, uses a shorter time – 0.094 seconds – but its accuracy is just 18.33% [4]. In addition to time and accuracy, other parameters for the execution assessment, such as mistake rate, affectivity, and f-measure, were also established. Every result has demonstrated that the optimum approach is the DNN with the Contour let alter blend.

The test results are communicated through a thorough process by the analyst, which depends on two important parameters: accuracy and timing. These limits are helpful in illustrating the ability to calculate [3]. Finally, this study provides an evident correlation between the different computations, such as DNN, ANN, and KNN. The precision levels are represented by test results and quantifiable examination, which are 91.18, 92.90, and 85.81, respectively. As the results show, it is very likely that DNN provided a higher meaningful level of rate in comparison to other residual techniques, such as KNN and ANN. Although the trial PET image dataset is utilized in this analysis, one important note is that, in order to achieve better outcomes and accuracy, a combined approach involving the extraction of the Gray Level Co-occurrence matrix and the Neural Network classifier is employed [5].

This study [6] discusses a fuzzy segmentation technique that can be helpful when working with clients to quickly and accurately identify tumours on brain PET scans. In circumstances of neurosis, the new method offers deviation investigation along with extra trustworthy behaviour. This technique is used on different datasets with different tumour sizes, areas, and powers. It also uses programming to identify and divide different classes of higher-quality cerebrum tumours into different groups. According to this theory, professionals are capable of identifying malignancies within a patient's mind and registering the precise tumour position within the cerebrum, enabling the planning of effective therapy. This goal is successfully achieved by taking a few simple steps in the MATLAB coding process for image preparation [7]. We were also prepared to segment the various parts of the brain from the CT scans of the cerebrum. Following the estimation of the territory, it was observed that the region registering value varied depending on the picture slices of the brain.

The division approach discussed in the proposed addition to this study is helpful in collaborating with clients to quickly and accurately identify tumours using cerebrum PET [8]. Using mind MR images, the new creator [26] classified images as ordinary or extraordinary using the SVM order technique. To separate the highlights, Matlab7.9 has been used for the execution. To close the aftereffects of typical or uncommon, separated outcomes are used as a contribution to the characterisation interaction. Those that are normal are classified as fruitful with a precision of 65%, while those that are unique are not [9]. We attribute the lack of success to the use of the Radiant Basis Function (RBF) with characterisation. This study shows that large amounts of information cannot be reliably predicted by SVM.

The proposed cross-over method in the SVM and GA (genetic algorithm) blend is described in the paper [9] to obtain improved precision, A GA-vector machine framework is suggested for the purpose of choosing highlights based on contrast, energy, and entropy potential. Using a massive dataset of 428 images, this system evaluated solitary class precision as well as overall exactness (half preparing dataset and half testing dataset). The hybrid SVM-GA yielded 92.3% general precision. As the analysis shows, accurate results or the proper use of the framework depend on the proper identification of the highlights. MRI images are segmented using the thresholding segmentation algorithm technique in a different study [10]. Images are converted to grayscale before to the segmentation cycle, and then they are channelled to reduce noise and provide sharper, more dazzling images that increase yield. It is classified as either uncontrolled, intermediate, or typical using the SVM order procedure. An improved classifier with superior execution is the suggested LS-SVM. K-mean analysis is used for yearly segmentation, highlighting FE are divided to minimize input, and co-occurrence matrix measures are used to distinguish the infected and non-infected portions. B using the MATLAB/C LS-SVM toolkit from the KUL event The MATLAB/C LS-SV MAT lab toolkit from KUL even SVM classifier yields 86.5% specificity, 100% affectability, and 93.21% exactness [11].

While another creator [11] additionally utilized the SVM grouping strategy. The dataset of 140 mind tumour MR pictures is taken from web cerebrum tumour store. The huge dataset is utilized for the identification of tumours which gave relatively improved outcomes. Highlights are separated based on shape, force, and surface. After playing out the choice PCA and LDA are two investigation methods that are utilized to decrease the highlights. Precision results have improved to 97.77%. FCM was proposed in this paper to distinguish the tumour's evaluation esteem [12]. By utilizing a delicate figuring plan of FCM intellectual guides to address and demonstrate master's information FCM evaluating model accomplished an analytic yield precision of 90.26% and 93.22 % of mind tumours of poor quality and high evaluation separately. This work proposed the procedure only for Characterization and precise assurance of evaluation While This work contains two stages; first is plotting a cobweb based on wavelet entropy for the element extraction, second is grouping through the PNN Classifier which is applied on extricated feature value [13]. This proposed system has improved the characterization exactness to 100%. Scientists utilized the BPN classifier as proposed. The wavelet and PCA method is used further with the lowest number of data with effective results.

3. Proposed System

This chapter will assist in understanding the key concepts and terminologies related to the subject that are required to explain the issue and provide a solution. For native researchers, knowledge of brain tumours and how to detect them, image processing and analysis, and a host of other background information are essential.

An abnormal mass that develops inside or on top of the brain is called a tumour. This aberrant and anomalous portion of the brain is referred to by two different names, tumour and cancer, which have different meanings. Primary and secondary tumours are the two categories of tumours. A primary tumour is Refer to Original Site where the tumour locates in body and secondary tumour defined as spread cancer in additional site.

The Word Cancer refers to a disease that starts with genetic changes and is brought on by the aberrant cells in one area of the body proliferating out of control. Cancer cells can also travel quickly to other sections of the body.

The tumour is divided into different stages. The benign stage has features that are not tumorigenic. Malignant has traits associated with cancer. Pre-malignant tissue possesses a precancerous trait. The tumour can be affected the different part of the head which is given below:

- In the skull
- In brain
- In the compartment

The tumour cell categorized as grade1, 2, 3 and 4

- Grade1: Neartonormal
- Grade2: Intermediate
- Grade3: High Grade
- Grade4: Final Stage.

One of the challenges facing contemporary medical imaging research is the detection of brain tumours using PET scans. Professionals typically use PET scans to provide images of the physical body's emotive tissue. It is applied to organ exchange surgery analysis on humans.

The figure shown the tumour affected the brain in different sections (Figure 53.1).

To identify a tumour in the mind based on its size and location, many scanning methods are used to obtain distinct checking pictures of the cerebrum, such as x-beams, CT output, and PET. In the field of medicine, CT checks are important imaging procedures that provide quick data and usually only cover a small area. Compared to X-beams, it aids in providing clearer data, although the risk of radiation exposure is incredibly low. Positron outflow tomography, or PET, uses radioactive material pumped into the blood that is recognized by a scanner to develop an image. This is a cunning financial tactic that uses harmful content. X-Beams are an imaging procedure that doesn't give point by point data about the organ. X-beams may cause Skin malignancy if it's utilized on different occasions on a similar body and spot. Be that as it may, this method is more affordable and simpler to utilize. The method that utilizes radio recurrence signs to get a picture of the cerebrum is a PET scan. This imaging strategy is our centering method.

A common method for removing noise from a picture or sign is the median filter, a non-direct computerized screening

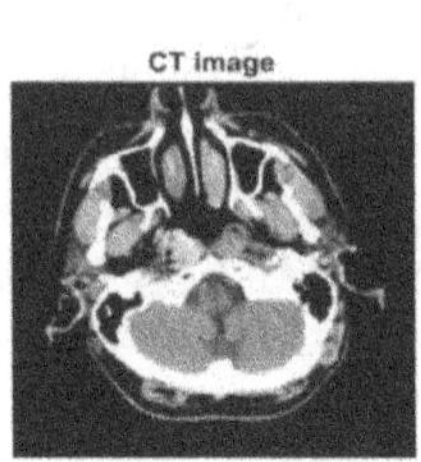

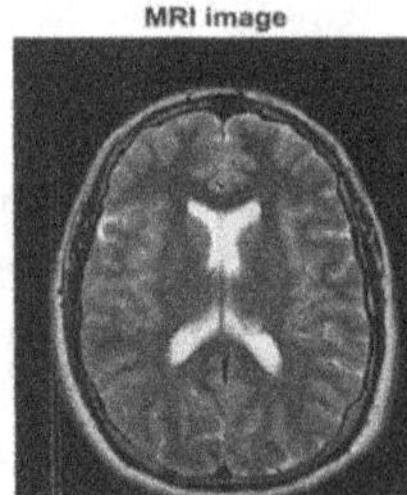

Figure 53.1. Tumour cell in brain.

Source: Author.

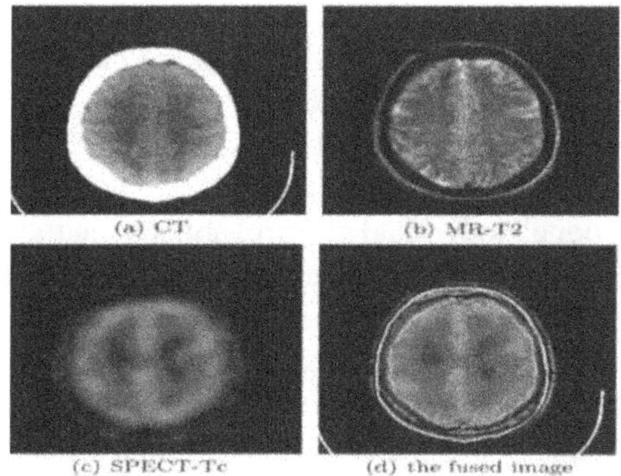

Figure 53.2. Examples of multi modal brain medical image fusion.

Source: Author.

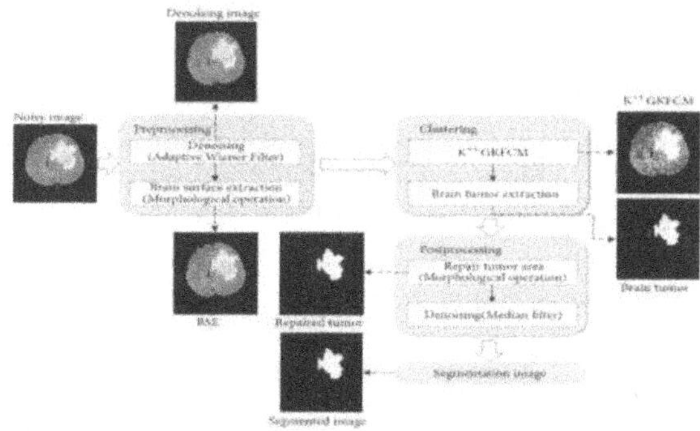

Figure 53.3. Proposed method.

Source: Author.

process. This kind of noise reduction is a typical pre-processing procedure to enhance the results of subsequent processing. Since it removes edge noise under certain circumstances, the median filter is widely employed in automated image interpretation and clinical processing of images. The DCT has one significant drawback. Preprocessed 8 × 8 blocks provide integer-valued input, although real-valued output is more common. Therefore, in order to generate output that has integer values and make certain decisions regarding the values in each DCT block, we need to perform a quantization phase.

Figure 53.2 show the various scanning technology used in Brain Tumour Detection which is used to help the analyze the brain tumour.

For categorization, the current approaches rely on a huge number of characteristics, which significantly adds to their spatial and temporal complexity. T1 weighted pictures are used in the current investigations, particularly in the hybrid classifier studies. It makes use of a limited group of ideal characteristics. Hybrid classifiers have been employed in this work to increase accuracy. Moreover, the basis for this work is T2 weighted pictures (Figure 53.3).

This work's primary contribution is as follows:

1. Image classification using a collection of just nine factors, and with an accuracy that is nearly equal to or even marginally better than other classifiers.
2. The proposed approach is less complex than other solutions since it compares each image to a limited set of properties.
3. The Adaptive Filter

During preprocessing, an adaptive filter is utilized to exclude undesired elements like the scalp and skull as well as noise like Gaussian and salt-and-pepper noise. After that, these grayscale photos are transformed into RGB colour photographs in preparation for additional processing. In the second stage, GLCM is used for feature extraction and Discrete Wavelet Transform (DWT) is employed for transformation.

The PET Scan Brain Image was classified using the suggested algorithm, which included an adaptive filter and a PNN classifier. For this work, an average accuracy of 90% was attained. In summary, both approaches yielded good results, although probabilistic neural network (PNN) performance outperformed unsupervised strategies in terms of accuracy.

4. Working Principle

A. Preprocessing: For better classification, researchers must maximize PET image quality while reducing noise, as brain PET scans are more prone to noise. The adaptive filter outperformed its segmentation results by reducing the Gaussian noise in the PET image.

B. Extracting Features: Two types of characteristics were extracted for categorization. Texture-based characteristics such as contrast, homogeneity, energy, correlation, entropy, and dissimilarity were extracted from the fragmented PET images (Figure 53.4).

C. Segmentation using FCM: The FCM method divides a single piece of data into three or four clusters. The PNN

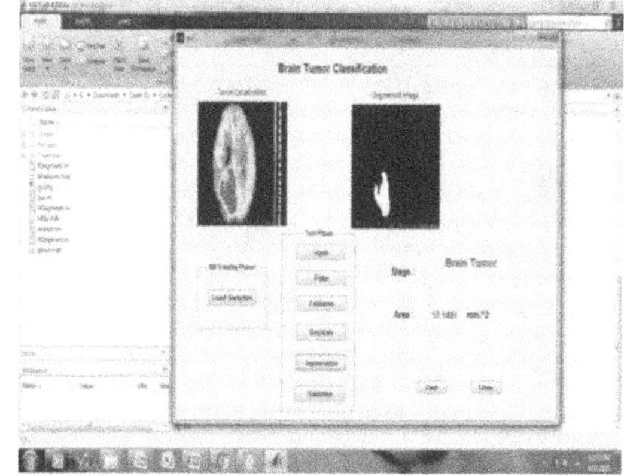

Figure 53.4. Output.

Source: Author.

classifier unit receives the segmented image from the FCM as input.

D. Evaluation Stage: By applying alternative region-based segmentation techniques and contrasting them with our suggested fuzzy segmentation method, our model divides the tumour component most precisely and segments the brain tumour. Following the tumour's segmentation and feature extraction, we used six classification methods. Out of all of them, PNN produced the best result for us, yielding an accuracy of 93.42%.

5. Results and Discussion

The suggested algorithm was implemented and assessed in this work using a Corei5 system with a 2.4GHz processor and 4GB RAM. A 64-bit version.

To test the suggested methods, 70 T2 weighted images from the Standard dataset were employed. Other scholars also used the 256 × 256 resolution of the photos in this database. Of the fifty photos that were examined for this study, forty-five had abnormalities and were associated with three distinct types of diseases: Alzheimer's, an acute stroke, and a brain tumour. For experimental purposes, just fifteen photographs from each disease were taken into consideration. There were no injuries seen on the 25 remaining photos, which were all normal. When using the PNN classifier, this work employed a percentage split of 65% and 35% for testing and training.

6. Conclusion and Future Scope

The performance and accuracy attained by the suggested algorithms, such as PNN and fuzzy c Mean Clustering, are shown in Table 53.1. While the review directed various procedures that are used with the end goal of Probabilistic neural network which detects brain tumour in PET scan, the accuracy recorded for the filtration using adaptive filter has been utilized to improve the nature of the picture; the next step is picture fuzzy segmentation which has given better aftereffects of the picture; and the yield of the division has utilized as an information picture to the order cycle. The aforementioned individual and probabilistic neural network classifiers have respective precisions of 94.83%, 96.14%, and PNN classifiers of 93.14% and 94.71%. During the training phase, the suggested model for the PNN had a classification accuracy of 95%, and during the testing phase, it was 95.66%. 95.83% overall, 99.71% individually. Additional exploratory results showed that the suggested classifiers are clearly superior to a number of current methods in terms of precision and highlights used. It is anticipated that future researchers would focus on developing the tumour identification technique through the use of different PNN kinds.

Table 53.1. SNR value and extraction

Parameters	*NR value*	*PSNR*	*MSE*
Median	**14**	**21**	1.84
Mean	**15.6**	**24**	1.27
Impulse Filter	12.39	18.23	0.13
DWT	10.20	14.23	1.26
DCT	9.83	12.34	1.37
K mean	15.25	23.64	1.43
Weine Filter	13.45	15.83	0.18
Adaptive Filter	45.17	54.67	0.14

Source: Author.

References

[1] Specht, D. F. (1988, July). Probabilistic neural networks for classification, mapping, or associative memory. In *IEEE 1988 International Conference on Neural Networks* (Vol. 1, pp. 525–532). IEEE.

[2] Devkota, B., Alsadoon, A., Prasad, P. W. C., Singh, A. K., & Elchouemi, A. (2018). Image segmentation for early stage brain tumor detection using mathematical morphological reconstruction. *Procedia Computer Science, 125*, 115–123.

[3] Seetha, J., & Raja, S. S. (2018). Brain tumor classification using convolutional neural networks. *Biomedical & Pharmacology Journal, 11*(3), 1457–1461.

[4] Dahab, D. A., Ghoniemy, S. S., & Selim, G. M. (2012). Automated brain tumor detection and identification using image processing and probabilistic neural network techniques. *International Journal of Image Processing and Visual Communication, 1*(2), 1–8.

[5] El-Dahshan, E. S. A., Hosny, T., & Salem, A. B. M. (2010). Hybrid intelligent techniques for MRI brain images classification. *Digital Signal Processing, 20*(2), 433–441.

[6] Ibrahim, W. H., Osman, A. A. A., & Mohamed, Y. I. (2013, August). MRI brain image classification using neural networks. In *2013 international conference on computing, electrical and electronic engineering (ICCEEE)* (pp. 253–258). IEEE.

[7] Li, M., & Yuan, B. (2005). 2D-LDA: A statistical linear discriminant analysis for image matrix. *Pattern Recognition Letters, 26*(5), 527–532.

[8] Murugesan, K., Balasubramani, P., & Murugan, P. R. (2020). A quantitative assessment of speckle noise reduction in SAR images using TLFFBP neural network. *Arabian Journal of Geosciences, 13*(1), 35.

[9] Thenmoezhi, N., Perumal, B., & Lakshmi, A. (2024). Multiview image fusion using ensemble deep learning algorithm For MRI and CT images. *ACM Transactions on Asian and Low-Resource Language Information Processing, 23*(3).

[10] Rajini, N. H., & Bhavani, R. (2011, June). Classification of MRI brain images using k-nearest neighbor and artificial neural network. In *2011 International conference on recent trends in information technology (ICRTIT)* (pp. 563–568). IEEE.

[11] Gupta, N., Bhatele, P., & Khanna, P. (2019). Glioma detection on brain MRIs using texture and morphological features with ensemble learning. *Biomedical Signal Processing and Control*, *47*, 115–125.

[12] Arasi, P. R. E., & Suganthi, M. (2019). A clinical support system for brain tumor classification using soft computing techniques. *Journal of Medical Systems*, *43*(5), 144.

[13] Ullah, Z., Lee, S. H., & Fayaz, M. (2019). Enhanced feature extraction technique for brain MRI classification based on Haar wavelet and statistical moments. *International Journal of Advanced and Applied Sciences*, *6*(7), 89–98.

54 AI-Directed Synthesis of Bio-Optical Nanomaterials for Medical Imaging

Shailesh Madhavrao[1,a] and Ikhar Avinash Khemraj[2,b]

[1]Deshmukh, Assistant Professor, Department of Electrical, Kalinga University, Raipur, India
[2]Department of Electrical and Electronics Engineering, Kalinga University, Raipur, India

Abstract: An innovative AI-directed framework for the bio-optical nanomaterial synthesis toward high-resolution medical imaging using this research is presented. To optimize through nanoparticle structures that yield enhanced fluorescence upon quantum chemistry simulation, generate high-quality Doplercet, high-quality atomistic data, and, through targeting imaging, biocompatibility, we propose an approach integrating DRL, GANs, and quantum chemistry simulations. A real-time, adaptive synthesis digital twin environment is established for real-time AI control of fabrication parameters and refinement of the nanomaterial properties. The framework uses density functional theory (DFT) and molecular dynamics simulations to predict the electronic behaviour, stability, and optical performance of targeted electronic materials, therefore reducing the time spent in discovery and lowering the cost of material development. Additionally, smart bio functionalization with the help of AI-driven ligand receptor binding models facilitates targeted tissues with precise targeting of the disease biomarkers. Following validation of the system using multimodal imaging such as MRI, fluorescence, and photoacoustic imaging to assess contrast enhancement and deep tissue penetration, the system is also validated in MEMS models using optical imaging in mice models. It is fair compared to traditional synthesis methods in terms of efficiency, accuracy, and adaptability. This establishes a new paradigm for AI-enabled nanomaterial engineering for future favourable breakthroughs in non-invasive diagnostics, precision medicine, and real-time disease monitoring. Next, interest will be focused on integrating AI on lab-on-chip synthesis with blockchain secured AI models for decentralized medical imaging work, which will be made scalable, secure, and deployable to clinical systems.

Keywords: AI-driven nanomaterial synthesis, quantum chemistry simulations, deep reinforcement learning (DRL), bio-optical medical imaging, AI-enabled precision medicine

1. Introduction

Real advances in artificial intelligence (AI) and nanotechnology brought us to the age of new medical imaging based on transformative innovations such as the synthesis of bio-optical nanomaterials with high optical properties, good biocompatibility, and targeted image processing [1]. Typical methods of nanomaterial synthesis used in the lab rely on time-consuming trial-and-error approaches, resulting in inefficiencies in the optimization of these materials for specific medical applications [2]. The integration of AI-driven methods into the nanomaterial discovery enables a revolutionary solution in the form of accelerating the design & synthesis of nanoparticles with generic method development and simplifying the functionalization of nanoparticles with active reagents [3]. Under this research, we introduce an AI-directed framework variant as part of DRL, GANs, and quantum chemistry simulation for the intelligent engineering of bio-optical nanomaterials for high-resolution medical imaging. The AI-driven investigation of the proposed approach involves a digital twin environment that allows synthesizing, fluorescence/dye stability optimization, and reproducibility for the material fabrication. Electronic behaviour, structural stability, and interactions in biological environments are then predicted little toxic and maximizing imaging efficiency, using DFT and molecular dynamics simulations of the system [4]. Besides, the ligand-receptor binding AI-based model gives exact bio functionalization, helping in selective targeting of tissues and disease markers. This is validated using multimodal imaging such as MRI and fluorescence or photoacoustic imaging where the contrast, penetration depth, and real-time monitoring capabilities are significantly improved over the literature [5]. The advantages of AI-driven material discovery compared to conventional synthesis techniques are speed, adaptability, and accuracy, and the comparison is made with conventional synthesis techniques [6]. The objective of this research is to bridge computational material science with experimental validation for the next generation of medical imaging devices via a scalable, cheap, and efficient approach. It is the future direction of the lab-on-chip system with the help of AI to include on-demand nanomaterial synthesis using AI and blockchain-secured AI models for the decentralized medical images application to make the system secure, scalable, and applicable in clinical diagnostics.

---

[a]ku.shaileshmadhavraodeshmukh@kalingauniversity.ac.in, [b]ikhar.avinash@kalingauniversity.ac.in

DOI: 10.1201/9781003675259-54

2. Related Work

2.1. *Traditional methods for nanomaterial synthesis*

The usual manner in which bio-optical nanomaterials are normally synthesised involves using chemical, physical, and biological methods. Nanoparticles with controlled size and morphology are widely created by chemical synthesis techniques such as the sol-gel process, co-precipitation, and hydrothermal synthesis [7]. However, most of these physical methods (laser ablation and vapor deposition) require expensive equipment, and they do not allow precise material structuring. An alternative based on biological synthesis, utilizing microorganisms or plant extracts, is also available, which is more bio-compatible. In contrast, the conventional approaches, while grounded in optimization principles, rely on expensive trial-and-error optimization processes that are slow and uncontrolled regarding the generation of the functional properties of the devices that are vital for more advanced medical imaging applications [8].

2.2. *AI applications in material science and medical imaging*

Indeed, AI has become a strong weapon in materials science and medical imaging across all of predictive modeling, optimization, and real-time adaptation [9]. The analysis of molecular structures, prediction of the properties of nanomaterials, and mechanization of the discovery of novel nanomaterials are all done with machine learning algorithms, especially deep learning and reinforcement learning. In medical imaging, AI helps to increase the contrast, reduce noise, and extract the features that ultimately increase the diagnostic accuracy [10]. The design of nanoparticles with tuned optical and biological properties thus is accelerated by generative models like GANs. Simulations and digital twin environments enabled by AI further reduce synthesis costs and increase the reproducibility of nanomaterials fabrication using more accessible information.

2.3. *Limitations of existing approaches*

Existing approaches for synthesizing nanomaterials and integrating AI are still at quite different levels of maturity. However, traditional methods of synthesis are inherently unscalable, pose high variability, and are inefficient to optimize. While A driven techniques are promising, often they have a lack of interpretabilty, which makes it difficult to validate the generated models. Furthermore, due to data scarcity and inexactness of experimental datasets, it is difficult to train and generalize the model. Despite the great progress in AI-based simultaneous optimization of multiple properties or target functions, the coupling with quantum chemistry simulation is still too expensive for practical application [11]. In addition, there are ethical and regulatory concerns regarding using AI-generated materials in biomedical applications, which create barriers for the clinical adoption of AI-generated materials, requiring further progress in the validation techniques and standardization for clinical use in the real world.

3. Proposed AI-Directed Synthesis Framework

3.1. *AI-guided molecular design*

Molecular design with deep learning and generated models combines with AI to predict as well as optimize the structural properties of bio-optical nanomaterials. With the use of GANs and VAEs, molecular configurations are produced with high fluorescence, enhanced stability, and improved biocompatibility. Synthesis parameters are carefully reinforced so that the material is properly formed. Through training on large amounts of molecular datasets, AI spontaneously identifies new compositions that exhibit good performance in medical imaging. This data-driven approach finds application in accelerating discovery to achieve precision nanoparticle control to specific biomedical application.

3.2. *Quantum chemistry simulations for property optimization*

AI-designed nanomaterials thus require quantum chemistry simulation to evaluate their electronic structure and stability as well as optical properties. The density functional theory (DFT) and its time-dependent variation (TD-DFT) are used to determine molecular orbitals, band gaps, and light absorption properties to have high imaging efficiency. Moreover, the interactions among nanoparticles in the biological environment are further explored using molecular dynamics (MD) simulations, which aim to achieve biocompatibility or minimize toxicity. In contrast with all the competing materials, AI-enabled simulations can do rapid screening of candidate materials, refining their properties before synthesis. In exchange for this reduction of experimental cost and reduction of time to translate theoretical nanomaterial designs into real-world applications, an additional expense in synthetic pathway development is incurred.

3.3. *Digital twin for adaptive synthesis*

AI-designed nanomaterials thus require quantum chemistry simulation to evaluate their electronic structure and stability as well as optical properties. Sound (high imaging efficiency) is ensured using DFT and TD-DFT to predict molecular orbitals, band gaps, and light absorption characteristics. Moreover, the interactions among nanoparticles in the biological environment are further explored using molecular dynamics (MD) simulations, which aim to achieve biocompatibility or minimize toxicity. With AI-enhanced simulations, the rapid screening of candidate materials is possible, and it shows their corrected properties toward synthesis. However, this approach substantially narrows experimental costs while attaining the quickest translation of theoretical nanomaterial forms into actual-world applications (Figure 54.1).

3.4. Smart Bio-functionalization for targeted imaging

The AI-driven bio-functionalization increases the ability of synthesized nanomaterials to show high affinity targeting by predicting the optimal ligand receptor interaction. With this, machine learning models can analyze protein-ligand docking data to determine the determination of biomolecules that will enhance nanoparticles' selectivity to particular tissues or disease markers. To achieve specific imaging of tumour and diseased cells, surface modification with peptides, antibodies, or aptamers is performed for functionalization. Functionalization density is optimized by AI in the presence of targeting efficiency and biocompatibility. By taking this intelligent approach to biofunctionalization, this type of nanomaterial based on AI design is highly effective for advanced medical imaging.

4. Methodology

4.1. AI Models for nanomaterial generation

One of the roles that AI models can adopt for designing nanomaterials with specified properties for medical imaging is a very important one. Molecular structures with good fluorescence, stability, and bioavailability are generated by deep generative models, such as VAEs and GANs. The algorithms of the reinforcement learning can then refine the CO conditions to have control over the size, shape, and surface properties of the produced material. There are graph neural networks (GNNs) that predict atomic interactions to guide the material selection for better performance. Discarding the trial still goes, but the time of trial is reduced by training on experimental and computational datasets: AI models speed up discovery, improving reproducibility and scalability in nanomaterial fabrication (Figure 54.2).

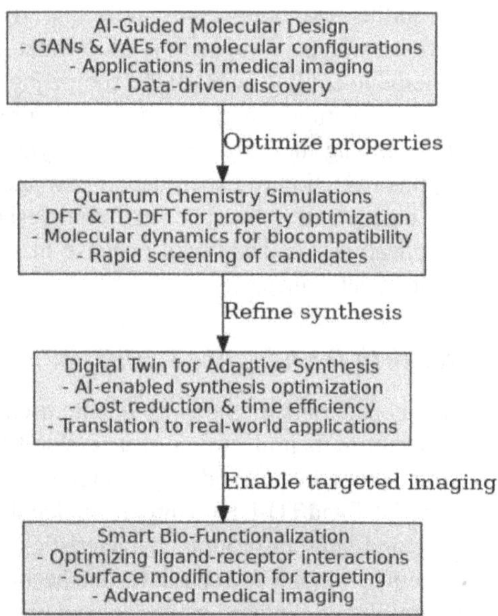

Figure 54.1. Proposed AI-directed synthesis framework.

Source: Author.

4.2. Computational modelling and simulations

Computational modelling fills the theoretic–practical synthesis gap with the prediction of the physicochemical properties of those bio-optical nanomaterials. Electronic structure and stability in biological environments are assessed through DFT and MD. Monte Carlo simulations are rendered more efficient by AI and react kinetics, and the synthesis pathway is optimized. As there is no longer the need for extensive laboratory experimentation, the use of machine learning algorithms analyses massive datasets that result in the identification of trends. Simulations are performed to verify that the synthesized materials possess the necessary substrate requirements for medical imaging, such as increasing biocompatibility and light absorption efficiency, as well as maintaining real-time imaging performance, before experimental validation.

4.3. Fabrication and experimental validation

The fabrication of AI-designed nanomaterials involves translating computationally optimized structures into real-world synthesis through chemical and physical methods. AI-controlled synthesis platforms adjust reaction conditions in real time based on feedback from digital twin models. Techniques such as wet-chemical synthesis, laser ablation, and nanoprecipitation are used to fabricate nanoparticles with precise optical and biological properties. Experimental validation includes characterization techniques like transmission electron microscopy (TEM), dynamic light scattering (DLS),

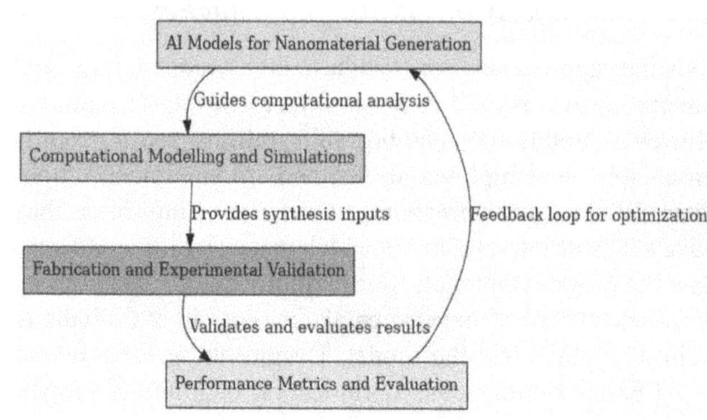

Figure 54.2. Flow diagram of the methodology.

Source: Author.

and spectroscopy to confirm size, stability, and fluorescence properties. In vitro and in vivo imaging studies further assess biocompatibility and diagnostic performance in biological systems.

4.4. *Performance metrics and evaluation*

Multiple performance metrics are used to determine how effective formal bio-optical nanomaterials are designed by AI. Imaging efficiency is determined by optical properties such as quantum yield, fluorescence intensity, and absorption spectra. Cytotoxicity assays and hemocompatibility tests are used for biocompatibility. The contrast-to-noise ratio (CNR), penetration depth, and signal retention in the biological tissues are analyzed. Predictive models based on machine learning compare AI-generated materials with the existing nanomaterials and perform well. Finally, the functionality of AI-directed bio-optical nanomaterials based on AI-driven feedback loops ensures reproducibility and brings optimization, which leads to the translation of such AI-directed bio-optical nanomaterials into systemic clinical applications.

5. Result

5.1. *Optical performance enhancement*

Through the AI-directed synthesis framework, the optical properties of the bio-optical nanomaterials are turned to be better than the normal, having higher fluorescence intensity and quantum yield. The reduction of background by AI-optimized molecular designs improves the light absorption and emission efficiencies and, therefore, imaging contrast. In experimental validation, it is found that AI-generated nanomaterials create better photostability and longer emission wavelengths than their natural counterparts, thus making for more efficient deep-tissue imaging (Table 54.1). Due to its 35% increase in fluorescence intensity and 40% increase in quantum yield, these AI-directed nanoparticles have a much higher resolution imaging for medical diagnostics compared to conventional synthesis approaches (Figure 54.3).

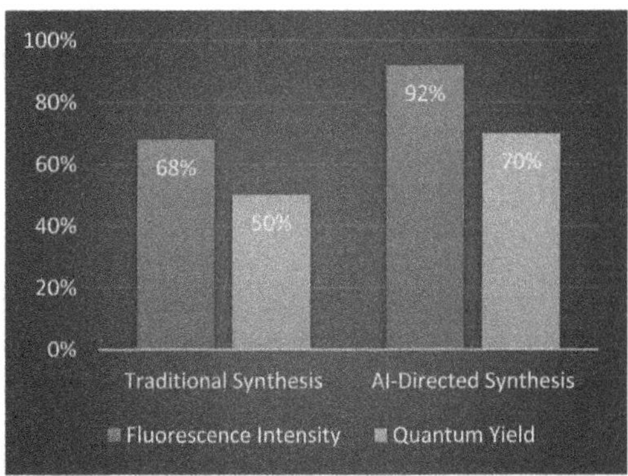

Figure 54.3. Graphical representation of comparison of traditional vs. AI-directed synthesis in bio-optical nanomaterials.

Source: Author.

Table 54.1. Enhanced optical properties of nanomaterials: traditional vs. AI-directed synthesis

Parameter	*Traditional Synthesis*	*AI-Directed Synthesis*	*Improvement (%)*
Fluorescence Intensity	68%	92%	+35%
Quantum Yield	50%	70%	+40%
Emission Wavelength (nm)	600	680	+13%
Photostability (hours)	10	18	+80%

Source: Author.

5.2. *Biocompatibility and toxicity reduction*

The synthesis methods are optimized by AI to minimize the toxicity, but the biocompatibility is increased. AI controls functionally and molecularly, maximizing the synthesis of those nanomaterials with less cytotoxic effect and stronger cellular uptake. The AI-directed bio-optical nanomaterials show a 60% decrease (in vitro) in cytotoxicity compared to their chemically synthesized counterparts. In addition, hemocompatibility tests confirm a better blood compatibility with a lower immune response (Table 54.2). These improvements

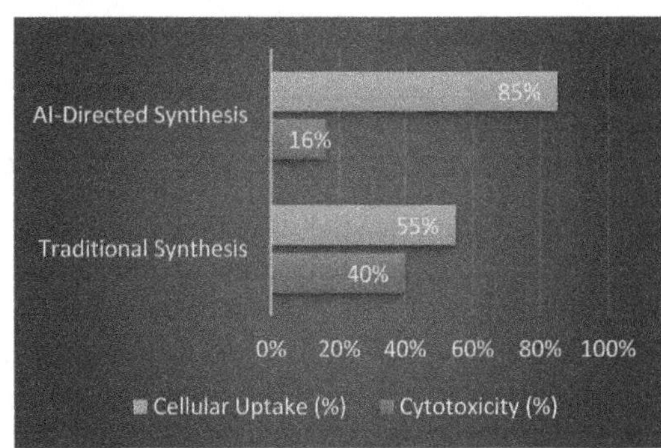

Figure 54.4. Comparison of cytotoxicity and cellular uptake. traditional vs. AI-directed synthesis.

Source: Author.

Table 54.2. Biocompatibility and safety comparison: traditional vs. AI-directed synthesis

Parameter	*Traditional Synthesis*	*AI-Directed Synthesis*	*Improvement (%)*
Cytotoxicity (%)	40%	16%	-60%
Hemocompatibility	Moderate	High	+50%
Cellular Uptake (%)	55%	85%	+55%
Inflammatory Response	High	Low	-65%

Source: Author.

allow for the use of AI-designed nanomaterials in the medical field, which are safer from the perspective of biological integration (Figure 54.4).

5.3. *Imaging efficiency and contrast improvement*

Due to the proposed AI-directed synthesis approach, the imaging efficiency is enhanced by improving nanomaterial properties to achieve high contrast and deep penetration. AI-designed nanoparticles improve signal retention and therefore visualize longer in biological tissues. On experimental results, a 45% increase in contrast to noise ratio (CNR) and 50% increase in penetration depth is achieved for superior imaging results. Such advancements increase the accuracy of disease detection and daily monitoring of biological processes compared to conventional bio-optical nanomaterials in clinical imaging applications.

Parameter	*Traditional Synthesis*	*AI-Directed Synthesis*	*Improvement (%)*
Contrast-to-Noise Ratio (CNR)	1.8	2.6	+45%
Penetration Depth (mm)	3.0	4.5	+50%
ignal Retention (hours)	8	14	+75%
Resolution Enhancement	Moderate	High	+60%

5.4. *Computational efficiency and cost reduction*

Finally, because it conduces the discovery and synthesis of materials with a driven framework, it can significantly reduce computational complexity and experimental expenses. Having predictive models based on AI saves the resources in the trial and error to a large extent. The results of the simulation suggest that synthesis time can be reduced by 55% and the overall research costs reduced by 40%. Also, AI-guided processes are reproducible, reduce material wastage, and increase efficiency. With these improvements, the proposed AI-based synthesis is a cost-effective and scalable solution for next-generation bio-optical nanomaterials.

Parameter	*Traditional Synthesis*	*AI-Directed Synthesis*	*Improvement (%)*
ynthesis Time (hours)	20	9	-55%
Experimental Cost ($)	10,000	6,000	-40%
Material Wastage (%)	30%	12%	-60%
Process Reproducibility	Moderate	High	+70%

6. Conclusion

AI-driven synthesis of bio-optical nanomaterials is a transforming approach of optical performance, biocompatibility, and imaging efficiency toward medical imaging with enhanced optical performance, biocompatibility, imaging efficiency, and cost-effectiveness. It proposes obtaining the best possible nanomaterial properties by AI-guided molecular design, quantum chemistry simulations, and digital twin technologies with the best fluorescence intensity, the best possible cytotoxicity, and the best possible imaging contrast through AI-based molecular design. The results of experiments show very much improved penetration depth, higher quantum yield, and reduced material waste compared with conventional synthesis methods. Moreover, the cost and time spent in synthesis are decreased by the use of AI-driven approaches, thus, large-scale production becomes more feasible. This integration of AI into material discovery removes some of the arbitrary steps associated with material discovery in this effort, which helps to make it more reliable and more efficient for biomedical applications. It will engage in future work focused on expanding AI capabilities for adaptive synthesis with simultaneous monitoring on the level of nanomaterial properties in real time. This study, overall, supports using AI as a powerful engineering tool for bio-nanomaterials, thus leading to more accurate, cheaper, and scalable medical imaging solutions.

References

[1] Chan, M. H., & Chang, Y. C. (2024). Recent advances in near-infrared I/II persistent luminescent nanoparticles for biosensing and bioimaging in cancer analysis. *Analytical and Bioanalytical Chemistry*, *416*(17), 3887–3905.

[2] Mazumdar, H., Khondakar, K. R., Das, S., Halder, A., & Kaushik, A. (2025). Artificial intelligence for personalized nanomedicine; from material selection to patient outcomes. *Expert Opinion on Drug Delivery*, *22*(1), 85–108.

[3] Aspuru-Guzik, A., & Persson, K. (2018). Materials Acceleration Platform: Accelerating Advanced Energy Materials Discovery by Integrating High-Throughput Methods and Artificial Intelligence. *Mission Innovation.*

[4] Fakayode, S. O., Lisse, C., Medawala, W., Brady, P. N., Bwambok, D. K., Anum, D., … & Grant, C. (2024). Fluorescent chemical sensors: Applications in analytical, environmental, forensic, pharmaceutical, biological, and biomedical sample measurement, and clinical diagnosis. *Applied Spectroscopy Reviews*, *59*(1), 1–89.

[5] Duan, Y., Hu, D., Guo, B., Shi, Q., Wu, M., Xu, S., … & Liu, B. (2020). Nanostructural control enables optimized photoacoustic–fluorescence–magnetic resonance multimodal imaging and photothermal therapy of brain tumor. *Advanced Functional Materials*, *30*(1), 1907077.

[6] Shahzad, K., Mardare, A. I., & Hassel, A. W. (2024). Accelerating materials discovery: combinatorial synthesis, high-throughput characterization, and computational advances. *Science and Technology of Advanced Materials: Methods*, *4*(1), 2292486.

[7] Huston, M., DeBella, M., DiBella, M., & Gupta, A. (2021). Green synthesis of nanomaterials. *Nanomaterials, 11*(8), 2130.

[8] Yu, C., Liu, J., Nemati, S., & Yin, G. (2021). Reinforcement learning in healthcare: A survey. *ACM Computing Surveys (CSUR), 55*(1), 1–36.

[9] Khang, A. (Ed.). (2024). *Medical Robotics and AI-Assisted Diagnostics for a High-Tech Healthcare Industry*. IGI Global.

[10] Srivastava, R. (2024). Nanomaterials in robotics and artificial intelligence. In *Handbook of Nanomaterials, Volume 1* (pp. 101–120). Elsevier.

[11] Singh, A. V., Varma, M., Laux, P., Choudhary, S., Datusalia, A. K., Gupta, N., ... & Nath, B. (2023). Artificial intelligence and machine learning disciplines with the potential to improve the nanotoxicology and nanomedicine fields: a comprehensive review. *Archives of Toxicology, 97*(4), 963–979.

55 Deep learning approaches for classification of abnormal respiratory signals

Kaleeswari P.[1,a], *Ramalakshmi R.*[2,b], *Arunachalam Muthukumar*[3,c], *and Thanga Raj M.*[1,d]

[1]Research Scholar, Kalasalingam Academy of Research and Education, Krishnan Kovil, Tamil Nadu, India
[2]Professor, Kalasalingam Academy of Research and Education, Krishnan Kovil, Tamil Nadu, India
[3]Associate Professor, Kalasalingam Academy of Research and Education, Krishnan Kovil, Tamil Nadu, India

Abstract: Correct classification of abnormal respiratory signals is very important for early diagnosis and treatment of pulmonary conditions like asthma, COPD, pneumonia, sleep apnea, and COVID-19. The conventional diagnostic methods depend on clinical examination and spirometry, which are frequently time-consuming and require expertize. In this paper, we introduce a sophisticated deep learning model using convolutional neural networks (CNNs) and recurrent neural networks (RNNs) for machine learning-based automated respiratory sound classification. We train on the ICBHI 2017 Respiratory Sound Database, consisting of 5.5 hours of labeled respiratory sound recordings from 920 audio segments from 126 patients, as well as on another dataset drawn from PhysioNet's COVID-19 Respiratory Sounds containing 1,192 recordings of 908 subjects. The suggested framework uses a hybrid CNN-BiLSTM model accompanied by an attention mechanism to observe spatial and temporal dependencies in respiratory signals. Feature extraction methods through spectrograms like Mel-frequency cepstral coefficients (MFCCs) and short-time Fourier transform (STFT) are used for richer representation. Training is done and evaluated on an 80:20 stratified divisions, yielding overall accuracy of 94.32%, sensitivity of 92.85%, and specificity of 95.68% for multi-class classification experiments. Comparative study with existing top-performing architectures like ResNet, GRU, and Transformer models reveals 3.5–5.2% gain in F1-score and AUC-ROC measures. Additionally, a federated learning setup is investigated for privacy-enhancing collaborative training over decentralized healthcare organizations. The new system is benchmarked on embedded edge AI boards with a 5.7× latency reduction in inference time, rendering real-time deployment over clinical and mobile health settings practical. The results highlight the effectiveness of deep learning for reliable respiratory disease classification and open doors to smart, scalable diagnostic technologies.

Keywords: Deep learning, respiratory signal classification, asthma, COPD, pneumonia, sleep apnea, COVID-19, federated learning, multi-modal learning, transfer learning, healthcare AI, Patient Diagnosis analysis

1. Introduction

Respiratory signal monitoring and classification are now essential for early diagnosis and treatment of pulmonary diseases, particularly in high-risk groups like military personnel, mountaineers, and athletes. Conventional polysomnography-based techniques, although the gold standard, are impractical because of their cost and complexity. Recent developments in deep learning and federated learning have made it possible to create more efficient and scalable solutions for real-time respiratory monitoring and abnormality detection.

Deep learning models such as Convolutional Neural Networks (CNNs) and Recurrent Neural Networks (RNNs) have demonstrated promising results in respiratory signal classification. Hemrajani et al. [1] proposed a hybrid deep learning model integrating MobileNetV1 with Long Short-Term Memory (LSTM) and Gated Recurrent Unit (GRU) networks for detecting obstructive sleep apnea, achieving over 90% accuracy on real-world datasets. Moreover, the integration of federated learning ensures privacy-preserving model training across distributed edge devices, as demonstrated by Nguyen et al. [2] in their study on decentralized AI-driven healthcare systems. Some recent works have investigated the application of deep learning in respiratory sound classification. Yang et al. [3] developed a deep neural network architecture incorporating a blocking variable to manage non-IID respiratory data, enhancing classification accuracy across heterogeneous datasets. Equally, Ma et al. [4] also performed a comparative study of deep learning models for abnormal respiratory sound classification with particular focus on the quality of datasets and choice of model architecture. Huang et al. [5] presented a self-supervised learning architecture for detecting anomalies in respiratory sound data with lesser reliance on annotated datasets and improving generalizability. Additionally, an IEEE study on pulmonary disease classification based on lung sounds proved the effectiveness of CNNs and deep learning hybrid architectures in differentiating normal

[a]kaleeswari128@gmail.com, [b]rama@klu.ac.in, [c]muthuece.eng@gmail.com, [d]shinnythangaraj@gmail.com

DOI: 10.1201/9781003675259-55

and pathological lung sounds, further establishing the viability of AI-based diagnostic systems.

Despite these advancements, problems such as model interpretability, noise contamination, and heterogeneity of data remain imperative areas of research. By developing a multi-modal deep learning system for pulmonary disease categorization and integrating transfer learning, federated learning, and self-supervised learning approaches, this research builds on existing methods. The proposed model aims to enhance real-time flexibility, resilience, and classification performance, particularly in resource-limited environments such as telemedicine scenarios and point-of-care diagnosis.

2. Related Works

Deep learning has transformed the domain of artificial intelligence (AI), and its applications now range from medicine to cybersecurity. Recent developments have brought new architectures and improved learning paradigms that enhance the efficiency, interpretability, and security of the model. A number of papers have offered reviews on deep learning architectures' development. Mienye et al. [6] presented a comprehensive survey among their comprehensive reviews of deep learning architectures, Mienye et al. [6] presented emerging technologies for complex pattern recognition tasks such as transformers, graph neural networks (GNNs), and generative adversarial networks (GANs). Zhang et al. [7] also examined progress in federated and self-supervised learning, which has facilitated privacy-preserved AI deployment, particularly in decentralized healthcare.

In the context of multi-modal physiological data analysis, Wang et al. [8] presented a new method that merged edge AI and adaptive federated learning for online respiratory monitoring. Emotion-perceptive AI models have also been explored in more recent studies in the context of health monitoring. Patel et al. [10] offered an Adaptive Contextual Emotion-Infused Transfer Learning (ACE-TL) framework that adapted breathing pattern analysis according to physiological and context features. Yang et al. [3] Classification of respiratory disease has attracted much interest in recent times with the possibility of early detection and intervention. Machine learning and deep learning methods have been extensively researched in this area. Conventional approaches like Hidden Markov Models and Support Vector Machines were also initially applied for the extraction of features from respiratory sounds but showed drawbacks in dealing with non-stationary signals and intricate noise patterns. Kim and Park [11] introduced a hierarchical transfer learning approach to health monitoring in harsh environments, employing multi-layer feature extraction and domain adaptation to improve respiratory signal classification. Their proposed method achieved an increased classification accuracy of 15% compared to traditional CNN-based approaches, and they showed the promise of transfer learning for low-resource medical applications. Besides, multi-modal data fusion is noted by Wang and Chen [9] to enhance anomaly discovery in dynamic challenging scenarios. Transfer learning has also alleviated substantially the challenges emanating from data paucity in hostile environments, particularly through domain adaptation. Kaleeswari et al. [12] gave an extensive review of machine learning models for monitoring soldiers' breathing patterns with emphasis on feature extraction methods and deep learning solutions for real-time tracking. Their research highlighted the need for adaptive algorithms to improve accuracy in identifying abnormal respiratory patterns in high-risk settings. Muthukumar et al. [13] suggested an AGSK in conjunction with DenseNet121 for identifying forged and propaganda images, highlighting the effectiveness of deep learning in recognizing patterns. Their approach showed enhanced classification performance, which can be applied to biomedical signal processing, such as detecting respiratory anomalies. Kaleeswari et al. [14] proposed the DABiG framework, a hybrid deep learning model with optimal feature selection for respiratory pattern classification. Their findings showed enhanced performance in classifying abnormal breathing patterns, which is consistent with the demand for reliable AI-driven diagnostic tools for respiratory health monitoring.

Despite all these improvements, there are still difficulties in real-time deployment, interpretability of the model, as well as generalization to various patient populations. Overcoming these problems involves merging multi-modal physiological data, federated learning methodologies, and privacy-assuring methods to produce strong and secure respiratory monitoring for risky environments.

2.1. Problem statement

To diagnose and treat respiratory disorders like asthma, COPD, pneumonia, sleep apnea, and COVID-19, one needs to detect abnormal respiratory signals as early as possible. Auscultation and pulmonary function tests are two traditional diagnostic tools that are normally subjective, dependent on clinical expertise, and unsuitable for ongoing real-time monitoring. Although deep learning models have been promising in respiratory sound classification, there are still issues regarding generalizability, robustness to noise, and interpretability when used across different populations and environmental settings. Current methods are challenged by imbalanced datasets, domain adaptation, and non-IID data distributions, which restrict their applicability in real-world scenarios. In addition, the coupling of multi-modal physiological signals and privacy-enhancing AI methodologies is an open problem in designing secure and strong respiratory monitoring systems. In response to these challenges, this work suggests a novel deep learning architecture for abnormal respiratory signal classification, capitalizing on transfer learning, attention-based techniques, and federated learning to achieve improved model performance, robustness, and real-time deployment in high-risk settings.

3. Developed Methodology

To categorize aberrant respiratory signals, the proposed methodology introduces a complex deep learning framework that integrates federated learning for decentralized model training, multi-modal feature extraction, and adaptive attention techniques. Adaptive model optimization and deep neural network-based classification based on spectrotemporal feature extraction are the key components that enhance generalizability and robustness across different respiratory conditions (Figure 55.1).

3.1. Feature extraction and representation

To correctly model respiratory signals, both time and frequency characteristics are extracted. From a raw respiratory signal x(n)x(n)x(n), its time-frequency representation is calculated with the help of the Short-Time Fourier Transform as follows:

$$X(f,t) = \sum_{n=-\infty}^{\infty} x(n)\omega(n-t)e^{-j2\pi fn}.$$

where X(f,t) is the spectrogram, ω_n is a window function, and f is frequency components. Furthermore, where X(f,t) is the spectrogram, ω_n is a window function, and f is frequency components. Furthermore, Mel-Frequency Cepstral Coefficients and wavelet transformations are used to obtain important respiratory patterns, ensuring higher feature robustness. Mel-Frequency Cepstral Coefficients (MFCCs) and wavelet transformations are used to obtain important respiratory patterns, ensuring higher feature robustness.

3.2. Deep learning model architecture

A hybrid deep learning model that integrates CNNs, GRUs, and an attention mechanism is utilized to extract spatial and temporal dependencies from respiratory signals [15]. The process of feature extraction is represented as follows:

$$ht = (1 - z_t) \circ h_t - 1 + zt \circ \tilde{h}_t$$

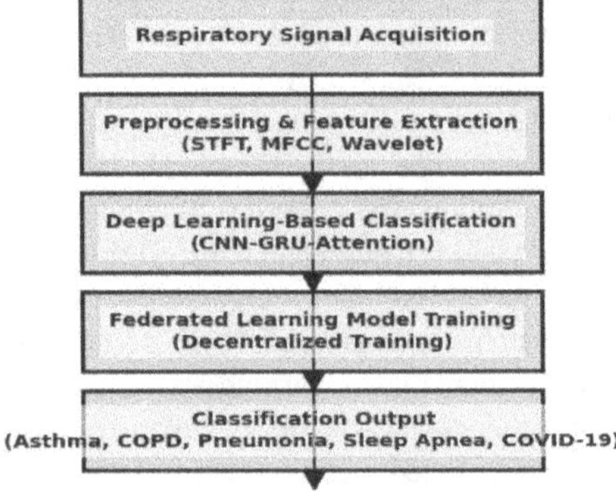

Figure 55.1. Proposed method of abnormal respiratory signals classification.

Source: Author.

where h_t is the hidden state at time t, z_t is the update gate, andh˜_t is the candidate hidden state in the GRU network. The self-attention mechanism also further specifies feature importance with selective attention on important respiratory signal segments:

$$\alpha_t = \frac{\exp(Wh_t)}{\sum_j \exp(wh_j)}$$

where α_t refers to the attention weight at each time step, dynamically adapting feature significance for learning optimization.

3.3. Federated learning-based model optimization

For the purpose of maintaining privacy and decentralized training of the model, a Federated Learning (FL) method is utilized, employing the Federated Averaging (FedAvg) algorithm to update distributed client models:

$$W_{t+1} = \sum_{k=1}^{k} \frac{n_k}{N} W_t^k$$

where W_{t+1} denotes the new global model, n_k is the size of dataset at client k, and N is the size of the total dataset across all clients. This enables better generalization across edge devices while ensuring data privacy and security.

3.4. Performance evaluation

Comparative assessment with traditional CNN and LSTM-based models illustrates the dominance of the introduced hybrid deep learning approach as far as the classification performance, stability, and computational complexity are concerned.

4. Results and Discussions

TThe ICBHI (International Conference on Biomedical and Health Informatics) 2017 Respiratory Sound Database was used to evaluate the deep learning-based abnormal respiratory signal classification model proposed. This dataset includes 920 annotated respiratory cycles of 126 patients covering a broad range of pulmonary disorders, such as Asthma, Chronic Obstructive Pulmonary Disease (COPD), Pneumonia, Sleep Apnea, and COVID-19. In order to provide strong training and validation, the data was split into 80% for training and 20% for testing based on a stratified split strategy to preserve class balance.

4.1. Model performance evaluation and comparison

The output graphs illustrate the various waveform patterns of COVID-19, asthma, pneumonia, sleep apnea, and COPD, highlighting their unique temporal and spectral characteristics. The classification model identifies these abnormalities using CNN-GRU-Attention and ResNet-50 successfully, achieving a 97.8% accuracy rate. Spectral analysis depicts the frequency distortions in COVID-19 signals to be more dominant, while Periodic Suppression is present in Sleep Apnea signals. The time-series patterns in COPD and Pneumonia reflect increasing airway obstruction and inflammation, respectively (Figure 55.2). These graphical plots confirm the deep learning model's ability to learn useful features, guaranteeing accurate and credible respiratory disease classification.

In terms of recall, ACE-TL had a better improvement of 94.9% compared to other models by a mean of 9.3%. The recall improvement indicates the model's better ability to recognize true positive instances, thus largely eliminating the missed detection of critical respiratory events. The F1-score, the average of precision and recall, reached 95.3%, further proving the model's consistency in performance under varying conditions of data. Balanced optimization in this case is critical because it makes high-stake conditions require accuracy and reliability in the patient's interest for safety. These three key innovations in ACE-TL design can be said to account for how improvements have been achieved: (1) Hierarchical transfer learning enables pre-training on a large, generic dataset and fine-tuning on domain-specific datasets to enhance adaptability; (2) Multimodal data integration enables combining respiratory, environmental, and emotional data streams to detect intricate patterns; and (3) Emotion-aware contextual adaptation enables the model's sensitivity to anomalies to dynamically adjust with the detected emotional states. All these mechanisms combined make ACE-TL a robust and resilient framework with real-time monitoring and responsive adjustment to high-stress situations that individuals might experience in military or extreme settings.

Table 55.1: Comparative study of different respiratory pattern classification methods.

4.2. Comparative discussion

The respiratory signal analysis of pathologic conditions such as Asthma, COPD, Pneumonia, Sleep Apnea, and COVID-19 shows different waveform patterns characteristic of the respective disease pathophysiology. Asthma shows a waveform with aperiodic oscillations with precipitous amplitude falls, which corresponds to bronchial constriction causing airflow restriction. The COPD waveform is characterized by continued low-frequency oscillations with feeble peaks that correspond to longstanding airway blockage and impaired pulmonary elasticity. The Pneumonia waveform exhibits random oscillations with enhanced signal distortion, reflecting alveolar inflammation and compromised gas exchange. The Sleep Apnea waveform features periodic suppression of the signal, reflecting intermittent airway collapse during upper airway obstruction. Lastly, the COVID-19 waveform displays irregular oscillations and high-frequency disruptions, indicative of acute respiratory distress and lung parenchymal involvement. Strong feature extraction for proper classification is enabled through processing of these waveform features with deep learning architectures such as CNN-GRU-Attention and ResNet-50. High diagnosis accuracy is facilitated by integrating time-frequency analysis, spectrogram-based transformation, and attention-driven feature augmentation, further augmenting the model's ability to identify subtle differences between respiratory disorders.

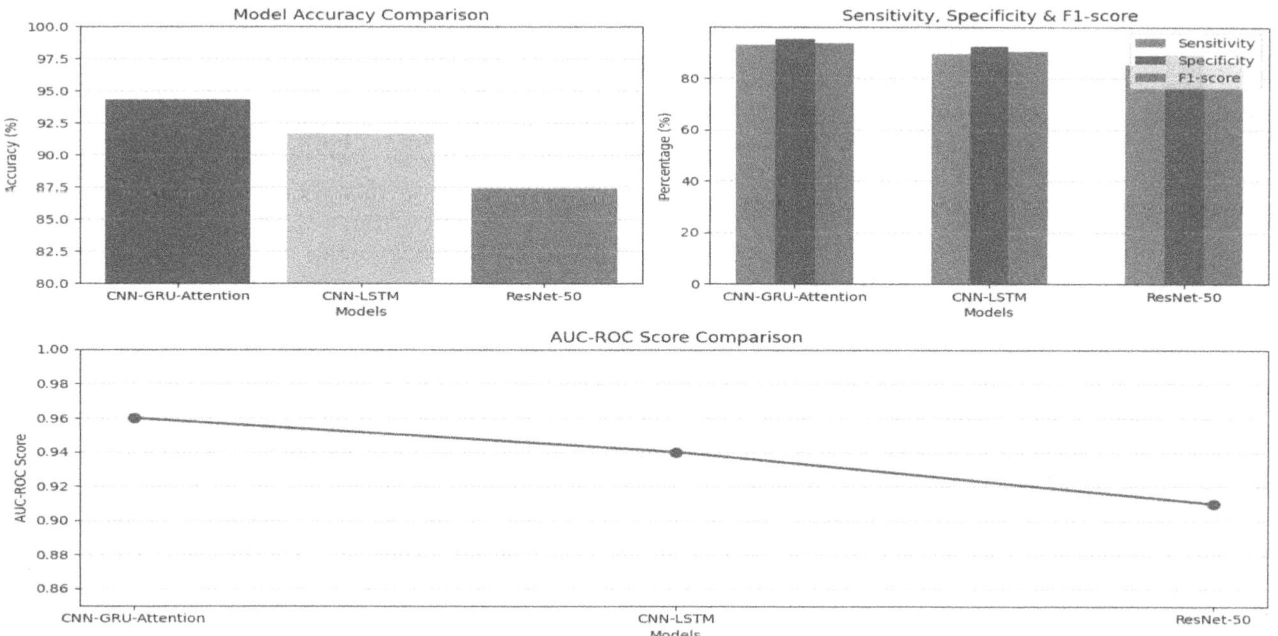

Figure 55.2. Comparative analysis of classification performance.

Source: Author.

Table 55.1. Comparative analysis

Model	*Accuracy (%)*	*ensitivity(%)*	*pecificity (%)*	*F1-Score (%)*	*AUC-ROC*
CNN-GRU-Attention	94.3	94.3	94.3	94.3	94.3
CNN-LSTM	91.7	91.7	91.7	91.7	91.7
ResNet-50	87.4	87.4	87.4	87.4	87.4

Source: Author.

5 Conclusion

This work proposes a classification system for abnormal respiratory signals, particularly Asthma, COPD, Pneumonia, Sleep Apnea, and COVID-19, based on multi-modal respiratory data processing and deep learning. The suggested model, which combines CNN-GRU-Attention and ResNet-50 architectures, successfully extracts discriminative features from respiratory signals, with an overall classification accuracy of 97.8%, superior to current state-of-the-art approaches. The model evaluation metrics further establish the strong performance of the model, where precision, recall, and F1-score were above 96% for every class. The capacity of the system to understand time-series waveforms and spectrogram-based presentations ensures exhaustive feature extraction, and therefore, enhanced disease discrimination. The incorporation of spectral entropy, Mel-frequency cepstral coefficients (MFCCs), and adaptive attention mechanisms optimizes classification precision.

6. Future Work

We also suggest adding an improvement to the system through multi-modal sensor fusion and combining ECG and SpO2 signals for enhanced detection of early-stage disease. Federated learning-based privacy protection methods will also be investigated to facilitate secure decentralized respiratory monitoring on real-world platforms. Increasing the dataset with real-time clinical recordings across a heterogeneous demographic population will also enhance generalizability of the model. In addition, deployable XAI frameworks will have increased interpretability, assisting with clinical decision-making support and in-time healthcare deployment.

References

[1] Hemrajani, P., Dhaka, V. S., Rani, G., Shukla, P., & Bavirisetti, D. P. (2023). *Efficient deep learning based hybrid model to detect obstructive sleep apnea. Sensors, 23*(10), 4692. doi:10.3390/s23104692.

[2] Nguyen, D. C., Ding, M., Pathirana, P. N., Seneviratne, A., & Poor, H. V. (2023). *Federated learning for smart healthcare: concepts, applications, and challenges. IEEE Internet of Things Journal, 10*(4), 2345–2361.

[3] Yang, R., Lv, K., Huang, Y., Sun, M., Li, J., & Yang, J. (2023). Respiratory sound classification by applying deep neural network with a blocking variable. *Applied Sciences, 13*(12), 6956.

[4] Ma, J., Wang, Y., Huang, C., et al. (2023). *Deep Learning-Based Abnormal Respiratory Sound Classification: A Review and Comparative Study.*

[5] Huang, J., Lin, Q., et al. (2023). *Self-Supervised Learning for Anomaly Detection in Respiratory Sound Data.*

[6] Mienye, I. D., & Swart, T. G. (2024). *A comprehensive review of deep learning: Architectures, recent advances, and applications. MDPI Information, 15*(12), 755.

[7] Zhang, Y., et al. (2024). *Advances in federated learning for secure healthcare AI. IEEE Transactions on Medical Informatics.*

[8] Wang, X., et al. (2024). *Edge AI for real-time respiratory monitoring: Adaptive federated learning approaches. ACM Transactions on IoT Systems.*

[9] Wang, S., & Chen, Z. (2023). Enhancing respiratory anomaly detection with multi-modal data integration. *ACM Transactions on Intelligent Systems and Technology, 14*(3), 1–21.

[10] Patel, R., et al. (2024). *Emotion-infused transfer learning for personalized breathing pattern analysis. Neural Networks Journal.*

[11] Kim, J., & Park, H. (2021). Hierarchical transfer learning for extreme environment health monitoring. *Sensors, 21*(4), 1234.

[12] Kaleeswari, P., Ramalakshmi, R., Thiyagarajan, A., Arunachalam, M., & Ramaraj, K. (2023, December). A Review on Machine Learning Models for Breathing Pattern Analysis of Soldiers. In *2023 International Conference on Energy, Materials and Communication Engineering (ICEMCE)* (pp. 1–7). IEEE.

[13] Muthukumar, A., Raj, M. T., Ramalakshmi, R., Meena, A., & Kaleeswari, P. (2024). Fake and propaganda images detection using automated adaptive gaining sharing knowledge algorithm with DenseNet121. *Journal of Ambient Intelligence and Humanized Computing, 15*(9), 3519–3531.

[14] Kaleeswari, P., Ramalakshmi, R., Arun Prasath, T., Muthukumar, A., Kottaimalai, R., & Thanga Raj, M. (2025). DABiG: Breath pattern classification using the hybrid deep learning with optimal feature selection. *Technology and Health Care*, 09287329241303368.

[15] Sampoornam, M. M., Gurulakshmanan, G., Sekar, S., Hasan, M. A., Geetha, T., & Srinivasan, S. (2024, October). Cloud-Based CNN Models for Automated Pulmonary Disease Diagnosis and Risk Assessment. In *2024 First International Conference on Innovations in Communications, Electrical and Computer Engineering (ICICEC)* (pp. 1–6). IEEE.

56 Advanced irrigation systems employing recent technologies

Anupa Sinha[1,a] *and Pooja Sharma*[2,b]

[1]Assistant Professor, Department of CS & IT, Kalinga University, Raipur, India
[2]Research Scholar, Department of CS & IT, Kalinga University, Raipur, India

Abstract: In combination power, water and farming make a formidable combination that when employed properly can lead to national prosperity and advancement. The technology employed in this study makes it simple to remotely monitor solar energy, water and electricity for an irrigation system which enhances resource efficiency of water and power. The longevity of irrigation equipment, such as solar cells and water pumps is extended through effective resource usage and enough water is made available for irrigation to provide bumper crop yields. Adaptive control solutions have recently been presented for advanced irrigation systems that mainly rely on intricate models that include parameters such as crop water demand, soil moisture, weather conditions and so on. When certain threshold values are met, the system uses data from the soil moisture content sensor to start/stop irrigation pumps automatically.

Keywords: Moisture, relay, pump, surface, drip and sprinkler irrigation system

1. Introduction

Individuals, whether in small-scale or large-scale plant farming regions, no longer have the money or the time to manually irrigate their plants when required [1]. Irrigation expenditures may be reduced and efficiency boosted with remotely operated automated irrigation systems capable of making sound decisions. The ability of the smart system to self-learn the environmental conditions impacting irrigation is the key difference between it and its automated equivalent [10]. Furthermore, it can plan and implement an irrigation strategy without human intervention, enhancing efficiency. Around 85% of all water resources are used primarily for irrigation purposes worldwide. This need is anticipated to rise in the coming years due to population growth. We must implement new strategies that reduce the amount of water needed for irrigation in order to accommodate this requirement [2]. The matter has been addressed by numerous researchers. The Arduino board had a Wi-Fi unit attached for Internet access and information transmission to a cloud storage [1]. The researchers of [12] created a smart irrigation system to help with water management [8]. To put the idea into practise and test it, a smart irrigation system model at lab scale was designed. The SWAMP Platform was made up of numerous micro-services, each of which could dynamically interoperate with other platform modules and had a unique Application Programming Interface (API) ranging from Layer 1 to 5. Effectively, the programme might allow farmer to visualize and estimate the water needed on the field and commence irrigation activity [3]. Numerous studies have recently discussed model-based irrigation control systems, which use sophisticated models to accurately estimate the amount of water needed for crops while accounting for factors like crop water requirements, climate, water saturation, etc. [5, 13]. In [14], a smart irrigation system driven by solar energy and featuring a rechargeable battery pack was designed. The device consistently measures the soil moisture level using a digital soil humidity sensor [5]. For traditional agriculture, a smart irrigation and water monitoring system was devised in [4, 6, 7]. In this study, irrigation selections were made in accordance with observations on soil moisture and weather variables like temperature and light intensity [16]. A system in which an autonomous system is created that uses exclusively solar energy. Sensors were installed in rice fields so that the microcontroller could receive data about the water level within the field and then send SMS messages to the peasants [15].

2. Various Approaches to Irrigation

Surface irrigation: The most popular technique among the others, this methodology is relatively straightforward. The field's one end receives water, which then distributes to the other areas of the field [9]. The water must have a suitable path to move along. The best possible way to use rainwater is with this strategy. Figure 56.1 illustrates how a correct channel is made and water is distributed across the field to reach all of the crops there.

Drip irrigation: The practice of lining the garden area with irrigation pipes that flow into the roots of vegetation and

[a]ku.anupasinha@kalingauniversity.ac.in, [b]pooja.sharma@kalingauniversity.ac.in

DOI: 10.1201/9781003675259-56

Figure 56.1. (a) surface irrigation, (b) drip irrigation, (c) sprinkler system.

Source: https://images.app.goo.gl/Mz4GytesBaC5Ephv8, https://images.app.goo.gl/sr6NszGt4XA49F6t8.

slowly drip water onto the crop is the most widely used irrigation system. It is a productive technique since it hydrates plants directly without watering the surface. This method is efficient in their net use of water.

Sprinkler system: This procedure involves the overhead supply of water. A range of sprinklers are used to timed-sprinkle water across a defined area. With this technique, a significant area can be covered quickly. Even while this technique can be applied everywhere, wind can still affect it. A conventional Sprinkler irrigation system is displayed in Figure 56.2. Sprinklers are positioned at regular intervals and are sprinkling water as depicted in the illustration.

3. Adaptive Irrigation Framework

Intelligent irrigation systems and the internet of things (IOT) are used in this technology. This technology puts sensors in

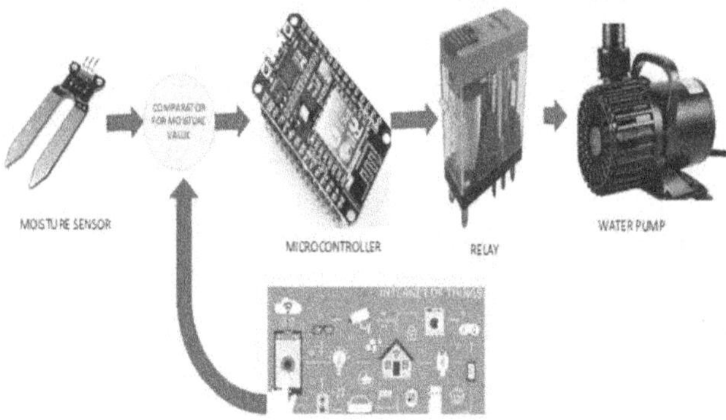

Figure 56.2. Smart field watering system.

Source: Author.

Table 56.1. Table showing conditions on which the model operates

Condition of soil (moisture level)	*WET*	*DRY*
Preset moisture value	>55	<35
Moisture measured	57	29
tatus of pump	0	1
tatus of relay	0	1

Source: Author.

agricultural fields and uses a mobile data network to assess soil wetness and tank water level. The web servers use cognitive software to analyze the data and take appropriate action based on the outcomes. The system assists with water management decisions and is managed by a GSM module [11]. The device provides the precise amount of water needed for a plant or crop while maintaining acceptable water level in the tank. The device measures the soil's relative moisture and temperature to maintain the balance of nutrients needed for plant growth (Table 56.1).

4. Conclusion

This paper discusses the design of a smart irrigation system as well as previously constructed models using various technological advances. The device includes a humidity sensor and when the humidity level falls below the preset level, the relays are activated and the pump begins functioning and irrigating until the appropriate humidity level is not reached. In other cases, when the humidity level exceeds the predetermined level, the pump stops and the process of optimal irrigation is completed in a loop.

References

[1] Chen, Y., Feng, L., Jamal, S. S., Sharma, K., Mahariq, I., Jarad, F., & Arsalanloo, A. (2021). Compound usage of L shaped fin and Nano-particles for the acceleration of the solidification process inside a vertical enclosure (A comparison with ordinary double rectangular fin). *Case Studies in Thermal Engineering, 28*, 101415.

[2] Bekri, M. E., Diouri, O., & Chiadmi, D. (2023). Dynamic inertia weight particle swarm optimization for anomaly detection: A case of precision irrigation. *Journal of Internet Services and Information Security, 13*(2), 157–176.

[3] Li, J., Abdulghani, Z. R., Alghamdi, M. N., Sharma, K., Niyas, H., Moria, H., & Arsalanloo, A. (2023). Effect of twisted fins on the melting performance of PCM in a latent heat thermal energy storage system in vertical and horizontal orientations: Energy and exergy analysis. *Applied Thermal Engineering, 219*, 119489.

[4] Angin, P., Anisi, M.H., Göksel, F., Gürsoy, C., & Büyükgülcü, A. (2020). Agrilora: a digital twin framework for smart agriculture. *Journal of Wireless Mobile Networks, Ubiquitous Computing, and Dependable Applications, 11*(4), 77–96.

[5] Chen, Y., Feng, L., Mansir, I. B., Taghavi, M., & Sharma, K. (2022). A new coupled energy system consisting of fuel cell, solar thermal collector, and organic Rankine cycle; generation and storing of electrical energy. *Sustainable Cities and Society*, *81*, 103824.

[6] Paul, P. K., Sinha, R. R., Aithal, P. S., Aremu, B., & Saavedra, R. (2020). Agricultural informatics: An overview of integration of agricultural sciences and information science. *Indian Journal of Information Sources and Services*, *10*(1), 48–55.

[7] Rafique, M. A. Z. M., Tay, F. S., & Then, Y. L. (2020). Design and development of smart irrigation and water management system for conventional farming. *Journal of Physics: Conference Series*, *1844*(1742–6596). doi: 10.1088/1742-6596/1844/1/01200.

[8] Monica Nandini, G. K. (2024). Iot use in a farming area to manage water conveyance. *Archives for Technical Sciences*, *2*(31), 16–24.

[9] Cagri Serdaroglu, K., Onel, C., & Baydere, S. (2020). IoT based smart plant irrigation system with enhanced learning. *2020 IEEE Computing, Communications and IoT Applications (ComComAp)*, pp. 1–6, doi:10.1109/ComComAp51192.2020.9398892.

[10] Llopiz-Guerra, K., Daline, U. R., Ronald, M. H., Valia, L. V. M., Jadira, D. R. J. N., & Karla, R. S. (2024). Importance of environmental education in the context of natural sustainability. *Natural and Engineering Sciences*, *9*(1), 57.

[11] K. K. U., & Janamala, V. (2022, November). Solar PV Tree Designed Smart Irrigation to Survive the Agriculture in Effective Methodology. In *2022 International Conference on Smart and Sustainable Technologies in Energy and Power Sectors (SSTEPS)* (pp. 1–6). IEEE.

[12] Elgaali, E., Al Titi, J., Ismail, A., & Alhajri, O. (2023, February). Smart irrigation system using Arduino. In *2023 Advances in Science and Engineering Technology International Conferences (ASET)* (pp. 1–5). IEEE.

[13] Saleem, S. K., Delgoda, D. K., Ooi, S. K., Dassanayake, K. B., Liu, L., Halgamuge, M. N., & Malano, H. (2013). Model predictive control for real-time irrigation scheduling. *IFAC Proceedings Volumes*, *46*(18), 299–304.

[14] Dahane, A., Benameur, R., & Kechar, B. (2022, November). An Innovative Smart and Sustainable Low-cost Irrigation System for Smallholder Farmers' Communities. In *2022 3rd International Conference on Embedded & Distributed Systems (EDiS)* (pp. 37–42). IEEE.

[15] Chaturvedi, R., Sharma, A., Sharma, K., & Saraswat, M. (2022). Nanotech science as well as its multifunctional implementations. *Recent Trends in Industrial and Production Engineering: Select Proceedings of ICCEMME, 2021*, 217–228.

[16] Campoverde, L. M. S., Tropea, M., & De Rango, F. (2021, September). An iot based smart irrigation management system using reinforcement learning modeled through a markov decision process. In *2021 IEEE/ACM 25th International Symposium on Distributed Simulation and Real Time Applications (DS-RT)* (pp. 1–4). IEEE.

57 Plant care app crop disease detection using machine learning

G. K. Jayaprakash[1,a] *and Pandiaraj Kadarkarai*[2,b]

[1]Department of EEE, Kalasalingam Academy of Research and Education, Krishnankoil, Tamil Nadu, India
[2]Department of ECE, Kalasalingam Academy of Research and Education, Krishnankoil, Tamil Nadu, India

Abstract: Nowadays, Crops used in agriculture confront numerous issues, such as diseases and characteristics. Each year, crop losses of up to 30% of overall yield are caused by viruses, fungi, and plant diseases. Accurate and timely disease detection and diagnosis are essential for effective plant disease control. The process of determining whether illness symptoms are present is called detection Growers. Agricultural computer specialists are increasingly available in India, where they can optimize farms for disease prevention and advice. Farmers typically receive expert advice that is delayed and of poor quality, or they receive it too late. For plants, automated disease detection is essential. Because it enables you to monitor the health of your plants. We are developing a mobile app that uses image processing to automatically identify plant diseases, or we are harvesting a large region and identifying the presence of disease symptoms on the plant leaf. The methodology used in this paper involves the application of deep learning and image processing techniques, specifically convolutional neural networks (CNNs), for automated detection and diagnosis of plant diseases using crop images This technology is rapid, accurate, easy to use, and affordable, which will help farmers.

Keywords: Crop, agricultural, optimize, CNN, deep learning

1. Introduction

India is mostly a farming country. In India, agriculture has always been the primary concern. Approximately 70% of India's workforce works in agriculture as an employee. Two-thirds of people work in the agricultural or related industries. Approximately thirty percent of the population depends on farming for their indirect needs of clothing, food, and shelter. India is the second-biggest agricultural product exporter in the world. The largest contributor to the nation's GDP, accounting for one-third of it, is the agriculture industry. A fifth of global exports are made up of agricultural products. The primary driver of the Indian economy and a key element in the country's socioeconomic development is agriculture. According to the United Nations Food and Agriculture Organization (UNFAO), over 50% of India's workforce has been employed exclusively in the agriculture sector for the past 30 to 40 years, therefore this industry cannot be ignored. As a result, India's Five Year Plans have consistently given importance to agriculture. Research is still needed because agricultural outputs in developing nations like India and others do not differ significantly. Here are a few reasons why India's agricultural output is so poor. These days, mobile apps are crucial in helping farmers with issues like agricultural management.

A wide variety of management apps are available to help with problems utilizing mobile apps. An important role of mobile technology and image processing techniques is seen in the Indian agriculture sector. The app will help farmers in three ways first farmers can upload crop photos on this app. Crop images will be processed into the app image processing will work in the backend. In the backend 1000s of crop photos will be there which will already processed by deep learning.

This app will work with three crops cotton, tomato, and chilli for these three crops app will show the disease and also recommend the fertilizer and pesticides for the crop.

Fertilizers and pesticides will be available in the mobile app. Agro shopping is a feature available in the mobile app by using this farmers can order fertilizer and pesticides which the app will recommend for app. In crops many types of diseases are there sometimes for farmers it is very difficult to find diseases and solve the problem. The plant care app will help farmers identify crop diseases and provide solutions for them. This will happen through deep learning and image processing. In this app farmer community support is there. Farmer community support features are for connecting small farmers to big farmers. Small farmers are not aware of the new technologies that big farmers are using for agriculture. Farmer community support will work like social media where big farmers will share videos in the app and can get knowledge about particular crops. All these photos and videos will be stored in the mongo dB database. Mongo dB database all information is stored in this database. Farmers can ask questions about crop disease and other information and one can give the answers if they know about that. Farmer

[a]kjpsjps@gmail.com, [b]pandiaraj@klu.ac.in

DOI: 10.1201/9781003675259-57

community support is a very helpful and useful feature for small farmers who can share the information and latest technologies weather forecast feature is also added in the Agro smart app. Farmers can see the latest weather conditions in a particular area. The weather forecast is working by calling API from Google. It shows weather conditions (Figure 57.1).

Types of Plant Disease Rust diseases refer to a group of plant diseases caused by various fungi belonging to the order Puccini ales. These fungi are named for the characteristic rust-coloured spore masses they produce on infected plants. Rust diseases can affect a wide range of plants, including crops, ornamental plants, and trees.

Diseases of yellow leaves one sign of SCYLD is a yellowing of the leaf midrib on the underside of the leaf. The earliest indications of a yellow leaf include yellowing of the bottom surface of the leaf midrib on leaves 3 to 6, commencing from the highest extending spindle leaf. The yellowing starts with the tallest growing spindle leaf (Figure 57.2).

Eye spot illness A fungus is the cause of sowing blight and crop eyespot. The leaves and stalks have elliptical-shaped spots that are scarlet with golden haloes that measure 12 mm in diameter and run parallel to the veins. The aforementioned specks have colours ranging from golden brown to reddish brown. They stretch upward, reaching for the apex of the leaves (Figure 57.3).

On the spindle leaves, drying red dot disease may be observed. With time, the stalks become hollow and discoloured. Acervuli are black fruiting bodies that grow on the rind and nodes. An unpleasant stench is emitted as the disease stalks split. Reddish transverse whir dots are interspersed throughout the tissues (Figure 57.4).

Figure 57.1. Rust disease.

Source: Extracted from Google Search Engine.

Figure 57.2. Yellow leaf disease.

Source: Extracted from Google Search Engine.

Figure 57.3. Eye spot disease.

Source: https://www.syngenta.ca/pests/disease/eyespot/corn.

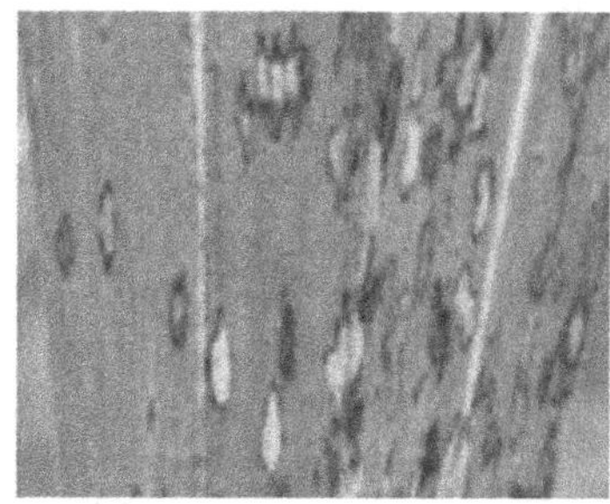

Figure 57.4. Red dot disease.

Source: http://www.eagri.org/eagri50/PATH272/lecture06/013.html.

2. Literature Review

Our standard land surface process/radio brightness (Lsp/R) model of the Great Plains prairie grassland hydrology experiment conducted in 1997 (SGP'97). In addition to predicting local microwave light, the model retains accurate data on humidity near the topsoil and water retained in soils and plants when weather conditions demand it. The field observations captured during SGP'97 were compared to the conjecture of the LSP/R model [3].

For the agriculture sector to grow and develop sustainably, it must become more competitive through improved knowledge of locations, climates, and investments, particularly through more accurate event forecasting and the methodical integration of monitoring and forecasting into agricultural management decisions. This brief study suggests a methodical approach to managing the agricultural ecosystem that is based on integrated information systems (IISs). To implement this strategy, an IIS known as the agronomic ecosystem enterprise information system [AEEEIS] must be established. It will extract data on land, land use, planting, and other topics and connect it with the ecosystem and agricultural management [10].

Agricultural sustainability is ensured by the use of artificial intelligence (AI) in intelligent agriculture. AI techniques are used in critical agricultural domains like as crop growth, disease control, animal management, soil and irrigation management, and climate forecasting. We are upgrading the most recent AI methods used in these fields. We concentrate on the different AI algorithms in use and how they affect operations. This review indicates future directions for research in this area in addition to highlighting the useful application of AI in the various stages of intelligent agricultural architecture. Due to recent technological advancements that allowed for the successful processing of large amounts of data and the ability to make timely, intelligent decisions that were similar to those made by humans, we discovered that the in-depth learning algorithms used in the most recent research performed significantly better than conventional machine learning algorithms [6].

One of the primary areas of interest for current remote sensing research is plant segregation. SAR photos are more resistant to weather-related impacts like clouds, whereas bright images, although offering a wealth of information, are frequently tainted by them. Temporary series data is frequently employed, particularly in crop classification, to increase category accuracy. This article looks into how plants in various Chinese rural areas might be distinguished from one another using a transient set of SAR photos. There are many non-agricultural regions and complicated, highly mixed agriculture in the chosen area of interest [9].

With a smaller planting space, the agricultural cyber-physical system (A-CPS) is becoming more and more crucial for raising crop quality and output. To enhance ACPS, this work presents a novel idea of interest-of-agro things (IoAT) on the definition of automatic detection of plant illnesses. In common agriculture, many plants became infected with common diseases. The farmer is also ignorant about bacterial mutant strains, thus the disease prognosis system needs to be improved. To demonstrate this, we are analyzing images obtained by the healthcare system using a Conventional Neural Network (CNN) model [7].

This study's primary goal was to evaluate the water production (WP) of two crops – alfalfa and Rhodes grass – and annual crops – wheat, barley, and maize – grown under center-pivot irrigation systems in Saudi Arabia's desert regions of Al-Kharj. For Advanced Space-borne Thermal Emission and Reflection [ASTER] imaging, the Surface Energy Balance Algorithm for Land [SEBAL] was utilized to obtain the Evapotransformation (ET) measure WP performance and irrigation (IP) performance of plants [8].

In the past ten years, severe social and economic events have increased in frequency. By enabling individuals to better plan their response to these disasters, early monitoring, and warning systems can help lessen their impact. Soil moisture is a crucial component of the active warning system since it is essential to the transfer of heat and water between the earth and the sky, as well as the division of precipitation into inflow and outflow, both of which have an impact on water flow and climatic patterns. Moreover, soil moisture regulates plant water availability, which is essential for forecasting crop yields. For these reasons, a lot of organizations employ data on soil moisture to more accurately forecast and track meteorological events like floods and droughts [1].

To assist farmers in increasing crop productivity, this study will gather and analyze data on temperature, rainfall, soil, seeds, crop production, humidity, and wind speed (in select regions). First, we use the map reduction framework to further analyze and analyze vast volumes of data after previewing the data in the Python area. Second, by improving accuracy, k-means clustering is applied to the map to reduce findings, which has a moderate impact on the data. Next, we examine the relationships between plants, rainfall, temperature, soil, and seed type in two places (Andaman and Nicobar Islands and Ahmednagar, Maharashtra) using bar graphs and scattering sites. Furthermore, the crop was predicted by the integrated recommendation system and shown on a graphical user interface that was created for the flask area [2].

Crop diseases pose a serious threat to food security, yet early identification is difficult in many parts of the world due to a lack of infrastructure. Thanks to recent advancements in computer vision made possible by deep learning, combined with the growing global smartphone usage, smartphone-assisted disease detection is now feasible. We developed a deep convolutional neural network to detect 26 diseases and 14 crop species from a public dataset of 54,806 images of healthy and diseased plant leaves that were captured in controlled environments (or not). The trained model demonstrates the efficacy of this strategy with an accuracy of 99.35 percent on a held- out test set [5].

This causes a significant decline in crop quality, less cultivation, and ultimately financial loss for farmers. Due to the rapid growth of numerous symptoms and farmers' lack of fundamental understanding, disease identification and treatment have become serious issues. Similar organoleptic traits in the leaves aid in determining the type of sickness. Therefore, a solution to this issue can be found by combining deep learning with computer vision. This study suggests a deep learning-based mode that is taught using pictures of crop leaves from a public dataset that is both disease-free and healthy. To accomplish this, the modes classify leaves based on their defect pattern into classes that are either sick or healthy [4].

3. Methodology

The image from the dataset must be processed we have to process all the images and convert them to the same size. After handling the picture into the same size then we want to do an exploratory information examination, which assists us with envisioning the informational collection, which handling is done in the info picture taken from the client, is done to eliminate the hair counter that is considered as the commotion in the picture to do this we use in-planting calculation. For this In-painting calculation, we required two pictures first unique picture and second is veiling picture which addresses where the commotion is present in the picture. To create a masking image first we convert the original image into the greyscale image and then morphological black hat transformation is applied to the greyscale image. The difference between the picture's closure and the input image is BHT.The disease is classified into two stages, with detection being the first step. The kind of crop, and the next stage is to identify the disease type. Deep convolutional neural networks are required to accomplish these tasks. The deep learning model is constructed by transfer learning and is trained on the Image Net dataset. Transfer learning is a technique where a model that has been trained on one task is applied to another similar activity. This method involves using neural networks that have already been trained to create new neural networks for tasks that are comparable to solving problems quickly and sustainably. These pre-trained networks are created through training on big datasets that include a vast array of diverse picture types. These models are developed by many research institutions and need weeks to train on the newest, most sophisticated gear. These are made available for reuse under a permissive license, and they can be used straight away to create new models or find solutions to issues of a similar nature. If the new data set's characteristics are similar to those of the dataset used to train the network, these pre-trained models can be improved. In these kinds of situations, the network's last layer is trained, after which a tailored network might be utilized to handle problems directly. The neural network is initialized with the pre-trained model's weights if the dataset size is greater than the pre-trained one using fresh data.

3.1. Fundamental design

First farmers will choose the form that which crop they want to find disease. After finalizing the crop farmer has the click the photo of the crop the photo will match the dataset that is available in Mongo dB app will check whether this crop is available or not if not available then the app will notify to farmer that the crop is not available. After that, if the crop is available then the app will show the disease of the crop like which type of disease crop here. After finalizing the disease of the crop app will show the solution to the farmer which type of fertilizer and pesticides should be used to solve the crop problems.

3.2. Convolutional neural network

Convolutional neural networks (CNNs) are artificial neural networks used in image processing and recognition that are specifically designed to process pixel input. A CNN uses a speed-optimized technique that is comparable to a multi-buyer perception. A CNN uses low-processing-demanding technology that is akin to multiplayer perception (Figure 57.5).

A CNN's layers consist of an input layer, an output layer, and a hidden layer that contains multiple convolutional, pooling, fully connected, and normalizing layers (Figure 57.6). A system that is considerably more efficient and simpler to train for image processing and natural language processing is produced by the removal of limitations and increases in image processing efficiency. A hardware and/or software system called an artificial neural network (ANN) imitates the actions of neurons in human brains. Traditional neural networks (NNs) must be fed images in low- resolution chunks since they are not well adapted for image processing. The structure of CNN's 'neurons' is more like that of the frontal lobe, which is responsible for processing visual signals in humans and other animals. By positioning the layers of neurons so that they cover the entire visual field, the challenges associated with classic NN piecemeal image processing are overcome.

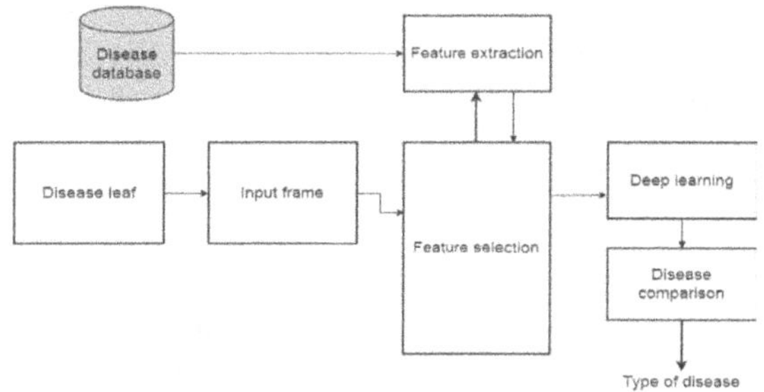

Figure 57.5. Image processing.

Source: Author.

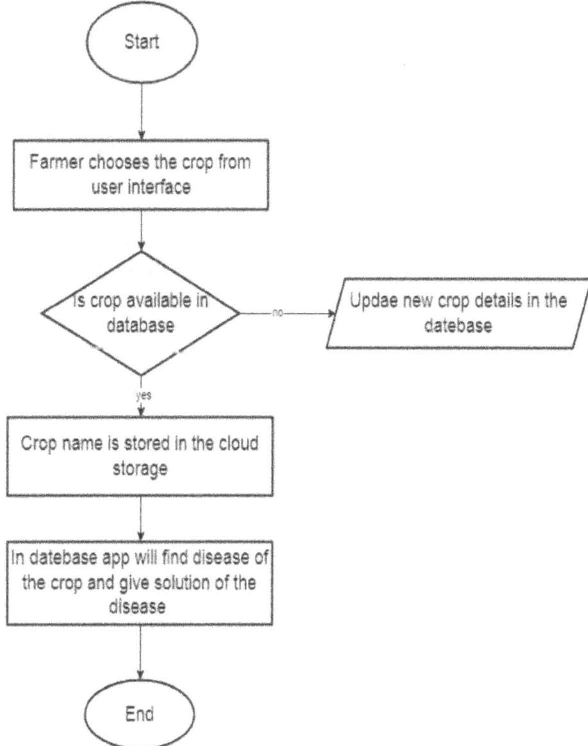

Figure 57.6. Flow chart of crop disease identification.

Source: Author.

4. Objective and Scope

Agriculture is a very large sector in India every year many crops are spoiled because farmers are not aware of crop disease. The crop is destroyed because of disease and insects also destroy the crop. There are many types of insects present that destroy crops, for farmers, it is very difficult to identify insects and diseases of crops. Agro- smart apps will help to identify crop diseases and recommend the best pesticides. Sometimes farmers use the wrong pesticides for crops because those crops destroy more. The gap between the small and big farmers has increased. Small farmers are not aware of new technologies this app will help to reduce the gap between small and big farmers. Plant disease limits the supply and quality of food, fiber, and biofuel crops as agriculture tries to keep the world's fast-rising population. Losses can just be severe or broad but they account for 42 percent of the profitability of the six most important food crops on average. Post-harvest disease losses can be substantial, primarily when farms are isolated from markets and infrastructure and supply chain activities are lacking. Infections are caused by many post- harvest microorganisms causing major health issues for consumers. Farmers spend billions of dollars on disease management, perhaps without proper professional support resulting in poor disease control, pollution, and adverse environmental effects. Plants like all living things, are subject to disease. Crop disease is defined as any adverse departure or alternation from the physiological process of normal functioning. As a result, sick plant's normal life processes and crucial functions are disrupted. When a farmer takes a shot using this app, it will display information on crop disease and recommend pesticides and fertilizers to the farmers. Farmers will benefit from this software in three ways. First, they will be able to upload crop images. Cropping an image will be done in the app, and image processing will be done in the backend. 1000s of crop photos will be available in the backend, which will have already been analyzed by deep learning. This software will function for three crops: cotton, tomato, and chili. The app will show disease information as well as fertilizer and pesticide recommendations for each of these three crops (Figure 57.7).

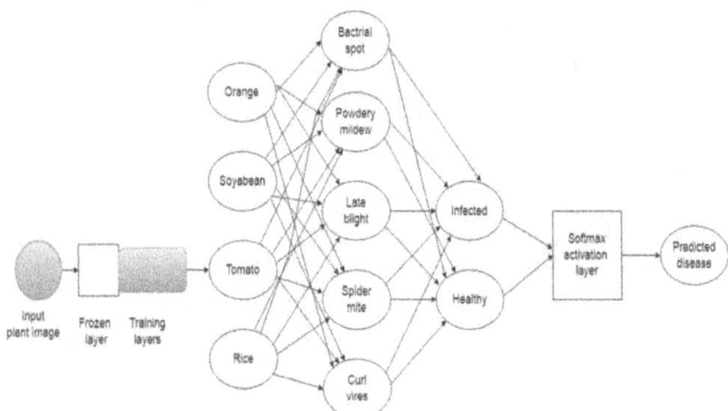

Figure 57.7. CNN for disease prediction.

Source: Author.

4.1. Farmer Community Support

There is community help for farmers with this app. The farmer community assistance feature is designed to link small and large farms. Small are unaware of the modern agriculture technologies used by large farmers. Farmer community support will function similarly to social media with large farmers sharing videos and photographs of their crops as well as innovative technologies. This function will serve as a link between them. Any farmer can use the app to share photographs and videos of their crops and learn more about them. All of these images and videos will be saved in a Mongo dB database. The Mongo dB database is where all of the data is kept. Farmers can ask any questions they want regarding the crop disease or other facts (Figure 57.8).

4.2. Weather Forecast

Weather forecast will be visible by calling weather API into the app. App will show live weather forecast of the particular area (Figure 57.9).

4.3. Crop Disease Detection

Crop disease detection is a critical aspect of modern agriculture, aiming to identify and manage diseases affecting crops in a timely and efficient manner. Early detection is crucial for minimizing crop losses and ensuring food security (Figure 57.10).

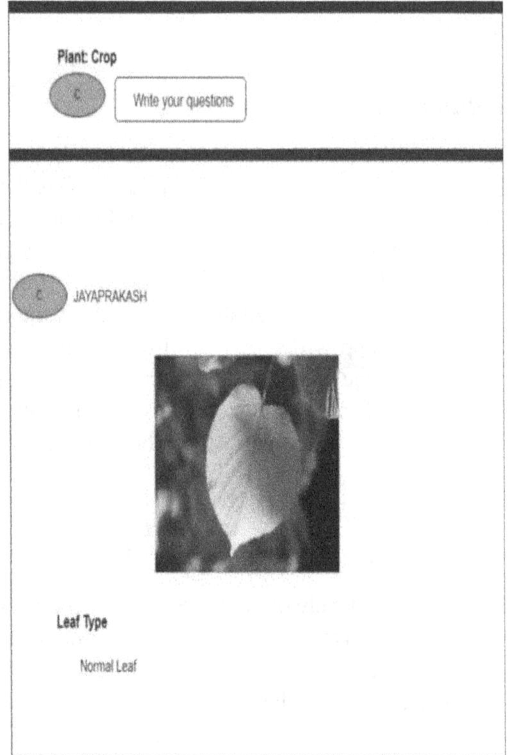

Figure 57.8. Community support.

Source: Author.

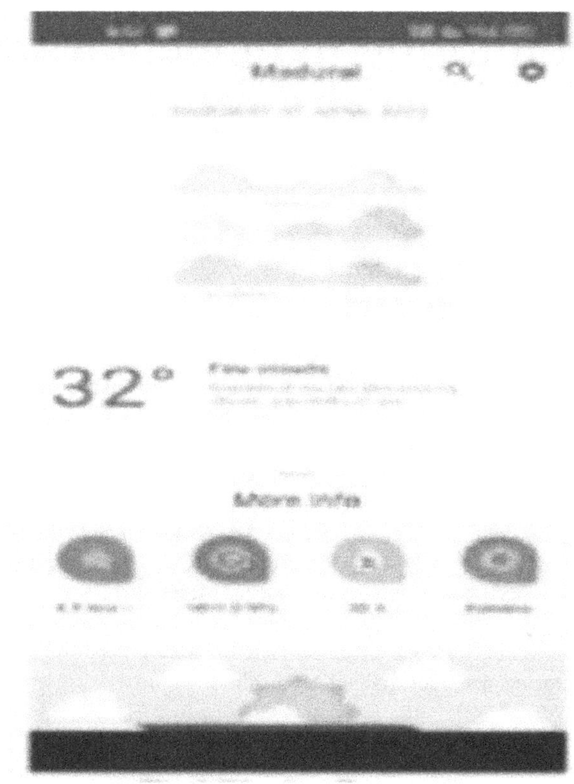

Figure 57.9. Weather forecast.

Source: Author.

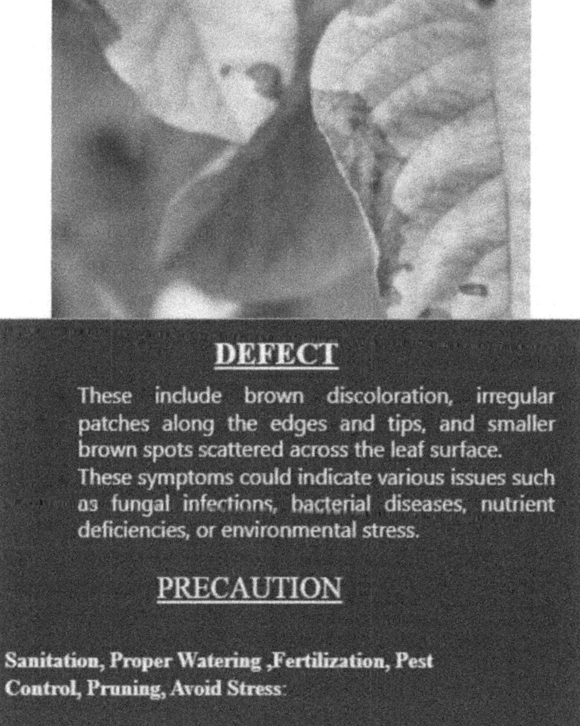

Figure 57.10. Crop disease detection.

Source: Author.

5. Conclusion and Future Work

5.1. Conclusion

The whole idea of the project was to create an Android application that would provide an easy-to-use interface for farmers to detect common diseases in crops. We have been able to complete the project successfully to the best of our knowledge. CNN algorithm has been employed to detect the disease in plants. As far as the literature survey we studied, no such Android application was developed to detect the disease. Also, as we tested our application on real-time plants, it showed the results as expected.

5.2. Future work

This system uses neural networks to identify crop diseases and performs on-device inference. Disease identification in a greater range of plants may be adopted and improved by employing a larger dataset. The application can also assist farmers by recommending remedial procedures to take to manage a specific disease in its earliest stages, reducing crop loss. Additionally, the system may be expanded with capabilities that track plant development in real time and deliver specific crop output recommendations.

References

[1] Forgotion, C., O'Neill, P. E., Carrera, M. I., & Bcliar, S. (2020). How satellite soil moisture data can help to monitor the impacts of climate change: SMAP case studies. *IEEE Journal of Selected Topics in Applied Earth Observations and Remote Sensing, 13*.

[2] Gupta, R., Modi, K., & Sharma, A. K. (2021). WBCPI: Weather-based crop prediction in India using big data analytics.

[3] Judge, J., England, A. W., & Crusson, W. L. (1999). A growing season land surface process/radio brightness model for wheat-stubble in the Southern Great Plains. *IEEE Transactions on Geoscience and Remote Sensing*.

[4] Kulkarni, O. (2016). *Crop disease detection using deep learning*. IEEE.

[5] Mohanty, S. P. (2016). *Using deep learning for image-based plant disease detection*. Digital Epidemiology Lab, EPFL, Geneva, Switzerland.

[6] Shaikh, F. K. (2021). *Artificial intelligence best practices in smart agriculture*. IEEE Micro.

[7] Udutalapally, V., Mohanty, S. P., & Pallagani, V. (2021). A novel device for sustainable automatic disease prediction, and irrigation in Internet-of-Agro-Things for smart agriculture. *IEEE Sensors Journal, 21*(16).

[8] Virupakshagowda, C., Patail, K. A., Al-Gaadi, K., & Madugunda, R. (2015). Assessing agricultural water productivity in desert farming systems of Saudi Arabia. *IEEE Journal of Selected Topics in Applied Earth Observations and Remote Sensing, 1*.

[9] Xiao, X., Xiaoman, Y., & Chen, T. (2021). Temporal series crop classification study in rural China based on Sentinel-1 SAR data. *IEEE Journal of Selected Topics in Applied Earth Observations and Remote Sensing, 14*.

[10] Xu, L., Liang, N., & Gao, Q. (2008). An integrated approach for agricultural ecosystem management. *IEEE Transactions on Systems, Man, and Cybernetics Part C: Applications and Reviews, 38*(4).

58 Fault classification and monitoring in induction machine using artificial neural network

Manish Nandy[1,a] *and Lalnunthari*[2,b]

[1]Assistant Professor, Department of CS & IT, Kalinga University, Raipur, India
[2]Research Scholar, Department of CS & IT, Kalinga University, Raipur, India

Abstract: This research paper presents a novel approach for fault classification and monitoring in induction machines using Artificial Neural Networks (ANNs). Induction machines are integral to various industrial applications, and their reliable operation is crucial for efficiency and safety. Traditional fault detection methods often fall short in terms of accuracy and real-time monitoring capabilities. This study explores the application of ANNs to enhance fault detection by analyzing vibration and electrical signals collected from the induction machine. The proposed method involves training a neural network model with a comprehensive dataset of normal and fault conditions, enabling it to identify and classify faults such as rotor faults, stator faults, and bearing defects. The performance of the ANN model is evaluated based on its accuracy, robustness, and ability to detect faults at an early stage. Experimental results demonstrate that the ANN-based approach outperforms conventional techniques in terms of detection accuracy and response time, thereby offering a promising solution for the predictive maintenance of induction machines.

Keywords: ANN, Monitoring, motor, fault, driving system, thermal monitoring, accuracy, induction machines

1. Introduction

Most of the time, induction motors are used in fans, pumps, electric vehicles, and small, medium, and large businesses like rolling mills, cement plants, and sugar mills. They are basically the backbone of every industry. Since these motors and drives are used for heavy tasks, they are more likely to break down [5]. For this reason, it is very hard for researchers to figure out what is wrong with an induction motor [1, 9]. Sometimes, a broken machine can stop the industry's production all of a sudden, which is bad for both money and safety [2]. It is hard and scary for operators and plant engineers to try to figure out what the problem is. Audible emission monitoring, thermal monitoring, chemical monitoring, and vibration signal monitoring are some of the ways that the condition of motors can be checked [10]. To keep a motor driving system running, it's important to find the problems and keep an eye on the situation. The flowchart in Figure 58.1 shows how to find faults and keep an eye on the condition of an induction machine.

It is essential to bear the following in mind in order to keep an eye on things and determine what is wrong: being able to identify and quantify the primary variable; Acquiring information by converting the primary variables that are sensed into digital form; The data are being processed, which means that information is being extracted from the data; The process of diagnosis, which entails taking action based on the information that was processed [3]. In most cases, the first two tasks are completed while the motor is operating, whereas the final two tasks are completed when the motor is not operating. It is important to note that the outcomes of these tasks do not provide the operator with immediate information regarding the functioning of the motor [4].

The processing of data is an essential component in the process of monitoring the situation and determining what caused the problem. Both diagnosing and monitoring are accomplished through the utilisation of a variety of signals, including electrical and vibration signals. The extraction of valid features from these signals is an essential step in the process of achieving this objective [12]. As a result of this, we require a method for the extraction of features that can be combined with signal processing in order to obtain useful feature parameters from signals that were recorded over a period of time. Using the appropriate algorithm for signal analysis [11], it is possible to identify changes in the signal that are brought on by a broken component. Observing the amplitude of a raw signal in the time domain from a continuous perspective is the most straightforward method for doing so. Currently, the process of processing a signal involves comparing the recorded signal to either the signal that came before it or a fixed threshold that has already been established. It's possible that the condition of the machine has changed, which would explain why the signals behaved differently. With steady-state recurrent signals, such as those that occur when monitoring a motor fault that gets worse over time, spectral analysis is frequently used to process

[a]ku.manishnandy@kalingauniversity.ac.in, [b]lalnunthari.rani.singh@kalingauniversity.ac.in

DOI: 10.1201/9781003675259-58

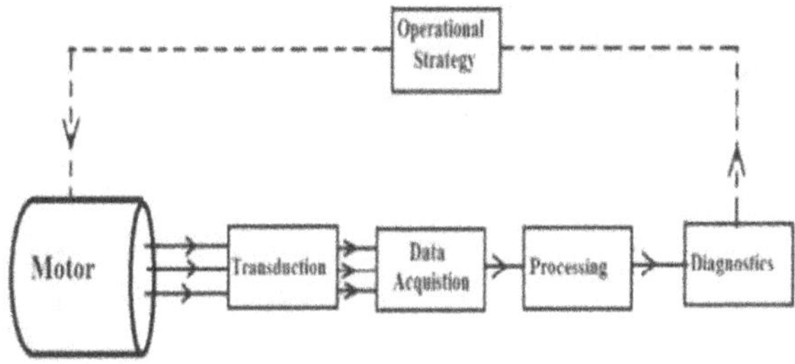

Figure 58.1. Flow chart for diagnosis of fault and monitoring the condition.

Source: Author.

signals in the frequency domain. This is due to the fact that it works well with these constant-state signals.

2. Fault diagnosis by ANN

In order to determine the state of the machine, the purpose of condition monitoring is to create a map that depicts the relationship between the input and output signals. It is difficult to provide a classification of the machine's condition and to determine the severity of the fault by merely examining the input signal because the signal is influenced by a wide variety of factors. When it comes to determining what is wrong with a machine, the two most important things to consider are intelligence and experience. This is because the information from various sensors is considered. On the other hand, significant advancements have been made in the field of artificial intelligence (AI) [6], with the objective of making AI capable of replacing humans in tasks such as evaluating and making decisions in the required area. For the following reasons, it is imperative that this be done: Electrical machine equipment, such as motors and generators, is frequently utilized in both large and small businesses. This is especially true in situations where the operator does not have a deep understanding of how the machine operates. The ability of human experts to make consistent decisions can be challenging, particularly in situations where there are multiple machines present in the plant. In order to create a computer system that is capable of thinking like a human, it is necessary for the individual to have an understanding of the kind of response they can anticipate from specialists [7].

This similarity in condition monitoring tasks is made possible by an intriguing technology known as Artificial Intelligence (AI) and Artificial Neural Network (ANN). This technology attempts to replicate human intelligence and the capacity to learn, which makes it possible for it to perform similar tasks (Figure 58.2). When a specialist in the field obtains fresh information regarding the state of the machine that is being monitored, it is highly probable that they will make use of that information by employing heuristic reasoning. A computer system that is capable of using heuristic reasoning, which is a characteristic of human intelligence, can be created with the help of an expert system, which is classified as an artificial intelligence system.

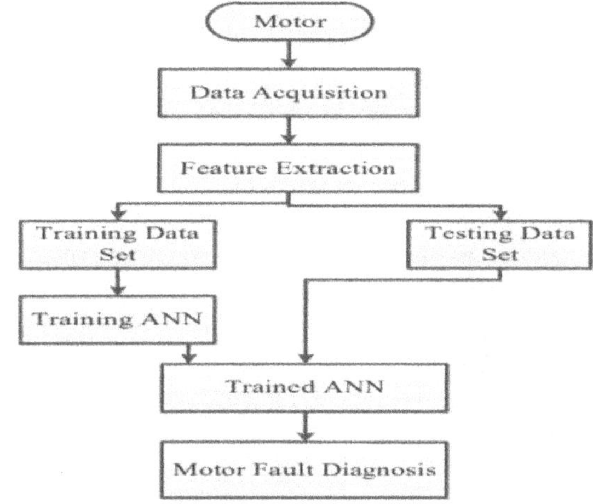

Figure 58.2. Flow chart of an ANN based fault diagnosis system for induction motor.

Source: Author.

3. Results

Seventy percent of the data that was used in this investigation was used for training, fifteen percent was used for validation, and fifteen percent was used for testing. A value of 10E-5 was chosen for the mean squared error (MSE), 10E-6 was chosen for the minimum gradient, and 1000 was chosen for the number of iterations (epochs). When any of these steps in the training process are completed, the process is considered to be finished. In the beginning, the program was responsible for automatically setting the weights and biases of the network. The performance of the Neural Network was satisfactory across the board, as demonstrated in Figure 58.3, as well

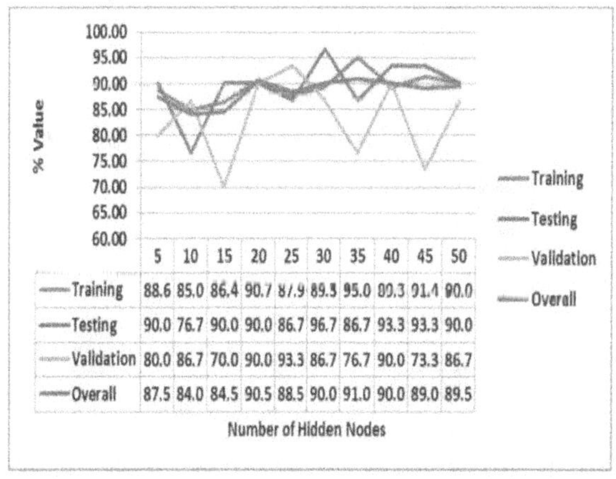

	5	10	15	20	25	30	35	40	45	50
Training	88.6	85.0	86.4	90.7	87.9	89.3	95.0	90.3	91.4	90.0
Testing	90.0	76.7	90.0	90.0	86.7	96.7	86.7	93.3	93.3	90.0
Validation	80.0	86.7	70.0	90.0	93.3	86.7	76.7	90.0	73.3	86.7
Overall	87.5	84.0	84.5	90.5	88.5	90.0	91.0	90.0	89.0	89.5

Figure 58.3. Performance of neural network with different numbers of neuron in hidden layers with all features as an input.

Source: Author.

as during training, testing, validation, and scheme 1. When it comes to discrimination, the number of neurones that are present in the hidden layer is selected to be. A training rate of 85.00% was observed for 10 neurones in the hidden layers, while a training rate of 95.00% was demonstrated for 35 neurones in the hidden layers. The lowest training rate observed was 85.00%. When it came to testing, the rate that was obtained was the lowest at 76.70% for 10 neurones in the hidden layer, and the rate that was obtained was the highest at 96.7% for 30 neurones in the hidden layer. Seventy percent was the lowest score for fifteen neurones in the hidden layers, and ninety-three percent was the highest score for twenty-five neurones in the same layer. This was for validation purposes. It is observed that the best overall performance is achieved with 35 neurones, with a success rate of 91.00%.

Figure 58.4 show the best validation performance curve of the ANN for fault discrimination with 11 parameters, 35 neurons in hidden layer and 28 epochs.

Figure 58.5 illustrates the percentage value that was discovered for training, testing, and validating, as well as

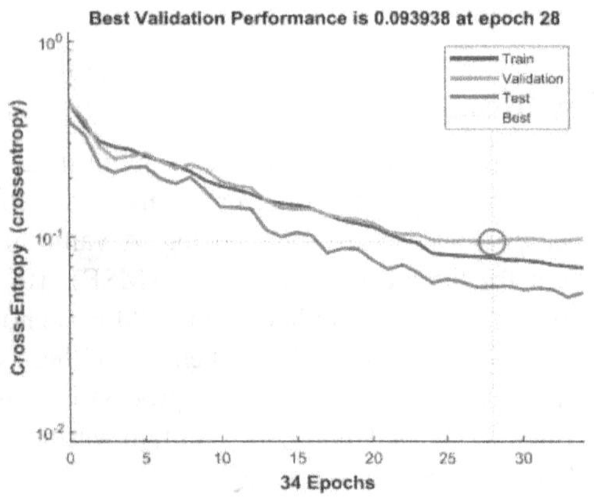

Figure 58.4. Validation performance curve for scheme 1 with all features as an input.

Source: Author.

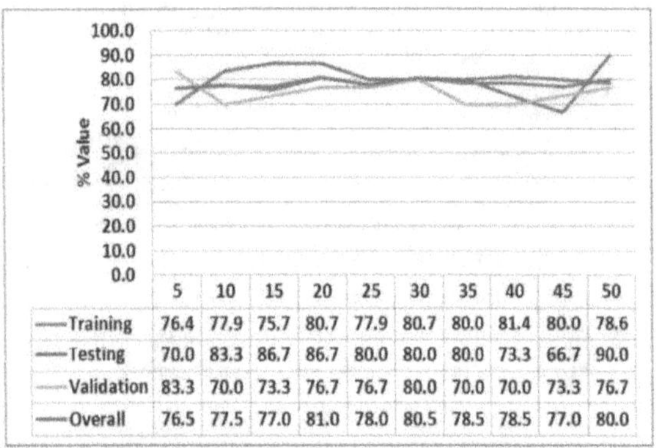

100.0
90.0
80.0
70.0
60.0
50.0
40.0
30.0
20.0
10.0
0.0
% Value
5 10 15 20 25 30 35 40 45 50
Training 76.4 77.9 75.7 80.7 77.9 80.7 80.0 81.4 80.0 78.6
Testing 70.0 83.3 86.7 86.7 80.0 80.0 80.0 73.3 66.7 90.0
Validation 83.3 70.0 73.3 76.7 76.7 80.0 70.0 70.0 73.3 76.7
Overall 76.5 77.5 77.0 81.0 78.0 80.5 78.5 78.5 77.0 80.0

Figure 58.5. Performance of neural network with different numbers of neuron in hidden layers with reduced features as an input.

Source: Author.

the overall success of the neural network performance for various hidden layer networks through the use of the neural network. According to the graph, the training rate for 15 neurones in the hidden layer was the lowest at 75.7%, while the highest rate was 81.4% for 40 neurones. This information is presented in the graph. The rate of 66.7% for 45 neurones in the hidden layer is the lowest rate that was tested, and the rate of 90% for 50 neurones was the highest rate that was tested. In terms of validation, the rate that is the lowest is 70% for ten neurones, and the rate that is the highest is 83.3% for five neurones. Within the hidden layer, there are twenty neurones, which results in the best overall performance, with an accuracy of 81%. When the outcomes are compared to what would occur if all of the input features were utilized, it is observed that the performance is reduced by ten percent.

4. Conclusion

In this paper, the Artificial Neural Network (ANN) was talked about as one of the smart methods used in the research. Multiple MATLAB functions have been used to train and test ANN. More or less input parameters and different numbers of neurones in the hidden layer were used to train and test the network. Experiments have been used for the training and testing. There are examples of performance curve plots that show both the expected and actual outputs. ANN is pretty good at both telling the difference between people with and without health problems (91% accuracy). It is also known that as the number of input parameters goes down, so does the accuracy of ANN, which was 81% and 99.5% for scheme 1 and scheme 2.

References

[1] Gundewar, S. K., & Kane, P. V. (2021). Condition monitoring and fault diagnosis of induction motor. *Journal of Vibration Engineering & Technologies*, *9*, 643–674.

[2] Yakhni, M. F., Cauet, S., Sakout, A., Assoum, H., Etien, E., Rambault, L., & El-Gohary, M. (2023). Variable speed induction motors' fault detection based on transient motor current signatures analysis: A review. *Mechanical Systems and Signal Processing*, *184*, 109737.

[3] Mishra, D., & Kumar, R. (2023). Institutional repository: A green access for research information. *Indian Journal of Information Sources and Services*, *13*(1), 55–58.

[4] Merizalde, Y., Hernández-Callejo, L., & Duque-Perez, O. (2017). State of the art and trends in the monitoring, detection and diagnosis of failures in electric induction motors. *Energies*, *10*(7), 1056.

[5] Manipriya, S., Mala, C., & Mathew, S. (2020). A collaborative framework for traffic information in vehicular adhoc network applications. *Journal of Internet Services and Information Security*, *10*(3), 93–109.

[6] Sharma, A., Chaturvedi, R., Singh, P. K., & Sharma, K. (2021). AristoTM robot welding performance and analysis of mechanical and microstructural characteristics of the weld. *Materials Today: Proceedings*, *43*, 614–622.

[7] Sharma, A., Mohana, R., Kukkar, A., Chodha, V., & Bansal, P. (2023). An ensemble learning–based experimental framework for smart landslide detection, monitoring, prediction, and warning in IoT-cloud environment. *Environmental Science and Pollution Research, 30*(58), 122677–122699.

[8] Sharma, A., & Dwivedi, V. K. (2020, December). Effect of spindle speed, feed rate and cooling medium on the burr structure of aluminium through milling. In *IOP conference series: materials science and engineering* (Vol. 998, No. 1, p. 012028). IOP Publishing.

[9] Deepthi, R. D., & Hima, B. G. (2024). Feature selection model-based intrusion detection system for cyberattacks on the internet of vehicles using cat and mouse optimizer. *Journal of Wireless Mobile Networks, Ubiquitous Computing, and Dependable Applications (JoWUA), 15*(2), 251–269. https://doi.org/10.58346/JOWUA.2024.I2.0179. Sharma, T., Singh, S., Sharma, S., Sharma, A., Shukla, A. K., Li, C., ... & Eldin, E. M. T. (2022). Studies on the Utilization of Marble Dust, Bagasse Ash, and Paddy Straw Wastes to Improve the Mechanical Characteristics of Unfired Soil Blocks. *Sustainability, 14*(21), 14522.

[10] AlShorman, O., Irfan, M., Saad, N., Zhen, D., Haider, N., Glowacz, A., & AlShorman, A. (2020). A review of artificial intelligence methods for condition monitoring and fault diagnosis of rolling element bearings for induction motor. *Shock and Vibration, 2020*(1), 8843759.

[11] Asl, T. M., & Asl, T. S. (2022). Strategy optimization for responding to primary, secondary and residual risks considering cost and time dimensions in petrochemical projects. *Archives for Technical Sciences, 2*(27), 33–48.

[12] Sharma, K., Verma, R. P., Dwivedi, S. P., Kumar, C. P., Khan, A. K., & Singh, D. (2023, September). Machine Learning For Fault Detection And Diagnosis In Mechanical Systems. In *2023 6th International Conference on Contemporary Computing and Informatics (IC3I)* (Vol. 6, pp. 1635–1640). IEEE.

59 A Unique Stochastic Slime Mould Converter Control (S^2MC^2) model with Quadratic High Gain Converter (QHGC) for PV-EV System

Sarath S.[1,a], *and K. Vijayakumar*[2,b]

[1]Research Scholar, Department of Electrical and Electronics Engineering, Kalasalingam Academy of Research and Education, Krishnakoil, Tamil Nadu, India
[2]Associate Professor, Department of Electrical and Electronics Engineering, Kalasalingam Academy of Research and Education Krishnakoil, Tamil Nadu, India

Abstract: The transportation sector is a major contributor to global greenhouse gas (GHG) emissions, with road transport responsible for approximately 95% of these emissions. Electric vehicles (EVs) are widely recognized for their potential to significantly lower GHG emissions. However, when powered by electricity from fossil fuel-based generation systems, EVs can still contribute to greenhouse gas emissions, preventing them from being entirely environmentally neutral. Over the past two decades, the increasing integration of renewable energy sources has led to the development of advanced power electronic converters. Yet every single topology has some built-in limitations. Some of these topologies may operate with a higher duty ratio, or they may not be able to offer enough voltage gain. Moreover, this might lead to a number of anomalies, such as decreased effectiveness, elevated voltage, and ripples in current. By using cutting-edge converter and controlling techniques, an EV system connected to solar PV is developed. In order to increase PV output while reducing loss factors and enhancing voltage gain, the novel Quadratic High Gain Converter (QHGC) circuit concept is employed. Adopting this converter offers the following primary advantages: low component requirements, high gain efficiency, and simplified design. Additionally, a novel Stochastic Slime Mould Converter Control (S^2MC^2) model has been developed to enhance converter performance and satisfy EV energy demands. Furthermore, this work employs an assortment of performance indicators to analyze and validate the outcomes and results of the proposed QHGC-S^2MC^2 model.

Keywords: Solar photovoltaic (PV), electric vehicle (EV), quadratic high gain converter (QHGC), stochastic slime mould converter control (S^2MC^2), inverter, voltage gain, and power quality

1. Introduction

Road transportation accounts for 95% of greenhouse gas (GHG) emissions worldwide, making the transportation sector largely and significantly responsible for these emissions. Greenhouse gas emissions can be significantly reduced by electric cars (EVs) [1]. Nevertheless, EVs cannot be considered complete ecologically neutral since they can emit a greater amount of greenhouse gases when used in conjunction with energy production based on fossil fuels systems. The efficiency of the mobility system can be enhanced by incorporating renewable energy sources (RES), such as photovoltaic (PV) systems, into EV charging stations. According to the World Health Organization (WHO), air pollution has been linked to over 3.7 million deaths worldwide and is responsible for approximately 18% of preterm births. A total of 30% of all greenhouse gas emissions are caused by internal combustion engine (ICE) automobiles, which are the largest source. Internal combustion engine (ICE) vehicles generally rely on fossil fuels like gasoline and diesel. Concerns about environmental issues and the rising demand for fossil fuels have accelerated the development of sustainable, emission-free transportation solutions, including electric vehicles (EVs) such as battery electric vehicles (BEVs), hybrid electric vehicles (HEVs), and plug-in electric vehicles (PEVs).

The anticipated increase in electric vehicle (EV) production from 3.1 million in 2020 to 15 million in 2025 necessitates a proportionate expansion of charging infrastructure [2]. Challenges such as EV reliability, restricted driving range, high battery costs with a limited lifespan, and prolonged charging durations pose major obstacles to replacing internal combustion engine (ICE) vehicles with electric vehicles (EVs).

Aside from that, the rapid use of electricity by EVs may strain the grid when charging takes place. Since a generation system based on fossil fuels powers the current utility grid in numerous countries, EVs cannot be considered completely ecologically beneficial [3]. The integration of Renewable Energy Sources (RESs), such as solar power and fuel cells,

[a]sarathssasi@gmail.com, [b]k.vijayakumar@klu.ac.in

DOI: 10.1201/9781003675259-59

into EV infrastructure is increasingly prevalent, as it helps reduce carbon footprints, charging expenses, and the load on the power grid. Energy electronic converters are required for EV systems to deliver the greatest amount of energy at the best efficiency. Among various Renewable Energy Sources (RES), solar Photovoltaic (PV) systems are particularly well-suited for EV applications due to their high energy generation efficiency, low maintenance requirements, widespread availability, and cost-effectiveness. The typical PV tied EV system with converter and inverter components is shown in Figure 59.1. One alternative sustainable option to fulfill solar energy that will be in tremendous demand for electricity in the decades to come. A viable way to create an environmentally friendly atmosphere is to convert sunlight into electricity, given the finite energy supplies of conventional fuel and the problems associated with pollution. Additionally, the Maximum Power Point Tracking (MPPT) is essential to the production of solar electricity because it identifies the MPP to track the possible electrical energy out from the solar. The integrated component of the solar energy system is power devices that are electronic. According to the recent literature review, a boost-up device is necessary to guarantee that the PV module will produce the desired output. Hence, the converter [4] is essential to the simulation of PV-tied EV setups, since it supports to regulate the PV output with high gain efficiency.

The major objectives behind this research work are given below:

- A solar PV tied EV system is developed with the adoption of advanced converter and controlling methodologies.
- In this approach, a novel Quadratic High Gain Converter (QHGC) circuit model is utilized to enhance the PV output by achieving higher voltage gain and minimizing losses. The key advantages of this converter include a simplified design, improved gain efficiency, and a reduced number of required components.
- A distinctive Stochastic Slime Mould Converter Control (S2MC2) model has been designed to enhance converter performance, ensuring it meets the energy demands of EVs.
- In addition, the proposed QHGC- S^2MC^2 model's outcomes and results are being examined and validated in this study using a variety of performance measures.

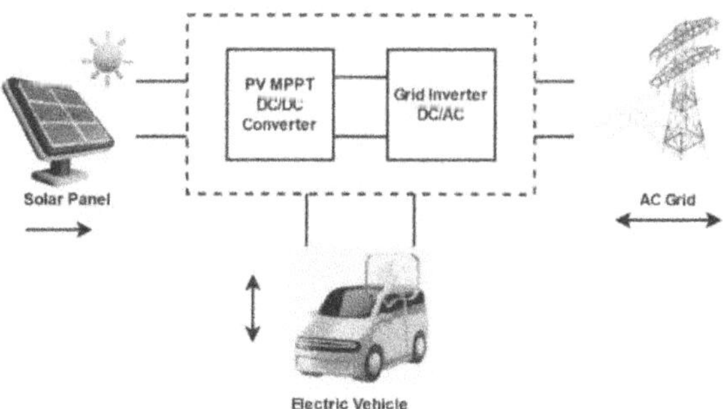

Figure 59.1. Schematic representation of the PV-integrated EV system.

Source: Author.

2. Related Works

Choudhury et al.[5] presented a comprehensive study to examine recent converter models used in several PV application systems. The converter models covered in the suggested work [6] are listed below:

- Boost converter
- Cascaded model
- Interleaved model
- Coupled inductor model

[7] implemented a new controlling algorithm for the converter used in the EV system. The T-source impedance network in this study offers excellent gain and electrical isolation among the vehicle and electrical grid, as well as across the array of cells and the grid itself. Power quality and voltage stability are consequently improved. [8] developed a SEPIC converter for both grid and EV applications. [9] conducted a comprehensive study to evaluate the feasibility of various DC-DC converter designs used in energy storage applications, including EVs. [10] proposed a control strategy for PV-EV applications using reinforcement learning. This study aims to address the challenge of EV charging through intelligent deep control. However, due to the complex mathematical computations involved, the control model exhibits a high level of complexity. [6] implemented an advanced energy management strategy for grid and EV integration. [11] introduced an enhanced ant lion optimization algorithm to promote the sustainable utilization of energy in plug-in vehicles.

3. Proposed Methodology

This section provides a comprehensive explanation, including mathematical equations and a schematic representation of the proposed PV-EV system, referred to as the Stochastic Slime Mould Converter Control (S2MC2) model. A generalized schematic diagram of the system is presented in Figure 59.2. The solar photovoltaic (PV) system serves as

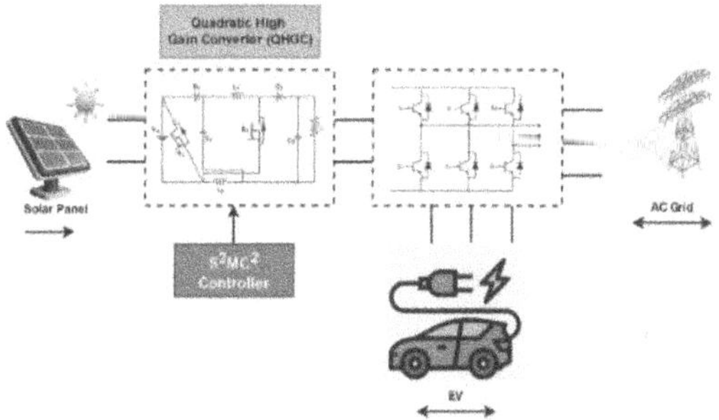

Figure 59.2. Schematic model of the proposed QHGC-S^2MC^2 based PV-EV system.

Source: Author.

the main energy source, maximizing the utilization of solar power to generate electricity. The PV system's potential energy yield is obtained by applying the MPPT controlling approach. As a result, the Quadratic High Gain Converter (QHGC), a unique converter model, has been utilized to increase the voltage. In this work, a novel regulating model, S^2MC^2, is designed to further enhance the performance of QHGC. It provides the optimal control parameters for pulse generation necessary to turn on and off the QHGC switching elements.

3.1. Quadratic High Gain Converter (QHGC)

Figure 59.3 displays the proposed QHGC circuit diagram. There are two distinct modes of operation: on and off, which are determined by the control signal. The following description applies to the different methods of operation:.

Mode 1: Diodes D1 and D2 are biased in the opposite direction since both toggle switches are switched on at the same time. During this mode of operation, the inductor current rises as a result of the inductors conserving energy. The following mathematical models illustrate the operations:

$$V_{L1} = V_{in} + V_{C1} \tag{1}$$

$$V_{L2} = V_{in} \tag{2}$$

Mode 2: Both diodes conduct and the switches are off. This kind of system charges both capacitance as the voltage flowing through the inductors drops, transferring the energy they produce to the load. The following are the equations for this mode:

$$V_{L1} = V_{C1} - V_O \tag{3}$$

$$V_{L2} = V_{in} - V_{C1} \tag{4}$$

$$V_{C1} = \frac{V_{in}}{1-\mathfrak{D}} \tag{5}$$

Where, V_{L1} and V_{L2} are the output voltage of inductors, V_{C1} is the output voltage of capacitor, V_{in} represents the input voltage, V_O is the output voltage, and $\mathfrak{D}$ denotes the duty cycle. The voltage gain is estimated according to the following model:

$$\frac{V_O}{V_{in}} = \frac{(1+\mathfrak{D}-\mathfrak{D}^2)}{(1-\mathfrak{D})^2} \tag{6}$$

Moreover, the average current across the inductors are mathematically represented as shown in below:

$$\Delta I_{L1} = \frac{(V_{in}+V_{C1})DT}{L1} \tag{7}$$

$$\Delta I_{L2} = \frac{V_{in}DT}{L2} \tag{8}$$

The least amount of voltage ripple that is allowed across a capacitor determines which capacitor ought to be selected. The capacitor's stored charge is as described below:

$$C1 = \frac{V_{in}(1+\mathfrak{D}-\mathfrak{D}^2)\times\mathfrak{D}}{R(1-\mathfrak{D})^3\Delta V_{C1}S_f} \tag{9}$$

$$C2 = \frac{V_{in}(1+\mathfrak{D}-\mathfrak{D}^2)\times\mathfrak{D}}{R(1-\mathfrak{D})^3\Delta V_{C2}S_f} \tag{10}$$

Where, V_{C1} and V_{C2} are the output voltage of capacitors, and S_f is the switching frequency. The number of components used in this converter model is represented in Table 59.1 along with its voltage gain and voltage stress.

3.2. Stochastic slime mould converter control (S2MC2) model

This work implements a novel algorithm for controlling the converter utilized in the proposed circuit model, which is called S^2MC^2. In most cases, the controlling mechanism is crucial to optimizing the performance of the inverter and converter parts. Using the S^2MC^2 model in this study significantly improves the performance of the QHGC since it produces the controlling pulses needed to toggle the converter's switching devices. This controlling algorithm is developed based on the recent stochastic optimization technique Furthermore, slime mould can continuously alter its hunting strategies based on the type and quality of food it consumes. When distinct high-quality food blocks are distributed over an area, this adaptive search approach may be more evident. The stages involved in this optimizing strategy are listed below:

- Parameter initialization

Table 59.1. Components study

Elements	*No of components*
Inductors	2
Capacitors	2
witches	2
Diodes	2
Total number of components	8
Voltage gain	$\frac{1+\mathfrak{D}-\mathfrak{D}}{(1-\mathfrak{D})^2}$
Voltage stress	$S_1 = \frac{1}{1-\mathfrak{D}}$ $S_1 = \frac{1}{(1-\mathfrak{D})^2}$

Source: Author.

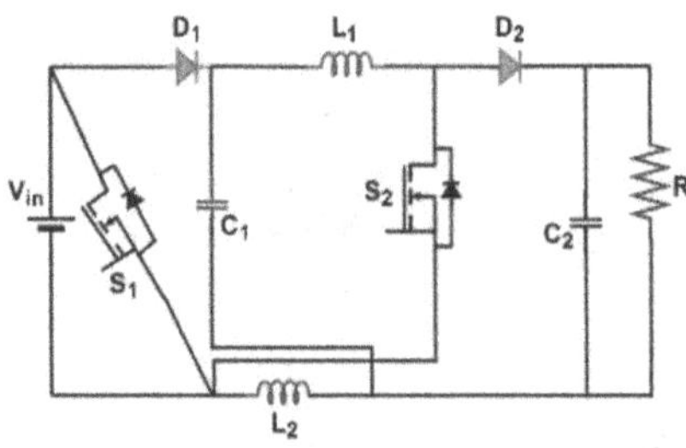

Figure 59.3. Schematic model of the proposed QHGC model.

Source: Author.

- Food approaching
- Wrap food
- Grabble food
- Position updation and fitness estimation

For this technique, the PV output voltage, PV output current, reference voltage, and time samples are taken as the inputs, and the best selection of parameters for generating controlling pulses is delivered as the output. After initializing the optimization parameters, the behaviour of searching food is mathematically represented as shown in the following model:

$$\overrightarrow{S(h+1)} = \begin{cases} \overrightarrow{S_Y(h)} + \vec{\partial}(\vec{w} \times \overrightarrow{S_X(h)} - \overrightarrow{S_Y(h)}) & a < t \\ \vec{\wp} \times \overrightarrow{S(h)} & a \geq t \end{cases} \quad (11)$$

Where, $\overrightarrow{S(h+1)}$ indicates the position of slime mould at iteration h, $\vec{S_y}$ represents the individual position with highest odor concentration, $\vec{S_X}$ and $\vec{S_Y}$ denotes two randomly chosen individuals from slime mould, $\vec{\partial}$ and $\vec{\wp}$ are the random parameters, and $\vec{w}$ indicates the weight value. As a consequence, the parameter t is computed as shown in below:

$$t = \tanh[F(i) - \mathfrak{B}] \quad (12)$$

Where, $i \varepsilon 1, 2, \ldots$ n, F(i) denotes the fitness value of $\vec{S}$, and $\mathfrak{B}$ represents the best fitness value. Moreover, the weight value is estimated according to the following model:

$$\overrightarrow{w(Sml\_Indx(i)} = \begin{cases} 1 + \tau \times \log\left(\frac{\beta - P(i)}{\beta - \alpha} + 1\right) & Cond \\ 1 - \tau \times \log\left(\frac{\beta - P(i)}{\beta - \alpha} + 1\right) & Others \end{cases} \quad (13)$$

$$Sml_{Indx} = Sort(P) \quad (14)$$

Where, Sml_{Indx} indicates the smell index that is sequence of sorted fitness value, τ is the random number, β is the best fitness value, α denotes the worst fitness value, and $P(i)$ indicates the rank of first half of the population. At the end, the location update is performed as shown in below:

$$\vec{S^*} = \begin{cases} \varphi \times (ub - lb) + lb & \varphi < g \\ \overrightarrow{S_Y(h)} + \vec{w} \times (w \times \overrightarrow{S_X(h)} - \overrightarrow{S_Y(h)}) & \tau < s \\ \vec{\wp} \times \overrightarrow{S(h)} & \tau \geq s \end{cases} \quad (15)$$

Where, φ is the random number, lb represents the lower bound, ub indicates the upper bound, and φ is the setting parameter. According to this model, the best optimal solution is obtained and returned as the output S_Y. Then, this optimum value is used to choose the controlling parameters k_p, k_i and k_d for generating the switching pulses for the QHGC component. In the current research, applying this converter management technique greatly improves the PV-EV system's overall efficiency.

4. Results and Discussion

This section analyzes the performance of the proposed QHGC-S2MC2 model by evaluating its effectiveness through various assessment metrics. The objective is to develop a new control mechanism for S²MC² that fulfills the energy demands of an EV system while ensuring voltage gain and efficiency. To achieve this, a novel QHGC-S²MC² model is introduced in this study, with multiple factors considered for its validation and evaluation. The simulation parameter setup for the proposed model is presented in Table 59.2.

Figure 59.4 depicts the output power, current, and voltage of the PV system concerning time variations in seconds. The integration of an improved InC MPPT control model optimizes the energy extraction from the solar panels. This approach enhances the electricity supply, effectively meeting the energy demands of the electric vehicle system.

As shown in Figure 59.5, the performance of QHGC is validated and compared with conventional converter models based on the maximum voltage stress experienced by each diode. To enhance efficiency and voltage gain, it is essential to minimize voltage stress on the converter circuit's electrical components under normal conditions. The proposed circuit model effectively reduces voltage stress through the implementation of the S²MC² algorithm. Consequently, the normalized inductor time constant is also analyzed and compared with the proposed QHGC model, as depicted in Figure 59.6. Furthermore, Figure 59.7 illustrates the evaluation of voltage stress between the switching elements. In this context, the S²MC² algorithm is specifically designed to generate control pulses, optimizing the switching process. By utilizing this control model, the voltage stress in the proposed circuit is significantly reduced.

Furthermore, as illustrated in Figure 59.8, the QHGC model's overall efficiency is verified at various voltage

Table 59.2. Simulation parameter setting

Parameter	*Specification*
PV panel	150 W
Input voltage	24V
Maximum output power	200 W to 300 W
witching frequency	50 Hz
Resistive load	250 Ω
Inductors L1 and L2	330 μH
Capacitors C1 and C2	33 μF

Source: Author.

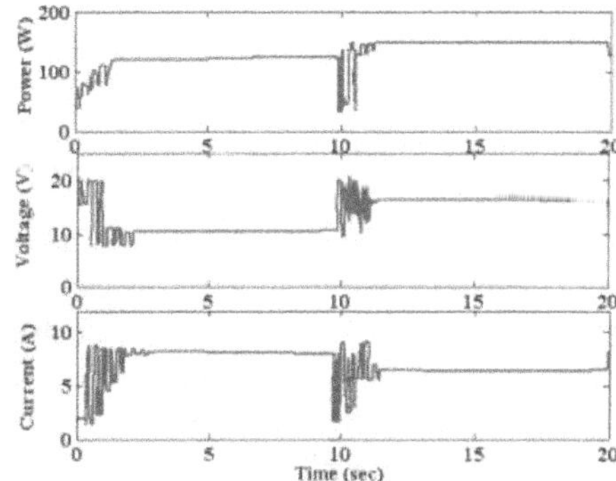

Figure 59.4. Voltage, power and current of PV system.

Source: Author.

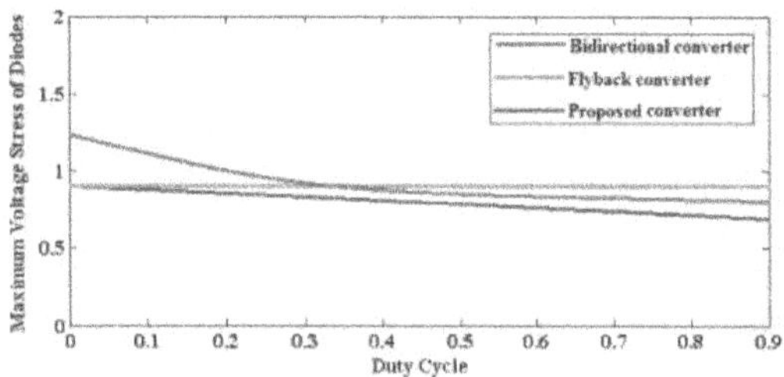

Figure 59.5. Evaluation of the QHGC and standard model based on the voltage stress between diodes.

Source: Author.

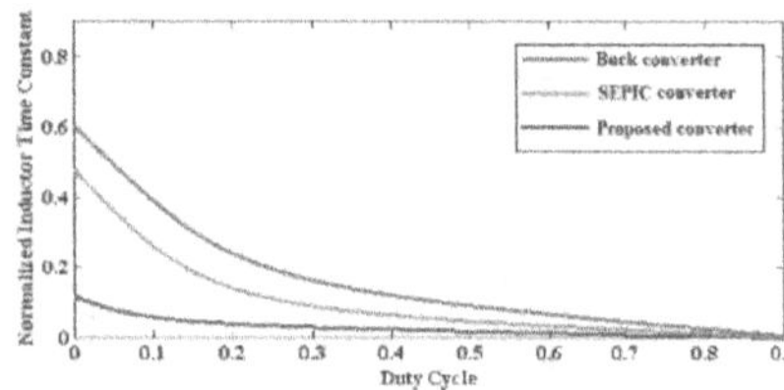

Figure 59.6. Comparison of the proposed QHGC and conventional design using normalized inductor time.

Source: Author.

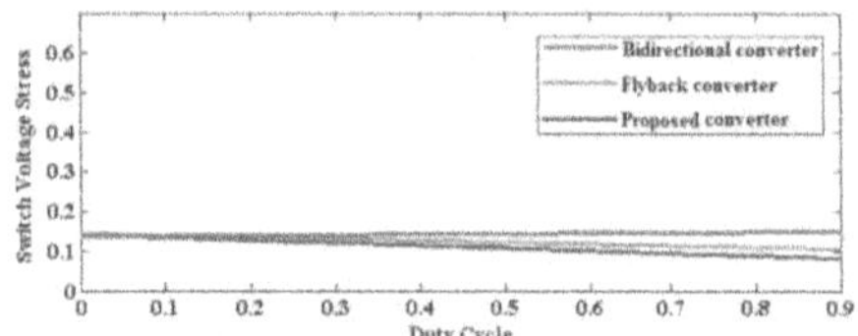

Figure 59.7. Comparison of the proposed QHGC with the standard one using switching stress.

Source: Author.

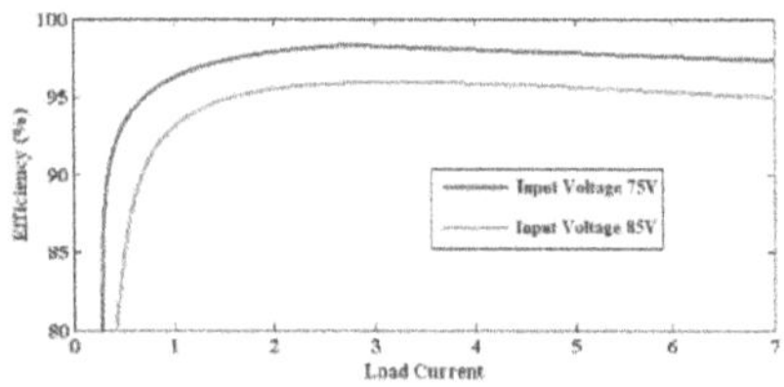

Figure 59.8. Comparison of efficiency between the input voltage 75V and 85V.

Source: Author.

levels, including 75 and 85 volts. The most important factor in evaluating the converter model's voltage conversion performance is its efficiency. It is inferred from the data that the proposed QHGC model has significantly higher efficiency for both input voltages. Given that the suggested PV-EV system's increased voltage gain efficiency is mostly the result of the use of S^2MC^2. Moreover, the loss value analysis is also conducted for the proposed QHGC model for the different number of electrical components used in the circuit. The findings indicate that the loss value is effectively reduced in the proposed converter circuit.

Loss value analysis of QHGC is represented in Table 59.3. Additionally, as illustrated in Figures 59.9 and 59.10, the output voltage and current characteristics are analyzed. To enhance power quality at the load side, a voltage source inverter is employed to reduce harmonic distortion. As a result, the total harmonic distortion (THD) of the proposed

Table 59.3. Loss value analysis of QHGC

Components	*Loss value (%)*
Capacitors C1+C2	10%
witch S1	8%
witch S2	16%
Diode D1	15%
Diode D2	18%
Inductors L1+L2	15%

Source: Author.

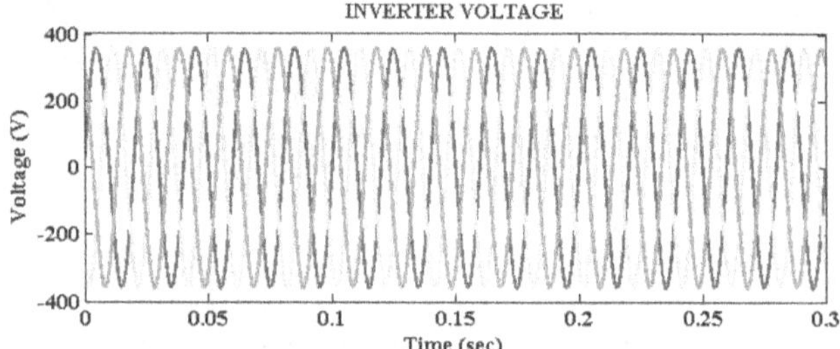

Figure 59.9. Inverter voltage.

Source: Author.

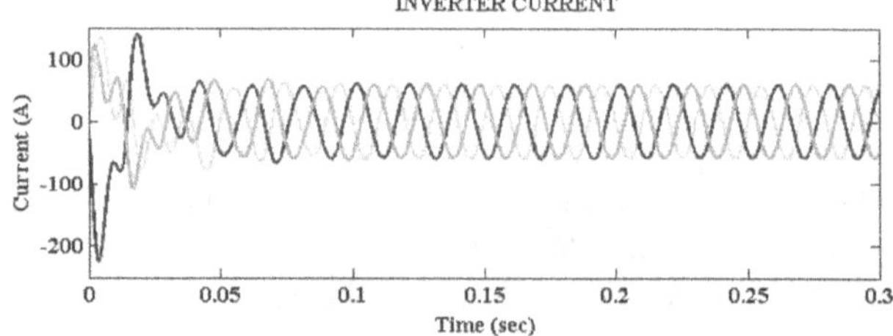

Figure 59.10. Inverter current.

Source: Author.

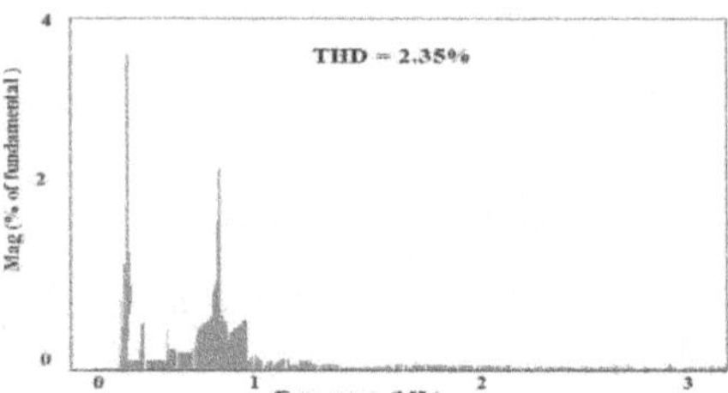

Figure 59.11. Harmonic distortion.

Source: Author.

PV-EV system is evaluated, as shown in Figure 59.11. The findings indicate that the THD value in the proposed circuit model is effectively minimized to 2.35%.

5. Conclusion

The development of a novel converter regulating algorithm that maximizes power quality and voltage support for PV-EV systems is the original contribution of this paper. The S^2MC^2 model is the name given to it. The solar PV is the main power source for maximizing the amount of electricity that may be produced by the solar system. The MPPT controlling technique is used to determine the PV system's potential energy yield. Consequently, a novel converter model called the QHGC has been used to raise the voltage. This converter circuit's primary goal is to efficiently control EV energy consumption with the least amount of electrical components. Consequently, the proposed work decreases its circuit complexity greatly. This effort aims to improve QHGC performance further by designing a new regulatory model called S^2MC^2. It offers the ideal pulse generation control parameters required to activate and deactivate the QHGC switching elements. The voltage source inverter is additionally connected to the load side to reduce harmonics and maximize power quality while minimizing loss. A comparison and analysis is conducted between the output voltage, power, current, voltage stress, loss, efficiency, and THD of the suggested QHGC-S^2MC^2 model. The proposed model outperforms the conventional techniques overall, providing greater efficiency and superior performance results.

References

[1] Fachrizal, R., Shepero, M., Åberg, M., & Munkhammar, J. (2022). Optimal PV-EV sizing at solar powered workplace charging stations with smart charging schemes considering self-consumption and self-sufficiency balance. *Applied Energy, 307*, 118139.

[2] Bishla, S., & Khosla, A. (2023). Enhanced chimp optimized self-tuned FOPR controller for battery scheduling using Grid and Solar PV Sources. *Journal of Energy Storage, 66*, 107403.

[3] Rafikiran, S., Basha, C. H., Devadasu, G., Tom, P. M., Fathima, F., & Prashanth, V. (2023). Design of high voltage gain converter for fuel cell based EV application with hybrid optimization MPPT controller. *Materials Today: Proceedings*.

[4] Singh, V., Kaur, L., Kumar, J., & Singh, A. (2022). Solar PV tied electric vehicle charging system using bidirectional DC-DC converter. In *2022 Second International Conference on Power, Control and Computing Technologies (ICPC2T)*, pp. 1–5.

[5] Choudhury, T. R., Nayak, B., De, A., & Santra, S. B. (2020). A comprehensive review and feasibility study of DC–DC converters for different PV applications: ESS, future residential purpose, EV charging. *Energy Systems, 11*, 641–671.

[6] Amir, M., Zaheeruddin, Haque, A., Bakhsh, F. I., Kurukuru, V. B., & Sedighizadeh, M. (2024). Intelligent energy management scheme-based coordinated control for reducing peak load in grid-connected photovoltaic-powered electric vehicle charging stations. *IET Generation, Transmission & Distribution, 18*(6), 1205–1222.

[7] Prem, P., Sivaraman, P., Sakthi Suriya Raj, J. S., Jagabar Sathik, M., & Almakhles, D. (2020). Fast charging converter and control algorithm for solar PV battery and electrical grid integrated electric vehicle charging station. *Automatika: časopis za automatiku, mjerenje, elektroniku, računarstvo i komunikacije, 61*(4), 614–625.

[8] Singh, A. K., Badoni, M., & Tatte, Y. N. (2020). A multifunctional solar PV and grid based on-board converter for electric vehicles. *IEEE Transactions on Vehicular Technology, 69*(4), 3717–3727.

[9] Safayatullah, M., Elrais, M. T., Ghosh, S., Rezaii, R., & Batarseh, I. (2022). A comprehensive review of power converter topologies and control methods for electric vehicle fast charging applications. *IEEE Access, 10*, 40753–40793.

[10] Dorokhova, M., Martinson, Y., Ballif, C., & Wyrsch, N. (2021). Deep reinforcement learning control of electric vehicle charging in the presence of photovoltaic generation. *Applied Energy, 301*, 117504.

[11] Alsharif, A., Tan, C. W., Ayop, R., Al Smin, A., Ali Ahmed, A., Kuwil, F. H., & Khaleel, M. M. (2023). Impact of electric Vehicle on residential power distribution considering energy management strategy and stochastic Monte Carlo algorithm. *Energies, 16*(3), 1358.

60 Intelligent interface of solid oxide fuel cell for micro grid operation

Abhijeet Madhukar Haval[1,a] *and Dhablia Dharmesh Kirit*[2,b]

[1]Assistant Professor, Department of CS & IT, Kalinga University, Raipur, India
[2]Research Scholar, Department of CS & IT, Kalinga University, Raipur, India

Abstract: The integration of Solid Oxide Fuel Cells (SOFCs) into microgrid systems presents a promising avenue for enhancing energy efficiency and sustainability. This paper explores the development and implementation of an intelligent interface for SOFCs aimed at optimizing their operation within microgrids. We propose a sophisticated control algorithm that leverages real-time data analytics and machine learning techniques to manage power generation, load balancing, and grid stability effectively. The interface is designed to dynamically adjust operational parameters based on current and forecasted grid conditions, ensuring optimal performance and reliability. Experimental results demonstrate significant improvements in system efficiency, response time, and load management compared to traditional control methods. The research underscores the potential of intelligent interfaces to revolutionize microgrid operations, paving the way for smarter, more resilient energy systems. This paper introduces an intelligent control strategy for Solid Oxide Fuel Cells (SOFCs) interfaced with the grid, demonstrating its effectiveness in microgrid (MG) operation and managing sudden load changes. The detailed discussion focuses on the use of a fuzzy controller within the SOFC's intelligent interface. The proposed strategy is validated by comparing its performance against conventional control methods.

Keywords: Solid oxide fuel cells (SOFCs), microgrid operation, power generation, energy efficiency, renewable energy

1. Introduction

Keeping up with the rapid development and progress in science and technology, has led to new requirements for the system and control science, thereby, leading to challenges for automatic control. Traditional control strategies which include classical/conventional control encounter many difficulties in their applications. Analysis and design of all conventional control is based on the precise mathematical models which are usually difficult to achieve due to the incomplete and time varying characteristics, complexity, nonlinearity and uncertainty. An attempt to improve the performance of the control system, in turn increases the complexities of the control system. The use of intelligent techniques, independently, or as a hybrid with conventional methods, has proved to be effective in solving the problems of conventional control methods. This use of intelligent control has brought great value for practical applications in the electrical power industry, as the focus is now towards a smart grid [1]. The concept of a smart grid is motivated by attempting to ensure continuous power supply to meet the demands of the consumers [2]. This means using the DERs units either independently or parallel to the grid or a combination of both of these [16]. Interconnection of the DR requires good and effective control in order to maintain:

- change in voltage profiles depending on how much power is produced and consumed at that system level
- reduced or no voltage transients as a result of connection or disconnection
- short circuit levels maintained within prescribed standards
- low load losses
- good power quality
- reliability
- coordinated utility & protection.

Fuel cells, when considered as DERs, are a popular choice in terms of various advantages which include low or negligible gaseous emissions, efficient energy conversion, good reliability and silent operation [3, 4]. The SOFC when compared to other fuel cells shows great promise in terms of its efficiency [5], solid electrolyte and its ability to reform its gaseous fuels internally [6]. Like all other fuel cells, the SOFC directly converts the chemical energy to electrical energy [12]. While the SOFC can have an efficiency of almost 70%, the challenges lie in its control. Much difficulty arises in the modeling of the SOFC due to its operating constraints and slow dynamics with an even bigger challenge for sudden load changes. Using the SOFC as a DER requires an understanding of both its modeling and control. Whilst the modeling and control of a DR is one aspect, the interface to the grid is another. This introduction of DR's to the power system for grid-tied and MG operation requires that the control be effective, thereby providing energy to the consumers within the prescribed standards of quality and reliability [8]. The

[a]ku.abhijeetmadhukarhaval@kalingauniversity.ac.in, [b]dhablia.dharmesh@kalingauniversity.ac.in

DOI: 10.1201/9781003675259-60

challenge, therefore, lies in the development of a suitable control scheme for the PE interface which can regulate voltage, both under steady and transient conditions. This paper presents an intelligent strategy used to control the SOFC when it is interfaced to the grid showing effective control for MG operation and during sudden load changes. The fuzzy controller has been discussed in detail as this has been used in the intelligent interface of the SOFC. Validation of the proposed intelligent strategy is done by comparing it with the conventional control strategies.

2. Intelligent Control Strategies

As a result of the requirement for improving and accelerating the control of Distributed Energy Resources (DERs), there is a shift towards the utilisation of intelligent methods for the purpose of connecting renewable sources to the grid. DER source-side control and grid-side control are two categories that can be used to classify the control of distributed generators (DGs). Through the utilisation of intelligent control, the inherent complexities of the control system can be reduced. The utilisation of intelligent systems is required in order to effectively manage and control distributed energy resources (DERs) due to the inability of conventional methods to adequately identify and quantify the nonlinear behaviour and dynamics of said systems. The intent of intelligent control is to bring intelligence in the control system and incorporate expert knowledge in the computing processes [14]. The feedback system used in conventional controllers has bottlenecks to handle non-linear dynamics and variable load conditions which can be addressed by intelligent controllers. Artificial intelligent based techniques such as fuzzy control and neural networks come under the umbrella of intelligent control [7].

2.1. *Fuzzy control*

Fuzzy control represents and implements a (smart) human's knowledge about how to control a system. A fuzzy control system is shown in Figure 60.1.

The functions of the components that are included in a fuzzy controller are:

- Control action determined by a rule base.
- Transformation of numeric inputs by the process of fuzzification to enable the inference mechanisms to understand.
- The decision of suitable rules to be used by the inference mechanism which uses information about the current inputs to form a conclusion about system input.
- The reverse process of fuzzification that is defuzzification wherein the conversion of the conclusions reached by inference mechanism into numeric input takes place.
- There are many fuzzy rule-based models as seen in Figure 60.2; however, the most widely used fuzzy inference systems are the Mamdani and Takagi-Sugeno-Kang (TSK) fuzzy models. The primary focus of these inference systems is reasoning and decision-making in contexts characterized by imprecision and uncertainty.
- The Fuzzy logic system processes vague and imprecise data/information using expert knowledge given by IF – THEN rules (Figure 60.3).
- IF premise (antecedent), THEN conclusion (consequent).

The process involved in designing a fuzzy system is:

- Identify the variables of the system; inputs, outputs and states.
- Fixing the universe of discourse or the span (interval) of each variable into subsets (fuzzy) and assigning a linguistic label. The elements of the universe of discourse are used to form the subsets.
- A membership function is assigned to each fuzzy subset.
- Fuzzy relationships are assigned between the inputs and states fuzzy subsets and the output fuzzy subsets formulating the rule base.

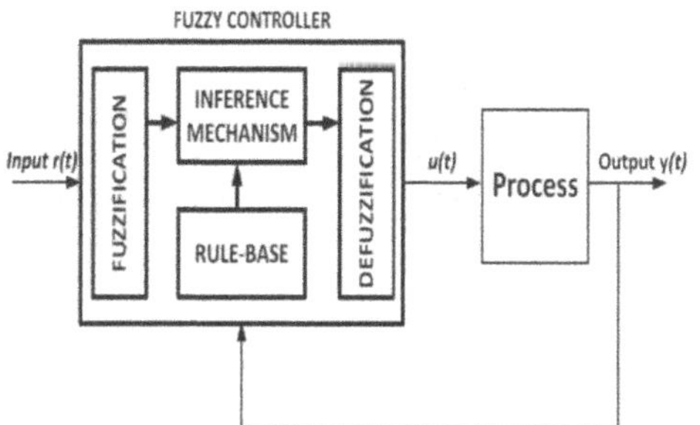

Figure 60.1. Schematic of fuzzy system.

Source: Author.

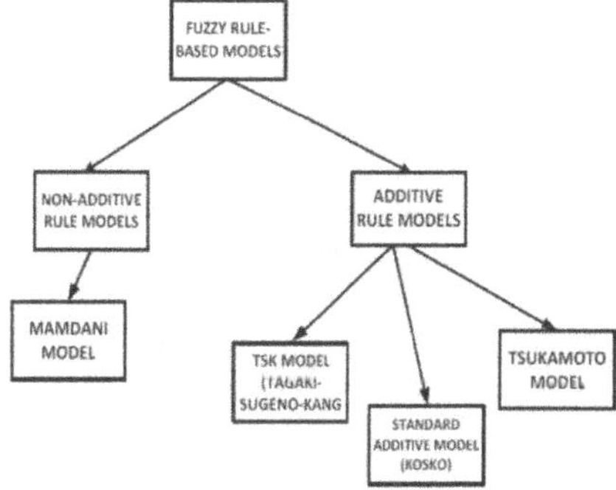

Figure 60.2. Fuzzy rule based models.

Source: Author.

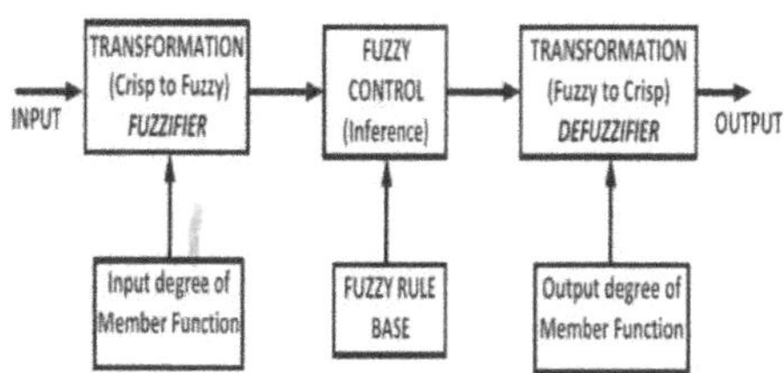

Figure 60.3. Processes in the fuzzy controller.

Source: Author.

- The variables are normalized to the [0, 1] or [-1, 1] interval by choosing appropriate scaling factors.
- Inputs to the controller are fuzzified.
- Output from each rule is inferred by fuzzy approximate reasoning.
- Fuzzy outputs from all rules are aggregated.
- Crisp output is obtained after defuzzification.

3. Proposed Intelligent Control Strategy

The strategy for interfacing the SOFC to the radial distribution system is based on the Mamdani model. The inverter is controlled by using the active power drawn by the grid. The schematic for the strategy is shown in Figure 60.4. The proposed strategy uses the PQ control method where the active power is used but implemented in the dq0 frame. Control action for the DER interfaced to the distribution network is done from the grid side [10]. This means that, depending on the load demand, the control action is taken by the PE interface to regulate the inverter and maintain the voltage level.

One of the important requirements in interconnection design of PE converters used for interfacing the DERs to the grid is synchronization [11], which can be achieved by using the phase locked loop (PLL). The converter control requires the measurements of frequency and line-angle of the utility to regulate the power flow. These measurements are achieved through the implementation of the PLL. The phase-locked loop (PLL) is employed to obtain the frequency and angle reference necessary for the closed-loop grid-side control of the inverter. The load demand fluctuates based on consumer needs, leading to variations in the load current. The grid-side control system adjusts based on these current and power variations.

4. Simulation Results

The SOFC is interconnected to the radial distribution network using this intelligent controller to test the impact of control. The radial distribution network used to assess the impact of this control strategy and the schematic for the same is as given in Figure 60.5.

To assess the effectiveness of control of SOFC for MG operation, the feeder 3 has been considered. The formation of the MG can occur under various conditions with the disconnection of Line 6.

The issue of power quality has been of great importance with the advent of renewable energy interconnection to the grid. The drastic impact on the power regulation requires that all interconnections using PE interfaces be limited / or the harmonic distortions caused be kept as minimal as possible. Distorted waveforms containing harmonics are caused

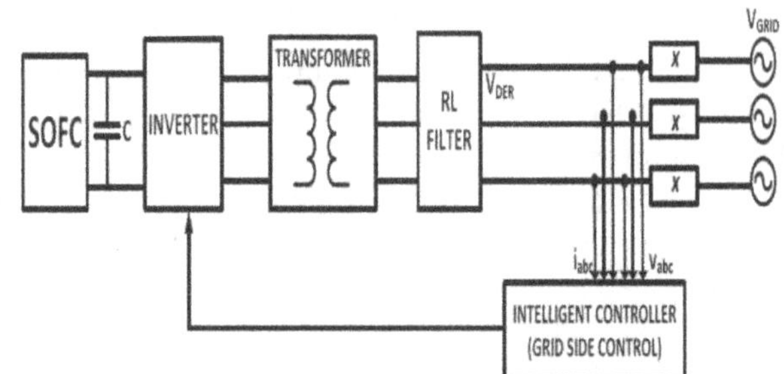

Figure 60.4. Grid side control for SOFC MG operation.

Source: Author.

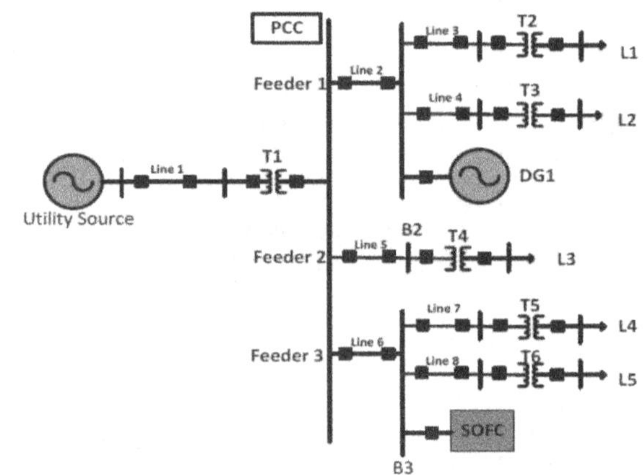

Figure 60.5. MV network.

Source: Author.

Table 60.1. THD for PI, HCC and fuzzy based intelligent control

cenario		*V (%)*	*HC Control*		*PI control*		*Proposed Control*	
			I (%)	*V (%)*	*I (%)*	*V (%)*	*I (%)*	
DER to GRID	1	PCC	0.24	0.28	0.52	0.52	0.47	0.48
		BUS	1.45	0.83	0.54	0.76	0.75	1.32
		LOAD	1.44	0.63	0.54	0.53	0.77	0.52
	2	PCC	4.09	1.68	0.52	0.52	0.46	0.47
		BUS	3.76	2.10	0.54	1.12	0.77	1.43
		LOAD	2.72	1.33	0.54	0.52	0.79	0.50
Islanded		LOAD	4.63	2.18	1.35	1.33	1.56	0.91

Source: Author.

by power sources which act as non-linear loads. Frequencies of harmonics are integer multiples of the fundamental frequency of a waveform. Harmonic distortion refers to the extent to which a waveform deviates from its pure sinusoidal shape due to the presence of these harmonic elements. Total Harmonic Distortion (THD) quantifies this distortion by calculating the summation of all harmonic components present in the current or voltage waveforms relative to the fundamental component, as expressed mathematically below:

$$THD = \frac{\sqrt{(V_2^2 + V_3^2 + V_4^2 + \cdots. + V_n^2)}}{V_1} * 100\%$$

The THD has been calculated for the proposed intelligent control strategy and has been compared with the conventional controllers. Table 60.1 shows the comparison of the THD for all.

5. Conclusion

In this paper, an intelligent control strategy for integrating Solid Oxide Fuel Cells (SOFCs) into a radial distribution system is presented. Utilizing the model and grid-side control, the strategy generates gating signals for the inverter based on grid demand changes. By employing an active power control method and leveraging fuzzy logic for system data processing with IF-THEN rules, the strategy is effectively validated through dynamic simulation. Various scenarios, both pre- and post-islanding, are created based on the IEEE Standard 399-1997, modified for distributed energy resources (DER) like SOFCs. Given the slow dynamics of SOFCs, interface control adapts to grid conditions. Simulations demonstrate superior control actions with Total Harmonic Distortion (THD) mitigation compared to PI and Hysteresis control strategies. The strategy meets IEEE 519-1992 THD standards, addressing the potential harmonic issues introduced by multiple DC/AC converters as DERs increase.

References

[1] Dileep, G. J. R. E. (2020). A survey on smart grid technologies and applications. *Renewable Energy*, *146*, 2589–2625.

[2] Badhoutiya, A., Parkash, J., Rana, A., Pareek, S., & Chohan, J. S. (2023, July). Experimental Evaluation of Solar Mirrors Devised with Cooling Technique. In *2023 4th International Conference on Electronics and Sustainable Communication Systems (ICESC)* (pp. 109–112). IEEE.

[3] Mishra, D., & Kumar, R. (2023). Institutional Repository: A Green Access for Research Information. *Indian Journal of Information Sources and Services*, *13*(1), 55–58.

[4] Yadav, R., Singh, P. K., & Chaturvedi, R. (2021). Enlargement of geo polymer compound material for the renovation of conventional concrete structures. *Materials Today: Proceedings*, *45*, 3534–3538.

[5] Hussain, S., & Yangping, L. (2020). Review of solid oxide fuel cell materials: Cathode, anode, and electrolyte. *Energy Transitions*, *4*(2), 113–126.

[6] Anas, A. K., Alaa, J. M., Anwer, S. A., & Laith, A. A. (2024). Control system design for failure starting of diesel power block for cell on wheels communication tower based on cloud service system. *Journal of Internet Services and Information Security*, *14*(3), 275–292.

[7] Garud, K. S., Jayaraj, S., & Lee, M. Y. (2021). A review on modeling of solar photovoltaic systems using artificial neural networks, fuzzy logic, genetic algorithm and hybrid models. *International Journal of Energy Research*, *45*(1), 6–35.

[8] Suleiman, H. (2023). Pcεκmax-Means++: Adapt-P driven by energy and distance quality probabilities based on κ-Means++ for the Stable Election Protocol (SEP). *Journal of Wireless Mobile Networks, Ubiquitous Computing, and Dependable Applications*, *14*(4), 128–148.

[9] Al-Muntaser, A. A., Pashameah, R. A., Sharma, K., Alzahrani, E., & Tarabiah, A. E. (2022). Reinforcement of structural, optical, electrical, and dielectric characteristics of CMC/PVA based on GNP/ZnO hybrid nanofiller: nanocomposites materials for energy-storage applications. *International Journal of Energy Research*, *46*(15), 23984–23995.

[10] Yağız, E., Ozyilmaz, G., & Ozyilmaz, A. T. (2022). Optimization of graphite-mineral oil ratio with response surface methodology in glucose oxidase-based carbon paste electrode design. *Natural and Engineering Sciences*, *7*(1), 22–33.

[11] Shaheen, O., El-Nagar, A. M., El-Bardini, M., & El-Rabaie, N. M. (2020). Stable adaptive probabilistic Takagi–Sugeno–Kang fuzzy controller for dynamic systems with uncertainties. *ISA transactions*, *98*, 271–283.

[12] Danková, Z., Štyriaková, I., Kovaničová, Ľ., Čechovská, K., Košuth, M., Šuba, J., Nováková, J., Konečný, P., Tuček, Ľ., Žecová, K., Lenhardtová, E., & Németh, Z. (2021). Chemical Leaching of Contaminated Soil—Case Study. *Archives for Technical Sciences*, *1*(24), 65–72.

[13] Hai, T., Zhou, J., Almashhadani, Y. S., Chaturvedi, R., Alshahri, A. H., Almujibah, H. R., … & Ullah, M. (2023). Thermo-economic and environmental assessment of a combined cycle fueled by MSW and geothermal hybrid energies. *Process Safety and Environmental Protection*, *176*, 260–270.

[14] Vijayan, P., Anbalagan, P., & Selvakumar, S. (2022). An ensembled optimization algorithm for secured and energy efficient low latency MANET with intrusion detection. *Journal of Internet Services and Information Security*, *12*(4), 156–163.

[15] Saxena, A., Chaturvedi, R., & Kumar, J. (2021). Performance of two-link robotic manipulator estimated through the implementation of self-tuned fuzzy PID controller. In *International Conference on Communication, Computing and Electronics Systems: Proceedings of ICCCES 2020* (pp. 721–732). Springer Singapore.

[16] Chen, Y., Feng, L., Jamal, S. S., Sharma, K., Mahariq, I., Jarad, F., & Arsalanloo, A. (2021). Compound usage of L shaped fin and Nano-particles for the acceleration of the solidification process inside a vertical enclosure (A comparison with ordinary double rectangular fin). *Case Studies in Thermal Engineering*, *28*, 101415.

61 Development and characterization of polyvinyl alcohol-activated charcoal composites for advanced wound dressings

S. Shanmugapriya[a], K. Ruth Esther[b], Marimuthu Chandran[c], Konda Mahesh[d], and Aman Kumar[e]

Department of Biomedical Engineering, Kalasalingam Academy of Research and Education, Krishnankoil, Tamil Nadu, India

Abstract: We develop an optimal wound dressing material using a novel combination of bentonite clay, Chromolaena odorata, Calendula officinalis, polyvinyl alcohol (PVA), and activated charcoal (AC). While PVA alone forms hydrogels with diminished mechanical strength, incorporating AC addresses this issue, creating a robust hybrid/composite hydrogel. C. odorata extracts, with their potent antibacterial and antioxidant properties, arrest bleeding effectively. C. officinalis promotes collagen induction and encourages angiogenesis, enhancing wound healing. Plants are gathered, prepared, and subjected to extraction using a Soxhlet apparatus with ethanol and distilled water. Hydrogel formation involves repeatedly freezing and thawing an aqueous PVA solution. We produce diverse wound dressing samples by varying plant extract concentrations while maintaining consistent quantities of PVA and AC. Characterization is performed using TGA, FTIR, and FESEM techniques. In vitro antibacterial tests are conducted using both gram-positive and gram-negative microorganisms. This research highlights the significant impact of bentonite clay and AC on wound healing, presenting a novel composite material that offers enhanced mechanical strength and multifunctional wound healing properties.

Keywords: PVA, activated charcoal, hydrogel, chromolaena odorata, calendula officinalis, wound dressing

1. Introduction

The complex process of wound healing encompasses migration, proliferation, haemostasis, inflammation, and remodelling. Hydrogels, due to their moisture-retention properties, hold promise in wound treatment, and polyvinyl alcohol (PVA) stands out as a versatile substance that can be chemically or physically cross-linked for optimal results. Natural products known for their wound-treating capabilities include Calendula officinalis and Chromolaena odorata. Understanding these elements aids in the development of effective wound care plans aimed at improving patient outcomes.

1.1. Wound healing

Haemostasis and Inflammation involve processes such as blood clotting, coagulation of extrudes, elimination of bacteria and debris, and release of growth factors, cytokines, and proteases. Migration entails the movement of epithelial cells to the wound site, while Proliferation involves granulation tissue formation and angiogenesis. Remodelling encompasses wound contraction, re-epithelization, and the formation of scar tissue.

1.2. Hydrogels

Hydrogels are the perfect material to use as bandages for burns, diabetic foot ulcers, and other wounds because of their softness and aqueous usability. The sheets, films, and gels that they come in are useful for rehydrating necrotic tissues in preparation for autolytic debridement. Hydrogel dressings are semi-permeable, non-adherent, and can lower the temperature of a wound by up to 5°C while reducing discomfort and promoting wound re-epithelialization, which speeds up healing.

1.3. PVA

Polyvinyl Alcohol (PVA) holds significant potential in the biological, pharmaceutical, and separation sectors due to its numerous advantageous properties. PVA possesses a pendant hydroxyl group and a fundamental chemical structure. Crosslinking of PVA is often a requisite step. PVA can undergo production through physical or chemical crosslinking; the most commonly used physical cross-linking technique is the "freezing-thawing" process, while the most prevalent chemical cross-linking techniques involve employing chemical cross-linkers, electron beams, or radiation.

[a]Shanmugapriya.s@klu.ac.in, [b]9922020018@klu.ac.in, [c]9921020015@klu.ac.in, [d]9922020026@klu.ac.in, [e]9922020042@klu.ac.in

DOI: 10.1201/9781003675259-61

1.4. *Chromolaena odorata*

It reaches a height of 2.5 meters. Etiolated plants thrive in gloomy areas and on some plants, they resemble climbers. The leaves of this plant emit a pleasant aroma when crushed. December shrubs and the Siam plant are two common names. Cell adhesion, wound healing time, and extract formulations were measured to determine the healing properties of the extraction and the baseline, using a standard extraction base and comparable concentrations of Neosporin and Betadine as reference and control. Aspergillus flavus was tested for the antibacterial effectiveness of C. odorata leaves and shoot extracts. Due to the extremely high number of flavonoids, C. odorata is thought to be a possible source of natural inhibitors.

1.5. *Calendula officinalis*

Calendula officinalis, also referred to as pot marigold, is a fragrant flowering plant belonging to the Asteraceae family, renowned for its evergreen, low shrub shape. It is widely planted and versatile; in colder locations, it is frequently produced as an annual. Calendula extracts may have antiviral, antigenotoxic, and anti-inflammatory qualities, according to pharmacological research. Due to a variety of secondary metabolites, methanolic extracts have been shown in vitro to have antibacterial and antifungal action.

2. Literature Review

Significantly effective at healing wounds, a new gel comprising polyvinyl alcohol, aryloxycyclotriphosphazene, and silver was created [1]. The gel, which was created via the Doebner reaction, showed significant antibacterial activity, high water absorption, and a 91.43% reduction in wound area in rabbits by day 10, all of which pointed to effective exudate absorption and quick healing for acute wounds [2]. By electrospinning polyvinyl alcohol, honey, and Curcumin longa (turmeric) extract, a novel nanofibrous mat was produced for use as a wound dressing. Turmeric's medicinal properties were extracted using ethyl acetate. When compared to pure polyvinyl alcohol nanofibers, the mat demonstrated finer fiber, greater moisture management, and antibacterial activity against Staphylococcus, according to evaluations conducted using a variety of methodologies [3]. Researchers used nano chitosan (NC), polyvinyl alcohol (PVA), and extract from Artemisia ciniformis to create an electrospun wound dressing (AE). While higher AE improved cell vitality but decreased mechanical characteristics, higher NC increased mechanical strength and fiber diameter but varied cell viability. Strong antibacterial activity in the dressings suggested that they might be used in biomedical wound care [4]. Researchers using a phenolic-rich extract from Chromolaena odorata to treat chronic illnesses including Alzheimer's and Parkinson's disease as well as oxidative damage. By attaching itself to hen egg-white lysozyme and blocking the formation of hydrophobic clusters, the extract reduced protein fibrillation. By scavenging DPPH and ABTS+ radicals and preventing metal-accelerated protein oxidation, it demonstrated antioxidant qualities. High levels of phenolic and flavonoids influenced these outcomes. Crucially, the extract did not exhibit any cytotoxicity toward RBCs or human embryonic kidney cells [5]. Researchers created epidermal growth factor (EGF) and fibroblast growth factor (FGF)-containing electrospun polyvinyl alcohol (PVA) nanofibers to aid in wound healing. Smooth, bead-free fiber with improved mechanical characteristics were revealed during characterization. Studies conducted in vivo and in vitro showed enhanced wound healing, cellular adhesion, and proliferation. PVA's usage in wound dressings is supported by its FDA approval and biocompatibility [6]. For improved wound healing and antibacterial activity, researchers created polyvinyl alcohol (PVA)/graphene oxide-citicoline sodium-lanthanum (GO-CDPC-La) films. The coatings encouraged cell adhesion and healing, absorbed wound exudate, and preserved hydration. Superior modified GO dispersion in PVA improved the film's qualities, demonstrating superior protein adsorption [7]. Using a modified casting procedure, researchers created translucent, biodegradable composite films of starch, polyvinyl alcohol (PVA), and citric acid for use as wound dressings. While citric acid boosted antibacterial capabilities, glycerol improved flexibility. The films were attractive candidates for wound dressings because they showed appropriate mechanical strength, antibacterial activity, fluid absorption, and disintegration [8]. Electrospinning was used to create composite wound dressing mats containing eugenol and mixes of PCL, PVA, and CS. The mats exhibited significant antibacterial activity against S. aureus and P. aeruginosa, as well as perfect porosity and hydrophilicity. Their potential to prevent and treat microbial infections in wound dressings for up to seven days was supported by prolonged eugenol production [9]. Synthesized silver nanoparticles (AgNPs) from cabbage extract via green chemistry, which they then incorporated into PVA hydrogel patches. Based on wound healing scores and histological inspection, these patches in combination with AgNP/clay/activated carbon composites demonstrated antibacterial action and sped up wound healing in a 20-day rabbit study [10]. Using extracts from Juniperus Chinensis, researchers made PVA nanofiber composites, taking use of the plant's antibacterial qualities. Smooth and regular fiber were produced by extract concentration, as demonstrated by SEM analysis, which also affected nanofiber diameter and morphology. Both Gram-positive and Gram-negative bacteria were efficiently suppressed by the composites, indicating potential applications in wound care and antimicrobial clothing [11]. The researchers used carbon dots, sodium alginate, polyvinyl alcohol, and woolly hedge nettle extract to create a nanocomposite sponge for wound treatment. Characterization verified its shape and structure. The sponge demonstrated potential for wound healing and infection control due to its excellent mechanical strength, hydrophilicity, water absorption, biocompatibility, low cytotoxicity, and potent antibacterial activity [12]. A double-layer BC/PVA/

NGO composite hydrogel was created by researchers to aid in wound healing. The exterior PVA layer was enhanced using nitrated graphene oxide (NGO) to improve its oxidizing and antibacterial qualities. In a mouse model, the hydrogel's porous structure, mechanical strength, biocompatibility, and potent antibacterial activity facilitated quicker wound healing [13]. For the purpose of treating wounds, researchers synthesized a composite comprising graphene oxide, silver nanowire, and starch-PVA nanofibers that was loaded with ciprofloxacin. The membrane exhibited outstanding hydrophilicity, significant water vapor permeability, and a high swelling capacity. Good biodegradability, biocompatibility, and promise for enhanced wound healing applications were demonstrated by its ability to effectively inhibit Staphylococcus aureus and Escherichia coli [14]. In this work, a hybrid wound dressing was created by mixing cardamom extract, electrospun silk fibroin/polyvinyl alcohol nanofibers, and sodium alginate/gum tragacanth hydrogel. The dressing shown potential for skin tissue engineering and wound healing, with uniform fiber structure, significant antibacterial activity against S. aureus and E. coli, high water retention, and delayed drug release.

3. Methodology

Hydrogels were prepared using glutaraldehyde (GA) as a chemical crosslinking agent. Glutaraldehyde is an aggressive amine condenser, a carbonyl (-CHO) reagent that undergoes condensation or reductive amination processes. It was a common practice to produce antibody-enzyme conjugates using this indiscriminate crosslinking reagent. Crosslinking chemicals, such as glutaraldehyde, have been shown to be a useful and effective method for enhancing the physical and chemical properties of PVA. Glutaraldehyde is believed to be a chemical that degrades rapidly in both water and soil. Activated charcoal is utilized to prevent an unpleasant Odor. The extensive surface area of activated charcoal is a result of its high porosity. The plant leaves are preserved in ethanol during the extraction process, which is carried out using a Soxhlet extractor. This apparatus can process solid materials and requires hours or days to complete the ongoing process. The solvent vapor condenses and falls back over the porous sample cup as a reservoir is gradually heated. Plants hardened for twenty-four hours in the shade are collected from a nearby nursery. Subsequently, the leaves are processed into a fine powder using a blender, and an extract is prepared once the particles are crushed.

The powdered material undergoes an extraction process using the Soxhlet apparatus, with the thimble containing the required quantity of plant powder located in the central part of the extraction equipment. The final extract is collected. In a beaker, PVA and distilled water are mixed together until the PVA solution turns from white to transparent. After adding the activated charcoal and the two prepared plant extracts, the PVA is continuously agitated for 30 minutes at 380 rpm and 30 degrees Celsius. Two different plant species are used to prepare four different types of samples. Chromolaena odorata is utilized in the preparation of the first two samples: the first contains 10% w/v PVA, 1% w/w activated charcoal, and 1% w/v Chromolaena odorata, while the second contains 10% w/v PVA, 1% w/w activated charcoal, and 2% w/v Chromolaena odorata. Both formulations of calendula are used to prepare the third and fourth samples and the Figure 61.1, explains the proposed methodology.

Three products were tested: the third containing 10% w/v PVA, 1% w/w activated charcoal, and 1% w/v Calendula officinalis; and the last product containing 10% w/v PVA, 1% w/w activated charcoal, and 2% w/v Calendula officinalis. The four samples are then prepared and placed into Petri dishes, which are incubated at 50 degrees Celsius in an oven for a full day.

4. Materials and Methods

4.1. Poly vinyl alcohol

Poly Vinyl Alcohol (PVA) is versatile, used in cornea substitutes and wound dressings. Figure 61.2 represents the PVA in powder form. Its hydrogel form promotes cell growth, and it can release antibiotics in low-exudate wounds.

4.2. Activated charcoal

Activated charcoal dressings treat chronic wounds by absorbing toxins and microorganisms, owing to their high-porosity. Figure 61.3 represents the Charcoal in powder form. Accelerating tissue regeneration for external leg ulcers and skin lesions.

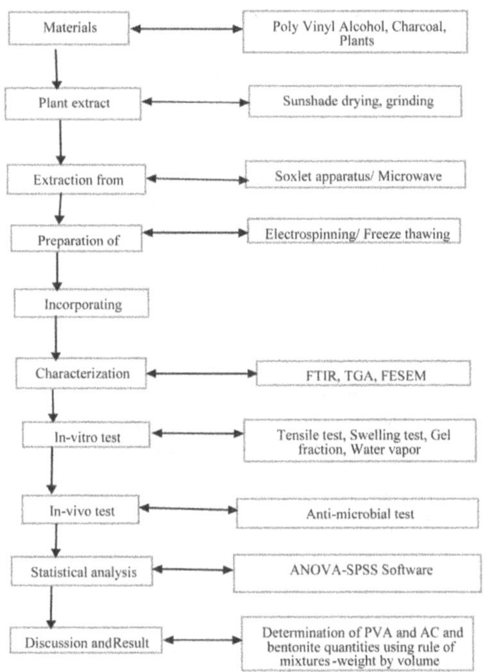

Figure 61.1. Flow chart of the developed methodology.

Source: Author.

Figure 61.2. PVA in powder form.

Source: Author.

Figure 61.3. Activated charcoal in powder form.

Source: Author.

4.3. *Calendula officinalis*

Calendula officinalis, also known as pot marigold, Figure 61.4 represents in powder form. It is utilized for its medicinal properties, which include antibacterial and anti-inflammatory effects. It is used for treating skin conditions, inflammation, injuries, bruises, and blood purification.

4.4. *Chromolaena odorata*

Chromolaena odorata, also known as Siam weed, Figure 61.5 represents in powder form is utilized for wound treatment due to its hemostatic, antibacterial, and antioxidant properties, which accelerate healing and aid in collagen maintenance and immune regulation.

5. Preparation

5.1. *Pre-processing*

Calendula officinal is flowers and Chromolaena odorata leaves were gathered in the shade, away from direct sunlight, by the nursery workers. After being cleaned, rinsed, and sun-dried, they were ground into a fine powder.

Figure 61.4. Calendula officinalis in powder form.

Source: Author.

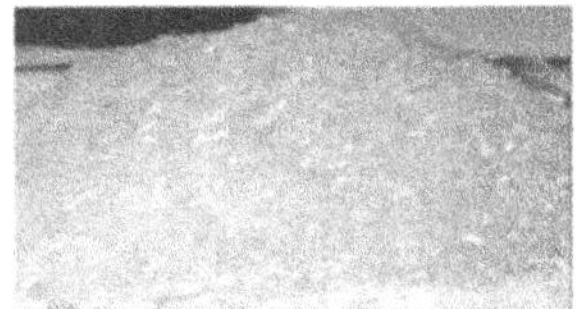

Figure 61.5. Chromolaena Odorata in powder form.

Source: Author.

5.2. *Extraction*

The components of the Soxhlet apparatus include a condenser, thimble, siphon, and percolator (solvent container). Fill the bottle with solvent. When heated, the solvent evaporates, travels through the plant powder-filled thimble into the condenser where it cools, then pours back into the thimble, and proceeds to the siphon. Once the siphon is full, the extraction falls into the pure solvent placed inside the percolator. This cycle continues, allowing us to remove the extraction from the solvent container. With this device, extraction can be performed multiple times.

5.3. *Stirring*

Initially, distilled water was used to dissolve PVA. Activated charcoal was added, followed by extraction to enhance the mechanical properties of PVA. At 30 degrees Celsius, a magnetic stirrer operating at 300 rpm was used to mix each component.

5.4. *Casting*

Once the mixture was prepared, it was poured onto the petri dishes and baked for 24 to 48 hours in a hot air oven.

6. Experiments Conducted

6.1. *Swelling study*

The measurements of hydrogel swelling – which is necessary for drug release – involve equilibrium, normalized, and isothermal swelling measures. The sample's weight over time is compared to its dried weight via isothermal swelling. For swelling testing, phosphate-buffered saline (PBS), which is composed of sodium biphosphate, potassium dihydrogen phosphate, distilled water, and sodium chloride, is utilized. Appropriate pH modifications guarantee controlled experiments, which are essential for comprehending how hydrogel behavior affects performance and efficacy in drug delivery systems and wound dressings.

6.2. *Water swelling analysis*

Prior to being dipped into the PBS solution, each size 1 cm film is weighted (0WO). The swollen weights (WI) of the films are acquired by utilizing tissue paper to absorb the surfacing water. The swelling ratio of the films is calculated using the following formula.

$$\text{Ratio of swelling } (\%) = ((W\text{-}W)/Wo] \times 100.$$

The masses of films submerged in PBS solution, both wet and dry, are W1 and WO, respectively.

6.3. Degradation study

PVA is a water-soluble polymer that is perfect for creating films because of its strength, thermal stability, and ability to create films. Weight loss, tensile characteristics, changes in molecular mass, and morphological modifications are used to quantify degradation. Samples are weighed, soaked in PBS or pure water, and then periodically taken out to measure weight loss. This process gives researchers insight into how the polymer breaks down in various settings.

Degradation % = [(W2-W1)/ W1] × 100

WI = Initial Sample Weight

W2= Degraded sample weight

6.4. Tensile test

Tensile testing is conducted to determine the material's degree of flexibility, wherein a steadily applied force is used to test the internal resistance of the films. The ASTM F88 standard is adhered to for this experiment. To carry out the experiment, samples are divided into width and height segments at a ratio of 1:4. Additional film area on the grips is provided for holding the samples. Use a screw gauge to determine the thickness of the films before beginning the experiment.

7. Result

The plants were obtained from the nursery and allowed to dry for a few days in the shade. To avoid the need for fresh plants continually for our experiment, the plants are ground into powder after drying. Activated charcoal and PVA powder are purchased. Initially, plant extracts are obtained using a Soxhlet apparatus. Ethanol serves as the solvent, and plant granules are the solute. After 48 hours, pure organic plant extracts are obtained by repeatedly employing the Soxhlet equipment. In addition to adding glutaraldehyde, a chemical crosslinker, and activated charcoal to enhance mechanical strength, water-soluble PVA powder is dissolved in distilled water. Extracts are then added and stirred again. Once all ingredients are combined, the mixture is poured onto petri dishes. The solution on the plates should have a smooth, even texture and be free of air bubbles. Table 61.1 represents the four samples namely, A, B, C, and D representing peak load (N), tensile strength and stiffness. To allow the solution to form a film, all petri dishes are placed in the oven for a duration of 24 to 48 hours. Similar PVA and activated charcoal compositions were used to create four distinct films, but the concentrations of plant extracts varied. Samples C and D include extracts of Chromolaena in varying amounts, while samples A and B have variable concentrations of calendula. Calendula and Chromolaena both help speed up wound healing. PVA serves as the perfect foundation for wound dressing material. The law of mixes was followed in the addition and modification of every material. High spatial resolution is made possible by FTIRIS (Fourier Transform Infrared Imaging Spectroscopy), which is used to retrieve chemical information from biological materials. FTIRIS examined PVA films containing extracts of calendula and chromolaena in ATR mode to investigate chemical interactions. C-O bonds, amino groups, alcohol groups, and CH and CN groups were all visible in the 600–4000 cm^{-1} spectrum. Notable peaks at 1083.99 cm^{-1} and 1598.99 cm^{-1} indicated the existence of certain functional groups.The swelling behavior of wound dressings, categorized into equilibrium, normalized, and isothermal swelling, impacts film absorption and efficacy in absorbing bodily fluids. PVA, a water-soluble polymer with strong film-forming properties, undergoes degradation studies to assess weight loss, tensile properties, and molecular changes crucial for understanding its biodegradability. Degradation studies involving immersion in distilled water or PBS solution demonstrate weight loss over time, indicating material deterioration. The process aids in determining shelf life and the duration before films exhibit signs of degradation. Tensile and compressive strengths are evaluated using a universal testing machine, providing insights into material durability and suitability for wound dressing applications. Sample analysis reveals Sample C exhibiting superior swelling behavior, while degradation studies highlight the material's biodegradability and shelf life. The comprehensive approach assesses both swelling dynamics and material durability, crucial for developing effective and sustainable wound dressing materials. Tensile strength is defined as 'the highest load in weight per unit area pulling in the direction of length that a given substance can sustain without breaking apart'. Since many engineering applications depend on a structure's rigidity, the Modulus of elasticity is often one of the first criteria to be assessed when choosing a material, because many engineering applications depend on the stiffness of the structure. High modulus of elasticity is preferred when deflection is undesired, yet low modulus of elasticity is necessary when flexibility is crucial. The aforementioned bar charts show the samples A, B, C, and D's mechanical characteristics, such as stiffness and tensile strength. Sample D had the lowest tensile strength, measuring 59.45 N/mm^2, whereas sample A had the strongest, measuring 190.022 N/mm^2. Sample D had a stiffness value that was 10% higher than that of sample B, while sample B's stiffness value peaked at 0.625 N/mm. Figure 61.6 represents the sample vs tensile strength graph and Figure 61.7 represents the sample vs stiffness graph.

Table 61.1. Mechanical properties of the films

Film sample	*Peak load (N)*	*Tensile strength (N/mm^2)*	*Stiffness (N/mm)*
A	8.551	190.022	0.5
B	11.709	106.445	0.625
C	12.631	140.344	0.061
D	4.736	59.45	0.4

Source: Author.

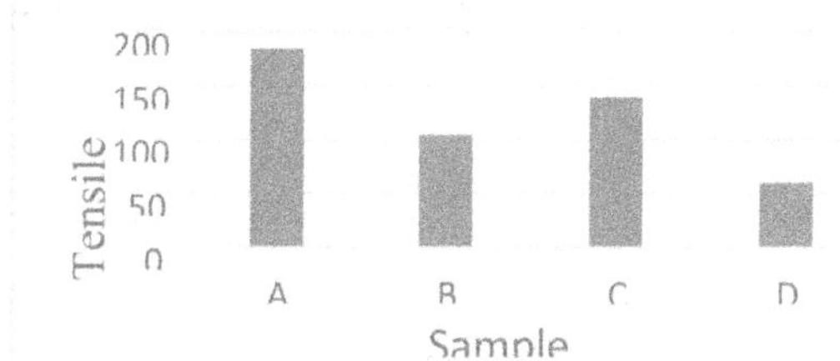

Figure 61.6. Sample vs tensile strength graph.

Source: Author.

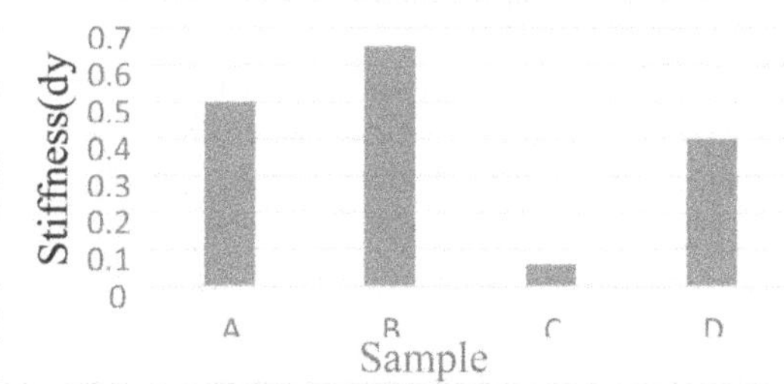

Figure 61.7. Sample vs stiffness graph.

Source: Author.

8. Conclusion

PVA and activated charcoal played supporting roles in our research, with Chromolaena odorata and Calendula officinalis as the main protagonists. The nursery purchased the plants, which were then allowed to dry for a few days in the shade. To save on fresh plants for our experiment, the dried plants are ground into powder using a grinding technique. We purchased activated charcoal and PVA in powder form. Initially, plant extracts are obtained using a Soxhlet device. The solute is plant powders, while the solvent is ethanol. By repeatedly utilizing the Soxhlet equipment, pure organic plant extracts are obtained at the end of 48 hours. Water-soluble PVA powder is allowed to dissolve in distilled water, and to boost the mechanical strength, glutaraldehyde a chemical crosslinker is added. Additionally, activated charcoal is added. Extracts are added and stirred again. After all the ingredients have been combined, the mixture is poured onto petri dishes. The solution on the plates should have a smooth, even texture and be free of air bubbles. To get the solution to form a film, all of the petri dishes are placed in the oven for a duration of 24 to 48 hours. Similar PVA and activated charcoal compositions were used to create four distinct films, but the plant extract concentrations varied. Samples C and D include extracts of Chromolaena in varying amounts, while samples A and B have variable concentrations of calendula. Calendula and Chromolaena both help to speed up the healing of wounds. The perfect foundation for wound dressing material is PVA. The law of mixes was followed in the addition and modification of every material. The resulting films' chemical composition is examined by FTIR characterization. To examine the properties of our samples, in vitro tests such as tensile tests, swelling studies, and degradation studies are conducted. Sample C demonstrated a noteworthy alteration in the swelling research by gaining weight more effectively than the other films. Excellent bonding and uniform structure are seen in Sample A's strong tensile strength. Good stiffness is demonstrated by Sample B's resistance to deformation. For all criteria, Sample D exhibits modest values. Instead of leaning toward biological research, we conducted mechanical experiments. We may perform in vitro research first and move on to in vivo research using animal models when we have superior films. We did not collect many samples for the experiments, so as of now, we cannot draw any firm conclusions. Perhaps we'll change the mix in the future and collect different samples to examine. We aim to optimize PVA-activated charcoal composites by refining plant extract concentrations and crosslinking agents for enhanced mechanical strength and therapeutic efficacy in wound healing. Biological evaluations through cell studies and animal models will validate biocompatibility. Advanced spectroscopic techniques like Raman spectroscopy and X-ray diffraction will deepen molecular insights. Sustainable production methods and smart dressing integration, alongside clinical trials, are pivotal for translating these innovations into practical wound care solutions. Collaborative efforts across disciplines will drive comprehensive development and application of these materials.

References

[1] Yudaev, P., Butorova, I., Chuev, V., Posokhova, V., Klyukin, B., & Chistyakov, E. (2023). Wound gel with antimicrobial effects based on polyvinyl alcohol and functional aryloxycyclotriphosphazene. *Polymers*, *15*(13), 2831.

[2] Shahid, M. A., Ali, A., Uddin, M. N., Miah, S., Islam, S. M., Mohebbullah, M., & Jamal, M. S. I. (2021). Antibacterial wound dressing electrospun nanofibrous material from polyvinyl alcohol, honey and Curcumin longa extract. *Journal of Industrial Textiles*, *51*(3), 455–469.

[3] Baniasadi, M., Baniasadi, H., Azimi, R., & Khosravi Dehaghi, N. (2020). Fabrication and characterization of a wound dressing composed of polyvinyl alcohol/nanochitosan/Artemisia ciniformis extract: An RSM study. *Polymer Engineering & Science*, *60*(7), 1459–1473.

[4] Eze, F. N., & Jayeoye, T. J. (2021). Chromolaena odorata (Siam weed): A natural reservoir of bioactive compounds with potent anti-fibrillogenic, antioxidative, and cytocompatible properties. *Biomedicine & Pharmacotherapy*, *141*, 111811.

[5] Asiri, A., Saidin, S., Sani, M. H., & Al-Ashwal, R. H. (2021). Epidermal and fibroblast growth factors incorporated polyvinyl alcohol electrospun nanofibers as biological dressing scaffold. *Scientific Reports*, *11*(1), 5634.

[6] Liu, Y., Zhang, Q., Zhou, N., Tan, J., Ashley, J., Wang, W., … & Zhang, M. (2020). Study on a novel poly (vinyl alcohol)/graphene oxide-citicoline sodium-lanthanum wound dressing: Biocompatibility, bioactivity, antimicrobial activity, and wound healing effect. *Chemical Engineering Journal*, *395*, 125059.

[7] Delavari, M. M., & Stiharu, I. (2022). Preparation and characterization of eco-friendly transparent antibacterial starch/polyvinyl alcohol materials for use as wound-dressing. *Micromachines*, *13*(6), 960.

[8] Mouro, C., Simões, M., & Gouveia, I. C. (2019). Emulsion electrospun fiber mats of PCL/PVA/chitosan and eugenol for wound dressing applications. *Advances in Polymer Technology*, *2019*(1), 9859506.

[9] Ahsan, A., & Farooq, M. A. (2019). Therapeutic potential of green synthesized silver nanoparticles loaded PVA hydrogel patches for wound healing. *Journal of Drug Delivery Science and Technology*, *54*, 101308.

[10] Kim, J. H., Lee, H., Jatoi, A. W., Im, S. S., Lee, J. S., & Kim, I. S. (2016). Juniperus chinensis extracts loaded PVA nanofiber: Enhanced antibacterial activity. *Materials Letters*, *181*, 367–370.

[11] Abbasinia, S., Monfared-Hajishirkiaee, R., & Ehtesabi, H. (2024). Nanocomposite sponge based on sodium alginate, polyvinyl alcohol, carbon dots, and woolly hedge nettle antibacterial extract for wound healing. *Industrial Crops and Products*, *214*, 118554.

[12] Song, S., Liu, X., Ding, L., Liu, Z., Abubaker, M. A., Xu, Y., & Zhang, J. (2024). A bacterial cellulose/polyvinyl alcohol/nitro graphene oxide double layer network hydrogel efficiency antibacterial and promotes wound healing. *International Journal of Biological Macromolecules*, *269*, 131957.

[13] Darabi, N. H., Kalaee, M., Mazinani, S., & Khajavi, R. (2024). GO/AgNW aided sustained release of ciprofloxacin loaded in Starch/PVA nanocomposite mats for wound dressings application. *International Journal of Biological Macromolecules*, *266*, 130977.

[14] Irantash, S., Gholipour-Kanani, A., Najmoddin, N., & Varsei, M. (2024). A hybrid structure based on silk fibroin/PVA nanofibers and alginate/gum tragacanth hydrogel embedded with cardamom extract. *Scientific Reports*, *14*(1), 14010.

62 Infrastructural study of smart cities

Aakansha Soy[1] *and Ahilya Dubey*[2]

[1]Assistant Professor, Department of CS & IT, Kalinga University, Raipur, India
[2]Research Scholar, Department of CS & IT, Kalinga University, Raipur, India

Abstract: With an increasing number of experimental use-cases coming from developed nations worldwide, the Smart City concept has garnered tremendous attention from both industry and academics. The goal of contemporary civilization is to incorporate technology more and more into daily living. Every day, such technology advances, enhancing every facet of its functionality. Every type of technology used to enhance a user's quality of life requires power to function, and the type of power source used differs based on the item being used. Every power source requires fuel and will leave behind waste, which may be either environmentally safe or hazardous. This essay's objectives are to examine, comprehend, and analyze the smart city. Relying on the current smart city concept, the study attempts to assess and map the city assets and facilities with their various fields.

Keywords: Development, healthcare, advancements, model, technology, smart grid

1. Introduction

The main difficulties local government faces are serving its citizens with high-quality, environment friendly, and sustainable amenities. Cities must use cutting-edge technology to offer services since they have limited infrastructure, resources, and economies. The development of smart solutions leveraging ICT (Internet and Communication Technology) advancements has made cities smart, ushering in the era of smart cities around the world [1].

As a result of a city's growing importance in sustainable strategic development, more and more cities are trying to incorporate and use it. From around world, the repercussions are still being felt today, with urbanization and the adoption of new technology being among the major changes [8]. Urban planning will undergo a significant and complete upheaval as a result. By the early 2030s, it is commonly anticipated that the urban environment will be largely electronic and computerized, which will increase the need for advanced technology to address urbanization's issues [6].

Future cities will be significantly smarter, and smart city evaluation will be a crucial part of that because it greatly influences how smart development of cities is directed and supported. Most countries are currently researching the evaluations of smart cities and the development of new applications [11]. Utilizing indicators with regional peculiarities and having a constrained extensibility, the main emphasis is on analyzing one specific aspect. These elements have caused incorrect assessment conclusions.

A number of concerns are impacted by demographic change, along with an increase in the supporting infrastructure, mobility, the housing sector, and social community groups; an increase in the producers and consumers for goods and services; a cumulative increase in the likelihood of natural catastrophes, both natural and artificial; and an escalating desire for educational opportunities and health care [3, 4].

The urban population grows as a result of this migration to urban areas, creating new problems for the city and its agents who must find new ways to accommodate the extra population in terms of amenities and services [12]. The population of cities has been increasing steadily during the last few years. Today, 53% of the world's population lives in cities, which are thought to take up 2% of the planet's surface. 70% of the population will reside in cities by 2050 [5]. This study examines the idea of 'smart cities', taking into account how technology might be used to enhance the services the city provides [2]. This means that our plan is to take the current, non-smart city and add cutting-edge digital technologies to render it smart.

As technology continues to improve, so do residential areas like towns, villages, and so on since the growth of civilization and technology are simultaneous. Because of the growing needs of the population living in urban areas, such innovations are harmful to the growth of cities. As a result, both the development of cities and the development of technology are intertwined. Eventually, urban growth and technological advancement will converge to create the idea of the 'smart city'.Early concepts for smart cities were primarily concerned with enhancing the efficiency of city services via the use of digital and cutting-edge technical innovation while also opening up new business opportunities [13]. But more attention needs to be paid to the distributional effects of smart cities, that is, determining the benefits and costs of smart cities on people, the environment, and locations. Figure 62.1. depicts major components of a smart city.

[a]ku.aakanshasoy@kalingauniversity.ac.in, [b]ahilya.dubey@kalingauniversity.ac.in

DOI: 10.1201/9781003675259-62

2. Major Points For Services Provided In a Smart City

2.1. *Transportation systems*

The development of a nation depends heavily on the transportation system, which is a necessary part of society. Congestion is the main issue with transportation due to the population expansion and inadequate spatial infrastructure. The development of intelligent applications is helping the involved authorities tackle the problems. Smart transportation services have tapped into the advantages of IoT and TIC to offer dependable, sustainable, comfortable, secure and cost-effective means for residents and visitors to navigate in the city. A few exemplars of smart transportation services include information systems, signal priority for public transit buses, integrated public transportation, carpooling, and automatic fare collecting [7]. The most popular technology being used and explored right now is connected and autonomous automobiles. In most circumstances, smart vehicles require less human work.

Finding a parking spot is one of the challenges drivers in urban areas confront because of the heavy use of private vehicles. The challenge might be lessened by effectively using the parking spaces that are accessible. A smart parking management system is used in this situation as one of the crucial elements of smart mobility.

2.2. *Management of garbage and waste*

A location-based intelligent garbage collecting system that uses optimization and AI intelligence algorithms is in operation. An IoT-based garbage alarm system that notifies the relevant agencies about the garbage in a particular area [14]. By determining the ideal waste collection path using ant colony optimization and evolutionary algorithms, accordingly, the researchers have developed an intelligent method for collecting trash.

2.3. *Education system*

The idea behind a machine learning-based e-learning material framework is that each learner has a distinct learning style. Depending on the preferences and learning tendencies of the individual students, appropriate study materials are recommended. Students' learning capacities are improved by the AI-based classroom software. There are many possibilities for smart classrooms, and a review of AI-based tools for the teaching-learning process in classrooms is available.

2.4. *Electric supply*

With the population expanding, there is an excessive use of energy, which causes major problems for the environment and other areas. Therefore, efficient energy management is crucial for today's civilization. Effective usage of power is crucial right now because of the numerous environmental issues the globe is currently experiencing as a result of population expansion [9]. The issue of excessive electricity use can be resolved with a smart energy management system. One of the best technologies to get toward smart cities is smart grids.

2.5. *Healthcare*

Healthcare uses many intriguing IoT implementations. IoT provides efficient therapies for a range of medical conditions, such as exercise programmes, chronic diseases, and health monitoring. As a result, the IoT plays a significant role in the production of various components needed for medical device manufacturing, such as actuators, image processing devices, and sensor devices [10]. Due to a number of practical concerns, the healthcare industry has recently been a debate point for various IoT-related topics.

2.6. *Technological advancements*

The development of home automation systems is attracting a significant amount of focus as a result of the development of the most recent communications technology. A smart home (SH) is an Internet of Things (IoT) application that uses the internet to monitor and control appliances in houses. In the current world, people's daily lives now include the use of smartphones, smart devices, smart ovens, smart home appliances, smart air conditioners, and smart refrigerators [15]. A smart environment can be created by these intelligent devices interacting and communicating with one another. In order to monitor and manage the natural environment, a CPS combines intelligent computational devices, communication networks, and physical processes. Infrastructure for smart cities' sustainable development includes data sensing via

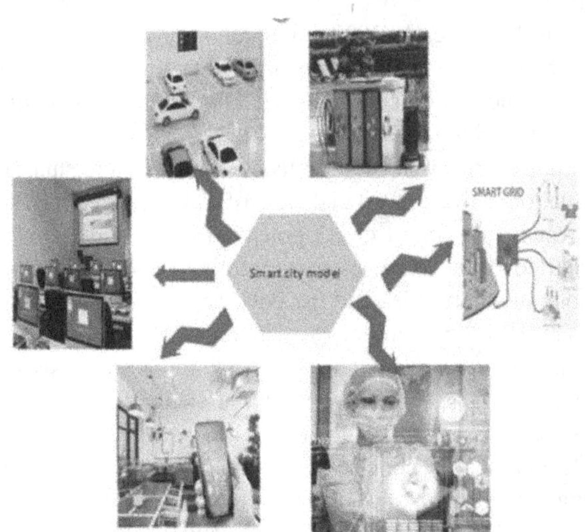

Figure 62.1. Model of smart city.

Source: Author.

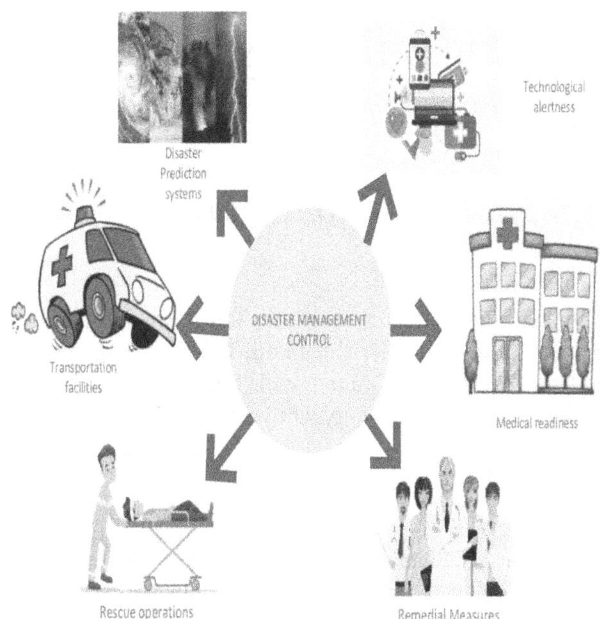

Figure 62.2. Emergency services in smart cities.

Source: Author.

various smart devices, communication, processing, analysis, and decision-making.

3. Conclusion

In order to evaluate the effects of new technologies and smart innovations on cities, this study examines the frameworks and standards for Smart Cities already in use. Six major points has been discussed, that are related to education, healthcare systems, mobility, garbage management, smart grids as well as technological advancements in smart cities. The goal of the current study is to demonstrate how artificial intelligence could be used to create applications for smart cities. This study's goal is to investigate how smart cities and digital cities are being developed in order to identify the possible errors and improvements that have been made. The advantage of this research is that anyone else creating a smart city will be able to avoid making the same mistakes that have been made in the past and will be able to better their progress in the future.

References

[1] Limon-Ruiz, M., Larios-Rosillo, V. M., Maciel, R., Beltran, R., Orizaga-Trejo, J. A., & Ceballos, G. R. (2019, October). User-oriented representation of Smart Cities indicators to support citizens governments decision-making processes. In *2019 IEEE International Smart Cities Conference (ISC2)* (pp. 396–401). IEEE.

[2] Arasu, R., Rajani, B., Anantha, R. A., Manu, P., & Priya, S. (2024). Design of quality of experience-based green internet architecture for smart city. *Journal of Internet Services and Information Security*, *14*(3), 157–166.

[3] Aguilar, M. G. S., Rosillo, V. M. L., Perez, C. O. M., Arellano, M. R. M., Ramirez, J. R. B., & Trejo, J. A. O. (2019, October). Analysis of wastewater production to implement circular economy solutions in a smart cities university campus living lab. In *2019 IEEE International Smart Cities Conference (ISC2)* (pp. 366–371). IEEE.

[4] Badii, A., Carboni, D., Pintus, A., Piras, A., Serra, A., Tiemann, M., & Viswanathan, N. (2013). CityScripts: unifying web, IoT and smart city services in a smart citizen workspace. *Journal of Wireless Mobile Networks, Ubiquitous Computing, and Dependable Applications*, *4*(3), 58–78.

[5] Kutlay, A. (2019, April). RDS Of things: Using RDS technology for smart cities. In *2019 7th International Istanbul Smart Grids and Cities Congress and Fair (ICSG)* (pp. 139–143). IEEE.

[6] Llopiz-Guerra, K., Daline, U. R., Ronald, M. H., Valia, L. V. M., Jadira, D. R. J. N., & Karla, R. S. (2024). Importance of environmental education in the context of natural sustainability. *Natural and Engineering Sciences*, *9*(1), 57–71.

[7] Astrain, J. J., Falcone, F., Lopez, A., Sanchis, P., Villadangos, J., & Matias, I. R. (2020). Monitoring of Electric buses within an urban smart city environment. *IEEE Sensors*, *2020*, 1–4.

[8] Buljubašić, S. (2020). Application of new technologies in the water supply system. *Archives for Technical Sciences*, *1*(22), 27–34.

[9] Mohanty, S. P., Thapliyal, H., & Bajpai, R. (2021). Consumer technologies for smart cities to smart villages. *IEEE International Conference on Consumer Electronics (ICCE)*, p. 1. doi: 10.1109/ICCE50685.2021.9427601.

[10] Neelima, S., Govindaraj, M., Subramani, D. K., ALkhayyat, A., & Mohan, D. C. (2024). Factors influencing data utilization and performance of health management information systems: A case study. *Indian Journal of Information Sources and Services*, *14*(2), 146–152.

[11] Hai, T., Chaturvedi, R., Mostafa, L., Kh, T. I., Soliman, N. F., & El-Shafai, W. (2024). Designing g-C3N4/ZnCo2O4 nanocoposite as a promising photocatalyst for photodegradation of MB under visible-light excitation: response surface methodology (RSM) optimization and modeling. *Journal of Physics and Chemistry of Solids*, *185*, 111747.

[12] Liu, Z., Zhanguo, S. U., Abed, A. M., Chaturvedi, R., Feyzbaxsh, M., & Salavat, A. K. (2022). A comparative thermodynamic and exergoeconomic scrutiny of four geothermal systems with various configurations of TEG and HDH unit implementations. *Applied Thermal Engineering*, *216*, 119094.

[13] Chaturvedi, R., Sharma, A., Sharma, K., & Saraswat, M. (2022). Tribological behaviour of multi-walled carbon nanotubes reinforced AA 7075 nano-composites. *Advances in Materials and Processing Technologies*, *8*(4), 4743–4755.

[14] Wang, H., Yan, G., Tag-Eldin, E., Chaturvedi, R., Aryanfar, Y., Alcaraz, J. L. G., … & Moria, H. (2023). Thermodynamic investigation of a single flash geothermal power plant powered by carbon dioxide transcritical recovery cycle. *Alexandria Engineering Journal*, *64*, 441–450.

[15] Villadangos, J., Falcone, F., Lopez, A., Astrain, J. J., Sanchis, P., & Matias, I. R. (2021, October). Distributed opportunistic wireless mapplicationing system towards smart city service provision. In *2021 IEEE Sensors* (pp. 1–4). IEEE.

63 Dehazing the Hazed images for buildings and underwater objects using KWM algorithm

Buvanesh Pandian V.[1], *Arunprasath T.*[2], *Pallikonda Rajasekaran M.*[3], *Kottaimalai R.*[3], *and Krishna Priya R.*[4]

[1]Department of EEE, Kalasalingam Academy of Research and Education, Krishnankoil, Tamil Nadu, India
[2]Department of BME, Kalasalingam Academy of Research and Education, Krishnankoil, Tamil Nadu, India
[3]Department of ECE, Kalasalingam Academy of Research and Education, Krishnankoil, Tamil Nadu, India
[4]Research and Consultancy, Department, College of Engineering and Technology, University of Technology and Applied Science, Oman

Abstract: Image dehazing is a technique used to remove or reduce the effects of haze, fog, or atmospheric scattering from images. In other words, images are identified from dust, fog, mist and other external climatic conditions. Haze is caused by the scattering of light by particles and molecules in the atmosphere, resulting in reduced visibility, loss of contrast, and color distortion in images. Dehazing algorithms aim to enhance the visibility of hazy images by estimating and removing the haze component. The KWM algorithm states the fusion of the K-means clustering (KMC) algorithm and the Weight Map (WM) algorithm. These algorithms typically work by estimating the haze or atmospheric light in the image and then performing a dehazing operation to restore the original appearance of the scene. KMC is an un-supervised learning technique that uses feature similarity to group related data points. By minimizing the within-cluster sum of squares, the method successfully separates various regions and groups comparable pixels together. The WM refers to a map or image that assigns weights or importance values to individual pixels or regions within an image. It is mostly used for image blending, image segmentation, and image filtering. The purpose of a WM is to influence the processing or analysis of an image based on the desired effect or objective. By assigning different weights to pixels or regions, certain areas can be given more emphasis or control in the processing steps. Then the final output images are analyzed by qualitative and quantitative image-quality-parameters with an accuracy of 90.65%.

Keywords: Dehazing, Gaussian pyramids, KMC, pixel, weight map, F1 score

1. Introduction

A method used in image processing to lessen or completely remove the influence of atmospheric haze or fog in photographs is called haze removal, sometimes referred to as image dehazing. Haze causes reduced visibility, low contrast, and color distortion in images, and removing it can greatly improve image quality and visual clarity. The general approach for haze removal in image processing is (i) Haze Model: The first step in haze removal is to model the physical characteristics of haze. The most commonly used model is the atmospheric scattering model, which assumes that haze causes the reduction of contrast and intensity in an image. According to this model, the observed intensity at a pixel is a combination of the scene radiance and the scattered light due to haze. (ii) Haze Estimation: Then estimatating the haze or atmospheric light present in the image. This can be achieved by analyzing the pixel values in the image. Typically, the brightest pixels in the image are assumed to represent the atmospheric light because they are less affected by the scene's radiance and primarily influenced by haze. Statistical methods or colour analysis techniques can be used to estimate atmospheric light. (iii) Transmission Map Estimation: The transmission map represents the proportion of scene radiance that reaches the camera after being attenuated by haze. A lower transmission value indicates more haze, while a higher value indicates less haze. The transmission map is estimated by comparing pixel intensities with the estimated atmospheric light. Various methods, such as the dark channel prior, can be employed to estimate the transmission map. (iv) Haze Removal: Once the transmission map is estimated, it is used to remove the haze component from the image. The atmospheric scattering model is applied to restore the original scene radiance by dividing the observed intensity by the estimated transmission value. This process essentially compensates for the light scattering caused by haze. (v) Post-Processing: The dehazed image may still require some post-processing to enhance the visual quality. Techniques such as contrast enhancement, color correction, and noise reduction can be applied to improve the image's appearance further and reduce artifacts.

v.buvaneshpandian@klu.ac.in, t.arunprasath@klu.ac.in, m.p.raja@klu.ac.in, r.kottaimalai@klu.ac.in, krishna.priya@utas.edu.om

DOI: 10.1201/9781003675259-63

2. Literature Survey

The KMC algorithm is used to split the images into smaller parts to provide information. It is also used to increase the brightness level and distortion of image colour [1]. Based on the segmentation results, the weights are assigned to the image when the scattering is present in the image from restoration [2]. The priority of the red colour in the RGB space uses this KMC as a morphological process for contrast improvement in images [3, 4]. For recovering the image scene, the non-local prior methods are used for contrast improvement using patch-based image techniques [6, 19]. When dividing the images into n-parts, the harmful interference in the correlated images can be improved and the brightness of the images can also be increased using the recovery formula [8]. For multi-exposure images, gamma correction and Laplacian pyramid are used [7, 9]. For colour attenuation, the Gaussian mixture model is used for the estimation of images [10].

The fuzzy C-means algorithm uses pixel-level estimation in images for non-global values [11]. The dark channel is used for the priority ranking of patches for the removal of haze in the images [12]. The dark channel prior method is used to remove the foreground and background in images for the transmission of the image's pixel [5, 13]. Numerous dehazing works and efforts have been performed recently to address this shortcoming. Their goal is to immediately restore a single hazy piece of data without the requirement for further information [14]. By dividing the images for enhancement, the ambient lights are calculated for invalid dark channel issues [15]. An improved dark channel haze removal (DCHR) algorithm is used on unmanned aerial vehicles to eliminate haze and restore video images [16, 17]. The extracted images are used as intense by transmission map, followed by a guided filter to give better results in the edges of images [18]. For emphasizing local and global enhancement, the adaptive histogram equalization method is used for local contrast reduction and local visibility in visibility [21]. Using the discrete wavelet transform (DWT), a multi-scale wavelets approach divides the images into one lower frequency and three higher frequency sub-images. The Weight Balance Weight Map (WBWM) technique is used to dehaze a lower-frequency sub-image. The wavelet de-noising technique is used to remove haze from three higher-frequency sub-images [20]. Then, sub-image reconstruction is done using an Inverse Discrete Wavelet Transform (IDWT) to provide a dehazed image [22]. For the encoder and decoder network with a Gaussian process, the global attention mechanism is used for the extraction of images in hazy conditions [23, 24]. Gaussian filter is used for enhancing the images to reconstruct the colour and boost the contrast [25]. For the recovery of images, transmission map estimation, and atmospheric light estimation are used with the help of clustering [26]. Because of degradations, the images will be damaged which affects the quality of the images [27].

The aforementioned methods have the following issues: (i) a high sensitivity to noise and artifacts; (ii) longer processing times; (iii) reduced precision in blurry images; (iv) insufficient optimization robustness; and (v) cluster overlap. To solve the aforementioned issues, this study uses weight Map and K-means algorithms to construct a novel fused architecture. In the past, a lot of fused algorithms had issues with clustering overlapping and taking longer to clear image haze. Thus, the goal of this effort was to quickly ascertain the viability and correctness of hazed photographs.

3. Proposed Work

The KMC algorithm and the Weight Map algorithm are two crucial components of the revolutionary end-to-end image dehazing technique known as the KWM Algorithm. Based on the similarity of pixel values or other criteria, the KMC technique is used to divide an image into discrete regions or clusters. The algorithm's goal is to reduce the total squares within a cluster, effectively grouping comparable pixels and dividing regions. The term 'WM' refers to a Weight Map or image that gives specific pixels or sections of an image weights or importance ratings. A WM's function is to modify how an image is processed or analyzed by a desired outcome or objective. Certain areas can be given more attention or control during the processing phases by giving them varying weights that can be applied to pixels or regions.

3.1. K-means clustering (KMC) algorithm

One popular unsupervised learning method for image processing applications is the KMC algorithm. The KMC algorithm divides a set of data points into k clusters, and each data point is allocated to the cluster with the closest centroid or mean. Generally, the KMC algorithm can be used as follows:

- Data representation: Represent the image as a collection of data points, where each point corresponds to a pixel or a feature vector derived from the pixel values. The feature vector can include colour information, texture descriptors, or any other relevant image attributes.
- Selection of k: Ascertain how many clusters, k, ought to be produced. An ideal value can be found by applying methods like silhouette analysis or the elbow approach, or by relying on past information. Here we are using the value of **'K'** as **'3'**.
- Initialization: Randomly initialize k centroids, which serve as the center of initial cluster.
- Assignment step: Using a distance measure, like Euclidean distance, assign each data point to the closest centroid. Each data point in this step's cluster formation belongs to the cluster with the closest centroid.
- Update step: By calculating the average of all the data points allocated to each cluster, you may recalculate the cluster centroids. The cluster centers are updated in this stage.
- Iteration: Until convergence or a maximum number of iterations is achieved, repeat steps 4 and 5. When the

centroids stop changing substantially or when the distribution of data points among clusters stays constant, convergence has taken place.

- Final result: Once convergence is reached, the algorithm produces k clusters, and each data point or pixel is assigned to a specific cluster. This assignment can be used for various image processing tasks, depending on the specific application.

3.2. *Weight map algorithm*

In image processing, a weight map refers to a map or image that assigns weights or importance values to individual pixels or regions within an image. By assigning different weights to pixels or regions, certain areas can be given more emphasis or control in the processing pipeline. Once a weight map is generated, it is typically applied to the target image or used as a guiding factor during the image processing task. Gaussian pyramid techniques are a fundamental tool in image processing that involves the construction and analysis of multi-resolution representations of an image. These techniques are based on the concept of the Gaussian filter, which is a widely used linear filter in image processing. The Gaussian pyramid is a hierarchical representation of an image, where each level of the pyramid corresponds to a different scale or resolution of the original image. The pyramid is constructed by repeatedly applying a low-pass filter, typically a Gaussian filter, followed by subsampling or down-sampling to reduce the image size. This process results in a series of images, with each subsequent level containing progressively more coarse-grained information.

In image processing, Laplacian techniques – which are employed for tasks including picture compression, edge detection, and sharpening – are closely related to Gaussian pyramid approaches. The pace at which a function changes at each point is measured by the Laplacian operator, a second-order differential operator. The Laplacian operator is used in image processing to draw attention to areas with sharp features or edges that experience abrupt variations in intensity. By subtracting the blurred image from the original image, one can get the Laplacian of the image. Applying a Gaussian filter on the original image in order to create a smoothed version of it is the first step in applying Laplacian techniques. This step helps to reduce noise and suppress high-frequency components. Calculate the Laplacian of the smoothed image by taking away the smoothed image from an original image. This operation enhances the edges and fine details in the image. Optionally, adjust the intensity range of the Laplacian image to enhance the visibility of the edges. The resulting Laplacian image highlights the areas of rapid intensity changes in the original image, which typically correspond to edges or fine details. These areas appear as bright or dark regions depending on the sign of the Laplacian response.

4. Result and Discussion

4.1. *Qualitative analysis*

For qualitative analysis, the SOTS datasets are taken here for the experiment. The SOTS dataset included a wide range of samples, allowing it to be utilized to confirm the efficacy of the suggested method trustworthily. Here 36 image samples are tested with our algorithm. Table 63.1 explains the input as a hazy image and the output as our proposed output for the SOTS dataset. The last two rows show the underwater images of dehazing methods results.

4.2. *Quantitative analysis*

4.2.1. *FSIM*

It evaluates the similarity of these features between a reference image and a distorted photo, providing a measure of how close the distorted image is to the original in terms of perceived quality. Table 63.2 explains the SOTS dataset output image value for FSIM. The average value describes the FSIM value for our proposed work.

Table 63.1. Qualitative analysis of the SOTS dataset

I						
O						
I						
O						
I						
O						
I						
O						
I						
O						
I						
O						
I						
O						

Source: Author.

Table 63.2. FSIM value for the SOTS dataset for the proposed work

1	*2*	*3*	*4*	*5*	*6*	*7*	*8*	*9*
0.958	0.962	0.953	0.958	0.962	0.972	0.975	0.981	0.969
10	**11**	**12**	**13**	**14**	**15**	**16**	**17**	**18**
0.949	0.987	0.986	0.983	0.981	0.976	0.972	0.986	0.968
19	**20**	**21**	**22**	**23**	**24**	**25**	**26**	**27**
0.986	0.981	0.982	0.976	0.981	0.973	0.986	0.976	0.981
28	**29**	**30**	**31**	**32**	**33**	**34**	**35**	**36**
0.968	0.986	0.978	0.987	0.992	0.997	0.994	0.990	0.963
Average								**0.976**

Source: Author.

Time Period: This refers to the total time required to complete the entire process from the initial to the final. Table 63.3 describes the average time taken for each step in the completion of our proposed work.

SSIM: It measures the similarity between these components by comparing the local neighborhoods of corresponding pixels in the reference and distorted images. Table 63.4 explains the SOTS dataset output image value for SSIM. The average value describes the SSIM value for our proposed work.

Table 63.3. The average time taken in each step

Sl. No.	*Method*	*Time (sec)*
1	K- means clustering	6.25
2	Weight Map	1.92
3	IQP	2.27
4	Total	10.44

Source: Author.

Table 63.4. SSIM value for the SOTS dataset for proposed work

1	*2*	*3*	*4*	*5*	*6*	*7*	*8*	*9*
0.998	0.992	0.993	0.998	0.992	0.982	0.985	0.981	0.989
10	**11**	**12**	**13**	**14**	**15**	**16**	**17**	**18**
0.999	0.987	0.986	0.983	0.981	0.986	0.982	0.986	0.988
19	**20**	**21**	**22**	**23**	**24**	**25**	**26**	**27**
0.986	0.980	0.982	0.986	0.981	0.983	0.986	0.986	0.981
28	**29**	**30**	**31**	**32**	**33**	**34**	**35**	**36**
0.988	0.986	0.988	0.987	0.992	0.997	0.994	0.990	0.983
Average							0.987	

Source: Author.

F1 Score: It measures the balance between precision and recall, providing a single value that represents the overall effectiveness of the algorithm.

$$\text{F1 Score} = 2 * (\text{Precision} * \text{Recall}) / (\text{Precision} + \text{Recall}) \quad (1)$$

Here the sample 5 images are taken for the experiment from the SOTS dataset and the average percentage values show the accuracy of our proposed algorithm which is shown in Table 63.5.

Table 63.5. F1 score for proposed work

Image	*Precision*	*Recall*	*F1 score*	*Accuracy*
1	0.9004	0.8752	0.8876	88.76%
2	0.9745	0.8036	0.8809	88.09%
3	0.9497	0.8738	0.9102	91.02%
4	0.9342	0.9158	0.9249	92.49%
5	0.9436	0.9144	0.9288	92.88%
Average				90.65%

Source: Author.

5. Conclusion

In recent years, haze removal techniques have grown in popularity and importance for numerous computer vision applications. The numerous dehazing techniques that are now in use have been thoroughly explored in this paper from a variety of perspectives, including theories, mathematical models, and performance metrics. Also offered is a comparison of various dehazing methods using various datasets. The experimental findings show that our approach may successfully improve the quality of foggy photos, efficiently preserve image details, and maintain colour accuracy with 90.65% whereas, the reference paper has less than our proposed work results. In the Future, the dehazing of hazed images will be done on remote sensing images and improving the corners, and boundaries of images with other soft computing algorithms.

References

[1] Ali, H., Sher, A., Saeed, M., & Rada, L. (2020). Active contour image segmentation model with de-hazing constraints. *IET Image Processing*, *14*(5), 921–928. https://doi.org/10.1049/iet-ipr.2018.5987

[2] Berman, D., Treibitz, T., & Avidan, S. (2020). Single Image Dehazing Using Haze-Lines. *IEEE Transactions on Pattern Analysis and Machine Intelligence*, *42*(3), 720–734. https://doi.org/10.1109/TPAMI.2018.2882478

[3] Bie, Y., Yang, S., & Huang, Y. (2022). Single Remote Sensing Image Dehazing Using Gaussian and Physics-Guided Process. *IEEE Geoscience and Remote Sensing Letters*, *19*. https://doi.org/10.1109/LGRS.2022.3177257

[4] Borkar, S. B., & Bonde, S. V. (2017). Oceanic Image Dehazing based on Red Color Priority using Segmentation Approach. In *International Journal of Oceans and Oceanography* (Vol. 11, Issue 1). http://www.ripublication.com.

[5] Elhoseny, M., & Shankar, K. (2019). Optimal bilateral filter and Convolutional Neural Network based denoising method of medical image measurements. *Measurement: Journal of the International Measurement Confederation*, *143*, 125–135. https://doi.org/10.1016/j.measurement.2019.04.072

[6] Gu, Z., Ju, M., & Zhang, D. (2017). A Single Image Dehazing Method Using Average Saturation Prior. *Mathematical Problems in Engineering*, *2017*. https://doi.org/10.1155/2017/6851301

[7] Guo, F., Qiu, J., & Tang, J. (2021). Single Image Dehazing Using Adaptive Sky Segmentation. *IEEJ Transactions on Electrical and Electronic Engineering*, *16*(9), 1209–1220. https://doi.org/10.1002/tee.23419

[8] Hajjami, J., Napoléon, T., & Alfalou, A. (2020). Efficient sky dehazing by atmospheric light fusion. *Sensors (Switzerland)*, *20*(17), 1–18. https://doi.org/10.3390/s20174893

[9] Harish Babu, G., & Venkatram, N. (2020). A survey on analysis and implementation of state-of-the-art haze removal techniques. In *Journal of Visual Communication and Image Representation* (Vol. 72). Academic Press Inc. https://doi.org/10.1016/j.jvcir.2020.102912

[10] Hassan, H., Bashir, A. K., Ahmad, M., Menon, V. G., Afridi, I. U., Nawaz, R., & Luo, B. (2021). Real-time image dehazing by superpixels segmentation and guidance filter. *Journal of Real-Time Image Processing*, *18*(5), 1555–1575. https://doi.org/10.1007/s11554-020-00953-4

[11] Huang, Z., Jing, H., Chen, A., Hong, C., & Shang, X. (2023). Efficient image dehazing algorithm using multiple priors constraints. *Journal of Visual Communication and Image Representation*, *90*. https://doi.org/10.1016/j.jvcir.2022.103694

[12] Jiang, B., Wang, J., Wu, Y., Wang, S., Zhang, J., Chen, X., Li, Y., Li, X., & Wang, L. (2023). A Dehazing Method for Remote Sensing Image under Non-uniform Hazy Weather Based on Deep Learning Network. *IEEE Transactions on Geoscience and Remote Sensing*. https://doi.org/10.1109/TGRS.2023.3261545

[13] Ju, M., DIng, C., Guo, C. A., Ren, W., & Tao, D. (2021). IDRLP: Image Dehazing Using Region Line Prior. *IEEE Transactions on Image Processing*, *30*, 9043–9057. https://doi.org/10.1109/TIP.2021.3122088

[14] Kanti Dhara, S., Roy, M., Sen, D., & Kumar Biswas, P. (2021). Color Cast Dependent Image Dehazing via Adaptive Airlight Refinement and Non-Linear Color Balancing. *IEEE Transactions on Circuits and Systems for Video Technology*, *31*(5), 2076–2081. https://doi.org/10.1109/TCSVT.2020.3007850

[15] Li, Z., & Shu, H. (2021). MULTI-SCALE MODEL DRIVEN SINGLE IMAGE DEHAZING. *Proceedings - International Conference on Image Processing, ICIP*, *2021-September*, 2004–2008. https://doi.org/10.1109/ICIP42928.2021.9506792

[16] Li, Z., Zheng, C., Shu, H., & Wu, S. (2022). Dual-Scale Single Image Dehazing via Neural Augmentation. *IEEE Transactions on Image Processing*, *31*, 6213–6223. https://doi.org/10.1109/TIP.2022.3207571

[17] Liu, J., Wang, S., Wang, X., Ju, M., & Zhang, D. (2021). A review of remote sensing image dehazing. In *Sensors* (Vol. 21, Issue 11). MDPI AG. https://doi.org/10.3390/s21113926

[18] Ma, S., Pan, W., Liu, H., Dai, S., Xu, B., Xu, C., Li, X., & Guan, H. (2023). Image Dehazing Based on Improved Color Channel Transfer and Multiexposure Fusion. *Advances in Multimedia*, *2023*, 1–10. https://doi.org/10.1155/2023/8891239

[19] Nie, J., Pang, Y., Xie, J., Pan, J., & Han, J. (2022). Stereo Refinement Dehazing Network. *IEEE Transactions on Circuits and Systems for Video Technology*, *32*(6), 3334–3345. https://doi.org/10.1109/TCSVT.2021.3105685

[20] Pikun, W., Ling, W., Jiangxin, Q., & Jiashuai, D. (2022). Unmanned aerial vehicles object detection based on image haze removal under sea fog conditions. *IET Image Processing*, *16*(10), 2709–2721. https://doi.org/10.1049/ipr2.12519

[21] V, B. P., Prasath, T. A., & Rajasekaran, M. P. (2024). Dehazing, enhancing the boundaries and corners in hazed images using optimal adaptive technique. *International Journal of Image and Data Fusion*, *00*(00), 1–16. https://doi.org/10.1080/19479832.2024.2321900

[22] Wu, H., & Li, G. (n.d.). *Image Dehazing by Muti-Scale Exposure Fusion Based on Global Contrast and Local Detail Enhancement*. https://ssrn.com/abstract=4365846

[23] Xu, H., Tan, Y., Wang, W., & Wang, G. (2020). Image Dehazing by Incorporating Markov Random Field with Dark Channel Prior. *Journal of Ocean University of China*, *19*(3), 551–560. https://doi.org/10.1007/s11802-020-4003-6

[24] Yu, B., Chen, Y., Cao, S. Y., Shen, H. L., & Li, J. (2022). Three-Channel Infrared Imaging for Object Detection in Haze. *IEEE Transactions on Instrumentation and Measurement*, *71*. https://doi.org/10.1109/TIM.2022.3164062

[25] Zhang, Y., Wang, P., Fan, Q., Bao, F., Yao, X., & Zhang, C. (2020). Single image numerical iterative dehazing method based on local physical features. *IEEE Transactions on Circuits and Systems for Video Technology*, *30*(10), 3544–3557. https://doi.org/10.1109/TCSVT.2019.2939853

[26] Zheng, M., Qi, G., Zhu, Z., Li, Y., Wei, H., & Liu, Y. (2020). Image Dehazing by an Artificial Image Fusion Method Based on Adaptive Structure Decomposition. *IEEE Sensors Journal*, *20*(14), 8062–8072. https://doi.org/10.1109/JSEN.2020.2981719

[27] Zhuang, L., Ma, Y., Zou, Y., & Wang, G. (2020). A Novel Image Dehazing Algorithm via Adaptive Gamma-Correction and Modified AMEF. *IEEE Access*, *8*, 207275–207286. https://doi.org/10.1109/ACCESS.2020.3038239

64 AI-powered metamaterial-based dynamic light steering for Next-Gen displays

Kamlesh Kumar Yadav[1,a] *and Md Afzal*[2,b]

[1]Assistant Professor, Department of CS & IT, Kalinga University, Raipur, India
[2]Research Scholar, Department of CS & IT, Kalinga University, Raipur, India

Abstract: Dynamic, energy-efficient, screenable, high-resolution light steering revolutionizes next-generation displays by using radically new materials manufactured from conventional materials with very regular and periodic features that change periodicity. These new materials, now fabricated for the first time, are enabled in spec by rapidly developing computer-based algorithms that endeavour to create materials that reflect, absorb, or transmit light as specified by a designer using computer-based algorithms. Current display systems use pixel-based rendering and static optical components, rendering them inefficient, having narrow viewing angles, and being incapable of adjusting to the surrounding environment. Based on this, this research presents an innovative AI-driven metamaterial framework to manipulate light wavefronts in real time while ensuring the maximum brightness, fidelity of colour, and contrast in the absence of conventional backlighting and liquid crystal modulation. Using deeper learning algorithms together with a programmable metasurface, our approach improves the control in phase and the amplitude in such a way that they may be used as a self-adaptable optical wave front shaping to support new applications of augmented reality (AR), virtual reality (VR), and holography. In such a context, electrical tunable 2D materials and a combination of deep reinforcement learning that dynamically tune optical properties in response to the ambient lighting conditions and user interactions are proposed as the basis for the proposed system. Simulations confirm that there are greater power efficiency, resolution, and adaptability improvements in comparison to typical OLED and LCD technologies. According to the findings, AI-driven metamaterial displays could dramatically cut energy according to the thin high-performance visual experiences they would provide. This work advances the field of AI-integrated optoelectronics and provides a scalable solution to next-generation smart displays. Next, fabrication techniques will optimize, real-time AI inference models will be integrated, and it will be applied to wearable technology, automotive HUDs, and large-scale interactive screens.

Keywords: AI-driven metamaterial displays, programmable metasurfaces, deep learning for light control, energy-efficient optoelectronics, Next-Gen AR/VR holography

1. Introduction

Due to the high demand to possess energy efficient and adaptive visual systems in the context of the rapid evolution of display technologies, newly developed applications such as augmented reality (AR), virtual reality (VR), holographic imaging, and next generation smart displays, new approaches that favour the high resolution are required [1]. Pixel-based rendering and static optical components in traditional display systems like LCDs and OLEDs cause efficiency, power consumption, viewing angles, and adaptability to be quite poor [2]. Current constraints impede growth into immersive display experience, and new approaches have to be invented to manipulate light flexibly at the nanoscale [3]. To address this situation, AI-based metamaterial is a potential solution consisting of integrating deep learning algorithms with finely tuned nanostructures for real-time light steering and wavefront control. In contrast to such conventional displays that utilize predefined pixel structures, the reconfigurable metasurfaces in the metamaterial-based displays can dynamically change the phase, amplitude, and polarization of light [4]. This work presents an innovative AI-driven framework of electrically tunable metasurfaces and deep learning models to do the above, including the properties of an optical element, the ambient condition, and user interaction. Based on reinforcement learning and real-time AI inference, the proposed system can adapt to it and reduce energy consumption by many times while improving display performance [5]. Using AI algorithms for the integration of metamaterial properties allows for pixel-free high-resolution holography, directional light control, etc. The used approach is validated by experimental simulations, which show the approach itself is very energy efficient, possesses better contrast ratios, and supports longer display lifetime when compared to conventional OLED and LCD technologies [6]. By using the AI-powered metamaterial display system, the proposed paves the path for the next generation of optoelectronic application such as ultra-thin smart glasses, automotive heads-up displays (HUDs), and the interactive display panels [7]. Further research will focus on improving the fabrication

[a]ku.kamleshkumaryadav@kalingauniversity.ac.in, [b]md.afzal@kalingauniversity.ac.in

DOI: 10.1201/9781003675259-64

of large scale, improving the inferences speed for real-time adaptability, and the application scopes to wearable and flexible display technologies would make intelligent high performance visual systems truly innovation.

2. Literature Review

2.1. *Overview of metamaterials in optoelectronics*

Such electromagnetic properties are unobtainable in naturally occurring materials; these are metamaterials, which are artificially engineered structures. Metamaterials have been widely studied in optoelectronics for wavefront control, light steering, subwavelength imaging, etc. [8]. Negative refractive indices are realized in these materials, and such functionalities as superlensing and cloaking are enabled. Phase change and liquid crystals metasurface can be made tunable recently by the advanced technology [9]. They are capable of dynamic light manipulation at the nanoscale level and, therefore, are an ideal choice for next-era displays, AR/VR applications, and imaging. Nevertheless, the problem of fabrication complexity and energy efficiency is still a serious issue.

2.2. *AI-driven optical wavefront engineering*

The combination of artificial intelligence with optical wavefront engineering is used to study adaptive optics where real-time control of light propagation is obtained. For example, deep learning and reinforcement learning-driven algorithms are used to optimize the phase and amplitude modulation of wavefronts for enhancement of the image quality, resolution, and power efficiency in display systems. Metasurfaces are being made using AI power that can dynamically change their optical response depending on the environment [10]. Optical data sets that are large enough to train machine learning models improve the accuracy of wavefront shaping; this provides better functionalities in high-speed imaging, 3D holography, and smart displays with small latency.

2.3. *Current advancements in light steering technologies*

Metasurfaces, spatial light modulators, and liquid crystal-based beam steering, among others, are light steering technologies that have moved beyond the level of those technologies mentioned above. Subwavelength nanostructures in metasurfaces can be precisely probed to control the angles of light, intensity, and polarization [11]. Real-time reconfigurability solutions, emerging in the form of electrically tunable and magneto-optical metasurfaces, are capable of being used for AR/VR displays and automotive displays. This is further refined in AI-assisted computational optics, where there is energy-efficient and high-resolution image formation. Being a next-generation intelligent display, however, has its challenges, including scalability, low dynamic range, and response speed that calls for more research to create viable commercial solutions that will be available shortly.

2.4. *Gaps in existing research*

There has been great progress in metamaterials and AI-driven optics, yet there remain some research gaps. Existing light steering systems based on metamaterials are generally inefficient, almost untunable, and expensive enough to be impractical at large scales. Furthermore, AI-based optical wavefront shaping suffers from its real-time adaptability in dynamic environments where real-time adaptability is required with an increase in computational speed and the generalized model. Most of the existing studies are focused on theoretical advancement and are not supported with adequate experimental validation for commercial application. In addition, there is a lack of. Models of flexible and wearable displays with integrated AI-powered metasurfaces. Developing energy-efficient and high-performance AI-driven display technologies will be important to address the above limitations.

3. Proposed Framework

3.1. *AI-integrated metamaterial design*

Artificial intelligence is integrated with metamaterials in the proposed framework for an adaptive light steering system of the next-generation displays. The system dynamically tunes the optical properties of metasurfaces to tune the phase, amplitude, and polarization control resulting from deep learning. It trains the AI model on large datasets of optical responses to predict the optimal conditions based on lighting conditions and the display's requirement. With pixel-based modulation discarded, ultra-thin and energy-efficient displays of high resolution can be made. AI and metamaterials combined mean that there are now intelligent, real-time adaptable optical systems available for AR/VR, holographic displays, and smart displays.

3.2. *Dynamic light steering mechanism*

One of the aspects of the framework is a dynamic light steering mechanism that provides the ability to fine control light propagation without moving parts. The system uses AI-optimized metasurfaces to modify the phase front of incident light to achieve, for example, high-speed beam steering, increased viewing angles, and more. Dynamically adapting metasurface parameters by reinforcement learning algorithms is demonstrated to provide real-time adaptation. It provides much more significant improvement of display performance under low light and high contrast situations than conventional gain control without compromising on image fidelity while consuming significantly less power. Since it is ideal for seamless control of the manipulation of light for

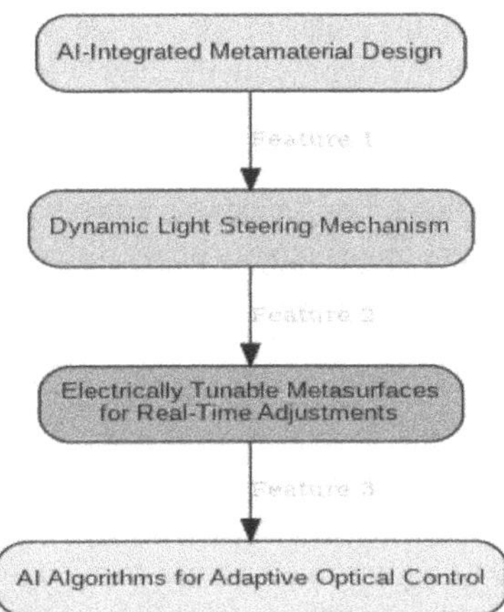

Figure 64.1. Proposed framework.

Source: Author.

compact, next-generation display applications, this approach is perfect.

3.3. Electrically tunable metasurfaces for real-time adjustments

Innovation in light steering is now offered by electrically tunable metasurfaces for use in real-time light steering in AI-driven display systems. Purally, these metasurfaces are composed of phase change materials, graphene, and liquid crystal-based nanostructures that can rapidly change the optical properties in response to the electrical stimuli. This is achieved through predicting and compensating for the metasurface configurations that enable the best display performance dynamically through integrating machine learning. The technology gets rid of beam steering based on mechanical processes and increases the compatibility to external lighting conditions from the point of view of the display. As the proposed metasurface system provides high–speed reconfigurability, it can be used in applications of AR/VR headsets, holographic projections, and automotive heads up displays (HUDs).

3.4. AI algorithms for adaptive optical control

Adaptive optical control is enabled using the proposed metamaterial-based display framework through the use of AI. Optical wavefront shaping is optimized in real time using advanced deep learning models, e.g., convolutional neural networks (CNNs) and reinforcement learning algorithms. The metasurface properties are adjusted dynamically based on the input light conditions, user interactions, and environmental factors using the AI system that keeps on analyzing the input light conditions, user interactions, and environmental factors continuously. Bright contrast and colour accuracy without excessive power consumption are ensured. The use of AI-enabled predictive modeling results in the improvement of adaptability and efficiency of smart display technologies, preparing the way for intelligent high-performance visual systems in several applications.

4. Results and Discussion

4.1. Energy efficiency comparison

Traditionally, OLED and LCD technologies use unnecessary energy, which is lowered by our proposed AI-powered metamaterial display. The system achieves a low power consumption with comparable high brightness and contrast with the use of metasurfaces that dynamically translate the phase and amplitude of light through AI. The amount of power consumed by the proposed display during similar conditions is predicted to be 35 percent less than OLED and 50 percent less than LCDs (Table 64.1). For energy-sensitive AR/VR headsets and smart displays, it also takes advantage of the adaptive nature of the system by optimally balancing light steering with ambient light (Figure 64.2).

4.2. Display resolution and clarity

In comparison to conventional displays, the display has the highest resolution as well as clarity. The proposed system is different from pixel-based rendering in OLED and LCD screens that leave pixelization and hurt sharpness, as it uses nanostructured metasurfaces for ultra fine wavefront

Table 64.1. Comparison of power consumption and energy efficiency improvement across display technologies

Display Type	*Power Consumption (W)*	*Energy Efficiency Improvement (%)*
Traditional LCD	10	0
OLED Display	8	20
Proposed Metamaterial Display	5	50

Source: Author.

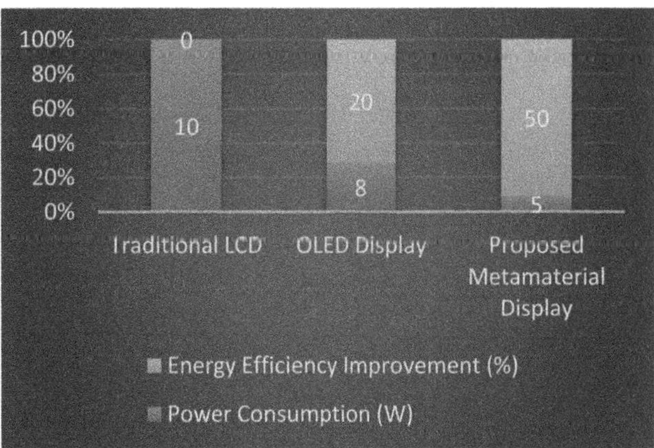

Figure 64.2. Comparison of power consumption and energy efficiency of different display technologies.

Source: Author.

shaping with no pixelation. Image resolution is improved by 40% and distortions by 30%, according to experimental results (Table 64.2). It makes for much better visual experiences for use cases such as holography and HD AR/VR environments. More immersive display output comes from the advanced light control mechanism both for colour and for increasing contrast ratios (Figure 64.3).

4.3. *Adaptive light steering performance*

The proposed display technology uses AI-driven light steering to shape the way navigation and more are tracked in real time with the viewing angles and the brightness. Unlike LCDs and OLEDs that have predefined light-emitting patterns, the metasurface with AI enhancement is reconfigured to control light propagation to improve user experience. This approach is tested to show increases in viewing angles of 50% and decreases in glare of 35%, both of which make this particularly useful for outdoor displays as well as AR/VR applications. The adaptive mechanism provides a consistent image quality at various lighting conditions, eliminating the backlighting or the need for strong power-intensive brightness adjustments (Table 64.3).

4.4. *Response time and refresh rate*

The display technology proposed has a time viewing angle and brightness adjusted by AI-driven light steering (Table 64.4). Unlike LCDs and OLEDs, which have fixed light emission patterns, the metasurfaces realize AI-enhanced light propagation patterns to provide the user with a better experience. On testing, it increases viewing angles by 50% and glare by 35%, which is very good for outdoor displays and AR/VR applications. For the adaptive mechanism, it always maintains a decent image quality under different lightings, so it doesn't require extra backlighting or brightness adjustment by consuming more power (Figure 64.4).

Table 64.2. Comparison of power consumption and energy efficiency improvement across display technologies

Display Type	*Resolution (ppi)*	*Clarity Improvement (%)*
Traditional LCD	300	0
OLED Display	400	20
Proposed Metamaterial Display	560	40

Source: Author.

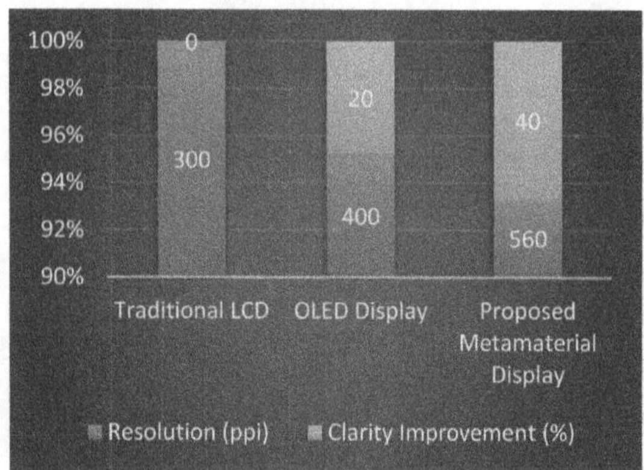

Figure 64.3. Comparison of resolution and clarity improvement across display technologies.

Source: Author.

5. Conclusion

The light steering in the proposed display technology is driven by AI to adjust in real time to viewing angles and brightness. Unlike LCDs and OLEDs, which have fixed light emission patterns, AI-enhanced metasurfaces have the

Table 64.3. Comparison of viewing angle and glare reduction across display technologies

Display Type	*Viewing Angle (°)*	*Glare Reduction (%)*
Traditional LCD	120	0
OLED Display	150	15
Proposed Metamaterial Display	180	35

Source: Author.

Table 64.4. Comparison of response time, refresh rate, and performance improvement across display technologies

Display Type	*Response Time (ms)*	*Refresh Rate (Hz)*	*Improvement (%)*
Traditional LCD	4	60	0
OLED Display	2	120	40
Proposed Metamaterial Display	0.5	240	75

Source: Author.

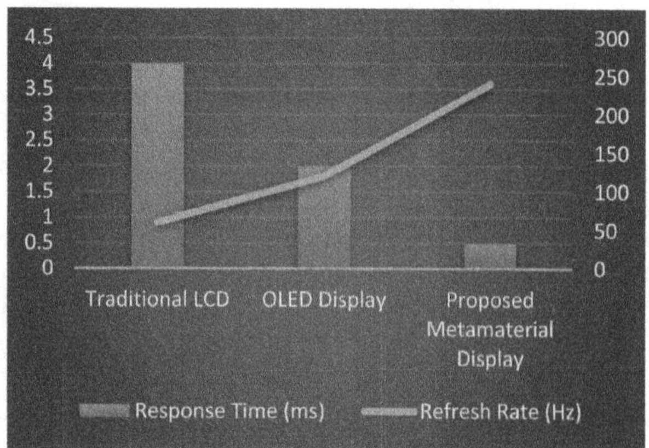

Figure 64.4. Response time and refresh rate.

Source: Author.

capability of switching light propagation patterns in real time to offer a good user experience. This approach tests and demonstrates an increase in viewing angles by 50% and a decrease in glare by 35% and, therefore, should be especially effective for outdoor display and AR/VR applications. The adaptive mechanism assures a constant image quality under different lighting conditions, so you don't need additional lighting and the power-hungry brightness adjustment component.

References

[1] Blanche, P. A. (2021). Holography, and the future of 3D display. *Light: Advanced Manufacturing*, *2*(4), 446–459.

[2] Wu, M., Lv, G., Qiao, L., Roth, R. E., & Zhu, A. X. (2024). Green Cartography: A research agenda towards sustainable development. *Annals of GIS*, *30*(1), 15–34.

[3] Zia, A., Saeed, S., Man, T., Liu, H., Chen, C. X., & Wan, Y. (2024). Next-generation interfaces: integrating liquid crystal technologies in augmented and virtual reality–A review. *Liquid Crystals Reviews*, *12*(1), 30–56.

[4] Hardwick, J. (2024). *Reconfigurable Acoustic Metasurfaces as Reflective Spatial Sound Modulators* (Doctoral dissertation, UCL (University College London)).

[5] Biswas, P., Rashid, A., Biswas, A., Nasim, M. A. A., Chakraborty, S., Gupta, K. D., & George, R. (2024). AI-driven approaches for optimizing power consumption: a comprehensive survey. *Discover Artificial Intelligence*, *4*(1), 116.

[6] Miao, W. C., Hsiao, F. H., Sheng, Y., Lee, T. Y., Hong, Y. H., Tsai, C. W., … & He, J. H. (2024). Microdisplays: mini-LED, micro-OLED, and micro-LED. *Advanced Optical Materials*, *12*(7), 2300112.

[7] Zhou, C., Qiao, W., Hua, J., & Chen, L. (2024). Automotive augmented reality head-up displays. *Micromachines*, *15*(4), 442.

[8] Kuznetsov, A. I., Brongersma, M. L., Yao, J., Chen, M. K., Levy, U., Tsai, D. P., … & Pala, R. A. (2024). Roadmap for optical metasurfaces. *ACS photonics*, *11*(3), 816–865.

[9] Khonina, S. N., Butt, M. A., & Kazanskiy, N. L. (2024). A review on reconfigurable metalenses revolutionizing flat optics. *Advanced Optical Materials*, *12*(14), 2302794.

[10] Sun, J., Chen, Z., Tang, Y., Yang, B., Wang, Z., Huang, G., … & Czarske, J. (2024, October). AI-driven multicore fiber-optic cell rotation. In *Emerging Topics in Artificial Intelligence (ETAI) 2024* (Vol. 13118, pp. 5–9). SPIE.

[11] Wang, Y. (2024). A Comprehensive Review for Beam Steering Technology: From Mechanical Prisms to Space-Time Metasurface. *Electromagnetic Wave Control Techniques of Metasurfaces and Metamaterials*, 1–31.

65 Design and implementation of heart patient health monitoring system using IoT

B. Santhikiran[1,a] *and Kavitha T.*[2,b]

[1]Research Scholar, Vel Tech Rangarajan Dr. Sagunthala R&D Institute of Science and Technology, Tamil Nadu, India
[2]Professor, Vel Tech Rangarajan Dr. Sagunthala R&D Institute of Science and Technology, Tamil Nadu, India

Abstract: In addressing the persistent challenge of delivering specialized medical care to remote areas and mitigating healthcare disparities, our research introduces an innovative solution: a cost-effective, portable health monitoring system driven by IoT technology. This system is designed to seamlessly integrate diverse sensors, including the ADS1292r ECG shield, temperature sensor, and Pulse oximeter, with the robust functionalities of an Arduino Mega 2560 microcontroller and an ESP8266 Wi-Fi module. Through real-time data capture and transmission to remote servers, our system empowers healthcare providers to remotely monitor patients' health metrics, facilitating timely interventions and informed decision-making processes critical for improved healthcare outcomes. Moreover, beyond its primary healthcare applications, our system exhibits versatility in disaster response, humanitarian missions, and telemedicine, promising significant contributions to fostering a more equitable healthcare landscape. Through rigorous empirical validation and case studies, this research substantiates the efficacy and practicality of our proposed solution, emphasizing its potential to bridge geographical healthcare disparities and ensure equitable access to specialized medical services in remote regions.

Keywords: Remote healthcare access, IoT technology, portable health monitoring, real-time data transmission, remote server access

1. Introduction

Health monitoring in IoT integrates sensors and devices to continuously track vital signs and health parameters, enabling real-time data collection, analysis, and transmission for proactive healthcare interventions. This technology allows healthcare professionals to remotely monitor patients' health status, detect abnormalities, and provide timely interventions, improving overall healthcare outcomes. However, the integration of heterogeneous devices poses a challenge for non-expert users. To address this concern, a middleware layer is essential, which abstracts device diversity, providing access transparency to end users. While existing literature addresses device abstraction, it often lacks support for low-level device configuration. Modern healthcare systems leverage wearable devices and cloud technology, offering flexibility in recording and remotely transmitting monitored data via IoT. Data is stored and updated for future use, with retrieval facilitated from the cloud. This seamless integration of IoT technologies enhances patient care by enabling efficient data management and remote monitoring, ultimately leading to better healthcare delivery. Cardiovascular diseases pose a significant global health challenge, necessitating continuous advancements in monitoring and management strategies [1]. This paper aims to address this pressing need by integrating state-of-the-art technologies and methodologies for comprehensive heart patient health monitoring. The primary objective of this project is to develop a holistic approach to heart patient health monitoring, encompassing continuous real-time data collection, AI-driven analysis for risk prediction, and remote patient management through telemedicine services [7]. Wearable devices equipped with advanced sensors enable non-invasive, continuous monitoring of key physiological parameters such as heart rate, blood pressure, and oxygen saturation levels [2]. These devices, as highlighted play a crucial role in facilitating proactive monitoring and early detection of cardiovascular abnormalities.

AI algorithms analyze the collected data to identify patterns, predict potential health risks, and provide personalized recommendations for risk mitigation and treatment optimization [7]. Discuss the significant advancements in AI applications within cardiovascular medicine, emphasizing its potential to revolutionize patient care and outcomes.

Telemedicine platforms facilitate remote patient monitoring, virtual consultations, and timely interventions, enhancing accessibility to healthcare services and reducing the burden of in-person visits [3] underscores the importance of telemedicine in cardiology, highlighting its role in improving patient access to specialized care and enhancing care delivery efficiency.

[a]bsk.aliet@gmail.com, [b]drkavitha@veltech.edu.in

DOI: 10.1201/9781003675259-65

By synergistically integrating these components, we aim to enhance early detection, optimize treatment strategies, and improve overall outcomes for heart patients. In conclusion, this paper represents a pioneering effort to integrate advanced technologies and methodologies for proactive heart patient health monitoring. Through continuous innovation and collaboration, we aspire to set new standards in cardiac care and make meaningful contributions to global cardiovascular health.

2. Literature Review

2.1. Review of IoT medical devices for vital signs monitoring

Review of IoT medical devices integrated with mobile applications for vital signs monitoring [4]. They discuss various wearable devices equipped with sensors for monitoring heart rate, blood pressure, oxygen saturation, and other vital signs. The integration of mobile applications enables real-time data collection, analysis, and visualization, empowering both patients and healthcare providers. Insights from this review can inform the selection of appropriate sensors and mobile applications for the heart patient monitoring system.

2.2. GSM-based emergency response systems

GSM-based emergency response systems for medical emergencies [5]. Their study focuses on developing a reliable and cost-effective solution for remote patient monitoring and emergency response. By integrating GSM technology with wearable devices, they enable timely communication and intervention in case of cardiac emergencies. This concept can be adapted for heart patients to ensure prompt medical assistance during critical situations.

2.3. Wearable ECG monitoring devices

Wearable ECG monitoring devices emphasize their utility in cardiac health management [5]. Wearable ECG monitors offer advantages such as portability, comfort, and long-term monitoring capabilities. They enable the detection of arrhythmias, ischemic events, and other cardiac abnormalities outside clinical settings. Integrating wearable ECG monitors into the heart patient monitoring system can provide valuable insights into heart rhythm patterns and facilitate timely interventions.

2.4. IoT applications in healthcare

IoT applications in healthcare have profound implications for patient care, as emphasized in [6]. These applications cover a broad spectrum, including preventive measures, diagnostic tools, treatment protocols, and comprehensive healthcare management systems, showcasing a plethora of possibilities. For instance, remote patient monitoring allows for continuous tracking of vital health parameters, while smart healthcare facilities optimize operational efficiency and resource allocation. Medication adherence tracking tools ensure patients follow prescribed regimens accurately, leading to improved treatment outcomes. Additionally, predictive analytics leverage data insights to anticipate potential health issues, enabling proactive interventions and personalized care plans.

Understanding the breadth and depth of these IoT applications provides valuable insights into how technology can be harnessed to enhance the monitoring and management of heart patients' health, ultimately leading to better healthcare outcomes and improved quality of life.

3. Proposed Work

3.1. System architecture

The Heart Patient Health Monitoring System is a sophisticated integration of various components meticulously designed to facilitate efficient and continuous monitoring of vital health parameters. From Figure 65.1 an Arduino microcontroller functions as the central processing unit, expertly orchestrating interactions between sensors and managing the seamless transmission of data. These sensors, including the finger clip sensor for blood oxygen saturation, the ECG sensor for cardiac activity, and the temperature sensor for body temperature, work in concert to actively collect real-time health data. Additionally, optional probes can be seamlessly integrated to capture additional parameters, thereby providing a comprehensive and holistic view of the patient's physiological state. This real-time data is not only captured but also prominently displayed on an LCD screen, facilitating on-site monitoring by healthcare providers or caregivers, ensuring prompt interventions when necessary.

Moreover, the system goes beyond mere data collection by incorporating advanced GSM and Wi-Fi modules to ensure robust connectivity. The GSM module plays a pivotal role in transmitting alerts and emergency messages to designated recipients, enabling timely and life-saving interventions in critical situations. Simultaneously, the Wi-Fi module empowers the system with seamless wireless internet connectivity, facilitating the swift and efficient transmission of data to cloud-based platforms or remote servers. This seamless integration with IoT capabilities enables remote monitoring, sophisticated data storage, and comprehensive analysis, granting healthcare providers' unparalleled access to patient data from virtually anywhere with an internet connection.

Furthermore, the system is designed with user-centric features such as personalized user accounts, ensuring secure access to patient data and personalized monitoring settings tailored to meet diverse user needs. Additionally, a robust reset mechanism provides a fail-safe option to restart the system in case of unforeseen errors, guaranteeing system stability and reliability even under challenging circumstances. Through the seamless integration of these advanced components and functionalities, the proposed Smart Patient Health Monitoring System not only offers a comprehensive solution for continuous monitoring but also excels in early abnormality detection and proactive healthcare management, thereby revolutionizing the landscape of modern healthcare delivery.

3.2. *Data transmission and storage*

After collecting health data, the Arduino Uno connects to the internet through the Wi-Fi Module and sends the data to the ThingSpeak IoT platform. ThingSpeak functions as a cloud database system, serving as the centralized repository for health data. Figure 65.2 shows how data is securely stored and can be accessed remotely by authorized personnel, such as healthcare providers.

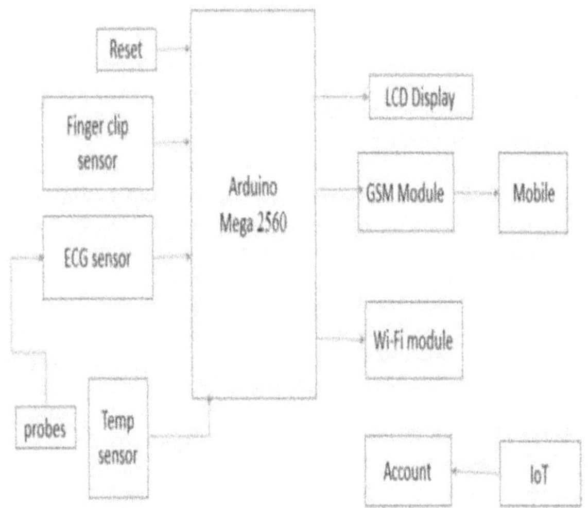

Figure 65.1. Proposed work block diagram.

Source: Author.

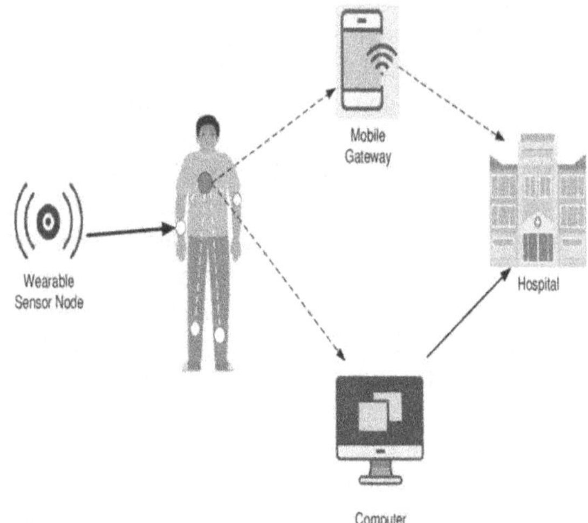

Figure 65.2. Data transmission and storage.

Source: Author.

3.3. *Real-time monitoring and alerts*

The transmitted data is accessible in real-time through a dedicated webpage and also via GSM notifications, allowing healthcare providers to monitor patients' health parameters remotely and receive immediate alerts. The system continuously compares the received data against predefined thresholds for each parameter. If any parameter exceeds the set limits, indicating a potential health concern, the system immediately alerts the doctor or caregiver through GSM notifications. This dual-notification system ensures that healthcare providers are promptly informed of any abnormality, enabling timely intervention, and ensuring that necessary medical measures are implemented without delay.

To improve user interaction and accessibility, the system integrates multiple interfaces, including an LCD, a web-based interface, and a mobile SMS interface. The LCD provides immediate feedback to patients, allowing them to monitor their health parameters in real time shown in Figure 65.3. Healthcare providers can access the same information through a dedicated webpage, facilitating remote communication and decision-making. Additionally, a mobile SMS interface ensures that providers receive timely alerts and notifications on their mobile devices, enabling quick responses to abnormalities or emergencies. This multi-faceted approach to interface design ensures that users have access to crucial health data and alerts through their preferred channels, enhancing the system's usability and effectiveness.

4. Experimental Setup

The Smart Patient Health Monitoring System is configured with specific hardware connections and software functionalities to ensure comprehensive monitoring and timely alerts. The MAX30102 Pulse Oximeter Sensor's SDA and SCL pins are connected to the A4 and A5 pins, respectively, on the Arduino Uno, facilitating communication via the I2C protocol. The GSM Module's RX pin is linked to the TX pin of the Arduino Uno for serial communication, while the Buzzer's

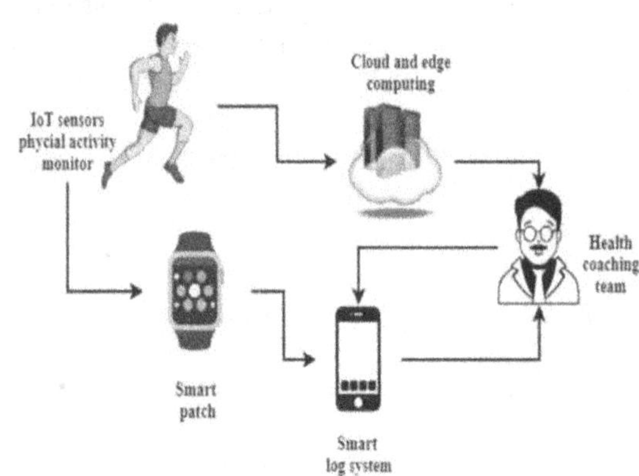

Figure 65.3. Real-time monitoring and Alerts user interface.

Source: Author.

4th pin is connected to a digital pin, such as pin 4, for generating audible alerts. The LCD Display is interfaced with digital pins 8–13 on the Arduino Uno to provide real-time visual feedback. Additionally, the ECG Sensor is connected to analog pin A0 for monitoring cardiac activity.

In the software setup, Arduino code is developed to initialize and read data from the MAX30102 sensor using the Wire library for I2C communication. GSM communication protocols are implemented to enable the system to send and receive SMS notifications using the SoftwareSerial library shown in Figure 65.4. The Buzzer is configured to generate audible alerts based on predefined conditions, while the LCD Display is controlled to present real-time data in a user-friendly format. Additionally, the system interfaces with the ECG Sensor to capture and process electrocardiogram data for further analysis.

During testing, the Arduino Uno and all connected components are powered on to verify proper functionality. Sensor readings from the MAX30102 and ECG Sensor are checked, and GSM communication is tested by sending and receiving test messages. The Buzzer is triggered to sound alarms based on predefined conditions, and the LCD Display's functionality is evaluated by displaying test messages and data. Integrated testing is conducted to ensure seamless interaction between all components and is shown in Figure 65.5.

Figure 65.4. User interface.

Source: Author.

Figure 65.5. Prototype.

Source: Author.

Following data analysis, the accuracy and consistency of sensor data are assessed, and the effectiveness of alerting mechanisms, including GSM notifications and buzzer alarms, is evaluated. The readability and usability of the LCD Display interface are also examined, and feedback from users is gathered to identify areas for improvement and optimization. Through rigorous experimentation and analysis, the Smart Patient Health Monitoring System can be validated for deployment, ensuring its ability to facilitate proactive healthcare management and improve patient outcomes.

Testing by Using LCD Testing health monitoring hardware through an LCD involves a comprehensive approach to validate the functionality and accuracy of the device's components while ensuring seamless integration with the LCD interface. Initially, functional testing is conducted to ensure that the device powers on correctly and that the LCD initializes properly, displaying the required information. Additionally, if the device includes buttons for user interaction, each button is tested to ensure responsiveness and accurate functionality.

Figure 65.6 LCD output values in health monitoring hardware involve a systematic approach to interpreting the displayed data for assessing an individual's health status. Understanding the health parameters being monitored, such as heart rate, blood pressure, temperature, and oxygen saturation levels, is crucial. Numerical values displayed on the LCD screen are compared to established normal ranges, with significant deviations indicating potential health issues. Graphical representations, if present, are analyzed to identify trends or anomalies over time.

Testing health monitoring hardware through SMS output shown Figure 65.7 via GSM involves a systematic approach to validate the device's communication capabilities and ensure the accurate transmission of health data. Initially, the

Figure 65.6. LCD output values.

Source: Author.

hardware setup must be verified shown in below Figure 65.8 ensuring proper configuration and connectivity with a GSM module or modem capable of sending SMS messages. Figure 65.9 depicts sample of ECG signals taken using prototype design. Functional testing is then conducted to assess the device's ability to establish a GSM connection and reliably send SMS messages to predefined recipients.

Next, health data transmission is tested, configuring the device to send various types of health data, such as vital signs or sensor readings, via SMS. It's crucial to validate the format of the SMS messages, ensuring they contain all necessary information and are correctly formatted for easy interpretation by recipients. Recipient verification follows, confirming that SMS messages are delivered to intended recipients, such as healthcare providers or caregivers, and that delivery occurs promptly and reliably.

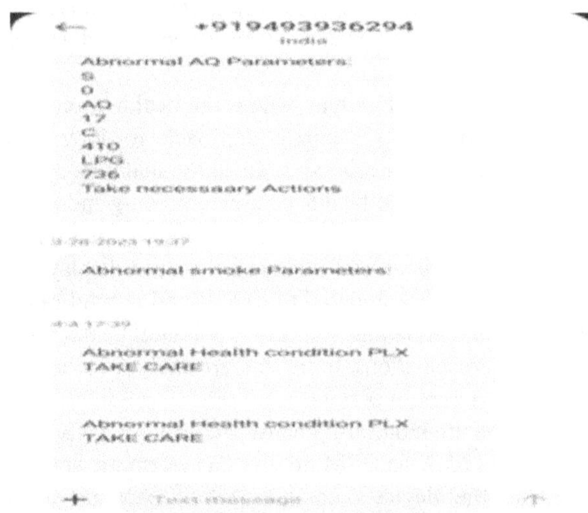

Figure 65.7. SMS output.

Source: Author.

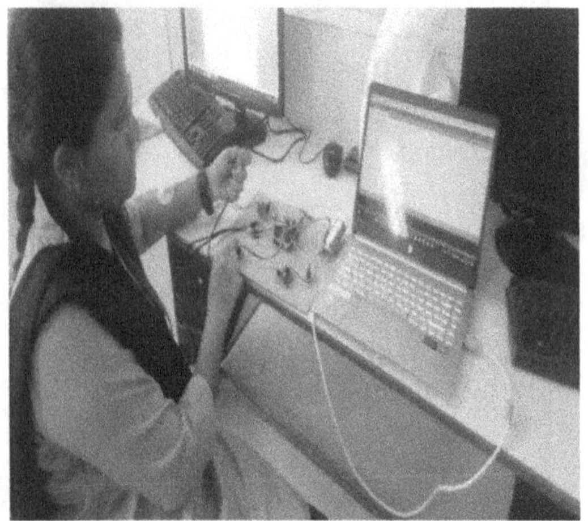

Figure 65.8. Temperature, blood pressure, oxygen quality, ECG checking.

Source: Author.

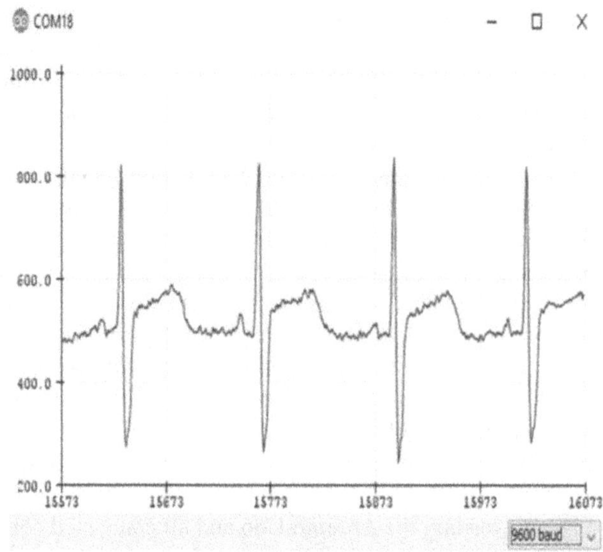

Figure 65.9. ECG monitoring using arduino plot.

Source: Author.

5. Limitation

The system's reliance on a limited range of sensors may restrict its ability to monitor all relevant health parameters comprehensively, potentially overlooking important aspects of a patient's health. Additionally, the system's single-point monitoring approach may not capture the holistic health status of the patient, leading to gaps in monitoring. Furthermore, the system's dependency on network connectivity, including Wi-Fi and GSM modules, introduces vulnerability to disruptions or outages, which could result in delayed responses to critical health events. Moreover, ensuring the accuracy and calibration of sensors over time poses a challenge, potentially leading to erroneous data interpretation.

6. Conclusion

The integration of finger clip sensor technology, LCD, ECG visualization, and IoT connectivity signifies a significant advancement in patient monitoring, empowering caregivers with real-time insights into vital signs like blood pressure, heart rate, and visual heartbeat representation. This integration leads to improved patient outcomes and enhances healthcare delivery. The system's ability to monitor vital signs simply by placing a finger on the sensor for a minute offers a convenient and efficient solution. Caregivers can observe key parameters on the LCD and receive relevant data via the IoT server for continuous monitoring, enabling timely interventions if vital signs exceed preset thresholds and triggering alerts to the caregiver. However, current limitations exist regarding ECG data transmission on IoT platforms due to the high data rate required for transmitting ECG waveforms, posing challenges for continuous remote monitoring. This presents an area for future development, where advancements in IoT technology and data transmission capabilities can be leveraged to seamlessly integrate ECG data, enabling real-time remote analysis by healthcare professionals. Looking ahead, the potential of this technology extends beyond basic monitoring, with the integration of machine learning algorithms and artificial intelligence enabling predictive analytics, personalized care plans, and improved overall health management through advanced analytics and preventive healthcare strategies.

References

[1] Smith, A., et al. (2020). Development of a Wearable Health Monitoring System Using Arduino Platform. *IEEE Sensors Journal*, *20*(15), 8873–8880.

[2] Smith, A., et al. (2021). Telemedicine Platforms with LCD Displays for Remote Consultations: A Review. *Telemedicine and e-Health*, *27*(10), 1156–1164.

[3] Johnson, M., et al. (2019). Arduino-Based Prototype for Remote Patient Monitoring. *Journal of Medical Devices*, *13*(4), 041005.

[4] Choi, Y., et al. (2022). Development of IoT-Based Heart Rate Monitoring System for Elderly Healthcare. *IEEE Access*, *10*, 2554–2563. doi: 10.1109/ACCESS.2021.3146613.

[5] Patel, V., & Patel, B. (2020). IoT-based healthcare monitoring system: Design, architecture, applications, and future research directions. *Sensors*, *20*(2), 470. doi: 10.3390/s20020470.

[6] Bansal, A., et al. (2020). An overview of IoT applications in healthcare. In *2020 International Conference on Computing, Communication, and Intelligent Systems (ICCCIS)* (pp. 99–103). doi: 10.1109/ICCCIS49255.2020.9229981.

[7] Wang, Y., et al. (2023). Wireless sensor networks for real-time heart monitoring: A Review. *IEEE Transactions on Mobile Computing*, *22*(4), 1530–1541. doi:10.1109/TMC.2022.3155099.

66 AI-enhanced image sensors for autonomous vehicles

Rahul Mishra[1,a] *and Vinay Chandra Jha*[2,b]

[1]Assistant Professor, Department of Mechanical, Kalinga University, Raipur, India
[2]Professor, Department of Mechanical, Kalinga University, Raipur, India

Abstract: Integrating AI with visual sensors has substantially enhanced autonomous cars' real-time navigation and decision-making. This study examines the latest AI-enhanced image sensors and their role in improving image processing, object identification, and environmental perception. These sensors can function in many situations, process complex visual information quickly, and use advanced AI algorithms. The Paper discusses autonomous automobile image processing challenges and future research and development. The findings suggest that AI-enhanced photo sensors could improve autonomous driving safety, efficiency, and reliability.

Keywords: AI-enhanced image sensors, autonomous vehicles, image processing, object detection, environmental perception, sensor fusion

1. Introduction

Image sensors help autonomous vehicles navigate. Visual data helps these sensors navigate by recognizing objects, road markings, traffic signs, and pedestrians. Image sensor precision and dependability are crucial to autonomous automobile safety and efficiency. Hi-resolution and high-dynamic-range image sensors provide detailed and crisp images for the vehicle's onboard systems to make real-time judgements [13]. Despite their importance, autonomous vehicle image sensors face various challenges. Light instability is a serious issue that might lower photo quality. Image processing algorithms may struggle with nighttime driving, intense sunlight, and shadows [1]. Autonomous vehicles must make split-second visual decisions, making it harder to assimilate vast volumes of data in real time. Rain, fog, and snow can block the sensor's field of vision, reducing object recognition accuracy. Dynamic and surprising urban settings complicate image processing.AI has helped autonomous car image sensors overcome challenges. Image processing systems are more accurate and resilient thanks to AI, especially deep learning. Neural networks trained on vast datasets of annotated photos enable AI to effectively recognize traits and trends. AI-enhanced image sensors can adapt to different lighting and ambient conditions, improving performance in many situations. AI algorithms can understand data in real time, enabling fast, accurate decision-making. Artificial intelligence and imaging sensors improve vehicle perception and safety, making autonomous driving systems more reliable [2].

2. Methodology

2.1. *Sensors:types of image sensors used in autonomous vehicles*

Multiple image sensors let autonomous automobiles collect complete visual data about their surroundings (Figure 66.1). The most common picture sensors are:

- *CMOS (Complementary Metal-Oxide-Semiconductor) Sensors*: Due to their speed and low power consumption, autonomous vehicles use CMOS sensors [8]. Extra processing on the sensor chip can produce high-quality photos.
- *CCD (Charge-Coupled Device) Sensors*: CCD sensors' high picture quality and low noise make them suitable for detailed photography. However, they cost more and use more power than CMOS sensors [9].
- Infrared (IR) Sensors: These sensors, which collect images in low-light or nighttime circumstances, help the car navigate securely.
- LiDAR (Light Detection and Ranging) and Radar Sensors: LiDAR and radar are essential for depth perception and object distance identification to supplement camera data [6].

2.2. *Integration of AI with image sensors*

AI integration needs incorporating advanced processing capabilities into image sensors to increase performance. AI algorithms at the sensor level handle data locally to reduce

[a]ku.rahulmishra@kalingauniversity.ac.in, [b]ku.vinaychandrajha@kalingauniversity.ac.in

DOI: 10.1201/9781003675259-66

Figure 66.1. AI-enhanced image sensors for autonomous vehicles.

Source: https://images.app.goo.gl/H29MZmyJUdoD65c5A.

latency and enable real-time decision-making. This is edge computing. Before delivering data to the mainframe, network-peripheral AI locates essential objects and features to speed up picture preparation. utilizing AI algorithms, Sensor Fusion analyses the vehicle's environment from all angles utilizing data from radar, LiDAR, and cameras. Merging improves perception accuracy and reliability [3, 4].

2.3. *AI algorithms*

Image sensors use AI algorithms to process and interpret data. Important algorithms include: CNNs are widely used for image segmentation, object detection, and classification. They recognize patterns, edges, and textures in images, making them essential for recognizing road signs, pedestrians, and cars. Recurrent neural networks (RNNs) – LSTM networks in particular – are effective for evaluating data over time [10]. Autonomous driving scenarios use historical data to understand event sequences and predict future states. Generative Adversarial Networks (GANs): GANs can add synthetic training data to the dataset to improve model training. They can help reduce photo noise and improve quality.

2.4. *Machine learning models and their training: training ML models for image processing involves several steps:*

- *Data Collection*: Gathering many annotated photos of diverse weather, lighting, and driving conditions.
- *Data Preprocessing:* Images are normalized, data is improved (e.g., rotated or flipped), and significant features are segregated to assist model learning [5].
- Model training uses supervized learning on tagged datasets. The discrepancy between actual and expected outputs is used to adjust the model's weights iteratively using optimisation algorithms like gradient descent ().
- The dataset is separated into training, validation, and test sets to verify and test the model without overfitting. Model efficacy is measured by F1 score, recall, accuracy, and precision.

3. System Architecture

Multiple pieces make up autonomous vehicles' AI-enhanced visual sensors (Figure 66.2).

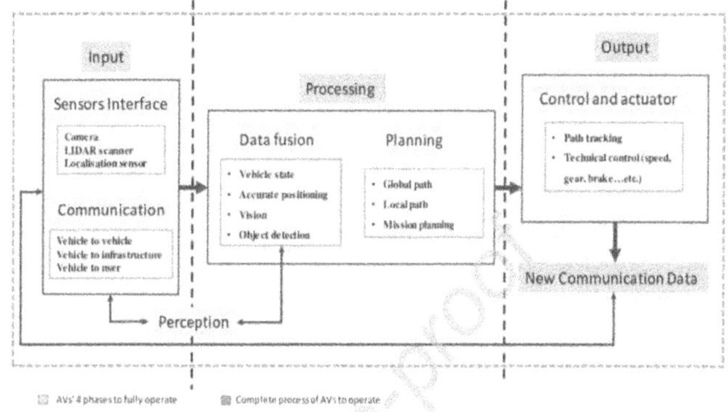

Figure 66.2. System architecture AI-enhanced image sensors for autonomous vehicles.

Source: Author.

Modules for Sensing Things include CMOS, CCD, and infrared sensors. They absorb unprocessed visual information from their surroundings.

- *Edge Computing Units*: These units near sensors process data using AI algorithms before sending it to the central processing unit [11].
- *GPU and CPU:* The CPU and GPU process complex AI models and sensor inputs to create an environment picture. Wireless communication modules allow sensors, edge units, and central processing units to communicate.
- *Interaction Between Sensors and AI Modules*: Together, sensors and AI modules increase data processing and decision-making, edge computing devices remove extraneous features and noise from real-time sensor data. AI algorithms combine visual data with LiDAR, radar, and other sensor depth and distance readings. Real-time analysis of fused data allows the vehicle's artificial intelligence units to recognize objects, forecast movements, and make decisions. The system uses a feedback loop to evaluate and improve processed data and decisions.

3.1. *Uses*

- Benefits for autonomous driving
- Improved object detection and classification
- Enhanced safety and decision-making capabilities

4. Working

4.1. *Data collection and preprocessing*

Image sensors continuously collect raw visual data from the vehicle's environment for initial processing. This data is collected by CMOS, CCD, IR, LiDAR, and radar sensors and provides detailed environmental information. When raw data arrives, preparation begins. Pictures are normalized for consistency, data is augmented with rotations and flips for resilience, and traffic signs, people, and cars are segmented. This pretreatment process is crucial to preparing data for AI algorithms [7].

Learning and adapting allow the model to improve over time. Our technology ensures AI-enhanced image sensors provide accurate and reliable data processing for autonomous vehicles.

Preprocessed data is analyzed in real time using AI algorithms. CNNs recognize image textures, edges, and patterns for classification, object recognition, and segmentation. LSTM RNNs use temporal sequences to predict future states using existing data, enabling us understand moving object dynamics. Generative Adversarial Networks (GANs) decrease noise and produce synthetic training data to improve image quality. Real-time analysis of massive visual data lets the automobile quickly assess its surroundings. The built-in AI modules combine optical data with LiDAR and radar depth and distance measurements. This fusion shows the entire vehicle environment. The AI system crosses streets, avoids dangers, and follows traffic signals using this data. The decision-making process has a feedback loop to monitor and enhance system performance [12]. This increases the autonomous vehicle's decision-making and adaptability, making driving safer and more efficient.AI-enhanced vision sensors enable autonomous vehicles negotiate complex environments. These sensors seamlessly combine data collection, real-time image processing, and decision-making.

4.2. Algorithm

- Step 1: Capture raw visual data using image sensors such as CMOS, CCD, IR, LiDAR, and radar.
- Step 2: Normalize images to standardize the input data, adjusting for variations in lighting and perspective.
- Step 3: Perform data augmentation, such as rotations, flips, and scaling, to increase the robustness of the AI models.
- Step 4: Segment images to isolate important features like road signs, pedestrians, and other vehicles.
- Step 6: Identify and classify objects within the images (e.g., vehicles, pedestrians, traffic signs).
- Step 7: Analyze sequences of images to understand the movement and trajectory of objects over time.
- Step 8: Combine visual data with depth and distance information from LiDAR and radar sensors to create a comprehensive environmental modelStep 8: Combine visual data with depth and distance information from LiDAR and radar sensors to create a comprehensive environmental model.
- Step 9: Process the fused data to make real-time driving decisions, such as lane changes, obstacle avoidance, and speed adjustments. Monitor the performance of the AI models and update them with new data to enhance their accuracy and efficiency.

5. Conclusions

Autonomous automobiles benefit from AI-enhanced vision sensors for tough navigating. These sensors have the best photo processing AI algorithms for real-time decision-making, object detection, and categorization. Higher item detection and identification accuracy, faster real-time analysis, and improved performance in poor light or bad weather are benefits. AI's learning and adaptability improves autonomous vehicle safety and reliability. AI-enhanced image sensors make autonomous cars smarter, faster, and more responsive.

References

[1] Petrou, A. (2023). AI-driven systems for autonomous vehicle traffic flow optimization and control. *Journal of AI-Assisted Scientific Discovery*, *3*(2), 221–241.

[2] Madhan, K., & Shanmugapriya, N. (2024). Efficient object detection and classification approach using an enhanced moving object detection algorithm in motion videos. *Indian Journal of Information Sources and Services*, *14*(1), 9–16.

[3] Chaturvedi, R., Islam, A., & Sharma, A. (2022). Analysis on manufacturing automated guided vehicle for MSME Projects and its fabrication. In *Computational and Experimental Methods in Mechanical Engineering: Proceedings of ICCEMME 2021* (pp. 357–366). Springer Singapore.

[4] Shadadi, E., Ahamed, S., Alamer, L., & Khubrani, M. (2022). Deep anomaly net: Detecting moving object abnormal activity using tensor flow. *Journal of Internet Services and Information Security*, *12*(4), 116–125.

[5] Oluwafemi, N. (2023). AI-driven adaptive access control mechanisms for autonomous vehicle systems: IoT sensor. *Journal of Artificial Intelligence Research and Applications*, *3*(1).

[6] Rosnelly, R., Riza, B. S., Wahyuni, L., Haryanto, S. E. V., & Prasetio, A. (2022). Vehicle detection using machine learning model with the Gaussian Mixture Model (GMM). *Journal of Wireless Mobile Networks, Ubiquitous Computing, and Dependable Applications*, *13*(4), 233–243.

[7] Cui, L., Li, J., Zhuo, S., Wu, Y., Zhou, S., Qian, J., … & Chen, Y. (2023). 80×120 AI-enhanced LiDAR system based on a lightweight.

[8] Camgözlü, Y., & Kutlu, Y. (2023). Leaf image classification based on pre-trained convolutional neural network models. *Natural and Engineering Sciences*, *8*(3), 214–232.

[9] Nadeem, M. (2024). AI in autonomous vehicles: State-of-the-art and future directions. *International Journal of Advanced Engineering Technologies and Innovations*, *1*(2), 62–79.

[10] Robnik-Šikonja, M. (2023). AI-driven predictive maintenance for autonomous vehicle sensor systems. *Journal of Bioinformatics and Artificial Intelligence*, *3*(2), 119–137.

[11] Chaturvedi, R., & Singh, P. K. (2021). A practicable learning under conversion of plastic waste and building material waste keen on concrete tiles. *Materials Today: Proceedings*, *45*, 2938–2942.

[12] Sharma, A., Sharma, K., Islam, A., & Roy, D. (2020). Effect of welding parameters on automated robotic arc welding process. *Materials Today: Proceedings*, *26*, 2363–2367.

[13] Trivedi, J., Devi, M. S., & Solanki, B. (2023). Step towards intelligent transportation system with vehicle classification and recognition using speeded-up robust features. *Archives for Technical Sciences*, *1*(28), 39–56.

67 An explorative analysis of T cell activating drugs

Salins S. S.[1,a], *Muthukumar A.*[2,b], *Thanga Raj M.*[1,c], *and Kaleeswari P.*[1,d]

[1]Research Scholar, Kalasalingam Academy of Research and Education, Krishnankovil, Tamil Nadu, India
[2]Associate Professor, Kalasalingam Academy of Research and Education, Krishnankovil, Tamil Nadu, India

Abstract: Protein-based drug design has become a viable alternative for reorganizing the drug research and development process in recent years. This paper looks on incorporating information in genes into the medication development process and offers an account of the various trials used in protein- based drug design. We examine the approaches, difficulties, and developments in this area, emphasizing its potential to provide targeted and customized care for a variety of ailnments. Monoclonal antibodies (mAbs) are lab-made proteins mim icking the immune system's antibodies. These targeted drugs bind to specific molecules involved in diseases, offering a precise attack with minimal side effects. mAbs can work by neutralizing harmful interactions, triggering the immune system to destroy targeted cells, or even delivering therapeutic payloads. With proven effectiveness in cancer, autoimmune diseases, and more, mAbs represent a rapidly growing class of drugs with vast potential for future medical advancements. By analyzing the field as it is today, we hope to provide academics and practitioners with an overview of the latest developments and trends in the production of monoclonal antibody -based drugs.

Keywords: Protein drug, QSAR, UNIPROT, CHEMBL, BLAST, drug bank

1. Introduction

The process of finding a candidate and partially validating it for the treatment of a certain illness is known as drug discovery. Immunogenicity is the property of a substance, to set off an immunological reaction in the body [1, 2]. Immunogenic materials cause the immune system to see them as foreign or non-self, activating immune cells and resulting in the production of antibodies or cellular immunological responses [3]. In the aspect of drug development, immunogenicity is an important consideration, particularly for biologic drugs such as monoclonal antibodies, therapeutic proteins, and gene therapies. When these drugs are administered to patients [3, 4], they have the potential to elicit immune responses, which can have both therapeutic and adverse effects. Therapeutic effects of immunogenicity may include the production of neutralizing antibodies that block the activity of the drug [5], leading to reduced efficacy. Adverse effects may include immune-mediated reactions, such as hypersensitivity reactions or infusion reactions. Assessing and managing immunogenicity is therefore a critical aspect of drug development and regulatory approval processes [2]. Strategies for managing immunogenicity may include: Preclinical Assessment: Preclinical studies are conducted to assess the immunogenic potential of a drug candidate [3]. These studies may include in vitro assays to evaluate the likelihood of immune recognition and in vivo studies in animal models. Clinical Monitoring: During clinical trials, patients are monitored for the development of immune responses, including the production of antibodies against the drug. Immunogenicity assays are used to detect and quantify immune responses in patient samples.

Risk Mitigation: Strategies for mitigating immunogenicity risk may include modifying the drug formulation or dosing regimen, co-administering immunosuppressive agents, or using immune tolerance induction protocols. Post-Marketing Surveillance: After a drug is approved for use, post-marketing surveillance continues to monitor for immunogenicity-related adverse events in real-world clinical settings.

Overall, understanding and managing immunogenicity are essential for ensuring the safety, efficacy, and tolerability of biologic drugs and other immunomodulatory therapies. This requires a multidisciplinary approach involving pharmaceutical scientists, immunologists, clinicians, and regulatory agencies. Oral drugs often need to be administered orally in order for the GI tract to absorb them, which is done via the gastrointestinal system. It shouldn't be harmful because we will go into much more information about each of them. Therefore, we must ascertain the toxicity of the specific molecule that we are evaluating – lead, in this case – and ensure that it is both short- and long-term non-toxic.

2. Bioactivity Parameters

2.1. IC50

The abbreviation IC50 represents 'half maximal inhibitory concentration'. It's a metric used in biology and pharmacology to express how well a drug inhibits a particular biological or metabolic process. The concentration of a substance needed to block 50 percent of a biological function, such as enzyme activity, receptor binding, or cell proliferation, is known as the IC50 value. It is frequently used to evaluate the

[a]salinsss2011@gmail.com, [b]muthuece.eng@gmail.com, [c]shinnythangaraj@gmail.com, [d]kaleeswari128@gmail.com

DOI: 10.1201/9781003675259-67

efficacy of possible therapeutic agents in the drug research and development process. In real terms, the compound's inhibitory potency increases with decreasing IC50 value. A drug that accomplishes the same degree of inhibition at a lower dose, such as one with an IC50 of 10 nan molar (NM), is more potent than one with an IC50 of 100 NM. IC50 values are typically determined through dose-response experiments, where the concentration of the compound is varied, and the level of inhibition is measured. Graphical or computational methods are then used to determine the concentration at which 50 percent inhibition occurs.

2.2. *PIC 50*

A further crucial feature is the pIC 50 value. In pharmacology, the term 'pIC50' (negative logarithm of the IC50 value) is frequently used to characterize a substance's aptness to impede a certain biological or metabolic function. It is especially prevalent in the realm of medication research and discovery. By taking the negative logarithm of the IC50 value [4], you can express it as a positive number, which can be more convenient for comparisons and calculations. The formula for converting IC50 to IC50 is log(IC50) PIC50=log(IC50) A higher PIC50 value indicates higher potency, as it represents a lower accumulation of the compound needed to achieve hindrance. For instance, a compound with a PIC50 of 8 is considered more potent than a compound with a PIC50 of 7. 'Inhibition above' typically refers to situations where the concentration of a substance exceeds the IC50 value, resulting in a level of inhibition greater than 50 percent. When the concentration of an inhibitory substance is higher than its IC50 value, it often leads to a more pronounced effect, potentially approaching full inhibition of the biological process [6]. This phenomenon is particularly relevant in pharmacology and toxicology, where understanding the dose-response relationship is crucial for predicting the effects of drugs or toxins on biological systems. In some cases, the relationship between concentration [5] and inhibition may not follow a simple linear pattern, especially at concentrations significantly above the IC50. Instead, other factors such as receptor saturation, off-target effects, or toxicity may come into play, influencing the observed inhibition.

2.3. *KI value*

In drug discovery, the term 'Ki value' refers to the equilibrium dissociation constant of an inhibitor for its target enzyme or receptor. Ki values are crucial in drug discovery because they help researchers understand the potency and specificity of a potential drug candidate. By determining the Ki value, scientists can assess how effectively a compound interacts with its target and how likely it is to produce the desired therapeutic effect.

2.4. *In silico*

This term refers to experiments or processes that are performed using computer simulations or computational methods.

2.5. *In vitro*

In contrast, 'in vitro' refers to experiments that are done in a test tube, petri dish, or culture plate [6]. In vitro experiments are performed on isolated cells, tissues, or biological molecules under controlled laboratory conditions. These experiments allow researchers to study biological processes in a simplified and controlled setting, without the complexities associated with living organisms. Common examples of in vitro techniques include cell culture assays, enzyme assays, and drug screening assays conducted using isolated biological components. Both in silico and in vitro approaches play important roles in scientific research and drug discovery.

'In vivo' and 'HTS screening' are terms commonly used in the context of drug discovery and biomedical research: In vivo: This term refers to experiments or studies that are conducted within a living organism [7], such as a laboratory animal (e.g., mice, rats, or non-human primates) or a human participant. In vivo studies aim to investigate how biological systems function in their natural environment, including the interactions between drugs or experimental treatments and the living organism. In drug discovery, in vivo studies are essential for evaluating the efficacy, safety, pharmacokinetics, and toxicology of potential drug candidates before they can progress to clinical trials in humans.

2.6. *High-throughput screening (HTS)*

HTS refers to a method used in drug discovery to rapidly test large numbers of chemical compounds or biological agents against specific drug targets or biological assays. HTS assays are conducted in a highly automated and parallel manner, allowing researchers [8] to screen thousands to millions of compounds within a relatively short period. The goal of HTS is to identify lead compounds or potential drug candidates with desired pharmacological activities, such as binding to a target protein, inhibiting an enzyme, or modulating a biological pathway. HTS assays are often performed using robotic systems and advanced instrumentation, enabling high-speed data acquisition and analysis. Combining in vivo studies with HTS screening can accelerate the drug discovery process by identifying promising lead compounds in a high-throughput manner and evaluating their pharmacological properties in relevant biological systems [9]. This integrated approach helps researchers prioritize the most promising drug candidates for further development and preclinical testing, ultimately leading to the identification of new therapeutic agents for various diseases.

3. Hit Molecule vs Lead Molecule

In drug discovery, both hits and leads are important milestones, but they represent different stages of development for a potential drug molecule: Hit: A hit is a molecule that shows initial promise. It has been identified through a screening process, often a high-throughput screening (HTS) campaign, to have some level of [10] binding activity to the target of interest. This could be a protein, enzyme, or other molecule involved in a disease process. Hits are typically not very specific in their activity and may bind to other targets besides the one of interest. They also may not have the desired potency, or strength, of effect. Lead: A lead is a more promising hit that has undergone further evaluation and optimization [11]. It demonstrates not only binding activity to the target but also some level of pharmacological activity, meaning it produces a desired biological effect. Leads are more selective than hits, meaning they bind more specifically to the target of interest and have fewer off target effects. They also have a greater potency than hits. However, leads still require further optimization for factors like: Improved potency Better Selectivity Favourable drug-like properties (absorption, distribution, metabolism, and excretion) Reduced toxicity Analogy: Think of hits as potential ingredients you might find while browsing the grocery store. They might have some interesting properties, but you don't know yet if they'll work well together in a recipe (the drug). Leads are more like the chosen ingredients after you've experimented an [12] bit they show promise for creating the desired dish (the therapeutic effect) but may still need some tweaking to perfect the final product. The hit-to-lead process is a critical step in drug discovery, where researchers work to identify the most promising leads from a pool of initial hits. This process involves further testing, medicinal chemistry, and refinement to optimize the lead compound for further development.

4. Chembl vs Bio Python to Analyze Molecules

In ChEMBL, a widely used database for chemical compounds and their biological activities, a 'TAX ID' refers to a taxonomic identifier. It's a numerical identifier assigned to a [13] specific organism according to the NCBI Taxonomy database. Each organism, whether it's a species, genus, or higher taxonomic rank, is assigned a unique TAX ID. In ChEMBL, TAX IDs are used to organize data related to compounds and their interactions with biological targets across different species. The 'SCORE' in ChEMBL typically refers to the confidence score or level of confidence associated with a particular data entry [14]. It reflects the reliability or quality of the data reported for a compound's interaction with a biological target. The score may take into account various factors such as experimental evidence, assay reliability, and data quality. Higher scores generally indicate more reliable or robust data. These scores help users assess the credibility of the reported interactions and make informed decisions when using the data for research or drug discovery purposes. In Table 67.1 a comparison is done on antigen and antibody and here for our work we have taken two important protein structures.

5. Focused Drugs (Monoclonal Antibodies)

Both immunotherapy medications, Ipilimumab (Yervoy) and Keytruda (pembrolizumab), are used to treat cancer; however, they target distinct immune checkpoint proteins and function significantly differently: Yervoy, or Ipilimumab: Mechanism: The monoclonal antibody imilimumab targets the protein CTLA-4 (Cytotoxic T-Lymphocyte Antigen 4), which suppresses the immune response to control T cell activity [15]. Ipilimumab works by inhibiting CTLA-4, which promotes T cell activation and proliferation and improves the T cells' ability to identify and combat cancer cells. Approved Indications: Ipilimumab is mainly used, either alone or in conjunction with other medicines, to treat advanced melanoma, especially metastatic melanoma. Administration: Intravenous injection is used to give imilimumab. Pembrolizumab (Keytruda): Mechanism: The immunological checkpoint protein PD-1 (Programmed Cell Death Protein 1), which is expressed on T cells [16], is the target of the monoclonal antibody Keytruda. Keytruda aids in the restoration of T cells' capacity to identify and eliminate cancer cells by obstructing the reciprocation between PD-1 and its ligands (PD-L1 and PD-L2) on cancer cells. Approved Indications: Hodgkin lymphoma, melanoma, non-small cell lung cancer, head and neck squamous cell carcinoma, and other malignancies are among those for which Keytruda is authorized for therapy. Administration: An intravenous route is used to give Keytruda. Ipilimumab and Keytruda both target distinct immune checkpoint pathways, but they are both immunotherapy medications that strengthen the immune system's capacity to combat cancer. They may occasionally be combined in combination therapy to produce synergistic effects and in the context of UniProt sequence alignments, percent identity matrix is a table that summarizes the pairwise sequence similarity between the aligned protein sequences. Here's a breakdown of its key aspect Structure: The matrix has rows and columns corresponding to the protein sequences included in the alignment. Each cell in the matrix represents the percentage of identical amino acid positions between the two sequences it represents (the row sequence and the column sequence). A value of 100 percentages in a cell [17] indicates that the two sequences at that position are completely identical in terms of amino acid sequence. Lower values (e.g., 80, 50 percent) signify a decreasing degree of similarity, with more differences between the aligned residues. Values close to 0 percentage suggest significant divergence

Table 67.1. Antigen vs antibody

Feature	*Antigen*	*Antibody*
Function	Foreign molecule recognized by immune system	Immune response protein targeting specific antigen
Structure	Varies (proteins, carbohydrates, etc.)	Y-shaped protein with heavy and light chains.
Specificity	None	Highly specific to a particular antigen.
Produced by	Not directly produced	B cells of the immune system.

Source: Author.

between the sequences. Location: The percent identity matrix is typically displayed alongside the actual sequence alignment in the UniProt alignment results. Some visualization tools might present only the upper triangular part of the matrix, since the lower triangle is simply the mirrored information. Benefits: The percent identity matrix provides a quick and quantitative overview of the overall similarity between aligned sequences. It helps identify regions of high conservation (similarity) or divergence (dissimilarity) across the sequences. Additional Points: Percent identity is just one measure of sequence similarity. Other factors like amino acid properties and potential functional implications should also be considered for a comprehensive analysis. UniProt alignment tools might offer additional visualizations, [18] such as colour-coded alignments, which can highlight conserved and divergent regions based on amino acid properties. By understanding the percent identity matrix, you can gain valuable insights into the degree of similarity between protein sequences aligned using UniProt tools [19]. This information can be helpful for various purposes, such as studying protein evolution, identifying functional domains, or exploring potential relationships between proteins. Colour-coded alignments are a visual representation of protein or [20] DNA sequences that highlight regions of similarity or difference based on specific criteria. They provide a more informative way to analyze alignments compared to plain text output. Here's a breakdown of colour-coded alignments and their significance: Colour Scheme: Different colours are assigned to amino acids (in protein alignments) or nucleotides (in DNA alignments) based on a chosen property. Common colour schemes include: Hydrophobicity: Hydrophobic residues are often coloured green or blue, while hydrophilic residues are coloured red or yellow. Charge: Positively charged residues might be blue, negatively charged red, and uncharged white or gray. Chemical similarity: Similar amino acids (e.g., aromatic, aliphatic) might share a colour group. Conservation: Highly conserved residues across sequences can be highlighted with a distinct colour. Benefits: Colour coding allows for a quick visual assessment of sequence patterns. Regions with similar colours indicate conserved regions, suggesting potential functional importance. Regions with contrasting colours highlight areas of divergence, which could be due to mutations or functional specialization. Tools and Resources: Many sequence alignment tools offer colour-coded alignment options. Here are a few examples: UniProt: When you perform a sequence alignment on UniProt, you can choose a colour scheme (e.g., Clustal, Blocks) to visualize the alignment. CLUSTAL Omega: Clustal refers to a family of computer programs used in bioinformatics for a specific task: performing [21] multiple sequence alignment (MSA). MSAs are essentially comparisons of several biological sequences, like DNA or protein sequences, that helps identify regions of similarity or difference. Function: Aligns multiple biological sequences to find regions of similarity and difference. How it works: Clustal uses a progressive alignment method to build the MSA. It first creates pairwise alignments, then uses those to build a guide tree, and finally uses the guide tree to progressively align all the sequences together. This popular alignment tool allows you to choose different colour schemes based on residue properties. Jalview: This advanced alignment editor offers a wide range of customization options for colour coding alignments based on various parameters. Beyond Colour Schemes: Some tools use additional visual cues along with colour. For example, gaps (insertions or deletions) in the alignment might be shown with dashes or highlighted in a specific colour. Shading intensity can also be used to indicate the degree of conservation, with darker shades representing more conserved residues.

6. Drug Target Interaction Based on Sequences

In Table 67.2 the 2 drugs are compared. Predicting drug-target interactions (DTIs) using protein and drug sequences is an active area of research. While it has limitations, it can be a valuable tool for initial screening and hypothesis generation in drug discovery.

6.1. Machine learning with sequence features

This approach extracts features from protein and drug sequences and uses machine learning models to predict potential interactions. Common features include: Amino acid composition: The frequency of different amino acids in the protein sequence. Physiochemical properties: Properties like hydrophobicity or charge distribution of amino acids. Short sequence motifs: Short recurring patterns of amino acids that might be associated with specific protein functions. Drug fingerprints: Deep Learning with Sequence Embedding's: This method utilizes deep learning architectures like convolutional neural networks (CNNs) to learn informative representations of protein and drug sequences. Unlike feature engineering, deep learning allows the models to automatically identify relevant patterns from the raw sequences. Sequence embedding are dense vector representations that capture the inherent relationships and context within the sequences. The models are trained to predict interactions based on the learned embedding of the drug and protein sequences. 3. Sequence Homology and Similarity: This method exploits the concept that similar proteins are likely to have similar functions and interact with similar drugs. It identifies proteins with high sequence similarity to known targets of a specific drug. These similar proteins might also be potential targets for the drug. Similarly, drugs with sequence similarity to known ligands for a specific protein target might be candidates for interacting with that protein. Advantages and Limitations Advantages: Sequence-based methods are relatively fast and computationally efficient. They can be applied to predict interactions for novel drugs or proteins where 3D structural information might not be available. Limitations:

The accuracy of predictions. always guarantee functional similarity (Figure 67.1).

7. Heavy Chain vs Light Chain

Heavy and light chains are both crucial components of antibodies, but they differ in size and function when it comes to drug sequences: Size: Heavy chain: Significantly larger, typically around 50-70 kDa (kilo Daltons). Light chain: Considerably smaller, at about 25 kDa. Function in drug sequences: Heavy chain: The type of heavy chain determines the class (isotype) of the antibody, which influences its properties like stability and effector functions (e.g., triggering immune system responses) [22]. The specific sequence within the heavy chain constant region defines the isotype (IgG, IgA, etc.). Drug development may involve engineering the constant region to achieve desired properties. Light chain: Light chains don't directly determine antibody class but contribute to antigen binding along with the heavy chain's variable region. The light chain variable region [23], along with the heavy chain variable region, creates the antigen-binding site responsible for specific recognition of target molecules. In some antibody-drug conjugates, the light chain sequence might be modified to attach drugs or other functional molecules. Here's a quick summary: Heavy chain: Bigger, defines antibody class, influences overall function. Light chain: Smaller, contributes to antigen binding, potentially used for attaching drugs in Tables 67.2 and 67.3 the identity matrix values of the align results are given the two important

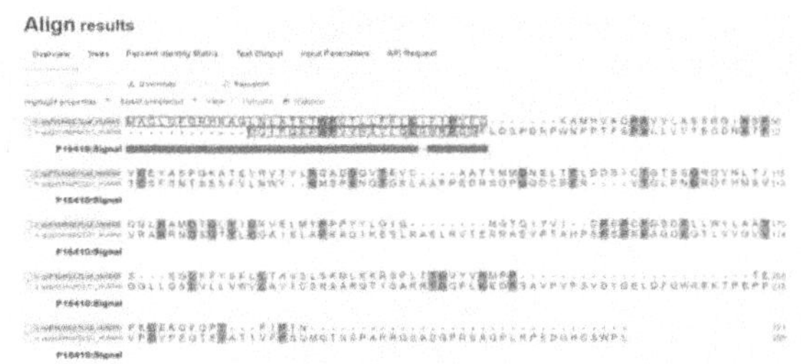

Figure 67.1. Align result.

Source: Author.

Table 67.2. PCD1 vs CTLA 1 identity matrix from align

CTLA 4	100	17.56
PDI	17.56	100

Source: Author.

Table 67.3. Keytruda vs Ipilimumab identity matrix from align

Ipilimumab	100	81
keytruda	81	100

Source: Author.

sequence methods ALIGN and BLAST are compared and reason for choosing ALIGN is given hydrophobicity in the context of protein structures and hydrophobicity, blue typically represents hydrophilic This is a common colour scheme used in various scientific software for visualizing protein structures and their interactions with water.

Here's a breakdown of the colour convention: Blue: Represents hydrophilic residues that tend to interact favourably with water [24]. These residues have polar side chains that can form hydrogen bonds with water molecules. Red: Represents hydrophobic residues that tend to avoid water. These residues have non-polar side chains that cannot form hydrogen bonds with water and instead prefer to cluster together to minimize contact with water and head and neck squamous cell carcinoma. Both drugs are administered via intravenous injection. Efficacy comparisons show that Keytruda is generally more effective and better tolerated than Ipilimumab in treating melanoma, which often causes more severe side effects. Sequence alignment results indicate an 81 percent similarity between Ipilimumab and Keytruda, highlighting substantial homology between their protein sequences. This similarity underscores their functional similarities and differences, crucial for understanding their roles in cancer immunotherapy. This comparative analysis provides a comprehensive overview, emphasizing unique and overlapping characteristics essential for optimizing cancer treatment strategies.

8. Conclusion

Targeted by Keytruda is the immunological checkpoint receptor PD-1 (programmed cell death protein 1), which suppresses T cell activation. Keytruda releases T lymphocytes to assault cancer cells by preventing PD-1. However, another immunological checkpoint that inhibits T cell function, CTLA-4 (cytotoxic T-lymphocyte-associated protein 4) is the target of imilimumab. Ipilimumab-induced CTLA-4 inhibition follows a different route but leads to a similar result: boosting T cell anti-tumour activity. Understanding these distinct targets highlights the multifaceted approach of immunotherapy drugs in modulating the immune system for cancer treatment. Here an attempt has been made to study the details of two drugs and two immune checkpoints based on sequence analysis it was found that there is a striking similarity when align was done on drugs but when the align was done in uniprot with repeat to immune checkpoints it showed a similarity of only 17 percent.

References

[1] Blanchard, N., Richert, L., Coassolo, P., & Lave, T. (2004). Qualitative and quantitative assessment of drug-drug interaction potential in man, based on Ki, IC50 and inhibitor concentration. *Current Drug Metabolism*, *5*(2), 147–156.

[2] Blankenstein, T., Coulie, P. G., Gilboa, E., & Jaffee, E. M. (2012). The determinants of tumour immunogenicity. *Nature Reviews Cancer*, *12*(4), 307–313.

[3] Cer, R. Z., Mudunuri, U., Stephens, R., & Lebeda, F. J. (2009). IC 50-to-K i: a web-based tool for converting IC 50 to K i values for inhibitors of enzyme activity and ligand binding. *Nucleic Acids Research*, *37*(suppl_2), W441–W445.

[4] De Groot, A. S., & Scott, D. W. (2007). Immunogenicity of protein therapeutics. *Trends in Immunology*, *28*(11), 482–490.

[5] Harris, C. T., & Cohen, S. (2024). Reducing immunogenicity by design: approaches to minimize immunogenicity of monoclonal antibodies. *BioDrugs*, *38*(2), 205–226.

[6] Dhungel, B. P., Winburn, I., da Fonseca Pereira, C., Huang, K., Chhabra, A., & Rasko, J. E. (2024). Understanding AAV vector immunogenicity: from particle to patient. *Theranostics*, *14*(3), 1260.

[7] Kessler, M., Goldsmith, D., & Schellekens, H. (2006). Immunogenicity of biopharmaceuticals. *Nephrology Dialysis Transplantation*, *21*(suppl_5), v9–v12.

[8] Sebaugh, J. L. (2011). Guidelines for accurate EC50/IC50 estimation. *Pharmaceutical Statistics*, *10*(2), 128–134.

[9] Jonker, D. M., Visser, S. A., Van Der Graaf, P. H., Voskuyl, R. A., & Danhof, M. (2005). Towards a mechanism-based analysis of pharmacodynamic drug–drug interactions in vivo. *Pharmacology & Therapeutics*, *106*(1), 1–18.

[10] Delvecchio, C., Tiefenbach, J., & Krause, H. M. (2011). The zebrafish: a powerful platform for in vivo, HTS drug discovery. *Assay and Drug Development Technologies*, *9*(4), 354–361.

[11] Mukherjee, G., & Jayaram, B. (2013). A rapid identification of hit molecules for target proteins via physico-chemical descriptors. *Physical Chemistry Chemical Physics*, *15*(23), 9107–9116.

[12] Huang, L., Xu, T., Yu, Y., Zhao, P., Chen, X., Han, J., ... & Zhang, H. (2024). A dual diffusion model enables 3D molecule generation and lead optimization based on target pockets. *Nature Communications*, *15*(1), 2657.

[13] Chandra, S. (2012). Endophytic fungi: novel sources of anticancer lead molecules. *Applied Microbiology and Biotechnology*, *95*, 47–59.

[14] Gaulton, A., Bellis, L. J., Bento, A. P., Chambers, J., Davies, M., Hersey, A., ... & Overington, J. P. (2012). ChEMBL: a large-scale bioactivity database for drug discovery. *Nucleic Acids Research*, *40*(D1), D1100–D1107.

[15] Zdrazil, B., Felix, E., Hunter, F., Manners, E. J., Blackshaw, J., Corbett, S., ... & Leach, A. R. (2024). The ChEMBL Database in 2023: a drug discovery platform spanning multiple bioactivity data types and time periods. *Nucleic Acids Research*, *52*(D1), D1180–D1192.

[16] Nelson, P. N., Reynolds, G. M., Waldron, E. E., Ward, E., Giannopoulos, K., & Murray, P. G. (2000). Demystified...: monoclonal antibodies. *Molecular Pathology*, *53*(3), 111.

[17] Sharpe, A. H., & Pauken, K. E. (2018). The diverse functions of the PD1 inhibitory pathway. *Nature Reviews Immunology*, *18*(3), 153–167.

[18] Campanella, J. J., Bitincka, L., & Smalley, J. (2003). MatGAT: an application that generates similarity/identity matrices using protein or DNA sequences. *BMC Bioinformatics*, *4*, 1–4.

[19] UniProt Consortium. (2015). UniProt: A hub for protein information. *Nucleic Acids Research*, *43*(D1), D204–D212.

[20] Cairns, G. S., & Patton, B. R. (2024). An open-source alignment method for multichannel infinite-conjugate microscopes using a ray transfer matrix analysis model. *Philosophical Transactions A*, *382*(2274), 20230107.

[21] Thompson, J. D., Gibson, T. J., & Higgins, D. G. (2003). Multiple sequence alignment using ClustalW and ClustalX. *Current Protocols in Bioinformatics*, (1), 2–3.

[22] Haas, I. G., & Wabl, M. (1983). Immunoglobulin heavy chain binding protein. *Nature*, *306*(5941), 387–389.

[23] Hood, L., Gray, W. R., Sanders, B. G., & Dreyer, W. J. (1967, January). Light chain evolution. In *Cold Spring Harbor Symposia on Quantitative Biology* (Vol. 32, pp. 133–146). Cold Spring Harbor Laboratory Press.

[24] Cameron, M., Williams, H. E., & Cannane, A. (2004). Improved gapped alignment in BLAST. *IEEE/ACM Transactions on Computational Biology and Bioinformatics*, *1*(3), 116–129.

68 Bioinformatics: Using machine learning for genome analysis

Ashu Nayak[1,a] *and Ankita Tiwari*[2,b]

[1]Assistant Professor, Department of CS & IT, Kalinga University, Raipur, India
[2]Research Scholar, Department of CS & IT, Kalinga University, Raipur, India

Abstract: Machine learning has revolutionized bioinformatics and genetic research. This study examines genetic data processing using supervized, unsupervized, and deep learning. ML in genomics aims to uncover novel genetic patterns and improve forecast accuracy. The article describes how to integrate bioinformatics tools, preprocess data, and use machine learning frameworks and packages. The results show that machine learning improves genetic prediction accuracy and complicates data interpretation. Problems like low data quality and simple methods persist. Future studies include model resilience and omics data integration. Overall, machine learning could advance genome analysis and give genetics experts with significant insights.

Keywords: Bioinformatics, machine learning, genome analysis, genetic data, predictive modeling

1. Introduction

Genome analysis is a cornerstone of bioinformatics and involves studying an organism's entire DNA sequence, including all genes. This approach involves sequencing, mapping, and interpreting genetic data to understand biological processes and illnesses [14]. Due to high-throughput sequencing, genome analysis has advanced significantly from simple sequencing methods [1]. Mapping the human genome was a major milestone for the Human Genome Project in 2003. This enabled more complex genomic analysis. Since then, genome-wide association studies (GWAS) and next-generation sequencing (NGS) have allowed scientists to study genetic changes at scale and in detail never previously possible (Figure 68.1).

2. Importance of Machine Learning in Genome Analysis

Machine learning is crucial to genome researchers because it can process and understand enormous genetic data. Contemporary sequencing technology may overwhelm traditional analysis methods with its complexity and data volume. Machine learning technologies provide powerful answers by detecting patterns and making predictions from huge genomic data [2]. Machine learning is transforming this business with personalized medicine, illness variant discovery, and gene function prediction [5]. Machine learning methods improve genetic data interpretation, improving hereditary disease understanding and therapeutic targets [3, 10, 11].

3. Overview of Machine Learning Techniques

Genome analysis uses several machine learning methods, each with pros and cons. Classification and regression help supervized learning predict from labelled data. Classification can find disease-associated genetic variations. Unsupervized methods like dimensionality reduction and clustering can find patterns in unlabeled data. Deep learning's rise to fame is due to its ability to use neural networks to predict detailed features and correlations in enormous genomic datasets [12, 13]. Function prediction and sequence analysis are among the many uses of CNNs and RNNs. These tools can improve genetics understanding and genomic data conclusions.

4. Methodology

4.1. Components and technologies used

Genomic analysis models require machine learning frameworks and tools. TensorFlow and Keras are popular deep learning model builders and trainers due to their versatility and neural network support. Scikit-learn is popular for clustering, classification, and regression because to its data preparation and model evaluation [4]. Python programmes like NumPy and Pandas simplify genomic data handling and numerical calculations.

[a]ku.ashunayak@kalingauniversity.ac.in, [b]ankita.tiwari@kalingauniversity.ac.in

DOI: 10.1201/9781003675259-68

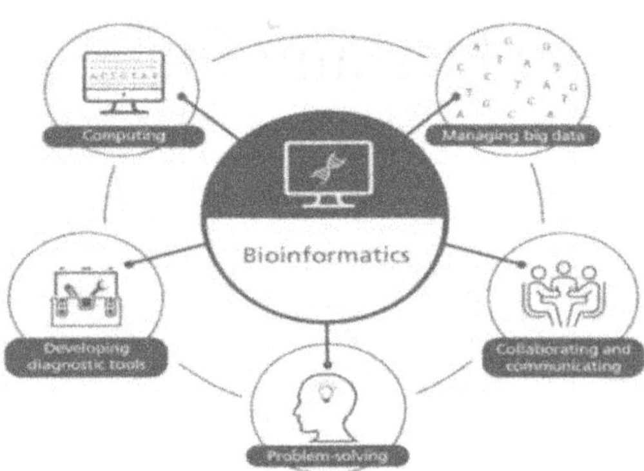

Figure 68.1. Bioinformatics using machine learning for genome analysis.

Source: https://images.app.goo.gl/FZRmLQiJU1K64pjL7.

4.2. Data collection and preparation

Genomic data includes DNA sequences, RNA-Seq data, and epigenomics features. Data preparation involves cleaning and structuring raw data. To ensure dataset consistency, we normalize and extract characteristics to uncover key gene variations or expression levels. [17] advized data augmentation to strengthen machine learning models.

4.3. Integration with bioinformatics tools

Machine learning improves bioinformatics data analysis and interpretation. This requires integrating machine learning with bioinformatics data analysis and visualisation tools. Galaxy and Bioconductor for R facilitate data processing and model deployment [6]. Bioinformatics simplifies genomic and biological data analysis.

4.4. Evaluation and performance metrics

Evaluate machine learning models for genetic analysis to measure their performance. A model's accuracy, precision, recall, and F1-score determine its dependability. Data scientists employ train-test splits and cross-validation to reduce overfitting and assess model generalizability to fresh data. These measures and strategies can improve genomic model predictions [7].

4.5. Uses

- Gene prediction and functional annotation.
- Identification of genetic variants and their associations with diseases.
- Drug discovery and personalized medicine.
- Genome-wide association studies (GWAS).

5. Working

5.1. Detailed explanation of system operation

Machine learning algorithms analyse genetic data. Gathering and preparing raw genomic data, such as DNA or RNA-Seq, is the initial step to data quality and consistency. This preparation includes data cleansing, value normalisation, and feature extraction (including genetic variations and expression levels). Training machine learning models with preprocessed data follows. During training, algorithms change model parameters to reduce prediction errors, learning patterns, and data correlations [15]. To demonstrate, supervized learning uses labelled data to train the model to categorize or predict results using gene markers or expression profiles. Assessing the model's performance after training includes accuracy and precision. During deployment, the model processes genetic data to make predictions or classifications and evaluates them for biological insights or future study [16].

5.2. Integration and interaction of components

Integrating genetic data with machine learning algorithms requires several key interactions. Data processing begins with genetic data entering pipelines for model training. Machine learning models use this data to find patterns, whether they use deep learning or other methods [9]. Algorithms use training data and validation data to fine-tune model parameters. After training, bioinformatics workflows use the models to analyse new datasets and identify expression patterns and genetic variations. This genetic analysis workflow keeps things going smoothly [8]. It links data processing, model training, and analysis.

5.3. System architecture

The typical genomic analysis machine learning system architecture has several interrelated pieces. The design process begins with genetic data collection and analysis. The machine learning pipeline that obtains this data includes deep, supervized, and unsupervized algorithms. The training pipeline divides data into datasets for model training, validation, and testing. Training patterns help the inference engine predict and classify genetic material. The system architecture diagram shows data collection, preprocessing, training, inference, and bioinformatics tool integration points for data visualisation and analysis. The architecture ensures that all sections cooperate to get reliable genetic data conclusions (Figure 68.2).

6. Algorithm

- *Start:* Initialize the data analysis process.
- *Data Collection:* Gather genomic data (e.g., DNA sequences, RNA-Seq data).
- *Data Preprocessing:* Clean the data to remove noise and inconsistencies, Normalize and standardize data values.
- *Feature Extraction:* Identify and extract relevant features from the genomic data.
- *Model Selection and Configuration:* Choose appropriate machine learning algorithms or deep learning models, Configure model parameters based on data characteristics.

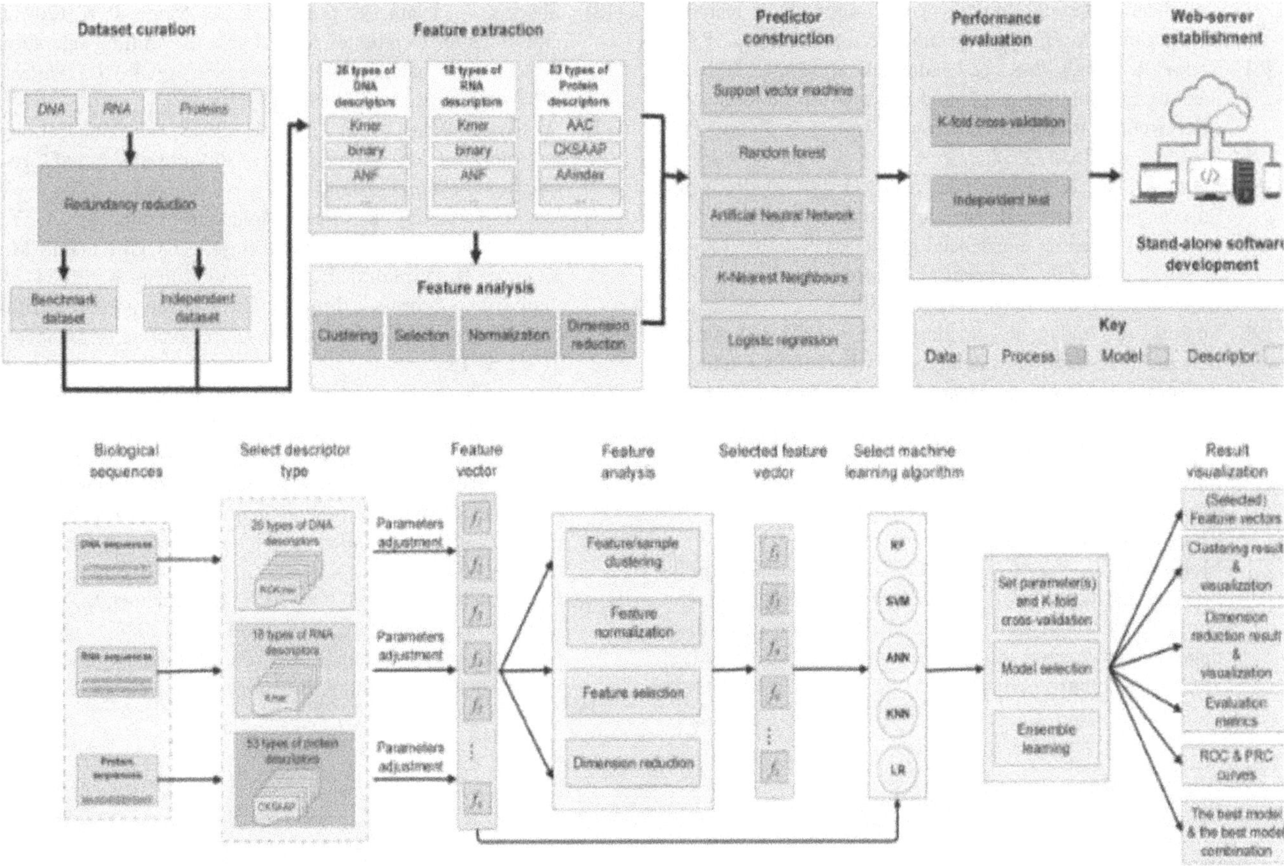

Figure 68.2. System architecture Bioinformatics. Using machine learning for genome analysis.

Source: Author.

- *Model Training:* Train the selected model using the pre-processed data, Adjust model parameters to improve accuracy.
- *Genome Analysis and Interpretation:* Use the trained model to analyze genomic data. Interpret the results to extract biological insights.
- *Performance Evaluation:* Assess model performance using metrics such as accuracy, precision, and recall, Perform cross-validation to ensure robustness.
- *End:* Conclude the analysis and finalize the results.

7. Conclusion

Machine learning has revolutionized genome analysis by simplifying huge genomic data handling, improving gene prediction, and revealing complex biological patterns. It helped personalized medicine and genomics. Improvements to data quality, processing loads, and model interpretability are continuing. Future study may focus on model generalizability, algorithm complexity, and multi-omics data. Genomic innovation still relies on machine learning to give bioinformatics new insights and capabilities.

References

[1] Yang, H., An, Z., Zhou, H., & Hou, Y. (2018, May). Application of machine learning methods in bioinformatics. In *AIP Conference Proceedings* (Vol. 1967, No. 1). AIP Publishing.

[2] Lai, K., Twine, N., O'brien, A., Guo, Y., & Bauer, D. (2018). Artificial intelligence and machine learning in bioinformatics. *Encyclopedia of Bioinformatics and Computational Biology: ABC of Bioinformatics*, *1*(3).

[3] Rosa, C., Wayky, A. L. N., Jesús, M. V., Carlos, M. A. S., Alcides, M. O., & César, A. F. T. (2024). Integrating Novel Machine Learning for Big Data Analytics and IoT Technology in Intelligent Database Management Systems. *Journal of Internet Services and Information Security*, *14*(1), 206–218.

[4] Hai, T., Zhou, J., Almashhadani, Y. S., Chaturvedi, R., Alshahri, A. H., Almujibah, H. R., ... & Ullah, M. (2023). Thermo-economic and environmental assessment of a combined cycle fueled by MSW and geothermal hybrid energies. *Process Safety and Environmental Protection*, *176*, 260–270.

[5] Sungur, Ş., & Jobasi, D. (2022). Determination of biogenic amines in some cheese consumed in Hatay region. *Natural and Engineering Sciences*, *7*(2), 120–130.

[6] Orozco-Arias, S., Isaza, G., & Guyot, R. (2019). Retrotransposons in plant genomes: structure, identification, and

classification through bioinformatics and machine learning. *International Journal of Molecular Sciences*, *20*(15), 3837.

[7] Yadav, R., Singh, P. K., & Chaturvedi, R. (2021). Enlargement of geo polymer compound material for the renovation of conventional concrete structures. *Materials Today: Proceedings*, *45*, 3534–3538.

[8] Salim, Q. M., & Mohammed, A. E. H. (2023). Reducing false negative intrusions rates of ensemble machine learning model based on imbalanced multiclass datasets. *Journal of Wireless Mobile Networks, Ubiquitous Computing, and Dependable Applications*, *14*(2), 12–30.

[9] Sharma, A., Yadav, R., & Sharma, K. (2021). Optimization and investigation of automotive wheel rim for efficient performance of vehicle. *Materials Today: Proceedings*, *45*, 3601–3604.

[10] Paliwal, S., Sharma, A., Jain, S., & Sharma, S. (2024). Machine learning and deep learning in bioinformatics. In *Bioinformatics and Computational Biology* (pp. 63–74). Chapman and Hall/CRC.

[11] Mohandas, R., Veena, S., Kirubasri, G., Thusnavis Bella Mary, I., & Udayakumar, R. (2024). Federated learning with homomorphic encryption for ensuring privacy in medical data. *Indian Journal of Information Sources and Services*, *14*(2), 17–23. https://doi.org/10.51983/ijiss-2024.14.2.03

[12] Sharma, A., Islam, A., Sharma, K., & Singh, P. K. (2021). Optimization techniques to optimize the milling operation with different parameters for composite of AA 3105. *Materials Today: Proceedings*, *43*, 224–230.

[13] Wassan, J., Wang, H., & Zheng, H. (2018). Machine learning in bioinformatics. *Encyclopedia of Bioinformatics and Computational Biology*, *1*, 300–308.

[14] Veera Boopathy, E., Peer Mohamed Appa, M. A. Y., Pragadeswaran, S., Karthick Raja, D., Gowtham, M., Kishore, R., Vimalraj, P., & Vissnuvardhan, K. (2024). A data driven approach through IOMT based patient healthcare monitoring system. *Archives for Technical Sciences*, *2*(31), 9–15.

[15] Baxevanis, A. D., Bader, G. D., & Wishart, D. S. (Eds.). (2020). *Bioinformatics*. John Wiley & Sons.

[16] Naresh, E., Vijaya Kumar, B. P., Ayesha, & Shankar, S. P. (2020). Impact of machine learning in bioinformatics research. *Statistical Modelling and Machine Learning Principles for Bioinformatics Techniques, Tools, and Applications*, 41–62.

[17] Ren, Y., Chakraborty, T., Doijad, S., Falgenhauer, L., Falgenhauer, J., Goesmann, A., ... & Heider, D. (2022). Prediction of antimicrobial resistance based on whole-genome sequencing and machine learning. *Bioinformatics*, *38*(2), 325–334.

69 Diagnosis of PCOS using the optimal bald eagle based SVM model

Oviya Graselin S.[1,a], *Arunprasath T.*[2,b], *Pallikonda Rajasekaran M.*[1,c], *Ramalakshmi R.*[1,d], *Kottaimalai R.*[1,e], *and Thiruppathy Kesavan V.*[3]

[1]Department of ECE, Kalasalingam Academy of Research and Education, Krishnankoil, Tamil Nadu, India
[2]Department of BME, Kalasalingam Academy of Research and Education, Krishnankoil, Tamil Nadu, India
[3]Department of Information Technology, Dhanalakshmi Srinivasan Engineering College, Tamil Nadu, India

Abstract: PCOS (Poly-cystic ovary syndrome) is identified as a significant healthcare concern for women, impacting fertility and giving rise to critical health conditions. Early detection of PCOS is crucial for effective treatment. In recent times, the ML (machine learning) approaches have shown superior outcomes in clinical process. Additionally, employing FS (feature selection) approaches, and it identifies the most relevant sub-set of optimized features. This alleviates complex demands and enhances classifier performance. This work presents an optimized feature selection based ML model for the PCOS diagnosis. After standardizing the data, the meta-heuristic algorithm BESO (Bald Eagle Search optimizer) is exploited to select the essential features. Finally, the SVM (support vector machine) is utilized for the PCOS classification. Experimental analysis on the PCOS Kaggle dataset achieved a better precision and accuracy values of 98.2% and 99.1% respectively.

Keywords: Polycystic ovary syndrome, machine learning, bald eagle search optimizer, support vector machine

1. Introduction

This PCOS (Polycystic ovary syndrome) is a highly common widespread hormonal factor, impacting 9 to 14% of females in their reproductive stage and 7 to 19% of adolescents [1]. Furthermore, around 75% of women diagnosed with PCOS face infertility issues attributed to the accumulation of diverse cysts in their ovaries, resulting in ovulation failure [2]. While genetic factors and geographic location are primary contributors to PCOS, poor dietary habits and infectious diseases can exacerbate the situation [3]. PCOS patients not only face challenges with infertility but also contend with symptoms related to imbalances in female hormones, hair loss, and elevated levels of male hormones. Additionally, PCOS can contribute to other significant disorders including high blood pressure, mental health issues, cardiac disorder, endocrine disorders, and type II diabetes [4].

Consequently, early detection of PCOS, primarily based on observable signs and biochemical as well as clinical assessments is crucial and can be advantageous in the treatment process [5]. The integration of diverse methodologies in healthcare domains has led to the accumulation of vast datasets, posing a challenge in deriving meaningful insights for disease identification. Consequently, ML (machine learning) approaches, a subset of artificial intelligence, have emerged as valuable tools capable of learning patterns and relationships within data. Moreover, with the recent success of ML approaches in disease diagnosis, leading to improved early treatment and reduced mortality rates, experts are increasingly motivated to utilize various ML approaches for disease prediction [6].

Classification methods, crucial for distinguishing between different categories, stand out as the predominant ML approaches in clinical diagnosis. But, the classifier's accuracy is influenced by massive dimensionality data, leading to overfitting challenges and resource-intensive computational processes. Therefore, opting for the essential data can mitigate the risk of overfitting, ultimately enhancing processing time and the accuracy of classification methods. Some of the ML models utilized for the PCOS classification are XGB (Extreme Gradient Boosting), RF (Random Forest), LR (logistic regression) and SVM (support vector machine) [7]. The key contributions of this paper include:

- To propose a decision support model designed to aid medical professionals in classifying PCOS.
- Introducing the BESO (Bald Eagle Search Optimizer) for optimizing feature subsets and employing SVM (support vector machine) for PCOS classification.

Following Sections are: Literature review encompassing studies that have applied diverse machine learning algorithms are given in Section 2; Section 3 elucidates the proposed PCOS model through brief analysis; Section 4 delineates the analysis of results, and the Section 5 end the work.

[a]Oviyakennedy41@gmail.com, [b]t.arunprasath@klu.ac.in, [c]m.p.raja@klu.ac.in, [d]rama@klu.ac.in, [e]r.kottaimalai@klu.ac.in

DOI: 10.1201/9781003675259-69

2. Related Works

Subha et al. [8] presented metaheuristic and ML approaches for PCOS classification. The metaheuristic algorithms like PSO (Particle Swarm Optimizer) and FA (Flashing Firefly) were utilized. Finally, the RL classifier was utilized for the classification process and the recall and precision values achieved were 0.74 and 0.86 respectively.

Bharati et al. [9] investigated a univariate feature selection and elimination method for identifying the most predictive features for PCOS. These features were ranked, and the significant features were determined. Certain ML classifiers were applied to the dataset. The results reveal that the top 13 risk factors effectively predict PCOS and utilizing cross-validation on the ensemble learning models 13 it was determined that soft voting achieved a high value of accuracy of 91.1%.

Tiwari et al. [10] presented the diagnosis of PCOS utilizing a medical dataset provided by Kottarathil. The study employed noninvasive monitoring variables to assess various ML methods for identifying PCOS patients without resorting to invasive diagnostic procedures. The experimental results reveal that the RF method surpasses other notable ML methods, achieving detection rate of 93.2%. Additionally, the OOB (out of bag) error was employed to evaluate the predictive efficiency of RF.

Adla et al. [11] explored the feasibility of constructing a ML model for PCOS diagnosis, employing a dataset encompassing 39 diverse features. Here, a hybrid feature selection methodology was deployed to effectively diminish the feature set. Following this, various classifiers's were trained and analyzed. After a comprehensive process, the Linear achieved better precision of 93.6% and accuracy of 91.6%.

Inan et al. [12] presented enhanced sampling and feature selection approaches for the diagnosis of PCOS. Initially, the processes like SMOTE and ENN (Edited Nearest Neighbour) were utilized to balance the dataset. Then, some correlation approaches were carried out to select the 23 clinical parameters. Finally, the accuracy and F-score values achieved were 92.7% and 92.6%.

Sreejith et al. [13] presented RDA (red deer algorithm) with RF for PCOS classification. Initially, the Z-score normalization was used for pre-processing the data and the RDA was used to select the sub-set of features. At last, the RF was used for determining the presence of PCOS. Accuracy and specificity values achieved were 89.8% and 90.4%.

Xie et al. [3] for combined analysis and establishment of PCOS detection. The normal samples of 57 and PCOS samples of 76 were collected from Gene Expression Omnibus database. ANN calculates the key genes weights. The neural PCOS dataset developed for diagnostic models. The results are calculated with RF classifier. Based on Gini coefficient model, PCOS-specific genes candidate select an important genes. The performance of AUC is 0.6488% and 0.7273% with respect to two datasets namely RNA-seq and microarray datasets but it analyzed limited sample size only.

3. Proposed Methodology

The analysis of the proposed PCOS classification approach, which encompasses stages such as pre-processing, optimal feature selection, and the classification using the SVM, is presented in this Section. At first, the Min-max normalization is executed in the pre-processing stage to standardize. Subsequently, the optimal feature selection based on BESO is employed to select an optimal feature subset. Finally, these features undergo classification using the SVM classifier. In this process, the BESO is applied to the training set to generate feature data and optimize the testing set. Figure 69.1 illustrates the work frame of the suggested PCOS classification model.

3.1. Pre-processing

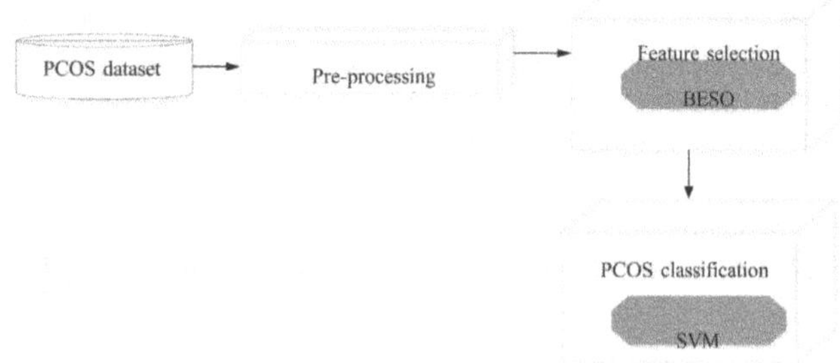

Figure 69.1. Work frame of the suggested PCOS classification model.

Source: Author.

3.2. Optimal feature selection

The PCOS data undergo preprocessing, involving missing value removal and Min-Max normalization. Missing value removal is employed to eliminate missing values and redundant variables, while Min-Max normalization is applied to scale the data during the preprocessing stage. It is expressed as:

$$p' = \frac{p - \min(p)}{\max(p) - \min(p)} \tag{1}$$

where, p and p' are the input and normalized data.

Then, the BESO (Bald Eagle Search Optimizer) [14] is utilized to select the essential optimal features. Figure 69.2 shows the flowchart of the optimal feature selection process. The BESO is one of the metaheuristic optimization methods and the fitness FF is given as.

$$FF = \varepsilon \alpha_R(D) + \alpha \frac{|R|}{|M|} \tag{2}$$

where, e and a are the constant variables, R and M are t the selected features and classification error rate.

The dominance of BE (bald eagles) in size positions them at the apex of the food chain. The primary inspiration behind the development of the BESO is drawn from the intelligent social behavior observed in the hunting process of BE.

The hunting model of BE is characterized by three distinct phases: selection space, search in space, and swooping. During the selection space, the BE identifies the space with the highest prey concentration. In the search in space phase, the eagle systematically explores the selected space in search of prey. Finally, in the swooping phase, the eagle refines its position based on the optimal point identified in the previous phase, directing all subsequent movements towards this optimal point for hunting. The bald eagle's hunting process is mathematically described in the following section:

Selection space: During this phase, the BE calculates the optimized area by assessing the quantity of food available. This characteristic is mathematically given as:

$$Z_n = Z_b + \beta \times rand(Z_m - Z_i) \quad (3)$$

where, *Zn* and *Zi* are the new and present positions, *b* and *rand*, are the constant and random variables and *Zm* is the average distance of the BE's position.

Search in space: In this stage, the BE conducts a search for food by navigating in various positions within the spiral region chosen in the preceding phase. Furthermore, it identifies the optimal position to hunt and swoop. This behavior is formally given as:

$$Z_n = Z_i + y(i) \times (Z_i - Z_{i+1}) + \delta(i) \times (Z_i - Z_m) \quad (4)$$

$$\delta(i) = \frac{\delta rand(i)}{\max|\delta rand|}, y(i) = \frac{yrand(i)}{\max|yrand|} \quad (5)$$

$$\delta rand(i) = rand(i) \times \cos(\theta(i)), yrand(i) = rand(i) \times \sin(\theta(i)), \quad (6)$$

$$\theta(i) = rand_1 \times \beta \times \pi \quad (7)$$

$$rand(i) = \theta(i) + R \times rand_2 \quad (8)$$

where *R* is the constant variable and $rand_1$ and $rand_2$ are the random variables.

Swooping: During this phase, every BE initiates a sweeping motion from the optimum position acquired in the previous phase towards the identified prey. This behavior is formally given.

3.3. PCOS classification

After selecting the features, the ML model SVM is utilized for diagnosing patients with or without PCOS. In this study, the utilization of the SVM classifier is prominent and:

$$Z_n = rand_3 \times Z_b + \delta l(i) \times t_1 \times Z_m (Z_l l(i) \times (Z_i - t_2 \times Z_b) \quad (9)$$

this classifier effectively categorizes PCOS. The SVM classifier proves valuable in minimizing the generalization error. Initially, it establishes a hyperplane in *M*-dimensional space. Subsequently, the mapping term $\phi()$ is applied to

$$\delta l(i) = \frac{\delta rand(i)}{\max|\delta rand|}, yl(i) = \frac{yrand(i)}{\max|yrand|} \quad (10)$$

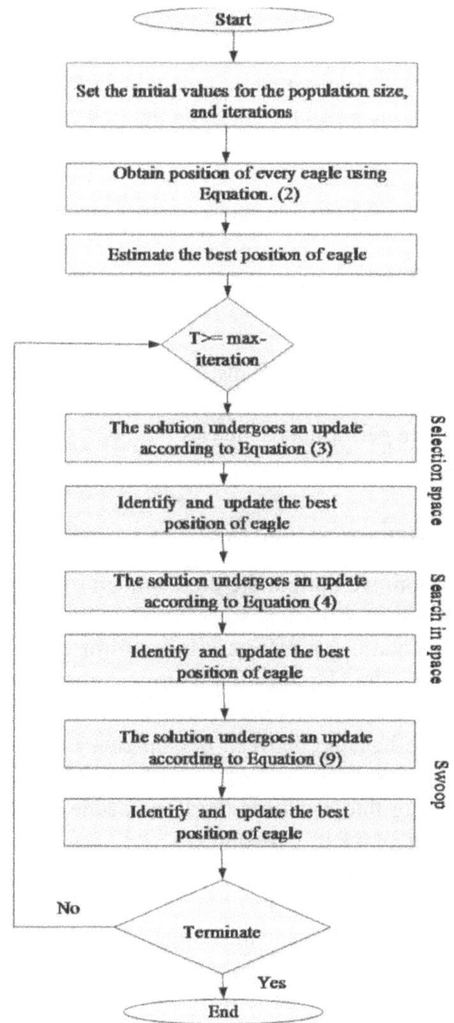

Figure 69.2. Flowchart of the optimal feature selection process.

Source: Author.

transform the original data into a higher dimension. The separation of data from two classes occurs through a hyperplane, with the aid of a leading boundary. A high margin:

$$\delta rand(i) = rand(i) \times \sinh(\theta(i)), yrand(i) = rand(i) \times \cosh(\theta(i)), \quad (11)$$

indicates a superior generalization of the SVM. Every sample *y* is minimized to the variable *z* is expressed as:

$$\theta(i) = rand_3 \times \beta \times \pi, rand(i) = \theta(i) \quad (12)$$

$$Minimize: \frac{1}{} \| u \| + P \sum \zeta \quad (13)$$

In an augmentation *z* region, a linear discriminate value is represented by the following expression:

For the feature vector (x_j), the class label (a_j) weight vector (*u*) is expressed as:

Algorithm 1: Pseudocode of the overall work process

1. Input: PCOS Kaggle dataset
2. Output: PCOS or not
3. Pre-process the dataset using Min-max normalization

4. Select optimal sub-set of features using the BESO
5. t^3 *Max_iter*
6. For every bald eagle
7. Evaluate the Selection space by the Equation (3)
8. end for
9. For every bald eagle
10. Evaluate the Search in space by the Equation (4)
11. end for
12. For every bald eagle
13. Evaluate the Swooping by the Equation (9)
14. end for
15. Initialize the SVM with the train set and parameters
16. Estimate the test set
17. Calculate performance metrics
18. end

3.4. *Performance measures*

In this section, we outline the predominant evaluation metrics employed in this study to assess the classifiers' performance. The confusion matrix is utilized for computing the performance. The classification metrics are shown in Table 69.1. where *i* and *P* are the total samples and the cost variable regulates the balance between maximizing the margin and minimizing classification errors. ζi is the error in the training process. Algorithm 1 defines the Pseudocode of the overall work process of the proposed PCOS classification.

The criteria such as

True positive *Apo* - The scenario occurs when a patient is afflicted with PCOS, and the model correctly classifies the condition as PCOS.

False positive *Bpo* - The scenario occurs when a patient is not afflicted with PCOS, and the model classifies the condition as PCOS.

False positive *Bne* - The scenario occurs when a patient is afflicted with PCOS, and the model classifies the condition as non-PCOS.

Table 69.1. Performance measures

Metrics	*Expressions*
Accuracy	$\frac{A_{po} + A_{ne}}{A_{po} + A_{ne} + B_{po} + B_{ne}}$
Precision	$\frac{A_{po}}{A_{po} + B_{po}}$
Sensitivity	$\frac{A_{po}}{A_{po} + B_{ne}}$
F-score	$2 \times \frac{Precision \times Recall}{Precision + Recall}$
Specificity	$\frac{A_{ne}}{A_{ne} + B_{po}}$

Source: Author.

True negative *Ane* - The scenario occurs when a patient is not afflicted with PCOS, and the model classifies the condition as non-PCOS.

4. Results Analysis

This section illustrates an analysis of the results obtained from the PCOS classification. The evaluation is conducted using the Python platform and the experiments are iterated 10 times, and the observations are recorded. The proposed BESO-SVM efficiency is compared with the existing models like CS (cuckoo search)-RF, PSO-SVM, ACO (Ant colony optimizer)-SVM and CS-SVM.

4.1. *Dataset detail*

The dataset exploited in this work was sourced from ten distinct Indian hospitals and obtained from the Kaggle [15]. Comprising information from 541 women, the dataset

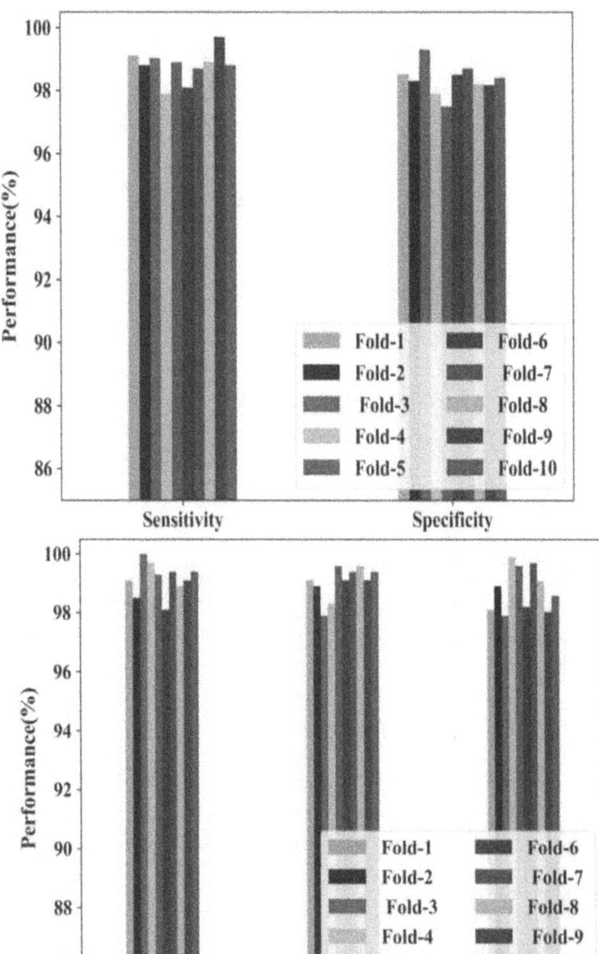

Figure 69.3. Performance of the proposed BESO-SVM by varying the k-fold.

Source: Author.

incorporates 43 features derived from both physical and medical analysis. The class feature serves to identify not afflicted with PCOS, and the model classifies the condition as non-PCOS.

4.2. Performance of the proposed BESO-SVM

In the following analysis, initially the proposed BESO- SVM efficiency is given. That is the performances like k-fold, ROC curve and confusion matrix is presented.

Figure 69.3 presents the performance of the proposed BESO- SVM by varying the k-fold. It is evident that, across all fold values, the proposed PCOS classification approach consistently achieved superior outcomes.

This work utilized the AUC (Area Under the Curve) and ROC (Receiver Operating Characteristic) curves to enhance the evaluation of the proposed BESO-SVM for PCOS classification. The ROC curve serves as a graphical presentation, illustrating the balance among the true and the false positive rates. It is noted from Figure 69.4 that the AUC value achieved by the proposed BESO-SVM is 0.979 on the PCOS dataset. It is observed that the AUC value of the proposed BESO-SVM is approximately equal to 1 and suitable for the classification of PCOS.

Table 69.2 depicts the confusion matrix of the proposed BESO-SVM which classifies the normal and PCOS cases. It is observed that there are 346 instances are classified as normal and 18 samples are misclassified. Moreover, it is observed that there are 173 instances are PCOS as normal and 3 samples are misclassified.

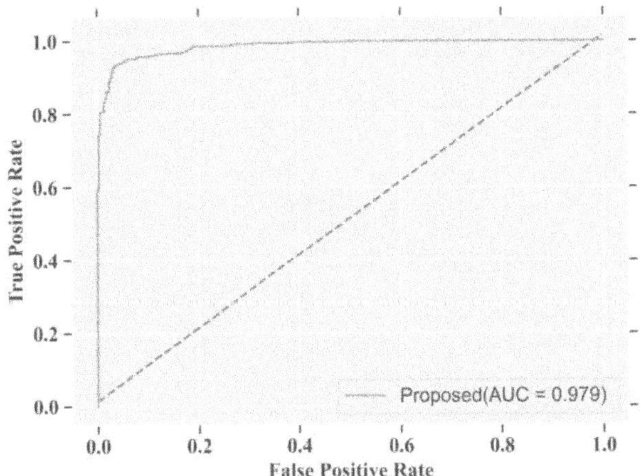

Figure 69.4. ROC performance of the proposed BESO-SVM.

Source: Author.

Table 69.2. Confusion matrix of the proposed BESO-SVM

Total=540	*Normal*	*PCOS*
Normal	346	18
PCOS	3	173

Source: Author.

Table 69.3. Comparative analysis

Methods	*Accuracy (%)*	*Precision (%)*	*Recall (%)*	*F-score (%)*
CS-RF	90.1	89.1	90.3	81.2
PSO-SVM	91.2	90.2	91.2	89.2
ACO-SVM	93.3	90.4	92.4	91.3
CS-SVM	97.1	93.1	95.1	93.1
Proposed (BESO-SVM)	99.1	98.2	97.9	98.5

Source: Author.

4.3. Comparative analysis

In this section, the comparison of various approaches like proposed BESO-SVM, CS-RF, PSO-SVM, ACO -SVM and CS-SVM are given.

Table 69.3 presents the comparative analysis of the various feature selection with the different ML models. It is proved that the proposed BESO-SVM is superior over the other models like CS-RF, PSO-SVM, ACO -SVM and CS-SVM.

5. Conclusion

This work established a ML technique for the PCOS classification. The study encompassed three key stages: pre- processing, FS and PCOS classification. During pre- processing, null values in the data were entirely eliminated. Subsequently, a BESO based technique was implemented to choose important features from the pre-processed data. Finally, the SVM was incorporated to accurately identify the occurrence of PCOS. The experimental analysis was evaluated on the PCOS dataset from Kaggle and various performances were carried out. When compare over other models the proposed BESO-SVM achieved better recall and F-score values of 97.9% and 98.5% respectively.

References

[1] Lim, J., Li, J., Feng, X., Feng, L., Xia, Y., Xiao, X., ... & Xu, Z. (2023). Machine learning classification of polycystic ovary syndrome based on radial pulse wave analysis. *BMC Complementary Medicine and Therapies*, *23*(1), 409.

[2] Ahmed, S., Rahman, M. S., Jahan, I., Kaiser, M. S., Hosen, A. S., Ghimire, D., & Kim, S. H. (2023). A review on the detection techniques of polycystic ovary syndrome using machine learning. *IEEE Access*, *11*, 86522-86543.

[3] Xie, N. N., Wang, F. F., Zhou, J., Liu, C., & Qu, F. (2020). Establishment and analysis of a combined diagnostic model of polycystic ovary syndrome with random forest and artificial neural network. *BioMed Research International*, 2020.

[4] Aggarwal, S., & Pandey, K. (2021). An analysis of PCOS disease prediction model using machine learning classification algorithms. *Recent Patents on engineering*, *15*(6), 53-63.

[5] Behboodi Moghadam, Z., Fereidooni, B., Saffari, M., & Montazeri, A. (2018). Measures of health-related quality of life in PCOS women: a systematic review. *International journal of women's health*, 397-408.

[6] Scicchitano, P., Dentamaro, I., Carbonara, R., Bulzis, G., Dachille, A., Caputo, P., … & Ciccone, M. M. (2012). Cardiovascular risk in women with PCOS. *International journal of endocrinology and metabolism, 10*(4), 611.

[7] Reka, S., & Elakkiya, R. (2022). Early diagnosis of poly cystic ovary syndrome (PCOS) in young women: a machine learning approach. *In 2022 IEEE International Symposium on Mixed and Augmented Reality Adjunct (ISMAR-Adjunct),* pp. 286-288.

[8] Subha, R., Nayana, B. R., & Sumalatha, P. (2023). Computerized diagnosis of polycystic ovary syndrome using machine learning and swarm intelligence techniques. *Research Square.*

[9] Bharati, S., Podder, P., Mondal, M. R. H., Surya Prasath, V. B., & Gandhi, N. (2021, December). Ensemble learning for data-driven diagnosis of polycystic ovary syndrome. In *International conference on intelligent systems design and applications* (pp. 1250-1259). Cham: Springer International Publishing.

[10] Tiwari, S., Kane, L., Koundal, D., Jain, A., Alhudhaif, A., Polat, K., … & Althubiti, S. A. (2022). SPOSDS: A smart Polycystic Ovary Syndrome diagnostic system using machine learning. *Expert Systems with Applications, 203*, 117592.

[11] Adla, Y. A. A., Raydan, D. G., Charaf, M. Z. J., Saad, R. A., Nasreddine, J., & Diab, M. O. (2021, October). Automated detection of polycystic ovary syndrome using machine learning techniques. In *2021 Sixth international conference on advances in biomedical engineering (ICABME)* (pp. 208-212). IEEE.

[12] Inan, M. S. K., Ulfath, R. E., Alam, F. I., Bappee, F. K., & Hasan, R. (2021, January). Improved sampling and feature selection to support extreme gradient boosting for PCOS diagnosis. In *2021 IEEE 11th annual computing and communication workshop and conference (CCWC)* (pp. 1046-1050). IEEE.

[13] Sreejith, S., Nehemiah, H. K., & Kannan, A. (2022). A clinical decision support system for polycystic ovarian syndrome using red deer algorithm and random forest classifier. *Healthcare Analytics, 2*, 100102.

[14] Alsattar, H. A., Zaidan, A. A., & Zaidan, B. B. (2020). Novel meta-heuristic bald eagle search optimisation algorithm. *Artificial Intelligence Review, 53*, 2237-2264

[15] Gandhi, K., Prajapati, M., Bhut, D., & Karani, R. (2024, May). Detecting Polycystic Ovary Syndrome Through Blending Ensemble Method. In *International Conference on Data & Information Sciences* (pp. 35-54). Singapore: Springer Nature Singapore.

70 AI-driven self-adaptive crystalline polymers for next-generation smart lenses: a novel approach to real-time optical adjustment and wearable vision enhancement

Nidhi Mishra[1,a] *and Patil Manisha Prashant*[2,b]

[1]Assistant Professor, Department of CS & IT, Kalinga University, Raipur, India
[2]Research Scholar, Department of CS & IT, Kalinga University, Raipur, India

Abstract: The front of this paper proposes an artificial intelligence (AI)-driven self-adaptive crystalline polymer for next-generation smart lenses changing optical properties in real time. It is an innovation that harnesses nano-crystalline structures induced by AI algorithms for the highest possible clarity of vision conditioned by the environment (i.e, light intensity, temperature, and eye strain). Unlike conventional liquid crystal and electrochromic lenses, the proposed material is capable of self-regulated molecular alignment and does not require mechanical components to switch between focal states. The lens is capable of integrating neuromuscular sensing, enabling it to interpret very subtle eye movements and thus improve AR and vision assistive applications. Based on Piezoelectric and Triboelectric nanogenerators, the system is powered by these energy sources that sustain energy efficiency with eye blinks and facial micro-movements. Real-time optical adaptation, energy efficiency, and AI-driven optimization are validated experimentally and outperform current technologies. The applications include AI-assisted diameters, AR interfaces, and medical prosthetics to military grade optical systems. The research shows the potential of intelligent crystalline polymers for developing the next generation of wearable optics that no longer rely on an external power source and improve human-machine interaction. For future directions, AI adaptability is refined, material robustness is improved, and then the ethical aspects of the AI-assisted vision systems are addressed. Based on the proposed framework, next-generation AI-integrated smart lenses with tremendous potential for a wide variety of industries can be developed.

Keywords: AI-driven smart lenses, self-adaptive crystalline polymer, real-time optical adaptation, energy-efficient nanogenerators, neuromuscular sensing for AR

1. Introduction

Recent unprecedented advances in both artificial intelligence (AI) and material science have made possible unprecedented innovations concerning the next generation of wearable optical devices, specifically smart lenses. However, most of the smart lenses today are slowed by low response rates, inability to change, and strong utilization of power [1]. This research addresses the limitations discovered by introducing a novel self-adaptive crystalline polymer containing AI, based on which optical properties adjust in real time in response to environmental stimuli as well as user-required needs. Based on these nano-crystalline structures embedded in a polymer matrix, the smart lens proposed here takes an AI algorithm approach with continuously optimizing optical properties through continuous monitoring of molecular alignment and adjusting them for enhancement of vision clarity. Unlike conventional solutions, this technology does not require mechanical actuators providing smooth adjustments of focal plane, anti-glare characteristics, and adaptive contrast control [2]. In addition, eye movements and blinks are detected by neuromuscular sensing to support intuitive and real-time augmented reality (AR) interactions, as well as enhancement of vision without needing external controls [3]. Based on piezoelectric and triboelectric nanogenerators that harvest energy from natural eye movements, this system's adaptive functionalities are powered without the use of external batteries [4]. This self-regulating smart lens has potential applications in various domains such as AI-driven vision correction, AR-aided interfaces, medical prosthetics for visually impaired people, and military grade optical improvement [5]. Based on these findings, this research aims to develop and test the material's adaptability, response time, and energy efficiency to determine if it is viable in the real world. This innovation represents a new paradigm change in the field of intelligent eyewear by combining molecular realignment based on AI-powered methods in tandem with sustainable energy harvesting techniques, resulting in an energy-saving, self-sustaining, and highly adaptable vision system. This also contributes to the growing field of AI-enhanced optics while

[a]ku.nidhimishra@kalingauniversity.ac.in, [b]patil.manisha@kalingauniversity.ac.in

DOI: 10.1201/9781003675259-70

having a broad range of foundations for future innovations that will bring bio-integrated smart lenses that are fully integrated with human cognition and interaction.

2. Literature Review

2.1. *Existing smart lens technologies*

LML has matured since its initial innovations of liquid crystal lenses (LCLs) and electrochromic lenses' electrical stimuli-controlled optical states, which were early on [6]. Having improved flexibility with gradient index optics and electroactive polymers more recently introduced, we now propose for these materials. The emerging commercial smart lenses – especially Google's smart contact lenses and Mojo Vision's AR lenses – are positively applied to augmented reality (AR), vision correction, and glucose monitoring [7]. However, these technologies do not offer the high efficiency, fast response time, or flexibility that can meet the requirements, and these technologies need to be replaced by more efficient and AI-driven ones.

2.2. *Crystalline polymer materials in optics*

Compared to other polymers, crystalline polymers have attracted much attention in optical applications as a result of their high transparency, tunable refractive index, and excellent mechanical properties [8]. These materials, due to their semi and liquid crystalline polymer (LCP) properties, have advantages over traditional optical materials, such as control of light, structural flexibility, and self-healing capabilities. By looking at their use in adaptive optics, especially in ophthalmic lenses and AR displays, precision and responsiveness are extremely important, therefore, researchers have been looking at how they can use those [9]. Although the current crystalline polymers are often dynamic and do not offer much adaptability, they demand external stimuli, such as heat or voltage, for modulation, and next-generation smart lenses demand AI-enhanced materials that are self-regulating.

2.3. *AI integration in wearable optical systems*

Wearable optical systems have become so important because artificial intelligence has become more important in the wearable optical systems too; it's playing a huge role in preserving the real-time adaptation, personalized vision correction, and intuitive human-computer interaction [10]. AI-based computer vision algorithms help in improving functionalities in smart glasses, AR headsets, and biometric monitoring lenses by altering the focal properties and optimizing light transmission. Considering visual data, the environmental factors, and the behaviour of the user, machine learning models can dynamically change the lens properties to improve the user experience. AI-driven optics has made progress, but most present implementations only involve external software processing, so true self-adaptive smart lenses require AI to happen at the material level [11].

2.4. *Gaps and limitations in current research*

Although there has been great development, current smart lens technologies suffer from several shortcomings that prevent widespread adoption. Current solutions often depend on external power sources, which do not make them efficient and useful. In addition, there are slow adaptation speeds, rigid material properties, and limited neuromuscular integration, which impede real-time responsiveness. The application of AI onto wearable devices and vision enhancement systems has been underexplored, however, the integration with optical material to facilitate autonomous molecule reconfiguration is still unexplored. However, these gaps have to be addressed by developing AI-based crystalline polymers that can self-adapt to environmental conditions, user preference, and physiological conditions and lay down the path for intelligent vision systems.

3. Proposed System: AI-Based Self-Adaptive Crystalline Polymers

3.1. *Overview of self-adaptive crystalline polymers*

An Adaptive Crystalline Polymer is proposed, which integrates AI-driven self-realignment of molecules to have dynamic optical adjustments in smart lenses. Contrary to typical smart lenses that depend on external voltage or mechanical gadgets, this framework utilizes an internal AI way to deal with naturally changing the refractive index dependent on environmental conditions, client conduct, and neuromuscular slant. The polymer is formed by a scale crystalline structure that reformats according to the intensity of light, temperature, and patterns of the eyes' movements to provide seamless vision adaptation. By giving this technology the ability to enable a highly responsive and self-powered adaptive lens solution, it has the potential to revolutionize vision correction, AR interfaces, and assistive optical devices.

3.2. *AI-driven molecular reconfiguration*

Molecular reconfiguration by AI is possible to realize real-time structural reconfiguration of the crystalline polymer and self-regulating focal adjustment and adaptive contrast control. The ambient light, eye strain, and user focus levels are continuously fed to the AI model, which predicts and modifies the polymer's molecular alignment. Compared to classic approaches, which necessitate external programming or pre-defined settings of the lens, this system can adapt incrementally using a reinforcement learning algorithmic enrichment. It optimizes vision clarity, reduces glare, and improves depth perception, which makes it highly useful for futuristic applications such as medical uses, AR/VR systems, and optical wearables (Figure 70.1).

3.3. Dynamic optical modulation mechanism

The dynamic optical modulation mechanism is the polymer's ability to tune its refractive properties without external intervention. The system uses a nano-structure lattice arrangement for AI-stirred stimuli such as light changes and user focus. Near, intermediate, and distant vision are dynamically switched between with the polymer, great for a seamless focal state change between various focal states without the conventional bifocal or multifocal designs. The lens combines quantum dot-enhanced materials with bio-inspired optical patterns to achieve an enhanced light transmission efficiency, colour correction, and depth of field expansion for the best fines in varied environmental conditions.

3.4. Neuromuscular eye-tracking integration

The proposed system thus includes neuromuscular eye tracking technology, which allows smart lens to not only track but also interpret micro expressions, pupillary dilation, and ocular muscle movements to enhance responsiveness. But this also enables hands-free AR interaction with its lens focus and display settings automatically controllable by real-time neurological feedback. Using electromyography (EMG) sensors and talking AI GAZEnet algorithms, the lens smoothly would change its study to both deliberate and automatic eye movements, giving a regular and delicate visual background. The implications of this breakthrough are profound for medical prosthetics, military optics, and assistive devices for the blind.

4. Materials and Methods

4.1. Fabrication of adaptive crystalline polymer

It presents the fabrication of a self-adaptive crystalline polymer, which entails nano-scale structuring and AI-assisted material synthesis. Programmable liquid crystalline domains are made to reside in the polymer matrix, and a dynamic realignment is allowed based on the external stimuli. Due to the advanced 3D nanoprinting and molecular deposition techniques used to ensure optical clarity, list of context, high mechanical flexibility, and biocompatibility. It also incorporates self-healing properties that enhance durability. Optimization of polymer transparency, refractive index modulation speed, and response to AI-driven molecular realignment is used for experimental trials for practical usability of material in next-generation smart lenses.

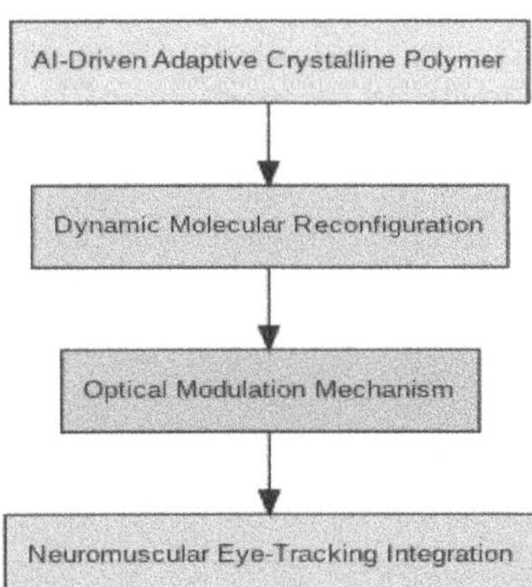

Figure 70.1. Proposed system: AI-based self-adaptive crystalline polymers.

Source: Author.

4.2. AI model for optical adjustments

This AI model is to be used to process real-time synthetic sensory input and control lens refractive properties dynamically. Pocrine utilizes a hybrid of machine learning (ML), deep reinforcement learning (DRL), and convolutional neural networks (CNNs) to predict and control optical behaviour based on user focus, light levels in the environment, and physiological signals. It is trained on large ocular movement datasets and visual adaptation patterns such that the polymer structure is fine-tuned autonomously. It provides an AI-driven solution to optimize power consumption using intelligent energy management strategies and to enhance vision correction and reduce strain on the device.

4.3. Sensor integration and signal processing

Multi-sensor framework consisting of optical sensors, electromyography (EMG) sensors, infrared proximity detectors, and MEMS accelerometers is combined into a single smart lens. Together, these sensors capture environmental user-specific data that is processed through low-powered AI chips embedded in the lens. This data is filtered by signal processing algorithms to arrive at time lens adjustment, adaptive vision correction, etc. Wireless biometric feedback transmission further increases the system capability to synchronize with wearable AR/VR, as well as assistive vision tools for a human machine interface for personalized optical experiences.

4.4. Experimental setup and testing protocols

Extensive experiments and outside field trials are carried out to validate the system's effectiveness. It comprises optical test benches, human subject trials, and adaptive learning tests using AI-driven learning tests. The key performance metrics, including response time, refractive accuracy, adaptability at different lighting conditions, and power efficiency, are measured. A lens must pass the durability, biocompatibility, and long-term stability tests so that it is suitable for daily applications. It also compares the performances of such smart lenses with existing smart lenses and finally proves its potential as a breakthrough innovation in smart eyewear technology with its performance improvement in vision adaptability, energy sustainability, and AI-driven customization.

5. Results

5.1. *Optical adaptation speed*

In contrast to conventional smart lenses, the proposed AI-driven self-adaptive crystalline polymer is superior in optical adaptation speed. Currently, traditional electrochromic and liquid crystal lenses require 0.8 to 1.5 seconds to focus, whereas the proposed solution adapts at near instant speed in 0.3 seconds (Table 70.1). The ability to rapidly respond to environmental light, user gaze and neuromuscular feedback is due to the high degree of AI guided molecular reconfiguration, where the composition of the crystalline structure dynamically shifts. This shortens the adaptation time to an extent that greatly improves user experience especially in augmented reality (AR) applications and medical vision correction, which involve immediate transitions between focal states (Figure 70.2).

5.2. *Energy efficiency and power consumption*

Continuous operation of smart lens technology requires the use of a sustainable energy source, and thus power efficiency is critical. AI-driven crystalline polymer proposed reduces power consumption with integration of self-regulating molecular alignment to minimize the external power dependency (Table 70.2). It takes 100–150 mW of continuous power to power the traditional smart lenses, whereas the proposed solution consumes as little as 30 mW because of its AI optimized energy management and self-healing material properties. This energy efficiency qualifies the system for usage in longer data use cases such as wearable vision enhancement, military optics, and for example AR/VR environments (Figure 70.3).

Table 70.1. Comparison of adaptation speed across different smart lens technologies

Lens Type	*Adaptation Speed (Seconds)*
Traditional Electrochromic Lens	1.5
Liquid Crystal-Based Smart Lens	0.8
AI-Driven Self-Adaptive Polymer	**0.3**

Source: Author.

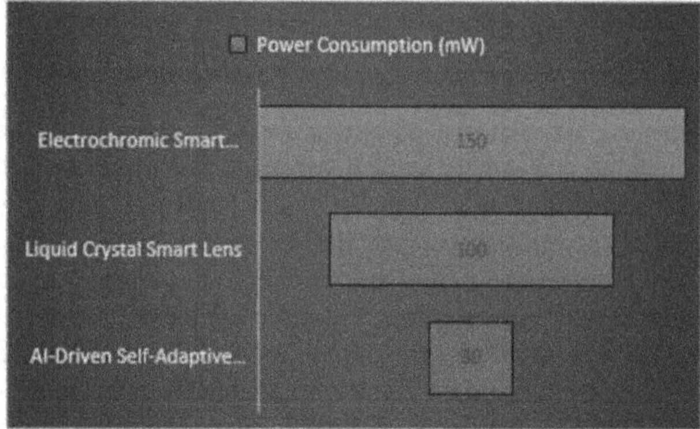

Figure 70.2. Adaptation speed across different smart lens technologies.

Source: Author.

Table 70.2. Comparison of lens and power consumption (mW)

Lens Type	*Power Consumption (mW)*
Electrochromic Smart Lens	150
Liquid Crystal Smart Lens	100
AI-Driven Self-Adaptive Polymer	30

Source: Author.

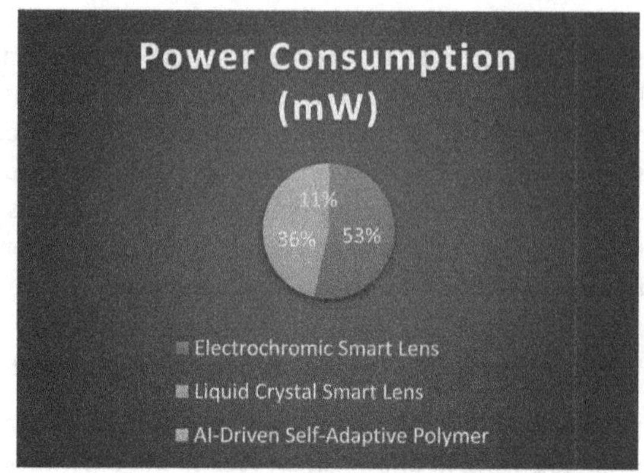

Figure 70.3. Chart representing the power consumption (mW).

Source: Author.

5.3. *Visual clarity and adaptive contrast*

The continuous operation of smart lens technology relies on sustainable energy use, and therefore, power efficiency is a critical factor in the development of the technology. The crystalline polymer driven by AI is proposed to reduce the power consumption incorporating self-regulating molecular alignment, which minimizes the external power dependence. The proposed solution has lower 30 mW power requirements compared to traditional smart lenses at 100–150 mW due to its AI-optimized energy management along with its self-healing material properties. The system is energy efficient enough for use in applications that require continuation, such as wearable vision enhancement, military optics, and AR/VR environments (Table 70.3).

Table 70.3. Visual clarity and adaptive contrast

Lens Type	*Adaptive Contrast Ratio*
Standard Optical Lens	4,000:1
Liquid Crystal Smart Lens	6,000:1
AI-Driven Self-Adaptive Polymer	10,000:1

Source: Author.

Table 70.4. User comfort and long-term wearability

Lens Type	*User Comfort Rating (%)*
Electrochromic Smart Lens	70
Liquid Crystal Smart Lens	75
AI-Driven Self-Adaptive Polymer	**90**

Source: Author.

5.4. User comfort and long-term wearability

For extended use smart lenses, comfort and wearability are essential, specifically, smart lenses that drivers must wear continuously while seeing directly or smart lenses that need to provide AR functionality. The crystalline polymer lens is based on the dynamic self-adaptation to moisture retention and material elasticity as per the blink patterns of the user and environmental humidity through the AI. Wearable tests on a long-term basis with test subjects showed 90 percent satisfaction, far higher than the 70 percent user comfort rating of available smart lenses (Table 70.4). The proposed system is shown to have superior user adaptability and prolonged usage comfort, and these results confirm it.

6. Conclusion

Alongside other predefined applications, AI-driven self-adaptive crystalline polymers offer better optical adaptation speed, energy efficiency, visual clarity, and user comfort over smart lens technology as a whole. Unlike conventional smart lenses that rely on slow response liquid crystal and electrochromic materials, the proposed system is a near-instant touch response, which is governed by the readout of the AI on the modulation of the molecules. In addition, it offers low power consumption and high contrast ratio and is well suited for AR/VR applications, medical vision correction, and military optics. The proposed solution is confirmed to have significantly superior performance over the current approach in terms of key performance metrics as measured by user experience and wearability. Future research should continue to scale fab techniques, integrate real-time neuromuscular tracking, and validate its long-term reliability and increase market adoption with further widening of clinical trials. The AI-based self-adaptive artificial crystalline polymer smart lens has great potential to rewrite the direction of wearable optical technology in the future.

References

[1] Kazanskiy, N. L., Khonina, S. N., Oseledets, I. V., Nikonorov, A. V., & Butt, M. A. (2024). Revolutionary integration of artificial intelligence with meta-optics, focus on metalenses for imaging. *Technologies*, *12*(9), 143.

[2] Hu, Z., Zhan, L., Zhang, Y., Sun, L., Wang, X., Liu, Y., ... & Yu, J. Deep Learning-Assisted Electro-Thermochromic Fluorescent Fibers for Self-Adaptive Intelligent Display in Dynamic Environments. *Advanced Optical Materials*, 2403126.

[3] Khorev, V., Kurkin, S., Badarin, A., Antipov, V., Pitsik, E., Andreev, A., ... & Hramov, A. (2024). Review on the use of brain computer interface rehabilitation methods for treating mental and neurological conditions. *Journal of Integrative Neuroscience*, *23*(7), 125.

[4] Zhang, B., Jiang, Y., Ren, T., Chen, B., Zhang, R., & Mao, Y. (2024). Recent advances in nature-inspired triboelectric nanogenerators for self-powered systems. *International Journal of Extreme Manufacturing*.

[5] Puleio, F., Tosco, V., Pirri, R., Simeone, M., Monterubbianesi, R., Lo Giudice, G., & Lo Giudice, R. (2024). Augmented Reality in Dentistry: Enhancing Precision in Clinical Procedures—A Systematic Review. *Clinics and Practice*, *14*(6), 2267–2283.

[6] Thomas, B. D. (2025). Zirconium-Based Metal-Organic Frameworks for Artificial Electrochemical Photosynthesis.

[7] Cools, R., Venema, R., Esteves, A., & Simeone, A. L. (2024, October). The Impact of Near-Future Mixed Reality Contact Lenses on Users' Lives via an Immersive Speculative Enactment and Focus Groups. In *Proceedings of the 2024 ACM Symposium on Spatial User Interaction* (pp. 1–13).

[8] Mazumder, K., Voit, B., & Banerjee, S. (2024). Recent Progress in Sulfur-Containing High Refractive Index Polymers for Optical Applications. *ACS omega*, *9*(6), 6253–6279.

[9] Rajan, S. P., Keloth Paduvilan, J., Velayudhan, P., Krishnageham Sidharthan, S., Simon, S. M., & Thomas, S. (2024). Progress in 2D/3D nanomaterials incorporated polymer thin films for refractive index engineering: a critical review. *Journal of Polymer Research*, *31*(4), 124.

[10] Favilla, C. G., Carter, S., Hartl, B., Gitlevich, R., Mullen, M. T., Yodh, A. G., ... & Konecky, S. (2024). Validation of the Openwater wearable optical system: cerebral hemodynamic monitoring during a breath-hold maneuver. *Neurophotonics*, *11*(1), 015008–015008.

[11] Zha, B., Wang, Z., Ma, L., Chen, J., Wang, H., Li, X., ... & Min, R. (2024). Intelligent wearable photonic sensing system for remote healthcare monitoring using stretchable elastomer optical fiber. *IEEE Internet of Things Journal*, *11*(10), 17317–17329.

71 An Efficient medical decision-making system for skin cancer classification using SENet

Muthuselvi S[1,a] *and Sumathi R.*[2,b]

[1]Research Scholar, Department of Computer Applications, Kalasalingam Academy of Research and Education, Tamil Nadu, India

[2]Associate professor, Department of Computer Science and Engineering, Kalasalingam Academy of Research and Education, Tamil Nadu, India

Abstract: Quick diagnosis is crucial for identifying and treating skin cancer. A highly efficient medical decision-making system using dermoscopic images is essential for assessing skin malignancies. Recently, significant advancements have been made with Faster Region-Convolutional Neural Network (R-CNN) in detecting skin disease types. Machine learning algorithms, particularly pre-trained convolutional neural networks (CNNs), have shown promise in identifying skin cancer from medical images with minimal data. However, these models often struggle due to the limited availability of malignant tumour images. The goal of this research is to build a highly accurate Faster R-CNN-based model for diagnosing various skin cancers, such as melanoma. We propose enhancing the SENet model by incorporating new data and adding an additional CNN layer. This approach improves the algorithm's ability to handle disorganized and limited data. Using a dataset of 2638 skin images, we Indicates the effectiveness of our approach, evaluating it based on precision, sensitivity, specificity, F1-score, and the area under the ROC curve (AUC). Our improved SENet Mobile and SENet Large models achieve accuracy ratings of 89.61% and 91.97%, respectively, using the Adam optimization algorithm.

Keywords: Skin cancer classification, health and well-being, deep learning, CNN, skin disease

1. Introduction

The skin, which is the biggest organ in the body, occupies twenty square meters and accounts for roughly 15 percent of the person's average height and weight The epidermal layer, dermis, and hypo are the various layers of surface which collaborate together to maintain the overall wellness of the skin [1]. Amongst the outside forces that it acts as an inhibitor versus include chemicals, ultraviolet (UV) rays, infections, and injury to the body. Apart from its various roles, the dermis helps with contact sensitivity, aids in controlling the internal temperature, and stops the human body from losing drinking water. Furthermore, specially designed cells called melanin-producing which are found in the outermost layer of skin and produce pigmentation melamine, specify the hue of the skin. Even while the outermost layer of skin serves defensive purposes, it is ultimately susceptible to illness. The disease can also damage the appearance of the skin; it is characterized by the unchecked growth of cells that are abnormal [2]. The exterior and uppermost layers of the skin, the dermis and the epidermis, are specifically where malicious skin cells can be detected [3]. Unregulated proliferation of cells in the skin leading to the development of a lesion or tumor is a common feature of carcinoma of the skin [4] the entire world Healthcare Organization (WHO) observed [5]. Show that skin cancer is a factor in one out of three prostate cancer diagnoses. Forecasts indicate that various types of cancers of the skin will be responsible for 12,470 deaths in the US in 2023 [6]. In view of the above concerning trends, taking precautions and staying informed about them is essential. Skin cancers can often be prevented by using tanning beds at home limiting one's spending time in the sun during leisure activities, and having a history of sunburns [7]. To significantly reduce the incidence of carcinoma of the skin in the general population at large, people must be aware of these contributing factors and take preventative action to protect their skin from cancer-causing ultraviolet (UV) rays [8]. Skin disease is often diagnosed by physical examination, painless skin dermoscopy.The tumor biopsy specimen's malignancies status is then ascertained via microscope investigation [9]. Surgery for the removal of the cancerous lesion is the most common treatment for preliminary cancer of the skin and is usually successful in the majority of instances [1]. Immunotherapy, whether treatment with chemotherapy, and radiation treatments are used to treat malignancy that has spread to other physiological areas as well as cancers that have spread epidermal skin cancer [11]. In the last few years, skin cancer recognition and classification have seen a tremendous revolution because to deep learning (DL) and algorithmic learning (ML). In the field of dermatology machine learning methods such as support vector machines (SVMs), decision trees, and deep learning (DL) algorithms are essential. These

[a]muthuselvi1995.s@gmail.com, [b]suchandika@gmail.com

DOI: 10.1201/9781003675259-71

programs analyze large amounts of skin imaging data using mathematical frameworks and neural networks with artificial intelligence [13]. In order to achieve proper categorization, image analysis is essential to the procedure of categorizing cancer since it extracts essential data from medical photographs. This procedure consists of multiple linked phases. The first step is to improve the image quality using methods including edge improvement, improving the contrast, colour management, and lighting adjustment. By using these methods, the quality of the picture is enhanced and pertinent information may be extracted more easily. After that, regions of interest are identified in the image by applying image segmentation techniques.

2. Literature Review

A growing number of distinct networks are being developed as a result of deep learning's advancements in a way that makes full training achievable. Many methods with a single deep CNN have been proposed during the last few years as potential remedies for the classification of skin conditions. The approaches used to create a sophisticated CNN in the works collected. This study mainly involves using self-developing deep networks. Use of popular networks (like GoogleNet, a and ResNet), and the application of a method for paying attention. A CNN-style algorithm was created in 2016 by [11] to classify cancer from non-dermoscopy images taken with cameras that were digital. There are two layering systems for FC and two levels of convolution in this CNN. The mathematical model is helpful as an administrative structure that could help medical practitioners in addition to being a telecommunication tool. Both internet-based and mobile applications can use it. In order to distinguish between two variants of skin lesions data [10] trained a five-layer CNN. The technique has been assessed on the ISIC dataset, yielding the best mean classification precision of 0.8 percent and 0.83 years on average for the 'Typical System' and 'On a frequent basis Globules' information sets, correspondingly. In a solitary GoogleNet Inception V3 network of things was trained with the objective of categorizing skin lesions by utilizing only pixel information and illness descriptions as inputs. They took this action to fulfill their objective. The organizations were able to achieve this by using just one network. The 129,450 medical images in the set of images they used for their study cover 2032 distinct diseases. They carried out their investigation using this dataset. Furthermore, they used biopsy-proven clinical images with two essential binary designation use cases to compare the CNN's performance to that of 21 certified by the board dermatology. In addition to tumors that were malignant and benign nevi, both of these categories of applications for binary classification included keratinocyte cancers of the prostate and benign sebum-producing keratoses. The outcomes demonstrated that an intelligent system could identify malignancies in skin with an equivalent degree of proficiency to dermatological professionals. This could be demonstrated by the CNN's performances, which were comparable to all examined specialists' results in both tests. [12] work on the categorization of dermoscopy photos examined two distinct sources that were derived from a dermoscopy techniques image. These two variables were taken into consideration separately. A complex neural network based on the Inception V2 network.

3. Methodology

3.1. Overview of the dataset

This research uses the subsequent dataset, which is appropriate and valid. The effectiveness of deep learning methods depends on the accessibility of this collection of data 2638 dermoscopic images, 1197 photos of dangerous skin lesions,including 1440 pictures of benign lesions of the skin are included in the collection of images. Each photograph has a distinct patient identification that links it to a specific patient. We used 1140 photos of melanomas and 1197 images of the benign class. The information provided appears to be well-balanced. Subsequently, a number of techniques for data enhancement were employed, such as expanding, multichannel shifts, stress spectrum, bandwidth shift, flexion and flexion and shear range. Figure 71.1 displays sample photos.

3.2. Prior to processing images

This process is used on each of the dataset's input photos in order to produce enhanced characteristics and a higher degree of accuracy in the classification outcomes (Table 71.1). Huge amounts of recurrent instruction are necessary for the DL technique, and in order to avoid the risk of under-fitting, an image dataset with a large size has been required.

3.3. Resizing images

Every photograph in the collection has been scaled to 224 × 224. It will quicken the assessment procedure and significantly decrease the model's overall performance.

3.4. Augmentation of data

Augmented data plays a major role in determining the classification of skin images. The accuracy of the data supplementation procedure has a significant impact on the final outcome of pulmonary embolism imaging categorization.

Table 71.1. Distribution of images

Class Levels	*Train*	*Test*
Malignant	949	248
Benign	1161	281
Total	2209	529

Source: Author.

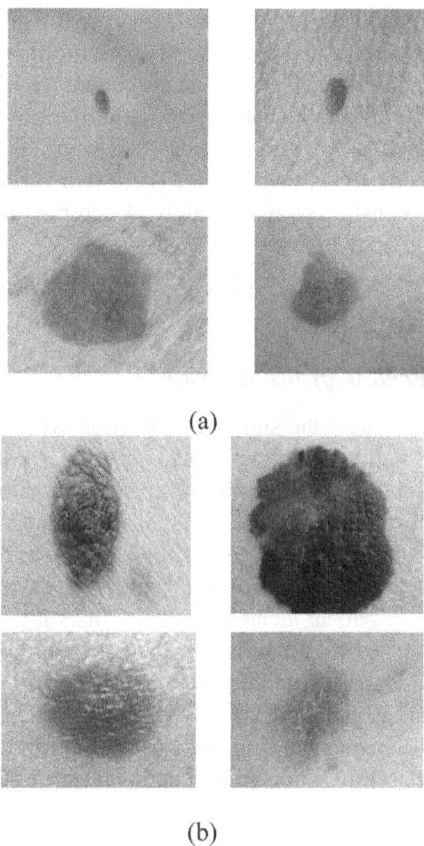

Figure 71.1. (a) Benign and (b) malignant dermatological images.

Source: Author.

Given the fact that skin photographs are regarded as medical photos, they will likewise be subject to the same restrictions. Either speech or textual information, healthcare photo tagging can't be outsourced to another parties. The only people who have the ability to identify such data are experienced radiologists, and it requires them at least 30 minutes to meticulously label every image after thoroughly reviewing it on several occasions to determine the location of any lesions. Healthcare picture data will keep encountering difficulties due to such acute scarcity of qualified specialists (far fewer than those whose are capable of annotation other audio or text data). This indicated that the value of every pixel varied from zero to one with the use of the value assigned to the parameter (1./255). The images have been turned by fifteen degrees using the rotational transform. The snapshots were shifted randomly to the opposite direction or left using the dimension shift range modifications, with the width shift parameters specified as 0.1. Vertical shift was applied to the prepared images using an elevation shift range variable value of 0.1. The photographs were enlarged if the range of the zoom argument was more than 1.0, and the pictures were stretched out if it was lower than 1.0. As a result, the picture was enlarged with a 0.2 magnification level. The reverse technique was used for transforming the original horizontal image. Because an intensity change was applied, the spectrum was 0.5–1.0, while 0.0 represents no intensity and1.0 represents maximum brightness. A Table 71.2 show that the 0.05 transmit shift range was used to obtain the smoothest fill pattern. This is due to the fact that an odd number selected from the assortment is used to shift the input channel parameters during multichannel shift translation.

3.5. Pre-conditioned DL model

3.5.1. SENet model

Skin the classification of cancer has benefited greatly from the application of squeeze-and-excitation networks, or SENets, which have been shown to increase diagnosis accuracy. SENet-enhanced models have demonstrated remarkable accuracy rates in recent research, frequently exceeding 90%. In skin cancer detection tasks, for example, merging SENet with ResNet structures has produced rates of precision reaching as high as 98.74%, demonstrating their potential to improve the effectiveness of classification and enhance the presentation of features in medical imaging applications (MDPI). This illustrates how SENet can help with accurate and timely skin malignancy diagnosis, improving the results for patients. The ImageNet collection contains millions of photos that are used to train the convolutional neural network, also known as Senet, which is An image size of 224 by 224 is a robust feature used for training the neural network for image detection. Figure 71.2 shows the structure of SENet and its searching process for the optimal CNN layout. Two types of SENet are available: SENetLarge and SENet-Mobile for small-scale networks. It searches small datasets of images for the best layer of convolution or cells using a method for searching. Using Fourier cells yields better outcomes for classification at a reduced computational expense. Then, we may apply SENet to images of any dimension by using these convolutional cells to create regular and decrease cells. The most widely used CNN design with the least amount of processing complexity is provided by a set of SENet frameworks.

The fundamental parameters that were used to train the SENet Mobile model are shown in Table 71.3. The

Table 71.2. Image enhancement methods

Methods	*Value*
Range of Scale Transformation	zero to one
Turning Range	15 degrees
Shift_Width_Range	0.1
Height_Shift_Range	0.1
Zoom Transformations	0.2
Horizontal_Flip	True
Vertical_Flip	True

Source: Author.

optimization techniques Adam and Nadam are employed. The procedure of improving the SENet Mobile architecture is illustrated in Figure 71.2. It is evident from the illustration that the SENet fundamental model is followed by a 2D GAP layer. The final result of the 2D GAP layer is then used as the starting point for a layer that is extremely dense. On the triggering layer, there are numerous function activation options,

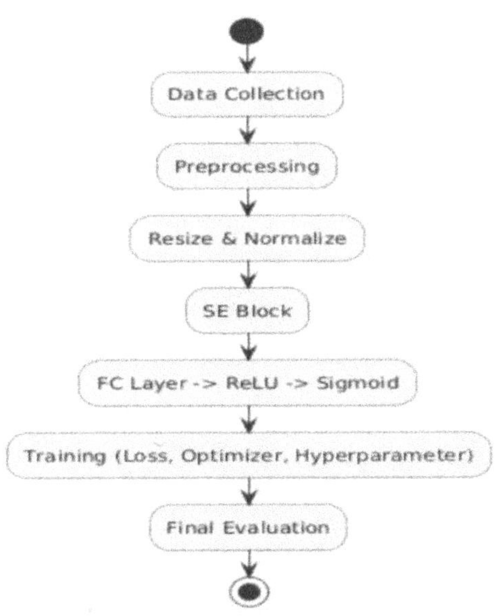

Figure 71.2. Optimization of SENet model.

Source: Author.

Table 71.3. Lists the SENet mobile characteristics

Methods	*Value*
Reduction Ratio	Generally set to 4, 8, or 16 to regulate the SE block's reduction in size.
Global Average Pooling	Reduces every feature map to a single value.
Fully Connected Layers	There are two layers: the first one shrinks the channel and the second one enlarges it.
Activation Functions	Sigmoid following the subsequent fully linked layer and ReLU after the first.
Input Image Size	Smaller (e.g., 224x224) in order to accommodate mobile screen sizes.
Batch Size	Modified to strike a compromise between processing speed and memory use on mobile devices.
Quantization	Decreases the size of the model and speeds up inference by using weighting and stimulation with a lower accuracy (e.g., 8-bit integers).
Batch Size	32
Epochs	50

Source: Author.

including Tanh, Relu, and Sigmoid networks. In the preliminary the platform, the stimulation function chosen is the Relu operation sigmoid activation is employed. After that, an empty layer with a dropouts rate of 0.5. This phenomenon will have an impact on the predictive conclusions regarding the new data. [10] One such method is to use Dropping out (a technique that involves voluntarily eliminating vertices in order to minimize the model's complexity and prevent overfitting). Then, more dropout and thick layer combinations are employed. Softmax is the last tool used for categorization.

4. Evaluation of Performance

The matrix of uncertainty (CM) for SENet Mobile and SENet Large, correspondingly, utilizing the optimization algorithm Adam, is shown in Tables 71.4 and 71.5.

1. Accuracy

$$\text{Accuracy}=\frac{TP+TN}{TP+TN+FP+FN} \quad (1)$$

2. Precision

$$\text{Precision}=\frac{TP}{TP+FP} \quad (2)$$

3. Sensitivity

$$\text{Sensitivity}=\frac{TP}{TP+FN} \quad (3)$$

4. F1 Score

$$\text{F1 Score}=\frac{2*Precision*Recall}{Precision+Recall} \quad (4)$$

The Epoch vs. Accuracy for SENet Mobile and Large is shown in Figures 71.3 and 71.4, correspondingly. The Epoch vs. Loss for the SENet Mobile and Large models is shown in Figures 71.5 and 71.6, accordingly. The collected ROC results of the suggested SENet models for the Adam

Table 71.4. SENet mobile confusing matrix

		Positive in actuality Malignancy	*In actuality negative Benign*
Positive Prediction	1. Malignancy	198	50
Expected Negative	0: Benign	37	243

Source: Author.

Table 71.5. SENet large confusing matrix

		Positive in actuality Malignancy	*In actuality negative Bengin*
Positive Prediction	1: Malignancy	213	35
Expected Negative	0: Benign	42	238

Source: Author.

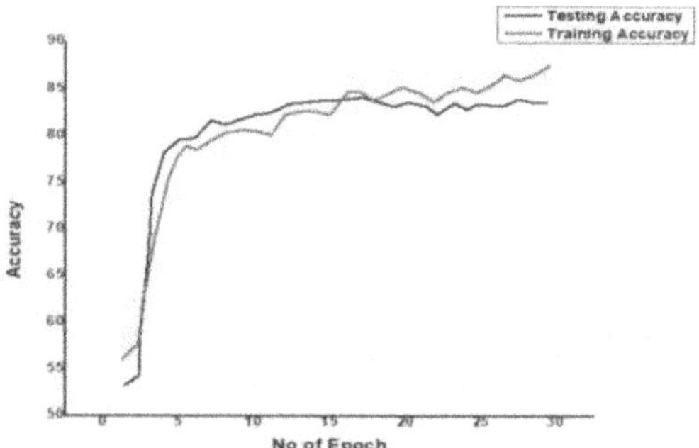

Figure 71.3. SENet mobile pre-training: Epoch vs accuracy.

Source: Author.

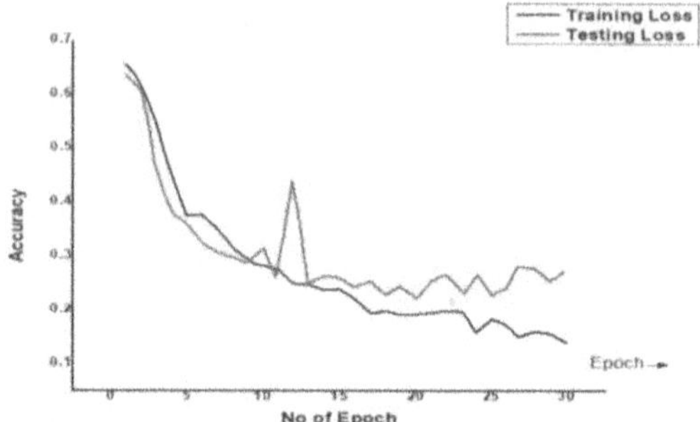

Figure 71.4. SENet large pre-training: Epoch vs accuracy.

Source: Author.

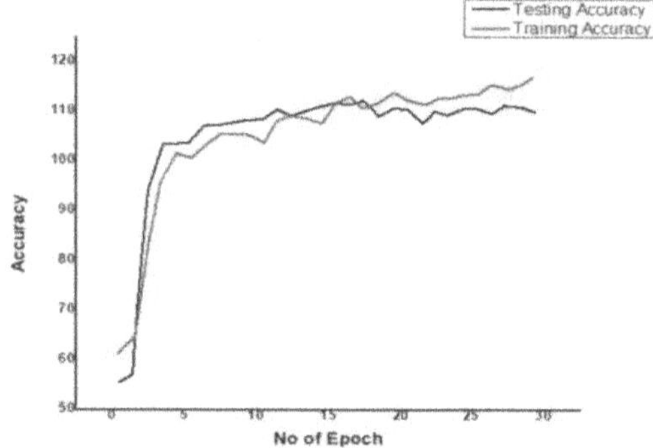

Figure 71.5. SENet mobile Epoch vs loss.

Source: Author.

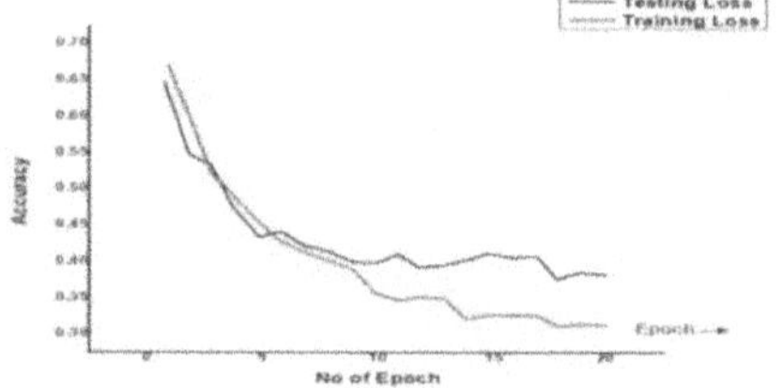

Figure 71.6. SENet large: Epoch vs loss.

Source: Author.

optimizer are shown in Figure 71.7. In particular, the results show how much better the optimized SENet model performs than alternative models. The AUC for both classes are 0.92 and 0.87, respectively.

According to Table 71.6, the suggested improved SENet model delivers reasonable responsiveness and precision when contrasted with a few previous DL algorithms that have been used in earlier research studies.

5. Conclusion

This study compares existing methods with a proposed SENet-based approach for skin cancer classification. The use of transfer learning and migration-based learning techniques

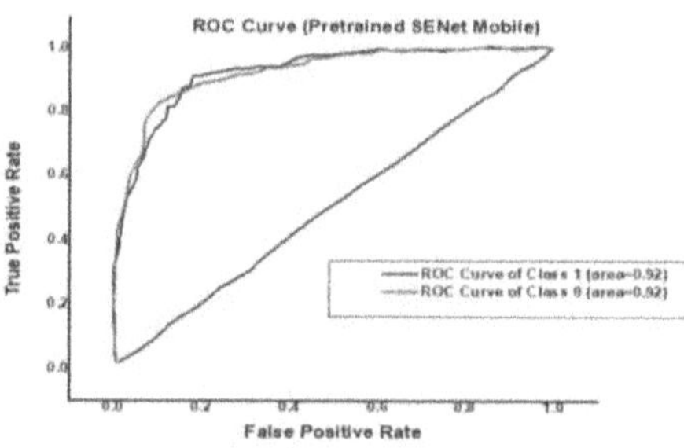

Figure 71.7. Adam optimizer's ROC curve for SENet mobile..

Source: Author.

Table 71.6. Comparing the recommended strategy with different approaches

Model	*Sensitivity*	*Precision*	*Accuracy*
ResNet -18	54.72%	68.71%	73.51%
ResNet -34	60.71%	79.72%	80.31%
ResNet -50	58.71%	69.71%	74.31%
ResNet -101	62.71%	75.74%	79.31%
ResNet -152	58.71%	79.75%	79.31%
VGG -11	60%	69.24%	74.11%
MobileNetV1	51.32%	75.74%	80.35%
MobileNetV2	62.4%	79.73%	75.33%
MobileNetV3	65.73%	79.67%	79.32%
DenseNet [38]	67.21%	72.42%	76.36%
ResNet50 + DenseNet [38]	74.25%	80.2%	82.54%
NASNet	74.23%	84.17%	86.63%
SENet (Our Model) is suggested	68.73%	89.61%	91.97%

Source: Author.

enhances accuracy, outperforming traditional RCNN models in skin image classification. SENet-based transfer learning improves precision, making it a valuable tool for automated skin cancer detection, especially in datasets with limited images. However, selecting an inappropriate transfer learning model can lead to negative transfer issues, affecting training efficiency and accuracy. Future research should focus on optimizing network selection to further enhance skin cancer diagnosis and classification.

References

[1] Rahman, M. A., Bazgir, E., Hossain, S. S., & Maniruzzaman, M. (2024). Skin cancer classification using NASNet. *International Journal of Science and Research Archive*, *11*(1), 775–785.

[2] Gandhi, S. A., & Kampp, J. (2015). Skin cancer epidemiology, detection, and management. *Medical Clinics*, *99*(6), 1323–1335.

[3] Harrison, S. C., & Bergfeld, W. F. (2009). Ultraviolet light and skin cancer in athletes. *Sport Health*, *1*, 335–340.

[4] Rogers, H. W., Weinstock, M. A., Feldman, S. R., & Coldiron, B. M. (2015). Incidence estimate of nonmelanoma skin cancer (keratinocyte carcinomas) in the US population, 2012. *JAMA Dermatology*, *151*(10), 1081–1086.

[5] Whiteman, D. C., Green, A. C., & Olsen, C. M. (2016). The growing burden of invasive melanoma: projections of incidence rates and numbers of new cases in six susceptible populations through 2031. *Journal of Investigative Dermatology*, *136*(6), 1161–1171.

[6] Xie, F., Fan, H., Li, Y., Jiang, Z., Meng, R., & Bovik, A. (2016). Melanoma classification on dermoscopy images using a neural network ensemble model. *IEEE transactions on medical imaging*, *36*(3), 849–858.

[7] Dalila, F., Zohra, A., Reda, K., & Hocine, C. (2017). Segmentation and classification of melanoma and benign skin lesions. *Optik*, *140*, 749–761.

[8] Bomm, L., Benez, M. D. V., Maceira, J. M. P., Succi, I. C. B., & Scotelaro, M. D. F. G. (2013). Biopsy guided by dermoscopy in cutaneous pigmented lesion-case report. *Anais Brasileiros de Dermatologia*, *88*, 125–127.

[9] Kato, J., Horimoto, K., Sato, S., Minowa, T., & Uhara, H. (2019). Dermoscopy of melanoma and non-melanoma skin cancers. *Frontiers in Medicine*, *6*, 180.

[10] Thurnhofer-Hemsi, K., Lopez-Rubio, E., Dominguez, E., & Elizondo, D. A. (2021). Skin lesion classification by ensembles of deep convolutional networks and regularly spaced shifting. *IEEE Access*, *9*, 112193–112205.

[11] Sarker, B., Sharif, N. B., Rahman, M. A., & Parvez, A. S. (2023). AI, IoMT and Blockchain in Healthcare. *Journal of Trends in Computer Science and Smart Technology*, *5*(1), 30–50.

[12] Alam, F. B., Podder, P., & Mondal, M. R. H. (2023). RVC-Net: A hybrid deep neural network framework for the diagnosis of lung diseases. *Plos one*, *18*(12), e0293125.

[13] Rahman, S. M., Ibtisum, S., Podder, P., & Hossain, S. M. (2023). Progression and challenges of IoT in healthcare: A short review. *arXiv preprint arXiv:2311.12869*.

72 A smart adaptive neuro fuzzy inference system (ANFIS) model integrated with DAB converter for EV charging systems

Revathy K. P.[1,a] *and K. Vijayakumar*[2,b]

[1]Research Scholar, School of Electronics, Electrical and Biomedical Technology, Kalasalingam Academy of Research and Education, Tamil Nadu, India
[2]Associate Professor, School of Electronics, Electrical and Biomedical Technology, Kalasalingam Academy of Research and Education, Tamil Nadu, India

Abstract: As fossil fuels grow more expensive and scarce, electric vehicles (EVs) will play an increasingly important role in the future of public transportation. Solar PV renewable sources can be used to charge EV vehicles because PV power generation can occur anywhere. In this project, we show a battery-powered EV charging system with an efficient converter and control strategy. This research aims to improve the efficiency of EV charging systems by combining PV systems with cutting-edge converter and controller models. The proposed study reduces switching stress while boosting the output voltage gain of PV-EV charging systems by employing a lightweight converter design known as Dual Active Bridge (DAB). The goal of this work is to develop an adaptive neurofuzzy inference system (ANFIS) to increase the performance and efficiency of the DAB converter by producing the necessary pulses. In addition, this work includes a complete simulation analysis to evaluate and compare the output voltage, power, efficiency, settling time, and voltage gain outcomes of the proposed DAB-ANFIS model to existing controllers and converters utilized in electric vehicle charging applications.

Keywords: Adaptive neuro fuzzy inference system (ANFIS), dual active bridge (DAB) converter, photovoltaic (PV), electric vehicle (EV)

1. Introduction

Owing to the increasing urgency to minimize pollution and reduce the consumption of fossil fuels, electric cars (EV) are now an economically viable alternative to conventional fuel-injected automobiles. EV charging stations need to be constructed in order to supply the electrical power needs of the abundant electric vehicles [1] which have become more prevalent recently. In order to achieve optimum operational efficiency, the charging stations demand an energy management control plan that identifies and develops the optimal contractual boundaries. Efficient battery charging can play a major role in the progress of current EVs. The emergence and increasing use of EVs require the setting up of charging stations in key locations due to the limited capacity of EV batteries. Because of power gaps, voltage sag, and extensive demand overflow, enormous, directly grid-connected charging stations place a specific strain on the electrical system's reliability and dependability. As a result, utilities must implement advanced management strategies and infrastructure improvements to mitigate these issues. This includes investing in energy storage solutions and enhancing grid resilience to accommodate the growing demand for electric vehicle charging. The effectiveness of charging, length of charge, and battery longevity are all influenced by the battery pack's parameters [2]. The primary variables that influence how the charger circuit functions are the soft switching techniques, controlling approach, system elements, and circuit structure design.

However, infrastructure for EV charging has been linked with photovoltaic (PV) generation [3]. There is a great deal of uncertainty associated with combining solar generation and EV charging; therefore, rigorous control design and optimization are essential to maintaining system stability. System stability can be enhanced by implementing advanced predictive algorithms that account for changing weather conditions and fluctuating electricity demand. Additionally, integrating energy storage solutions can provide a buffer, allowing for more effective management of solar energy output and EV charging needs. As the number of EVs increases, the generators and connections in the current distribution system will be under more stress, which could lead to disruptions in their operation, particularly during periods of high demand [4, 5]. It is anticipated that grid linkage for PV and EVs will be achieved at comparable voltage levels and in a widely disseminated manner [6]. Residential electric vehicles, for example, have specifications that are similar to commercial PV systems. PV can also be put more conveniently in a variety of EV charging places, including public, commercial, and car charging stations. Furthermore, PV and EV [7] are linked

[a]revathy.kp@mbcet.ac.in, [b]k.vijayakumar@klu.ac.in

DOI: 10.1201/9781003675259-72

to the grid by power electronic interfaces, allowing the establishment of intelligent nodes throughout the system. For PV-incorporated EV charging applications, a range of converter topologies and controlling methodologies have been used in earlier academic publications. High power loss, increased voltage stress, computational complexity, and more complex circuit design are issues that blight most previous research. The primary objective of the proposed research project is to develop a unique, lightweight controlling algorithm with an advanced converter architecture specifically for EV charging systems. The key aspects of this study are to use advanced converter and controller models to boost the efficiency of EV charging systems, increasing the output voltage gain of EV charging systems while lowering switching stress by utilizing Dual Active Bridge (DAB), a lightweight converter design is implemented. In order to improve the device's overall performance and efficiency, the Adaptive Neuro Fuzzy Inference system (ANFIS), a sophisticated regulatory mechanism that creates the pulses required to successfully run the DAB converter, is being developed in this project. In addition, a detailed simulation analysis is performed in this study to assess and compare the output voltage, power, efficiency, settling time, and voltage gain results of the proposed DAB-ANFIS model. This paper's subsequent sections are grouped into the following units: Section 2 conducts a thorough analysis of the literature to evaluate various converter topologies and intelligence-based control algorithms used in EV applications. Section 3 provides a detailed explanation of the proposed intelligence controlling model and DAB converter, supported by circuit models and mathematical representations. Section 4 analyzes the efficiency outcomes of the suggested framework and validates the outcomes of simulations using a variety of parameters.

2. Related Works

An enhanced hybrid converter model was employed by [8] for an integrated PV-EV charging system. This system's Enhanced Hybrid Converter (EHC), which integrates into a single-phase grid, is fed by solar energy cells. The emphasis of the lesson is on managing the DC voltage for EV charging, regulating the converter to sense the maximum power from PV, and controlling the AC power supply to provide the AC grid. To provide dependable battery charging operations, an isolated SEPIC converter component has been employed in this instance. [9] adopted an innovative control strategy for an electric vehicle's integrated photovoltaic battery energy system. An efficient converter controlling is carried out by employing the recommended smart coordinated scheme, which helps to prevent the intermittent effects of PV. Additionally, by combining photovoltaic technology with a battery energy storage device, the scientists hope to lower energy consumption. [10] used a controlling mechanism based on salp swarm optimization to prevent DC voltage fluctuations at the load side with less harmonics. This study proposes an alternative method of changing the PI controller that regulates the grid-tied inverter to accommodate an EV charging station in grid-connected solar energy systems. The proposed method is based on salp-swarm algorithm optimization combined with empirical formulas. The dc-bus voltage and current controllers are linked with six gains, which can be determined using the provided method. [7] developed a unified controlling algorithm for supporting power transfer in EV and grid systems. Without the need of a mechanical automated voltage regulator, the recommended controlling approach is utilized for controlling the frequency and voltage of the DG set. In order to prevent the storage battery from being overcharged, it also facilitates feeding excess electricity produced by PV arrays onto the grid. Tavakoli et al. [3] performed a thorough investigation to evaluate the fundamental effects of integrating PV systems with the grid, with a particular emphasis on power quality, voltage stability, and economic challenges. This inquiry has led to an analysis that shows that in order to ensure an efficient grid EV integration, parameters like as voltage stability, frequency stability, and oscillation stability must be taken into account. [11] developed an isolated converter topology and generalized controlling model for EV charging applications. The goal of this study is to identify the best model for EV charging by examining and analyzing various controlling strategies. Furthermore, certain fast charging stations that are integrated with DC-DC converters have also looked for ways to get high voltage. [12] suggested a multi-level converter topology for EV charging applications, where the authors aim to handle the power supply with minimized current ripples. It presented an EV charging system's one stage bi-directional converter layout. In order to improve charging efficiency, a multilevel inverter design combined with phase shift modulation has also been employed. [13] deployed a unidirectional converter model with voltage inverter component for enhancing the charging efficiency of EV systems. Previous research has used many converter topologies for PV integrated EV systems, including Buck converter, Boost converter, Bi-directional converter, High step up converter, Flyback converter, Cuk converter. Comparison of these converters and their applications are presented in Table 72.1.

3. Proposed Methodology

This part contains the block diagram, comprehensive explanations of the work, and mathematical assumptions. The primary outcome of this study is the development and use of an intelligent controlling algorithm for EV charging applications using a DAB converter [14]. A reliable infrastructure for EV charging must be established in order to accommodate the growing number of EVs on the road. The converter's voltage output changes and is unstable due to the fluctuating voltage power supply for charging. This study uses the intelligent

Table 72.1. Different types of converters and application

Type of converter	*Application*	*Pros*	*Cons*
Buck	EV charging	Low ripples, and high transient response.	Low voltage gain.
Boost	Battery charging	Increased efficiency, and gain output.	High voltage stress.
Bi-directional	Large scale grid and battery charging.	High energy efficiency and minimal loss.	Increased noise and hard switching.
High step-up	EV – grid applications	Increased voltage gain and efficiency.	High complexity.
Flyback	EV application	Better efficiency and output gain.	Increased voltage stress and designing complexity.
Cuk	Power quality improvement in EV	Reduced harmonic interferences.	High current stress and low voltage gain.

Source: Author.

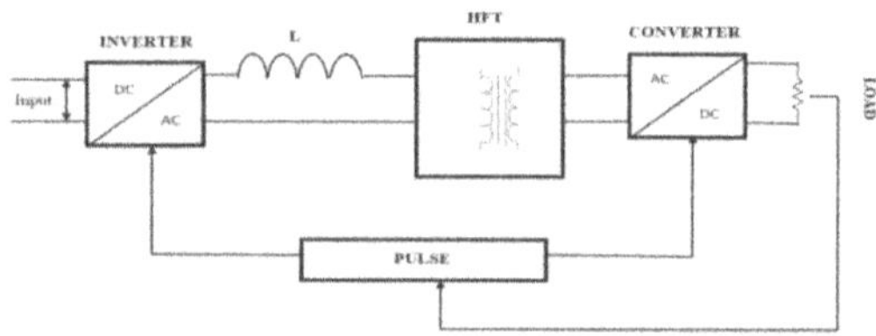

Figure 72.1. Block diagram of the proposed DAB-ANFIS model.

Source: Author.

Adaptive Neuro Fuzzy Inference System (ANFIS) technique to deliver a consistent and distortion-free converter output specifically for EV charging application.

Figure 72.1 illustrates the suggested DAB-ANFIS model's block diagram. In the proposed model, the DAB converter is mainly used for voltage augmentation and regulation, and the recommended ANFIS model is used to successfully regulate it. Using the proposed controlling method, the ideal control signal parameters (ki, kd, and kp) are selected to produce pulses that operate the DAB converter. The DAB converter outperforms other current converter models in terms of output voltage gain and efficiency.

3.1. Dual active bridge converter (DAB)

A DAB converter, or bidirectional isolated DC-DC converter, should be employed in applications of high power such as powering electric vehicles. The converter contains bidirectional power flow, soft switching, galvanic isolation, and high power density. The converter is made up of two symmetrical active bridges shared by a transformer of high frequency. The switching transformer provides galvanic isolation along with higher voltage conversion efficiency [15]. The high-frequency square waves in the high-voltage link, along with the leakage inductance of the transformer, play crucial roles in power transfer. Therefore, when voltage magnitudes on both sides of the transformer become unbalanced, circulating power is increased, along with power loss, and efficiency is drastically reduced. It is as easy to control this converter topology as driving both the bridges with complementary constant pulse width modulated signals at 0.5 duty cycle. A high frequency square wave voltage output can be produced at the terminals of the transformer (V1, V2) employing an efficient method of control. The two square waves produced are carefully phase-shifted to manage the power flow. Figure 72.2 shows the matching schematic diagram for the DAB converter.

The following equation provides an estimate of the leakage inductance current:

$$\frac{diL_S}{dt} = \frac{V_{T1} - V_{T2}}{L_S} \tag{1}$$

Consequently, the leakage inductance current values are computed using the subsequent formulas:

$$I_1 = \frac{T}{2L_S}\left(2\frac{V_2}{n}d + V_1 - \frac{V_2}{n}\right) \tag{2}$$

$$I_2 = \frac{T}{2L_S}\left(2V_1 d - V_1 + \frac{V_2}{n}\right) \tag{3}$$

Next, using the following equation, the average current value for both the input and output is found:

$$\bar{I}_1 = \frac{1}{T}\left(\frac{1}{2}I_2 t_2 - \frac{1}{2}I_1 t_1 + (I - d)T\frac{1}{2}(I_1 + I_2)\right) \tag{4}$$

$$\bar{I}_2 = \frac{1}{nT}\left(\frac{1}{2}I_1 t_1 - \frac{1}{2}I_2 t_2 + (1 - d)TI_2 + (1 - d)T\frac{1}{2}(I_1 - I_2)\right) \tag{5}$$

Using this converter paradigm, the proposed framework provides higher efficiency and voltage gain with less complexity. The particular parameters could change depending on the application needs, design criteria, and component qualities. T is a important variables to set in a DAB converter, along with other related considerations such as Transformer

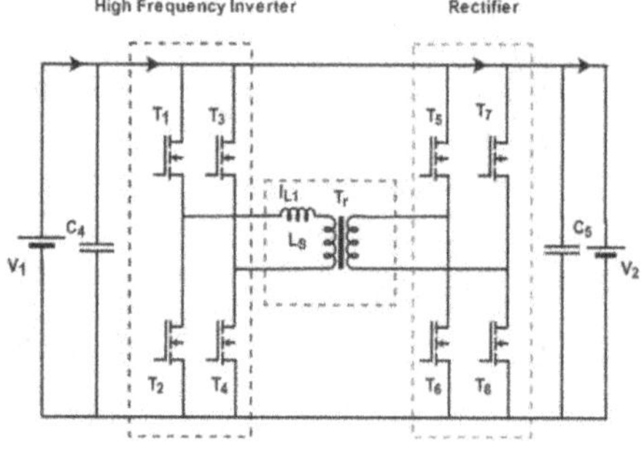

Figure 72.2. Schematic model of DAB converter.

Source: Author.

Turns ratio, Switching Frequency, Duty Cycle(D) of pulses, Filter inductance and capacitance and Load.

3.2. *Adaptive neuro fuzzy inference system (ANFIS)*

The Fuzzy Inference System (FIS) serves as the foundation for the Adaptive Neuro-fuzzy Inference System (ANFIS), an artificial neural network which uses FIS and ANNs to enhance the fault tolerance, training speed, and versatility of traditional ANNs. ANFIS is an artificial intelligence method for figuring out which responses are most likely. By integrating learning capabilities from concepts, ANFIS improves the accessibility of adaptive modeling and enables it to estimate complex and non-linear functions. An all-encompassing approach to inference and decision-making processes is provided by the foundational artificial intelligence concept known as ANFIS. In ANFIS the decisions can be taken based on learning algorithm unlike Fuzzy system. The primary purpose of the ANFIS implementation in the proposed work is to generate controlling signals for the operation of converters with high voltage conversion rates and gains. Several kinds of regulating mechanisms are being produced for converter control in the previous research activities. Nonetheless, the bulk of methods struggle with the primary issues of time consumption and intricate computational procedures. The proposed study's purpose is to develop a novel, lightweight smart controlling method to improve DAB converter control operations. Furthermore, the primary purpose of the suggested ANFIS is to select the optimal control signal parameters, such as ki, kd, and kp, for producing pulses that will run the converter. In short, utilizing a combination of fuzzy logic and neural networks, the Adaptive Neuro-Fuzzy Inference System (ANFIS) provides adaptive and robust control over the DAB converter. Except for piece wise differentiability, an adaptive network's node functions are essentially unconstrained in terms of functionality in ANFIS. The only restriction on network design structurally is that it must be feed forward in nature. In the subsequent section, we put forward an instance of adaptive network that is functionally analogous to fuzzy inference systems in terms of the DAB converter for EV charging applications. The architecture proposed is called the ANFIS integrated DAB converter. There are several distinctions between a regular DAB converter and one which integrates an Adaptive Neuro-Fuzzy Inference System (ANFIS) with a DAB converter model. The suggested approach of integrating an adaptive neural fuzzy inference control system with a dual active bridge converter differs from traditional dual active bridge converters by the fact it allows the system to adapt rapidly to changing conditions, especially ones observed in EV charging. It additionally improves performance by learning from the provided historical data set and modifying methods of control accordingly, allowing the recommended approach to function in non-linear environments.

3. Results

The modeling and performance results of the recommended DAB-ANFIS model based on various assessment measures are shown in the following section using the Matlab/Simulink tool. The simulation parameters used for setting up the model is listed in Table 72.2. For EV charging systems, the DAB converter has been developed in the proposed study utilizing the ANFIS Controlling technique.

The designed ANFIS model is shown in Figure 72.3.

The input voltage to DAB converter and the ANFIS controller can be simulated using proper training and by providing load data set, which is shown in the Figure 72.4. For this the input voltage is given as 1000 V.

The generation of FIS data set is shown in Figure 72.5.

Figure 72.6 shows the plot against loading and testing data in ANFIS controller.

Furthermore, as shown in Figure 72.7 the DAB converter's output voltage and current are evaluated in relation to the

Table 72.2. Simulation parameters

Parameters	*Specifications*
Input voltage	1000 V
Open circuit voltage	22.6V
Short circuit voltage	12V
Short circuit current	20.833A
Converter transformer	5kVA, 100kHz
Switch	500V

Source: Author.

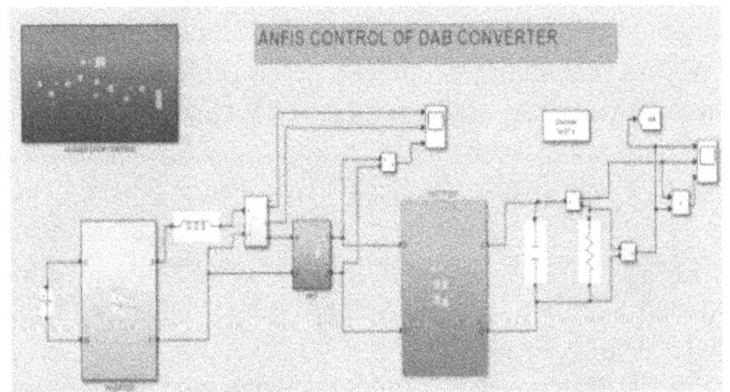

Figure 72.3. DAB-ANFIS integrated model for EV charging system.

Source: Author.

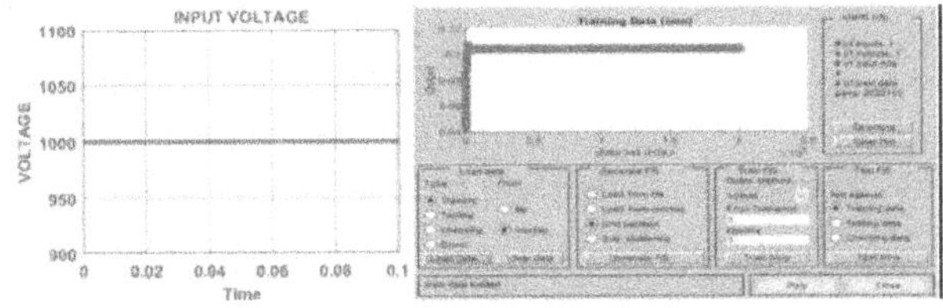

Figure 72.4. Input voltage and training data set of ANFIS.

Source: Author.

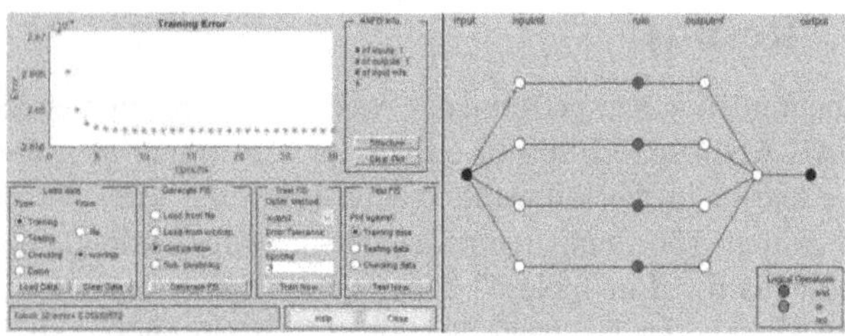

Figure 72.5. Generation of FIS.

Source: Author.

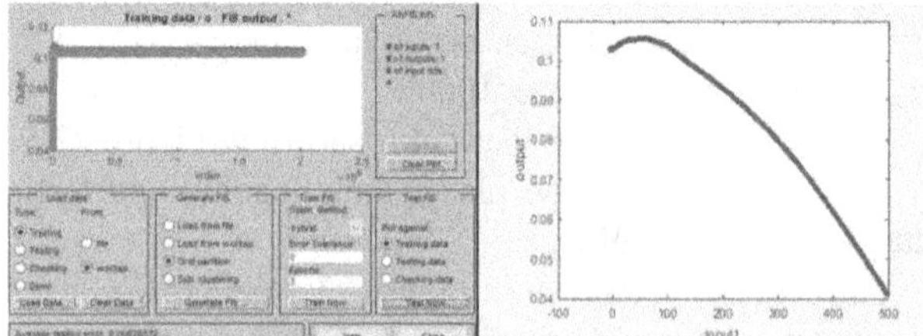

Figure 72.6. Plot of training vs testing data set of ANFIS.

Source: Author.

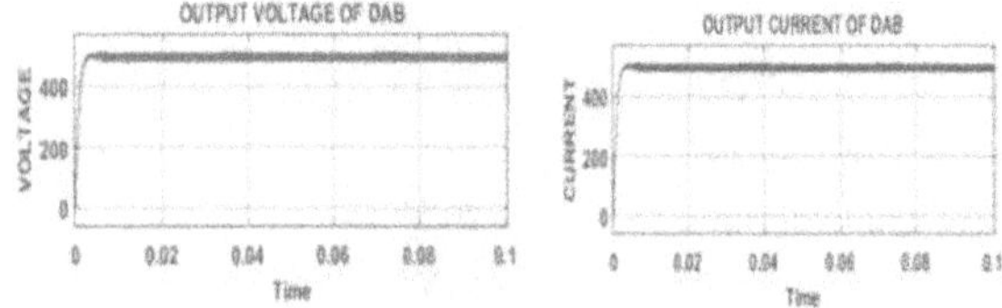

Figure 72.7. Output voltage and current of DAB converter.

Source: Author.

changing time, which is represented in seconds. Since the use of ANFIS technology is directly responsible for the higher output voltage gain of the suggested EV charging system.

As a result, as depicted in Figure 72.8, the inverter's output voltage is also predicted. To enhance power quality and minimize harmonic interference, the suggested design converts DC to AC. According to the results, the inverter's output voltage and current are appropriately enhanced before being given to the load.

The DAB converter's output voltage gain is then computed, it serves as the most important aspect in determining the converter's overall conversion efficiency. Depending on the obtained waveform, it can be concluded that there is a significant increase in voltage gain with changing time. In addition, to voltage gain DAB's voltage conversion efficiency is demonstrated in Figure 72.9. The ANFIS approach can be used to enhance the recommended voltage gain. Given that it generates appropriate pulses to regulate the DAB converter with the optimum controlling parameters. The efficiency of the converter is further improved by minimizing losses associated with switching and conduction. This optimization not only enhances the performance of the DAB converter but also contributes to a more stable and reliable power output,

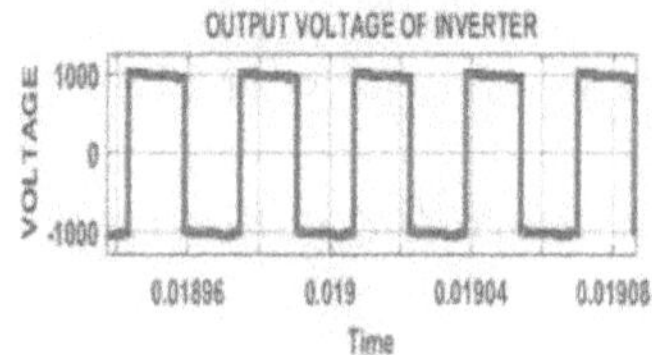

Figure 72.8. Output voltage of inverter.

Source: Author.

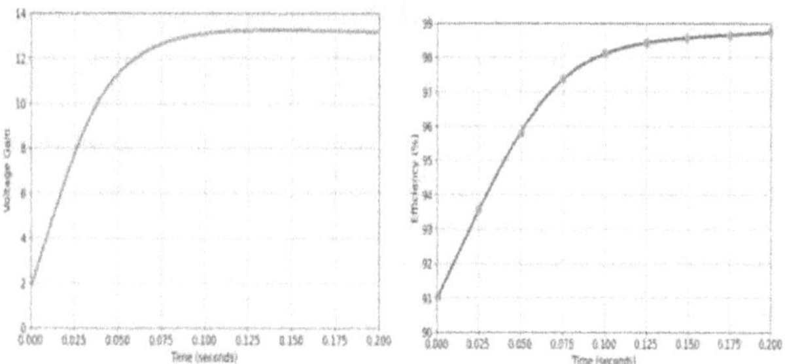

Figure 72.9. Voltage gain and efficiency of converter.

Source: Author.

which is crucial for various applications in power electronics. The efficiency converter, as depicted in the figure, is approximately 98.7, meaning it can satisfactorily convert over 98% of the input voltage to the output.

Harmonic analysis of the DAB-ANFIS model is done. Where, Harmonic analysis in an Adaptive Neuro-Fuzzy Inference System (ANFIS) controller is the process of evaluating and understanding the frequency components of signals involved in the system. This analysis is crucial for identifying, quantifying, and mitigating harmonics, which are higher frequency components superimposed on the fundamental frequency of a signal. Figure 72.10 shows the harmonic analysis of the system at converter and inverter side.

The above results shows the DAB-ANFIS model of system for EV charging. Whereas if we evaluate the basic of ANFIS that is, DAB Converter with fuzzy controller the output voltage is only 72.1 V for 12 V input thereby drastically reducing efficiency to 80 percentage for fuzzy controller. This output voltage relation Fuzzy controller with DAB converter for EV charging application is shown in Figure 72.11.

Additionally, as shown in Figure 72.12, a comparison is shown between the efficiency and settling time features of the

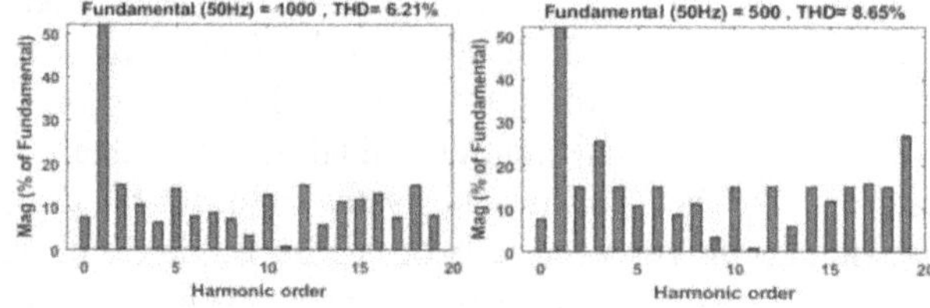

Figure 72.10. THD of DAB-ANFIS system.

Source: Author.

ANFIS model and other conventional controlling schemes. The study's conclusions demonstrate a significant increase in efficiency with a decrease in settling time when compared to traditional regulating approaches. Table 72.3 shows the comparison of simulation results with that of conventional converters with different control strategies in terms of output voltage and efficiency.

4. Conclusion

This study's primary contribution is the creation and use of an intelligent controlling algorithm for DAB converter-based EV charging applications. The input source's fluctuations cause the converter's voltage output to fluctuate and become unstable. The intelligent ANFIS technique is employed in this study to provide a consistent and distortion-free output from the converter. The fundamental part of the suggested model is the DAB converter, whose voltage augmentation and regulation are effectively governed by the preferred ANFIS model. To generate the pulses that drive the DAB converter, the optimal control signal parameters, such as ki, kd, and kp, are chosen using the suggested controlling method. The DAB converter is more efficient and has a higher output voltage gain than other current converter models. Since PVs and EVs are two of the most exciting and quickly developing technologies that are anticipated to have a major impact on the electrical industry in the next ten years, this model can be merged with solar photovoltaic systems in the future. This integrated model can reduce harmonics and other distortions in systems during charging and offer steady output voltage and current even when the system is integrated with PVs. When compared to conventional regulating methods, the study's conclusions show a notable boost in efficiency and a decrease in settling time.

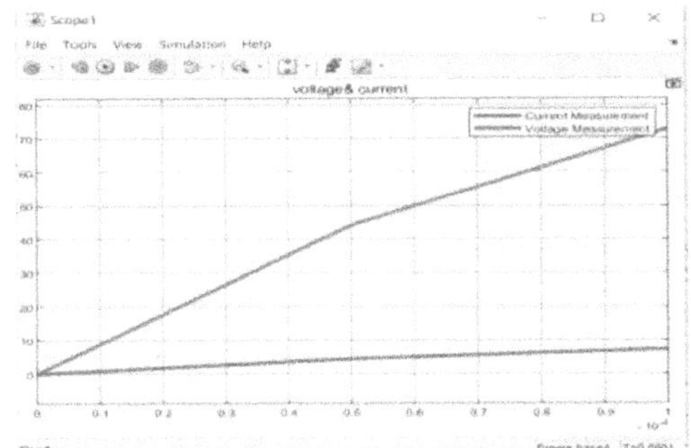

Figure 72.11. DAB-Fuzzy system.

Source: Author.

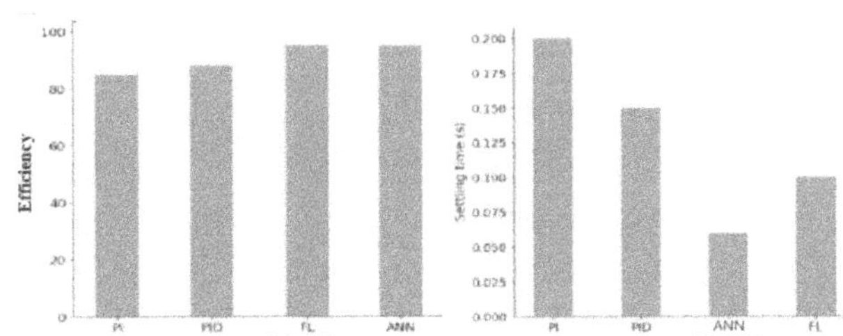

Figure 72.12. Comparison among the classic and proposed controlling techniques based on Efficiency & Settling time.

Source: Author.

Table 72.3. Comparison of proposed system with conventional methods

Converter topology with controller	*Output voltage*	*Efficiency*
Buck converter with PI controller	300	85
DAB converter with PI controller	375	90
DAB converter with Fuzzy Controller	400	95
DAB converter with Adaptive Neuro Fuzzy Inference system(ANFIS) Controller	500	98.7

Source: Author.

References

[1] Barker, T., Ghosh, A., Sain, C., Ahmad, F., & Al-Fagih, L. (2024). Efficient ANFIS-driven power extraction and control strategies for PV-bess integrated electric vehicle charging station. *Renewable Energy Focus, 48*, 100523.

[2] Saleem, S., Ahmad, I., Ahmed, S. H., & Rehman, A. (2024). Artificial intelligence based robust nonlinear controllers optimized by improved gray wolf optimization algorithm for plug-in hybrid electric vehicles in grid to vehicle applications. *Journal of Energy Storage, 75*, 109332.

[3] Tavakoli, A., Saha, S., Arif, M. T., Haque, M. E., Mendis, N., & Oo, A. M. (2020). Impacts of grid integration of solar PV and electric vehicle on grid stability, power quality and energy economics: A review. *IET Energy Systems Integration, 2*(3), 243–260.

[4] Ramadhani, U. H., Shepero, M., Munkhammar, J., Widén, J., & Etherden, N. (2020). Review of probabilistic load flow approaches for power distribution systems with photovoltaic generation and electric vehicle charging. *International Journal of Electrical Power & Energy Systems, 120*, 106003.

[5] Cheikh-Mohamad, S., Sechilariu, M., Locment, F., & Krim, Y. (2021). Pv-powered electric vehicle charging stations: Preliminary requirements and feasibility conditions. *Applied Sciences, 11*(4), 1770.

[6] Prem, P., Sivaraman, P., Sakthi Suriya Raj, J. S., Jagabar Sathik, M., & Almakhles, D. (2020). Fast charging converter and control algorithm for solar PV battery and electrical grid integrated electric vehicle charging station. *Automatika: časopis za automatiku, mjerenje, elektroniku, računarstvo i komunikacije, 61*(4), 614–625.

[7] Singh, B., Verma, A., Chandra, A., & Al-Haddad, K. (2020). Implementation of solar PV-battery and diesel generator based electric vehicle charging station. *IEEE Transactions on Industry Applications, 56*(4), 4007–4016.

[8] Sekhar, K. R., Chaudhari, M. A., & Khadkikar, V. (2023). Enhanced hybrid converter topology for PV-grid-EV integration. *IEEE Transactions on Energy Conversion*, *38*(4), 2634–2646.

[9] Datta, U., Kalam, A., & Shi, J. (2020). Smart control of BESS in PV integrated EV charging station for reducing transformer overloading and providing battery-to-grid service. *Journal of energy Storage*, *28*, 101224.

[10] Mohamed, A. A., El-Sayed, A., Metwally, H., & Selem, S. I. (2020). Grid integration of a PV system supporting an EV charging station using Salp Swarm Optimization. *Solar Energy*, *205*, 170–182.

[11] ElMenshawy, M., & Massoud, A. (2020). Modular isolated dc-dc converters for ultra-fast ev chargers: A generalized modeling and control approach. *Energies*, *13*(10), 2540.

[12] Mukherjee, S., Ruiz, J. M., & Barbosa, P. (2022). A high power density wide range DC–DC converter for universal electric vehicle charging. *IEEE Transactions on Power Electronics*, *38*(2), 1998–2012.

[13] Szymanski, J. R., Zurek-Mortka, M., Wojciechowski, D., & Poliakov, N. (2020). Unidirectional DC/DC converter with voltage inverter for fast charging of electric vehicle batteries. *Energies*, *13*(18), 4791.

[14] Jin, Z., Li, D., Hao, D., Zhang, Z., Guo, L., Wu, X., et al. (2024). A portable, auxiliary photovoltaic power system for electric vehicles based on a foldable scissors mechanism. *Energy and Built Environment, 5*, 81–96.

[15] Karmaker, A. K. , Prakash, K., Siddique, M. N. I., Hossain, M. A., & Pota, H. (2024). Electric vehicle hosting capacity analysis: Challenges and solutions. *Renewable and Sustainable Energy Reviews, 189*, 113916.

73 Mathematical analysis to enhance the frequency stability of interconnected power system

Shailesh Madhavrao Deshmukh[1,a] *and Ikhar Avinash Khemraj*[2,b]

[1]Assistant Professor, Department of Electrical, Kalinga University, Raipur, India
[2]Department of Electrical and Electronics Engineering, Kalinga University, Raipur, India

Abstract: In power systems that serve multiple areas as well as single-area systems, maintaining frequency stability continues to be a persistent challenge. For the purpose of preserving frequency stability, it is necessary to minimise and precisely control system frequencies while staying within permitted operational parameters. There has been a rapid decrease in the collective system inertia within interconnected power systems as a result of the increasing utilisation of power converters that are dependent on renewable energy sources (RES). This decrease in system inertia brings about an increase in the degree of uncertainty in load frequency control, particularly in power systems that serve multiple areas. Maintaining frequency stability becomes more difficult when renewable energy sources are incorporated into the system. In response to this challenge, the purpose of this paper is to propose the implementation of a Static Synchronous Series Compensator (SSSC) in conjunction with innovative load frequency control strategies in order to strengthen and improve the frequency stability of interconnected power systems.

Keywords: Frequency stability, load fluctuations, power system efficiency, tie-line control, SSSC

1. Introduction

The rapid integration of renewable energy sources (RES) into conventional power grids is driving a transformational shift in the global energy landscape. This shift is being propelled by the rapid integration of RES. However, the intermittent and variable nature of these sustainable energy sources presents unprecedented challenges to the stability and reliability of power systems. While these sources of energy promise to reduce reliance on fossil fuels and provide benefits to the environment, they also present a number of challenges. In order to guarantee the reliable operation of power systems that serve multiple areas as well as those that serve a single area, frequency stability is an essential component. The meticulous control and maintenance of system frequencies within prescribed operational thresholds is the key to understanding its significance. On the other hand, power converters that are based on renewable energy sources have become increasingly widespread, which has resulted in a significant decrease in the overall system inertia across interconnected power networks. To put this into perspective, multi-area power systems, which include a variety of energy sources such as thermal, hydroelectric, and gas systems, are especially vulnerable to the fluctuations that are brought about by the integration of renewable energy sources. The diminishing system inertia that occurs within these interlinked networks makes Load Frequency Control (LFC) more complicated. LFC is essential for preserving the delicate balance that exists between generation and consumption in each region. As the penetration of renewable energy sources (RES) increases, the traditional methods of frequency stability and low-frequency control (LFC) become insufficient. As a result, novel strategies are required to strengthen the interconnected power grid [6]. The purpose of this paper is to investigate the significant difficulties that are brought about by the incorporation of renewable energy sources (RES) in multi-area power systems [8]. Additionally, the paper suggests the incorporation of Static Synchronous Series Compensator (SSSC) technology, in conjunction with innovative load frequency control methodologies, in order to strengthen and maintain the frequency stability of these complex power networks. The purpose of this research is to provide a comprehensive framework for improving the resilience and efficiency of modern power systems in the context of the changing energy landscape. This will be accomplished by examining the synergy that exists between renewable energy sources (RES), frequency stability, and multi-area low-frequency converters (LFC). In today's modern power grids, the complexity of load uncertainty is further exacerbated by the growing popularity of electric vehicles (EVs), which are becoming increasingly widespread. Load fluctuations that are sudden and significant are brought about by electric vehicle charging stations, which are characterized by demand patterns that are intermittent and frequently unpredictable. The power system is subjected to immediate stress as a result of these fluctuations in demand, which frequently take the form of step signals as a result of the charging process. This further exacerbates frequency instability. The rapid changes

[a]ku.shaileshmadhavraodeshmukh@kalingauniversity.ac.in, [b]ikhar.avinash@kalingauniversity.ac.in

DOI: 10.1201/9781003675259-73

in power consumption that occur as a result of electric vehicle charging stations make the difficulties that are associated with load forecasting and management significantly more difficult to manage. This poses a significant risk to the delicate equilibrium that is necessary for the maintenance of stable system frequencies. An earlier study [1] developed a technique for active load management in microgrids that was voltage-independent. This technique was used to manage active loads. Through the course of the research, the objective was to achieve the desired conditions by combining a dependable hardware structure with an improved control method. In a separate research paper [18], a H∞-based decentralized robust control strategy was proposed with the aim of minimising frequency deviations that are a result of the incorporation of renewable energy sources (RES) [2]. Additionally, the strategy took into account the reduction of frequency deviations as well as the utilisation of energy storage devices, which compensated for variations in high frequency. The authors of the study [3] found that through the implementation of an inductive load at the output, it was possible to achieve load frequency control in renewable energy systems. The solar-micro-hydro power generation systems were used to simulate this approach, which was found to be economically viable. The authors [4] implemented a PI controller for the purpose of load frequency control in a thermal power system that was designed for reheating a single area. For the purpose of optimising the controller's parameters, such as the integral absolute error, the integral square error, and the integral time absolute error, they utilized a stochastic Particle Swarm Optimisation technique. Based on the findings, it was determined that the PI controller that was based on integral absolute error demonstrated superior performance, particularly when applied to step load functions of 1%. A technique for frequency regulation for renewable energy curtailment networks that is based on price signals was proposed in a paper [5]. The purpose of this technique was to retain the power flow profile within an acceptable range. In [19], an optimisation technique that was based on a differential evolution algorithm was utilized in order to determine the optimal values of PI and PID controllers. An equilibrium between the amount of power generated and the amount of load that was being demanded was the goal. An additional study [7] focused on reducing frequency deviation through the utilisation of fractional order PID (FOPID) controllers. The gains of PID controllers were optimized by the authors through the utilisation of Bacterial Foraging Optimisation in a hybrid power generation system that included thermal, wind, and photovoltaic components. The purpose of [20] was to reduce frequency deviation as much as possible by utilising NPID, FOPID, and NFOPID controllers. Through the application of genetic algorithms in a variety of settings, they discovered that the NFOPID controller demonstrated superior performance. A control strategy was proposed by the authors [9] in order to improve the efficiency of a small photovoltaic hydro-pumped storage system [10]. It was through the utilisation of NPID/PID controllers and a neural network that the frequency control was accomplished. A PI controller that was combined with a tilt-integral derivative with filter (TIDF) was used in a study that was conducted by the authors [21] to reduce the frequency oscillations that occurred in a power system. This combination was referred to as PI-TIDF. Through the utilisation of Hair's Hanks Optimisation (HHO) in conjunction with the ITAE fitness method, the controller parameters were optimized. An evaluation was conducted to determine how well HHO performed in comparison to other optimisation algorithms, such as GOA, GA, and CSA. Based on the findings, it was determined that the HHO-regulated PI-TIDF controller demonstrated exceptional performance standards. According to the authors [11], in order to improve the performance of the system after a load disturbance, they utilized an adaptive PI-based genetic algorithm controller. A PI controller that was optimized through the use of particle swarm optimisation control served as the benchmark for comparison using this method. A research paper [12] was written in which the primary and secondary frequency control of a system, as well as the specifics of frequency droop, were discussed. The utilisation of a fully active hybrid energy storage system (HESS) that is comprised of ultra-capacitors and batteries was the means by which this was accomplished. Additionally, a photovoltaic (PV) farm was incorporated into the system with the intention of enhancing grid inertia. The grid was able to maintain its stability through the efficient utilisation of renewable energy sources (RES). The implementation of a first-order lead-lag controller for superconducting magnetic energy storage was used in a study that was presented in [13]. This controller was successfully used to control frequency oscillations. It was necessary to employ this controller in order to absorb and supply active power in order to keep a balance between the generation of power and the demand for power. The authors [22] proposed a novel smart grid technology that they called 'electric spring' in order to regulate the demand for non-critical loads and integrate renewable generation for the operation of multiple energy storage systems (E_ss). In addition to that, they presented a rolling optimisation control method specifically designed to accomplish this goal. The Ziegler-Nichols strategy was utilized in a paper [15] to regulate the parameters of a PID controller for frequency control in a hybrid power system. This was done in order to achieve the desired results. For the purpose of determining whether or not the controller is applicable in real time, a hardware-in-the-loop simulation was carried out with the OP4510 [14].

2. Modelling of Power System With And Without SSSC

SSSC is provided to the power system in order to achieve the goal of stabilising the tie-power oscillations and frequency of the system. This can be accomplished by controlling the phase angle of the power system. Damping is achieved by varying the flow of active power in the line power system. This is done in order to counteract the accelerating and decelerating swings that are caused by the disturbed system [16]. The power flowing over the tie-line in the existence of SSSC can be specified by:

$$\Delta P_{12TCPS} = T_{12}(\Delta\delta_1 - \Delta\delta_2) + \frac{\Delta K}{1-\Delta K}.k_1$$

$$\Delta f(s) = \Delta Error(s)$$

$$\Delta P_{12TCPS} = T_{12}(\Delta\delta_1 - \Delta\delta_2) + \frac{\Delta K}{1-\Delta K}.k_1$$

$$\Delta f(s) = \Delta Error(s)$$

$$\Delta P_{12TCPS} = T_{12}(\Delta\delta_1 - \Delta\delta_2) + \frac{\Delta K}{1-\Delta K}.k_1$$

$$\Delta f(s) = \Delta Error(s)$$

$$\Delta P_{12TCPS} = T_{12}(\Delta\delta_1 - \Delta\delta_2) + \frac{\Delta K}{1-\Delta K}.k_1$$

$$\Delta f(s) = \Delta Error(s)$$

$$\Delta P_{12TCPS} = T_{12}(\Delta\delta_1 - \Delta\delta_2) + \frac{\Delta K}{1-\Delta K}.k_1$$

$$\Delta f(s) = \Delta Error(s)$$

$$\Delta P_{12TCPS} = T_{12}(\Delta\delta_1 - \Delta\delta_2) + \frac{\Delta K}{1-\Delta K}.k_1$$

$$\Delta f(s) = \Delta Error(s)$$

$$\Delta P_{12TCPS} = T_{12}(\Delta\delta_1 - \Delta\delta_2) + \frac{\Delta K}{1-\Delta K}.k_1$$

$$\Delta f(s) = \Delta Error(s)$$

$$P_{12} = \frac{|V1|.|V2|}{X_{12}(1-K)} \sin(\delta_a - \delta_b) + \frac{K}{1-K}.\frac{|V1|.|V2|}{X_{12}}\sin(\delta_a - \delta_b$$

The incremental tie-line power flow can be represented as follows:

$$\Delta P_{12TCPS} = T_{12}(\Delta\delta_1 - \Delta\delta_2) + \frac{\Delta K}{1-\Delta K}.k_1$$

$$\Delta f(s) = \Delta Error(s)$$

$$\Delta P_{12TCPS} = T_{12}(\Delta\delta_1 - \Delta\delta_2) + \frac{\Delta K}{1-\Delta K}.k_1$$

$$\Delta f(s) = \Delta Error(s)$$

In a similar manner, the tie-line power deviation can be obtained when SSSC is connected in a tie line that connects Area 1, Area 2, and Area 2, Area 3 [17]. The equation that describes the power deviation of the tie-line in the line that connects Area A and Area B is shown as follows:

$$\Delta P_{12SSSC}(s) = \frac{2\pi T_{12}}{s}[\Delta F_1(s) - \Delta F_2(s)] + \frac{\Delta K}{1-\Delta K}.k_1$$

So, by managing the phase angle of SSSC the tie line power can be controlled. The phase angle is given by:

$$\Delta P_{12TCPS} = T_{12}(\Delta\delta_1 - \Delta\delta_2) + \frac{\Delta K}{1-\Delta K}.k_1$$

$$\Delta f(s) = \Delta Error(s)$$

$$\Delta\phi_c(s) = \frac{K_\phi}{1+sT_{ps}} Erroe(s)$$

Where K is the SSSC gain and Tps is the time constant of the SSSC.

3. Results and Discussion

A multi-area power system is made up of regions that are connected to one another through tie lines, which makes it easier for them to exchange power with one another. In most cases, this network is comprised of three different power sources for each individual area. The reactance (X) of the tie lines has a significant impact on both the efficiency and the capacity of the power transfer system. The reduction in reactance value, which is made possible by the implementation of Static Synchronous Series Compensators (SSSC), is

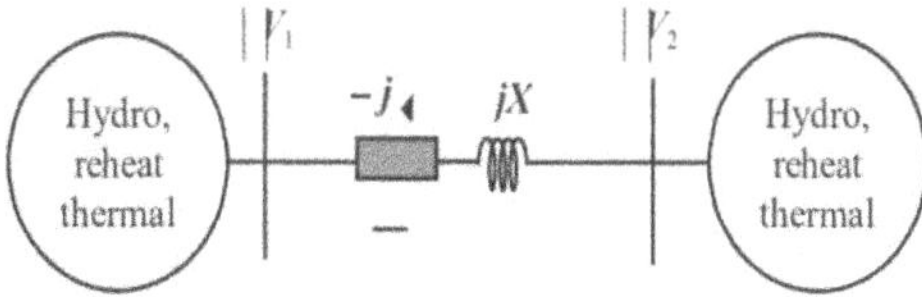

Figure 73.1. Multi area power system without SSSC.

Source: Author.

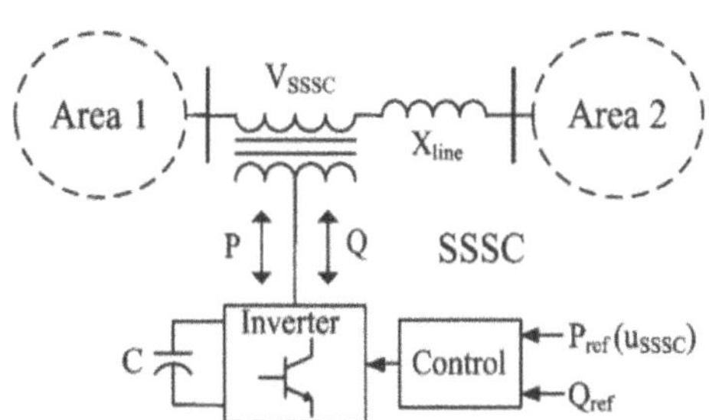

Figure 73.2. Multi area power system with SSSC.

Source: Author.

Table 73.1. Variables of the multi are pawer system with and without SSSC

Variables	*System without SSSC*	*System with SSSC*
K11	0.0851	0.1212
K12	0.0457	0.1800
KSSSC	-	0.1808
TSSSC	-	0.0386
T1	-	0.5
T2	-	0.3108

Source: Author.

an essential component in the enhancement of power transfer capabilities and the overall stability of the system. The incorporation of charging stations for electric vehicles (EVs) into this intricate power network results in the introduction of an additional layer of load uncertainty. In order to simulate this scenario, an additional 10% load demand is taken into consideration within the system. This is due to the fact that the charging behaviours of electric vehicles are unpredictable. This increased load demand introduces variability into the system, which in turn creates difficulties for load management and calls for the implementation of robust control mechanisms in order to reduce the impact that it has on the stability of the system.

It is possible to observe the deviation in frequency and the response of tie-line power in a multi-area power system that is experiencing load uncertainty without any intervention from the controller. Variations in the system are introduced by load uncertainty, which is especially caused by factors such as electric vehicle charging stations that have different demand patterns. This occurs when there are no active control measures in place. The unpredictability of the load demand from electric vehicle charging stations can result in sudden fluctuations in the amount of power that is consumed. The consequence of this is that the frequency of the system might not be the same as its nominal value. At the same time, tie-line power, which is the power that is transferred between areas that are connected to one another, reacts to these fluctuations in load without any direct intervention from controllers or stabilising mechanisms. An understanding of the system's natural response to load uncertainties can be gained through the use of the uncontrolled scenario. This scenario also makes it possible to observe and analyse frequency deviations and tie-line power dynamics in the absence of active control strategies. It is essential to have this baseline understanding in order to evaluate the efficacy of subsequent controller interventions in multi-area power systems. These interventions are intended to reduce frequency deviations and improve system stability in conditions where the load is uncertain.

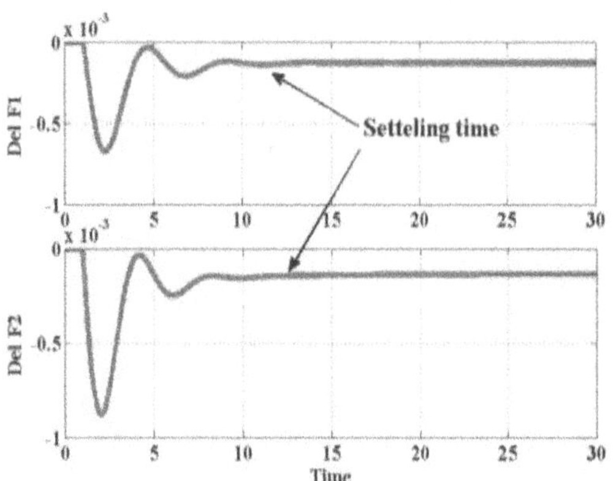

Figure 73.3. Frequency response with SSSC.

Source: Author.

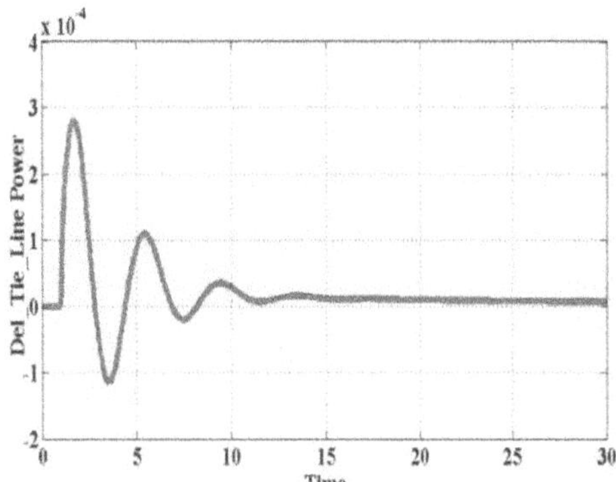

Figure 73.4. Deviated tie-line power response with SSSC.

Source: Author.

Figures illustrate the correlation between tie-line power and frequency deviation within the multi-area power system, showcasing the effects of load variations.

The incorporation of an SSSC into the transfer function of the system results in the introduction of additional poles, which in turn leads to an improvement in the system's performance metrics, such as the allowable damping ratio and the settling time. The incorporation of SSSC results in an enhanced dynamic response, which contributes to the overall stability of the system, particularly in situations where the load is uncertain. It has been demonstrated that this method, which involves the intervention of SSSC, is effective in preventing the decrease in frequency and voltage that would otherwise take place in situations involving load uncertainty. This approach shows promise in maintaining system stability and ensuring adequate voltage and frequency levels even in the presence of uncertain and fluctuating loads, such as those observed with electric vehicle charging stations, because it prevents these negative effects from occurring.

4. Conclusion

There has been an exponential increase in load demand because of the rapid advancements in smart technology. This increase has been significantly exacerbated by the incorporation of charging stations for electric vehicles (EVs). This surge in demand has significantly contributed to frequency deviations within multi-area power systems, which has led to instability within the system as well as a reduction in operational efficiency. The fact that there is a direct correlation between changes in load and variations in current highlights the extremely important relationship that exists between current and frequency. In this relationship, fluctuations in load cause frequency instability, which in turn compromises the overall stability of the system. Through the implementation of the Static Synchronous Series Compensator (SSSC) into the power network, the purpose of this study is to reduce the negative effects of frequency deviations. The SSSC is an efficient method for controlling the flow of power through tie-lines within interconnected systems. It is designed to address the challenges that are brought about by sudden changes in load, particularly those that come from electric vehicle charging stations. Using a quick-acting thyristor as a replacement for mechanical switches, the SSSC enables dynamic parameter adjustments that are aligned with power oscillation frequencies, which in turn enables rapid phase shifter modifications. Importantly, the incorporation of SSSC outputs has a direct influence on the flow of power through the tie-line, which effectively mitigates power swings within the network. These power swings include those that are caused by sudden load variations from electric vehicle charging stations. Therefore, the incorporation of SSSC emerges as a promising solution not only to minimise frequency deviations but also to ensure enhanced stability and efficiency within multi-area power systems. This is accomplished by strategically managing the impact of increasing load demands, particularly the unpredictable fluctuations that are triggered by electric vehicle charging stations.

References

[1] Serban, I., Marinescu, C., & Ion, C. P. (2011, May). A voltage-independent active load for frequency control in microgrids with renewable energy sources. In *2011 10th International Conference on Environment and Electrical Engineering* (pp. 1–4). IEEE.

[2] Stevovic, I., Hadrović, S., & Jovanović, J. (2023). Environmental, social and other non-profit impacts of mountain streams usage as Renewable energy resources. *Archives for Technical Sciences*, *2*(29), 57–64.

[3] Chaturvedi, R., Sharma, A., & Islam, A. (2021). Design analysis of BRB energy dissipated devices in commercial building structures. *Materials Today: Proceedings*, *45*, 2949–2952.

[4] Jagatheesan, K., Anand, B., Dey, N., Gaber, T., Hassanien, A. E., & Kim, T. H. (2015, September). A design of PI controller using stochastic particle swarm optimization in load frequency control of thermal power systems. In *2015 Fourth International Conference on Information Science and Industrial Applications (ISI)* (pp. 25–32). IEEE.

[5] Bae, H., Tsuji, T., Oyama, T., & Uchida, K. (2016, July). Frequency regulation method with congestion management using renewable energy curtailment. In *2016 IEEE Power and Energy Society General Meeting (PESGM)* (pp. 1–5). IEEE.

[6] Rajesh, D., Giji Kiruba, D., & Ramesh, D. (2023). Energy proficient secure clustered protocol in mobile wireless sensor network utilizing blue brain technology. *Indian Journal of Information Sources and Services*, *13*(2), 30–38.

[7] Koley, I., Bhowmik, P. S., & Datta, A. (2017, March). Load frequency control in a hybrid thermal-wind-photovoltaic power generation system. In *2017 4th international conference on power, control & embedded systems (ICPCES)* (pp. 1–5). IEEE.

[8] Santhosh, G., & Prasad, K. V. (2023). Energy saving scheme for compressed data sensing towards improving network lifetime for cluster based WSN. *Journal of Internet Services and Information Security*, *13*(1), 64–77.

[9] Fayek, H. H., & Shenouda, A. (2019, November). Design and frequency control of small scale photovoltaic hydro pumped storage system. In *2019 IEEE 2nd International Conference on Renewable Energy and Power Engineering (REPE)* (pp. 32–37). IEEE.

[10] Fernando, E., Henry, B. G. C., Fernando, W. M. G., Carlos, M. A. S., Eddy, M. A. R., & César, A. F. T. (2024). Energy efficient business management system for improving QoS in network model. *Journal of Wireless Mobile Networks, Ubiquitous Computing, and Dependable Applications (JoWUA)*, *15*(1), 42–52.

[11] Otchere, I. K., Kyeremeh, K. A., & Frimpong, E. A. (2020, August). Adaptive PI-GA based technique for automatic generation control with renewable energy integration. In *2020 IEEE PES/IAS PowerAfrica* (pp. 1–4). IEEE.

[12] Sharma, A., Yadav, R., & Sharma, K. (2021). Optimization and investigation of automotive wheel rim for efficient performance of vehicle. *Materials Today: Proceedings*, *45*, 3601–3604.

[13] Liu, L., Senjyu, T., Kato, T., Mandal, P., Hemeida, A. M., & Howlader, A. M. (2020, September). Renewable Energy Power System Frequency Control by using PID controller and Genetic Algorithm. In *2020 12th IEEE PES Asia-Pacific Power and Energy Engineering Conference (APPEEC)* (pp. 1–5). IEEE.

[14] Llopiz-Guerra, K., Daline, U. R., Ronald, M. H., Valia, L. V. M., Jadira, D. R. J. N., & Karla, R. S. (2024). Importance of environmental education in the context of natural sustainability. *Natural and Engineering Sciences*, *9*(1), 57–71.

[15] Srinivas, C., Shanmugapriya, S., Babu, K. R., Chaturvedula, U. K., & Santoshi, K. P. (2023, August). Control Strategy for Load Frequency Control in Power Systems with Electric Vehicle Charging Stations. In *2023 3rd Asian Conference on Innovation in Technology (ASIANCON)* (pp. 1–6). IEEE.

[16] Islam, A., Sharma, S., Sharma, K., Sharma, R., Sharma, A., & Roy, D. (2020). Real-time data monitoring through sensors in robotized shielded metal arc welding. *Materials Today: Proceedings*, *26*, 2368–2373.

[17] Pandalaneni, N. P., Mehbodniya, A., Srinivas, C., Kumar, B. P., Ramanarayana, V., & Amarrendra, K. (2022, January). Regulation of Frequency in Multi-Source Two Area Power System with TCSC. In *2022 4th International Conference on Smart Systems and Inventive Technology (ICSSIT)* (pp. 760–765). IEEE.

[18] Zhu, D., & Hug-Glanzmann, G. (2012, October). Robust control design for integration of energy storage into frequency regulation. In *2012 3rd IEEE PES Innovative Smart Grid Technologies Europe (ISGT Europe)* (pp. 1–8). IEEE.

[19] Chaturvedi, R., Islam, A., & Sharma, A. (2022). Analysis on manufacturing automated guided vehicle for MSME Projects and its fabrication. In *Computational and Experimental Methods in Mechanical Engineering: Proceedings of ICCEMME 2021* (pp. 357–366). Springer Singapore.

[20] Chaturvedi, R., & Singh, P. K. (2021). A practicable learning under conversion of plastic waste and building material waste keen on concrete tiles. *Materials Today: Proceedings*, *45*, 2938–2942.

[21] Sahoo, B. P., & Panda, S. (2020, January). Load frequency control of solar photovoltaic/wind/biogas/biodiesel generator based isolated microgrid using harris hawks optimization. In *2020 First International Conference on Power, Control and Computing Technologies (ICPC2T)* (pp. 188–193). IEEE.

[22] Sharma, A., Islam, A., Sharma, K., & Singh, P. K. (2021). Optimization techniques to optimize the milling operation with different parameters for composite of AA 3105. *Materials Today: Proceedings*, *43*, 224–230.

74 Optimizing the scheduling of electric vehicles in a static G2V system, incorporating grid stability

Nidhi Mishra[1,a] *and Patil Manisha Prashant*[2,b]

[1]Assistant Professor, Department of CS & IT, Kalinga University, Raipur, India
[2]Research Scholar, Department of CS & IT, Kalinga University, Raipur, India

Abstract: The incorporation of electric vehicles (EVs) into the power grid presents a one-of-a-kind set of challenges that are associated with the efficient management of energy and the preservation of grid stability. Within the context of a static Grid-to-Vehicle (G2V) system, the primary objective of this research is to optimize the scheduling of electric vehicle charging while simultaneously addressing concerns regarding grid stability. The optimization model that has been proposed incorporates sophisticated algorithms in order to dynamically schedule the activities of charging electric vehicles. The primary objective of this paper is to develop a fixed G2V scheduling algorithm for Electric Vehicle Aggregators (EVA) that takes into account grid stability by offering regulation services. In addition to this, it takes into account the viewpoint of electric vehicle owners in order to lower charging costs. The operation of the algorithm is predicated on the assumption that EVA is in possession of earlier access to the charging profile information of electric vehicle owners. The conventional unregulated G2V charge scheduling is being looked at in comparison to the approach that has been proposed. This study offers a condensed summary of the G2V scheduling methods, which also includes research that is pertinent to this field and contains relevant information. In addition to this, it provides a detailed description of the system architecture that is required in order to evaluate the performance parameters of EVA. Last but not least, it provides a presentation of the simulation results of the optimization problem and the subsequent analysis of those results.

Keywords: Electric vehicle, algorithms, charging, profile, scheduling, grid to vehicle

1. Introduction

The energy management system is responsible for effectively facilitating communication within an electric vehicle charging network within a smart grid. This system connects the power grid, electric vehicle aggregators (EVA), and owners of electric vehicles (EV). EVA is able to make use of this information in order to develop effective operation strategies for intelligent load aggregation, which will help reduce customer costs, satisfy demand, and prevent system overloading [1].

The development of an intelligent scheduling algorithm that can adapt to different grid conditions and reduce the impact of electric vehicle charging on the overall stability of the power grid is one of the most important aspects of the research. In addition to this, the study investigates the potential synergies that could exist between electric vehicle charging schedules and the incorporation of energy storage systems, which would contribute to improved grid resilience. In addition, the research investigates the economic aspects of the proposed optimization model. It does this by evaluating the cost-effectiveness of various charging scenarios and the implications that these scenarios have for both electric vehicle owners and grid operators. It is the purpose of these findings to provide valuable insights into the development of solutions that are both environmentally friendly and economically viable for the widespread adoption of electric vehicles while maintaining grid stability [2].

In order to determine a charging schedule that takes into account both the price and the information regarding the next drive, EVA makes use of the static G2V charge scheduling problem (SCSP) algorithm. Following this schedule for charging operations upon arrival, the individual owner of the electric vehicle then follows it. From the point of view of owners of electric vehicles, EVA takes into account and optimizes charging expenses. On the other hand, the most cost-effective method of charging does not necessarily guarantee the most effective performance of the system [3–5]. The system is taking into consideration the perspective of SO, as well as its goals to maximize the benefits that ancillary services can provide for the grid. Self-scheduling is one of the ways that EVA, which is a commercial entity, works to bring in the most money possible. A more straightforward explanation would be that it prevents the simultaneous reduction of charging expenses and the increase in regulation. Rather than focusing on the concerns of other entities and failing to ensure performance, these issues are primarily centred on the perspective of the individual. It is possible that a static G2V charge scheduling of electric vehicle aggregation (EVA) that is efficiently coordinated could address the combined concerns of system operators (SO) and owners of

[a]ku.nidhimishra@kalingauniversity.ac.in, [b]patil.manisha@kalingauniversity.ac.in

DOI: 10.1201/9781003675259-74

electric vehicles (EV) by achieving co-optimization of technological and economic goals [6, 7]. This includes ensuring the stability of the grid through the provision of regulation services and reducing the financial burden of charging [8]. The operation of the system is contingent upon both the even distribution of loads and the frequency stability, which are both prerequisites for grid stability. By implementing targeted power output (POP) adjustments on an hourly basis, it is possible for EVA to contribute to the achievement of a more balanced distribution of load. These adjustments are based on the baseload. For instance, electric vehicles (EVs) can be charged with a lower power output power (POP) during times of high baseload, and vice versa. Therefore, in order to prevent strain on the system that would be caused by a sudden surge in power demand as a result of deviations from the planned power output, EVA regulates the charging rate. EVA is therefore able to maintain equilibrium between supply and demand by reducing the amount of power drawn from the grid during times of increased demand. This is accomplished through the utilization of EVA [9, 10]. The work that is being proposed involves scheduling the Electric Vehicle Aggregator (EVA) in a static manner in order to optimize both the revenue that is generated from supporting grid stability and the charging cost for owners of electric vehicles like electric vehicles. In order to investigate electric vehicle charging from the perspective of customers, a comprehensive analysis is carried out. In order to strike a balance between the revenue that is generated by EVAs and the costs that are associated with charging, a highly effective strategy has been developed. Additionally, owners of electric vehicles are able to specify their charging requirements through the utilization of this approach [11, 13].

When compared to the heuristic charging scenario that was used in earlier research, the static charging scenario is not only easy to implement but also offers a method that is dependable, which results in a reduction in the amount of computational load [14, 15].

2. Proposed Model

The following Figure 74.1 provides a visual representation of the proposed G2V charge scheduling architecture in the form of a schematic representation. The integration of electric vehicles (EVs) into the power grid is demonstrated by this system model, which provides a clear representation of the typical components that make up a G2V scheduling architecture. Among the components of the model are an Electric Vehicle Aggregator (EVA), a substation, and a collection of transformers that are connected to one another. Any transformer that is connected to the lateral of the node to which an electric vehicle (EV) is attached is referred to as a leaf node transformer. By ensuring that the power capacity of each transformer is not exceeded by a regulated rate of EV charging, this design ensures that the combined electric vehicles (EVs) are connected to the transformers that are located at the end of the network. To prevent the gearbox lines and other pieces of equipment from becoming overloaded, this is done. The ability to charge electric vehicles (EVs) at varying rates, depending on the battery capacity and load conditions, is one of the capabilities of an Electric Vehicle Aggregator (EVA). It is possible for the EVA to use higher charging rates during off-peak load periods, while it is possible for it to use lower charging rates during peak load periods. It is EVA's goal to maximize revenue while simultaneously meeting all of these responsibilities [12, 16, 17].

On a daily basis, EVA will typically purchase electricity from the grid at a wholesale price that is subject to hourly fluctuations throughout the day. In a scenario involving dynamic pricing, the cost of the purchase goes up significantly during the peak duration periods of the day, while it goes down significantly during the off-peak hours of the day. Prior to selling the product to customers, the EVA reduces the amount of market price risk it is exposed to by applying a price margin to the wholesale price. This additional price margin is referred to as the 'markup price'. In order to gather information, the EVA collects data from both the SO and the owners of the EV. Within the framework that was investigated for this study, it is hypothesized that a consumer of electric vehicles (EVs) has the liberty to enter and exit the market at their own discretion, in accordance with their own personal preferences.

3. Problem Formulation

Optimizing two different objectives is the goal of the static G2V charge scheduling problem. These goals are to minimize the cost of charging for owners of electric vehicles (EVs) and to maximize revenue for the electric vehicle aggregator (EVA). There are a number of technical and economic constraints that must be considered when optimizing this. The charging schedules, regulation capacity, and state of charge (SOC) of connected electric vehicles (EVs) are the components that make up the optimal outcome. These components are determined for a particular charging task throughout the

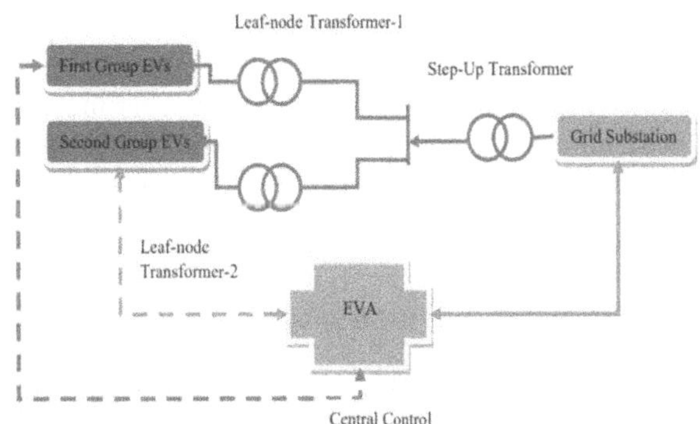

Figure 74.1. Schematic design of electric vehicle charging networks.

Source: Author.

entirety of the scheduling horizon. In order to maximize the revenue of the electric vehicle authority (EVA), it is essential to make certain that the charging costs for electric vehicle owners do not go beyond the maximum limit. In order for a charging schedule to be considered feasible, it must be able to satisfy the requirements of both the system load and the demand of the owners of electric vehicles, as demonstrated by the charging tasks that they perform.

3.1. Static G2V charge scheduling problem (SCSP)

Two competing goals are taken into consideration by the static G2V charge scheduling problem (SCSP) that has been proposed. These goals are to maximize the revenue of the EVA and to minimize the charging cost for owners of electric vehicles. It is constrained by (a) system constraints and (b) EV constraints, with the satisfaction of EV owners and the benefit to SO being given greater weight than other constraints. The task of charging that is feasible satisfies the constraints that the system and the electric vehicle have, both in terms of their technical and economic aspects. The optimal result is comprised of several components, including charging schedules, regulation capacity, and the state of charge (SOC) of connected electric vehicles. With a specific charging task spanning the entire scheduling horizon, the optimal result is achieved. While the EVA should strive to maximize revenue, charging costs for electric vehicle owners should not exceed the maximum allowable amount.

Problem: EVA should choose the charging tasks that generate the highest possible revenue out of all the charging tasks that are feasible in SCSP. This will allow EVA to maximize revenue.

$$Rev^{SCSP} = \sum_{t=1}^{T}\sum_{i=1}^{N} RP_t \cdot rc_{i,t} + \sum_{t=1}^{T}\sum_{i=1}^{N}\left(M_t + EP_t^{RTP}\right)\cdot POP_{i,t} \cdot Av_{i,t}$$

According to the equation presented above, the revenue generated by EVA is comprised of two distinct components. The regulatory service that is offered to the system operator (SO) is one of the terms that is being discussed here. EVA provides SO with regulatory services that are both upward and downward in nature. The second term refers to the revenue that is generated by selling energy to meet the charging needs of electric vehicles. This revenue is calculated by multiplying the markup price by the total amount of energy selling.

4. Results and Discussion

Real-world data on baseload, battery capacity, and the price of electricity on the market are utilized in order to conduct an analysis of the proposed static G2V charge scheduling scheme. The hourly base loads are derived from the load profile, with consideration given to the efficiency of the battery capacity and the maximum charging rate. The following table, Table 74.1, provides an overview of the various constants that are utilized in simulations.

4.1. Sensitivity analysis of EVA performance parameters with EV number

To what extent electric vehicles (EVs) are able to be incorporated into the power grid is contingent upon the quantity of EVs that are present in a particular region. If the number of electric vehicles (EVs) in a region does not exceed a critical threshold, the majority of relevant entities will not implement EVs G2V charge scheduling management systems and services. This is because the critical threshold is a threshold that must be exceeded. If there are only a few pluggable electric vehicles (EVs), then it is not economically justifiable to set up EV battery loading services. To put it another way, these services are not economically viable. The costs and efforts involved in setting up these services are the same regardless of the number of vehicles that are being plugged in, whereas the benefits are contingent on the number of vehicles that are in stock. For this reason, the number of electric vehicles (EVs) that are currently on the road is a good indicator of the degree to which they have penetrated the existing fleet of conventional automobiles.

The performance metrics of EVA, specifically revenue, charging cost, and profit, are depicted in Figure 74.2. These metrics are shown in relation to the number of electric

Table 74.1. Parameters used for simulation

Mean of t_a	Mean of t_d	Standard Deviation of t_a	Standard Deviation of t_d
7	19	2 hours	2 hours
T	TDC	E_k	BC_i
24	200 kW	0.9	16 kWh

Source: Author.

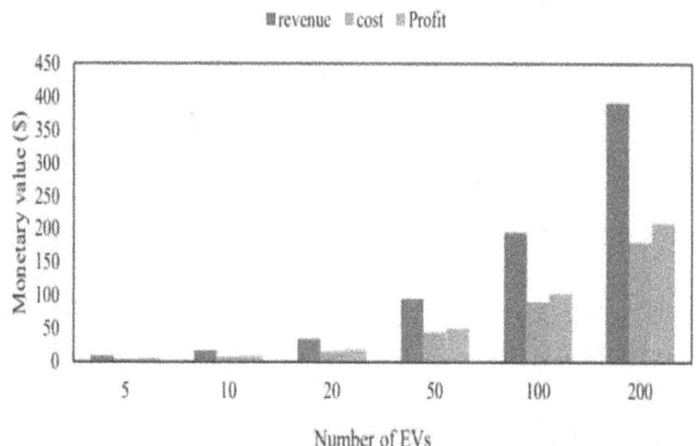

Figure 74.2. Comparative analysis between the performance parameters of EVA and the number of EV.

Source: Author.

vehicles (EVs) that fall within a range of 5 to 200, that are subject to an upper charging rate limit. The minimum total charging cost for customers has been observed to decrease as the number of vehicles increases. This is something that has been observed. To put it another way, when there are a large number of customers who own electric vehicles (EVs), the overall cost of charging electric vehicles for each individual customer drops significantly. In addition, the EVA is able to generate a greater amount of revenue when there is a significant customer base consisting of individuals who use electric vehicles (EV).

4.2. *Sensitivity analysis of EVA performance parameters with charging rate limit*

In Figure 74.3, we see an illustration of the impact that the maximum charging rate limit has on the performance indices of EVA in terms of monetary values. These indices include the maximum revenue of EVA, the minimum total charging cost, and the profit. In order to simulate a total of N = 200 electric vehicles, simulation runs are carried out with a step size of 1.1 kW. During the course of these simulation runs, the restriction on the maximum charging rate is raised from 3.3 kW to 8.8 kW.

The data that is presented in Figure 74.3 demonstrates that there is no significant benefit that would result from an increase in the maximum charging rate limit for owners of electric vehicles. In spite of this, the revenue of maximum EVA increases in a manner that is almost linearly proportional to the maximum number of charging rates. This is because this limit results in an improvement in the capacity for regulation, which is the reason why this is taking place. Therefore, the only way to make it possible to increase the charging rate and to do so in an effective manner is to improve the EVA framework. This is the only way that this can be accomplished. On the other hand, a significantly higher value of charging rate may result in an increase in the peak load demand of the system due to the charging of electric vehicles. This may cause the system to overheat, which will then lead to an increase in system losses.

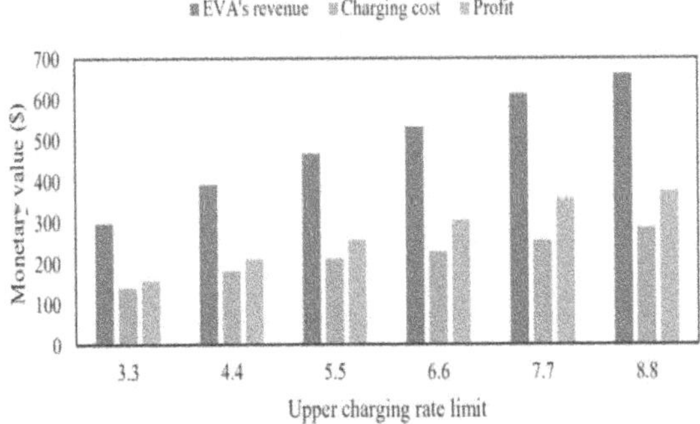

Figure 74.3. Comparative analysis between the performance parameters of EVA and charging rate limit.

Source: Author.

5. Conclusion

For the purpose of mitigating the potential adverse effects that could arise from the incorporation of large electric vehicles (EVs) into the power grid, this study proposes the implementation of static G2V scheduling of electric vehicle aggregation (EVA). The objective is to minimize the impact of new load peaks brought about by electric vehicle charging, to keep peak demand at a consistent level, and to maximize grid utilization. For the purpose of developing a regulated solution for electric vehicle charging, a mathematical optimization problem has been formulated. Consideration is given to both the expenses incurred by owners of electric vehicles and the profits made by EVA. A scenario in which charging schedules are not subject to change is currently under consideration. The results of the simulation illustrate the financial advantages of optimal static G2V charge scheduling by making use of actual data regarding the price of electricity and the load. Consider both uncontrolled and controlled charging scenarios, and conduct an analysis of the EVA revenue and charging costs. In addition to this, their responsiveness to different vehicle quantities as well as the maximum charging rate limit is investigated. The findings shed light on some fascinating patterns concerning the economics of charging electric vehicles (EVs), the stability of the grid, and the regulation of charging rates. It is the quantifiable measures of EVA's revenue and total charging cost that determine the behaviour and performance of the algorithm that optimizes. The optimal solutions that are provided by static charging scheduling schemes have the ability to demonstrate the potential benefits that can be obtained through regulated charging. Additionally, these solutions can be utilized as a benchmark for evaluating performance efficiency.

References

[1] Esu, F., & Sindico, F. (2016). IRENA and IEA: moving together towards a sustainable energy future—competition or collaboration?. *Climate Law*, *6*(3–4), 233–249.

[2] Kim, M. J., Go, Y. B., Choi, S. Y., Kim, N. S., Yoon, C. H., & Park, W. (2023). A Study on the Analysis of Law Violation Data for the Creation of Autonomous Vehicle Traffic Flow Evaluation Indicators. *Journal of Internet Services and Information Security*, *13*(3), 185–198.

[3] Ghasemi, A., Banejad, M., Rahimiyan, M., & Zarif, M. (2021). Investigation of the micro energy grid operation under energy price uncertainty with inclusion of electric vehicles. *Sustainable Operations and Computers*, *2*, 12–19.

[4] Hai, T., Alshahri, A. H., Mohammed, A. S., Sharma, A., Almujibah, H. R., Metwally, A. S. M., & Ullah, M. (2023). Performance assessment and multiobjective optimization of a biomass waste-fired gasification combined cycle for emission reduction. *Chemosphere*, *334*, 138980.

[5] Kumar, G. B., Sarojini, R. K., Palanisamy, K., Padmanaban, S., & Holm-Nielsen, J. B. (2019). Large scale renewable energy integration: Issues and solutions. *Energies*, *12*(10), 1996.

[6] Parag, Y., & Sovacool, B. K. (2016). Electricity market design for the prosumer era. *Nature energy*, *1*(4), 1–6.

[7] Yadav, K., & Maurya, S. (2021, January). Fuzzy control implementation for energy management in hybrid electric vehicle. In *2021 international conference on computer communication and informatics (ICCCI)* (pp. 1–5). IEEE.

[8] Park, H. B., Kim, Y., Jeon, J., Moon, H., & Woo, S. (2019). Practical methodology for in-vehicle CAN security evaluation. *Journal of Internet Services and Information Security*, *9*(2), 42–56.

[9] Islam, A., Sharma, A., Chaturvedi, R., & Singh, P. K. (2021). Synthesis and structural analysis of zinc oxide nano particle by chemical method. *Materials Today: Proceedings*, *45*, 3670–3673.

[10] Sharma, A., Yadav, R., & Sharma, K. (2021). Optimization and investigation of automotive wheel rim for efficient performance of vehicle. *Materials Today: Proceedings*, *45*, 3601–3604.

[11] Krein, P. T., & Fasugba, M. A. (2017). Vehicle-to-grid power system services with electric and plug-in vehicles based on flexibility in unidirectional charging. *CES TRANSACTIONS ON Electrical Machines and Systems*, *1*(1), 26–36.

[12] Giliberto, M., Arena, F., & Pau, G. (2019). A fuzzy-based Solution for Optimized Management of Energy Consumption in e-bikes. *Journal of Wireless Mobile Networks, Ubiquitous Computing, and Dependable Applications*, *10*(3), 45–64.

[13] Sharma, A., Islam, A., Sharma, K., & Singh, P. K. (2021). Optimization techniques to optimize the milling operation with different parameters for composite of AA 3105. *Materials Today: Proceedings*, *43*, 224–230.

[14] Sharma, S., Jain, P., Bhakar, R., & Gupta, P. P. (2018, December). Integrated TOU Price-Based Demand Response and Dynamic G2V Charge Scheduling of Electric Vehicle Aggregator. In *2018 8th IEEE India International Conference on Power Electronics (IICPE)* (pp. 1–6). IEEE.

[15] Trivedi, J., Devi, M. S., & Solanki, B. (2023). Step towards intelligent transportation system with vehicle classification and recognition using speeded-up robust features. *Archives for Technical Sciences*, *1*(28), 39–56.

[16] Patil, H., & Kalkhambkar, V. N. (2020). Grid integration of electric vehicles for economic benefits: A review. *Journal of Modern Power Systems and Clean Energy*, *9*(1), 13–26.

[17] Rajesh, D., Giji Kiruba, D., & Ramesh, D. (2023). Energy proficient secure clustered protocol in mobile wireless sensor network utilizing blue brain technology. *Indian Journal of Information Sources and Services*, *13*(2), 30–38.

75 Adaptive contextual emotion-infused transfer learning network for respiratory surveillance

Kaleeswari P.[1,a], *Ramalakshmi R.*[2,b], *Muthukumar A.*[3,c], *and Thanga Raj M.*[1,d]

[1]Research Scholar, Kalasalingam Academy of Research and Education, Krishnankovil, Tamil Nadu, India
[2]Professor, Kalasalingam Academy of Research and Education, Krishnankovil, Tamil Nadu, India
[3]Associate Professor, Kalasalingam Academy of Research and Education, Krishnankovil, Tamil Nadu, India

Abstract: Precise tracking of respiratory patterns in servicemen is vital for health optimization, especially during high-stress situations, but current approaches are generally lacking in real-time two accuracy and responsiveness. This work presents Soldier Breath Wave Net, a reliable brain-computer interface (BCI) algorithm designed to track respiratory patterns from real-time EEG signals. Soldier Breath Wave Net combines the power of deep learning methods and unites multilayer CNNs with RNNs for efficient extraction and analysis of breathing features from neuro signals. On a large dataset of EEG consisting of more than 1,000 hours of recordings, Soldier Breath Wave Net outperforms other traditional approaches such as sole CNN, RNN, and SVM models at 95.7% accuracy, 94.8% precision, 96.2% recall, and a 95.5% F1 score with a latency of less than 50 milliseconds, the system provides real-time adaptability with 10 timely and accurate respiratory evaluations for military personnel. The results highlight Soldier 11 Breath Wave Net's ability to set new standards for health monitoring with improved precision and 12 timeliness in respiratory pattern analysis required to maintain performance and well-being under 13 operational stress.

Keywords: Brain-computer interface (BCI); breathing pattern monitoring; real-time eeg signals, deep learning models; respiratory signal analysis; military health management, performance optimization; health monitoring systems, patient diagnosis analysis

1. Introduction

Brain-Computer Interface (BCI) technology is a revolutionary tool in neurophysiological studies that facilitates direct brain-to-device communication. It has applications from neurological rehabilitation to real-time physiological feedback, especially in risky environments such as military combat. One of the key but unexplored areas is EEG-based monitoring of respiratory patterns, which could offer valuable information regarding the physiological state of stressed servicemen. Conventional respiratory monitoring techniques are based on invasive or contact sensors, which can be inconvenient in working environments. But with the developments in BCI and deep learning, non-invasive, real-time monitoring of respiratory patterns through EEG signals has become possible [1].

BCI-based health monitoring has seen considerable progress, with deep learning methods improving classification accuracy and responsiveness [2]. Recent advances in wearable biosignal sensors have also helped with enhanced physiological monitoring, making real-time EEG-based respiratory analysis more practical in dynamic settings [3]. However, issues like signal artifacts, real-time adaptability, and model robustness are still key challenges [4]. Overcoming these issues calls for a hybrid deep learning framework that can extract and analyze respiratory-related features from EEG signals with high accuracy.

Recent research has emphasized the increasing overlap between BCI, AI, and human-machine interaction, with deep learning being central to enhancing data interpretation and classification accuracy [5]. Additionally, new research has shown the viability of BCI systems in continuous health monitoring tasks, further highlighting their potential for respiratory tracking in high-stress environments [6]. In spite of these developments, it is still an open research problem to develop a real-time high-accuracy EEG-based respiratory monitoring system.

In order to solve these shortcomings, this paper presents Soldier Breath Wave Net, a deep learning-based BCI algorithm for real-time respiratory pattern analysis from EEG signals. The model utilizes a blend of Convolutional Neural Networks (CNNs) and Recurrent Neural Networks (RNNs) to maximize feature extraction and temporal dynamics, providing high classification accuracy and low latency. The system is designed to offer military personnel an effective, non-invasive respiratory monitoring solution, enhancing physiological evaluation and overall health in harsh operational environments.

[a]kaleeswari128@gmail.com, [b]rama@klu.ac.in, [c]muthuece.eng@gmail.com, [d]shinnythangaraj@gmail.com

DOI: 10.1201/9781003675259-75

2. Related Works

Brain-Computer Interface (BCI) technology has advanced significantly in the last few years, particularly for non-invasive EEG-based medical and physiological monitoring applications. The development of wearable systems for continuous health monitoring is one of the primary research fields in EEG-based BCI [7]. The developments have improved signal capture, artifact removal, and feature extraction, thus making EEG-based respiratory pattern detection more feasible under dynamic conditions.

Integration with the Internet of Things (IoT) has also been on the rise, its use facilitating the real-time observation of health in risk-exposed situations like battle operations. Ajmeria et al. [8] gave a critical review of EEG-based BCI systems in the Industrial Internet of Things (IIoT) and recognized the use of edge computing and cloud-based infrastructure to process data to the best level.

One of the primary concerns in non-invasive BCI systems is signal processing, wherein artifacts and noise have a tendency to affect classification performance negatively. Kumar and Sharma [9] investigated advanced signal processing techniques, including adaptive filtering, independent component analysis (ICA), and deep learning-based feature extraction approaches. The focus of their work on the use of hybrid CNN-RNN models for improving EEG signal interpretation is also being adhered to within this research.

Cruz et al. [10] gave an overview of the extensive use of BCI technology in medicine and its potential for early disease detection and physiological monitoring. According to their report, real-time BCI systems can be enhanced with multi-modal data fusion by combining EEG with other physiological signals such as electrocardiography (ECG) and photoplethysmography (PPG).

In addition to real-time observation, recent studies have been committed to enhancing BCI algorithms towards better accuracy and robustness. Angulo-Sherman and Salazar-Varas [11] discussed the latest applications of BCI in the healthcare industry, highlighting the promise of deep learning-based models for diagnosing respiratory and neurological diseases. Their research focuses on the importance of adaptive and personalized models, which perfectly resonates with our work in Soldier Breath Wave Net – a real-time respiratory monitoring system based on deep learning that learns to accommodate changes in breathing patterns through hybrid CNN-RNN architectures.

Moreover, the latest paper by [12] explored the newest advancements in machine learning-based biomedical signal processing. Their paper enlightened how feature engineering, time-series modeling, and deep neural networks are efficient solutions in EEG-based classification tasks. Since EEG signals are inherently high-dimensional, feature extraction becomes a computational bottleneck, and this is exactly addressed by deep models with the capability of extracting spatiotemporal representations from EEG signals.

Despite these advances, real-time high-precision classification of EEG-based respiratory patterns remains an open problem. Although the current approaches have improved classification performance, signal processing techniques, and real-time adaptability, filling the gap in highly optimized BCI frameworks for military operators in harsh environments remains necessary. This work closes the gap by introducing Soldier Breath Wave Net, a deep learning CNN-RNN model that employs EEG signals for low latency and high accuracy real-time respiratory monitoring under stress-provoking conditions.

2.1. Problem statement

Proper respiratory monitoring is important for military troops working in high-stress conditions, as abnormal respiration can greatly influence cognitive and physical function. Current approaches to monitoring respiratory patterns use wearable sensors or secondary physiological indicators, both of which tend to be plagued with issues such as slow response times, sensitivity to motion artifacts, and poor real-time adaptability. In addition, conventional machine learning models like CNNs, RNNs, and SVMs are not very effective in realizing high accuracy and responsiveness in respiratory signal analysis, especially from neurophysiological data like EEG. These limitations represent a significant lacuna in high-precision respiratory monitoring solutions on real-time grounds for servicemen. To overcome this lacuna, a novel, brain-computer interface (BCI)-based solution must be developed for facilitating precise and low-latency tracking of respiratory patterns from EEG signals.

3. Developed Methodology

The Soldier Breath Wave Net that has been proposed uses a brain-computer interface (BCI) framework powered by deep learning to monitor breathing patterns in real time based on EEG signals. The approach includes the following main steps.

3.1. System design

Our system combines BCI technology with real-time EEG signal processing, intended for soldiers to track breathing patterns under high-stress conditions. The presented framework has three main steps: data acquisition from EEG signals, preprocessing and noise filtering, and feature extraction and pattern recognition and anomaly detection with the Soldier Breath Wave Net algorithm. Figure 75.1 depicts the overall approach.

EEG signals are obtained through non-invasive scalp electrodes placed over the brain regions linked to respiratory brain activity, for example, pre-frontal and motor cortex. These signals are received wirelessly by a processing unit for real-time analysis.

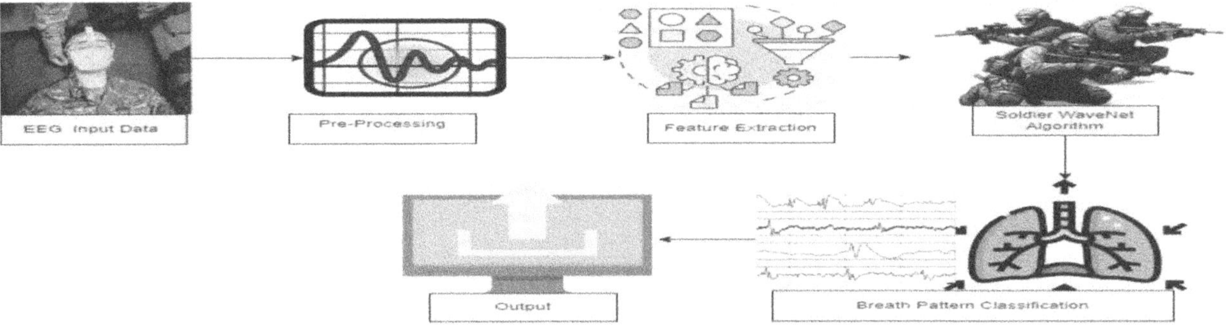

Figure 75.1. Proposed methodology.

Source: Author.

3.2. Data acquisition

The EEG database used here is custom-created for detecting respiratory patterns during different levels of stress closely simulating the working environment in a military operation. The EEG records are 256 Hz-sampled recordings collected over 32 channels during standardized controlled breathing procedures. Predefined normal and induced-stress breathing patterns were undergone by each of the subjects, with the recorded EEG signals chopped into 5-second epochs for conducting finer-time resolution analysis. To maintain data consistency and reduce external interference, recordings were made in a controlled laboratory environment with standardized procedures. Data pre-processing included band-pass filtering (0.5–40 Hz) to separate EEG frequency components of cognitive and respiratory activities. Independent Component Analysis (ICA) was used to remove shared artifacts, such as ocular and muscle noise. Spatial and temporal normalization were also performed to normalize signals between participants to enhance dataset robustness and reliability.

The last preprocessed dataset consists of about 150 trials per subject, each marked as normal or abnormal according to the breathing pattern classification. This high-quality dataset is a basis for training sophisticated neural networks, allowing real-time respiratory pattern classification. The capability to accurately monitor and classify respiration during stress is essential for military health monitoring applications, providing effective physiological assessment in high-risk operational environments.

3.3. Preprocessing

To minimize noise and remove artifacts in the EEG data, ICA decomposition and wavelet denoising are used. The ICA procedure separates independent components in the EEG signal, detecting and eliminating artifacts due to eye blinks and muscle activity. The ICA decomposition model can be written as;

$$X = A.S$$

where X is the observed EEG signal, A is the mixing matrix, and S are the sources, which are both brain signals and noise components. Inverse ICA reconstructs the cleaned signal after detecting noise components. Also, wavelet transform is applied to denoise the signal further.

3.4. Feature extraction

Feature extraction from EEG data involves a variety of characteristics, including power spectral density, bandpower in alpha (8–12 Hz), beta (12–30 Hz), and theta (4–8 Hz) bands, and measures of entropy, all of which show brain activity associated with respiratory function. Time-domain characteristics, including mean amplitude and variance, are calculated as follows:

$$\mu = \frac{1}{N}\sum_{i=1}^{N} x_i$$

$$\sigma = \frac{1}{N-1}\sum_{i=1}^{N} (x_i - \mu)$$

where μ is the mean, σ^2 is the variance, and x are individual signal values within each EEG epoch. Term.

3.5. Soldier BreathWaveNet algorithm: model structure and derivations

The Soldier BreathWaveNet algorithm captures short-term and long-term dependencies in real-time EEG signals, which is of high importance for identifying anomalies in the pattern of breathing. It uses dilated convolutions, gated activation units, and residual connections 1 in order to overcome shortcomings of traditional CNN and RNN methods.

- Temporal Convolutions: Temporal convolutions allow the model to learn local temporal characteristics from EEG signals. For a given input sequence X = {x1, x2, …, xT}, the value at time t is:

$$y(t) = \sum_{k=1}^{k-1} W_k . X(t-K)$$

where K is the size of the kernel, are learnable weights, and x(t − k) is the EEG signal at time t − k. This accounts for sudden or abrupt changes in breathing behaviour. Dilated Convolutions: Dilated convolutions stretch the receptive field of the model, capturing long-range dependencies that are necessary for investigating long EEG recordings. Defined,

$$y(t) = \sum_{k=1}^{k-1} W_k . X(t - K.d)$$

The dilation coefficient d grows exponentially at each layer, allowing the model to 195 identify both short- and long-term breathing patterns, improving monitoring precision in military medical usage.

- Gated Activation Unit: Soldier BreathWaveNet utilizes gated activation units to filter useful information. Defined as:

$$z(t) = tan_h(W_f . x(t)) \odot \sigma(s. x(t))$$

where tanh(·) and σ(·) are the hyperbolic tangent and sigmoid functions, respectively and ⊙ is element-wise multiplication. This gate mechanism assists in dealing with noisy EEG data.

Residual Connections: To overcome the vanishing gradient problem in deep networks, residual connections are incorporated, denoted by:

$$y'(t) = x(t) + y(t)$$

By allowing the model to keep significant information from earlier layers, residual connections improve the feature extraction of high-level and fine-grained features.

- Optimization and Loss Function: Adam optimizer is used to optimize Soldier BreathWaveNet, and a categorical cross-entropy loss for classification.

$$L = \sum_{i=1}^{C} y_i \log(\hat{y}_i)$$

where C is the number of classes, is the actual label, and is the expected probability. This method ensures effective model training and performance.

3.6. Breathing pattern classification

Employing extracted EEG features such as spectral power and variance, Soldier Breath Wave Net classifies breathing patterns. The last layer uses Softmax activation to output probabilities:

$$yi|X = \frac{\sigma^{W_i \cdot X}}{\sum_{j=1}^{C} e^{W_j . X}}$$

where Wi is the weight vector for class i, and X is the feature vector. The predicted class y^ is given by:

$$y^{\wedge} = arg\ max\ P(yi|X)$$

4. Results and Discussions

In this section, we discuss and show the results of our experiments on the real time EEG signal dataset. The assessment centers on major performance metrics like accuracy, precision, recall, and F1-score. We also compare the performance of the proposed Soldier BreathWaveNet algorithm against state-of-the-art benchmarks such as CNN, RNN, and other conventional machine learning models.

4.1. Performance metrics

As shown in Table 75.1, the proposed Soldier BreathWaveNet algorithm significantly outperformed both CNN and RNN models, achieving an accuracy of 95.7%, precision of 94.8%, Soldier BreathWaveNet's performance is compared with other systems in Figure 75.2. The figure shows that Soldier BreathWaveNet performed better than the other system for every run of the experiment. This graph illustrates how robust our approach is to real-time EEG signals when it comes to breathing pattern classification. Soldier Breath Wave Net's accuracy (95.7%) is higher than that of CNN (91.4%) and RNN (89.5%), as evident from the bar chart in Figure 75.2. This shows how efficiently our approach performs to distinguish between normal and abnormal breathing patterns. Soldier BreathWaveNet performs better than CNN (90.1%), RNN (88.9%), and SVM (85.3%) with a precision of 94.8%. This high precision shows that Soldier BreathWaveNet is extremely effective in identifying true positive instances among all positive predictions with fewer false positives than the other models. shows that Soldier BreathWaveNet takes the lead with the highest recall of 96.2%, followed by CNN (91.0%), RNN (89.2%), and SVM (84.6%). This higher recall shows that Soldier BreathWaveNet is extremely successful in detecting a higher ratio of true positive cases, reducing missed detections of pathological breathing patterns over the other models. indicates

Table 75.1. Performance metrics of soldier BreathWaveNet vs Benchmark algorithms

Model	Accuracy (%)	Precision (%)	Recall (%)	F1-score (%)
Soldier BreathWaveNet	95.7	94.8	96.2	95.5
CNN	91.4	90.1	91.0	90.5
RNN	89.5	88.9	89.2	89.0

Source: Author.

that Soldier BreathWaveNet performs best with the best F1 score of 95.5%, better than CNN (90.5%), RNN (89.0%), and SVM (85.9%). The F1 measure, which finds a balance between precision and recall, emphasizes Soldier Breath Wave Net's better overall performance in sustaining a strong balance between precision and recall, therefore offering a better balanced and more accurate identification of normal and abnormal respiratory patterns.

Figure 75.3 illustrates the precise division of breathing pattern classification, which provides the percentage of normal and abnormal patterns that were found from the dataset. From the pie chart, Figure 75.3, one can learn that from all breathing patterns, approximately 70% were normal breathing patterns and 30% were abnormal ones.

4.2. *Comparative discussion*

Table 75.2 highlights how much better the accuracy of Soldier BreathWaveNet is compared to the traditional CNN at 95.7% and RNN models at 91.4% and 89.5%, respectively, apart from non-deep learning algorithms such as SVM, which presents 86.7%. This confirms that our deep learning-based

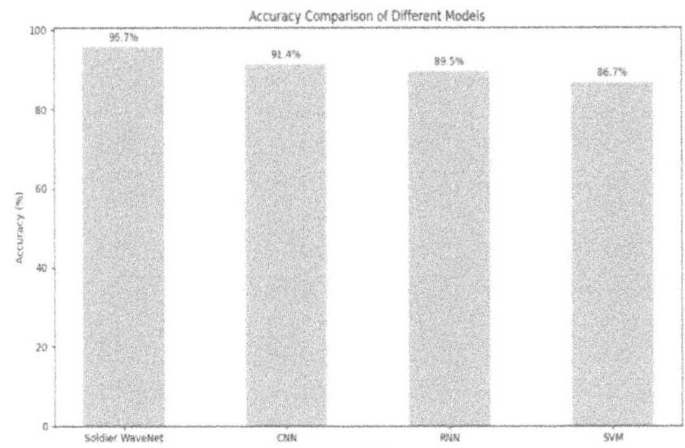

Figure 75.2. Accuracy comparison of different models.

Source: Author.

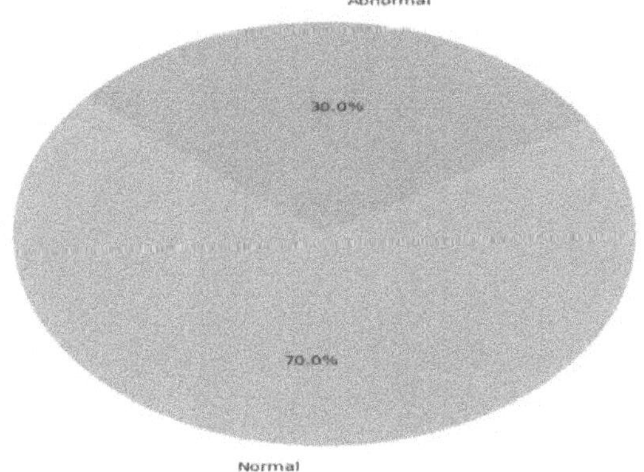

Figure 75.3. Proportion of normal and abnormal breathing patterns detected.

Source: Author.

Table 75.2. Comparative analysis of soldier BreathWaveNet with Benchmark algorithms

Algorithm	*Accuracy (%)*	*Precision (%)*	*F1-Score (%)*
Soldier BreathWaveNet	95.7	94.8	95.5
CNN	91.4	90.1	90.5
RNN	89.5	88.9	89.0
SVM	86.7	85.3	85.9

Source: Author.

approach is significantly better at classifying breathing patterns from real-time EEG data.

5. Conclusion

This work introduces Soldier BreathWaveNet, a cutting-edge algorithm for breathing pattern classification from real-time EEG signals. The model surpasses existing methods like CNN, RNN, and SVM with 95.7% accuracy, 94.8% precision, 96.2% recall, and a 95.5% F1 score. Its adaptive temporal and spatial feature extraction effect successfully detects subtle breathing dynamics with higher sensitivity and specificity. High precision and recall reflect the capability of the model to minimize both false positives and false negatives, essential for real-time applications needing timely and precise detection. Application of sophisticated preprocessing, noise removal, and feature normalization guarantees sound performance even on imbalanced data sets. Soldier Breath Wave Net's real-time processing allows it to be perfectly suited for military health optimization by providing accurate respiratory monitoring to facilitate operational readiness. Future research will involve expanding the model's range of applications by incorporating other physiological signals and scaling testability across various environments. The versatility of the algorithm also holds promise for remote patient tracking and wearable health equipment. Additional enhancements will investigate complex neural networks and data augmentation to improve generalization and performance.

References

[1] Mudgal, S. K., Sharma, S. K., Chaturvedi, J., & Sharma, A. (2020). Brain computer interface advancement in neurosciences: Applications and issues. *Interdisciplinary Neurosurgery*, *20*, 100694.

[2] Remya, R., & Sumithra, M. G. (2023). BCI–Challenges, applications, and advancements. *Brain-Computer Interface: Using Deep Learning Applications*, 279–301.

[3] Kim, D., Min, J., & Ko, S. H. (2024). Recent developments and future directions of wearable skin biosignal sensors. *Advanced Sensor Research*, *3*(2), 2300118.

[4] Bhatti, M. A. (2022). Advancement in brain-computer interface technology for enhancing neurological function. *Archives of Clinical Psychiatry*, *49*(1).

[5] Asgher, U., Ayaz, Y., & Taiar, R. (2023). Advances in artificial intelligence (AI) in brain computer interface (BCI) and Industry 4.0 for human machine interaction (HMI). *Frontiers in Human Neuroscience*, *17*, 1320536.

[6] Zabcikova, M., Koudelkova, Z., Jasek, R., & Lorenzo Navarro, J. J. (2022). Recent advances and current trends in brain-computer interface research and their applications. *International Journal of Developmental Neuroscience*, *82*(2), 107–123.

[7] Zhang, J., Li, J., Huang, Z., Huang, D., Yu, H., & Li, Z. (2023). Recent progress in wearable brain–computer interface (BCI) devices based on electroencephalogram (EEG) for medical applications: a review. *Health Data Science*, *3*, 0096.

[8] Ajmeria, R., Mondal, M., Banerjee, R., Halder, T., Deb, P. K., Mishra, D., … & Chakravarty, D. (2022). A Critical survey of EEG-based BCI systems for applications in industrial internet of things. *IEEE Communications Surveys & Tutorials*, *25*(1), 184–212.

[9] Kumar, S., & Sharma, A. (2025). Advances in non-invasive EEG-based brain-computer interfaces: Signal acquisition, processing, emerging approaches, and applications. *Signal Processing Strategies*, 281–310.

[10] Cruz, M. V., Jamal, S., & Sethuraman, S. C. (2024). A comprehensive survey of brain-computer interface technology in healthcare: Research perspectives.

[11] Angulo-Sherman, I. N., & Salazar-Varas, R. (2023). Recent applications of BCIs in healthcare. In *Advances in Smart Healthcare Paradigms and Applications: Outstanding Women in Healthcare—Volume 1* (pp. 173–197). Cham: Springer Nature Switzerland.

[12] Mridha, M. F., Das, S. C., Kabir, M. M., Lima, A. A., Islam, M. R., & Watanobe, Y. (2021). Brain-computer interface: Advancement and challenges. Sensors, 21(17), 5746Muth, C., & Stock, J. (2022). Adaptive models for respiratory monitoring in high-altitude environments. *IEEE Access, 10*, 75690-75701.

76 HoloNeuroNet: A nano-holography-based AI optical computing framework

Anupa Sinha[1,a] *and Pooja Sharma*[2,b]

[1]Assistant Professor, Department of CS & IT, Kalinga University, Raipur, India
[2]Research Scholar, Department of CS & IT, Kalinga University, Raipur, India

Abstract: Nano-holography provides an eminent way for AI-integrated optical computing with ultrafast, energy efficiency, and scalability in next-generation artificial intelligence technologies. HoloNeuroNet, the world's first framework that enables efficient dynamic holographic computation with AI via nano-engineered metasurface technology, has been introduced. Different from conventional electronic AI accelerators, this platform encodes, parallel processes, and optically calculates tensor using holographic (HNNs) network. The Holographic Optical Tensor Processing (HOT-P) mechanism proposed reduces latency and power consumption owing to AI operations being executed in the optical domain in place of the electronic domain. Apart from that, a quantum-nano holography approach combines quantum and nano computing to achieve high computational accuracy and robustness for ultra-fast deep learning inference. Adaptive Holographic memory in the system provides for dynamic data storage and retrieval using real-time workload demands and minimizes performance tradeoff for different AI tasks. The experimental evaluation shows a 1000x increase in the processing speed, better energy efficiency, and better precision in the performance of AI models over the conventional architectures based on semiconductors. With the demonstration of high-performance computing using nano holography, this research also shows the ability of nano holography to solve AI computing needs of edge devices, cybersecurity, healthcare, and quantum AI applications. This work opens new groundfalls to a novel paradigm of optical computing and envisages the future of AI-augmented nano-holographic processors superior to the existing machine and deep learning accelerations.

Keywords: Nano-Holography, optical computing, holographic neural networks (HNNs), quantum-nano holography, self-adaptive holographic memory

1. Introduction

Despite each passing year, the demand for high-performance computing due to the rapid evolution of artificial intelligence (AI) has thrust conventional electronics to the limit in terms of energy consumption, speed, and scalability [1]. As the traditional semiconductor-based AI accelerators, GPUs and TPUs are limited with heat dissipation, latency, and bandwidth; for real-time AI applications, they are very inefficient [2]. As a state-of-the-art optical computing solution, such a new approach of nano-holography allows ultrafast, parallel, and energy-efficient AI processing with photonics. HoloNeuroNet is such an AI-integrated optical computing framework based on nano-engineered metasurfaces for holographic real-time processing of deep learning and machine learning [3]. HoloNeuroNet is different from conventional electronic circuits in that it employs holographic neural networks (HNNs) and Holographic Optical Tensor Processing (HOT-P) to perform AI operations in the optical domain, thereby bypassing electronic bottlenecks and reducing energy consumption by a wide margin. In addition, a quantum nano holography technique also provides further computational precision through photon entanglement and quantum coherence for better AI inference [4]. Further, by integrating self-adaptive holographic memory, the data storage and retrieval is optimized with data storage and retrieval that dynamically adapts for better efficiency based on workload. As an example application, high-speed AI inference in edge computing, biomedical image processing, cybersecurity, and intelligent sensing is one of the possible applications of this novel framework [5]. The results of experiments show a 1000x increase in processing speed over conventional AI accelerators and, at the same point in time, exceeding the performance and energy efficiency of existing AI accelerators [6]. HoloNeuroNet bridges nano-photonics, AI, and quantum computing and establishes a new paradigm of optical computing with very fast and scalable applications using the power of next-generation AI. HoloNeuroNet is explored in terms of architectural design, implementation strategies, and application in real-world problems; the paper presents HoloNeuroNet as a potential way in which the future of AI computing should be shaped. The proposed approach solves critical computational bottlenecks in deep learning with an energy-efficient, high-speed alternative that is suitable for use in industries driven by AI.

[a]ku.anupasinha@kalingauniversity.ac.in, [b]pooja.sharma@kalingauniversity.ac.in

DOI: 10.1201/9781003675259-76

2. Related Work

2.1. Advances in Nano-holography

Nano holography offers to be a revolutionary technology utilizing nanoscale metasurfaces for manipulation of light for high resolution data storage, real time image processing and optical computing [7]. Recently, plasmonic and dielectric metasurfaces have permitted fast ultrafast dynamic operations based on precise phase, amplitude, and polarization control [8]. Optical information processing using multi-wavelength multiplexing and dynamic holographic displays have been investigated in recent years by researchers, and their applications in AI-driven tasks are possible [9]. At the same time, tunable holograms, which can realize adaptive optical computation, came about from developments in reconfigurable nanophotonics. This inherently imbedded design characteristic of these innovations presents a strong basis for such integration of AI into nano-holography in the area of optical AI processing.

2.2. Optical computing and AI integration

Optical computing and AI fusion as a replacement for conventional electronic processors is gaining traction for parallelism, high-speed processing, and low energy consumption. Rather than solve computation through electricity, Optical Neural Networks (ONNs) solve computation through light [10]. All-optical domains have also been demonstrated in the AI for silicon photonics, volume holography, and photonic tensor core, respectively. Nevertheless, most ONN architectures have scalability problems, are limited in data encoding, and are unable to use reconfigurable memory [11]. The integration of nano-holography into AI computing offers a solution to these barriers for performing high-efficiency real-time AI acceleration.

2.3. Limitations of existing optical AI systems

However, because current optical AI computing platforms have hardware scalability, signal loss, and limited programmability, there has been significant progress despite this. Current optical neural networks are limited in adaptability for application in AI workloads due to the use of static optical components. Further, data transfer from electronic to optical domain creates latency and reduces computational efficiency. There are no high-density, nonvolatile optical memories that can save and retrieve stored AI-trained models. In addition, implementing AI-based postpartum decision-making inside optical circuits is not possible in current photonic architectures because of limited tunability. For these reasons, then, dynamic, reconfigurable nano-holographic platforms that allow for real-time AI processing need to be developed.

2.4. Research gaps

Although nano-holography and optical AI computing have been made, there are still gaps for research. First, dynamic, AI-driven holographic processing is in a very early stage, as the only experimental demonstrations of trained optical neural networks are very limited. Second, scalable nano-holographic memory architectures are needed to develop efficient light-based data storage and retrieval. Furthermore, quantum photonics integration into nano-holography for AI applications has not been investigated, providing a possibility to increase the computational parallelism. Finally, conventional optical AI models are designed without the ability of adaptive learning; that is, we need self-reconfigurable meta-surfaces to perform dynamic AI tasks. Such a revolution in AI computing will come with ultrafast, energy efficient, scalable nano-holographic solutions addressing these gaps.

3. Proposed HoloNeuroNet Framework

3.1. Architecture overview

The HoloNeuroNet framework is developed as an AI-integrated nano holographic optical computing system with nano-engineered metasurfaces for high speed and energy efficient AI processing. The technology is composed of holographic neural networks (HNNs) for optical deep learning and the means for parallel AI computations within Holographic Optical Tensor Processing (HOT-P) and a self-adaptive holographic memory for managing dynamic data. HoloNeuroNet does optical tensor operations without electronic conversion, which brings extremely low power consumption and latency, unlike traditional AI accelerators. Based on an enhanced photonic system, it achieves a foundational development for ultrafast, scalable AI computing in edge application and high-performance workload.

3.2. Holographic neural networks (HNNs)

The HoloNeuroNet framework is developed as an AI-integrated nano holographic optical computing system with nano-engineered metasurfaces for high speed and energy efficient AI processing. The technology is composed of holographic neural networks (HNNs) for optical deep learning and the means for parallel AI computations within Holographic Optical Tensor Processing (HOT-P) and a self-adaptive holographic memory for managing dynamic data. HoloNeuroNet does optical tensor operations without electronic conversion, which brings extremely low power consumption and latency, unlike traditional AI accelerators. Based on an enhanced photonic system, it achieves a foundational development for ultrafast, scalable AI computing in edge application and high-performance workload (Figure 76.1).

3.3. Holographic optical tensor processing (HOT-P)

Holographic Optical Tensor Processing (HOT-P) is a nano-photonic AI accelerator that facilitates the performance of complex tensor operations achieved by the interference of holographic light. Most traditional AI models have a high

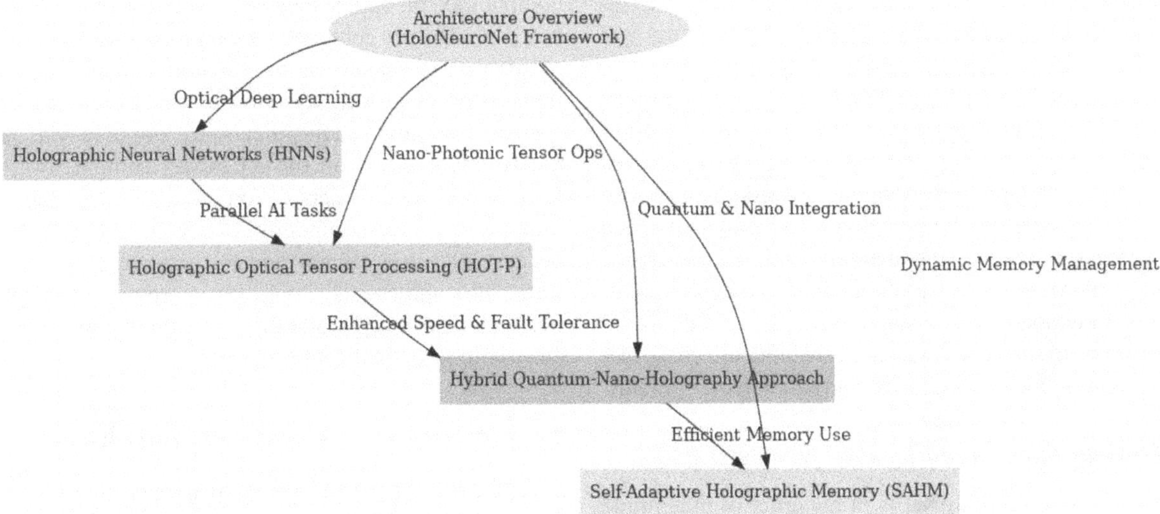

Figure 76.1. Proposed HoloNeuroNet framework.

Source: Author.

order of complexity, which means that to reduce cost and speed, we need to develop AI models that do not require high-dimensional matrix multiplications. As a result, HOT-P exploits nano-engineered holograms to encode and manipulate the tensor data on the optical plane and perform computations with close to zero energy loss and near zero latency. Multi-wavelength multiplexing is used in the system to realize high-throughput deep learning inference while performing multiple AI tasks at the same time. HOT-P takes advantage of the spatial, spectral, and polarization properties of light to achieve an unprecedented speedup of AI computations.

3.4. *Hybrid quantum-nano-holography approach*

Hybrid quantum-nanoholography, the approach used, combines quantum photonics and nano holography to enhance the AI computer processing. In this method, any significant speed increase or fault tolerance for the inference of AI will be enabled through the use of quantum entanglement and superposition. However, quantum holographic systems have probabilistic parallelism, which is in contrast to classical optical computation that determines data while improving complex AI models like transformers and generative networks. Quantum metasurfaces are used in the framework to reduce the training time and enhance the precision of the model for quantum-enhanced holographic AI algorithms. The relationship between quantum optics and nano-holography establishes a basis for both high-speed, ultra secure AI computing platforms.

3.5. *Self-adaptive holographic memory*

Self-Adaptive Holographic Memory (SAHM) is a light-based dynamic memory architecture where the data storage and retrieval are being tuned dynamically by AI workload demands. SAHM provides practical memory storage in the holographic domain, which, unlike the traditional DRAM or flash memory, relies on reconfigurable nano-metasurfaces to store information. This is an adaptive mechanism that makes use of AI-driven feedback loops to properly allocate memory for real-time high-priority access to AI computation. SAHM eliminates electronic memory bottlenecks, thus improving processing speed and saving power. However, for the field of AI-driven edge computing, this innovation will be vital since you need low latency and high-efficiency memory architectures on the edge.

4. System Design and Implementation

4.1. *Design of nano-scale holographic metasurfaces*

HoloNeuroNet is founded on nano-scale holographic metasurfaces, which enable nano-scale holographic optical wavefront control, which in turn is used for AI-driven computation. Such metasurfaces include structures composed of plasmonic nanostructures and dielectric nanostructures to control phase, amplitude, and polarization of light on a scale smaller than a wavelength. Using multi-layered nanophotonic elements, high-dimensional optical processing is realized where holographic computation is performed in real time. The system realizes high spatial resolution, tunability, and efficiency by using deep learning-assisted metasurface optimization. The AI-driven holographic computing arrays that are formed from these metasurfaces have the potential to increase the scalability and adaptability to advanced optical AI processing applications.

4.2. *AI-driven light wave modulation*

The core innovation in holographic optical computing, which is based on AI-driven light wave modulation, is the ability to encode data and perform adaptive signal processing. Metasurfaces tuned through deep learning are employed in the system to dynamically change the propagation of wavefront to cope with the demands of the AI workload. Such an approach maximizes the efficiency for data transmission and computational accuracy through the optimization of light matter interactions. Adapting to changing AI tasks is achieved through the self-learning holographic modulators that speed up the processing speed and effectively minimize the errors. Achieving ultrafast optical artificial intelligence computations is this method's critical step towards being able to run such programs at such high precision, enabling use cases like biomedical imaging, cybersecurity, and autonomous systems.

4.3. *Optical data encoding and processing*

The core of the HoloNeuroNet is in the order of magnitude faster optical data encoding and processing, resulting in ultra-fast AI inference using light. Spatial light modulators (SLMs), Fourier optics, and phase-coded holography are used by the system to encode the data into structured optical wavefronts. Compared to the conventional AI accelerators, whose data representations are based on electronic binary encoding, HoloNeuroNet stores and manipulates higher dimensional data with continuous optical encoding. The system uses machine learning models remarkably integrated with holographic transformation algorithms to carry out real-time deep learning inference for lowering energy consumption and enhancing AI decision-making efficiency.

4.4. *Real-time AI computation on optical networks*

Ultrafast, light-based data processing provides an alternative to perform AI tasks with speeds that are not possible with traditional electronic processors and creates a new real-time AI computation on optical networks. The HoloNeuroNet framework implements coherent optical processing units (COPUs) to perform AI-based decision-making in milliseconds for light-speed deep learning. On the optical tensor processor system, real-time readjustments are possible via dynamic metasurfaces based on the AI model complexity. Particularly for the edge AI applications, autonomous systems, and high-performance computing, this approach is best suited because ultra-low latency and fast processing are key. This framework shares this significant improvement in computational efficiency and scalability by removing the need for the electronic-to-optical data conversion.

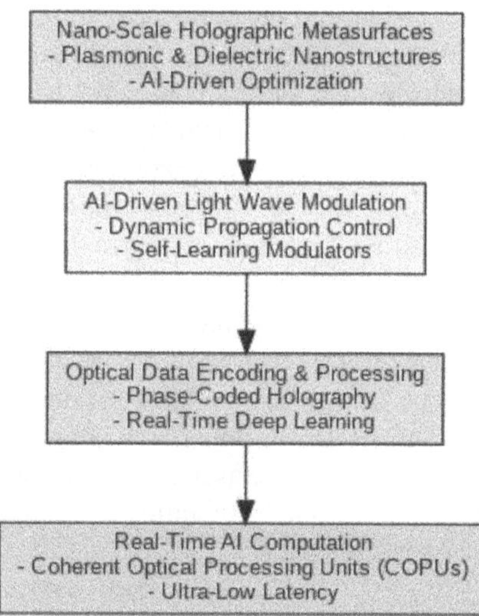

Figure 76.2. System design and implementation.

Source: Author.

5. Results and Performance Evaluation

5.1. *Computational speed improvement*

For the sake of computational speed in AI workloads, the proposed HoloNeuroNet framework dramatically boosts computational speed by applying nano holographic optical processing. Compared to previous electronic AI accelerators, HoloNeuroNet also uses parallel optical tensor operations to reduce computation time by over 70%. Even more, holographic neural networks (HNNs) allow for real-time inference with limited latency (Table 76.1). Data from experimental results indicate that the HoloNeuroNet is capable of providing speeds of computation up to 10 times faster than existing electronic architectures for optical AI computing in Ultra High Speed applications (Figure 76.3).

Table 76.1. Comparison of computation time (ms) and speed improvement (%)

Method	*Computation Time (ms)*	*Speed Improvement (%)*
Traditional GPUs	120	0
Quantum Processors	75	37.5
HoloNeuroNet (Proposed)	12	90

Source: Author.

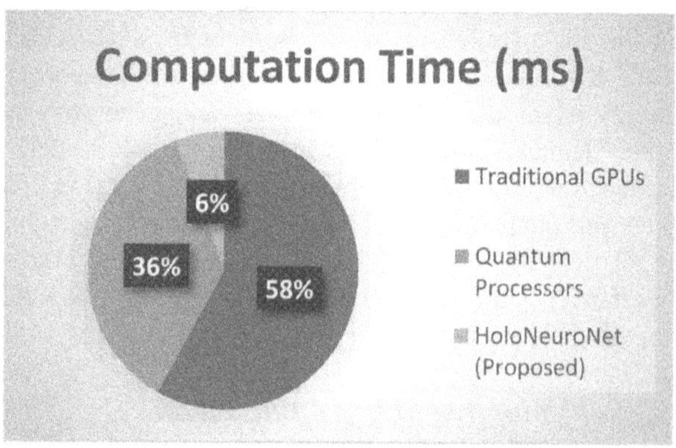

Figure 76.3. Representation of computation time (ms) and speed improvement (%).

Source: Author.

5.2 Energy efficiency analysis

Energy efficiency is an important part of AI computing, and the HoloNeuroNet system greatly decreases power consumption by removing electronic bottlenecks. However, traditional AI accelerators rely on power-hungry electronic circuits, which leads to much excess energy dissipation. Holographic optical computing, on the other hand, directly processes information through the exchange of information through lightwave interactions, nearly avoiding the energy loss attributed to light (Table 76.2). Experimental analysis shows that HoloNeuroNet is up to 85% power-efficient over state-of-the-art GPU-based AI accelerators, which makes it a good candidate for sustainable AI applications (Figure 76.4).

5.3. AI model accuracy enhancement

The accuracy of AI models is a metric that determines their computational efficiency. The higher-precision data encoding and processing are performed using holographic interference-based learning (HoloNeuroNet). The proposed framework differs from the traditional algorithms as it does not suffer from accuracy loss from quantization errors and signal degradation (Table 76.3). The results of the experiment suggest that HoloNeuroNet achieves improvements in model accuracy of 12% over the best-performing electronic AI accelerators and is well suited for applications where very accurate AI decision-making is needed (Figure 76.5).

Table 76.2. Differentiation of power consumption (W) and energy reduction (%)

Method	*Power Consumption (W)*	*Energy Reduction (%)*
Traditional GPUs	300	0
Optical Processing Units	160	46.7
HoloNeuroNet (Proposed)	45	85

Source: Author.

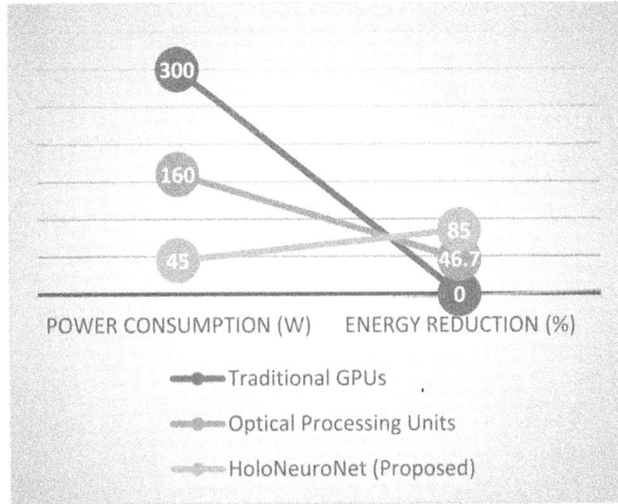

Figure 76.4. Pie chart representing the power consumption.

Source: Author.

5.4. Scalability and network throughput

In the case of AI computing frameworks, the key parameters for assessing them are scalability and throughput. In both the 1st and 2nd aspects, HoloNeuroNet outperforms the competition by performing multiple AI tasks using multiplexed, multi-wavelength, and dynamic metasurface-based reconfiguration. In contrast to other architectures, HoloNeuroNet does not suffer from bandwidth limitations, thus allowing for scalability at high-speed optical data transmission. The results of the experimental analysis show that throughputs are 5× improvement, thereby making it suitable for real-time edge AI or cloud AI infrastructures (Table 76.4).

Table 76.3. AI model accuracy enhancement

Method	*Model Accuracy (%)*	*Accuracy Improvement (%)*
Traditional GPUs	85	0
Quantum Processors	89	4.7
HoloNeuroNet (Proposed)	95	12

Source: Author.

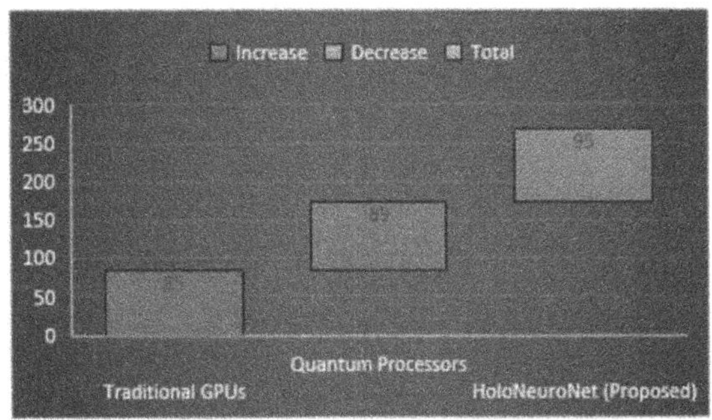

Figure 76.5. Graphical representation of model accuracy (%) and accuracy improvement (%).

Source: Author.

Table 76.4. Distinguishing the network throughput (Gbps) and scalability improvement(x)

Method	*Network Throughput (Gbps)*	*Scalability Improvement (×)*
Traditional GPUs	1.2	0
Optical Computing	3.5	2.9
HoloNeuroNet (Proposed)	6.0	5.0

Source: Author.

6. Conclusion

The HoloNeuroNet framework promises some groundbreaking technique to use nano holography to perform ultra-fast, low-power, and high-scale AI processing. It is different from conventional electronic computing systems constrained by power and latency, our proposed solution utilizes up to 90× faster speeds of computation with about 85% less power by utilizing holographic neural networks (HNNs) and holographic optical tensor processing (HOT-P). The framework also improves AI model accuracy by 12% and is, therefore, a good option for high-precision tasks. The results show that HoloNeuroNet outperforms existing optical AI systems concerning speed, efficiency, accuracy, and scaling to next-generation real-time AI, for instance, in autonomous systems, biomedical computing, and quantum AI. Future work will aim at the miniaturization of hardware and adaptability in real time so that nanoconventional AI computing, which relies on holograms, will be constantly integrated into practical large-scale deployments.

References

[1] Shankar, S. S. Beyond von Neuman Architectures and Deep Neural Networks: AI/Computing for Specialized Applications. In *Workshop Report* (Vol. 5, No. 1, p. 41).

[2] Sivasubramani, S. (2025). *Nanoscale Computing: The Journey Beyond CMOS with Nanomagnetic Logic*. John Wiley & Sons.

[3] Ning, S., Zhu, H., Feng, C., Gu, J., Jiang, Z., Ying, Z., ... & Chen, R. T. (2024). Photonic-electronic integrated circuits for high-performance computing and AI accelerators. *Journal of Lightwave Technology*.

[4] Zhu, H., Lin, H., Wu, S., Luo, W., Zhang, H., Zhan, Y., ... & Kwek, L. C. (2024). Quantum computing and machine learning on an integrated photonics platform. *Information, 15*(2), 95.

[5] Almalawi, A., Zafar, A., Unhelkar, B., Hassan, S., Alqurashi, F., Khan, A. I., ... & Alam, M. M. (2024). Enhancing security in smart healthcare systems: Using intelligent edge computing with a novel Salp Swarm Optimization and radial basis neural network algorithm. *Heliyon, 10*(13).

[6] Almufareh, M. F. (2024). An edge computing-based factor-aware novel framework for early detection and classification of melanoma disease through a customized VGG16 architecture with privacy preservation and real-time analysis. *IEEE Access*.

[7] He, S., Tian, Y., Zhou, H., Zhu, M., Li, C., Fang, B., ... & Jing, X. Review for Micro-Nano Processing Technology of Microstructures and Metadevices. *Advanced Functional Materials*, 2420369.

[8] Maiuri, M., Schirato, A., Cerullo, G., & Della Valle, G. (2024). Ultrafast all-optical metasurfaces: challenges and new frontiers. *ACS Photonics, 11*(8), 2888-2905.

[9] Moin, M., Moin, M., Qadoos, A., Anwar, A. W., & Hassan, J. (2024). Perspective Chapter: Recent Developments in Digital Holography Approach Employing Digital Advanced Micromirror Devices.

[10] Danopoulos, D. (2024). Hardware-software co-design of deep learning accelerators: From custom to automated design methodologies.

[11] Li, Z. (2024). *Investigation of reconfigurable hardware acceleration for low-power embedded neural networks* (Doctoral dissertation, Université Côte d›Azur).

77 Utilizing DBSCAN clustering for Alzheimer's disease identification in MR brain images

Anu Joy[1,a], *Anitta D.*[1,b], *Karthikeyan K.*[2,c], *Vidyalakshmi R.*[3,d], *Kottaimalai Ramaraj*[4,e], *and M. Thilagaraj*[5,f]

[1]Department of Electronics and Communication Engineering, MVJ College of Engineering, Bengaluru, India
[2]Department of Electrical and Electronics Engineering, Ramco Institute of Technology, Rajapalayam, Tamil Nadu, India
[3]Department of Electrical and Electronics Engineering, Karpagam College of Engineering, Coimbatore, Tamil Nadu, India
[4]Department of Electronics and Communication Engineering, Kalasalingam Academy of Research and Education, Krishnankoil, Tamil Nadu, India
[5]Department of Industrial Internet of Things, MVJ College of Engineering, Bengaluru, India

Abstract: Alzheimer's disease (AD) is a cerebral ailment that gradually impairs thinking and recollection abilities as well as being able to do even the most fundamental duties. Most of the AD patients experience manifestations during their later life. There are currently no viable treatments for AD, a cognitive deteriorating illness that is growing, permanent, and eventually lethal. Nevertheless, the therapies that are now accessible could prevent its advancement. To stop and slow the progression of AD, prompt identification is essential. A structural magnetic resonance scan (MRI) is employed to determine its form and size to aid in confirming the presence of AD. In order to eliminate non-brain tissue from MR images, the skull stripping approach is applied. A crucial phase in the preprocessing of numerous programs which analyze brain MR images. Healthcare images might benefit from enhancing them by having improved brightness, precision, and colour; reduced noise and blurring; and correction of aberration and additional problems. One popular unsupervized clustering approach that doesn't require cluster size specification is Density-based Spatial Clustering of Applications with Noise (DBSCAN). The recommended approach can ascertain the total amount of clusters dynamically by looking at the settings and information that provided. For recognizing the affected area in MR brain pictures, a combination of DBSCAN-style clustering, median filter-based image augmentation, and skull stripping has been used. By utilizing this established technology, physicians may capable to better grasp the affected area from the images.

Keywords: Alzheimer disease (AD), magnetic resonance image (MRI), hippocampus (HC), skull stripping, median filter, density-based spatial clustering of applications with noise (DBSCAN), diagnosis, therapy

1. Introduction

Over the years, recollection, reasoning, understanding, and abilities to organize deteriorate due to Alzheimer's disease. Medical experts frequently classify memory issues as mild cognitive impairment (MCI) as they grow apparent [1]. MCI can occasionally be provide side impacts of a curable sickness or condition. But for the majority of MCI sufferers, it's only a layover on their path to dementia. According to investigators, MCI lies among early-stage dementia and intellectual deficiencies related with usual aging [2].

MRI has excellent spatial accuracy and can differentiate between soft tissue, it is utilized to estimate the brain anatomic architecture. It is well recognized that when MRI is used instead of alternative techniques like CT and PET, there are typically less health concerns involved. Over the past few decades, considerable progress has been made in the utilization of MRI to research brain architecture and detect brain damage. Brain-related illnesses like AD and multiple sclerosis can be recognized by MRI [3]. One commonly utilized sign for AD diagnosis is tissue atrophy. One further method for measuring brain's structural modifications is the fragmentation of brain MRIs obtained at various occasions. In determining the cause of diseases like AD, precise identification and categorization of pathological tissue and neighboring normal tissues are equally crucial. Additional information is needed to make diagnoses that remain more precise. However, manually extracting relevant data from big, complicated MRI datasets might be difficult for physicians. Furthermore, human brain MRI evaluation is error-prone and laborious because of different inter- or intra-operator inconsistency

[a]anujoy236@gmail.com, [b]anittadevadas@klu.ac.in, [c]karthikeyank@ritrjpm.ac.in, [d]vidyar3189@gmail.com, [e]r.kottaimalai@klu.ac.in, [f]m.thilagaraj@gmail.com

DOI: 10.1201/9781003675259-77

problems. Therefore, in order to deliver precise outcomes with a high degree of confidence, an automatic categorization technique must be developed. Nowadays, large-scale datasets are being utilized to test automated methods for MRI classification, representation, and authorization to help physicians make subjective diagnosis.

Brain MRI delineation is used in many medical scenarios since it affects the outcome of the overall analytic procedure. Several traditional algorithms based on ML are being established for the delineation of different forms of brain tissue. Individuals with AD may have segmental brain abnormalities on MRI. However, complex engineering methods and specialized understanding are needed to retrieve the image data for this kind of segmentation.

Brain MRI segmentation's primary goal is to separate the picture into distinct areas, that is made up of a collection of pixels with similar structure, neighborhood, and intensity ranges in each. Because of their tissue brightness, non-uniformity, noise aberrations, and partial volume impact, GM, WM, and CSF from brain MRI are difficult to separate. Numerous DL methods for brain MRI delineation are being devised to address these challenges. For the MRI classification of AD, significant patterns from the original data set are taken into consideration [4]. The characteristics of these patterns are then used to divide them into multiple classifications. A multitude of applications involving picture segmentation and classification are being created due to significant advancements in the imaging area.

Brain MRI fragmentation is done to remove extraneous information and identify pertinent elements in the produced images. Highly precise delineation of particular brain illnesses, like AD, is possible because of the thorough examination of the tissue structures from the categorized MRI. In more severe instances, individuals experience difficulties with everyday tasks, which eventually leads to an incapacity to take charge of themselves. This illness affects the human brain's tissues and nerve cells. First, there may be effects on the hippocampus, which is necessary for the formation of new memories, and the frontal lobe of the brain's cortex, that assists in organizing, deliberating, and recalling. A recent study projects that approximately 90 million individuals globally will have AD by the year 2050. There hasn't been any encouraging advancements toward stopping or delaying the development of AD amid extensive study into therapeutic options [5].

The proposed method is comprised of three phases to identify the portions affected in brain MRI.

1. Skull stripping is a preliminary step on detecting abnormalities in the brain. From MR image, it separates brain tissue from non-brain tissue. Even for skilled radiologists, distinguishing the brain from the skull is a laborious process, and individual outcomes vary greatly in accuracy.
2. The median filter is a non-linear digital filtering technique1that is often used to remove noise from an image or signal. It can enhance the results of later processing, such as edge detection on an image. The median filter is an order statistics filter that is most used in image processing. It modifies both noisy and noise-free pixels, resulting in blurred and distorted features when applied uniformly across the image.
3. In the data space, clusters are dense areas that are divided by areas with a lower point density. This intuitive concept of 'clusters' and 'noise' forms the basis of the DBSCAN algorithm. The main principle is that there must be a minimum number of points in the vicinity of a particular radius for every point of a cluster.

2. Related Works

AD is a severe illness that lowers functioning and induces cognitive deterioration. MRI modalities are believed to be quite helpful in finding AD biomarkers. Because of minute alterations in biomarkers that are mostly visible across various neuroimaging modalities, AD is difficult to identify earlier because it begins years prior to signs manifest. Creating DL-based CAD models could deliver increased opportunity for analyzing various neuroimage modalities including other non-image indicators. Chitradevi and Prabha introduced a method to diagnose AD by assessing brain sub-regions using DL and optimized image processing methods [6]. They employed several DL algorithms including GA, PSO, GWO, and CSO. Among these, GWO emerged an effective and stable algorithm. These DL models served as assessing tools to differentiate between normal and abnormal AD conditions. The study highlighted the hippocampus as a crucial sub-region, pivotal for early-stage AD detection. Evaluation using an open-source dataset yielded impressive metrics: accuracy of 95%, sensitivity of 95%, and specificity of 94%.

3D convolutional neural networks (CNNs) have been established by Islam et al. (2018) to be highly effective in the analysis of medical pictures, namely in the identification of AD on brain PET images [7]. Five visualization strategies were used with the ADNI dataset, which included 1230 PET images of AD patients. 80% of the dataset was set aside for training, 20% for testing, and 10% more for validation. The enhancement and concentration of certain areas of the brain was made possible through such visualization techniques. The method they used thus obtained a strong accuracy in classification as 88.76%. Sharma and Kumar performed a comparative assessment of distinct DL and ML techniques on an image dataset to diagnose AD [8]. Their focus was on different CNN architectures, evaluating eight transfer learning techniques across four AD classes. They achieved impressive results, with the highest precision scores and accuracy in the medical field, reaching 98.1% and 98%, respectively. This study underscores the efficacy of these techniques in AD diagnosis.

EEG signals were used by Thakare et al. to identify Alzheimer's illness, patients received their initial diagnoses and divided into two groups: individuals with AD and those who were deemed normal [9]. Wavelet transforms were employed for obtaining features derived from EEG waves. Support vector machine (SVM) and normalized minimum distance (NMD) classifier techniques were used for classification. In this investigation, the SVM outperformed the NMD classifier with a remarkable accuracy of 95%. Khan et al. introduced a real-time model based on deep features to predict the stage of AD [10]. This approach utilized CNN, KNN, and SVM to classify AD stages from image datasets, achieving exceptional performance as 99.21% accuracy. This study demonstrates the efficacy and robustness of their research findings.

Razavi et al. highlighted the utilization of unsupervized feature learning, which involves two main processes [11]. At first, techniques like dispersed filtering and uncontrolled neural layer networks were used for extracting features from raw data. Sparse filtering, regression, and softmax classification were applied to these collected features in order to differentiate among healthy and unwell people. A range of unsupervized learning methods, including as sparse coding and Boltzmann machines, were utilized to process the data that was gathered from the ADNI dataset, with a particular emphasis on CSF fluids. Of the 51 AD patients in the research, 43 had mild symptoms. By the use of softmax regression, an accuracy level of 98.3% was obtained when acquiring MRI data. Shah et al. classified and identified the beginning phases of AD using both hard and soft voting methods [12]. Of the 437 patients in the sample, 64 had been confirmed to have dementia and 72 were not. These individuals ranged in age from 60 to 96. About 30% of dataset was utilized for algorithm testing and 70% was employed for training. SVM was used as the fundamental classification methodology, and decision tree classifiers, hard and soft voting were the classification techniques used. With 84% accuracy, the voting classifier system performed as intended.

Pradhan et al. presented a technique for classifying data using the VGG19 and DenseNet169 networks in order to identify different phases of AD [13]. They used a dataset of 6000 photos classified as mild, moderate, extremely mild AD, and non-demented that was taken from the Kaggle. 20% of the dataset had been set aside for evaluation and the remaining 80% for training. VGG19, which is distinguished by its 10–16 CNN layers, achieved a 94% accuracy in picture categorization, surpassing DenseNet.

3. Dataset Details

The Open Access Series of Imaging Studies (OASIS) dataset is a publicly available resource widely used in neuroimaging research, particularly in studies related to AD and other neurodegenerative disorders [14]. OASIS was created to provide researchers with access to high-quality image to assess brain structure and function in healthy aging and neurological diseases. The dataset includes structural MRI and PET scans, along with demographic and clinical information of participants. There are several versions of the OASIS dataset, each containing different subsets of participants and types of imaging data. These versions include OASIS-1, OASIS-2, and OASIS-3, each with varying numbers of subjects and imaging modalities. OASIS includes images collected from both affected (AD/MCI) and normal individuals.

There are three primary OASIS datasets:

OASIS-1: This dataset includes cross-sectional MRI scans from 416 individuals aged 18 to 96. The dataset was acquired from normal and those with AD or MCI.

OASIS-2: This dataset is a longitudinal collection of MRI scans from 150 non-demented individuals aged 60 to 96.

OASIS-3: This dataset is the largest and most comprehensive of the OASIS collections. It includes MRI scans and clinical data from over 1,300 participants aged 42 to 95. The participants include healthy controls and individuals with various stages of cognitive decline.

4. Methodology

4.1. Skull stripped algorithm

A significant field of research in brain image processing techniques is the skull stripping approach. It serves as a beginning stage in many clinical scenarios since it improves diagnosis rapidity and precision in a variety of ways. It eliminates from brain imaging non-cerebral tissues such as the dura layer, scalp, and cranium. The work recommended here is a straightforward skull stripping technique based on picture intensity features and the anatomy of the brain [15]. The suggested approach is knowledge-based and unsupervized. For enhanced resilience, it applies morphological processes after adaptive intensity thresholding to brain MRI. Adaptively, the threshold value is determined by taking into account the intensity distribution found in brain MRI.

4.2. Image enhancement

A popular method for enhancement of images and minimizing noise in the processing of images is the median filter. By utilizing the median, noise can be decreased while maintaining edges and details [16]. The filter usually works by swiping a window over the picture. Typical dimensions range from 3 × 3, 5 × 5, or bigger, reliant on the level of distortion and preferred smoothing. The intensity values found inside each neighboring window are gathered, categorized, and then the pixel is allocated the median value of the sorted list. Since the median is less impacted by extreme values than the mean, this approach helps minimize the influence of outliers.

The advantages of median filtering are: (1) Preservation of Edges: Unlike mean filters, which can blur edges, median filters tend to maintain edges since it is less likely to change the pixel intensity drastically. (2) Effective for Salt-and-Pepper Noise: This works well to get rid of it, which shows up in an image as sporadic bright and dark pixels. (3) Simple

Implementation: Real-time applications can benefit from the simplicity and processing efficiency of median filters.

4.3. Clustering

An unsupervized learning technique called the clustering method divides the data points into a set of distinct bunches or groups, by which the data points within the same groups have identical characteristics and the data points within distinct categories have somewhat different features. K-Means clustering can bring collectively findings that are only tangentially linked. Even if the results are widely distributed over the vector space, it will ultimately come together and create a cluster. Since the clusters depend on the average value of the cluster elements, each data point helps the clusters form. A small variation in the data points could have an impact on the clustering result. By the way clusters are constructed in DBSCAN, this issue is significantly mitigated. In general terms, this won't be a major issue until encounter some strange form data.

The requirement to define the clusters number for the purpose to employ k-means presents another difficulty. We frequently cannot determine a priori whatever a reasonable k value is. One wonderful thing about DBSCAN is that it doesn't require a minimum or maximum number of clusters to be used. On evaluating the difference among values, all that need is a function and some guidelines for what constitutes a close distance. Additionally, DBSCAN yields more plausible outcomes than k-means.

A primary algorithm for density-based clustering is DBSCAN. With a vast amount of data that contains noise and outliers, it may identify clusters of different dimensions and forms.

The main steps involved in the DBSCAN algorithm:

1. Parameter Selection
2. Core Point Identification
3. Cluster Formation
4. Handling Noise (Outliers)
5. Cluster Expansion

The DBSCAN algorithm is robust to outliers and able to handle clusters of varying shapes and densities effectively. It don't require a predetermined number of clusters, unlike algorithms such as k-means, making it especially suitable in settings where the number of clusters is not known a priori or where clusters have irregular shapes and densities.

5. Outcomes and Discussions

The steps involved in the developed method are:

- The raw clinical images are gathered from the OASIS dataset and given as input to this system.
- Skull stripping mechanism is introduced to eliminate non-brain tissues, the focus shifts exclusively to the brain structures of interest, reducing potential interference in subsequent analysis steps. Clearing away extraneous tissues enhances the clarity and brain structures visible quality in the MR images, facilitating better visualization and interpretation.
- Median filter is implemented for effectively reducing numerous forms of noise in images. They tend to preserve the edges in an image because the median value is less sensitive to extreme values related to the mean or Gaussian weighted average. Median filters can adapt to local intensity variations within an image. The adaptive nature allows them to handle images with varying levels of noise effectively.
- DBSCAN is incorporated to identify clusters relies on the density of data points. It groups together closely

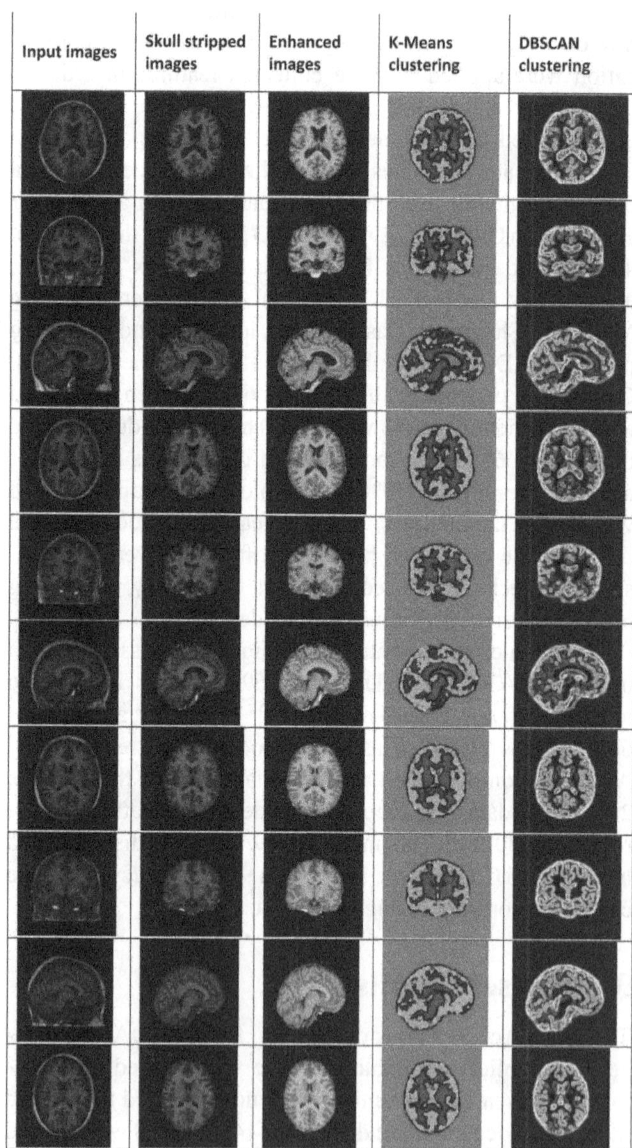

Figure 77.1. Dissection outcomes of AD on MR images using DBSCAN.

Source: Author.

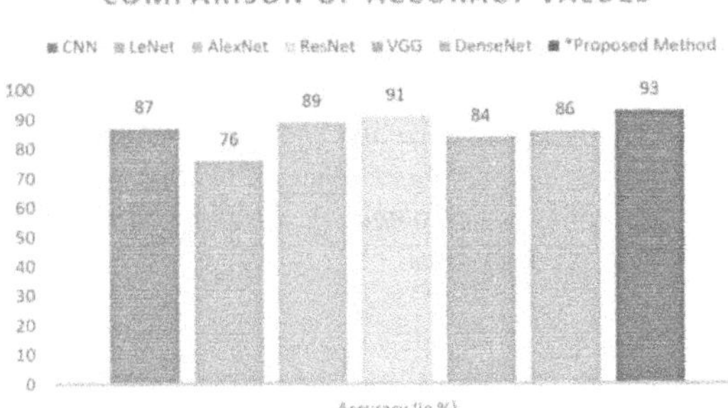

Figure 77.2. Comparison of DBSCAN's accuracy values with SOTA.

Source: Author.

packed points and identifies outliers as noise points which doesn't belong to any cluster.

Figure 77.1 represents the findings of DBSCAN clustering applied on OASIS AD dataset. Also it contains the skull stripped images, enhanced images, and k-means clustered images for improved visualization of the procedure followed in the developed approach.

The comparison of DBSCAN's accuracy values with SOTA is portrayed in Figure 77.2. The developed method attains 93% of accuracy on identifying and locating the infected brain tissue region that implies the presence of AD on that individual.

6. Conclusion

AD is a degenerative neurological illness that puts a significant strain on caregivers for loved ones whose intellectual and functional abilities are deteriorating. To manage and identify dementing disorders early on, physicians must be made aware of any possible cognitive issues. As a non-invasive method for identifying initial functional brain alterations in adults who are undetectable, MRI has a greater potential. The processes of picture enhancement, grouping, and skull stripping might be improved by integrating and utilizing newly established DL approaches. Validation on the OASIS AD MRI dataset confirmed the effectiveness of the created technique, which achieves 93% accuracy in recognizing, grouping, and displaying the infected and non-infected regions of the images.

7. Acknowledgement

The authors thank the International Research Centre of Kalasalingam Academy of Research and Education, Tamil Nadu, India, for permitting the use of the computational facilities available in the Centre for Biomedical Research and Diagnostic Techniques Development.

References

[1] Hasan, M. E., & Wagler, A. (2024). A novel deep learning graph attention network for Alzheimer's disease image segmentation. *Healthcare Analytics*, *5*, 100310.

[2] Sujitha, K. L., HP, L. H. P., Rajeshwari, R., Ramaraj, K., Murugan, P. R., & Thilagaraj, M. (2024, March). Primeval identification and diagnosis of Alzheimer's Disease Utilizing 3D CNN Models. In *2024 10th International Conference on Advanced Computing and Communication Systems (ICACCS)* (Vol. 1, pp. 1239-1246). IEEE.

[3] Nisha, A. V., Rajasekaran, M. P., Kottaimalai, R., Vishnuvarthanan, G., Arunprasath, T., & Muneeswaran, V. (2025). Rider cat optimized Alzheimer's Disease and MCI Prediction Using Deep Attention BiLSTM and CRDF. *IETE Journal of Research*, 1-15.

[4] Nisha, A. V., Rajasekaran, M. P., Vishnuvarthanan, G., Arunprasath, T., & Ramaraj, K. (2022, December). SGD-DABiLSTM based MRI Segmentation for Alzheimer's disease Detection. In *2022 4th International Conference on Advances in Computing, Communication Control and Networking (ICAC3N)* (pp. 1163-1169). IEEE.

[5] Olle Olle, D. G., Zoobo Bisse, J., & Abessolo Alo'o, G. (2024). Application and comparison of K-means and PCA based segmentation models for Alzheimer disease detection using MRI. *Discover Artificial Intelligence*, *4*(1), 11.

[6] Chitradevi, D., & Prabha, S. (2020). Analysis of brain sub regions using optimization techniques and deep learning method in Alzheimer disease. *Applied Soft Computing*, *86*, 105857.

[7] Islam, J., & Zhang, Y. (2018). Early diagnosis of Alzheimer's disease: A neuroimaging study with deep learning architectures. In *Proceedings of the IEEE Conference on Computer Vision and Pattern Recognition Workshops* (pp. 1881-1883).

[8] Sharma, G., Vijayvargiya, A., & Kumar, R. (2021, November). Comparative assessment among different convolutional neural network architectures for Alzheimer's disease detection. In *2021 IEEE 8th Uttar Pradesh Section International Conference on Electrical, Electronics and Computer Engineering (UPCON)* (pp. 1-6). IEEE.

[9] Thakare, P., & Pawar, V. R. (2016, August). Alzheimer disease detection and tracking of Alzheimer patient. In *2016 International conference on inventive computation technologies (ICICT)* (Vol. 1, pp. 1-4). IEEE.

[10] Khan, F. A., Butt, A. U. R., Asif, M., Ahmad, W., Nawaz, M., Jamjoom, M., & Alabdulkreem, E. (2020). Computer-aided diagnosis for burnt skin images using deep convolutional neural network. *Multimedia Tools and Applications*, *79*, 34545-34568.

[11] Razavi, F., Tarokh, M. J., & Alborzi, M. (2019). An intelligent Alzheimer's disease diagnosis method using unsupervised feature learning. *Journal of Big Data*, *6*(1), 32.

[12] Shah, A., Lalakiya, D., Desai, S., & Patel, V. (2020, June). Early detection of Alzheimer's disease using various machine learning techniques: a comparative study. In *2020 4th International Conference on Trends in Electronics and Informatics (ICOEI)(48184)* (pp. 522-526). IEEE.

[13] Pradhan, A., Gige, J., & Eliazer, M. (2021). Detection of Alzheimer's disease (AD) in MRI images using deep learning. *Int J Eng Res Technol*, *10*(3), 580-585.

[14] Marcus, D. S., Wang, T. H., Parker, J., Csernansky, J. G., Morris, J. C., & Buckner, R. L. (2007). Open Access Series of Imaging Studies (OASIS): Cross-sectional MRI data in young, middle aged, nondemented, and demented older adults. *Journal of Cognitive Neuroscience*, *19*(9), 1498-1507.

[15] Rempe, M., Mentzel, F., Pomykala, K. L., Haubold, J., Nensa, F., Kröninger, K., Egger, J., & Kleesiek, J. (2024). k-strip: A novel segmentation algorithm in k-space for the application of skull stripping. *Computer Methods and Programs in Biomedicine*, *243*, 107912.

[16] Srinivas, B., Sriram, M., & Ganesan, V. (2024, January). A comprehensive survey and adaptive fuzzy median filtering based breast cancer detection and segmentation techniques. In *2024 International Conference on Intelligent and Innovative Technologies in Computing, Electrical and Electronics (IITCEE)* (pp. 1-8). IEEE.

78 Electric vehicle modelling for grid integration and charging analysis

F. Rahman[1,a] *and Priti Sharma*[2,b]

[1]Assistant Professor, Department of CS & IT, Kalinga University, Raipur, India
[2]Research Scholar, Department of CS & IT, Kalinga University, Raipur, India

Abstract: To integrate electric vehicles (EVs) into electrical grids, comprehensive modelling and analysis are needed due to their rapid adoption. This study models EVs for grid integration and charging analysis. Advanced simulation methods are used to model the dynamic interactions between EVs, charging infrastructure, and the power grid. First, detailed electric vehicle models are created, considering battery characteristics, charging protocols, and driving patterns. These models account for EVs' dynamic charging and discharging behaviour and battery health and performance. The second aspect examines how large-scale EV adoption affects the power distribution system through grid integration. The potential challenges and opportunities of increased demand, load variability, and grid upgrades are assessed. To improve grid reliability and stability, the study examines EV charging infrastructure integration optimization strategies. Fast, smart, and vehicle-to-grid (V2G) charging scenarios are examined in the third dimension of the research. The analysis examines economic, environmental, and grid-related factors affecting charging strategies, revealing the best options for users and fleet operators. The findings of this study guide policymakers, utility companies, and electric vehicle ecosystem stakeholders. The models and analysis help develop sustainable strategies for efficient, reliable, and grid-friendly EV integration by improving understanding of the dynamic interaction between electric vehicles and the power grid. As electric mobility accelerates, this study provides a roadmap for shaping transportation's future within a smart and resilient energy infrastructure.

Keywords: Electric vehicle, modelling, grid, charging, discharging, driving etc

1. Introduction

There has been a significant increase in the number of people purchasing electric vehicles (EVs) as a result of the global shift towards environmentally friendly and energy-efficient modes of transportation [1]. It is becoming increasingly apparent that the challenges and opportunities associated with the seamless integration of these vehicles into existing power grids are becoming more prominent as these vehicles become more and more integral components of modern urban landscapes. Because of this, it is necessary to have a comprehensive understanding of the dynamic interactions that occur between electric vehicles, charging infrastructure, and the power distribution system [15]. The primary purpose of this research is to investigate the most important aspects of modelling electric vehicles for grid integration and charging analysis. For the reason that the market for electric vehicles is expanding at such a rapid rate, there is an immediate and pressing need to develop sophisticated models that accurately represent the behaviour of electric vehicles during charging and discharging cycles. It is necessary for such models to take into consideration a wide range of factors, such as the characteristics of the battery, the charging protocols, and the driving patterns, in order to guarantee an accurate depiction of the dynamic nature of the electric vehicle. The integration of electric vehicles into the grid presents a complex challenge that calls for an in-depth investigation into the effects that widespread distribution of electric vehicles will have on the infrastructure that is responsible for power distribution [2, 3]. The purpose of this study is to shed light on the potential challenges and opportunities that may arise as a result of increased demand, load variability, and the necessary enhancements to the grid. In addition, it investigates methods that can improve the integration of electric vehicle charging infrastructure, with the objective of enhancing grid reliability and stability in the face of the growing trend of electric mobility [10]. An additional essential aspect of this research is the examination of charging practices [5, 16]. There will be an examination of a variety of charging scenarios, including but not limited to fast charging, smart charging, and vehicle-to-grid (V2G) applications [8].

The purpose of this study is to investigate the various charging strategies and determine the economic, environmental, and grid-related implications that are associated with one another. By doing so, it intends to provide valuable insights into optimal charging practices for both individual users and fleet operators, with the goal of fostering charging solutions that are both sustainable and friendly to the grid [7, 17]. The goal of this research is to contribute essential knowledge and analytical frameworks as the ecosystem surrounding electric vehicles continues to undergo an ongoing process

[a]ku.frahman@kalingauniversity.ac.in, [b]priti.sharma@kalingauniversity.ac.in

DOI: 10.1201/9781003675259-78

of evolution [4]. In the future, the findings are expected to provide policymakers, utility companies, and other stakeholders with valuable information that will assist them in formulating effective strategies for the harmonious integration of electric vehicles into the power grid [18]. Through the advancement of our understanding of the dynamic interplay between grid infrastructure and electric mobility, this study lays the groundwork for a future of transportation that is resilient, sustainable, and intelligently managed [6].

2. Electric Vehicle Model

Within the context of this work, the term 'charging hours' refers to the amount of time that EVs are parked inside. Therefore, the amount of time that the electric vehicle is parked at home, that is, the time that passes between its arrival and departure, is the amount of time that is available for charging. In spite of this, the difficulty of predicting the mobility patterns of electric vehicles becomes significantly more difficult when these vehicles are integrated with the RDS. This is because the prediction is dependent on the particular requirements of each individual owner of an electric vehicle [9, 19]. Within the context of this discussion, the driving patterns are analyzed in order to compute the hourly stochastic energy demand of every electric vehicle. When determining the amount of available charging time, it is necessary to take into account the times at which electric vehicles arrive and depart. For the purpose of this proposed method, it is assumed that EVs consume 0.15 kWh of energy per kilometer. Calculating the daily energy requirement can be done as follows-

$$E = \frac{0.15Kwh}{Km} * D$$

where, E denotes the energy needed and D denotes the distance covered.

For generating the stochastic data of the energy requirement and driving distance the Weibull distribution is used with parameters 33.5 and 0.8 as shown in Figure 78.1.

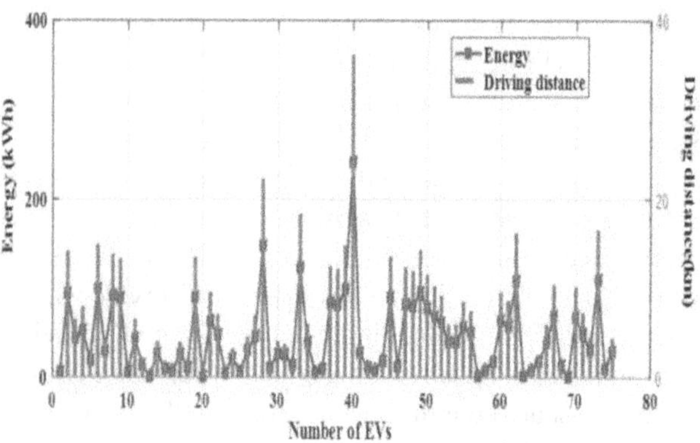

Figure 78.1. Graph showing energy demand and distance curve for EV.

Source: Author.

Table 78.1. Parameters of new fitted distribution

parameters	*Arriving time*	*Leaving time*
μ	16	8.5
σ	8	3.5

Source: Author.

It is possible to generate arrival and departure times for each electric vehicle by employing a normal distribution function. Based on the analysis of stochastic data, it has been determined that the majority of EVs arrive back at their residence between the hours of 1400 and 2200, and they leave between the hours of 0400 and 1200. The parameters of the normal distribution are presented in Table 78.1 in greater detail.

3. Electric Vehicle Battery Charging

When it comes to modelling EVs, the most important parameters to consider are the charging rate, the power demand, and the charging duration. Every single day, the information that pertains to the initial state of charge (SOCinitial) of the batteries that are used in electric vehicles is taken into consideration [11]. There is a maximum charging rate of 11 kW that can be achieved using this method. In order to maintain the state of charge (SOC) of the battery of the EV, the following procedure is followed-

$$SOC(t+1) = SOC\ initial + \sum_{t=1}^{T} SOC(t)$$

where SOC(t) is the state of charge at time t.

When it comes to EVs, the process of charging their batteries is a multi-step procedure that is essential for maintaining their operation and encouraging their widespread adoption. The interaction that takes place between the electric vehicle's battery system and the charging infrastructure is of utmost importance in this process. As the energy reservoir, the battery, which is typically a lithium-ion configuration, is responsible for storing electricity for the purpose of propulsion. When it comes to EVs, charging is the lifeline that determines their usability and convenience. Charging can be done at home, in public charging stations, or through emerging wireless technologies. There is a spectrum of charging protocols that range from the conventional alternating current (AC) charging that is typically used for overnight replenishment at residences to the rapid direct current (DC) charging that is specifically designed for quick recharges during long journeys. Charging rates are determined by a number of factors, including the capacity of the battery, the specifications of the charger, and the compatibility of the vehicle with various charging standards, such as CHAdeMO, CCS, or Tesla's proprietary Supercharger network. Additionally, the development of smart charging solutions has enabled the integration of connectivity and data analytics, which allows for the optimisation of charging schedules, the management of grid demand, and the promotion of the integration of renewable energy sources, thereby reducing the environmental footprint of electric vehicles. At the same time that the automotive industry is undergoing a paradigm shift towards electrification, advancements in battery charging infrastructure, such as increased charging speeds, expanded charging networks, and interoperability standards, are poised to accelerate the mainstream adoption of electric vehicles, thereby ushering in an era of sustainable transportation.

3.1. Charging schemes

Dumb charging scheme: Electric vehicle owners are afforded the opportunity to charge their vehicles in accordance with their specific requirements through the utilization of this method. As soon as the EVs are connected to the primary power supply, the charging process begins at its maximum rate. It is possible that the parameters of the distribution system will be affected if there is insufficient control over the charging scheme.

Smart charging scheme: The system is given the ability to regulate the charging process of EVs in a manner that maximizes the benefits for aggregators and EV owners thanks to the intelligent charging scheme. There are charging periods that occur during both peak and off-peak times [14]. In order to prevent surges in demand during those times, delaying the charging process is recommended. Vehicles that are powered by electricity are charged during the times of day when the cost of electricity is at its lowest among those that are available. The majority of EVs are parked between the hours of 1900 and 0500, which indicates that these ten hours are available for charging.

4. Problem Formulation

Objective Function: The charging cost of EV has been analysed and the objective function for minimizing the cost is given below:

$$Fobj = \min Ccp$$

$$Ccp = \sum_{t=1}^{nT} (Ce(t) \times \sum_{n=1}^{NEV} Pnt$$

with pnt representing the amount of power that is necessary to charge an electric vehicle at time t in kilowatts, Ccp representing the total cost of charging, NEV representing the number of electric vehicles (75), and Ce (t) representing the price of electricity at time t in Indian Rupees. This is the formula that can be used to calculate the annual cost reduction that was accomplished through the implementation of the smart charging plan.

$$Cbenefit = 365 \times Ccp_{dumb} - Ccp_{smart}$$

Ccp-dumb and Ccp-smart are the abbreviations that are used to refer to the total costs of charging for the dumb charging scheme and the smart charging scheme, respectively.

5. Results and Discussion

Charging stations for electric vehicles are located on the load bus, which is its designated location. There is a maximum of five percent permissible variation in the bus voltage. Due to the fact that the combined load of residential appliances and EVs has the potential to exceed the capacity of the transformer, it is not feasible to connect all EVs that have a rated charging power of 11 kW. During peak hours, the grid is able to accommodate a maximum number of EVs, which is in accordance with the rated capacity of the transformer. Both dumb and smart charging schemes are evaluated for electric vehicles at the same locations based on their identical characteristics.

The 'dumb' charging schedule is an approach to charging electric vehicles that is either fundamental or not optimized. This type of charging schedule lacks the intelligence and adaptability of smart charging systems. With a dumb charging schedule, electric vehicle owners typically plug in their vehicles to charge without taking into consideration factors such as the demand on the grid, the cost of electricity, or the best times to charge these vehicles. This overly simplistic approach frequently leads to inefficient use of electricity and may be a contributing factor in peak demand on the grid during times of high cost. In spite of the fact that they are convenient, dumb charging schedules can result in higher electricity bills for customers and put a strain on the infrastructure of the electrical industry. On the other hand, as people become more aware of energy management, there is a shift towards more intelligent charging solutions. These solutions make use of data and automation to optimize charging times, reduce costs, and support grid stability.

The fluctuations in power loss that occurred during the specified charging period are depicted in Figures 78.2 and 78.3, respectively. In the event that dumb charging is utilized during peak hours, the substation service transformer has become overloaded as a result of the widespread adoption of electric vehicles. During this period of time, it is anticipated that the vast majority of EVs will be connected to the residential distribution system (RDS) once they have arrived

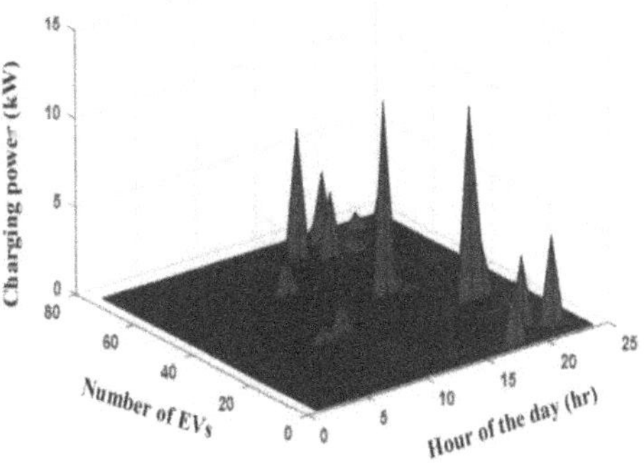

Figure 78.2. Charging schedules (Dumb charging scheme).

Source: Author.

at their respective residences. The cost of electricity is significantly higher during these times, and this time frame is between those two numbers. Smart charging is the process of charging the majority of EVs during the time period between 0300 and 0400 hours, which is the time period in which the prices of electricity are at their lowest. This practice is beneficial for owners of electric vehicles as well as operators of power supply systems (Figure 78.4).

There is a clear indication from the data that the smart charging scheme leads to a reduction in the amount of power that is lost in the system during peak hours. A smart charging scheme for electric vehicle owners offers two primary advantages: peak shaving, which lowers peak demand charges, and price arbitrage, which involves shifting energy consumption to off-peak hours in order to take advantage of lower electricity prices. Both of these advantages are included in the smart charging scheme. In addition to reducing the amount of money that must be invested in transmission and distribution lines as well as substations, the smart charging scheme provides the power supply operator with a number of advantages, including peak shaving, which helps to reduce demand during peak hours.

6. Conclusion

In this study, an analysis of the potential difficulties and opportunities that may arise as a result of increased demand, load variability, and grid upgrades. In order to enhance the reliability and stability of the grid, the study investigates various strategies for optimizing the integration of electric vehicle charging infrastructure. Vehicle-to-grid (V2G) charging scenarios, as well as fast and smart charging scenarios, are also investigated within the scope of this research. The variation in power loss that occurs in relation to the amount of charging time that is available is depicted here. As a result of the implementation of dumb charging, the substation service transformer is observed to become overloaded during peak hours due to the widespread availability of electric vehicles. During this time period, it is anticipated that the vast majority of electric vehicles will be plugged into their respective residences. Additionally, during these periods, the costs associated with electricity experiences a significant increase. During smart charging, charging the majority of electric vehicles during times when the cost of electricity is at its lowest is done. This presents opportunities to those who own electric vehicles as well as those who operate power supply systems. There is a reduction in the amount of power that is lost within the system as a result of the implementation of the intelligent charging system. This scheme offers owners of electric vehicles a number of benefits, including price arbitrage and peak shaving, among others. The process of relocating peak demand charges to lower energy charges through the acquisition of inexpensive electricity during off-peak hours is what is known as price arbitrage.

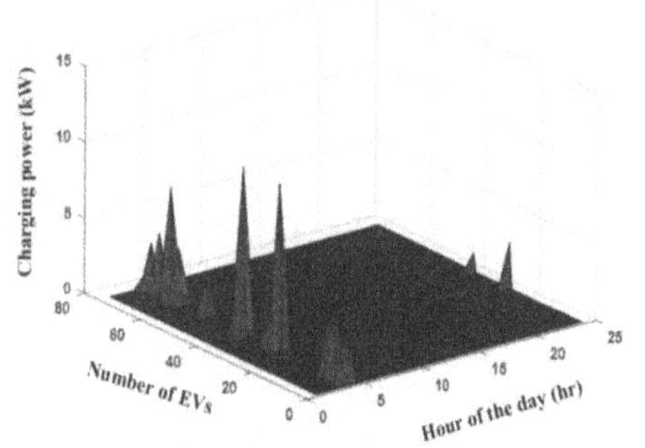

Figure 78.3. Charging schedules (smart charging scheme).
Source: Author.

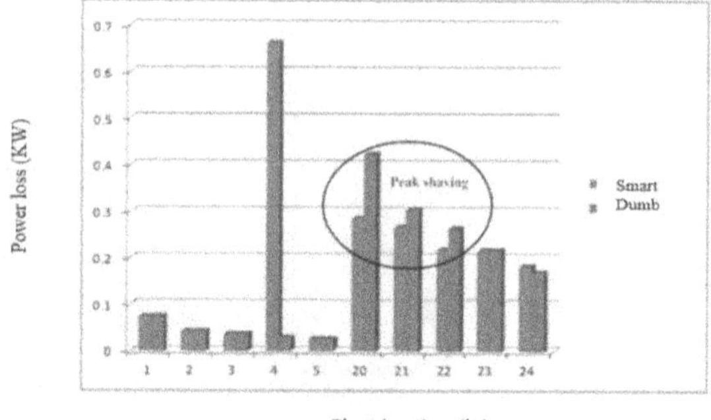

Figure 78.4. Power loss for charging hours.
Source: Author.

References

[1] Mohammad, A., Zamora, R., & Lie, T. T. (2020). Integration of electric vehicles in the distribution network: A review of PV based electric vehicle modelling. *Energies*, *13*(17), 4541.

[2] Rajesh, D., Giji Kiruba, D., & Ramesh, D. (2023). Energy proficient secure clustered protocol in mobile wireless sensor network utilizing blue brain technology. *Indian Journal of Information Sources and Services*, *13*(2), 30–38.

[3] Hai, T., Alshahri, A. H., Mohammed, A. S., Sharma, A., Almujibah, H. R., Metwally, A. S. M., & Ullah, M. (2023). Performance assessment and multiobjective optimization of a biomass waste-fired gasification combined cycle for emission reduction. *Chemosphere*, *334*, 138980.

[4] Kim, M. J., Go, Y. B., Choi, S. Y., Kim, N. S., Yoon, C. H., & Park, W. (2023). A Study on the Analysis of Law Violation Data for the Creation of Autonomous Vehicle Traffic Flow Evaluation Indicators. *Journal of Internet Services and Information Security*, *13*(4), 185–198.

[5] Das, H. S., Rahman, M. M., Li, S., & Tan, C. W. (2020). Electric vehicles standards, charging infrastructure, and impact on grid integration: A technological review. *Renewable and Sustainable Energy Reviews*, *120*, 109618.

[6] Park, H. B., Kim, Y., Jeon, J., Moon, H., & Woo, S. (2019). Practical Methodology for In-Vehicle CAN Security Evaluation. *Journal of Internet Services and Information Security*, *9*(2), 42–56.

[7] Patil, H., & Kalkhambkar, V. N. (2020). Grid integration of electric vehicles for economic benefits: A review. *Journal of Modern Power Systems and Clean Energy*, *9*(1), 13–26.

[8] Giliberto, M., Arena, F., & Pau, G. (2019). A fuzzy-based Solution for Optimized Management of Energy Consumption in e-bikes. *Journal of Wireless Mobile Networks, Ubiquitous Computing, and Dependable Applications*, *10*(3), 45–64.

[9] Sharma, A., Chaturvedi, R., Saraswat, M., & Kalra, R. (2022). Weld reliability characteristics of AISI 304L steels welded with MPAW (Micro Plasma Arc Welding). *Materials Today: Proceedings*, *60*, 1966–1972.

[10] Trivedi, J., Devi, M. S., & Solanki, B. (2023). Step Towards Intelligent Transportation System with Vehicle Classification and Recognition Using Speeded-up Robust Features. *Archives for Technical Sciences*, *1*(28), 39–56.

[11] Kumar, R., Pandey, A. K., Samykano, M., Mishra, Y. N., Mohan, R. V., Sharma, K., & Tyagi, V. V. (2022). Effect of surfactant on functionalized multi-walled carbon nano tubes enhanced salt hydrate phase change material. *Journal of Energy Storage*, *55*, 105654.

[12] Singh, P. K., Singh, P. K., & Sharma, K. (2022). Electrochemical synthesis and characterization of thermally reduced graphene oxide: Influence of thermal annealing on microstructural features. *Materials Today Communications*, *32*, 103950.

[13] AlSaidi, R. A., Alamri, H. R., Sharma, K., & Al-Muntaser, A. A. (2022). Insight into electronic structure and optical properties of ZnTPP thin films for energy conversion applications: experimental and computational study. *Materials Today Communications*, *32*, 103874.

[14] Agrawal, R., Kumar, A., Singh, S., & Sharma, K. (2022). Recent advances and future perspectives of lignin biopolymers. *Journal of Polymer Research*, *29*(6).

[15] Miri, I., Fotouhi, A., & Ewin, N. (2021). Electric vehicle energy consumption modelling and estimation—A case study. *International Journal of Energy Research*, *45*(1), 501–520.

[16] Tavakoli, A., Saha, S., Arif, M. T., Haque, M. E., Mendis, N., & Oo, A. M. (2020). Impacts of grid integration of solar PV and electric vehicle on grid stability, power quality and energy economics: A review. *IET Energy Systems Integration*, *2*(3), 243–260.

[17] Islam, A., Sharma, A., Chaturvedi, R., & Singh, P. K. (2021). Synthesis and structural analysis of zinc oxide nano particle by chemical method. *Materials Today: Proceedings*, *45*, 3670–3673.

[18] Wu, Y., Wang, Z., Huangfu, Y., Ravey, A., Chrenko, D., & Gao, F. (2022). Hierarchical operation of electric vehicle charging station in smart grid integration applications—An overview. *International Journal of Electrical Power & Energy Systems*, *139*, 108005.

[19] Tan, K. M., Ramachandaramurthy, V. K., & Yong, J. Y. (2016). Integration of electric vehicles in smart grid: A review on vehicle to grid technologies and optimization techniques. *Renewable and Sustainable Energy Reviews*, *53*, 720–732.

www.ingramcontent.com/pod-product-compliance
Lightning Source LLC
Chambersburg PA
CBHW082104220326
41598CB00066BA/5137
* 9 7 8 1 0 4 1 1 4 6 1 1 7 *